Environmental Electrochemistry
Fundamentals and Applications
in Pollution Abatement

Environmental Electrochemistry

Fundamentals and Applications in Pollution Abatement

KRISHNAN RAJESHWAR
Department of Chemistry and Biochemistry
The University of Texas at Arlington

JORGE G. IBANEZ
Department of Engineering and Chemical Sciences
Universidad Iberoamericana, Mexico City

ACADEMIC PRESS

San Diego   London   New York   Boston   Sydney   Tokyo   Toronto

Find Us on the Web! http://www.apnet.com

This book is printed on acid-free paper. ∞

Copyright © 1997 by ACADEMIC PRESS

All Rights Reserved.
No part of this publication may be reproduced or transmitted in any form or by any means, electronic or mechanical, including photocopy, recording, or any information storage and retrieval system, without permission in writing from the publisher.

Academic Press
A Division of Harcourt Brace & Company
525 B Street, Suite 1900, San Diego, California 92101-4495

United Kingdom Edition published by
Academic Press Limited
24-28 Oval Road, London NW1 7DX

Library of Congress Cataloging-in-Publication Data

Environmental electrochemistry / edited by Krishnan Rajeshwar, Jorge Ibanez.
    p.   cm.
  Includes bibliographical references and index.
  ISBN 0-12-576260-7 (alk. paper)
  1. Pollution. 2. Electrochemistry. 3. Photoelectricity.
I. Rajeshwar, Krishnan. II. Ibanez, Jorge.
TD192.2.E58 1996
628.5--dc20   96-10927
                                                  CIP

PRINTED IN THE UNITED STATES OF AMERICA
      98 99 00 01 MM 9 8 7 6 5 4 3 2

To Rohini C. Krishnan who made "everything" possible.

K. R.

To Dr. Wayne E. Wentworth and Fr. Carl B. Trutter, O. P.
for their guidance and friendship, and

To the Mexican Province of the Society of Jesus (the Jesuits)
for their unceasing work towards a better world.

J. G. I.

TABLE OF CONTENTS:

CHAPTER 1

1.1.   Introduction 1
1.2.   Some Definitions and Classification of Pollutants 4
1.3.   Environmental Media and Pollutant Transport 8
      1.3.1.   *Air* 8
      1.3.2.   *Water* 10
      1.3.3.   *Soil* 13
      1.3.4.   *Biota* 16
1.4.   Environmental Chemistry and Toxicology of Common Pollutants -
      A Primer 18
      1.4.1.   *Organics* 19
      1.4.2.   *Inorganics* 23
      1.4.3.   *Microorganisms* 29
1.5.   Current Methods for Pollutant Analyses 30
      1.5.1.   *Analysis of Organic and Inorganic Pollutants* 30
      1.5.2.   *Assay of Microorganisms* 35
1.6.   Current Methods for Pollutant Treatment 36
      1.6.1.   *Incineration and Pyrolysis* 37
      1.6.2.   *Air Stripping* 38
      1.6.3.   *Microbial Treatment* 39
      1.6.4.   *Precipitation and Coagulation* 41
      1.6.5.   *Chemical Treatment* 42
      1.6.6.   *Adsorption* 44
      1.6.7.   *Membrane Processes* 44
      1.6.8.   *Distillation* 45
      1.6.9.   *Advanced Oxidation Processes* 46
      1.6.10. *Treatment of Polluted Sites* 47
1.7.   Electrochemical Technology and the Environment 49
1.8.   Summary 50
      References 50
      Supplementary Reading 54

CHAPTER 2

2.1.   Introduction 57
2.2.   Current, Charge, and Potential 58
2.3.   Charge and Mass Transport 62
2.4.   Electrode/Electrolyte Interfaces and Electrochemical Cells 64

Table of Contents

- 2.5. Thermodynamics and Kinetics in Electrochemical Systems 69
  - 2.5.1. Use of Standard Reduction Potentials 69
  - 2.5.2. Redox Potentials and the Nernst Equation 70
  - 2.5.3. Electrode Reaction Kinetics 72
  - 2.5.4. Ohmic, Activation, and Concentration Polarization 73
  - 2.5.5. Pourbaix Diagrams 78
  - 2.5.6. Mixed Potentials 81
- 2.6. Electroanalytical Chemistry 83
  - 2.6.1. Potentiometric Methods 84
  - 2.6.2. Dynamic Methods 88
- 2.7. Electrolysis and Electrodeposition 97
- 2.8. Mass Transport Under Forced Convection in an Electrochemical Cell 100
- 2.9. Electrochemical Reactor Design 104
  - 2.9.1. Performance Figures of Merit of an Electrochemical Reactor 111
  - 2.9.2. Economic Criteria 112
- 2.10. Electrokinetic Phenomena 112
- 2.11. Semiconductor Electrochemistry 116
  - 2.11.1. Band Structure of a Semiconductor and Band Bending at a Semiconductor/Liquid Junction 117
- 2.12. Photoemission at Metal Electrodes 120
- 2.13. Instrumentation 122
- 2.14. Summary 123
- References 123
- Supplementary Reading 124

CHAPTER 3

- 3.1. Introduction 127
- 3.2. Electrochemistry of Organic Pollutants 129
  - 3.2.1. General Considerations 129
  - 3.2.2. Survey of the Electrochemistry of Environmentally Important Organics 136
- 3.3. Electrochemistry of Inorganic Pollutants 155
  - 3.3.1. Metals 155
  - 3.3.2. Nonmetals (Metalloids) 167
  - 3.3.3. Inorganic Anions and Gaseous Species 175
  - 3.3.4. Organometallic Compounds 180
  - 3.3.5. Organo Derivatives of Arsenic, Phosphorus, and Sulfur 181
- 3.4. Summary 189
- References 189
- Supplementary Reading 198

CHAPTER 4

- 4.1. Introduction 201
- 4.2. Flow Injection Analysis 204
- 4.3. Potentiometric Sensors 208
  - 4.3.1. Speciation Studies 212
  - 4.3.2. Solid State and Gas Sensors 213
  - 4.3.3. Flow Systems 215
  - 4.3.4. Examples of Environmental Applications of Potentiometric Sensors 216
- 4.4. Amperometric–Coulometric and Voltammetric–Polarographic Detection in Flow Systems 220
  - 4.4.1. Cell design: General Considerations 220
  - 4.4.2. Specific Cell Designs and Electrode Geometries 223
  - 4.4.3. Dual-Electrode and Array Detectors 226
  - 4.4.4. Electrode Materials and Electrode Surface Modification 231
  - 4.4.5. Dioxygen Removal, Derivatization, and Other Practical Considerations 247
- 4.5. Amperometric Sensors for Environmental Pollutants: Some Examples 250
- 4.6. Amperometric Gas Sensors 271
- 4.7. Stripping Analyses: Specialized Aspects 276
  - 4.7.1. Adsorptive Stripping Voltammetry 276
  - 4.7.2. Anodic Stripping with Collection 280
  - 4.7.3. Subtractive Stripping Voltammetry 280
  - 4.7.4. Potentiometric Stripping Analysis 282
  - 4.7.5. Miscellaneous Aspects of Stripping Analyses 287
  - 4.7.6. Trace Element Speciation in Water by Stripping Analyses 290
  - 4.7.7. Electrochemical Speciation Data on Environmentally Important Elements 299
  - 4.7.8. Other Studies on Environmental Pollutants by Stripping Analyses 304
- 4.8. Direct Voltammetric (or Polarographic) Determination of Pollutants 304
- 4.9. Electrochemistry as an Auxiliary Tool to Atomic Spectroscopies ... 307
- 4.10. Conductivity Detectors 310
- 4.11. Photoassisted Detection of Pollutants 312
  - 4.11.1. Photoelectrochemical Detection 314
  - 4.11.2. Photogalvanic and Luminescent Detection 315
  - 4.11.3. Spectroelectrochemical Detection 317
- 4.12. Summary 318
  - References 319
  - Supplementary Reading 358

CHAPTER 5

5.1.	Introduction	361
5.2.	Positive Features of Electrochemical Remediation	362
5.3.	Direct Electrolysis of Pollutants	363
	5.3.1. Anodic Oxidation of Sample Organic Pollutants	363
	5.3.2. Anodic Oxidation of Sample Inorganic Pollutants	374
	5.3.3. Cathodic Reduction of Pollutants: Electrode Materials	376
	5.3.4. Cathodic Reduction of Sample Organics	377
	5.3.5. Cathodic Reduction of Sample Inorganics	378
	5.3.6. Cathodic Reduction of Metal Ions	380
5.4.	Indirect Electrolysis of Pollutants	398
	5.4.1. Reversible Processes	399
	5.4.2. Irreversible Processes	405
	5.4.3. Fenton Reaction Chemistry	408
	5.4.4. Approaches Relying on pH Manipulation	409
5.5.	Electroflotation, Electrocoagulation, and Electroflocculation	410
5.6.	Electrochemical Remediation of Gaseous Pollutants	417
	5.6.1. CO_2 Reduction	419
	5.6.2. H_2S Treatment	423
	5.6.3. SO_2 Removal	426
	5.6.4. NO_x Removal	429
	5.6.5. Chlorine Removal and Concentration	431
5.7.	Membrane-Assisted Processes	432
	5.7.1. Ion Exchange Membranes: General Considerations	433
	5.7.2. Applications of Ion Exchange Membranes: Electrodialysis	436
	5.7.3. Membrane Cell Electrolysis	440
	5.7.4. Bipolar Membrane-Based Processes	442
	5.7.5. Electrochemical Ion Exchange	444
	5.7.6. Other Membranes and Adsorbents	447
5.8.	Electrokinetic Processing of Soil	449
5.9.	Emerging Materials for Electrochemical Treatment of Pollutants	453
	5.9.1. Electrode Materials	454
	5.9.2. Membranes and Electrode Surface Modification Agents	465
	5.9.3. Electrolytes	466
	Summary	470
	References	470
	Supplementary Reading	496

CHAPTER 6

6.1.	Introduction	499
6.2.	Photolysis of H_2O_2 and O_3 and Generation of e^-_{aq}	500
6.3.	Destruction of Organics as Mediated by ·OH and e^-_{aq}	504
6.4.	Direct Photodissociation of the Pollutant	506
6.5.	Reaction Kinetics, Mechanisms and Examples of Application of the UV–H_2O_2 System	509
6.6.	Reaction Kinetics, Mechanisms and Examples of Application of the UV–O_3 System	520
6.7.	The UV–H_2O_2–O_3 Process	527
6.8.	UV–H_2O_2 and UV–O_3 Systems: Practical Considerations	527
6.9.	Heterogeneous Photocatalysis	531
	6.9.1. Operating Principle and Scope of the Photocatalysis Method	532
	6.9.2. Thermodynamic Aspects	541
	6.9.3. Choice of Semiconductor Photocatalyst, Photocatalyst Configuration and Other Materials-Related Aspects	541
	6.9.4. Absorption of Light by the Photocatalyst and Quantum Yield	545
	6.9.5. Carrier Dynamics in Irradiated TiO_2 Particles	553
	6.9.6 Photocharging of the TiO_2 Particles	559
	6.9.7. Dynamics of Interfacial Electron and Hole Transfer at Irradiated TiO2 (and Other Semiconductor) Suspensions	563
	6.9.8. Kinetics and Mechanistic Aspects of Photocatalytic Reactions at TiO_2	575
	6.9.9. Reaction Intermediates	584
	6.9.10. Comparison of the Direct Photolysis, UV–H_2O_2, Fenton Reaction, and Heterogeneous Photocatalysis Approaches for the Destruction of Organic Pollutants	591
	6.9.11. Photocatalytic Oxidation of Organic Pollutants: Miscellaneous Aspects	592
	6.9.12. Photocatalytic Treatment of Inorganic Pollutants	595
	6.9.13. Gas-Phase Photocatalysis and the Treatment of Air-Borne Pollutants	595
	6.9.14. Photocatalytic Reactor Design Considerations	597
	6.9.15. The Photocatalytic Pollutant Treatment Approach: Prospects and Problems	599
6.10.	Summary	600
	References	600

CHAPTER 7

7.1.	Introduction	625
7.2.	Water Disinfection: Background and Principles	627
	7.2.1. General Considerations	627
	7.2.2. Chemical Disinfection	628
	7.2.3. Disinfection By-Products	635
	7.2.4. Taste and Odor Removal	636
	7.2.5. Indicator Organisms	637
7.3.	Electrochemical Disinfection of Water	638
	7.3.1. Introductory Remarks	638
	7.3.2. Electrosorption of Microorganisms and Direct Electron Transfer	638
	7.3.3. In situ Electrogeneration of Disinfection Agents	640
7.4.	Disinfection by High-Energy Radiation	643
7.5.	UV Disinfection of Water	644
	7.5.1. General Considerations	644
	7.5.2. UV Dose and Disinfection Kinetics	647
	7.5.3. The Role of Suspended Particles and Chemicals in UV Disinfection	652
	7.5.4. UV Dose Sensitivity of Various Microorganisms and Comparison with Chlorine and other Dininfection Methods	658
	7.5.5. Photoreactivation and Sublethal UV Damage Repair	659
	7.5.6. Flow Systems for UV Disinfection of Water	660
	7.5.7. Photolysis of Aqueous Chlorine	663
	7.5.8. UV–O_3 and Other Advanced Oxidation Processes for DBP Control and Water Treatment	664
7.6.	Photoelectrochemical Disinfection of Air and Water	666
7.7.	Electrochemical Detection and Enumeration of Microorganisms	668
7.8.	Summary	673
	References	673
	Supplementary Reading	673

CHAPTER 8

8.1.	Introduction	687
8.2.	Commercial Vendors of Electrochemical Equipment and Accessories	690
8.3.	Electrochemical Sensors of Environmental Pollutants: Commercialization Efforts	690
8.4.	Electrochemical Reactors for Pollutant Treatment and Chemical Recycling	691
	8.4.1. Water Treatment Systems	691

Table of Contents   xiii

      8.4.2. Commercial Processes for Water Treatment and
            Chemical Recycling 695
      8.4.3. Gas and Air Treatment Systems and Processes 705
8.5. **Advanced Oxidation Processes: Pilot and Field Tests and
     Commercialization Efforts** 708
      8.5.1. Supplies of Ozonators, UV Lamps, and Catalysts for UV–H_2O_2,
            UV–O_3, and UV–TiO_2 Technologies 708
      8.5.2. Companies Marketing AOP Technologies 710
      8.5.3. Companies Marketing Direct Photolysis Technology 716
      8.5.4. Pilot, Field, and Demonstration Tests of the AOPs 717
      8.5.5. Economic Considerations 726
8.6. **Water Disinfection Technology** 730
      8.6.1. Case Studies 730
      8.6.2. Economic Aspects 732
8.7. **Summary** 733
     References 734

Appendix A
Abbreviations, Acronyms, Symbols, and Notation 739

Appendix B
Physical Constants, Units and Unit Conversions,
and Reference Electrode Potentials 753

Appendix C
Selected Standard Reduction Potentials in Aqueous Media at 25°C 757

Appendix D
Companies Marketing Electrochemical Technologies and
Accessories for Pollution Sensors and Pollution Abatement 759

PREFACE

The field of environmental science and technology has made enormous strides in recent years. This rapid growth has been stimulated both by the concerted efforts of scientists and engineers trained in traditional environmental disciplines, as well as by the diffusion of ideas from other (non-environmental) fields. This book is about the positive role that electrochemical science and engineering can play in the detection, quantification, and treatment of environmental pollutants. It is targeted both at an environmental specialist audience and at the practicing electrochemist. In so designing the theme, compromise necessarily had to be made between depth, breadth of coverage, and the size of the book; it is our hope that an appropriate balance has been achieved.

The book is divided into eight chapters. The first two are introductory chapters. Those well versed in environmental problems will find little that is new in Chapter 1. Similarly, practicing electrochemists can safely skip Chapter 2. Chapter 3 provides a survey of the electrochemical data base on common types of environmental pollutants. Chapters 4 through 7 attempt to delve into the details of environmental electrochemical analyses (Chapter 4), electrochemical methods for pollution abatement (Chapter 5), photo-assisted methods for pollution control (Chapter 6), and water/air disinfection approaches (Chapter 7). The reference list is extensive in each case, and ought to facilitate easy entry into the specialized literature. Illustrations are liberally employed to illustrate a particular principle or approach.

We realize that this book describes an evolving discipline, and undoubtedly there are many shortcomings. We hope to hear from readers about the flaws - in detail, logic, or in other respects - that have escaped our notice. The purist may also find some of our discussion approaches to be rather unconventional. For example, the classification of environmental pollutants into neat little "boxes" (see Table 1.3) is completely arbitrary, and has been done only to facilitate a concise review. Similarly, many environmentalists may be taken aback by the virtual lack of material related to nuclear waste treatment. Unfortunately, neither of us are knowledgeable enough about this topic to write about it.

Many people made meaningful contributions to the preparation of this book. Gloria Madden managed to transcribe (with her usual efficiency) unreadable drafts into the final manuscript version, improving the organization along the way. Ivonne Konik, Flor Gomez Esparza and Yolanda Alegre also helped in the preparation of the manuscript. Sanjay Basak, AnnaLou Busboom and her staff at the Media Center, University of Texas at Arlington, and Alberto Sosa and Marco Antonio Villaseñor at the Universidad Iberoamericana, succeeded in translating primitive drawings into (hopefully more presentable) illustrations. Hannah Frieser and her staff, at the Office of University Publications, University of Texas at Arlington, spent countless hours preparing the text and illustrations for camera-ready production. The University of Texas at Arlington and the Universidad Iberoamericana both contributed time and resources toward the completion of this book. We gratefully acknowledge the many authors (too numerous to list!) who made preprints/reprints available to us, as well as companies that provided brochures and information on their products and services (Chapter 8). Last but not least, we thank David Packer and Jackie Garrett at Academic Press for their patience and encouragement.

Finally, we offer simple thanks to our wives Rohini and Luz Teresa, and our daughters Reena, Rebecca, Georgina and Lucia, for their encouragement, patience, and understanding during the writing of this book.

Krishnan Rajeshwar
Jorge G. Ibanez

CHAPTER ONE

1.1. INTRODUCTION

Spectacular advances in technology and improvements in the quality of everyday life, especially in the industrialized parts of the world, unfortunately have come at the expense of ravages to our resource base and the environment. Several human-made chemical disasters in recent history (Table 1.1) tell only part of the environmental story. Every day our atmosphere, water resources and soil are being contaminated with human-made pollutants at levels that are unnoticed, and thus far more environmentally potent in a cumulative sense. We understand fairly well the health hazards associated with the acute overdose of many chemicals, but the same cannot be said about the long-term consequences of chronic exposure to them.

Fortunately, however, environmental awareness also has grown dramatically, especially in the past few years. Several nations around the world are taking the lead in the implementation of new laws that regulate, and in many cases even ban, the use and disposal of hazardous chemicals. Figure 1.1 provides a perspective of how these legislative mandates have grown in an exponential manner in the United States.[1] Concomitantly, people and institutions are becoming increasingly aware that their actions can have not only local but also global environmental consequences; this bodes well for the future.

Table 1.1. Examples of Documented Instances of Environmental Disasters Attributable to Hazardous Chemical Exposure

Comments	Incident(s)	Period
Lead poisoning and the Franklin Expedition	1845	Members of this ill-fated expedition from England to discover the Northwest Passage through the Canadian Arctic are suspected to have died from lead poisoning from the lead solder which was used for the crew's provision storage.
Mercury poisoning in Minamata	1953–1960	About 200 people died from mercury poisoned fish in this Japanese fishing village off Minamata Bay.
Love Canal pollution	1942–1953	Approximately 23,000 tons of chemical wastes were dumped in this canal in Niagara Falls, New York. A health emergency was declared by the State in 1978, and cleanup efforts began.
The Reed Paper controversy	1962–1970	About 10 tons of mercury were lost from the chlor-alkali plant into the Wabigoon–English River system. This plant supplied the chemicals needed to bleach the pulp at the pulp mill in Dryden, Ontario.
The vinyl chloride episode	1971–1974	Exposure to this chemical —an integral component of the plastics (PVC) industry—is now regulated following studies of cause-and-effect relationship between vinyl chloride and human angiosarcoma of the liver.
PCB poisoning in Yusho and Yu-Cheng	1968, 1978	Contamination of rice oil with PCB occurred in Japan and Taiwan in which thousands of people were afflicted with skin problems.
TCDD pollution in Seveso and Times Beach, Missouri	1976	Although no human deaths were attributed to these incidents, entire neighborhoods were evacuated in these communities in Italy and in the United States.
Lekkerkerk	1986	The Rhine River drains a vast basin in four countries (Switzerland, Germany, France and the Netherlands) as it runs from the Alps to the North Sea. The basin is heavily industrialized and the river accumulates and transports to the Netherlands a heavy load of pollutants.
Bhopal explosion	1984	A chemical plant explosion in India released methyl isocyanate causing more than 3,000 deaths and blindness and other permanent injuries to thousands in the adjacent village community.
Castleford explosion	1992	An explosion in a distillation vessel associated with a mononitrotoluene plant killed five people and injured many others in this incident in the United Kingdom.

1.1. Introduction

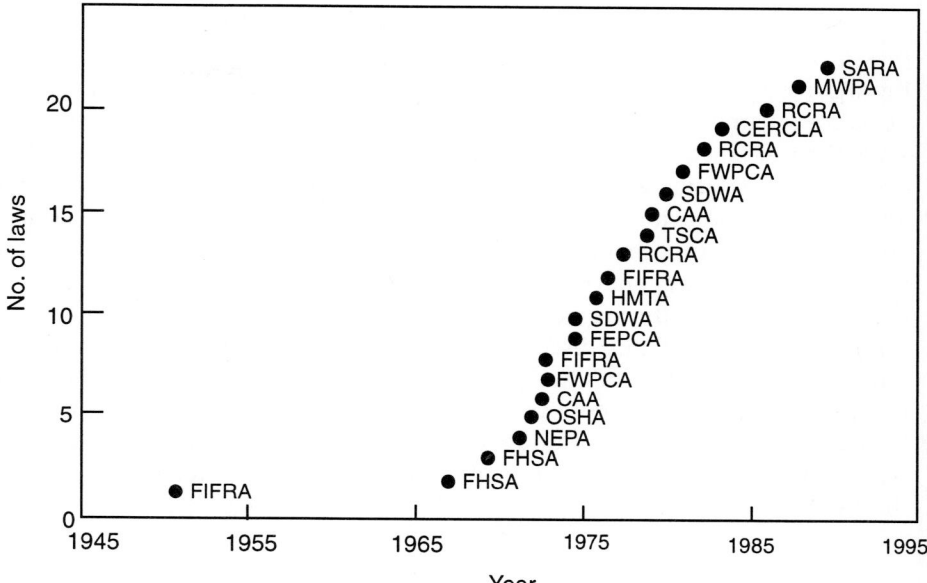

FIGURE 1.1. Growth of legislative mandates for environmental quality control in the United States. (Reproduced with permission from Grasselli.[1])

This chapter begins with a bird's-eye view of pollutants in the environment. A survey of existing technologies for pollutant sensing and treatment is given next. While there can possibly be no argument about the importance of developing new assay techniques, pollution control itself and an extensive discussion of this topic may be considered retrograde. In other words, is it not true that efforts to discover new approaches for controlling pollution is an admission that pollution is a necessary evil of growth? Indeed, it has been stated that the time has come to move away from "end-of-the-pipe" thinking toward pollution reduction (or even prevention) at the source.[2,3] While this is indeed a laudable approach, it is unlikely that we can dispense with the "command-and-control" strategy altogether, especially given the cumulative result of the widespread use of harmful chemicals over the past decades and the growing recognition that pollution is not a local but a global problem. Advocates of the new (pollution prevention) strategy also associate the command-and-control approach with the problem of moving pollution from air to land to water and back. This is undoubtedly true with many of the current pollution control strategies, as we shall see shortly. However, there are many important exceptions to this trend.

The strategy of treating potential pollutants at the discharge end of the pipe also is not incompatible with the concept of chemical recycling. Indeed, the argument could be made that a parallel strategy to the reduction (or abolition) of the use of harmful chemicals (and a search for potentially more costly alternative synthetic routes) could involve efficient recycling of chemicals back into the process. This latter approach is likely to be more universally applicable and, equally important, more economically viable in many cases. Many electrochemical approaches to pollution control are amenable to incorporation of chemical recycling, and this is well exemplified by the Cr(III) → Cr(VI) conversion system discussed later in this book.

In summation of this discussion, all three approaches to environmental remediation—namely, pollution prevention, pollution control, and chemical recycling—are likely to continue to play important roles in the foreseeable future. An increased level of understanding nature's capacity to repair itself from environmental damage is equally important but beyond the scope of this book.

1.2. SOME DEFINITIONS AND CLASSIFICATION OF POLLUTANTS

Environmental science and technology cuts across many disciplines. Thus, the definition of the underlying terminology and the "trade jargon" is a necessary prelude to undertaking an overview of the field.

We begin with the term pollutant. A reasonable definition of a pollutant is a substance present in greater than natural concentration as a result of human activity and that has a net detrimental effect upon its environment or upon something of value in that environment. Interestingly enough, time and place determine what may be called a pollutant. Thus, the phosphate that has to be removed from wastewater in a sewage treatment plant is chemically indistinguishable from the commodity that a nearby farmer buys at high prices as fertilizer! A contaminant causes a deviation from the normal composition of an environment. Contaminants are not normally classified as pollutants unless they have a deleterious effect on the environment. Every pollutant has a source from which it originates. This source can be either natural or anthropogenic (i.e., human-made).

The definition of waste has been problematical.[3] Waste has been defined by the U.S. EPA as any material emanating from a process that is not directly used in another process. Under this definition, many chemical products (even benign ones) could reasonably be considered as wastes and thus subjected to

1.2. Some Definitions and Classification of Pollutants

regulatory restrictions. A further conceptual roadblock to this definition is that it does not make sense to expect an output from a chemical process to be reduced or eliminated, but rather is to be increased in the interest of enhanced industrial productivity! An alternative and perhaps more reasonable definition would be to classify waste as something released into the environment. Hazardous wastes have been classified according to their source or their characteristics.[4] Thus, in the first category, the U.S. EPA has compiled a list of industrial sources that generate hazardous substances (mostly chemicals). As many as 31 lists of chemicals are similarly available in a computerized database.[5] The second category comprises waste materials not listed but that exhibit one or more of five hazardous characteristics: ignitability, toxicity, corrosivity, reactivity, or biological activity.

This brings us to another controversial issue; the definition of toxicity. The standard method for measuring the toxicity of a chemical is to perform bioassays on genetically sensitive animals. Unfortunately, these laboratory tests require relatively high chemical dosage—much higher than levels typical of the environment—to generate adverse effects in the test animals. Predictive models are then used to extrapolate to the lower doses to which people are usually exposed. However, whether it is even meaningful to extrapolate data from test animals to humans is far from settled, as are conclusions drawn from epidemiological studies.

Given all these, it is perhaps not surprising that there is disagreement between scientific experts and public opinion on the seriousness of risks from hazardous waste, relative to other environmental problems. Table 1.2 illustrates this priority gap in the United States.[6,7] Indeed, the controversial Superfund legislation in the United States quietly passed legislation for cleaning up abandoned hazardous waste sites in spite of concerns expressed about the economic viability of (and even the need for) such remedial action.[8]

In Table 1.3 we have attempted a classification of pollutants commonly found in air, water, soil, and biota. We shall discuss them in more detail in a subsequent section of this chapter. It must be noted that the examples listed in Table 1.3 are by no means comprehensive. The four types of environmental media provide the path that the pollutants take from the source to the sink, where they remain for a long time, though not necessarily indefinitely.

We review next the major transport pathways for the pollutants in environmental media, and then discuss the individual categories of the pollutants listed in Table 1.3 later in this chapter. Finally, examples of specific pollutants are provided, along with identification of their sources and their toxic effects.

Chapter 1

Table 1.2. A Comparison of the U.S. Public Environmental Concerns and the Priorities of the U.S. Environmental Protection Agency

The Public's Concerns[a] (ranked)	EPA's 12 Highest Concerns (not ranked)
1. Active hazardous waste sites (67%)	**Ecological risks**
2. Abandoned hazardous waste sites (65%)	Global climate change
3. Water pollution from industrial wastes (63%)	Stratospheric ozone depletion
	Habitat alteration
4. Occupational exposure to toxic chemicals (63%)[b]	Species extinction and biodiversity loss
5. Oil spills (60%)	**Health risks**
6. Destruction of the ozone layer (60%)[b]	Criteria air pollutants (e.g., smog)
7. Nuclear power plant accidents (60%)	Toxic air pollutants (e.g., benzene)
8. Industrial accidents releasing pollutants (58%)	Radon
	Indoor air pollution
9. Radiation from radioactive wastes (58%)	Drinking water contamination
	Occupational exposure to chemicals
10. Air pollution from factories (56%)[b]	Application of pesticides
11. Leaking underground storage tanks (55%)	Stratospheric ozone depletion
12. Coastal water contamination (54%)	
13. Solid waste and litter (53%)	
14. Pesticides risks to farm workers (52%)[b]	
15. Water pollution from agricultural run-off (51%)	
16. Water pollution from sewage plants (50%)	
17. Air pollution from vehicles (50%)[b]	
18. Pesticide residues in foods (49%)	
19. Greenhouse effect (48%)[b]	
20. Drinking water contamination (46%)[b]	
21. Destruction of wetlands (42%)	
22. Acid rain (40%)	
23. Water pollution from city runoff (35%)	
24. Non-hazardous waste sites (31%)	
25. Biotechnology (30%)	
26. Indoor air pollution (22%)	
27. Radiation from X-rays (21%)	
28. Radon in homes (17%)[b]	
29. Radiation from microwave ovens (13%)	

[a] Figures in parentheses represent the percentages of those surveyed who rated the problem as "very serious."
[b] Also appears on EPA's list of highest concerns.
Source: Kunreuther and Patrick[6] and Roberts.[7]

1.2. Some Definitions and Classification of Pollutants

Table 1.3. Classification of Pollutants in the Environment and Some Examples

Organics	Inorganics	Microorganisms
Herbicides	**Metals**	**Bacteria**
Alachlor	Lead	*Escherchia coli*
Butachlor	Cadmium	*Salmonella typhi*
Atrazine	Mercury	*Pseudomonas aeruginosa*
Cyanazine	Copper	*Salmonella enteritis*
Dioxin	Chromium	*Shigella dysenteriae*
		Shigella paradysenteriae
Insecticides	**Metalloids**	*Shigella flexneri*
Chlordane	Arsenic	*Shigella sonnei*
Dieldrin	Selenium	*Staphylococcus aureus*
Heptachlor		*Legionella pneumophilia*
		Vibrio cholerae
Solvents	**Anions**	**Viruses**
Acetone	Chloride	Poliovirus 1
Benzene	Cyanide	Coliphage
Toluene	Bromide	Hepatitis A virus
Ethylbenzene	Nitrate	Rotavirus SA 11
Xylene	Fluoride	
Trichloroethylene	Phosphate	
Chloroform		
Polycyclic Aromatic Hydrocarbons (PAHs)	**Gases**	**Protozoan cysts**
Benzo(a)pyrene	SO_x, NO_x	*Giardia muris*
Cyclopenta(cd)pyrene	Ozone	*Acanthamoeba castellanii*
	Carbon dioxide	*Cryptosporidium*
	Carbon monoxide	
	Ammonia	
Dyes and Surfactants		**Algae**
Other Industrial Organics		
Phenols		
Formaldehyde		
Polychlorinated biphenyls (PCBs)		
Chlorofluorocarbons (CFCs)		

1.3. ENVIRONMENTAL MEDIA AND POLLUTANT TRANSPORT

Traditional control of environmental pollution has focused primarily on the immediate vicinity of the pollution source. For example, tall stacks were once hailed as the panacea to local sulfate emissions from coal-fired power plants. However, acid rain is now recognized as a global environmental problem. Similarly, CO_2 emissions are leading to global climate warming, and chlorofluorocarbon (CFC) releases are resulting in the depletion of the earth's protective ozone layer.[9] Other dramatic examples for the global transport of pollutants include the finding of high lead levels in Greenland snow, and the detection of polychlorinated biphenyls (PCBs) and pesticides in the Arctic.[10] It is clear that multimedia environmental transport models[11] are needed to account for global distribution of pollutants. Concomitantly, pollutant regulation has gradually shifted from local governments to national level and even to an international scale. A recent example of the latter is the 1987 Montreal Protocol on Substances that Deplete the Ozone Layer. This unprecedented international regulatory action requires the signatory nations to severely curtail the production and use of five CFCs. This accord sets two important precedents. It recognizes the atmosphere as a limited, shared resource, and it curtails the right of individual countries to release wastes to the atmosphere. Regulation of other regional and global pollutants such as CO_2, NO_x, and SO_x on an international level, however, appears to be much more difficult.

We shall now discuss each environmental medium in turn.

1.3.1. Air

The atmosphere's composition has changed at a significantly faster pace in the past two centuries than it has at any time in human history. Yet the concentrations of the major constituent gases (N_2, O_2, and the inert gases) have remained nearly constant over a timespan much longer than human life on this planet. On the other hand, trace level constituents such as methane, CO, SO_x, and NO_x and pollutants such as the CFCs have largely undergone increases in their concentrations with rather disastrous environmental consequences including acid rain, photochemical smog, and climatic changes.

A good rule of thumb for understanding pollutant transport is the partitioning rule; that is, chemicals, once they are released into the environment, seek out the environmental medium in which they are most soluble. For example, volatile organics (see Table 1.3) such as benzene and trichloroethylene (TCE) are most soluble in air. Hence, they tend to partition into the vapor phase and inhalation is the principal means of human exposure.

1.3. Environmental Media and Pollutant Transport

Figure 1.2 shows that the fate of emissions in the atmosphere can vary.[12] A gas that is unreactive and insoluble in water (*a*) will spread through the troposphere (the lower 10–15 km from the earth's surface) and in some cases, into the stratosphere (10–50 km above the surface). A small fraction, however, may be assimilated by the soil and water (*b*). On the other hand, soluble gases dissolve in moisture on particulate matter (*c*) or in water droplets (*d*), mainly in clouds. The particles and droplets then carry the gas to the earth directly (*e*) or via rain, snow, fog, or dew (*f*). Most gases are reactive enough to undergo chemical transformation (*g*). The resulting gaseous products can sometimes be deposited dry on the earth (*h*). Most often, these products, being more soluble than their precursors, are also more readily incorporated into wetted particles (*i*) and, directly (*j*) or indirectly (*k*), into water droplets. Thus the gaseous products tend to be removed (*e*, *f*) quickly, and are much less likely than their predecessors to diffuse above the troposphere.

The oxidants present in the atmosphere such as ozone act as natural cleansing agents because they transform gases into water-soluble products, according to the aforementioned scheme. Another important "atmospheric detergent" is the hydroxyl radical, •OH, about which we will have a lot more to say later in this book. Chlorine atoms from CFCs play a central role in one of the most efficient ozone-destroying catalytic cycles in the stratosphere.[12]

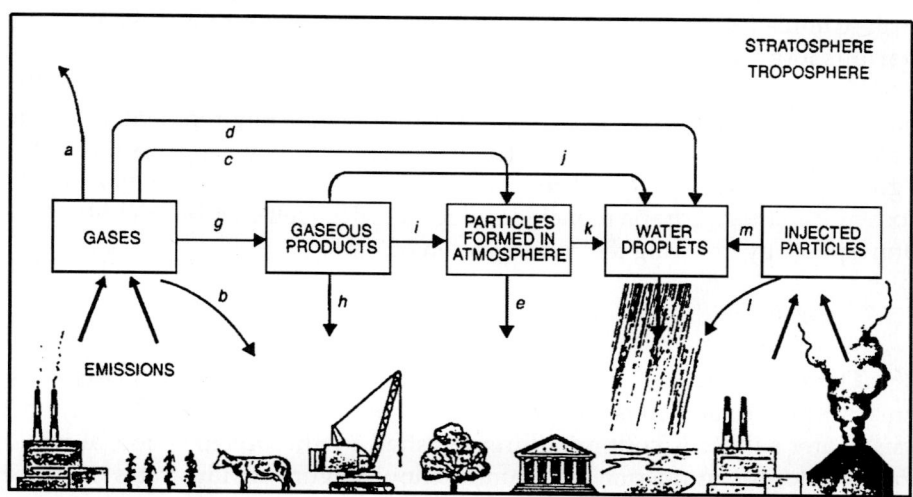

FIGURE 1.2. Fate of emissions in the atmosphere. (Reproduced with permission from Graedel and Grutzen.[12])

The protective ozone layer thus shields the earth's surface from the intense ultraviolet radiation via the photochemical Reaction (1.1).

$$Cl^{\bullet} + O_3 \rightarrow ClO^{\bullet} + O_2$$

$$ClO^{\bullet} + O^{\bullet} \rightarrow Cl^{\bullet} + O_2$$

$$O_2 \xrightarrow{h\upsilon} O^{\bullet} + O^{\bullet}$$

$$O^{\bullet} + O_2 \rightarrow O_3$$

$$O_3 \xrightarrow{h\upsilon} O^{\bullet} + O_2 \tag{1.1}$$

In general, the drastic increases in the atmospheric levels of other gases such as methane, CO and CO_2, NO_x, and SO_x in the 20th century must be attributed to the cumulative effects of fossil-fuel combustion, automobile exhaust and various forms and consequences of human activity such as farming, deforestation and biomass burning in tropical forests, microbial activity in municipal landfills, and so forth. The adverse effects of these gases on the environment appear to be rather well understood. Thus CO_2 and methane play a central role in the "greenhouse effect"; SO_x and NO_x are implicated in acid rain and NO_x in photochemical smog. Carbon monoxide undermines the self-cleansing ability of the atmosphere by lowering the concentration of $^{\bullet}OH$.

1.3.2. Water

Of the total store of water on this planet earth, the fresh water in the lakes, creeks, streams, and rivers amounts to only about 0.01 %.[13] Fortunately, this freshwater supply is continually replenished by the precipitation of water vapor from the atmosphere as rain or snow. Figure 1.3 illustrates the three major pathways in the global water cycle: precipitation, evaporation, and vapor transport. The precipitation route constitutes the major fraction (~385,000 cubic kilometers per year[13]) and this water falls into the oceans. Some flows from the land to the sea as runoff or groundwater. The reverse flow from the earth to the atmosphere is sustained by evaporation or transpiration.

1.3. Environmental Media and Pollutant Transport

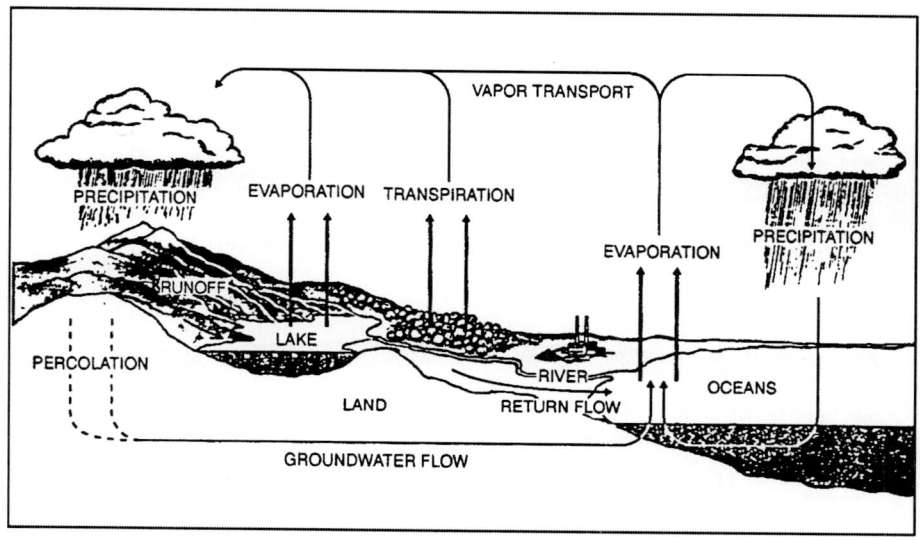

FIGURE 1.3 The global water cycle. (Reproduced with permission from Maurits la Riviere.[13])

Contamination of the water supply can occur along any or all of the three pathways illustrated in Figure 1.3 as well as during groundwater transport.

Surface water also is contaminated by traditional organic wastes (human and animal excreta, etc.), and from dumping of industrial process wastes. Wastes can enter lakes and streams in discharges from *point* sources or from *nonpoint* (diffuse) sources, as in the case of pesticides and fertilizers in runoff water. Groundwater is contaminated by pollutants leaching through the soil; this pollution front is further transported by advection and molecular diffusion. Normally, biodegradation is a natural self-cleansing process in nature. The extent to which hazardous wastes can be thus biodegraded depends on their half-lives. In other words, a hazardous substance that can be broken down easily to a nontoxic form by organisms poses less risk than does one that is difficult to destroy. Groundwater pollution tends to be much less reversible than the pollution of rivers and lakes. This is because most self-cleansing microbes are aerobic — they need dioxygen to do their job. Because groundwater is cut off from the atmosphere's dioxygen supply, its capacity for self-purification is very low. In general, metal drums containing hazardous chemicals are nothing less than time bombs that will go off when they rust through. The incidents at Lekkerkerk in the Netherlands and at Love Canal in the United States (Table 1.1) are cases in point. Unfortunately, little is known about the overall quality of the earth's vast groundwater reserves, except in those instances where particular aquifers are being actively exploited. Figure 1.4 contains a schematic of the movement of hazardous substances in groundwater.[6]

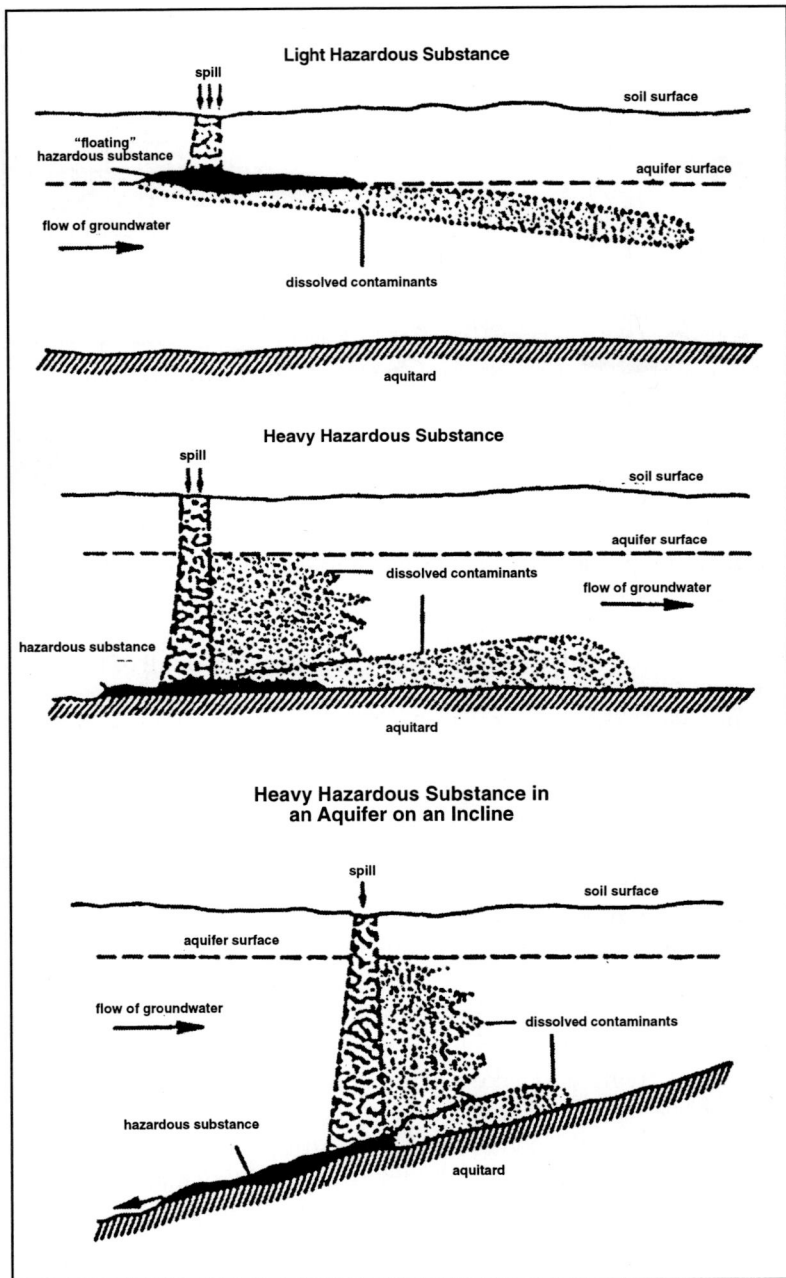

FIGURE 1.4. Movement of hazardous substances in groundwater. (Reproduced with permission from Cherry. [72])

1.3. Environmental Media and Pollutant Transport

Some pollutants enter the water cycle by way of the atmosphere. Best-known examples of this transport route include the acid generation from NO_x and SO_x emissions. An indirect effect of acid deposition is positive ion leaching out of the soil, that has been implicated in several cases of fish-kill in rivers and lakes. Ammonia from the atmosphere can also enter the groundwater through the soil. Microbial action converts it into soluble nitrates. Another deleterious effect of human activity is the introduction of large amounts of nutrients, especially phosphates, into the surface water reservoirs. This leads to excessive growth of algae. When these die, their microbial degradation consumes most of the dissolved dioxygen in the water, vastly reducing the water's capacity to sustain life. This process is called *eutrophication*. Water salinity is also increased, with disastrous agricultural consequences, as a result of a number of human activities. Table 1.4 contains a listing of some of the primary constituents of sewage from a city sewage system and their role in water pollution.

Two important measures of the water quality are the *biological oxygen demand,* or BOD, and *chemical oxygen demand,* or COD. The former measures the degree of microbially mediated oxygen consumption by contaminants in water. Although BOD is a reasonably realistic measure of water quality, it is a time-consuming and cumbersome test to perform. The COD of a water sample is more easily determined by the chemical oxidation of an organic material in water by dichromate ion in 50% H_2SO_4; the amount of unreacted dichromate is titrimetrically determined. For a given water sample, the COD and BOD values may differ appreciably especially when poorly biodegradable compounds are present. Yet another parameter, the *total organic carbon* or TOC is gaining in popularity because it is easily measured instrumentally. In this test, the carbon is catalytically oxidized and the evolved CO_2 is measured.

1.3.3. Soil

Soil receives large quantities of waste products each year. Much of the SO_x and NO_x emissions end up in the soil as sulfates and nitrates, respectively. Elevated levels of heavy metals (e.g., lead) are found in the soil near industrial and mining facilities. The soil receives enormous quantities of pesticides as an inevitable result of their application to crops. The degradation and eventual fate of these chemicals in soil largely determine their ultimate environmental impact. For example, humic substances in soils have a strong affinity for organic compounds with low water solubility such as atrazine, a widely used herbicide

Table 1.4. Major Constituents of Sewage from a City Sewage System

Constituent	Sources	Effects in Water
Oxygen-demanding substances	Mostly organic materials, particularly human feces	Consume dissolved dioxygen
Refractory organics	Industrial wastes, household products	Toxic to aquatic life
Viruses	Human wastes	Cause disease (possibly cancer); major deterrent to sewage recycle through water systems
Detergents	Household detergents	Esthetics; toxic to aquatic life
Phosphates	Detergents	Algal nutritients
Grease and oil	Cooking, food processing and industrial wastes	Esthetics; harmful to some aquatic life
Salts	Human wastes, water softeners, industrial wastes	Increase water salinity
Heavy metals	Industrial wastes, chemical laboratories	Toxicity
Chelating agents	Some detergents	Heavy metal ion transport, industrial wastes
Solids	All sources	Esthetics; harmful to aquatic life

Source: S. E. Manahan, "Environmental Chemistry." 3rd ed., Willard Grant Press, Boston, 1979.

(Table 1.5). Soil humic substances may contain levels of uranium more than 10^4 times that of the water with which they are in equilibrium. Further examples of the environmental partitioning effect were discussed earlier. The three primary ways in which chemicals (including pesticides) are decomposed in soil are biodegradation, chemical degradation, and in surface soil scenarios, photochemical reactions. Microbial degradation, however, remains the predominant decomposition pathway.

Table 1.5. Organic Chemicals Regulated by the U.S. EPA Drinking Water Standards

Pesticides and Herbicides
  Alachlor
  Atrazine
  Chlordane
  Dibromochloropropane (DBCP)
  Ethylene dibromide
  Heptachlor
  Heptachlor epoxide
  Lindane
  Pentachlorophenol
  Simazene
  Toxaphene

Chlorinated Solvents and Related Chemicals
  Carbon tetrachloride
  1,2-dichloroethane
  1,1-dichloroethylene
  Dichloromethane
  Tetrachloroethylene
  1,1,2-trichloroethane
  Trichloroethylene
  Vinyl chloride

Disinfection By-Products (DBPs)
(Note: the sum of all DBPs has a MCLG of 0.1 mg/liter)
  Chloroform
  Bromodichloromethane
  Dibromochloromethane
  Bromoform

Aromatic Hydrocarbons
  Benzene
  Benzo(a)pyrene

Other Chlorinated Synthetic Organic Compounds
  p-Dichlorobenzene
  Hexachlorobenzene
  Polychlorinated biphenyls (PCBs)
  Dioxin (1,2,7,8-TCDD)
  Epichlorohydrin
  Hexachlorocyclopentadiene

Other Nonchlorinated Synthetic Organic Compounds
  Acrylamide

Landfills have become the focal point for media scrutiny in recent years. Most of the hazardous materials that have been identified in landfills have been mixed in with many different waste substances as well as with the soil. The biodegradability of chemicals in landfills remains an issue of crucial concern to the experts and the public alike.

The soil constitutes an integral part of the human exposure pathway whereby the pollutant ultimately enters the food chain. This is because most lipophilic compounds reside mainly in the soil. The transfer chain of events is schematized in Figure 1.5. The biota obviously play an important role in mediating the transfer of pollutants from the environment to the humans; the role of this environmental medium is briefly considered next.

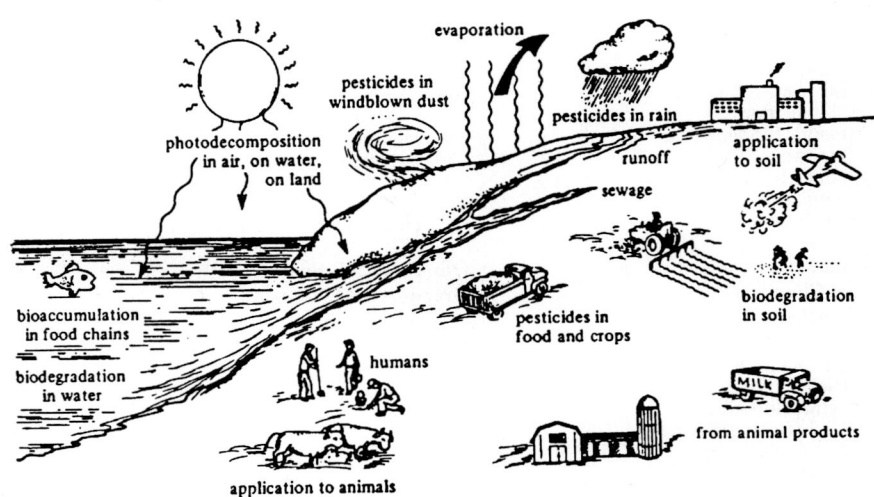

FIGURE 1.5. Pollutant cycles in the environment. (Reproduced with permission from S.E. Manahan, "Environmental Chemistry." 3rd ed., Willard Grant Press, Boston, 1979.)

1.3.4. Biota

Chemicals such as DDT and PCBs (see Table 1.3) are insoluble in water and tend to partition into organic matter. Thus, they bioaccumulate in vegetation, beef, milk, and fish (Figure 1.6), and the food chain becomes an important pathway of human exposure for most global pollutants. Atmospheric pollutants enter the food chain through three routes: direct deposition onto leaf surfaces, vapor phase air-to-leaf transfer, and root uptake. For many years, vegetative contamination was thought to result primarily from root uptake. However, the importance of the other two assimilative mechanisms is now recognized.[10] Herbivores (e.g., cattle) take up these pollutants by feeding on the contaminated vegetation. Ultimately, the pollutants are assimilated into the predators that feed on these herbivores and the cycle is completed.

1.3. Environmental Media and Pollutant Transport

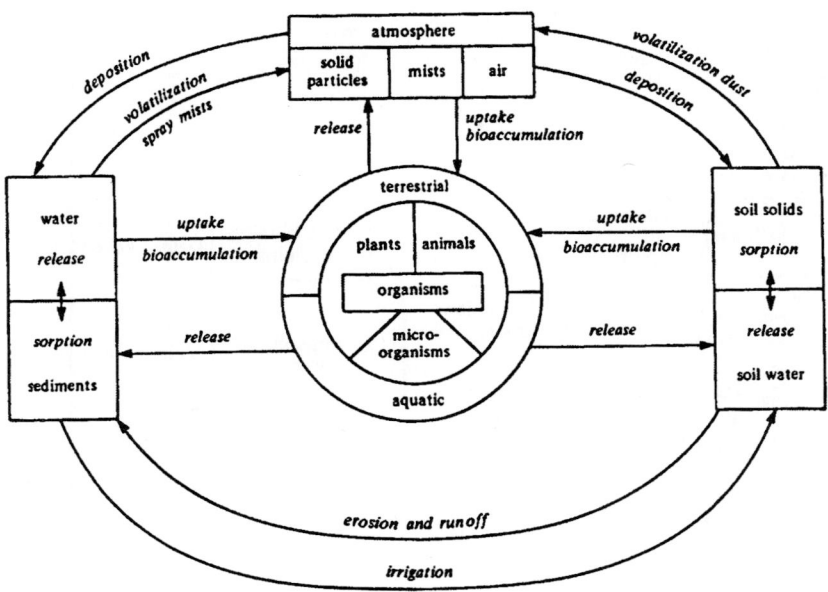

FIGURE 1.6. Multimedia transport of pesticides in the environment. (Reproduced with permission from S.E. Manahan, "Environmental Chemistry." 3rd ed., Willard Grant Press, Boston, 1979.)

Much research has been done in recent years on mechanisms of pollutant transport across biological membranes. It is now believed that there are three major transport routes (Figure 1.7): diffusion, carrier mediated pathways involving interactions with membrane components, and migration driven by the potential gradient and high electric fields developed across the membrane.[14] Lipophilic species undergo facile transport via diffusion; and methyl mercury, for example, appears to be bioassimilated in this manner. The varying response of biota to metal speciation is also rationalizable on the basis of transport mechanisms.[14] Therefore, orders-of-magnitude differences have been observed in fish for the bioaccumulation of lipophilic metal species vis-à-vis the ionic metal counterparts.

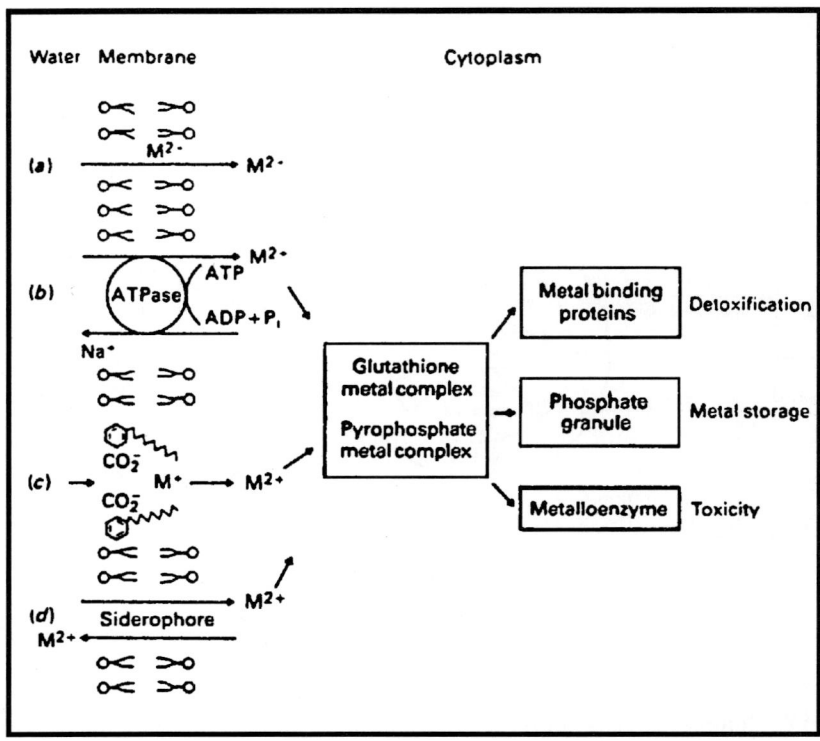

FIGURE 1.7. Metal uptake, transport, and cell interactions in a biological cell membrane. (Reproduced with permission from Morrison, et al.[14])

In summarizing the discussion in this section, it is important to reiterate that not only the environmental media in themselves are important in global pollution but also the *interchange* of pollutants among them. We hope it is clear from the preceding discussion that it is no longer valid to treat pollution (even from a point source) as a localized problem.

1.4. ENVIRONMENTAL CHEMISTRY AND TOXICOLOGY OF COMMON POLLUTANTS: A PRIMER

We now return to the three classes of pollutants in Table 1.3 and discuss sample candidates within each.

1.4. Environmental Chemistry and Toxicology of Common Pollutants

1.4.1. Organics

Pesticides. The introduction of the chlorinated hydrocarbon DDT during World War II marked the beginning of a period of very rapid growth in pesticide use, especially in Western countries. Pesticides include insecticides for insect control, herbicides for weed control, and fungicides for fungi control. Some 22 families of water-soluble pesticides have been encountered in the environment, constituting two major categories: chlorinated hydrocarbons and organic phosphates, the latter group being generally more biodegradable. A third category, carbamate pesticides which are derivatives of carbamic acid, are even more biodegradable; and their use has increased sharply in recent years.

The best-known pesticide is perhaps DDT. Although its toxicity is generally low, its persistence and accumulation in the food chain have led to a ban on DDT in the United States, although its use is continuing in other countries. Methoxyclor is a popular DDT substitute that is reasonably biodegradable and has a relatively low toxicity to mammals. Another herbicide, 2,4,5-T (2,4,5–trichlorophenoxyacetic acid and its salts and esters) was employed in very large quantities as the chemical defoliant "Agent Orange" by the U.S. Army in the Vietnam War. Interestingly enough, 2,4,5-T was initially suspected of causing birth defects, although later research showed the actual toxin to be a trace constituent, namely 2,3,7,8-tetrachlorodibenzo-p-dioxin (TCDD). Dioxins including TCDD and other forms of polychlorinated dibenzo-p-dioxins (PCDDs) and polychlorinated dibenzofurans (PCDFs) have captured widespread media attention.[15] Dioxins have been detected in practically all types of environmental media even though their lipophilicity would render their partitioning mainly in soil. However, a small fraction volatilizes and is transported globally.

The vapor phase, after uptake by vegetation, is primarily responsible for human exposure to the highly toxic TCDD. Subsequent air-to-vegetation transfer of atmospheric TCDD results in its entry into the food chain, especially meat and dairy products, and accounts for 99% of human exposure to TCDD.[16]

The available data are inconclusive as to the extent of toxicity of TCDD, especially in humans. Laboratory animals, however, manifest a wide variety of symptoms ranging from teratogenicity (birth defects) and skin lesions to porphyria. The severity of skin lesions (exemplified by chloracne) appears to correlate with TCDD levels in blood samples of human subjects, and occupational exposure to TCDD appears to be linked with increased risk of soft-tissue cancers.

Organic Solvents. The BTEX family (benzene, toluene, ethylbenzene, xylene) are of concern in wastewater, and particularly in sources of drinking water, because they are generally nonbiodegradable (biorefractory). The use of benzene as a solvent has been banned in the United States because of its suspected role in the incidence of leukemia in humans. Whether benzene is a direct causal agent for leukemia, however, is a controversial issue.[6] Benzene partitions mainly into the air (99%) with less than 1% partitioning into water, soil, sediment, and biota.[10] Because benzene is not very lipophilic, it does not accumulate greatly in the food chain. Inhalation is the primary exposure pathway for benzene and for most of the other organic solvents.

Polycyclic Aromatic Hydrocarbons. Polycyclic aromatic hydrocarbons (PAHs) have attracted a good deal of attention because of the known carcinogenicity of some of these compounds. These compounds are most likely to be found in urban atmospheres and appear to be transported via adsorption on particulate matter (e.g., soot). Coal furnace emissions constitute a particularly fertile source for these pollutants. Interestingly enough, these compounds undergo secondary chemical reactions (including photochemical reactions) in the atmosphere. For example, perylene is neither mutagenic nor carcinogenic. Under conditions simulating smog formation, however, perylene forms 3-nitroperylene, which is a known mutagen.[17]

As with the preceding group of pollutants, inhalation is the primary exposure pathway for PAHs.

Surfactants. Until the early 1960s, the most common synthetic detergent used was an alkyl aryl sulfonate, ABS. However, ABS is only very slowly biodegradable because of its branched-chain structure. Consequently, ABS has been replaced by α-dodecane benzenesulfonate. This surfactant is more easily biodegradable because its alkyl portion is not branched. It also does not contain the tertiary carbon (unlike ABS), which is detrimental to biodegradability. Most of the *current* environmental concerns associated with detergents, however, are not related to the surfactant portion, which constitutes only 10–30% of the total formulation. Most detergents contain polyphosphates that are added to complex calcium (a water hardness ion), to improve the detergent action. We shall return to a discussion of these inorganic anions later in the book.

Industrial Organics. Chlorinated phenols and polychlorinated biphenyls (PCBs) are two important categories of industrial organic pollutants. Pentachlorophenol (PCP) accounts for almost half of the total world production of the first group of compounds. It is used largely as a wood preservative. Volatilization of PCP into the atmosphere is an important transport route for this pollutant. Formulations of PCP are rarely pure and usually contain other congeners, which are often more toxic. In rats, purified PCP has been found to have a LD_{50}^* of ~150 mg/kg.

---

*LD_{50} is a parameter commonly used in toxicity studies and stands for "lethal dose, 50% level."

1.4. Environmental Chemistry and Toxicology of Common Pollutants

Chlorinated hydrocarbons were discussed earlier in connection with pesticide formulations. A related class of compounds is the PCBs, of which some 209 congeners are now known. These were primarily used as dielectric fluids in power transformers and capacitors because of their lack of flammability, coupled with excellent thermal and electrical insulation characteristics. Unfortunately, a lack of flammability translates to antioxidant behavior, which explains why the PCBs are so persistent when released into the environment. While PCB use, at least in North America, is largely curtailed and replaced by other alternatives (e.g., silicone fluids), disposal of PCBs from discarded electrical equipment continues to be a vexing environmental problem.

Polychlorinated biphenyls are highly lipophilic and thus human exposure to them occurs primarily via the food chain. Volatilization of PCBs from spills, landfills, road oils, and the like also forms another important transport pathway via atmospheric emissions. Indeed, atmospheric transport is now recognized as a primary mode of global PCB pollution.[10] The toxicity of PCBs continues to elicit much controversy.[18] As with the other chlorinated compounds discussed earlier, it is not clear to what extent the trace contaminants, invariably present in industrial formulations, are the real culprits.

Table 1.5 contains a list of regulated organic chemicals in drinking water with a maximum "contaminant" level goal (MCLG) equal to or less than 0.01 mg/liter (0.01 ppm). Table 1.6 contains a list of the hazardous organic substances regulated by the U.S. Clean Water Act.

Organic pollutants in general are classifiable as follows:

(1) Suspended solids: These can develop into sludge deposits and consequently generate anaerobic conditions. In wastewater of medium strength (i.e., with a typical total organic carbon, TOC, value of 160 mg/liter), some 75% of the suspended solids are organic in nature. Treatment methods include: screening and comminution (crushing), grit removal, sedimentation, filtration, flotation, polymer addition, coagulation/sedimentation, and landfill.

(2) Biodegradable organics: These include proteins (soluble and insoluble), carbohydrates (e.g., sugars, starches, cellulose, wood fibers), fats (which are normally less biodegradable due to their high stability), and surfactants. Their presence may lead to depletion of dioxygen from water with the resulting development of septic conditions and interference with biological action; some of these compounds can also cause maintenance problems as well as negative aesthetic impact (e.g., oils and dyes). Surfactants made earlier (ca. in the 1950s and 1960s) were not biodegradable and caused major foaming problems. Treatment methods include activated sludge, fixed-film reactor, lagoon treatment, sand filtration, physical–chemical systems, and natural systems.

(3) High-priority pollutants: These are selected based on their known or suspected behavior as carcinogens, mutagens, teratogens, or highly acute toxins. Typical examples include benzene, ethylbenzene, toluene, chlorobenzene, chloroethene, dichloromethane, tetrachloroethene as well as many pesticides, herbicides and insecticides.

(4) Refractory organics: These compounds are rather resistant to biodegradation. Treatment methods include carbon adsorption and ozonation.

(5) Volatile organic compounds (VOCs): These include organic compounds with boiling points below that of water or vapor pressures greater than 1 mm Hg at 25°C or both. When transferred to the vapor phase, their mobility increases greatly and can lead to increased health risks or the formation of photochemical oxidants. Treatment methods include: air stripping, off-gas treatment and carbon adsorption.

(6) Malodorous compounds: Some organic compounds, volatile or otherwise, may lead to a foul smell. The effect of odors often has stronger psychological and social consequences than physiological ones. Some typical malodorous organic compounds include amines, diamines, indole (C_8H_7N), mercaptans, organic sulfides, and skatole (C_9H_9N). Their odor detection thresholds can be as low as 1 ppb by volume (e.g., some mercaptans).

Table 1.6. Hazardous Organic Substances as Regulated by the U.S. Clean Water Act (Section 311)

Acetone	Dicofol	Methoxychlor
Cyanohydrin	Dieldrin	Mevinphos
Acrolein	Dinitrophenol	Neled
Aldrin	Dinitrotoluene	Parathion
n-butyl phthalate	Disulfoton	Pentachlorophenol
Captan	Endosulfan	Phosgene
Carbofuran	Endrin	Polychlorinated biphenyls (PCBs)
Carbon tetrachloride	Ethion	Propargite
Chlordane	Ethylene dibromide	Pyrethans
Chloroform	Guthion	Strychnine
Chloropyrifos	Heptachlor	Tetraethyl pyrophosphate
DDT	Hexachlorocyclopentadiene	Toxaphene
Diazinon	Kepone	Trichlorophenol
Dichlone	Lindane	
Dichlorvos	Mercaptodime	

1.4.2. Inorganics

With reference to Table 1.3, we can again group inorganic pollutants into several categories including the following.

Metals. The toxicity of metals in aqueous environments depends on their physicochemical form (i.e., speciation).[14] For example, copper ions are very toxic while copper bound to natural organic matter is less harmful. Table 1.7 lists possible forms of trace metals in polluted waters; both dissolved species and particulate matter can be present.[14] Of the metal species listed in this tabulation, two groups contribute to toxicity: simple ionic and lipid soluble (lipophilic) forms. Clearly, assays of *total* metal levels in a medium are inadequate from an environmental perspective; *speciation* measurements are needed to identify those metal species that have adverse effects on biota and human beings. This is because the interaction of metals with intracellular components is highly dependent on chemical form. We shall further consider speciation in Chapter 4.

The addition of metal residues in substantial amounts to the atmosphere, hydrosphere and lithosphere from anthropogenic sources dates back to ancient times. For example, lead production began some six millennia ago and lead concentrations four times higher than natural values have been analyzed to be present in Greenland ice from about 2,500 to 1,700 years ago.[18a] In modern times, several billions of liters of metal-containing wastes are produced every year in the United States alone.

Table 1.7. Inorganic Pollutant Forms

Form	Examples
Simple ions	$Cd(H_2O)_6^{2+}$
Complex ions	$CdCl^+$
Differing oxidation states	Cr(III), Cr(VI)
Lipid soluble complexes	CH_3HgCl
Organometallic complexes	$CH_3AsO(OH)_2$
Adsorbed species	Cu^{2+}/humic acid/Fe_2O_3
Particulates	Metals adsorbed onto or intercalated within clay particles

Source: Morrison, *et al.*[14]

The main sources today are industrial wastes from processes such as electroplating, nuclear fuel processing, photography, batteries, catalyst production and recovery, and other metal extraction, production, treatment, etching, cleaning, finishing, recovery, or refining procedures. Combustion engines contribute to lead emissions from leaded gasolines in those countries that still produce and use this type of fuel. Corrosion processes generate dangerous metal compounds (e.g., cadmium, chromate) from apparently innocuous materials such as galvanized iron and stainless steel. Other anthropogenic sources of toxic metals include paints, automobile and domestic metal wastes, and the like. Metal ion concentrations in effluents from chemical process industries can be as high as 10^5 ppm, whereas maximum permissible concentrations are typically in the range 0.05–5 ppm.[18b]

The effects of metals and their compounds on humans, animals, and plants are quite varied. Whereas some are essential nutrients, others are very toxic due to their interactions with life processes (such as enzymatic activity). Especially problematic are those capable of bioaccumulation; for example, concentrations ranging from ~1 to 78 mg of Pb or Cd per kg of dry organ weight, were found in the kidneys, lungs, livers, and adrenal glands, extracted from over 1,000 human autopsies in Hungary some years ago.[18c]

Some elements play both roles; that is, they are essential for life processes in small amounts but toxic in larger amounts or in other organisms. For example, zinc is required in humans for protein synthesis; its deficiency can cause anemia, hyperpigmentation, sexual dysfunction, or other problems. However, ingestion of zinc salt solutions also can cause nausea, vomiting, and purging. Likewise, copper is known to be a trace element fundamental for the activity of several enzymes and plays a key role in the catecholamine metabolism of the human brain. However, in concentrations as low as 10^{-6}–10^{-8} M, it induces high mortality rates in aquatic organisms such as *daphnia* and carp.[18c,d] In fact, permissible levels for Cu and Zn in drinking water have been set by the U.S. EPA at 1 and 5 mg/liter, respectively.

Metal ion intake is not only a health concern, but an economic one as well. For example, in the United States, an employer must remove an employee from exposure to lead every time the employee shows blood lead concencentrations above regulatory levels, and under certain circumstances the worker may be granted up to 18 months of benefits.[18e]

Human metal intake may occur primarily from contaminated foods, substances, or objects introduced in the mouth (e.g., toys, dust), drinking water, skin, and lung absorption; a high correlation has been observed worldwide between blood lead levels and household lead-containing dusts. Also, the World

1.4. Environmental Chemistry and Toxicology of Common Pollutants

Health Organization recommends a maximum concentration of this metal in drinking water of 0.05 mg/liter.[18f]

In addition to health and economic concerns, the presence of metal ions in solution can bring about side effects such as bad taste, increase in turbidity and coloration, or staining of fixtures in contact with water. Iron and manganese ions may also promote bacterial growth with its concomitant drawbacks, like foul odor. Some of these problems can be at least temporarily eliminated upon *sequestration* (i.e., selective solubilization) by complex formation. Of particular importance is the pollution of soils, since residence times of metal ions in these media are much longer than those in the gas or liquid phases. Soil restoration is considered in more detail in Chapter 5.

Application of the following point-source procedures has resulted in dramatic decreases in heavy metal ion emissions to water receiving bodies, like the Rhine River[18g]:

(1) Installation of water pollution control equipment
(2) Adoption of accepted waste management policies
(3) Increase in recycling of materials
(4) Replacement of old technologies with more environmental friendly ones (where possible, with a zero-emission design)

Landfill disposal of metal-bearing wastes has been commonly used in the past; however, increasing costs and decreasing site availability have prompted attention to alternative technologies. In addition, this method is not a ultimate solution since leachates are known to be produced in many (leaking) landfills, and a trend in legislation towards a cradle-to-after-grave responsibility will very likely be widely adopted.

The important metallic pollutants are listed in Table 1.3; next we shall discuss three of these candidates; namely, lead, mercury, and chromium.

Lead is a naturally occurring element that has no recognized essential biological function; yet it is toxic to almost all organisms. Recognition of its toxicity appears to date back to Hippocrates, and the clinical effects have been recognized for centuries (see also Table 1.1). Table 1.8 lists possible sources of lead exposure in everyday life.[19] The major routes of human exposure to lead are through ingestion and inhalation. Lead exposure is particularly toxic for children and the young of other species.[19] It is disturbing to note that the use of leaded gasoline continues even now in many parts of the world. A concerted effort is necessary to curtail such anthropogenic sources of global lead pollution. Adverse health effects of lead include impaired cognitive development, gastrointestinal distress, ataxia, and in extreme poisoning, even coma and death.

Table 1.8.	Sources of Lead Exposure

Occupational
    Plumbing, pipe fitting
    Lead and other mining
    Auto mechanics
    Lead smelting
    Battery manufacturers
    Gas stations
    Painting

Environmental
    Lead paint
    Soil or dust near lead industries
    Leaded gasoline

Hobbies
    Target shooting
    Lead soldering
    Car repair
    Home remodeling

Source: Taylor.[19]

Mercury is a highly volatile toxic metal that has generated the greatest concern among the class of heavy metal pollutants. The primary anthropogenic sources of mercury are the burning of fossil fuels, mining and extraction of mercury from cinnabar, the chloralkali industries, and, to a lesser extent, volatilization from paper pulp, paints, fungicides, electrical equipment, laboratory wastes, and municipal waste incineration. Evidence for global pollution from this metal is surfacing with increasing regularity. For example, large-scale mercury contamination of fish has been reported in lakes remote from anthropogenic sources. Practical examples for the high toxicity of mercury may be found in two instances: the Minamata Bay area of Japan during the period 1953–1960 and the Wabigoon–English River contamination near Dryden, Ontario during the period 1962–1970 (Table 1.1).

As with lead, both inorganic and organic forms of mercury are known. While mercury in metallic form is widely used in many applications (e.g., electrodes, thermometers), organic mercury compounds find wide application as fungicides. Alkyl mercury compounds tend to resist degradation and pose more of an environmental threat than either the aryl or inorganic compounds. Methylation of mercury also provides a facile pathway for bioaccumulation.

1.4. Environmental Chemistry and Toxicology of Common Pollutants

Toxicological effects of mercury include neurological damage, paralysis, blindness, and birth defects.[20] Milder symptoms of mercury poisoning include depression and irritability.

Chromium is a suspected carcinogen, particularly in the Cr(VI) state.[21] Anthropogenic sources of chromium include effluent discharge from cooling towers, electroplating, tanning, aerospace, electronics and photography industries, oxidative dyeing wastes, leaching from sanitary landfills, and leaks from timber treatment sites. Hexavalent chromium is notoriously mobile in nature, contributing to global pollution from this element.[22] On the other hand, Cr(III) is readily precipitated or sorbed onto a variety of substrates (including soil) at near-neutral pH.[23] Interestingly enough, trivalent chromium is an essential element to mammals as a regulator of glucose, lipid, and protein metabolism.

Table 1.9 lists permissible levels of lead, mercury, and chromium in drinking water. It must be borne in mind that these are only guidelines, as determined, for example, by the public health authorities. Many of the estimates in terms of environmentally acceptable levels continue to be updated as new knowledge evolves, and this is particularly true for elements such as arsenic, which is discussed next.

<u>Metalloids</u>. Arsenic occurs in the earth's crust at an average level of 2–5 ppm. The primary anthropogenic source of arsenic is fossil fuel (primarily coal) combustion. Mining provides for a secondary source, especially as a by-product of copper, gold, and lead refining.

Table 1.9. Drinking Water Standards in the United States

Species	Maximum Acceptable Level, ppm
Copper	1.0
Chromium	0.05
Cadmium	0.005
Lead	0.003
Mercury	0.001
Selenium	0.01
Arsenic	0.05[a]
Cyanide	0.2
Nitrate	45.0
Uranium	0.02
Fluoride	1.5

[a]Expected to be substantially lowered in the near future.

Table 1.10. Chemical Forms of Arsenic in Water Samples

Form	Chemical Formula
Arsenate [As(V)]	$H_2AsO_4^-$
Arsenite [As(III)]	$As(OH)_3$
Arsine	AsH_3
Monomethylarsenate (MMA)	$CH_3AsO_2OH^-$
Dimethylarsenate	$(CH_3)_2AsOO^-$
Dimethylarsine (DMA)	$(CH_3)_2AsH$
Trimethylarsine (TMA)	$(CH_3)_3As$
Trimethylarsine oxide (TMAO)	$(CH_3)_3AsO$
Arsenobetaine (AsBet)	$(CH_3)_3As^+CH_2COOH$
Arsenocholine (AsChol)	$(CH_3)_3As^+(CH_2)_2OH$

As with lead and mercury, discussed earlier, dissolved arsenic can occur in natural waters in both inorganic and organic forms. Table 1.10 contains a list of the chemical forms of arsenic observed in water samples and in biota.[24] The location of arsenic in the Periodic Table directly below phosphorus implies similar chemical behavior, and indeed arsenic and phosphorus coexist in natural sources. It also interferes with numerous biological mechanisms (e.g., phosphorylation) normally dependent on phosphorus.[25]

There is evidence that organisms have developed methylation mechanisms to isolate and detoxify arsenic as organoarsenicals,[25] and in this respect, arsenic is similar to mercury in its bioaccumulation mechanism. Additionally, the incorporation of arsenic into arsonium zwitterions such as arsenobetaine and arsenocholine (see Table 1.10) may serve the dual purposes of detoxification and osmoregulation analogous to some sulfur compounds.[26] The toxic action of arsenic resembles that of lead and mercury. Biochemical effects include protein coagulation, enzyme inhibition, and uncoupling of phosphorylation. Acute poisoning results from the ingestion of ~100 mg of the element, and much lower levels cause chronic poisoning. There is some evidence that arsenic is also carcinogenic.[25]

Selenium is reported to be the 69th most common element.[27] It is found in the earth's crust at levels of approximately 6 ppm. Major markets for selenium are the glass and photoreceptor industries. It is also used in various optoelectronic devices and for X-ray medical imaging. Selenium is an essential nutrient at trace levels, and in fact is added in trace amounts to chemical fertilizers, animal feeds, and various veterinary preparations. The beneficial role is reported to involve a synergistic relationship with vitamin E. Deficiency symptoms include muscle disease, embryonic mortality, *heptosis dietetica* in weaning pigs, and kwashiorkor in children. Levels of selenium in animal feeds higher than 5 ppm have been reported to result in chronic selenosis, which manifests as blind staggers or, in acute cases, death. The carcinogenicity of selenium at high levels remains controversial.[28]

1.4. Environmental Chemistry and Toxicology of Common Pollutants

Table 1.9 contains drinking water standards for arsenic and selenium.

<u>Anions</u>. The role of soil nutrients such as phosphates in eutrophication was mentioned earlier. However, it has been extremely difficult to find alternative detergent builders with performance qualities that match those of polyphosphates. Paradoxically enough, the present method of phosphate removal by precipitation may have the important benefit of also removing other undesirable impurities from the water stream via coprecipitation. This in itself may prove to be a cogent argument for continuing phosphate use in detergent formulations.

Nitrate in drinking water has been a source of increasing concern because of its toxicity, especially toward children and adults deficient in glucose-phosphate dehydrogenase.[29,30] Actually, it is not the nitrate ion itself that is toxic but rather the nitrite ion formed from it by the reducing action of intestinal bacteria such as *E. coli*. In adults, on the other hand, the nitrate is absorbed high in the digestive tract before reduction can take place. Instances of "blue baby syndrome" or methemoglobinemia have been recorded and related to the complexing action of nitrite on hemoglobin. Further, at stomach pH levels, nitrite can generate secondary amines. These N–nitrosamines have been found to cause cancer in many animal species; whether they are human carcinogens is still an unresolved issue.

Fluoridation of drinking water has been a public health matter of much debate.[31,32] The beneficial dental effects of fluoride are well documented, although high levels are not recommended (see Table 1.9). Epidemiological evidence for the toxicity of fluoride, however, has been inconclusive.[32]

Bromide has created recent interest in the drinking water industry in connection with disinfection by-products (DBPs). Chlorination or ozonation of water containing bromide generates the bromate ion, which is a suspected carcinogen as revealed by animal tests.[33]

<u>Gaseous Pollutants</u>. These were already discussed in Section 1.3.1.

1.4.3. *Microorganisms*

As with the majority of the pollutants discussed in the preceding sections, microorganisms play a positive role in the environment. These include a variety of natural detoxification mechanisms involving soil bacteria. Artificial bioremediation schemes (see later) also rely on the use of microorganisms for combating organic and inorganic pollutants. Unfortunately, however, pathogenic microorganisms have made headline news in the media. These include the outbreak of gastrointestinal disorders in the Milwaukee area, which affected thousands of people. Contamination of the municipal water supply with *Cryptosporidium* was found to be the source of this outbreak. Viruses (Tables 1.3 and 1.4) constitute another source of concern to the drinking water industry. Unlike bacteria, viruses cannot grow by themselves. They are of only fractional size relative to bacterial cells and are responsible for a variety of water-borne diseases. They are also notoriously resistant to disinfection agents, such as chlorine, as we shall see later in this book (Chapter 7).

1.5. CURRENT METHODS FOR ENVIRONMENTAL ANALYSES

1.5.1. Analysis of Organic and Inorganic Pollutants

It is again convenient to use the classification in Table 1.3 as a framework for this discussion. Therefore, Table 1.11 contains a summary of analytical techniques that are currently used for sensing (and quantifying) organic and inorganic pollutants. Only instrumental techniques are included in this list. Classical wet analytical methods are sometimes superior (especially in the hands of an experienced practitioner) but suffer from their "user unfriendly" and tedious nature, and consequently to their incompatibility with routine use scenarios. Amenability to automation is an important consideration, especially when hundreds of samples are to be analyzed. Automated procedures also have the virtue that the quality of the analytical results is not severely compromised by human error.

How far have we progressed in environmental analytical science? The charts contained in Figures 1.8 and 1.9 are reassuring. They show that we have indeed come a long way in improving sensitivity and detection limits. An interesting synergism has been pointed out[1] between these improvements in our analytical capability and the environmental revolution. That is to say, as our ability to detect trace amounts of pollutants has improved, so has the awareness of the societal impact of these chemicals. Perhaps, no other example illustrates this as vividly as lead. Concurrent with the analytical trend in Figure 1.9 has been the response from the society, such as the use of unleaded gasoline, substitutes for lead-based paints, replacement material for lead water pipes, and alternatives to lead solder.[1]

The analytical protocol for pollutant analyses is specified by regulatory agencies such as EPA, American Public Health Association, Occupational Safety and Health Administration, American Water Works Association in the United States, and also in other parts of the world. This is formulated as "recipes" that describe everything the analyst needs to know to complete a satisfactory analysis. The EPA methods include, for example, sample collection, preservation, shipment and storage, instrumentation, apparatus, glass or plastic ware for sample containment, reagents, analytical calibration, quality control, calculations, and reporting results. In most cases, the methods are directed toward specific environmental matrices; for example, industrial wastewater.[34] Examples of these methods include the "600 series" for the analysis of organics in wastewater and the "500 series" developed in response to the requirements of the Safe

1.5. Current Methods for Environmental Analyses

Table 1.11. Instrumental Analysis of Organic and Inorganic Pollutants

Pollutant(s)	Technique(s)
Organics	
Herbicides and pesticides	GC with electron capture detection
	GC-MS
Solvents	Reverse-phase HPLC
	GC-MS
Polycyclic aromatic hydrocarbons	Fluorescence
	Phosphorescence
Dyes	Spectrophotometry
Phenols	Colorimetry
Surfactants	Colorimetry
Formaldehyde	Colorimetry
	Fluorescence
Inorganics	
Metals	Atomic absorption spectroscopy
	ICP-MS
	X-ray fluorescence
Metalloids (arsenic and selenium)	Colorimetry
Anions	Ion chromatography
	Capillary electrophoresis
Gases	GC-MS
	FT-IR spectrometry
	Chemiluminescence
	Colorimetry

Abbreviations: GC = gas chromatography, MS = mass spectrometry, HPLC = high-performance liquid chromatography, ICP = inductively coupled plasma, FT-IR = Fourier transform infrared.

Drinking Water Act. Clearly, each type of environmental medium discussed earlier in Section 1.3—air, water, soil, and biota— requires its own set of sample acquisition, storage, and pretreatment procedures.

The importance of speciation in the assay methodology was mentioned earlier, especially with reference to metals and inorganic pollutants of a metalloid nature (e.g., arsenic). It must be noted that many of the older techniques do not offer this capability. A case in point is atomic absorption spectroscopy (AAS). When performed in the usual manner, this technique yields only the *total* metal content of a particular water or soil sample. This is an instance where hyphenated techniques play a crucial role. In this analytical approach, two or more techniques of a *complementary* nature are coupled together in a tandem combination. The progeny of this is the GC-MS system (see Table 1.11), where the *separation* power of the GC unit is combined with the *detection* capability of MS for the analysis of complex mixtures.

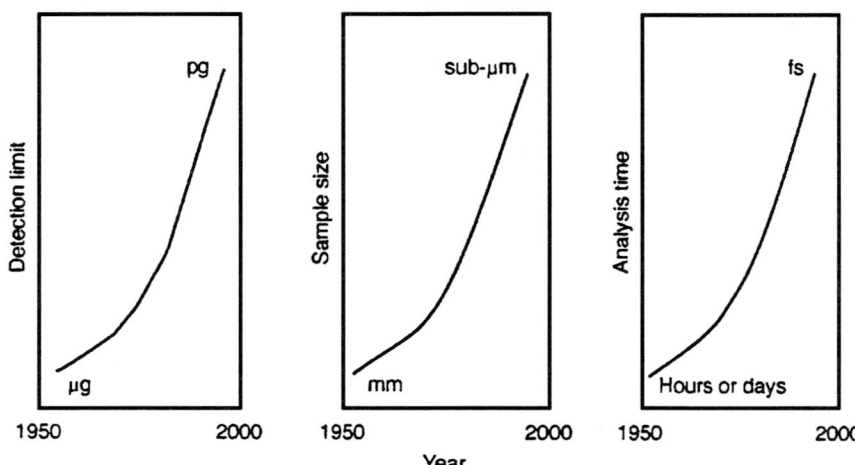

FIGURE 1.8. Progress in detection limits, sample size, and analysis time since World War II. (Reproduced with permission from Grasselli.[1])

Similarly, atomic spectroscopy can be combined with chromatography (e.g., HPLC) for speciation and quantitation of pollutants. Thus, AsBet, AsChol, AsO_2^- and AsO_4^{3-} (see Table 1.10) have been separated and determined on a reversed-phase C_{18} column by means of HPLC-AAS.[35] A $NaBH_4$ hydride generation unit is sometimes incorporated in such systems[36] for converting the HPLC eluents into volatile species that can be subsequently piped into the atomic spectroscopy detector. Electrothermal atomization has also been utilized for this purpose.[37] More recently, the detection capability has been improved with the advent of inductively coupled plasma–atomic emission spectroscopy (ICP-AES) and ICP-MS systems. Finally, the organometallics (e.g., organoarsenicals) are amenable to UV photolysis[38] prior to routing into the hydride generation unit. Hyphenated techniques clearly are the wave of the future, and Figure 1.10 contains an overall schematic and chromatographic traces from an HPLC-UV-HG-ICP-AES assembly for the analysis and speciation of arsenic.[39] Novel chromatographic methods are also being combined with advanced detectors such as the ICP-MS system.[40]

We shall see later in this book that a variety of electrochemical and photoelectrochemical detectors are compatible for use with a separation system (e.g., HPLC, capillary electrophoresis) for the speciation or analysis of environmental pollutants. An important virtue of electrochemical systems is their *portability*; we discuss this analytical issue next.

1.5. Current Methods for Environmental Analyses

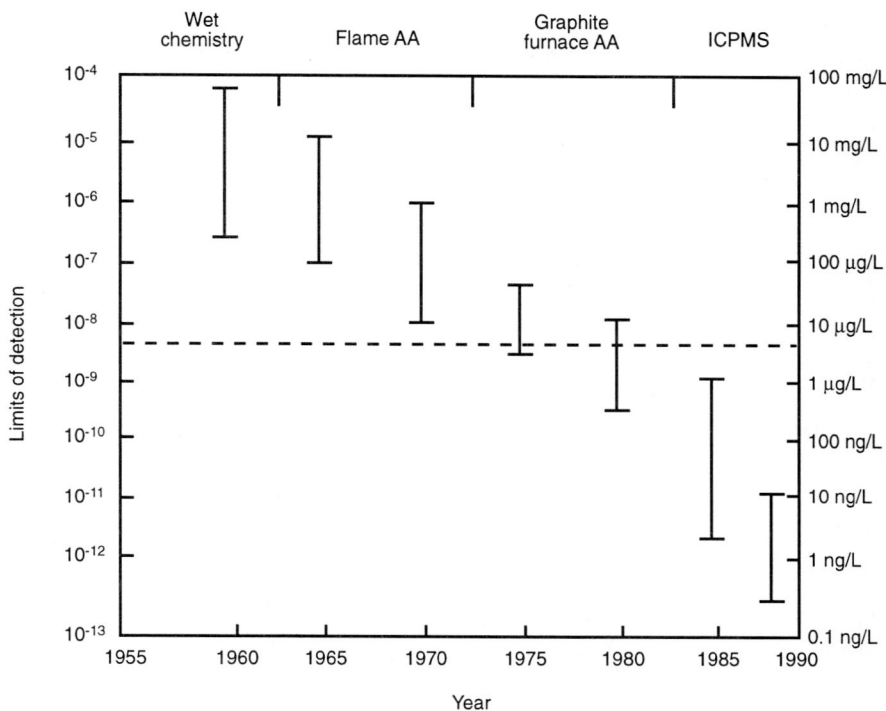

FIGURE 1.9. Detection limits for lead in routine water quality measurements from 1955 to 1990. The dashed line represents U.S. federal guideline for lead in fresh water. (Reproduced with permission from Grasselli.[1])

On-site analytical approaches have the obvious advantage of providing information in a timely manner. For example, consider an active cleaning operation site (e.g., a Superfund site in the United States). Should the personnel wear protective clothing and self-contained breathing apparati? It is obvious that laboratory tests will be inadequate in this scenario. An *in situ* sampling and analysis technique is the obvious solution here and, importantly, can be designed to provide the pollutant level information in *real time*. It is a fair assessment, at least at this juncture, to conclude that current on-site pollutant analysis is much further advanced for air relative to the other media. In particular, photoionization detectors and toxic gas analyzers are now available in a variety of portable configurations.[41,42] These instrumentation advances have also been undoubtedly spurred by requirements for field use by military personnel. The portable units are configured for use with a minimal amount of training.

(a)

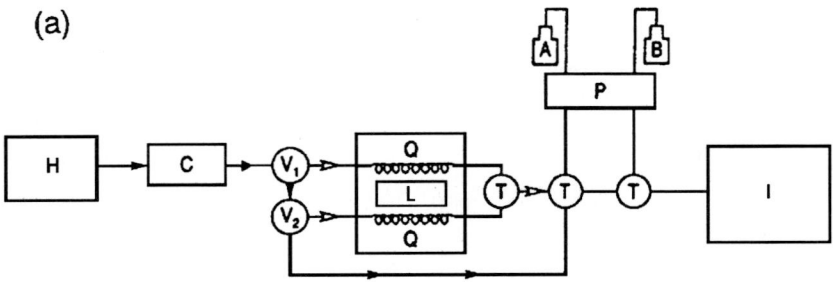

(b)

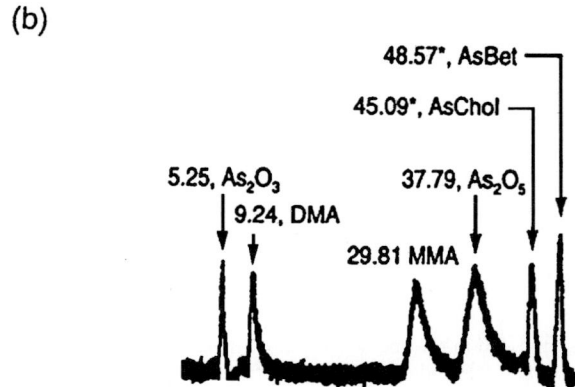

FIGURE 1.10. (a) Schematic diagram of a HPLC-UV-HG-ICP-AES assembly: A=3 M H_2SO_4; B= $NaBH_4$; C= columns; H=HPLC; I=HG/ICP-AES system; L=UV lamp; P=peristaltic pump; Q=quartz coils; T=tee connections; V_1 and V_2 = switch valves. The open arrows mark AsChol and AsBet pathways; the closed arrows indicate the DMA, MMA, As_2O_3, and As_2O_5 pathways. All connections were made using zero-dead-volume Teflon tubings. (b) Chromatographic pattern of the six arsenical species with the UV decomposition and HG system: As_2O_3: 5 ng; DMA, MMA, As_2O_5, AsChol, and AsBet: 50 ng; *= total time of treatment before detection. (Reproduced with permission from Violante et al.[39])

A more subtle aspect of on-site environmental analysis relates to legality issues. In particular, the current trend in air pollution legislation hinges on chain-of-custody steps in keeping accurate accounts of who handles which samples where and when. When field analysis is possible this chain becomes smaller because data collection is facilitated on site. Quality assurance and quality control requirements for on-site analysis promise to be quite stringent because of the "bubble concept" of emission testing. Specifically, if a plant emits less than the allowed amount of chemicals within a given period of time, it is afforded the opportunity to "sell" the residual amount to an adjoining industrial facility in the same area. Naturally, an emission trading situation like this requires a sophisticated analysis system, involving minimal labor and turnaround time. Other analytical requirements and problems have been reviewed recently.[43,44]

1.5.2. Assay of Microorganisms

A coliform count of water samples is performed by making appropriate dilutions and incubating them at 35°C for 48 hr in a nutrient broth. If the medium becomes turbid, the presence of bacteria is indicated, and it can be spread over an agar gel plate. After an additional 24 hr incubation, the number of bacterial colonies on the plate is counted under a microscope and expressed in colony forming units (CFUs). Table 1.12 lists the commonly accepted standards for coliform counts in drinking water.

Coliform bacteria, however, provide only a partial profile of the extent of pollution of the water with pathogenic micro-organisms. For example, approximately 50% of water-borne disease outbreaks are traceable to enteric viruses and protozoan cysts. *Giardia* and *cryptosporidium* are examples of biological pollutants in this category. Unfortunately, unlike bacterial assays, tests for detecting and quantitating such organisms are not so straightforward. The availability of gene probe technology, however, promises to open the way for inexpensive, sensitive and accurate monitoring

Table 1.12. Standards for Coliform Counts in Drinking Water

Counts[a]	Medium
<1	Drinking water
10–100	Normal, unpolluted inland lake water
1,000–5,000	Mild sewage pollution
5,000–10,000	Definite evidence of pollution, caution
10,000–100,000	Heavy pollution, dangerous
>100,000	Sewage

[a]Expressed per 100 cm^3 of water.
Source: Adapted from J. W. Moore and E. A. Moore, "Environmental Chemistry." Academic Press, New York, 1976.

1.6. CURRENT METHODS FOR POLLUTANT TREATMENT

Table 1.13 lists current methods for the treatment of organic, inorganic, and microbiological pollutants in air, water, and soil.

Note that electrochemical treatment has been traditionally applied "in the field" only for the treatment of inorganic (mainly metallic) pollutants. However, there are many manufacturers of electrochemical equipment for pollution control at present, as we shall see later in the book (Chapter 8). Electrochemical methods for the treatment of organics and to a lesser extent, microorganisms are reasonably well developed and appear to be poised for more widespread adoption. A similar situation holds for photoelectrochemical methods. In a limited number of instances, electrochemical technologies have been compared with the other alternatives in Table 1.13 in terms of performance and cost. We shall defer this comparative discussion to a later stage in the book in Chapter 8.

Another approach to a discussion of the current methods for pollutant treatment revolves around the host phase of the pollutant; namely, solid (soil), air, or liquid. However, a difficulty with this approach is that some wastes such as sludge from a wastewater treatment plant are semi–liquids (much like mud) with solids content ranging from 0.5–5%. Primary treatment of industrial

Table 1.13. Current Technologies for the Treatment of Organic, Inorganic, and Microorganism Pollutants in Air, Water, and Soil

Organics	Inorganics	Microorganisms
Incineration and pyrolysis	Precipitation/coagulation	Incineration
Air stripping	Membrane separation	High-energy (γ)-irradiation
Carbon adsorption	Distillation	Filtration
Microbial treatment	Chemical treatment	Carbon adsorption
	Electrochemical treatment	Direct UV irradiation
	Microbial treatment	Ozonation
		Chlorination

1.6. Current Methods for Pollutant Treatment

wastes involves physicochemical techniques such as screening, coagulation, flocculation, and sedimentation. The BOD is reduced at this stage. Secondary treatment involves aerobic oxidation, which removes the residual BOD. Concurrently, additional solids are removed by filtration or sedimentation. Large quantities of sludge are thus generated at both these stages of waste treatment. There are four principal alternatives for disposal of the sludge: dumping in the sea or rivers, disposal at landfill sites, incineration, and application to agricultural land as fertilizer. The landfill option will be severely limited in the future by land availability and stringent legislation. Some legislative control will also constrain the dumping and agricultural options, with the result that incineration or thermal destruction becomes the most viable disposal route in many scenarios.

1.6.1. Incineration and Pyrolysis

Combustion of oxidizable material is an effective way for reducing the mass and volume of hazardous wastes. An important side benefit of this technology is the resultant heat of combustion, which can be utilized to generate process steam. Thus the concept of using municipal refuse as part of the fuel and energy supply for a nearby electric power utility has found favor. The cement industry is another potential big user of hazardous waste; cement kilns can burn appreciable amounts of waste as part of their fuel mix. Unfortunately, however, incineration also can generate appreciable amounts of air pollution, particularly toxic degradation products and particulate matter. The public outcry to widespread adoption of this technology consequently has been vehement, and many incineration units have been forced to either shut down or significantly upgrade the quality of their output.

Three types of incinerator designs have been in use: rotary kiln, fixed or moving hearth furnaces, and fluidized bed. Of these, the use of the rotary kiln design has dominated incineration applications. A crucial parameter is the combustion temperature in all three cases. Ideally, the temperature must be high enough to ensure complete combustion (typically in the range 1050–1250 K) but not too high that slag formation occurs. Recent innovations in incinerator technology include nonslagging designs and the use of dioxygen (instead of air) as oxidizer. Other developments include electric vitrification, plasma degradation, and pyrolytic technology. The last approach differs from incineration in that dioxygen is excluded, and only enough combustion to supply heat for the process is permitted to occur. A major advantage of pyrolysis is that value–added products such as *liquid* hydrocarbon fuels are generated in the process. An innovative example of this approach is the generation of fuel oil from the liquefaction of scrap tire.[45]

Incineration has been widely used for the treatment of PCBs, municipal solid waste and medical waste. Unfortunately, the formation of PCDDs and

PCDFs in the combustion process has fueled public concern about this disposal approach. In general, incomplete combustion is a vexing problem with the thermal approach, and one that is common to almost all organics including plastics. For example, combustion of polyvinylidene chloride (a component of plastic wrapping film) affords chlorinated aromatics. Toxic emissions remain an active area of research in incineration and pyrolysis technology. Four mechanisms have been identified for these emissions:[46]

- Incomplete destruction of PCDD-PCDF in the waste
- Incomplete destruction of long-chain organics that convert to PCDD-PCDF
- Formation from precursors (e.g., soot)
- Low-temperature (catalyzed) reactions

Attempts are being made to correlate the thermal stability of the pollutant (especially for organics), the destruction and removal efficiency of the incinerator, and the products of incomplete combustion.[4]

Another source of concern with incineration technology is toxic metal (or inorganic) emissions. Thus pollutants such as arsenic, lead, chromium, and cadmium have been identified in the exhaust gas stream from the incinerator. Available data[47] suggest that cadmium and lead have a tendency for higher concentration at the multiple hearth design relative to the fluidized bed counterpart. The reverse trend appears to hold for chromium and nickel.

Despite the lack of public acceptance, there is sentiment that incineration will not become moribund. In particular, it is believed that this will continue to be the disposal technology of choice for the treatment of highly halogenated wastes and wastes of high calorific value. Along with the other innovations mentioned earlier, parallel advances are being made for the effective control of air emission from incinerators. Since requirements for this have much in common with air stripping, the two will be discussed together next.

1.6.2. Air Stripping

Organics may be removed from water by packed-tower aeration, in which air is used to "strip" the organics from the aqueous phase. This methodology depends on the Henry's law of partitioning of solutes between the aqueous and gas phases; and obviously works only for volatile organic compounds. After partitioning into the gas phase, the VOCs may be subsequently destroyed either by combustion or pyrolysis followed by scrubbing of the out-gases. Aside from wet scrubbers (to remove HCl, for example), electrostatic precipitators may also be added as auxilliary equipment for controlling particulate and metals emissions. A venturi-type scrubber is most commonly employed, and a quench section may precede the venturi to precool the exhaust gases. The effectiveness of the total unit is rated in terms of the total hydrocarbon (THC) emissions from it.

1.6.3. Microbial Treatment

Microorganisms—bacteria, fungi, and algae—are living catalysts that enable a vast number of chemical processes to occur in water and soil. In microbial remediation procedures, these processes are *induced* to the place via either the micro-organisms naturally present in the contaminated medium or those suitably innoculated into it. As an example of the first type, much of the organic matter in a sanitary landfill undergoes anaerobic decomposition catalyzed by the enzymes present in the soil bacteria. Microbial treatment is commonly done in (above-ground) bioreactors, pits, lagoons, and using soil biofilters, although *in situ* (subsurface) biorestoration of groundwater is emerging as a primary technology.[48] An important aspect of microbial treatment is that the contaminants are partially or even completely destroyed *with minimal impact on the environment*. This is in stark contrast to many remedial techniques that simply *transfer* the pollutant from one part of the environment to another. Perhaps the most widespread application of bioremediation has been in the treatment of oil spills, where marine bacteria and filamentous fungi, which thrive on petroleum, are used.

Heterotrophic bacteria, those types of bacteria that are dependent on a carbon source for energy, are commonly used. Those include *Micrococcus*, *Pseudomonas*, *Mycobacterium*, and *Nocardia*. Fungi are entirely heterotrophic, deriving carbon and energy from the degradation of organic matter. However, fungi do not grow well in water, although they play an indirect environmental remediation role through the humic material that they generate. This material, which is formed from the breakdown of wood cellulose and plant matter, enters the soil and, ultimately, the water stream. Humic material can bind metal ions, as can algae. These microorganisms thus provide a natural mechanism for the sequestering of toxic metals.[49]

Microbial degradation constitutes an important component of the secondary treatment of municipal sewage. The organic matter in the water is biologically degraded in the presence of added dioxygen until the BOD of the waste is reduced to acceptable levels. A commonly used treatment approach is the trickling filter process in which the wastewater is sprayed over rocks or other solid supports covered with deposits of microorganisms. A more versatile approach is the activated sludge process, which mimics the natural self-cleansing processes occurring in streams and other aquatic environments. Unlike in the latter case, however, the remediation is speeded up by continual recycling of the active organisms in an aeration tank.

Aerobic microbial processes require an electron acceptor, usually dioxygen. In many instances, especially in subsurface process scenarios, the bioremediation rate may be actually limited by the O_2 transport to the contamination zone. Other

Chapter 1

electron acceptors such as nitrate may be used under anaerobic conditions, although nitrate addition has other environmental consequences, as pointed out earlier. Nitrous oxide has been used in the biodegradation of *m*-xylene; unlike nitrate, nitrous oxide does not form toxic intermediates such as nitrite upon reduction.

The ubiquity of chlorinated aliphatic solvents (e.g., TCE) in aerobic aquifers suggests that they are biologically recalcitrant in the presence of dioxygen. However, enrichment of environmental samples with methane brings about their biodegradation through cometabolism. Cometabolism involves the biodegradation of an organic substance by a microbe that cannot use the compound itself for growth and hence must rely on other carbon and energy sources. Such methanotrophs have been used for the oxidation of several alkanes, alkenes, and halogenated methanes.[48,50] However, heavily chlorinated samples (e.g., CCl_4) are degraded only slowly and sometimes not at all even under these conditions.

Another synergistic effect involving an aromatic pathway has been identified in the aerobic biodegradation of TCE.[51] A strain, G4, tentatively ascribed to the genus *Acinetobacter*, was found to act via an enzyme, most likely a dioxygenase in an inducible aromatic pathway involving *meta* fission. Thus the presence of phenol, toluene, *o*-cresol, or *m*-cresol was found to be required in the water for the biodegradation to take place.[52] A similar mechanism has been invoked for TCE degradation by another organism, *Pseudomonas putida* PpF1.[53] The aromatics act as "enzyme inducers" in these cases.

Aromatic oxidation plays a key role in the biodegradation of PAHs. Among the microorganisms that attack aromatic rings is the fungus *Cunninghamella elegans*.[54] The prototype breakdown pathway is

$$\text{benzene} \xrightarrow{O_2} \text{catechol} \xrightarrow{O_2} \text{cis,cis-muconic acid} \tag{1.2a}$$

Similar oxidative cleavage has been used for the treatment of benzo(a)pyrene and benzo(a)anthracene by *Beijerincka*.[55]

The biodegradability of pesticides varies greatly. Oxidative routes are mediated by oxygenase enzymes in the treatment with microorganisms and include epoxide formation, aromatic ring hydroxylation (Reaction 1.2a), and β-cleavage of saturated hydrocarbon chains. Reduction routes usually involve the conversion of $-NO_2$ groups to $-NH_2$ groups and quinones to phenols. Hydrolysis is a third route in the microbial degradation of pesticides; for example, ester and amide groups in these compounds are

1.6. Current Methods for Pollutant Treatment

hydroxylable. A specific example of this pathway is the action of *Pseudomonas* on malathion. Dehalogenation reactions involve the replacement of a halogen atom with –OH and can be mediated by some bacteria:

$$\text{pesticide-Cl} \longrightarrow \text{pesticide-OH} \tag{1.2b}$$

Dealkylation is another important degradation route.

In some instances, the microbial degradation intermediates may be more toxic than the original compound. An example is the bioconversion of DDT:

$$\text{Cl-C}_6\text{H}_4\text{-C(CCl}_3\text{)-C}_6\text{H}_4\text{-Cl (DDT)} \longrightarrow \text{Cl-C}_6\text{H}_4\text{-C(CHCl}_2\text{)-C}_6\text{H}_4\text{-Cl} \tag{1.3}$$

In summary, microbial treatment of pollutants is an attractive approach and has the important virtue of mimicking natural self-cleansing processes in the environment. Seeding microorganisms into above-ground bioreactors and surface impoundments is a mature technology. The subsurface biorestoration approach shows promise, although it is as yet an undemonstrated technology. Finding indigenous microorganisms to degrade highly chlorinated aromatics remains a challenging problem. Nonetheless, the biodegradation of organics into mineralized products and biomass offers a better solution to hazardous materials disposal than many other methods, which simply transfer the pollutant from one environmental medium to another.

1.6.4. Precipitation and Coagulation

This approach has been classically used in the primary treatment of wastewater. Hence, lime is added after aeration to raise the pH and to precipitate the "water hardness" ions, Ca^{2+} and Mg^{2+}. Additional coagulants such as aluminum are added to remove the colloidal matter as gelatinous hydroxides. Activated silica or polyelectrolytes may also be added to stimulate coagulation and flocculation.

Pollutants such as Cr(VI) are also removed via hydroxide formation after first reducing them to a lower oxidation state, such as Cr(III). The commonly used reagent for this purpose is ammonium metabisulfite. Arsenic exists in natural waters primarily as arsenite (As(III))and arsenate (As(V)) species. Conventional coagulation with iron or aluminum salt is effective in removing 80–90% As(V) from an initial concentration of ~0.1 ppm at pH 7 or below. Lime softening removes 95% As(V) at pH above 10.5. Reduction of As(V) to lower oxidation states is sometimes done (via the addition of iron salts or sulfides) prior to precipitation.[56] On the other hand, increased removal of As(III) has been reported as well by prior oxidation, using agents such as ozone, chlorine, or permanganate.[57]

Coprecipitation is another mechanism for pollutant removal. Four mechanisms have been identified for coprecipitation of a pollutant by a precipitate[58]: (a) isomorphic inclusion, in which the impurity substitutes into the crystal lattice for a lattice ion of similar size and chemical characteristics; (b) nonisomorphic inclusion, in which the impurity appears to be dissolved in the precipitate; (c) occlusion, in which an impurity differing in size or chemical characteristics from the lattice ions is adsorbed at lattice sites as the crystals are growing, producing crystal imperfections; and (d) surface adsorption, in which the impurity is not incorporated into the internal crystal structure, but is adsorbed only on the outer surface of the precipitate.

Coagulation can also be effective in removing organic compounds from natural waters. The major factors here are the water pH, coagulant dose, and the molecular structure of the organics present. The important mechanisms have been identified as colloid destabilization, precipitation, and coprecipitation.[59] In the last, the organic material is adsorbed (as an impurity) onto a lattice site of the growing precipitate. Coprecipitation is the dominant mechanism for the removal of fulvic acid by lime softening.[60,61] Recent results suggest that lime softening is superior to low pH coagulation for lowering water turbidity and total organic carbon.

1.6.5. *Chemical Treatment*

We have seen several examples of chemical routes to the detoxification or removal of pollutants in the preceding two sections. Classical organic reaction routes can be employed for waste treatment. These include nucleophilic substitution, whereby hydroxyl ion substitution is induced in aryl chlorides (e.g., PCBs). Usually KOH is used as the OH⁻ source and polyethylene glycol is additionally employed for solubilizing the KOH in an aprotic medium.[62] Reductive dechlorination (hydrogenolysis) has also been effected at low temperatures (e.g., 100°C) using a mixed oxide catalyst. Aprotic media such as tetrahydrofuran have been used for dissolving active metals such as sodium:

$$Na + ArH \rightarrow Na^+ + ArH^{-\bullet} \tag{1.4}$$

Similar chemistry occurs with chlorinated aromatics, and the radical anions thus generated can expel Cl^-:

$$ArCl^{-\bullet} \rightarrow Ar^{\bullet} + Cl^- \tag{1.5}$$

The aryl radicals subsequently participate in typical free radical reactions including dimerization, abstraction of hydrogen from the solvent, or arylation of hydrogen from the solvent or arylation of another molecule of the aromatic. These chemistries are all variants of the Wurtz reaction:

$$2\,R\text{--}Cl + 2\,Na \rightarrow R\text{--}R + 2\,NaCl \tag{1.6}$$

Redox (i.e., electron transfer) reactions such as the one shown in Eq. 1.4 can also be used for treating inorganic pollutants. The treatment of Cr(VI) with sodium metabisulfite was mentioned in the preceding section. Another redox reagent for this purpose is ferrous sulfate; the electron transfer reaction involved here can be represented by the scheme

$$Cr(VI) + 3\,Fe(II) \rightarrow 3\,Fe(III) + Cr(III) \tag{1.7}$$

Such (redox) reactions are discussed in more detail in Chapter 5.

Aluminum has been suggested[63] as a reductant for the treatment of nitrate in water, the reaction proceeding stepwise as

$$3\,NO_3^- + 2\,Al + 3\,H_2O \rightarrow 3\,NO_2^- + 2\,Al(OH)_3 \tag{1.8a}$$

$$NO_2^- + 2\,Al + 5\,H_2O \rightarrow NH_3 + 2\,Al(OH)_3 + OH^- \tag{1.8b}$$

$$2\,NO_2^- + 2\,Al + 4\,H_2O \rightarrow N_2 + 2\,Al(OH)_3 + 2\,OH^- \tag{1.8c}$$

The compatibility of this process with conventional water treatment procedures (e.g., lime softening) has been pointed out.[63] Preliminary data show ammonia to be the principal product (60–95%) followed by nitrite and nitrogen gas via Reactions (1.8a) and (1.8c), respectively. The generation of nitrite, however, could pose a problem (see Section 1.4.2).

1.6.6. Adsorption

The role of adsorption in water treatment procedures was already discussed in Section 1.6.4. Another standard water treatment method consists of adsorption on activated carbon, a product obtained from a variety of carbonaceous materials including wood, peat, and lignite. Activated carbon comes in two types: granular activated carbon (GAC), consisting of particles 0.1–1 mm in diameter, and powdered activated carbon, in which the particles are predominantly in the 50–100 µm size range. The former is more commonly used, and in a water treatment system, a fixed GAC bed is employed through which the water flows downward. Accumulation of particulate matter requires periodic backwashing. A parameter for defining the adsorptive capacity of GAC is the phenol value. This is defined as the amount of carbon (in mg) required to reduce by 90% the phenol content of a 100 ppb solution. Over and above simple adsorption, bacterial growth on the carbon can provide an additional route to organics removal. Economics requires the spent carbon to be regenerated. This is normally done by heating to 950°C in a steam–air atmosphere, where the accumulated organics are burned off.

Adsorption at GAC or other surfaces can also remove metals at the ppm level. Sometimes a chelating agent is sorbed onto the adsorbent to enhance metal removal. Coal humic acids, for example, can remove ca. 80% of copper at the ppm level from acidic water; and these materials (which effectively mimic soil behavior) show promise as toxic metal and inorganics scavengers. Enhanced As(V) removal has been observed for activated carbon that had been pretreated with ferrous salts, which were also used to regenerate the adsorbent.[64]

In spite of its widespread use, it must be recognized that GAC adsorption merely transfers the pollutant from the process stream to the solid phase. Ultimately, the contaminated carbon poses a disposal problem. With the rapidly dwindling landfill disposal possibilities, combustion remains the only viable route, and even this is not without problems associated with thermal emissions (see Section 1.6.1).

Alternatives to activated carbon are being considered including the use of peat[65] and adsorbent synthetic polymers. The latter, for example, the Amberlite group of resins, remove virtually all nonionic organic solutes via hydrophobic interactions.

1.6.7. Membrane Processes

These include electrodialysis, reverse osmosis, and ion exchange. The common denominator here is the membrane (usually a polymer) phase. Water desalinization and softening form the bulk in terms of technology applications,[66] although these methods are now being considered for industrial wastewater treatment.

1.6. Current Methods for Pollutant Treatment 45

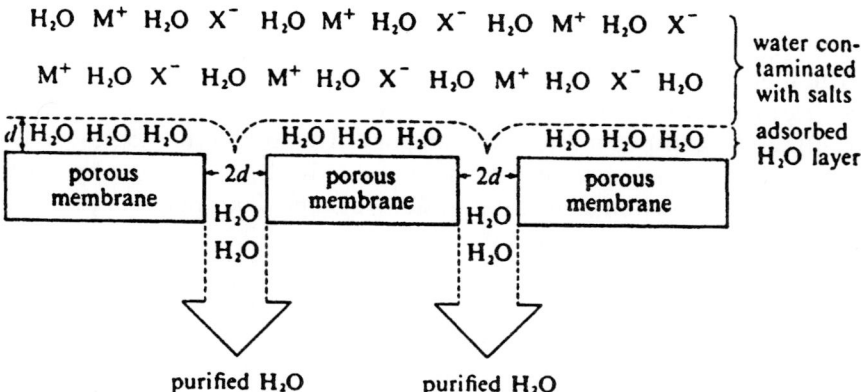

FIGURE 1.11. Schematic diagram of a reverse osmosis process. (Reproduced with permission from S. E. Manahan, "Environmental Chemistry." 5th ed., Lewis Publishers, Michigan, 1991.)

Electrodialysis consists of applying an electric field across water layers alternately separated by cation- and anion-selective membranes. This process shall be discussed in detail in Chapter 5. Membrane fouling is remedied by periodic reverse flushing. Generally, electrodialysis can remove up to 40–50% dissolved inorganics (see Chapter 5).

Reverse osmosis or ultrafiltration employs an osmotic membrane (usually made of cellulose acetate). This process is illustrated in Figure 1.11. The water flow in this process is entropy driven and occurs in a direction opposite to normal osmotic flow, hence the name. Rather strong pressure is required, however, to reverse the normal flow, especially to speed up the process, and the mechanical strength of the membrane becomes a crucial factor.

1.6.8. Distillation

An obvious method for removing inorganics from water is by distilling it. Unfortunately, this process is not energy efficient enough to be economically attractive, especially for large-scale applications. Further, volatile pollutants and odorous compounds (especially organics) are carried over to a large extent in the distillation process unless special precautions are taken.

Freezing is another related method for producing very pure water. Freeze–pump–thaw (FPT) cycles, for example, are commonly used for purifying solvents. Again this approach to water treatment does not appear to be economically attractive at present.

1.6.9. Advanced Oxidation Processes

The methods considered in Sections 1.6.6. and 1.6.7. together constitute the *tertiary* or *advanced* treatment of wastewater. Advanced oxidation processes (or AOP) can be grouped in the same category. The AOP technologies almost all rely on the generation of very reactive free radicals, such as the hydroxyl radical (•OH), to function as pollutant-killing agents. Four major approaches to AOP are under development at present[67]:

(1) Homogeneous photolysis (UV/H_2O_2 and UV/O_3): These processes employ UV photolysis of H_2O_2 and/or O_3 and other additives in homogeneous solution to generate •OH and other free radicals (Chapter 6).

(2) Radiolysis: A source of high-energy radiation (γ-rays) is used to irradiate the waste water. A variety of species, such as •OH, H•, e_{aq}^- (hydrated electrons) are created under these conditions (Chapter 7).

(3) Dark oxidation processes: These obviously do not employ UV light but instead the radicals are generated by other means—Fenton's reaction, ozone at high pH and O_3/peroxide (Chapter 5).

(4) Heterogeneous photolysis or photocatalysis: This uses a semiconductor catalyst and a light source to induce photoelectrochemical reactions at or near the catalyst particle surface. We shall discuss this approach in Chapter 6.

In concluding this section, the role of *natural* photochemical processes in "self-cleansing" of the environment must be mentioned. These processes in the atmosphere are all too familiar to us thanks to the publicity generated in the media about the ozone layer and the adverse effects of chlorofluorocarbons. Perhaps less well known are the processes that occur in aquatic environments. We discussed mainly the negative environmental role of microorganisms in Section 1.4.3. Algae, which are microscopic organisms that subsist on inorganic nutrients and produce biomass via photosynthesis, are largely the "good guys" in the environment. They bioaccumulate toxic species. For example, marine algae are known to concentrate arsenic as arsenic phospholipids and arsenolipids. This occurs principally in phosphate-deficient waters and works because of the chemical similarity of arsenic to phosphorus. We are learning more about the role of photochemical transformations in arsenic speciation in natural waters.[38] There are also indications for a diurnal cycle in the oxidation state of chromium in low-salinity estuarine water.[68] The interaction of sunlight with soil-particulate matter (presumably containing ferric ions, see Reaction (1.7), Section 1.6.5) appears to drive this cycle, resulting in Cr(VI) reduction during daylight hours. The Fe(II) ions needed to mediate the Cr(VI) reduction are apparently generated photolytically. This finding may have important implications in natural repair by the environment of Cr(VI) contaminated media.

1.6. Current Methods for Pollutant Treatment 47

1.6.10. Treatment of Polluted Sites

So far, in Section 1.6, we have discussed the important methods available for "end-of-the-pipe" treatment of organic and inorganic pollutants. We have said little about sites already contaminated with pollutants and how they are being environmentally restored at present. The polluted medium in these cases is either soil or groundwater (or both). The Superfund sites in the United States belong to this category; some 1,200 of these have been identified already at an estimated cleanup cost of ca. $90 billion.[69] However, the viability of restoring all these sites to the pristine state has been questioned, given the costs involved and the unavailability of effective and permanent remedial technologies. Nonetheless, progress is being made both in the refinement of relatively old technologies (e.g., incineration, "pump-and-treat" methods) and in the evaluation of novel methods (e.g., *in situ* biorestoration, see Section 1.6.3).

Compendia of organic plumes in sand and gravel aquifers in the United States and Canada are available.[70] Much, if not most, of the groundwater pollution is caused by leakage, spillage, or disposal of organics into the ground. Subsequent dissolution and transport of these by groundwater (see Section 1.3) are thought to generate these plumes, although vapor-to-soil water transfer is also thought to be an important transport mechanism.

Difficulties with the common "pump-and-treat" method have been discussed.[70] These may occur because at many sites of significance a relatively large mass of pollutants has been leaked, spilled, or disposed into the subsurface; and in comparison, the rate of pollutant mass removal by pumping wells is exceedingly slow. Another major difficulty is the range of pollution problems encountered; this range forces site-specific solutions. There are also troublesome antagonisms in the choice of a treatment candidate (e.g., biotreatment obviously will not work for a biotoxic waste). Other problems that can preclude treatment approaches include[69]:

- Mercury, lead, bromine, and reduced nitrogen compounds can cause problems with incinerator air emissions.
- Biotreatment of chlorinated organics produces vinyl chloride, a known human carcinogen.
- Conventional chemical stabilization tends to mobilize arsenic.
- The use of metal complexing agents (e.g., cyanides, ammonia) renders many stabilization processes ineffective.

Nonetheless, the future is not at all bleak. Advances are being made in both incinerator (see previously) and chemical stabilization technology. Technologies prescribed by the U.S. EPA for waste site cleanup are listed in Table 1.14. It is a fair assessment that many of these technologies are currently being pushed to their limits. Simultaneously, promising technology alternatives

Table 1.14. Conventional and Emerging Technologies for the Treatment of Hazardous Waste Sites

Technology	Comments
Conventional	
Incineration and thermal destruction	Technological advances will continue in spite of increasing concerns
Solidification, stabilization, and neutralization	Bans on landfill will spur increasing use of this technology
Volatilization and soil aeration	Methods effective for most volatile organics including PCBs
Soil washing and flushing	Because of the volume of wastewater generated and dissolved solids, applications may be limited
Biotreatment	Perhaps the most attractive *in situ* treatment technology available at present
Vacuum extraction	Has worked well for petroleum hydrocarbons
Emerging	
Supercritical oxidation	High-quality water can be produced from feeds with up to 10% organics and a range of inorganics
Evaporation with catalytic oxidation	Organic compounds are oxidized in the vapor phase over a solid catalyst; can handle feeds with up to 10% dissolved solids
High-pressure and supercritical fluid extraction	Can remove nonpolar contaminants from extraction wastewater as well as separating emulsions and dewatering sludges
Electrokinetic extraction	Discussed in Chapter 5

are also emerging; these are also listed in Table 1.14. Of these, supercritical oxidation and evaporation with catalytic oxidation appear to hold particular promise. These newer technologies can tackle both organic and inorganic pollutants and also deal with the sludge problem. Electrochemical candidates are discussed later in Chapter 5.

1.7. ELECTROCHEMICAL TECHNOLOGY AND THE ENVIRONMENT

This book is about the *positive* role that electrochemical science and technology can play in pollution abatement. However, the discussion would not have been complete without mention of some of the deleterious aspects of electrochemical technology and the manner in which electrochemistry has contributed to environmental pollution. Chlor-alkali and battery industries are two of the major culprits in this regard.

The chlor-alkali process is the electrolysis of brine to produce chlorine and sodium hydroxide:

$$2\,NaCl(aq) + 2H_2O(l) \rightarrow 2\,NaOH(aq) + H_2(g) + Cl_2(g) \qquad (1.13)$$

Layers of liquid mercury were used as the cathode in a flow cell. In the late 1960s, elevated levels of mercury were detected at sites around the chlor-alkali plants. Unfortunately, the requirement of cooling water in these plants necessitates their proximity to rivers. Poor maintenance and inventory control led to loss of mercury and subsequent contamination of the river sediments downstream. Attempts to combat this pollution problem have included stricter governmental regulation to ensure better housekeeping by the plants and, more important, the search for mercury-free cathode technology.

Spent or discarded batteries constitute another serious source of environmental pollution. Over half a million pounds of lead–acid batteries are estimated to require daily disposal in the United States.[71] As mentioned earlier in other contexts, municipal landfill is no longer a viable disposal alternative. The heavy-metal toxicity of incinerator ash, particularly lead and cadmium, is directly related to the indiscriminate manner in which lead and cadmium batteries have been disposed of in the past. Recycling of battery constituents is the soundest option in both environmental and cost terms. Reclamation of battery (and also fuel-cell) materials is an increasingly important topic, and the underlying technical, regulatory and safety issues are the focus of at least two annual symposia in the United States.[71]

1.8. SUMMARY

In this chapter, we have attempted to present a broad picture of environmental science and technology and, in particular, environmental chemistry. The emphasis, as with the remainder of this book, has been on pollution abatement. However, a discussion of pollutant types, environmental media, and pollutant transport cycles has been provided as a necessary framework for the discussion of specific pollution control technologies. An integral part of this background is a survey of environmental analysis methods, for it goes without saying that a pollutant must be first identified and quantified before it can be either treated or disposed of. The alert reader will note the omission from the background material in this chapter of drinking and wastewater treatment and disinfection practice. This rather specialized topic is deferred to Chapter 7 (which deals with electrochemical and photoelectrochemical disinfection of water and air) to provide for a more cohesive discussion.

REFERENCES

1. J. G. Grasselli, Analytical Chemistry—Feeding the Environmental Revolution? *Anal. Chem.* **64**, 677A (1992).
2. M. R. Deland, An Ounce of Prevention....After 20 Years of Cure. *Environ. Sci. Technol.* **25**, 561 (1991).
3. Pollution Reduction. *Chem. Eng. News*, November 16, p. 22 (1992).
4. Hazardous Waste Incineration Presents Legal, Technical Challenges. *Chem. Eng. News*, March 29, p. 7 (1993).
5. S. Miller, Where All Those EPA Lists Come From. *Environ. Sci. Technol.* **27**, 2302 (1993).
6. H. Kunreuther and R. Patrick, Managing the Risks of Hazardous Waste. *Environment* **33**, 12 (1991).
7. L. Roberts, Counting on Science at EPA. *Science* **249**, 616 (1990).
8. C. C. Travis and C. B. Doty, Superfund: A Program Without Priorities. *Environ. Sci. Technol.* **23**, 1333 (1989).
9. M. J. Molina and F. S. Rowland, Stratospheric Sink for Chlorofluoromethane: Chlorine Atom Catalyzed Destruction of Ozone. *Nature (London)* **249**, 810 (1974).
10. C. C. Travis and S. T. Hester, Global Chemical Pollution. *Environ. Sci. Technol.* **25**, 814 (1991).
11. D. Mackay, S. Patterson and B. Cheung, Evaluating the Environmental Fate of Chemicals: The Fugacity-Level III Approach as Applied to 2, 3, 7, 8 TCDD. *Chemosphere* **14**, 859 (1985).

12. T. E. Graedel and P. J. Grutzen, The Changing Atmosphere. *Sci. Am.* September, p. 58 (1989).
13. J. W. Maurits la Rivière, Threats to the World's Water. *Sci. Am.* September, p. 80 (1989).
14. G. M. P. Morrison, G. E. Batley and T. M. Florence, Metal Speciation and Toxicity. *Chem. Br.*, August, p. 791 (1989).
15. G. H. Eduljee, Dioxins in the Environment. *Chem. Br.* December, p. 1223 (1988).
16. C. C. Travis and H. Hattemer-Frey, Human Exposure to 2,3,7,8-TCDD. *Chemosphere* **16**, 2331 (1987).
17. J. N. Pitts, Jr., Atmospheric Reactions of Polycyclic Aromatic Hydrocarbons: Facile Formation of Mutagenic Nitro Derivatives. *Science* **202**, 515 (1978).
18. K. L. Idler, "PCBs—The Current Situation," Publ. P86-3E. Canadian Centre for Occupational Health and Safety, Hamilton, Ontario, 1986.
18a. S. Hong, J.P. Candelone, C. C. Patterson, and C. F. Boutron, Greenland Ice Evidence of Hemispheric Lead Pollution Two Millenia Ago by Greek and Roman Civilizations. *Science* **265**, 1841 (1994).
18b. K. Rajeshwar, J. G. Ibanez, and G. M. Swain, Electrochemistry and the Environment. *J. Appl. Electrochem.* **24**, 1077 (1994).
18c. S. Takacs and A. Tatar, Trace Elements in the Environment and in Human Organs. *Environ. Res.* **42**, 312 (1987).
18d. Y. M. Nor, Ecotoxicity of Copper to Aquatic Biota: A Review. *Environ. Res.* **43**, 274 (1987).
18e. K. Sweetland, Dealing with Lead in the Workplace. *Chem. Eng.* **100**, 127 (1993).
18f. P. E. Body, G. Inglis, P. R. Dolan, and D. E. Mulcahy, Environmental Lead: A Review. *Crit. Rev. Environ. Control* **20**, 299 (1991).
18g. W. M. Stigliani, P. R. Jaffé, and S. Anderberg, Heavy Metal Pollution in the Rhine Basin. *Environ. Sci. Technol.* **27**, 786 (1993).
19. D. Taylor, Lead: Nature's Ubiquitous Poison. *Lab. Q.* **5**, (4), 1 (1994).
20. Mercury Toxicity, *Am. Fam. Physician* **46**, 1731 (1992).
21. S. A. Katz and H. Salem, The Toxicology of Chromium with Respect to its Chemical Speciation: A Review. *J. Appl. Toxicol.* **13**, 217 (1993).
22. L. M. Calder, Chromium in the Natural and Human Environment. *Adv. Environ. Sci. Technol.* **20**, 215 (1988).
23. D. Rai, B. M. Sass, and D. A. Moore, Chromium(III) Hydrolysis Constants and Solubility of Chromium(III) Hydroxide. *Inorg. Chem.* **26**, 345 (1987).
24. L. C. D. Anderson and K. W. Bruland, Biogeochemistry of Arsenic in Natural Waters: The Importance of Methylated Species. *Environ. Sci. Technol.* **25**, 420 (1991).
25. S. Tamaki and W. T. Frankenberger, Jr., Environmental Biochemistry of Arsenic. *Rev. Environ. Contam. Toxicol.* **124**, 79 (1992).
26. P. H. Yancey, M. E. Clark, S. C. Hand, R. D. Bowlus and G. N. Somero, Living with Water Stress: Evolution of Osmolyte Systems. *Science* **217**, 1214 (1982).

27. E. Hoyne, "The Selenium and Tellurium Markets," Bull. Selenium-Tellurium Development Association, 1992.
28. R. W. Andrews and D. C. Johnson, Voltammetric Deposition and Stripping of Selenium(IV) at a Rotating Gold-Disk Electrode in 0.1 M Perchloric Acid. *Anal. Chem.* **47**, 294 (1975).
29. G. Walton, Survey of Literature Relating to Infant Methemoglobinemia Due to Nitrate-Contaminated Water. *Am. J. Public Health* **41**, 986 (1951).
30. B. C. Challis, Rapid Nitrosation of Phenols and its Implications for Health Hazards from Dietary Nitrites. *Nature (London)* **244**, 466 (1973).
31. B. Hileman, Fluoridation of Water. *Chem. Eng. News*, August 1, p.26 (1988).
32. E. Marshall, The Fluoride Debate: One More Time. *Science* **247**, 276 (1990).
33. R. J. Joyce, Determination of Bromate in Drinking Water Using Ion Chromatography. *Am. Environ. Lab.*, May, p. 1 (1994).
34. R. A. Hites and W. L. Budde, EPA's Analytical Methods for Water: The Next Generation. *Environ. Sci. Technol.* **25**, 998 (1991).
35. R. A. Stockton and K. J. Irgolic, The Hitachi Graphite Furnace-Zeeman Atomic Absorption Spectrometer as an Automated Element-Specific Detector for High Pressure Liquid Chromatography. *Int. J. Environ. Anal. Chem.* **6**, 313 (1979).
36. C. T. Tye, S. J. Haswell, P. O'Neill, and K. C. Bancroft, High Performance Liquid Chromatography with Hydride Generation/Atomic Absorption Spectrometry for the Determination of Arsenic Species with Application to Water Samples. *Anal. Chim. Acta* **169**, 195 (1985).
37. S. Beres, R. Thomas, E. Denoyer, and P. Brückner, The Benefits of Electrothermal Vaporization for Minimizing Interferences in ICP-MS. *Spectroscopy* **9**, 20 (1994).
38. C. I. Brockband, G. E. Batley, and G. K-C. Low, Photochemical Decomposition of Arsenic Species in Natural Waters. *Environ. Technol. Lett.* **9**, 1361 (1988).
39. N. Violante, F. Petrucci, F. La Torre, and S. Cardi, On-Line Speciation and Quantification of Arsenic Using an HPLC-UV-HG-ICP-AES System. *Spectroscopy* **7**, 36 (1992).
40. E. Blake, M. W. Raynor, and D. Cornell, Combined SFC-ICP-MS: A Solution for Organometal Speciation in Environmental Samples. *Am. Lab.*, June, p. 46 (1994).
41. R. E. Clement, C. J. Koester, and G. A. Eiseman, Environmental Analysis. *Anal. Chem.* **65**, 85R (1993).
42. J. N. Driscoll, Portable Instrumentation for On-Site Monitoring of Toxic Gases. *Am. Lab.*, May, p.37 (1993).
43. C. F. D'Elia, J. G. Sanders, and D. G. Capone, Analytical Chemistry for Environmental Sciences. *Environ. Sci. Technol.* **23**, 768 (1989).
44. W. Chudyk, Field Screening of Hazardous Waste Sites. *Environ. Sci. Technol.* **23**, 504 (1989).

45. D. Shaw, Scrap Tyres Can Just Melt Away. *Eur. Rubber J.*, May, p.12 (1994).
46. "State of the Art Assessment of Medical Waste Thermal Treatment," Report from the EER Corporation to the EPA Risk Reduction Engineering Laboratory, 1990.
47. M. B. Foisy, R. Li, A. Chattopadhyay, and M. Karell, Controlling Air Emissions from Incinerators. *Water Environ. Technol.*, April, p. 40 (1994).
48. J. M. Thomas and C. H. Ward, In Situ Biorestoration of Organic Contaminants in the Subsurface. *Environ. Sci. Technol.* **23**, 760 (1989).
49. B. Vokeski, ed., "Biosorption of Heavy Metals." CRC Press, Boca Raton, FL 1990.
50. C. L. Haber, L. N. Allen, and R. S. Hanson, Methylotrophic Bacteria: Biochemical Diversity and Genetics. *Science* **221**, 1147 (1983).
51. M. J. K. Nelson, S. O. Montgomery, E. J. O'Neill, and P. H. Pritchard, Aerobic Metabolism of Trichloroethylene by a Bacterial Isolate. *Appl. Environ. Microbiol.* **52**, 383 (1986).
52. M. J. K. Nelson, S. O. Montgomery, W. R. Mahaffey, and P. H. Pritchard, Biodegradation of Trichloroethylene and Involvement of an Aromatic Biodegradative Pathway. *Appl. Environ. Microbiol.* **53**, 949 (1987).
53. M. J. K. Nelson, S. O. Montgomery, and P. H. Pritchard, Trichloroethylene Metabolism by Microorganisms that Degrade Aromatic Compounds. *Appl. Environ. Microbiol.* **54**, 604 (1988).
54. C. E. Cerniglia and D. T. Gibson, Metabolism of Naphthalene by *Cunninghamella elegans*. *Appl. Environ. Microbiol.* **34**, 363 (1977).
55. D. T. Gibson, V. Mahadevan, D. M. Jerina, H. Yagi, and H. J. C. Yeh, Oxidation of the Carcinogens Benzo(a)pyrene and Benzo(a)anthracene to Dihydrodiols by a Bacterium. *Science* **189**, 295 (1975).
56. T. R. Harper and N. W. Kingham, Removal of Arsenic from Wastewater Using Chemical Precipitation Methods. *Water Environ. Res.* **64**, 200 (1992).
57. S. R. Qasim, The University of Texas at Arlington, private communication (1994).
58. D. A. Skoog, D. M. West, and F. J. Holler, "Analytical Chemistry," 6th ed., pp. 100-106. Saunders College Publishing, Philadelphia, 1994.
59. S. J. Randtke, Organic Contaminant Removal by Coagulation and Related Process Combinations. *J. Am. Water Works Assoc.* **80**, 40 (1988).
60. M. Y. Liao and S. J. Randtke, Removing Fulvic Acid by Lime Softening. *J. Am. Water Works Assoc.* **77**, 78 August (1985).
61. M. Y. Liao and S. J. Randtke, Predicting the Removal of Soluble Organic Contaminants by Lime Softening. *Water Res.* **20**, 27 (1986).
62. D. J. Brunelle, A. K. Mendlratta, and D. A. Singleton, Reaction/Removal of Polychlorinated Biphenyls from Transformer Oil: Treatment of Contaminated Oil with Poly(ethylene glycol)/ KOH. *Environ. Sci. Technol.* **19**, 740 (1985).
63. A. P. Murphy, Chemical Removal of Nitrate from Water. *Nature (London)* **350**, 223 (1991).

64. C. P. Huang and L. M. Vane, Enhancing As^{5+} Removal by a Fe^{2+}-treated Activated Carbon. *J. Water Pollut. Control Fed.* **61**, 1596 (1989).
65. D. Couillard, The Use of Peat in Wastewater Treatment. *Water Res.* **28**, 1261 (1994).
66. A. S. Michaels, Membranes, Membrane Processes, and Their Applications: Needs, Unsolved Problems and Challenges of the 1990's. *Desalination* **77**, 5 (1990).
67. J. R. Bolton and S. R. Cater, "Aquatic and Surface Photochemistry" (G. R. Helz, R. G. Zepp and D. G. Crosby, eds.), Chapter 33, pp.467-490. Lewis Publishers, Boca Raton, Florida 1994.
68. R. J. Kleber and G. R. Helz, Indirect Photoreduction of Aqueous Chromium(VI). *Environ. Sci. Technol.* **26**, 307 (1992).
69. P. S. Daley, Cleaning Up Sites with On-Site Process Plants. *Environ. Sci. Technol.* **23**, 912 (1989).
70. D. M. Mackay and J. A. Cherry, Groundwater Contamination: Pump-and-Treat Remediation. *Environ. Sci. Technol.* **23**, 630 (1989).
71. J. M. Fenton, Electrochemistry: Energy, Environment, Efficiency and Economics. *Interface*, Electrochem. Soc. **3**, 38 (1994).
72. J.A. Cherry, Contaminant Migration in Groundwater: Processes and Problems, in J. Wilson, ed., "The Fate of Toxics in Surface and Ground Waters," The Proceedings of the Second National Water Conference, p. 72, Academy of Natural Sciences, Philadelphia, 1984.

FURTHER READING

Books and Monographs

The following listing, by no means comprehensive, provides more details on the topics discussed in this chapter.
1. L. Hodges, "Environmental Pollution." Holt, New York, 1973.
2. H. C. Perkins, "Air Pollution." McGraw-Hill, New York, 1974.
3. H. S. Stoker and S. L. Seager, "Environmental Chemistry: Air and Water Pollution." Scott, Foresman, Glenview, IL, 1976.
4. J. W. Moore and E. A. Moore, "Environmental Chemisry." Academic Press, New York, 1976.
5. S. E. Manahan, "Environmental Chemistry." Willard Grant Press, Boston, 1979.
6. N. J. Bunce, "Environmental Chemistry." Wuerz Publishing, Winnipeg, Canada, 1991.
7. T. Godish, "Air Quality." Lewis Publishers, Chelsea, MI, 1991.
8. S. E. Manahan, "Fundamentals of Environmental Chemistry." Lewis Publishers, Chelsea, MI, 1993.
9. R. N. Reeve, "Environmental Analysis." Wiley, New York, 1994.
10. C. Vandecasteele and C. B. Black, "Modern Methods for Trace Element Determination." Wiley, New York, 1994.

11. M. W. Sigrist, editor, "Air Monitoring by Spectroscopic Techniques." Wiley, New York, 1994.
12. R. E. Sievers, editor, "Selective Detectors: Environmental, Industrial and Biomedical Applications." Wiley, New York, 1995.
13. K. D. Racke and A. R. Leslie, eds., "Pesticides in Urban Environments: Fate and Significance," ACS Symp. Ser. No. 522. American Chemical Society, Washington, DC, 1993.
14. J. J. Breen and M. J. Dellarco, eds., "Pollution Prevention in Industrial Processes: The Role of Process Analytical Chemistry," ACS Symp. Ser. No. 508. American Chemical Society, Washington, DC, 1992.

Journals

The following journals are specifically devoted to environmental chemistry and environmental science and technology. Again, the listing is representative rather than comprehensive.

1. *Journal of the American Water Works Association*
2. *Journal of the Water Pollution Control Federation*
3. *Water Research*
4. *Environmental Science and Technology*
5. *Environment*
6. *Applied and Environmental Microbiology*
7. *Environmental Technology Letters*
8. *Environmental Progress*
9. *International Journal of Environmental Analytical Chemistry*
10. *Environmental Remediation Technology*

The following are trade journals in this field.

1. *Pollution Engineering*
2. *Industrial Wastewater*
3. *Environmental Solutions*

Articles of environmental interest frequently appear in the following journals and magazines.

1. *Chemtech*
2. *Chemical and Engineering News*
3. *Chemistry and Industry*
4. *Nature*
5. *Science*
6. *Scientific American*
7. *American Laboratory*

8. *American Environmental Laboratory*
9. *Analytica Chimica Acta*
10. *Talanta*
11. *Analytical Chemistry*
12. *Spectroscopy*
13. *The Analyst*

Manuals and Handbooks

Many organizations provide comprehensive treatments, particularly of hazardous chemicals, sampling and analytical procedures. A selection follows.

1. American Public Health Association, American Water Works Association, Water Pollution Control Federation, "Standard Methods for the Examination of Water and Wastewater." American Public Health Association, Washington, DC, 18th ed., 1992.
2. American Water Works Association, Inc., "Water Quality and Treatment." McGraw-Hill, New York.
3. G. Weiss, ed., "Hazardous Chemicals Data Book." Noyes Data Corporation, Park Ridge, NJ, 1986.
4. M. Sittig, "Handbook of Toxic and Hazardous Chemicals and Carcinogens." Noyes Data Corporation, Park Ridge, NJ, 1991.
5. R. Lewis, "Hazardous Chemicals Desk Reference." Van Nostrand-Reinhold, New York, 1993.
6. H. Seiler and H. Sigel, eds., "Handbook on Toxicity of Inorganic Compounds." Dekker, New York, 1988.

Other Literature on Toxic Waste Treatment

1. J. H. Exner, "Detoxification of Hazardous Waste." Ann Arbor Science Publishers, Ann Arbor, MI, 1982.
2. B. B. Berger, ed., "Control of Organic Substances in Water and Wastewater." Noyes Data Corporation, Park Ridge, NJ, 1987.
3. S. D. Faust and O. M. Aly, "Chemistry of Water Treatment." Butterworth, Boston, 1983.
4. S. D. Faust and O. M. Aly, "Adsorption Processes for Water Treatment." Butterworth, Boston, 1983.
5. B. A. Bolto and L. Pawlowski, "Wastewater Treatment by Ion Exchange." Spon, London and New York, 1987.
6. J. N. Armor, ed., "Environmental Catalysis," ACS Symp. Ser. No. 552. American Chemical Society, Washington, DC, 1992.
7. G. F. Vandegrift, D. T. Reed, and I. R. Tasker, eds., "Environmental Remediation: Removing Organic and Metal Ion Pollutants," ACS Symp. Ser. No. 509. American Chemical Society, Washington, DC, 1992.

CHAPTER TWO

2.1. INTRODUCTION

Although we may not always recognize it, electrochemistry and electrochemical processes have an impact on our everyday life in many different ways. Batteries, corrosion, commodity chemicals, metallurgy, and electroplating are but a few of the important examples. Before we begin to examine how electrochemistry can play a useful role in environmental science and technology, it is first necessary to review some fundamental concepts that can then be used as building blocks for later chapters in this book.

We begin with an examination of charge and mass transport, then move to a discussion of electrode/electrolyte interfaces and electrochemical cells, review thermodynamic and kinetic aspects related to these systems, and finally focus on individual topics that form the prelude to more detailed and specific descriptions in later chapters. These include Pourbaix diagrams (Chapter 3), electroanalytical chemistry (Chapter 4), mass transport and electrochemical reactor engineering (Chapter 5), and semiconductor electrochemistry (Chapter 6), the companion material being shown in parentheses in each case. The reader may, therefore, wish to peruse each of these sections in conjunction with its detailed counterpart as a unit.

2.2. CURRENT, CHARGE, AND POTENTIAL

It is conceptually useful to treat electricity in a manner analogous to fluid flow. Thus the ability of a medium to conduct electricity is related to its electrical resistance. A continuous range of materials behavior exists from *insulators* to *conductors* with an important intermediate category comprising *semiconductors*. The unit of current in the SI (Systéme International d'Unités) system is the ampere (A). The current, I, is defined as the increment of charge, dQ flowing across a plane in unit time:

$$I = \frac{dQ}{dt} \tag{2.1}$$

The unit of electric charge in the SI system is the coulomb (C), and according to Eq. (2.1), 1 ampere equals 1 coulomb per second. Another important quantity is the *current density, i*:

$$i = I/A \tag{2.2}$$

which is the current flowing per unit cross-sectional area, A, of a plane (A/m²).

It must be noted that the fluid analogy is valid for current flow through a medium. That is to say, just as a liquid or a solid medium behaves almost incompressibly, the same current, I, flows across any given plane in a conductor and indeed through an electric circuit in which all the conductive elements are connected in series. Note, however, that the current *density* (Eq. (2.2)) will be much greater in an electrical wire than, for example, in a cylindrical conductor of greater dimensions, even though the same current may flow through both.

An *electrochemical cell* is unique in that it comprises the flow of both *electronic* and *ionic* charges. Recall that our definition of current, in general, encompasses the movement of electric charge. Thus, we have to differentiate an *electronic* current from an *ionic* current, and a given material often is an excellent ionic conductor but not an electronic conductor. We shall have opportunities to examine these variant types of behavior in the sections that follow. However, unless explicitly stated, a current or a conductor here refers to the *electronic* component.

In electrochemical terms, a useful measure of the quantity of charge is the Faraday constant, F. It is defined as the charge borne by one mole of electrons or equivalently, by 1 mole of univalent cations.

$$F = N_A e_o = (6.0220 \times 10^{23} \text{ mol}^{-1})(1.6022 \times 10^{-19} \text{ C}) = 96,485 \text{ C mol}^{-1} \quad (2.3)$$

In Eq. (2.3), N_A is the Avogadro number and e_o is the charge on an electron.

What is the driving force for flow of electronic charge from one location in space to another? Again, the "water-in-a-pipe" analog is useful. Pressure is the driving force for fluid flow. The electrical counterpart of pressure is the *electrical potential*, and we use the symbol ϕ for it. It is measured in volts (V), and we know from elementary physics that 1 volt equals 1 joule per coulomb (i.e., 1 V = 1 J/1 C). Unlike with pressure, however, it is not possible to measure an absolute potential. The measurement of absolute potential brings with it both conceptual and practical difficulties, and it is possible to measure only a *difference* between two potentials. We term this a *potential difference* or simply a *voltage*.

Electrons flow from a point of lower (i.e., more negative) electrical potential toward a more positive one. This is the "downhill" direction, much like a water column flowing downhill across a gradient of potential energy (Figure 2.1). Note that the *direction* of electric current is often defined counter to the flow of electronic charge. However, we shall adhere to the description in terms of the direction of electron (or, in some cases, ion) movement rather than the convention adopted, for example, by the engineering or physics community. A more negative potential also corresponds to a higher potential energy because the energy, E, is equal to $-Q_e V$. This leads to the useful unit of electron–volt (eV) for energy, which often is used in spectroscopy and semiconductor physics. The reference level for the physics and the semiconductor device communities is the vacuum level; that is, the electron energy in a vacuum and infinitely removed from bulk medium effects. This is referred to as the *zero energy level*. On the other hand, electrochemists and electrochemical engineers prefer to use a *standard reference electrode* such as the *standard hydrogen electrode* (SHE) and assign to it a potential of zero volts. All other potentials are expressed relative to this reference electrode. The location of the SHE on the vacuum scale has been the topic of much discussion and variant values have been quoted for it.[1] However, a value of -4.44 eV now appears to be the most tenable choice. Figure 2.2 provides a comparison of the two scales on this basis.

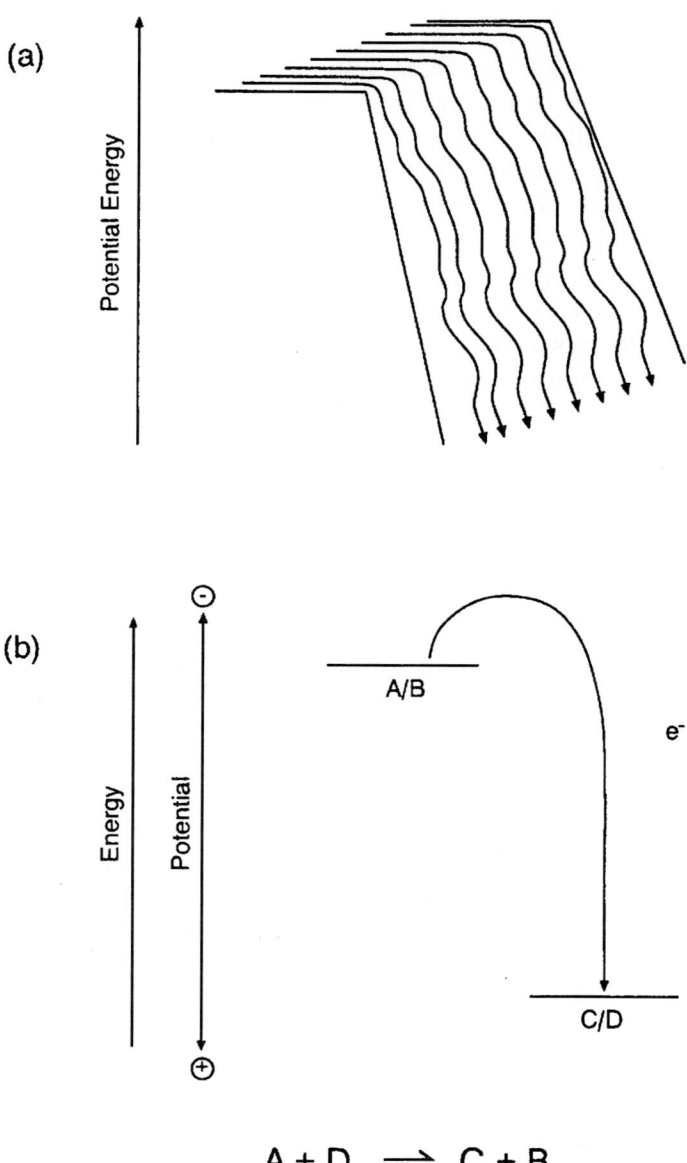

FIGURE 2.1. Analogy between fluid flow (a) and spontaneous electron transfer (b) from one redox couple, A/B, to another, C/D.

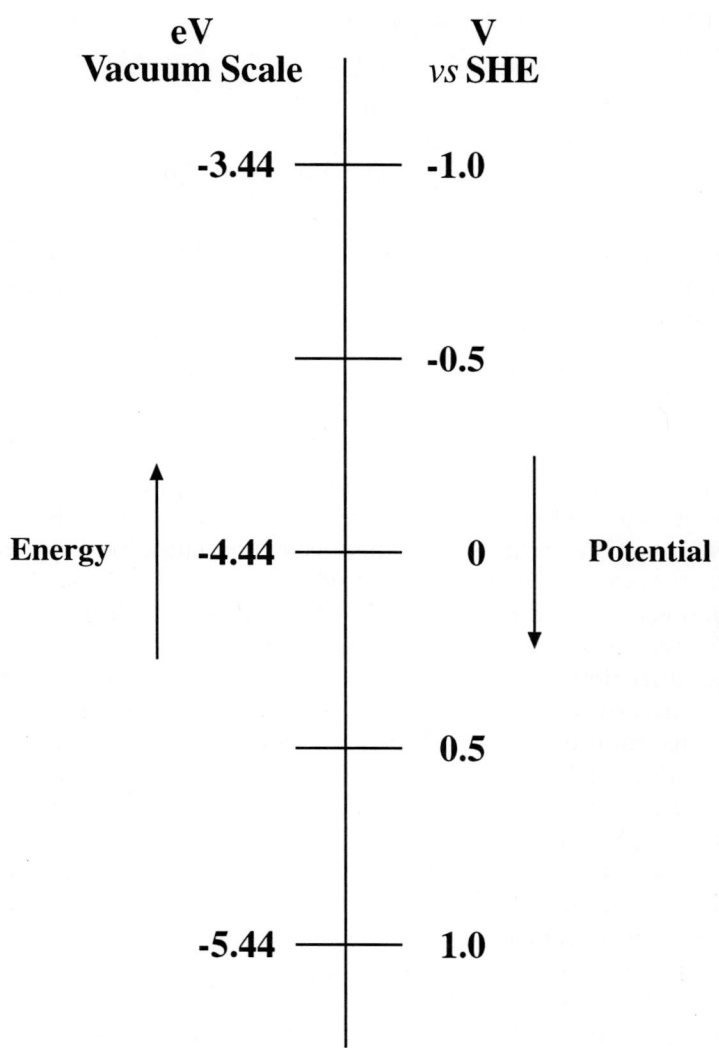

FIGURE 2.2. Comparison of the vacuum electron energy and the standard hydrogen electrode scales.

2.3. CHARGE AND MASS TRANSPORT

The flux, **J**, of any species, j, across a plane of cross-sectional area, A, is given by

$$\mathbf{J}_j = \underbrace{D_j \nabla C_j}_{\text{diffusion}} - \underbrace{\frac{z_j F}{RT} D_j C_j \nabla \phi}_{\text{migration}} + \underbrace{C_j \mathbf{u}}_{\text{convection}} \qquad (2.4)$$

The bold-faced quantities in Eq. (2.4) signify that they are vectorial; that is, they have both magnitude and direction. The flux has the units of mol m^{-2} s^{-1}. The Laplace operator ∇,

$$\nabla = \mathbf{i}\frac{\partial}{\partial x} + \mathbf{j}\frac{\partial}{\partial y} + \mathbf{k}\frac{\partial}{\partial z} \qquad (2.5)$$

models the system in a rectangular region of space with **i**, **j**, and **k** being the unit vectors along the x, y and z coordinates, respectively. In Eq. (2.4), D_j is the diffusion coefficient (m^2/s), C_j is the species concentration (mol/m^3), $\nabla \phi$ is the potential gradient (i.e., the *electric field*, V/m), z_j is the ionic charge, **u** is the solution velocity, and F, R, and T have their usual significance.

We see that the overall flux and mass transport is driven by three terms: diffusion, migration, and convection. The goal of electrochemical cell engineering is to maximize mass transport of the electroactive species to the electrode surface, and this is true regardless of whether the cell will be employed for pollutant sensing (Chapter 4) or pollutant degradation (Chapter 5). We shall explore strategies for doing this later in this chapter. But for now, let us simply note that the driving force for each of the three components in Eq. (2.4) is concentration, potential, and density gradients, respectively. Further, convection can be either natural or forced in an electrochemical system. The former arises even in an unstirred or stagnant solution because of the density gradients generated as a result of electrolysis in the electrochemical cell.

Two other concepts related to the flow of charge and current flow are pertinent before we move on to a discussion of electrochemical cells in general. The first is the definition of an *electrical conductance*, K:

$$K = \kappa (A/l) \qquad (2.6)$$

The value of K is directly proportional to the cross-sectional area, A, perpendicular to the electrical field vector and it is inversely proportional to the length (l) of the segment along the field. The unit of K is the Siemen (S) and the proportionality constant, κ (also called the *electrical conductivity*) is expressed in S m^{-1}. For a given species, κ is given by:

2.3. Charge and Mass Transport

$$\kappa_j = ne_o\mu = c_j |z_j| \mu_j \qquad (2.7)$$

That is, the conductivity intuitively may be expected to be governed by the number density (or concentration) of the species, its charge e_o (or z_j if the species of interest is an ion) and by its migration velocity, μ, which is termed the *mobility*. Mobility has the units of $m^2\ V^{-1}\ s^{-1}$. The mobilities of some species (both ionic and electronic) are contained in Table 2.1; they can be seen to vary over a wide range. As is to be expected, ionic mobilities are much lower than their electronic counterparts, and even among these, protons (or more correctly, the hydronium ions, H_3O^+) have very high mobility relative to some larger species. Note that Li^+ ions have rather low mobility (relative to K^+, for example) attesting to the effect of ionic solvation (i.e., hydration) in aqueous media. The mobility and the diffusion coefficient are related through the Nernst–Einstein law:

$$\mu_j = |z_j| \frac{F}{RT} D_j \qquad (2.8)$$

This equation underlines the fact that many factors that enhance one parameter (temperature, solvent viscosity, etc.) have a similar influence on the other.

Table 2.1. Ion and Electronic Carrier Mobilities in Various Media at 298 K

Specie	$\mu / m^2\ s^{-1}\ V^{-1}$
H^+	362.5×10^{-9}
K^+	76.2×10^{-9}
Na^+	51.9×10^{-9}
Li^+	40.1×10^{-9}
$(C_4H_9)_4N^+$	19.8×10^{-9}
OH^-	204.8×10^{-9}
SO_4^{2-}	82.7×10^{-9}
Cl^-	79.1×10^{-9}
Br^-	81.3×10^{-9}
e^-	6.7×10^{-3} (Ag)
	$\sim 10^{-4}$ (TiO_2, rutile)
	0.15 (Si)
	0.02 (ZnO)
h^+	0.06 (Si)

Sources: K. B. Oldham and J. C. Myland, "Fundamentals of Electrochemical Science." Academic Press, San Diego, CA, 1994; S. M. Sze, "Physics of Semiconductor Devices." Wiley (Interscience), New York, 1969.

Note: The values for ions are for extreme dilution and aqueous solutions. The medium for the electronic carrier in each case is listed in parentheses.

It is important to recognize the *chemical* significance of electric current as an indicator of *reaction rate*; that is, a higher current translates to a faster reaction. This can be seen by combining Eq. (2.1) with Faraday's law:

$$Q = nFN \qquad (2.9)$$

In Eq. (2.9), n is the electron stoichiometry and N is the number of moles undergoing electrochemical reaction. Thus,

$$I = \frac{dQ}{dt} = nF\frac{dN}{dt} \qquad (2.10)$$

We see that the current indeed has the rate dimension of moles per unit time. This equality will become important when we associate current flow with the rate of pollutant decomposition.

2.4. ELECTRODE/ELECTROLYTE INTERFACES AND ELECTROCHEMICAL CELLS

An electrode/electrolyte interface perhaps is the most important component of an electrochemical system. This is the junction between the electronic conductor (electrode) and the ionic conductor (electrolyte). The capacitive properties of this interface limit the temporal response (i.e., the transient behavior) of the electrochemical cell and also its function as an analytical sensor. We adopt the usual electrochemical practice of employing vertical slashes to denote phase boundaries within the cell. Thus the prototype Daniell cell may be schematically represented by

$$- \text{Zn} \mid \text{Zn}^{2+}_{(aq)} \parallel \text{Cu}^{2+}_{(aq)} \mid \text{Cu} +$$

The two slashes in the middle of the cell signify the presence of a *liquid* junction. As is usual with thermodynamic practice, standard states denote solution species at unit activity (approximately unit molarity) and gaseous species at one atmospheric partial pressure. Further, the SHE considered in a preceding section would correspond to the *half-cell*:

$$\text{Pt} \mid \text{H}_2 \text{ (1 atm), H}^+ \text{ (1 M)}$$

Note that the platinum in this case functions as an *inert* electrode.

Two half-cells compose an overall *electrochemical cell*, and the cell voltage is given by the difference between the two half-cell potentials. Particular attention must be paid to the sign of these potentials. We recommend the *exclusive* use of *reduction potentials* in the computation of cell voltages.

2.4. Electrode/Electrolyte Interfaces and Electrochemical Cells

To illustrate, let us return to the Daniell cell. The two relevant half-cell reactions are

$$Zn^{2+}_{(aq)} + 2e^- \leftrightarrows Zn$$
$$Cu^{2+}_{(aq)} + 2e^- \leftrightarrows Cu$$

The *standard reduction potentials* of these reactions (on the SHE scale) are -0.76 V and +0.34 V, respectively (see Appendix C). We have seen that the *spontaneous* direction of electron flow is from a negative terminal to the positive one. Therefore, when the Zn and Cu electrodes are short-circuited, electrons will flow from the Zn terminal to the Cu half-cell. In other words, the zinc electrode will corrode and copper will plate onto the copper rod. The cell is said to operate in the *galvanic* mode. All batteries operate in this manner in the *discharge* mode.

How long will the current flow through the external load? As with the water flow downhill (Figure 2.1), the situation cannot last forever, and the cell will "die" when the electrical pressure is equalized on the two half-cells, or alternatively when the cell voltage becomes zero. This will happen when the positive shift of the zinc half-cell potential is exactly counterbalanced by the negative shift undergone by the copper half-cell potential.

The cell voltage is easily computed geometrically as illustrated in Figure 2.3. Some textbook treatments adopt a reversal of the zinc half-cell reaction (as written previously) accompanied by *a switch in sign of the corresponding standard reduction potential,* followed by the addition of the two half-cell potentials; that is, 0.34 V + (+0.76 V) = 1.10 V. We prefer the alternative strategy of adhering to standard reduction potentials *regardless of whether an oxidation or reduction reaction is involved*.

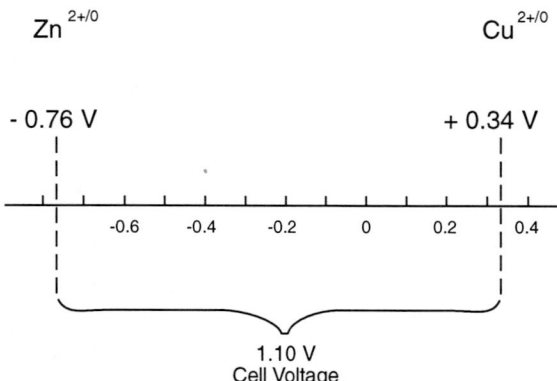

FIGURE 2.3. Geometric computation of the cell voltage in a Daniell cell. All solution species are assumed to be at unit activity.

In terms of relevance to the theme of this book, the contrathermodynamic direction is often more important. Thus the preceding reaction in the Daniell cell may be reversed (i.e., to plate Zn metal onto the zinc rod and to corrode the copper rod) by applying an external voltage of 1.10 V to the cell. The zinc half-cell attains a potential more positive than -0.76 V, and the Cu half-cell potential becomes more negative than 0.34 V. In other words, the zinc half-cell becomes the positive terminal and the copper rod becomes negative relative to it. When operated in this mode, the cell becomes an *electrolysis* unit; that is, the passage of electric current produces a chemical change.

The *sign* of the cell voltage becomes important in differentiating whether the cell is operating in the galvanic or electrolytic mode. We know from thermodynamics that the Gibbs free-energy change is negative for a spontaneous reaction system. Thus, for a galvanic cell,

$$\Delta G = -nFE \tag{2.11a}$$

or $$\Delta G° = -nFE° \tag{2.11b}$$

The voltage in an operating Daniell cell (for all cell participants in their standard states) thus will be +1.10 V.

In many instances, we are interested in the individual electrode/electrolyte interface (rather than the overall cell reaction). Two examples may be cited, both using the Cr(VI)/Cr(III) *redox* system as an example. When a spent chrome etchant is to be regenerated, we are interested in only the electrode at which the process

$$Cr^{3+} \rightarrow Cr^{6+} + 3\,e^-$$

occurs. In other words, as long as this process is not reversed at the other electrode, we really do not care what happens at that half of the cell! Second, we know that Cr(VI) is carcinogenic and environmentally of greater concern than Cr(III). Suppose the task is to design an electrochemical sensor for Cr(VI) assay. In this instance, we would focus our efforts on designing and optimizing an electrode/electrolyte interface capable of selectively and sensitively monitoring the $Cr^{6+/3+}$ redox process.

In both these examples, the second electrode/electrolyte interface serves only to complete the electrical circuit. Note, however, that the same current flows through both, and we must be careful that this counterelectrode is not rate-limiting in applications where the overall flux is of importance as, for example, in hazardous chemical treatment. In other words, the counterelectrode/electrolyte interface must be optimized to have fast reaction kinetics. The preceding discussion also underlines the importance of ensuring that the

2.4. Electrode/Electrolyte Interfaces and Electrochemical Cells

product generated in the counterelectrode half-cell does not reverse the targeted electrode reaction or otherwise interfere with our test interface. This can be done using two-compartment cell designs (Figure 2.4). Cell designs are discussed later in this chapter (Section 2.9).

Polarization is an inevitable consequence of current flow at an electrode/electrolyte interface. This phenomenon can have many underlying reasons including charge- and mass-transfer limitations. Notwithstanding the mechanism involved, the electrode potential shifts to an extent depending on the magnitude of the current flowing across the interface. Hence, unless the current flowing in the electrochemical cell is very small (as for example when the electrode area is very small or when potentiometric measurements are being made, see later), a two-electrode cell containing the test electrode and a reference electrode generally cannot be used. This is because the latter will polarize and its potential will start to shift such that the potential of the working (test) electrode relative to it will become of uncertain magnitude. Simply put, we have the untenable situation of two unknown (half-cell) potentials whose difference is being monitored! Therefore, three-electrode cell geometry is often the norm, especially in laboratory situations where the current-potential behavior at the test electrode/electrolyte interface must be characterized in a reliable fashion. This is done using a device called the potentiostat–galvanostat (Figure 2.5).

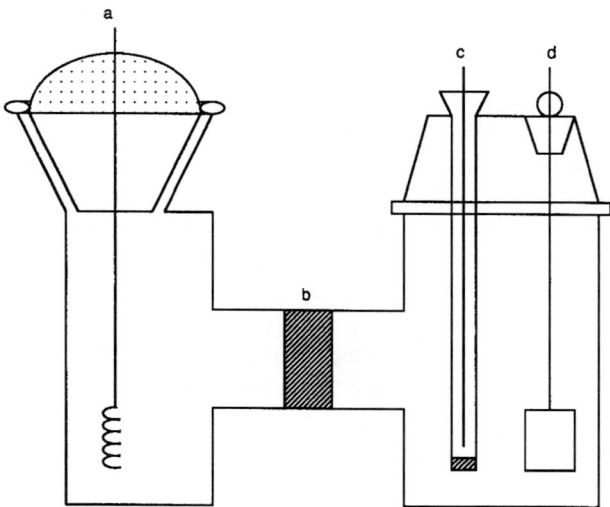

FIGURE 2.4. Schematic diagram of a two-compartment cell design: (a) Pt spiral counterelectrode, (b) porous glass or ceramic frit, (c) reference electrode, and (d) working electrode. Provisions for sample addition and cell purge (with inert gas) may be made and are not shown.

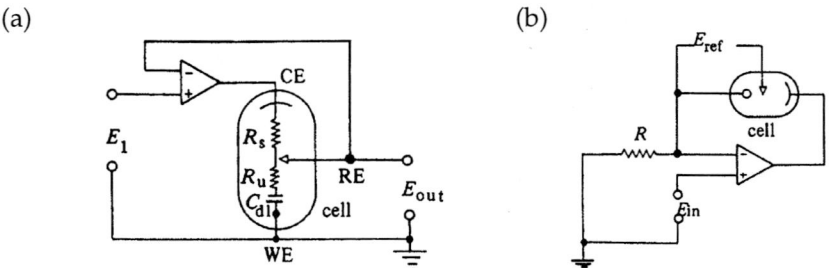

FIGURE 2.5. Schematic diagram of operational amplifier based circuitry for (a) potentiostat and (b) galvanostat: CE = counterelectrode, RE = refernce electrode, and WE = working electrode. Refer to text for other symbols. (Reproduced with permission from The Southampton Electrochemistry Group, "Instrumental Methods in Electrochemistry." Ellis Horwood, Chichester, 1985.)

In this arrangement, the current is passed between the working electrode and the counterelectrode. The potential of the working electrode is monitored relative to a separate reference electrode positioned with its tip close to the working electrode surface. A Luggin capillary positioned very close to the working electrode surface often is used for this purpose. Negligible current is drawn through the reference electrode in the potential measurement loop such that polarization of the latter is minimal. The placement of the reference electrode relative to the working electrode is important, especially in nonaqueous media and in flow cell geometries. This is because of the ohmic (IR_s) drop in the intervening electrolyte (see Section 2.5.4). This is not compensated for in the arrangements in Figure 2.5, although strategies exist for compensation of this voltage drop.[2]

Using the potentiostat/galvanostat, the potential of the working electrode can be precisely controlled to a set value relative to the reference electrode. Alternatively, a constant amount of current can be passed through the test electrode/electrolyte interface in the galvanostatic mode. A disadvantage with this approach is that the test electrode potential will sometimes drift to a value in an undesirable regime (e.g., solvent electrolysis) when the reactant (substrate) supply to the interface is depleted. Thus the potential will adjust to the value required to sustain the value set for the current. Finally, in conjunction with a programmer, the potential can be swept linearly in the so-called potentiodynamic mode. This is especially useful for analytical purposes such as in a *voltammetry* experiment (see Section 2.6).

2.5. Thermodynamics and Kinetics in Electrochemical Systems

The reader will note that, in the preceding discussion, we have avoided use of the terms *anode* and *cathode* for the working and counterelectrodes in the cell. This terminology is not especially useful, since at a given electrode/electrolyte interface, the reaction can proceed in either the forward or the backward direction. Thus the same electrode can function either as an anode or a cathode, depending on whether an oxidation or a reduction reaction is occurring on it. Indeed, in many situations (e.g., semiconductor particles, a corrosion system), both types of reactions occur simultaneously on the same surface! Of course, in these latter cases, there is no *net* current flow because the cathodic and anodic current components are exactly balanced. We shall treat such cases later in our discussion of mixed potentials.

2.5. THERMODYNAMICS AND KINETICS IN ELECTROCHEMICAL SYSTEMS

2.5.1. Use of Standard Reduction Potentials

Standard reduction potentials are a crucial aid to understanding and rationalizing phenomena such as electron transfer and catalysis. Thus, species with high (i.e., positive) standard reduction potentials are good oxidizing agents and are easily reduced. Examples of environmental import include Cr(VI) and H_2O_2. Conversely, chemical species with low (i.e., negative) standard reduction potentials are good reducing agents and are easily oxidized. Alkali metals and sulfite are examples of this group. When two redox couples with disparate standard reduction potentials coexist in solution, spontaneous electron flow will occur from the couple with a negative standard reduction potential to the couple with a more positive standard reduction potential (see Figure 2.1). Thus, permanganate is expected to oxidize Fe(II) to Fe(III), the two relevant standard reduction potentials being 1.512 V and 0.77 V on the SHE scale (Appendix C).

By the same token, MnO_4^- and $Cr_2O_7^{2-}$ ions are thermodynamically expected to oxidize water (to O_2). The good stability of these solutions in water (as all undergraduate chemistry students can attest to in the laboratory) is a result of sluggish electron transfer *kinetics*. Often a *catalyst* is required to *mediate* charge transfer and accelerate the reaction rate. Herein lies the limitation of thermodynamic predictions, thermodynamics tells us only whether a given process is feasible or not. It does not afford insights into how *fast* the process will occur.

Figure 2.6 contains an extension of the graphical method first shown in Figure 2.3 and is a useful aid for understanding and rationalizing the *chemical* significance of half-cell (or redox) potentials.

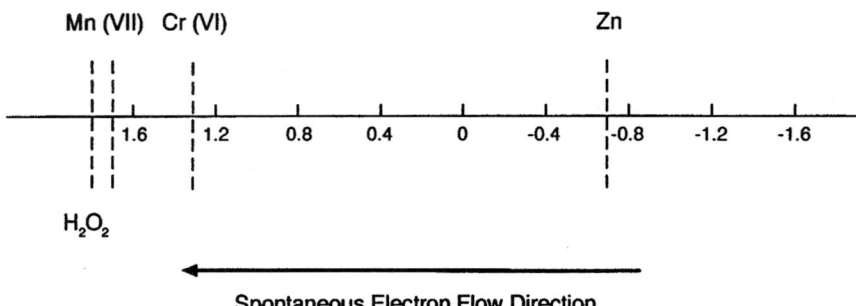

FIGURE 2.6. Oxidizing and reducing agents shown on a common scale based on the standard hydrogen electrode.

2.5.2. Redox Potentials and the Nernst Equation

So far we have mainly considered cases where all the cell participants are in their standard states. What happens when the concentration (or, more correctly, the activity) of a given ionic species is reduced below 1 M? We have the Nernst equation:

$$E_{eq} = E° - \frac{RT}{nF} \ln Q_R \tag{2.12}$$

to describe the nonstandard situation. This equation, which is equally applicable to both the overall cell reaction and a half-cell reaction, will be considered only for the latter case for reasons enumerated earlier. In Eq. (2.12), E_{eq} and $E°$ are the equilibrium potential and the standard reduction potential, respectively, and Q_R is the reaction quotient. Consider the half-reaction:

$$Zn^{2+}(aq) + 2\,e^- \leftrightarrows Zn(s) \tag{2.13}$$

which is the negative electrode in the Daniell cell. The reaction quotient for this system is

2.5. Thermodynamics and Kinetics in Electrochemical Systems

$$Q_R = a_{Zn(s)}/a_{Zn^{2+}(aq)} \tag{2.14a}$$

where a represents the activities. As is the thermodynamic norm, the activity of a pure state is taken to be unity. Thus,

$$Q_R = 1/a_{Zn^{2+}(aq)} \tag{2.14b}$$

Substitution into Eq. (2.12) yields

$$E_{eq} = E° + \frac{RT}{2F} \ln a_{Zn^{2+}(aq)} \tag{2.15}$$

At 25°C, the collection of constants preceding the logarithmic term in Eq. (2.15) can be evaluated, and along with the conversion of natural to base 10 logarithms, Eq. (2.15) becomes

$$E_{eq} = E° + \frac{0.0592}{2} \log a_{Zn^{2+}(aq)} \tag{2.16}$$

Thus the potential shifts by ca. 60 mV for a one-electron process for a decade change in the specie concentration. This shift will be negative for reduction and positive for an oxidation process. In other words, the process becomes more difficult as the substrate activity is lowered, which makes intuitive sense.

The Nernst equation has the general form

$$E_{eq} = E° + \frac{RT}{nF} \ln \frac{a_O}{a_R} \tag{2.17}$$

In Eq. (2.17), the subscripts O and R denote the oxidized and reduced forms of a redox couple.

In practical terms, molar concentrations are more convenient to use rather than activities. Since,

$$a_j = \gamma_j C_j \tag{2.18}$$

where γ_j is the activity coefficient, Eq. (2.17) becomes

$$E_{eq} = E° + \frac{RT}{nF} \ln \gamma_O/\gamma_R + \frac{RT}{nF} \ln \frac{C_O}{C_R} \tag{2.19}$$

This can be reexpressed as

$$E_{eq} = E°' + \frac{RT}{nF} \ln \frac{C_O}{C_R} \tag{2.20}$$

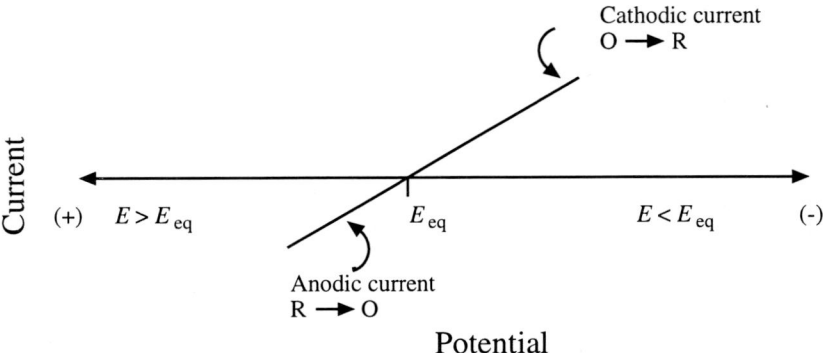

FIGURE 2.7. The electrochemical consequence of perturbing a redox couple poised at a potential, E_{eq} by application of an external bias potential.

where $E^{o\prime}$ is called the *conditional* or the *formal* potential. The key difference between $E°$ and $E^{o\prime}$ is that the latter is medium dependent. However, use of $E^{o\prime}$ has the advantage of operational convenience.

For a redox couple containing soluble components, Eqs. (2.17) and (2.20) underline that activity or concentration *ratio* controls $E°$ and $E^{o\prime}$, respectively. Thus, a Pt wire immersed in either 1 M each or 1 mM each of Fe^{3+} and Fe^{2+} ions should register the same $E^{o\prime}$ value (e.g., 0.77 V *vs.* SHE in 1 M HCl). The $E^{o\prime}$ values may differ slightly in the two instances and in another medium (e.g., 1 M H_2SO_4) because of activity differences (see Eq. (2.19)). This is the equilibrium null or the "rest" value for the interface. When the Pt electrode potential is made positive of 0.77 V (with a power source or a potentiostat), some of the Fe^{2+} species will be oxidized to Fe^{3+} and an anodic current will flow. Conversely, when the electrode potential is made slightly negative of 0.77 V, the reaction

$$Fe^{2+} \rightarrow Fe^{3+} + e^-$$

will occur and the concentration ratio will accordingly adjust to a new value as dictated by Eq. (2.20).

Figure 2.7 illustrates the two scenarios. We have in effect begun to map the current–potential behavior of a test electrode/electrolyte interface.

2.5.3. Electrode Reaction Kinetics

The implicit assumption in the preceding discussion was that the charge-transfer kinetics was infinitely fast, and the concentrations (or, more precisely, the activity ratio) were always equilibrium values as dictated by the

2.5. Thermodynamics and Kinetics in Electrochemical Systems

Nernst equation. Such redox couples are viewed as *electrochemically reversible* or *Nernstian* in behavior. But we know that the $Fe^{3+/2+}$ aquo complex does not have very fast reaction kinetics. In such cases, the redox couple often will obey the Butler–Volmer equation:

$$i = nF(r_{ox} - r_{red})$$
$$= nFk^{\circ\prime}\{C_R^s \exp[\frac{(1-\alpha)nF(E-E^{\circ\prime})}{RT}] - C_O^s \exp[\frac{-\alpha nF(E-E^{\circ\prime})}{RT}]\} \quad (2.21a)$$

where r_{ox} and r_{red} are the rates of the anodic and cathodic reaction branches (see Figure 2.7), α is called the symmetry factor, $k^{\circ\prime}$ is the *conditional* rate constant (or, alternatively, the formal standard heterogeneous rate constant), and the superscripts on the concentration terms, C_R and C_O, respectively for the reduced (e.g., Fe^{2+}) and oxidized (e.g., Fe^{3+}) forms of the redox couple, signify that these are the concentrations *at the electrode surface*.

Another useful measure of charge-transfer kinetics is the exchange current density, $i^{\circ\prime}$:

$$i^{\circ\prime} = nFC^\circ k^{\circ\prime} \quad (2.21b)$$

The concentration C° in Eq. (2.21) is for the particular (standard) case when $C_O^* = C_R^*$, where the asterisks denote *bulk* concentration. The $i^{\circ\prime}$ parameter may be regarded as an indicator of the "idling speed" of the redox reaction at a particular electrode/electrolyte interface. The higher the value of $i^{\circ\prime}$, the smaller is the magnitude of the "kick" necessary to initiate a *net* current flow in either direction.

Table 2.2 lists the $k^{\circ\prime}$, $i^{\circ\prime}$, and α values for selected redox couples. The effect of the parameters α and $i^{\circ\prime}$ on the shape of the current–potential curve is shown in Figures 2.8A and 2.8B, respectively. Now we have a more quantitative basis for ordering redox couples in terms of their kinetic facility. Thus, redox couples with $i^{\circ\prime}$ values greater than $\sim 10^4$ A/m² and $\alpha = 0.5$ qualify for being classified as reversible or Nernstian. On the other hand, a redox system with an $i^{\circ\prime}$ of 10^{-5} A/m² (Figure 2.8B) would be irreversible. Many redox couples fall somewhere between the two extremes in the *quasi-reversible* regime (see Table 2.2).

2.5.4. Ohmic, Activation and Concentration Polarization

The shift in the potential caused by the passage of current is called *polarization* as was briefly discussed in Section 2.4. Polarization has three underlying causes: ohmic, activation, and concentration. Their effects are more or less additive and the potential shift is called an *overpotential*, η:

$$E = E_{eq} + \eta \quad (2.22)$$

$$\eta_{tot} = \eta_{ohmic} + \eta_{act} + \eta_{conc} \quad (2.23)$$

Table 2.2. Kinetic Parameters for Selected Electrode Reactions

Reaction	Medium	Electrode	α	$k^{o\prime}$ m s^{-1}	$i^{o\prime}$ A/m^2
$Fe^{3+}(aq) / Fe^{2+}(aq)$	0.5 M $HClO_4$	Pt	0.50	9×10^{-8}	87 (25)
$Fe^{3+}(aq) / Fe^{2+}(aq)$	1.0 M HCl	C	0.59	1.2×10^{-6}	116 (21)
$MnO_4^-(aq) / MnO_4^{2-}(aq)$	1.0 M KOH	Pt	—	3.2×10^{-5}	3,088 (20)
$Ag^+(aq) / Ag(s)$	1.0 M $HClO_4$	Ag	—	2×10^{-4}	19,297 (25)
$Fe(CN)_6^{3-}(aq) / Fe(CN)_6^{4-}(aq)$	1.0 M KNO_3	Pt	0.49	6.6×10^{-4}	63,680 (35)

Source: Data from K. B. Oldham and J. C. Myland, "Fundamentals of Electrochemical Science." Academic Press, San Diego, CA, 1994.

Note: Values in parentheses denote the measurement temperature in °C.

Note that the magnitude of η depends on the current density, i. For ohmic polarization,

$$\eta_{ohmic} = I R_s \qquad (2.24)$$

In Eq. (2.24), R_s is the resistance of the solution between the working and reference electrodes. This expression shows why the placement of the reference electrode relative to a working electrode in a three-electrode arrangement (Figure 2.5) is critical especially in the (rather resistive) nonaqueous solution media.

The Butler–Volmer equation (Eq. 2.21a) may be rewritten to bring out the activation overpotential term:

$$i = i^{o\prime} \{ \exp[\underbrace{(1-\alpha) \tfrac{nF}{RT} \eta_{act}}_{\text{anodic}}] - \exp[\underbrace{-\alpha \tfrac{nF}{RT} \eta_{act}}_{\text{cathodic}}] \} \qquad (2.21c)$$

The fundamental equation in this form illustrates several useful points:
(1) When η_{act} is large and positive, the anodic branch dominates and $i_{ox} \gg i_{red}$.
(2) When η_{act} is large and negative, the cathodic current branch dominates and $i_{red} \gg i_{ox}$.

2.5. Thermodynamics and Kinetics in Electrochemical Systems

(A)

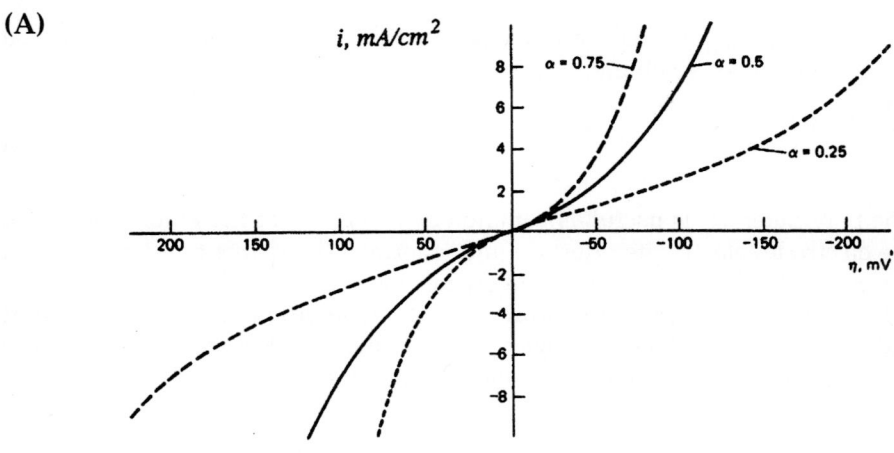

(B)

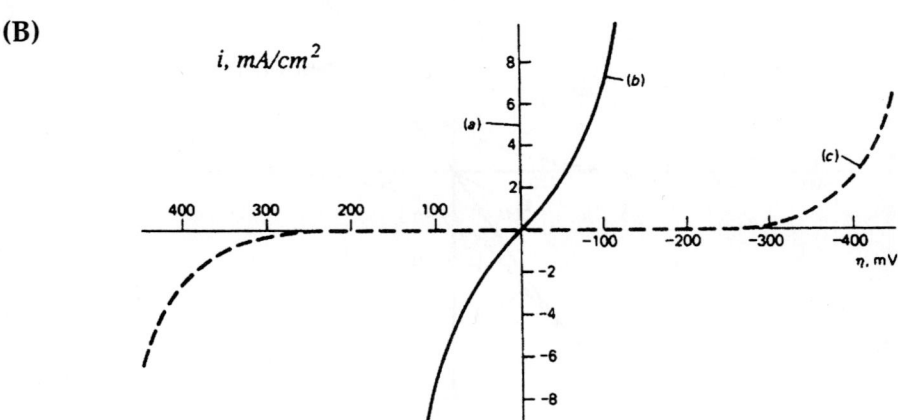

FIGURE 2.8. Effect of the transfer coefficient (α) and the exchange of current density (i_o) on the shape of current–potential curves. In (A) $n=1$, $T=298$ K, and $i^{o\prime}=10^{-6}$ A/cm^2; and in (B) $\alpha=0.5$, $n=1$, $T=298$ K, and $i^{o\prime}= 10^{-3}$, 10^{-6}, and 10^{-9} (A/cm^2) for curves (a), (b), and (c), respectively. (Reproduced with permission from A. J. Bard and L R. Faulkner, "Electrochemical Methods." Wiley, New York, 1980.)

(3) η_{act} takes the same sign as the applied potential; that is, positive for anodic processes and negative for cathodic reactions.
(4) When $i_{ox} = i_{red}$, $i = i^{o\prime}$.
(5) When η_{act} is small (a few mV) relative to RT/nF, Eq. (2.21c) may be linearized to obtain

$$R_{ct} = \frac{RT}{nFi^{o\prime}} \qquad (2.25)$$

The parameter, R_{ct} is useful as yet another descriptor of the kinetic facility of the electrode/electrolyte interface and is termed the *charge-transfer resistance*. It is accessible via AC electrical impedance measurements.

Figure 2.9 illustrates the additivity of the anodic and cathodic current branches and their relationship with respect to the overall current–voltage curve for the particular electrode/redox electrolyte interface.

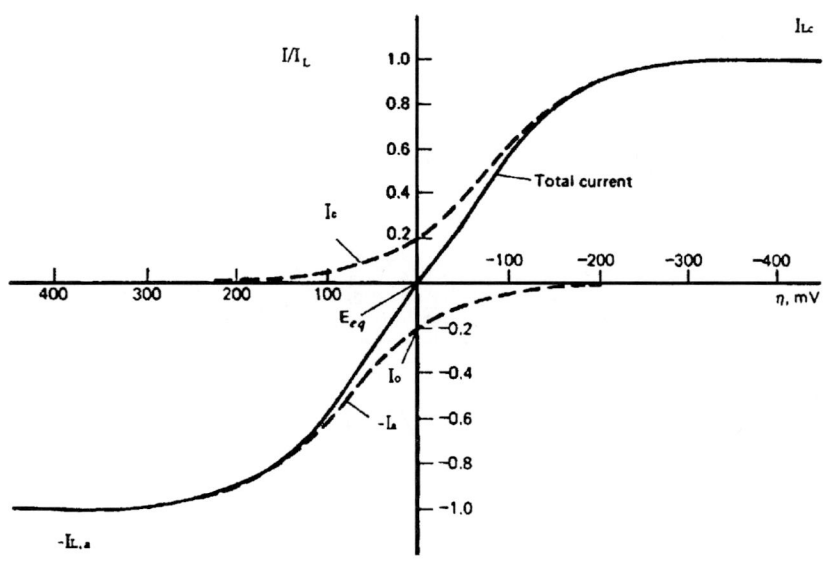

FIGURE 2.9. Current–potential curve illustrating the additivity of the component anodic and cathodic processes. The limiting currents, $I_{L,a}$ and $I_{L,c}$ are assumed to be equal in magnitude. Further, $I_O/I_L = 0.2$, $\alpha = 0.5$, $n = 1$, and $T = 298$ K. (Reproduced with permission from A.J. Bard and L.R. Faulkner, "Electrochemical Methods." Wiley, New York, 1980.)

2.5. Thermodynamics and Kinetics in Electrochemical Systems

Consider again the anodic current branch:

$$i \approx i_{ox} = i^{o'} \exp\{(1-\alpha)\frac{nF}{RT}\eta_{act}\} \text{ (when } i_{ox} \gg i_{red})$$
$$\ln i \approx \ln i^{o'} + (1-\alpha)\frac{nF}{RT}(E - E_{eq}) \tag{2.26a}$$

A similar treatment for the cathodic branch yields

$$\ln(-i) \approx \ln i^{o'} - \alpha \frac{nF}{RT}(E - E_{eq}) \tag{2.26b}$$

Equations (2.26a) and (2.26b) are known as Tafel equations. They illustrate a route to the analysis of current-potential data by plotting log i vs. potential. Such *Tafel plots* may be used to measure E_{eq}, $i^{o'}$, α, and n, as illustrated in Figure 2.10. The departure from linearity at low values of η_{act} illustrates that the assumptions $i_{ox} \gg i_{red}$ and $i_{red} \gg i_{ox}$ underlying Eqs. (2.26a) and (2.26b), respectively, become invalid in this regime. That is, the back reactions become important at low η_{act}.

Concentration polarization begins to limit the overall current flow at high values of η_{act}; that is, the electrode/electrolyte interface starts to become depleted of the substrate (reactant), $C_R^s < C_R^*$ and $C_O^s < C_O^*$ where the superscript s denotes concentrations at the interface. This is best illustrated using another form of the Butler–Volmer equation:

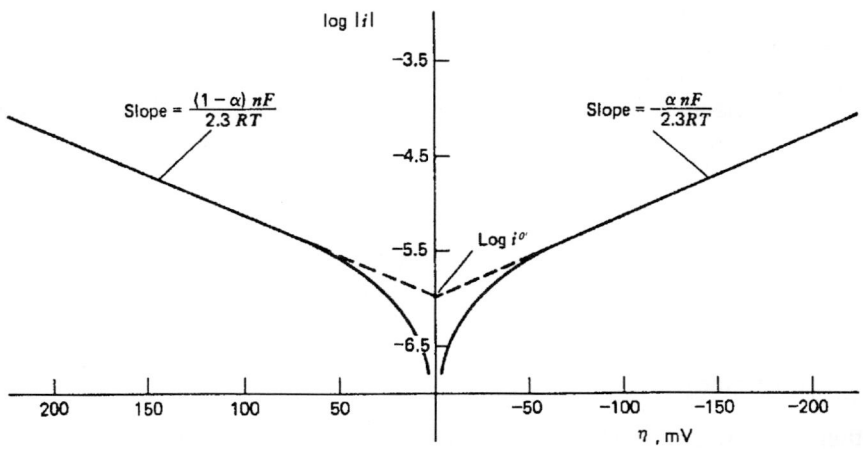

FIGURE 2.10. Tafel plots for anodic and cathodic branches of the current density–overpotential curve for $O + ne^- = R$ with $n = 1$, $\alpha = 0.5$, $T = 298$ K, and $i^{o'} = 10^{-6}$ A/cm². (Reproduced with permission from A.J. Bard and L.R. Faulkner, "Electrochemical Methods." Wiley, New York, 1980.)

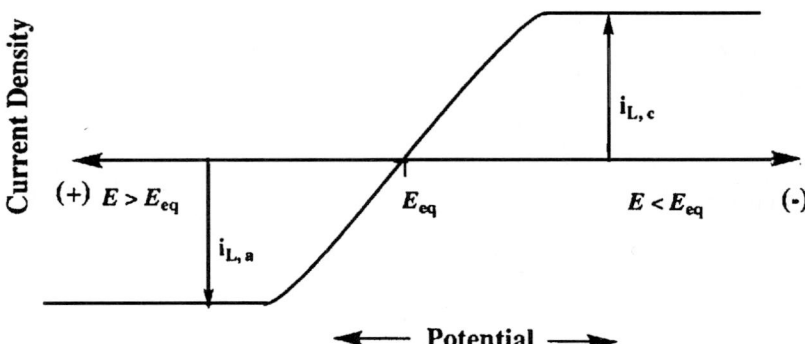

FIGURE 2.11. Limiting current flow in anodic and cathodic branches of the current-potential curve.

$$i \approx i^{o\prime} \left\{ \frac{C_R^{\,S}}{C_R^{\,*}} \exp\left[\frac{(1-\alpha)nF\eta_{act}}{RT}\right] - \frac{C_O^{\,S}}{C_O^{\,*}} \exp\left[\frac{\alpha nF}{RT}\right] \eta_{act} \right\} \quad (2.21d)$$

The exponential growth of the current–potential curve (in either direction) is ultimately kept under control by the factors $C_R^{\,S}/C_R^{\,*}$ and $C_O^{\,S}/C_O^{\,*}$, which describe the reactant supply to the interface. Thus the currents ultimately plateau at the two extremes to yield the corresponding *limiting currents* (Figure 2.11), so called because the $i_{L,a}$ and $i_{L,c}$ values are limited by $C_R^{\,*}$ and $C_O^{\,*}$ (the bulk concentrations of the redox species), respectively.

Quantitative treatments of concentration polarization are available, depending on whether the solution is quiescent or stirred (as in hydrodynamic voltammetry). We shall examine these in greater detail later in this chapter.

2.5.5. Pourbaix Diagrams

The discussion in Section 2.5.1 did not take into account the pH of the medium in which the electrochemical reaction is occurring. For example, consider the $Zn^{2+/0}$ redox couple. In addition to

$$Zn^{2+}(aq) + 2\,e^- \rightleftarrows Zn(s) \quad (2.13)$$

other equilibria are possible, including

$$Zn(OH)_2(s) + 2\,H^+ + 2\,e^- \rightleftarrows Zn(s) + 2\,H_2O \quad (2.27)$$

and

$$ZnO_2^{2-}(aq) + 2\,H_2O + 2\,e^- \rightleftarrows Zn(s) + 4\,OH^- \quad (2.28)$$

2.5. Thermodynamics and Kinetics in Electrochemical Systems

Reactions 2.13 and 2.28 are important in acidic and basic media respectively (see Figure 2.12a).

Predominance diagrams that take pH into account are known as Pourbaix diagrams. Figure 2.12 contains such diagrams for the Zn–H_2O (Figure 2.12a) and the H_2O solvent (Figure 2.12b) systems, respectively. These diagrams are plots of the standard reduction potential *vs.* the solution pH; all species are usually defined in their standard state unless otherwise noted. That is, the solution species are at unit activity (approx. 1 M), gaseous species are at 1 atm partial pressure, and the system is defined at 25°C. Similarly, pure phases and the solvent have unit activity.

Therefore, the zones in Figure 2.12 indicate the pH and potential regimes of predominance of a particular species. These diagrams are to be used much like phase diagrams. Consider Figure 2.12b, for example. Liquid water and the ions with which it is in equilibrium are stable only in the central rhombohedral zone. At a potential lying above the upper diagonal line, water should decompose according to

$$H_2O(l) \rightarrow O_2(g) + 4\,H^+(aq) + 4\,e^- \tag{2.29}$$

or

$$4\,OH^-(aq) \rightarrow O_2(g) + 2\,H_2O(l) + 4\,e^- \tag{2.30}$$

depending on the pH. On the other hand, at lower potentials (below the lower diagonal line), water can be reduced to H_2 via either

$$2\,H^+(aq) + 2\,e^- \rightarrow H_2(g) \tag{2.31}$$

or

$$2\,H_2O(l) + 2\,e^- \rightarrow H_2(g) + 2\,OH^-(aq) \tag{2.32}$$

At pH 0, these "critical" potentials lie at 1.23 V and 0 V, respectively, and they shift with pH at the rate of -59 mV/pH (at 25°C) according to the Nernst equation (2.12).

In practice, however, water is much more stable (especially toward oxidation) than suggested by Figure 2.12b. This is a major weakness of Pourbaix diagrams, in that they indicate only what is thermodynamically feasible. For example, the environmentally important specie, H_2O_2(aq) is completely absent from the Pourbaix diagram for dioxygen (Figure 2.12b) because it is thermodynamically labile toward water and O_2. However, H_2O_2 is usually the first product of the cathodic reduction of dissolved O_2:

$$O_2(aq) + 2\,H^+(aq) + 2\,e^- \rightarrow H_2O_2(aq) \tag{2.33a}$$

(a)

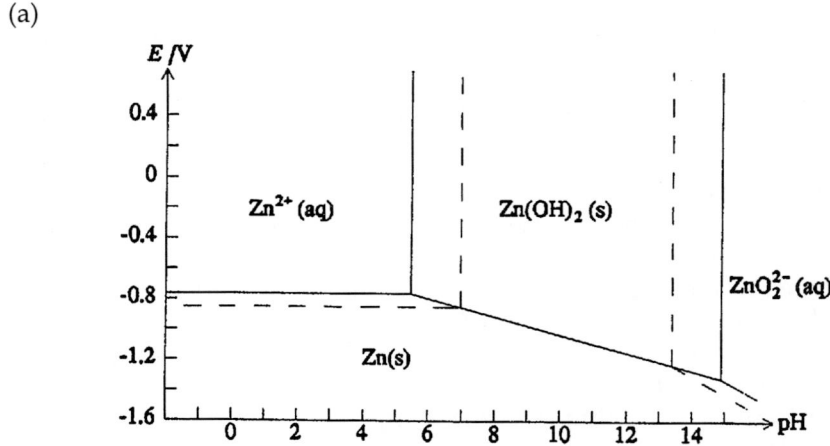

(b)

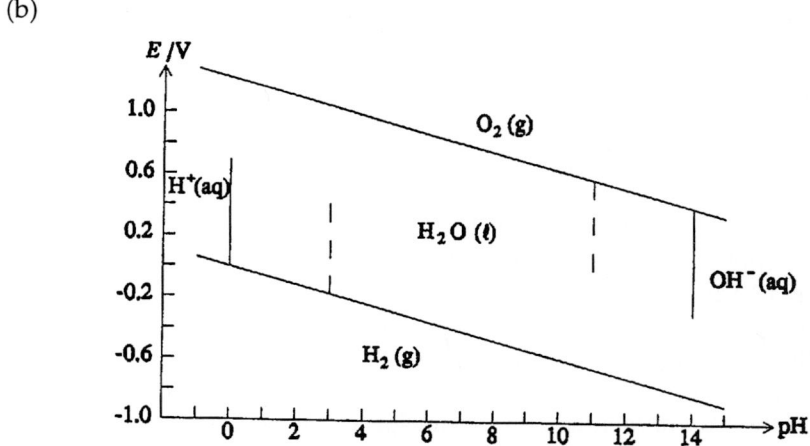

FIGURE 2.12. Pourbaix diagram of the Zn-H_2O (a) and the H_2O solvent (b) systems at T=298 K. Solid lines relate to unit activities of all species and dashed lines denote predominance zones when ions are at an activity of 10^{-3}. (Reproduced with permission from K. B. Oldham and J. C. Myland, "Fundamentals of Electrochemical Science." Academic Press, San Diego, CA, 1994.)

2.5. Thermodynamics and Kinetics in Electrochemical Systems

The subsequent reduction to water,

$$H_2O_2(aq) + 2 H^+(aq) + 2 e^- \rightarrow 2 H_2O(l) \tag{2.33b}$$

has a much more positive standard reduction potential.

Nonetheless, Pourbaix diagrams have a useful diagnostic role, and we shall consider them again in conjunction with the electrochemistry of inorganic pollutants in the next chapter.

2.5.6. Mixed Potentials

Many practical systems contain a mixture of redox couples or species capable of undergoing electrochemical reactions. In such cases, the concept of the *mixed potential* (a term borrowed from the corrosion field) is useful. Consider the $Cu^{+/0}$ redox couple in a strong chloride medium. (This stabilizes the Cu(I) redox state.) Figure 2.13 contains the current–potential curves for a copper rotating disk electrode. The anodic polarization curve for the dissolution of copper

$$Cu(s) \rightarrow Cu^+(aq) + e^- \tag{2.34a}$$

in the absence of appreciable amounts of Cu(I) in solution is shown as curve *a* in Figure 2.13. Note the absence of a mass-transport limited plateau (see Section 2.5.4) since the supply of the reactant, copper metal, is virtually unlimited.

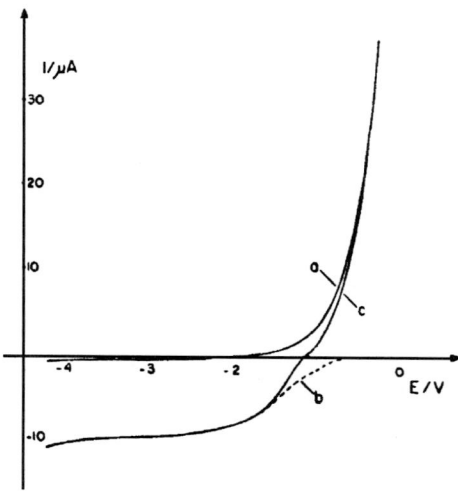

FIGURE 2.13. Current–voltage curves for copper rotating disk electrode in 3 *M* chloride. Disk rotation speed 1600 RPM, area 7.07×10^{-7} m². See text for description of curve symbols. (Reproduced with permission from Power and Ritchie.[3])

On the other hand, the cathodic polarization counterpart for the reverse reaction

$$Cu^+(aq) + e^- \rightarrow Cu(s) \tag{2.34b}$$

in the *absence of the metal* is clearly not directly accessible. Attempts to measure it will yield only the combined curve c. A polarization plot for the cathodic discharge of Cu(I), however, can be *synthesized* using the principle of *current additivity*[3]:

$$I_{net} = I_{ox} + I_{red} \tag{2.35}$$

Thus curve b in Figure 2.13 is obtained as the difference of I_{net} and I_{ox}. The potential at which curve c cuts the zero current axis, of course, is the reversible (equilibrium) potential.

Suppose now that two redox couples are present, $Fe^{3+/2+}$ and $Cu^{+/0}$, as before. The anodic reaction is the same as in Eq. (2.34a), and the corresponding polarization curve is shown as curve a in Figure 2.14. The polarization curve for the cathodic reaction

$$Fe^{3+} + e^- \rightarrow Fe^{2+} \tag{2.36}$$

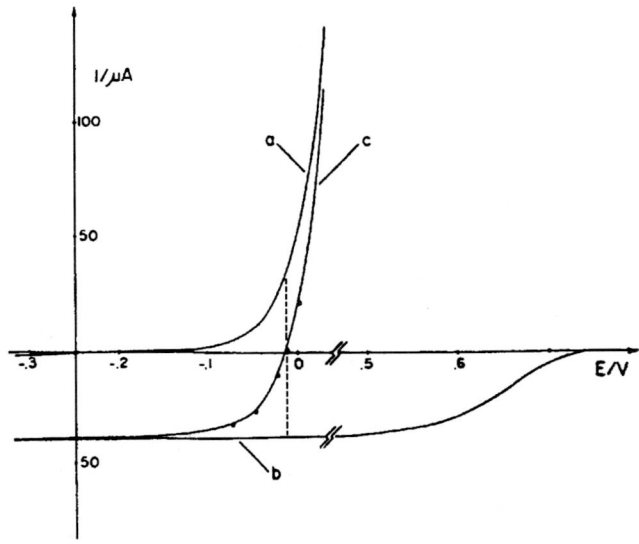

FIGURE 2.14. Current–voltage curves for the dissolution of copper in iron(III) (Cl⁻ = 3 M). Disk rotation speed 900 RPM, area 7.07×10^{-6} m²: (a) Polarization curve for the oxidation of copper(0). (b) Calculated polarization curve for the reduction of iron(III) on platinum. The points marked by dots were calculated by adding a and b. (Reproduced with permission from Power and Ritchie.[3])

is experimentally inaccessible. This is because of the location of the standard reduction potentials of the two redox couples: 0.34 V and 0.77 V for the $Cu^{+/0}$ and $Fe^{3+/2+}$ systems, respectively. Therefore, the reduction of Fe^{3+} cannot be studied independent of the simultaneous corrosion of copper metal. However, the reduction of Fe(III) on *platinum* (curve b, Figure 2.14) is similar to that on copper and is mass-transport controlled; and hence relatively independent of the nature of the electrode material. Curve c in Figure 2.14 is the measured current–voltage curve for a copper electrode dipped in an iron(III) solution. The points on curve c were calculated[3] as before, according to Eq. (2.35); good agreement is seen between the experimental data and the model prediction. The potential at which curve c cuts the zero current axis in Figure 2.14 defines the mixed potential of the system. This is the potential at which the anodic and cathodic current components exactly balance, as seen by comparing the magnitude of the "tie-lines" in Figure 2.14. The mixed potential for the reaction

$$Fe^{3+}(aq) + Cu(s) \rightarrow Fe^{2+}(aq) + Cu^{+}(aq) \tag{2.37}$$

has exactly the same significance as the reversible potential for the reaction considered earlier; namely,

$$Cu^{+}(aq) + Cu(s) \rightarrow Cu(s) + Cu^{+}(aq) \tag{2.38}$$

We shall use this concept in the modeling of semiconductor colloidal particles in Chapter 6 of this book.

2.6. ELECTROANALYTICAL CHEMISTRY

In this section, the necessary background for the use of electroanalytical techniques is provided as a prelude to the subsequent discussion in Chapter 4 on their use in the assay of pollutants. Techniques such as *voltammetry* are also useful for first defining the electrochemical behavior of a targeted pollutant. Based on these voltammetric data, "end-of-the-pipe" electrochemical pollution abatement procedures can be devised to either decompose the pollutant (to less harmful products) or to convert the spent effluent (e.g., Cr(III)) to the starting state (Cr(VI)), which can then be recycled back into the process input.

Electroanalytical techniques can be classified as *dynamic* (or *active*) or *passive*, depending on whether the process of measurement itself forces concentration (or activity) changes at the electrolyte interface. Thus, techniques such as voltammetry belong to the former category, whereas *potentiometry* is an example of a passive technique. These categories of techniques are now considered in turn.

2.6.1. Potentiometric Methods

The indicator electrode is of paramount importance in analytical potentiometry. The indicator electrode potential, E_{ind} is related to the cell voltage, E_{cell} via Eq. (2.39):

$$E_{cell} = E_{ind} - E_{ref} + E_{lj} \qquad (2.39)$$

E_{ref} and E_{lj} are the reference electrode and liquid junction potentials, respectively. If these are invariant in the measurement process, and if E_{ind} responds to variations in the analyte activity, then measurement of E_{cell} provides a route to quantification of the analyte activity (or concentration).

Perhaps the most common example of this approach is the use of the glass membrane electrode for measurement of solution pH. A wide variety of electrodes have been developed for potentiometry and will be discussed in Chapter 4. The connection between E_{cell} and the analyte activity is provided by the Nernst equation (2.12) re-expressed as follows for the specific case of an ionic analyte:

$$E_{cell} = K + \frac{RT}{nF} \ln a_o \qquad (2.40)$$

where K is a constant that contains E_{lj}. If the measurement conditions are such that the activity coefficients are approximately constant, then the activity terms such as a_o in Eq. (2.40) can be replaced by concentrations:

$$E_{cell} = K' + \frac{RT}{nF} \ln C_o \qquad (2.41)$$

Note that temperature fluctuations must be minimized because the *slope* of a calibration plot of E_{cell} vs. C_o will be dependent on temperature. In the ideal case, the slope will have a value of 0.0591 V for measurements made at 25°C (see Section 2.5.2).

Selectivity is an important consideration in all analytical procedures, and potentiometry is no exception. Equation (2.40) may be modified to explicitly show the selectivity of the indicator electrode to the target analyte:

$$E_{cell} = K + \frac{RT}{nF} \ln (a_i + k_{ij} a_j^{n/x}) \qquad (2.42)$$

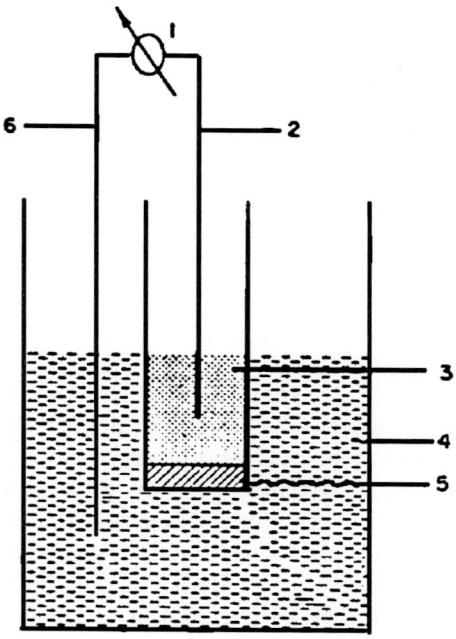

FIGURE 2.15. Schematic representation of membrane electrode cell assembly: 1 = membrane potential to be recorded; 2 = internal reference electrode; 3 = internal filling solution; 4 = sample solution to be measured; 5 = membrane; and 6 = external reference electrode. (Reproduced with permission from N. Lakshminarayanaiah, "Membrane Electrodes." Academic Press, New York, 1976.)

In Eq. (2.42), a_i is the activity of the analyte ion, a_j is the activity of the interferent ion, x is its charge, and k_{ij} is the selectivity constant (or alternatively called the *cross-sensitivity term*). Thus k_{ij} is a useful figure-of-merit for the selectivity of the potentiometric electrode. Small values of k_{ij} obviously are preferable. A selectivity constant of 0.01 translates to a situation where the interferent would have to be at an activity 100-fold that of the analyte to yield a comparable potentiometric response. Glass membrane electrodes for pH measurement, for example, have a selectivity order of $H^+ \ggg Na^+ > K^+, Rb^+, Cs^+ \ldots \gg Ca^{2+}$. Glasses containing less than about 1% Al_2O_3 yield good pH electrodes with little metal–ion response. Unfortunately, much of the compositional optimization is empirical, especially for membrane electrodes, since the fundamental interfacial phenomena underlying their response is not well understood at present.

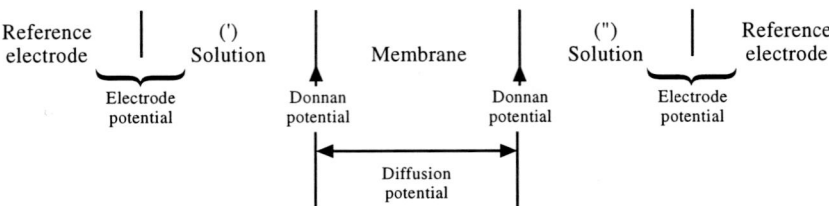

FIGURE 2.16. The components of a membrane potential measurement. (Reproduced with permission from N. Lakshminarayanaiah, "Membrane Electrodes." Academic Press, New York, 1976.)

Given that the membrane is the key component of a potentiometric sensor, it is perhaps not surprising that much effort has gone into gaining a fundamental understanding of membrane phenomena. Any phase that separates two other phases and affords passage with varying degrees of restriction of one or more species of these two external phases may be defined as a membrane. The membrane electrode in a potentiometric sensor could be either a solid or liquid, and the membrane phase may or may not be charged. Electrical potentials arising across membranes when they separate two electrolyte solutions are called *membrane potentials* (Figure 2.15). There are three components to this potential, as illustrated in Figure 2.16. First, there may exist a *diffusion* potential because of differences in the mobilities of the ions. A second source is the presence of charged species (i.e., ions) that cannot pass through the membrane because of its permselectivity. This is the Donnan potential. A third (rather trivial) source is the ohmic component because of the membrane resistance. This potential of course will be negligible at or near zero current as is typical of a potentiometric measurement. If the membrane has no fixed charge (e.g., a neutral carrier ionophore membrane), the membrane potential would be equivalent to a diffusion potential. In this case, the Nernst–Planck flux equation becomes applicable. The various theoretical approaches to the calculation of membrane potentials depart from one another in the nature of the flux equation used in the treatment. Thus, theoretical approaches may be classified into those based on the Nernst–Planck equation or its refinements, a second category based on the use of the principles of pseudo-thermostatics and irreversible thermodynamics, and a third category based on the concept of absolute rate processes.

The interested reader is referred to the literature listed at the end of this chapter for a more detailed discussion on the theories of membrane electrode potentials.

2.6. Electroanalytical Chemistry

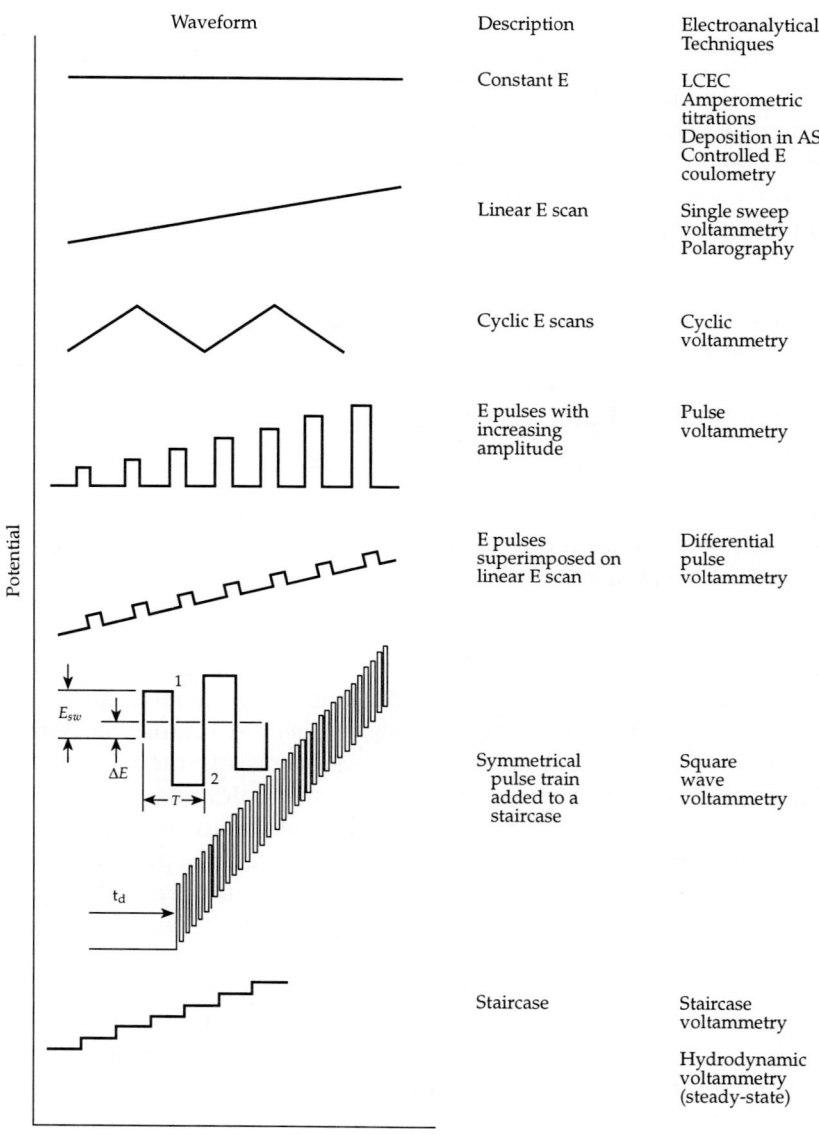

FIGURE 2.17. Potential excitation signals for various dynamic electroanalytical techniques. (Reproduced with permission from H.A. Strobel and W.K. Heineman, "Chemical Instrumentation: A Systematic Approach." Wiley, New York, 1989.)

2.6.2. Dynamic Methods

A variety of voltammetric techniques have been developed in which an external potential is applied to the electrochemical cell and the resulting current is measured. Then, a *voltammogram* is prepared, a plot of the current *vs.* the applied potential. Figure 2.17 contains potential excitation signals for some commonly used voltammetry techniques. The variant ways in which voltammetry may be implemented may be recognized as follows:

- The solution may be moving or quiescent with respect to the electrode.
- The waveform for the applied potential may be varied.
- The timing sequence of the current measurement with respect to the potential waveform may be varied.
- The electrode and cell geometry, which affect the current response, may be varied.

We shall now elaborate on four of these useful techniques in environmental electrochemistry; namely, cyclic voltammetry, polarography, pulse voltammetry, and stripping voltammetry.

<u>Cyclic Voltammetry.</u> This is perhaps the most versatile electroanalytical technique. The effectiveness of cyclic voltammetry (CV) results from its capability for rapidly observing redox behavior over a wide potential range. Indeed, CV has been termed *electrochemical spectroscopy*[4] because of broad similarities in the information content in the two types of measurements. Figure 2.18 contains a cyclic voltammogram for an organic substrate, *p*-aminophenol (PAP), in an aqueous medium. The data contained in this figure illustrate the power of this technique for mechanistic and diagnostic studies. In particular, coupled chemical reactions can be studied via CV. In the example in Figure 2.18, the starting compound generates benzoquinoneimine (BQI) on oxidation, which subsequently undergoes a chemical reaction with the solvent (water) to form quinone. This chemical reaction diminishes the amount of BQI available in the electrode vicinity for reduction back to PAP during the negative-going scan. This illustrates the utility of the return cycle. The diminution of the peak current ratio for the forward and reverse cycles (I_{pc}/I_{pa}) can be used for the estimation of the rate constant for the chemical reaction.

The potential scan rate also can be used as a variable to probe the kinetic facility of the redox couple since, at sufficiently fast scan rates, the effect of the chemical reaction becomes negligible and I_{pc}/I_{pa} approaches unity.

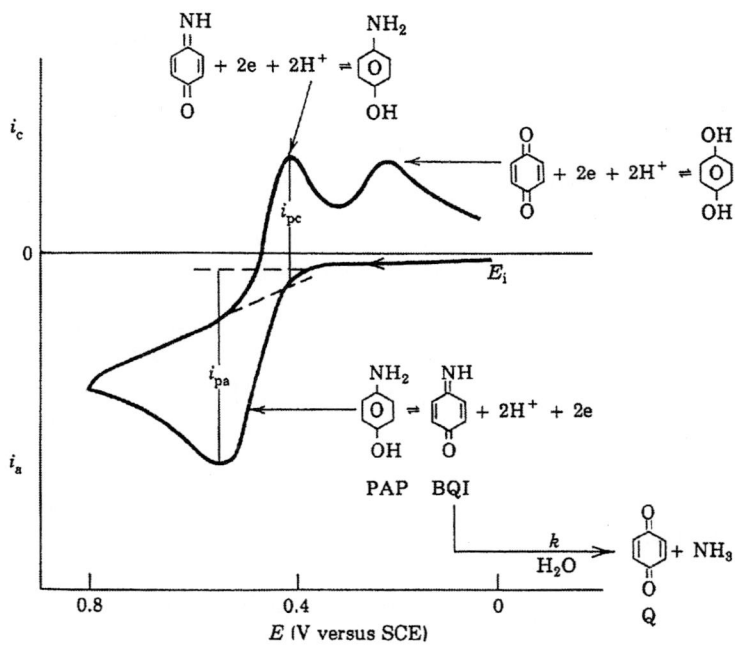

FIGURE 2.18. Cyclic voltammogram of *p*-aminophenol in 0.01 M H_2SO_4 at a carbon paste electrode. (Reproduced with permission from H. A. Strobel and W.K. Heineman, "Chemical Instrumentation: A Systematic Approach." 3rd ed., Wiley, New York, 1989.)

Cyclic voltammetry can also be used for exploring the kinetics of charge transfer at an electrode/electrolyte interface. For a reversible (fast) system, the current in the forward sweep of the first cycle is given by the Randles–Sevcik equation:

$$I_p = (2.69 \times 10^5)\, n^{3/2}\, A\, D^{1/2}\, C^*\, v^{1/2} \tag{2.43}$$

where v is the potential scan rate (in V s^{-1}). The formal reduction potential for a reversible couple is centered between the anodic and cathodic peak potentials:

$$E^{o'} = \frac{E_{pa} + E_{pc}}{2} \tag{2.44}$$

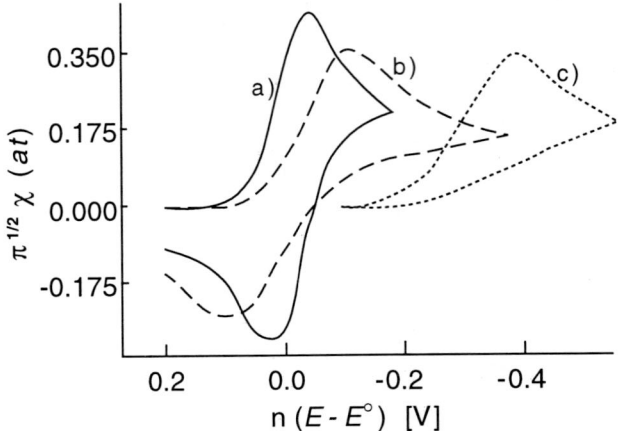

FIGURE 2.19. Cyclic voltammograms for (a) reversible (solid line, $k^{o\prime} = 1$ cm/s), (b) quasi-reversible (dashed line, $k^{o\prime} = 1.25 \cdot 10^{-3}$ cm/s), and (c) irreversible charge transfer (dotted line, $k^{o\prime} = 6.25 \cdot 10^{-6}$ cm/s) cases; $D = 1 \cdot 10^{-5}$ cm^2/s, $v = 0.1$ V/s, $\pi^{1/2} \chi$ (at): normalized current function. (Reproduced with permission from Heinze.[4])

The peak separation ΔE_p for a reversible couple is given by

$$\Delta E_p = E_{pa} - E_{pc} = \frac{0.059}{n} \qquad (2.45)$$

For sluggish redox couples, these relationships will not generally hold. A plot of the peak current *vs.* the square root of the scan rate will often still be linear but the slope will be lower than that dictated by Eq. (2.43). Similarly, ΔE_p will deviate from the reversible value, and this deviation can be used for an estimation of $k^{o\prime}$ (Section 2.5.3).[5]

Figure 2.19 contains examples of CV traces for electrochemically irreversible redox couples and contrasts these with the reversible case. Note that *thermodynamic* parameters such as $E^{o\prime}$ no longer are accessible from CV data in these cases, but instead kinetic parameters such as $k^{o\prime}$ can be extracted.

Redox couples immobilized on the electrode surface are less prone to effects of mass transport. Thus, plots of I_p *vs.* $v^{1/2}$ will not be linear in such cases; instead, I_p will exhibit a linear dependence on v. Similarly, ΔE_p will be close to zero for redox couples confined to electrode surfaces.

The advent of advanced software and computation has significantly enhanced the scope of CV for mechanistic studies.[6,7] Thus comparison of *simulated* voltammograms with experimental ones serves to validate a particular scheme for the test system. Computer simulation also is useful for gauging the sources of errors in data extraction from CV such as the magnitude of the (uncompensated) *IR* drop.

2.6. Electroanalytical Chemistry

Equation (2.43) suggests that the peak current, I_p should be linearly related to the analyte concentration, C. Alternatively, since the abscissa is a time-related parameter (recall that the potential is ramped in a linear manner), the area encompassed in a CV peak should afford the electroactive charge, Q, and thence the analyte concentration (Eq. (2.9)). Both these expectations are borne out; however, CV is hardly the technique of choice for quantifying analyte concentrations especially at trace (sub-mM) levels. This is because its sensitivity is seriously limited by the *capacitive* current that flows at the electrode/electrolyte interface.

Polarography. Polarography predates voltammetry, and the distinction between the two techniques relates to the working electrode material used — dropping mercury in the former and a solid electrode in the latter. The two main virtues related to the use of dropping mercury as an electrode material are as follows. First, its surface is continually renewable, and therefore a perennial problem with electroanalytical techniques, namely, that of electrode fouling, is circumvented. Second, mercury has a high overpotential for hydrogen evolution, and therefore its negative potential window can extend up to ~-2.0 V (*vs.* SCE). On the other hand, mercury has a very restricted positive window because of its proclivity to undergo oxidation. The environmental problems associated with the routine use and disposal of mercury in the laboratory have curtailed its popularity in recent years. Nonetheless, polarography is often the method of choice for the determination of many environmentally important reducible species and for speciation applications.

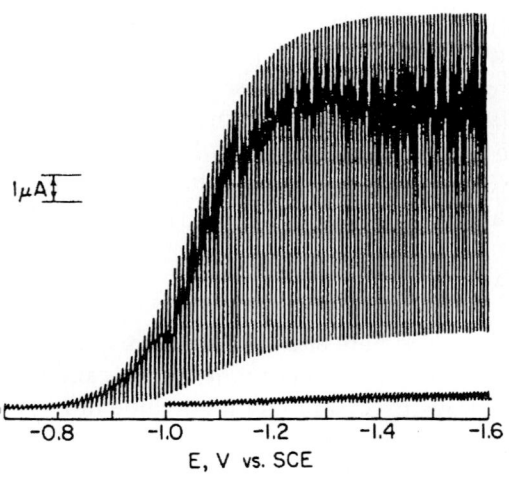

FIGURE 2.20. Polarogram for 1 mM CrO$_4^{2-}$ in deaerated 0.1 M NaOH. The lower curve is the residual current obtained in the absence of CrO$_4^{2-}$. (Reproduced with permission from A.J. Bard and L.R. Faulkner, "Electrochemical Methods." Wiley, New York, 1980.)

An expression analogous to Eq. (2.43) can be derived by taking into account the variation of the drop area with time. The result is the *Ilkovic equation* for the limiting diffusion current, I_d:

$$I_d = 708\, nD^{1/2}\, m^{2/3}\, t^{1/6}\, C^* \tag{2.46}$$

where I_d has the units of µA, m is the mass flow rate, and t is the drop life of the mercury in mg s^{-1} and s, respectively. A representative polarogram is contained in Figure 2.20. Quantitation of analytes may be made (usually at the ppm range) by using calibration standards, an internal standard, or by the method of standard additions. Alternatively, the electroactive species may be identified by the location of the *half-wave potential*, $E_{1/2}$ which is the potential at which the limiting current is half its maximum value. This characteristic potential is related to $E^{o\prime}$ according to the expression

$$E_{1/2} = E^{o\prime} + \frac{RT}{nF} \log (D_R/D_O)^{1/2} \tag{2.47}$$

where D_R and D_O are the diffusion coefficients of the reduced and oxidized species, respectively. Since the ratio D_R/D_O is usually close to unity for most redox systems, the two potentials are not very different.

As with its CV counterpart, polarography suffers from sensitivity problems associated with capacitive charging of the electrode/electrolyte interface. Substantial improvements are possible via the use of pulse excitation instead of a linear potential ramp. This class of voltammetric–polarographic techniques is reviewed next.

Pulse Voltammetry. The electrode–solution interface behaves like a capacitor, and a *double-layer capacitance*, C_{dl} can be defined for it. The capacitance is nominally on the order of 10–40 mF/cm. However, unlike a real capacitor, whose capacitance is independent of the voltage across it, C_{dl} is often a weak function of potential. The transient current, $I(t)$, to a potential step is given by

$$I(t) = \frac{E}{R_s} e^{-t/R_s C_{dl}} \tag{2.48}$$

Ohmic behavior is assumed here and the initial current, $I(o)$ is given by E/R_s. The product $R_s C_{dl}$ has the dimension of time, and is called the *time constant*, τ. Hence the current decays to 37% of its initial value at $t = \tau$, and to 5% of its initial value at $t = 3\tau$. For example, if $R_s = 1$ ohm and $C_{dl} = 20$ mF, $t = 20$ µs, and *double-layer charging* is 95% complete in 60 µs.

2.6. Electroanalytical Chemistry

(a)
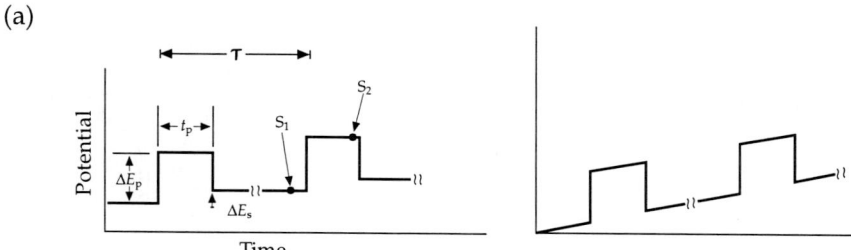

(b)
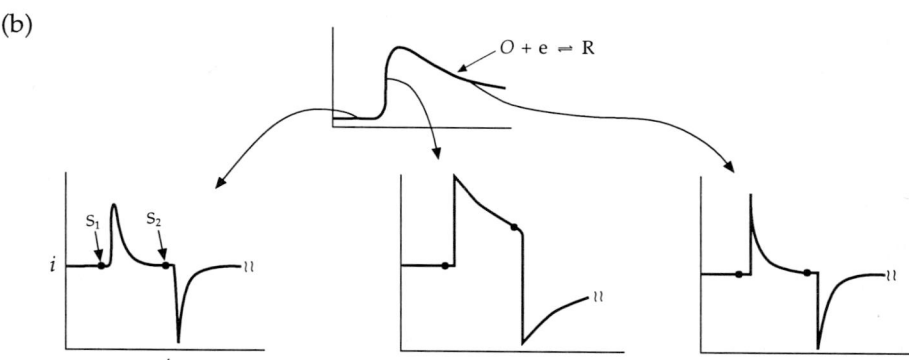

(c)
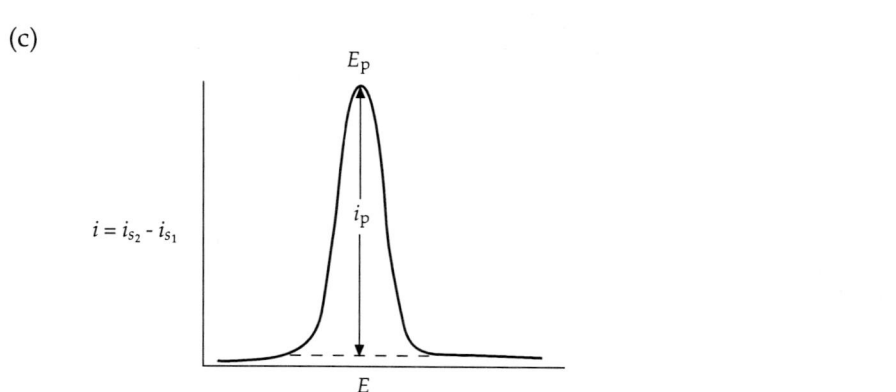

FIGURE 2.21. Excitation signals (a), current response signals obtained at different locations in a corresponding linear sweep voltammogram (b), and the voltammogram (c) for differential pulse voltammetry. (Reproduced with permission from H. A. Strobel and W. K. Heineman, "Chemical Instrumentation: A Systematic Approach." 3rd ed., Wiley, New York, 1989.)

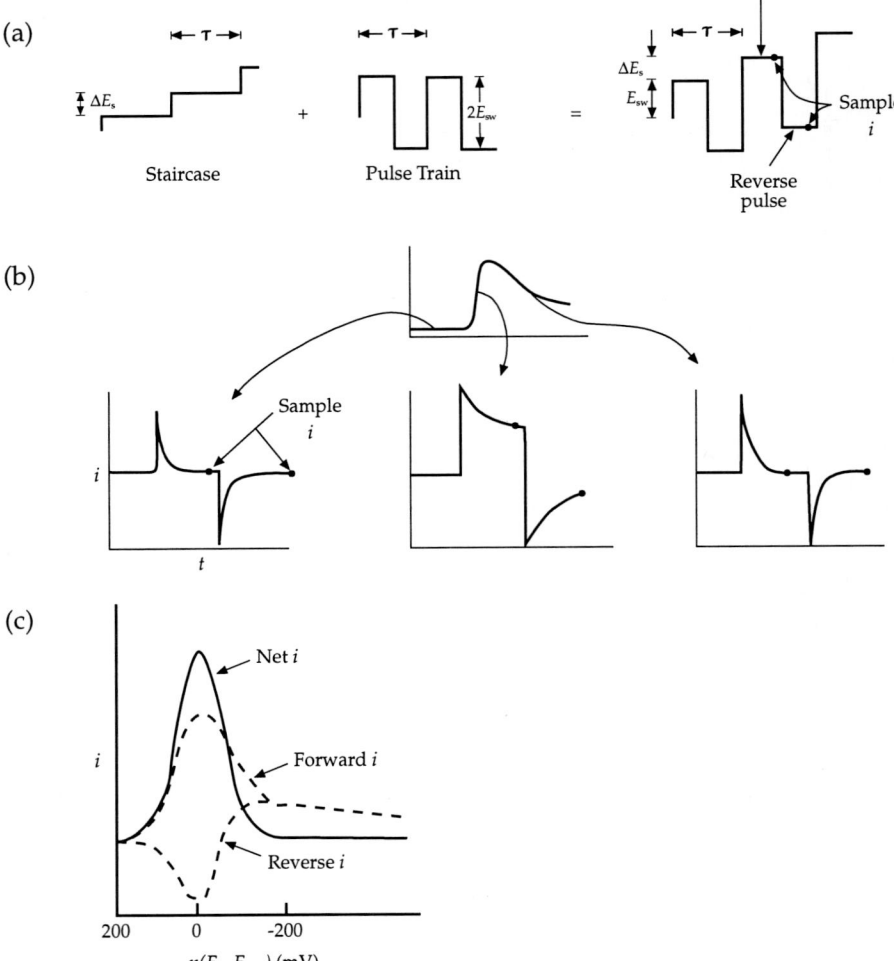

FIGURE 2.22. Same as in Figure 2.21 but for square voltammetry. (Reproduced with permission from H. A. Strobel and W. K. Heineman, "Chemical Instrumentation: A Systematic Approach." Third Edition, John Wiley and Sons, New York, 1989.)

For a potential sweep experiment (e.g. voltammetry), the current is given by

$$I = vC_{dl} + [(\frac{E_i}{R_s} - vC_{dl})\, e^{-t/R_s C_{dl}}] \tag{2.49}$$

where E_i is the initial potential. The second term on the right-hand side is the transient part, which dies away with a time constant, t.

2.6. Electroanalytical Chemistry

Equations (2.48) and (2.49) pertain only to the capacitive component (I_{nf}) of the total current. The latter is the sum of its Faradaic and capacitive components:

$$I = I_f + I_{nf} \qquad (2.50)$$

The time dependence of I_f to a step potential is given by the Cottrell equation:

$$I = \frac{nFAD^{1/2}C^*}{\pi^{1/2}t^{1/2}} \qquad (2.51)$$

Comparison of Eqs. (2.43) and (2.49) and Eqs. (2.48) and (2.51) shows why potential-sweep techniques such as cyclic voltammetry have poor analytical sensitivity, and time-resolved methods such as pulse voltammetry improve the analytical sensitivity. Specifically, Eq. (2.43) shows that $I_f \propto v^{1/2}$. On the other hand, $I_{nf} \propto v$ (Eq. (2.49)). Thus, at very low analyte concentrations and relatively high scan rates, I_{nf} will swamp the influence of I_f to the total current. By the same token, $I_{nf} \propto e^{-t/\tau}$ and $I_f \propto t^{-1/2}$. Thus, the Faradaic component persists at times longer than the capacitive current. If the current is *sampled* at a later time in the potential pulse, the ratio I_f/I_{nf} can be enhanced, and thereby the analytical sensitivity can be improved. This is the rationale behind *pulse voltammetry* methods.

A variety of pulse waveforms and timing/current sampling patterns have been devised; a few of these are illustrated in Figure 2.17. The two important pulse voltammetry methods are *differential pulse voltammetry* and *square-wave voltammetry*. Figures 2.21 and 2.22 illustrate the excitation signals, the current responses, and the voltammograms for the two techniques, respectively. With these innovations, detection limits in the range $10^{-7} - 10^{-8}$ M are easily achieved, compared with 10^{-5} M typical of cyclic voltammetry.

Stripping Voltammetry. This has the lowest detection limit ($\sim 10^{-10}$ M) of the commonly used electroanalytical techniques. Stripping voltammetry consists of two steps. In the first, the analyte is deposited at the electrode by *controlled-potential* electrolysis. This step *preconcentrates* the analyte. In the second step, the deposited analyte is "stripped" from the electrode by an appropriate potential scan. The resulting current signal is used to quantify the analyte. Any of a number of pulse voltammetry routines can be used for the stripping step. If the stripping is done anodically, the technique is called *anodic stripping voltammetry* (ASV). An example of this approach is the determination of metal ions at a hanging mercury drop electrode (HMDE) (Figure 2.23).

An alternative approach involves the deposition of anions at a mercury surface in the form of an insoluble mercury salt:

$$2\,Hg + 2\,X^- \rightleftarrows Hg_2X_2 + 2\,e^- \qquad (2.52)$$

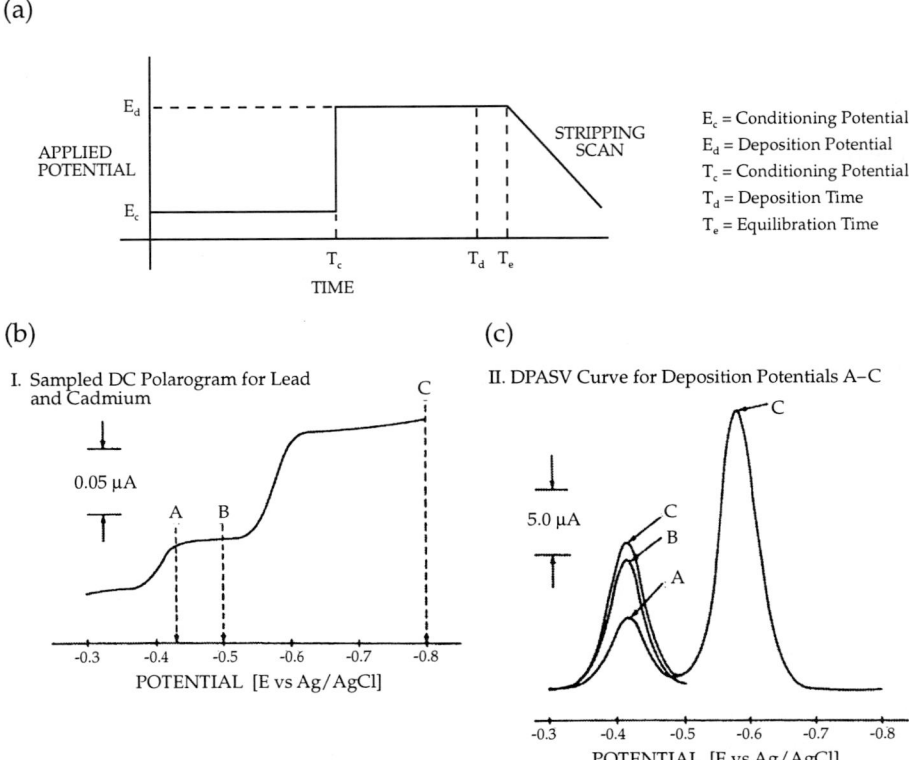

FIGURE 2.23. Excitation signal (a), sampled dc polarogram (b), and differential pulse anodic stripping voltammogram (c) for an analyte containing both lead and cadmium. The influence of choice of the deposition potential, A–C, on the DPASV response is also shown. (Reproduced with permission from W. M. Peterson and R. V. Wong, *Am. Lab*, November 1981.)

A negative potential scan (i.e., a cathodic scan) is then initiated for the stripping step, and the resultant cathodic current may be used for analyte quantification. This approach is called cathodic stripping voltammetry (CSV).

Several innovations in this area must be mentioned. First, the concerns with the use of Hg coupled with its anodic instability have prompted the search for alternative electrodes for stripping voltammetry. These include the use of solid electrodes such as graphite or glassy carbon. In some cases, these electrodes are also chemically modified with a polymeric film. Second, the preconcentration step has utilized analyte *adsorption* instead of *deposition*. This has extended the range of analytes that can be detected via stripping voltammetry. Third, the electrodes and the measurement system have been miniaturized and made user-friendly for routine, on-field, and portable use.

Fourth, to further improve the analytical sensitivity, and to avoid the problem of capacitive current flow altogether, the quantification step has utilized *potential* rather than *current* measurement. This approach is called *potentiometric stripping analysis* (PSA).[8]

In view of the importance of stripping analysis to environmental assay and speciation, we shall defer further details to Chapter 4.

We conclude Section 2.6 by noting that all electroanalytical techniques employ an excess of *supporting electrolyte* in the measurement medium. In many cases, especially in the analysis of real-world samples in environmental scenarios, the water samples may already contain an abundance of *natural* electrolytes in addition to the analyte. The electrolyte serves two important functions: (a) it improves the electrical (i.e., ionic) conductivity of the medium and thus minimizes undesirable *IR* drops, and (b) the electrolyte ions bear the burden of transporting charge via *migration* between the working and counterelectrodes in the electrochemical cell. Thus the usually small amount of analyte that is present in the cell moves primarily via *diffusion* (and in some cases, via convection) in the electrolytic medium. This condition ensures the applicability of diffusion models that underpin the development of Eqs. (2.43) and (2.51).

2.7. ELECTROLYSIS AND ELECTRODEPOSITION

Bulk electrolysis is a versatile route to the electrochemical treatment of pollutants and shall be discussed in Chapter 5. A related approach, which is especially suited to the removal of metal ion pollutants from effluents and process streams, is electrodeposition, where these ions are "plated out" onto a support electrode structure for subsequent recycling or disposal. We have also seen the usefulness of electrodeposition as a preconcentrating technique for trace-level analyses in the preceding section. In this section, we shall briefly consider some aspects of the fundamentals of bulk electrolytic/electrodeposition procedures.

Bulk (or exhaustive) electrolytic procedures (unlike the assay methods considered in Section 2.6) are characterized by large electrode area/cell volume cell conditions and as effective mass transfer conditions as possible. Therefore, the use of wire gauzes, foils, packed beds of powders, slurries, or fluidized beds as electrodes, rotating electrodes, and solution stirring is common. Proper orientation of the counterelectrode is also critical for providing a uniform current density across the working electrode surface. High cell resistances are very deleterious because large values of I^2R mean wasted power and undesirable heat generation in the cell. Finally, the use of separators that do not allow intermixing of *anolyte* and *catholyte* during the electrolysis procedure, and yet do not contribute appreciably to the cell resistance, is of paramount importance.

Consider the deposition of a solid product (e.g., Cu) from its solution precursor, O (e.g., Cu^{2+}):

$$O + ne^- \rightleftarrows R(solid) \tag{2.53}$$

When several monolayers of R are deposited on an inert electrode or an electrode made of R (Cu), the activity of R, a_R, is constant and equal to 1 at the completion of electrolysis. Hence, the Nernst equation takes the form

$$E = E° + \frac{RT}{nF} \ln [\gamma_o C_o (1-x)] \tag{2.54}$$

where C_o is the initial concentration of O, and x is the fraction of O reduced to R.

For less than one monolayer coverage, $a_R \neq 1$, and it can be assumed that a_R is proportional to the fractional coverage of the deposit on the electrode surface, θ:

$$a_R \approx \gamma_R \theta = \gamma_R \frac{A_R}{A} = \frac{\gamma_R n_R A_a}{A} \tag{2.55}$$

where A_R is the area occupied by R, A_a is the cross-sectional area of a molecule of R (cm^2), and n_R is the number of molecules of the deposit. At equilibrium,

$$n_R = V_s C_o N_A \tag{2.56}$$

where V_s is the solution volume and N_A is the Avogadro number. Combining Eqs. (2.54) – (2.56),

$$E = E° + \frac{RT}{nF} \ln \left(\frac{\gamma_o A}{\gamma_R N_A V_s A_a} \right) + \frac{RT}{nF} \ln \left(\frac{1-x}{x} \right) \tag{2.57}$$

Figure 2.24 contains a plot of x vs. the potential E referred to $E°'$ ($E - E°'$). The deposition begins at potentials more positive than values wherein deposition of *bulk R* ensues. This process is termed *underpotential deposition* (UPD). The preceding (simplified) treatment ignores several complications: (a) The deposition potential often depends on the nature of the support electrode material and pretreatment. (b) The deposition also is affected by any *adsorption* of O at the electrode surface. (c) We have assumed that growth of the second

2.8. Mass Transport under Forced Convection in an Electrochemical Cell

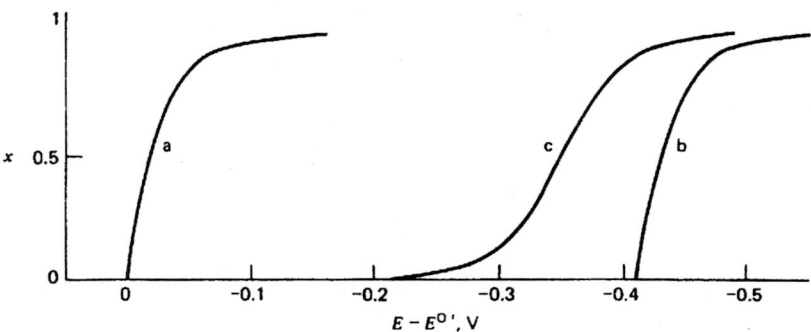

FIGURE 2.24. Fraction of a metal M (e.g., Ag) deposited (x) as a function of potential. Curve a: 1M Ag^+ on Ag. Curve b: 10^{-7} M Ag^+ on Ag. Curve c: 10^{-7} M Ag^+ on Pt. (Reproduced with permission from A.J. Bard and L.R. Faulkner, "Electrochemical Methods." Wiley, New York, 1980.)

layer does not start until the monolayer is completely formed; however, this is frequently not the case. There is often a tendency for agglomeration and *dendrite* formation. (d) Nucleation and growth phenomena are of variant types, and reviews of the thermodynamics and kinetics underlying the formation of new phases at electrode surfaces are available.[9]

Nonetheless, the simple treatment can be extended to the case of multimetal deposition from a solution phase. Consider the quantitative separation of two metals, M_1 and M_2. For 99.9% completeness of deposition of O, Eq. (2.55) becomes

$$E = E^{o'} + \frac{0.059}{n} \log\left(\frac{0.001}{0.999}\right) \simeq E^{o'} - \frac{0.18}{n} \tag{2.58}$$

Thus, nearly complete deposition of M_1 will occur when the potential is more negative than $\sim 0.18/n_1$ V of $E_1^{o'}$. Similarly, very little deposition of M_2 will occur at a potential $\sim 0.18/n_2$ V positive of $E_2^{o'}$. Therefore, the separation between the two formal potentials must be at least 0.18 ($n_1 + n_2$) for reasonably complete separation of the two metals M_1 and M_2. Often, however, the deposition potentials may be judiciously shifted via metal ion complexation. Complexation reduces the free metal ion activity in the solution phase, and thus shifts the deposition to more negative potentials relative to the uncomplexed metal ion case.[10]

Mercury cathodes were used for classical electroseparation procedures. For environmental applications, however, a variety of other electrode material choices exist, including reticulated vitreous carbon (RVC), carbon foams, and various forms of graphite. Other (solid) electrode material candidates are discussed in Chapters 4 and 5.

2.8. MASS TRANSPORT UNDER FORCED CONVECTION IN AN ELECTROCHEMICAL CELL

Since the goal in many industrial processes and electroanalytical procedures is to maximize the supply of electroactive species to the electrode surface, the solution is most often under forced convection. This is done either by maintaining solution flow within the electrochemical cell or by using electrode rotation. Since the concentration of the electroactive species in the cell will be changing due to the electrolytic reaction, concentration gradients will give rise to diffusive processes. Assuming that there is enough supporting electrolyte to minimize the migration of electroactive species, these systems will be under diffusion–convection and this combined mode of mass transport is termed *convective diffusion*.

At the electrode surface, the liquid will have a zero flow velocity (perpendicular to the surface), u, and this velocity will increase with distance to reach a limiting value, u_o. The velocity profile is shown in Figure 2.25. Likewise, a tangent drawn at the onset of the curve will cross a line with the maximum value of u (i.e., u_o) at a distance called the *thickness of the hydrodynamic layer*, δ_H (also called the *Prandtl layer*). Diffusion toward the electrode will gradually decrease the concentration from C_O to C_O ($x = 0$); this will occur within the Nernst diffusion layer, where $\delta_H < \delta$. In addition, this forced convection creates a steady-state distribution of C_O within the Nernst diffusion layer and so the expression for the rate of change of concentration here is given by

$$\frac{\partial C_O}{\partial t} = 0 = D_O \left(\frac{\partial C_O}{\partial t}\right)_{diff} - u \left(\frac{\partial C_O}{\partial t}\right)_{conv} \tag{2.59}$$

(The different sign accounts for the different direction of the flux induced by each mechanism.) In three dimensions this can be written as

$$\frac{\partial C_O}{\partial t} = 0 = D_O \left(\frac{\partial^2}{\partial x^2} + \frac{\partial^2}{\partial y^2} + \frac{\partial^2}{\partial z^2}\right) C_O - \left(u_x \frac{\partial}{\partial x} + u_y \frac{\partial}{\partial y} + u_z \frac{\partial}{\partial z}\right) C_O \tag{2.60}$$

2.8. Mass Transport under Forced Convection in an Electrochemical Cell

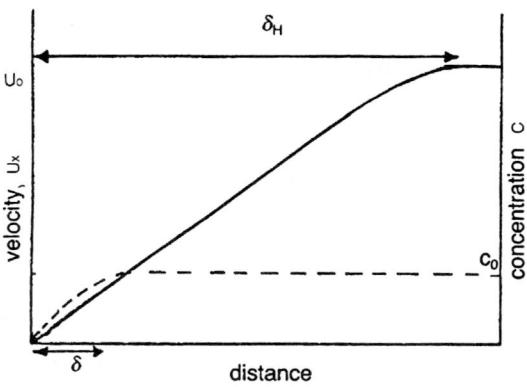

FIGURE 2.25. The velocity and concentration profiles at an electrode/electrolyte interface showing the Nernst diffusion layer (δ) and the Prandtl hydrodynamic layer (δ_H). (Adapted from J. Wang, "Analytical Electrochemistry." VCH, New York, 1994.)

where x, y and z are the normal Cartesian coordinates and u_x, u_y, and u_z are the flow velocities corresponding to these coordinates.

The relationship between the velocity of a viscous liquid and the forces acting upon it (forced and natural convection, and viscosity forces due to internal friction) is given by the Navier–Stokes equation. If the viscosity forces are considered dominant, this equation can be written in its simplified form:

$$\left[\left(u_x \frac{\partial}{\partial X}\right)+\left(u_y \frac{\partial}{\partial Y}\right)+\left(u_z \frac{\partial}{\partial Z}\right)\right] u_u = \left(\frac{\nu}{u_o L}\right) \nabla^2 u_u \tag{2.61}$$

where X, Y and Z are the Cartesian coordinates x, y and z divided by a characteristic length for the specific geometry analyzed, L; u_u stands for each velocity component ($u = x$, y or z) divided by a characteristic velocity within the Prandtl layer, u_o; ν is the kinematic viscosity of the liquid ($\nu = \eta/\rho$, where η is the viscosity and ρ is the liquid density); and ∇^2 is the Laplace operator $\left(\frac{\partial^2}{\partial x} + \frac{\partial^2}{\partial y} + \frac{\partial^2}{\partial z}\right)$. Assuming that the continuity equation for incompressible fluids holds,

$$\text{div } \vec{u} = 0 \tag{2.62}$$

The last three differential equations define the system in terms of convective mass transport; their analytical solution can be written as a relationship between dimensionless numbers. Therefore, any mass transport process is determined by its velocity, viscosity, diffusion coefficient, mass-transfer coefficient, and characteristic length and can be concisely described in terms of dimensionless numbers (π) as follows:

$$\pi = f(L, k, D, \nu, u) \tag{2.63}$$

where the exponents must be such that π is dimensionless. The following equation fulfills this condition:

$$\pi = \left(\frac{k_m L}{D}\right)^a \left(\frac{uL}{\nu}\right)^b \left(\frac{\nu}{D}\right)^{-(b+c)} \tag{2.64}$$

where a, b and c are characteristic exponents for each cell arrangement and the groupings in parentheses correspond to the following definitions of dimensionless numbers that simplify the number of variables to be handled for relating the geometry and hydrodynamics of the system to the process parameters.

(1) Sherwood number = Sh = $\left(\frac{k_m L}{D}\right)$. This gives the ratio between effective mass transport and mass transport by diffusion. It can also be defined as $\frac{J_{av} L}{D\left[C^* - C(x=0)\right]}$ = Sh, where J_{av} is the average value of the material flux or mass transfer rate.

(2) Reynolds number = Re = $\left(\frac{uL}{\nu}\right)$. This gives the ratio between the inertia and friction forces and characterizes the type of flow of the liquid: small Re values indicate laminar flow, whereas large Re indicates turbulent flow. The boundary value between these regimes depends on the characteristics of each system. For example, in a system comprising a stationary plate electrode in a flowing solution, the boundary value of Re is ~ 1.5×10^3. The thickness of the Prandtl layer is related to Re by

$$\delta_H \approx L\, \text{Re}^{1/2} \tag{2.65}$$

(3) Schmidt number = Sc = $\frac{\nu}{D}$. It gives the ratio between momentum

2.8. Mass Transport under Forced Convection in an Electrochemical Cell

transport and mass transport by diffusion; that is, it compares the rates of transport by convection and diffusion. Its inverse is proportional to the Nernst diffusion layer width:

$$\delta_N = \frac{1}{Sc} \tag{2.65a}$$

Sc is sometimes called the *Prandtl number* (see earlier).

(4) Peclet number $= Pe = Sc\, Re = \left(\frac{v}{D}\right)\left(\frac{uL}{v}\right) = \left(\frac{uL}{D}\right).$ (2.65b)

(5) Grasshof number $= Gr = \dfrac{gL^3 \Delta\rho}{v^2 \rho}$ (g = gravitational acceleration, ρ = density, $\Delta\rho$ = density difference between phases). (2.65c)

Table 2.3 lists the Sh values for some electrochemical cell configurations of importance to pollutant treatment scenarios. These empirical correlations are most often experimentally determined. Consider a packed-bed electrode flow reactor design. The current density (or applied potential) can be increased until a constant concentration profile is established for the reactant across the bed. With mass transfer control and plug flow in the bed, the following relation is applicable:

$$C/C_{in} = \exp(-A_e k_m L/u) \tag{2.66}$$

where A_e is the volumetric surface area (m^2/m^3), L is the height (at which the reactant concentration is C), and u is the linear flow velocity of the fluid (m s^{-1}). By monitoring the ratio C/C_{in} at various L, a semi-log plot can be constructed, the slope of which yields k_m. Thus, the parameter, Sh ($= k_m L/D$) can be computed for the specific reactor geometry and flow conditions. Since the correlations in Table 2.3 can be expressed as

$$Sh = (\text{Constant})\, Sc^x\, Re^y$$

a log – log plot of the ratio Sh/Sc^x vs. Re yields the exponent, y. The Schmidt number, Sc, is approximately constant for most dilute waste waters. The value of the exponent x is therefore of less practical interest and is usually taken to be 1/3 (Table 2.3).

The reader is referred to additional references listed at the end of this chapter for further details on solution hydrodynamics in electrochemical cells.

Table 2.3. The Average Sherwood Number for Some Electrode Configurations

Electrode configuration	Average Sh numbers
Rotating disk electrode	$Sh = 0.01\, Sc^{1/3}\, Re^{0.87}$ (turbulent flow) $Sh = 0.62\, Sc^{1/3}\, Re^{1/2}$ (laminar flow)
Laminar flow over a flat plate	$Sh = 0.646\, Sc^{1/3}\, Re^{1/2}$
Parallel plates	$Sh = 0.023\, Sc^{1/3}\, Re^{4/5}$ (turbulent flow) $Sh = 1.85\, Sc^{1/3}\, Re^{1/3}\, (d_e/L)^{1/3}$ (laminar flow)
Tubular electrodes	$Sh = 0.023\, Sc^{1/3}\, Re^{4/5}$
Packed bed	$Sh = 0.32\, Sc^{1/3}\, Re^{2/3}$ (for $100 \leq Re \leq 10{,}000$)
Fluidized bed	$Sh = 0.29\, Sc^{1/3}\, Re^{3/4}$ (for $100 \leq Re \leq 5{,}000$)
Rotating cylinder electrode	$Sh = 0.079\, Sc^{0.33}\, Re^{0.92}$
Rotating concentric cylinders	$Sh = 0.0791\, Sc^{0.356}\, (d_R/d_L)^{0.70}\, Re^{0.70}$
Vertical flat plate (free convection)	$Sh = 0.66\, (Sc\, Gr)^{1/4}$ (for $10^4 < Sc Gr < 10^{12}$) $Sh = 0.31\, (Sc\, Gr)^{0.28}$ (for $Sc Gr > 10^{12}$)

Notes: d_e = hydraulic diameter, L = length of plate, d_R = diameter of inner cylinder, d_L = diameter of the cylinder with the limiting current.

2.9. ELECTROCHEMICAL REACTOR DESIGN

Electrochemical cells vary widely in design depending upon the specific needs for a given process. Their shape, mode of operation, mode of electrolyte flow, temperature control, number of electrodes, type of electrical connections, degree of separation between anolyte and catholyte, and electrode structure and movement are the main variations. The decision of which cell to use will necessarily involve chemical, economic, engineering, technological, safety, and environmental criteria (Table 2.4).

2.9. Electrochemical Reactor Design

Usually, only one of the two reactions occurring in a cell is of interest; the following discussion assumes this to be the case. If both reactions are important, the reactor design might be somewhat more complicated. Likewise, if the product of the electron transfer at one electrode is capable of somehow affecting the electrochemistry at the other electrode, a separator (ion-selective membrane, felt, porous frit, etc.) would normally be included in the cell design. In addition, the process conditions may be such that a (deleterious) side reaction can occur at the site of the reaction of interest. Here, a judicious selection of the solution conditions as well as the electrode materials and applied voltage can normally lead to the minimization or even the disappearance of the side reaction. For example, dimensionally stable anodes (DSA) are used to facilitate the evolution of chlorine at the expense of O_2 during the electrolysis of a NaCl solution.

An electrochemical reactor can be classified according to its mode of operation as batch reactor (SBR), plug flow reactor (PFR), or continuous stirred tank reactor (CSTR). Corresponding models as well as some composite schemes are depicted in Figures 2.26 and 2.27. By using the appropriate reaction rate equation in conjunction with that for the current under mass-transport control, the equations that govern the design of electrochemical reactors as a function of the corresponding fractional conversions can be obtained (x = fraction of the initial amount that has been transformed at time t). These are shown in Table 2.5 for some relatively simple reactor designs and in the absence of complications associated with, for example: gas evolution, nonlimiting current flow, side reactions, or homogeneous chemical reactions.

Table 2.4. Important Decisions on Design Features in Electrochemical Reactors

Batch	Mode of operation	Continuous
Single	Number of electrode pairs	Multiple
Two-dimensional	Electrode geometry	Three-dimensional
Static	Electrode motion	Moving
Monopolar	Electrode connections	Bipolar
Moderate	Interelectrode gap	Capillary
External	Electrolyte manifolding	Internal
Undivided	Cell division	Divided
Open	Reactor sealing	Closed
Single	No. of electrolyte phases	Multiple
Liquid	Type of electrolyte	Solid polymer

Source: F. C. Walsh and G. Reade, *Analyst (London)*, 119, 794 (1994).

(a)

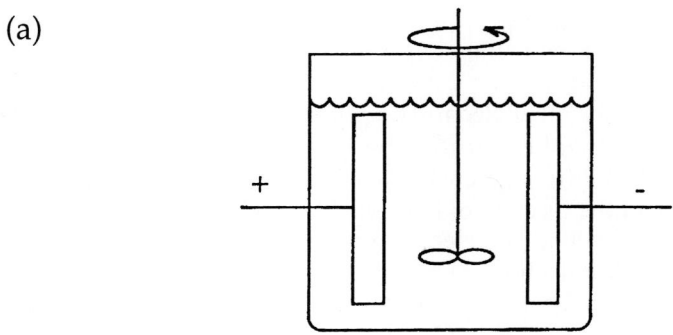

(b)

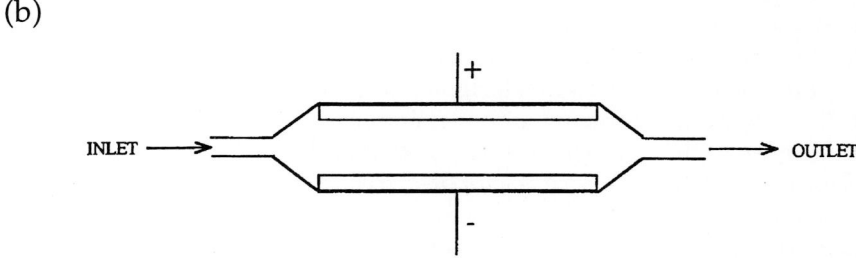

(c)

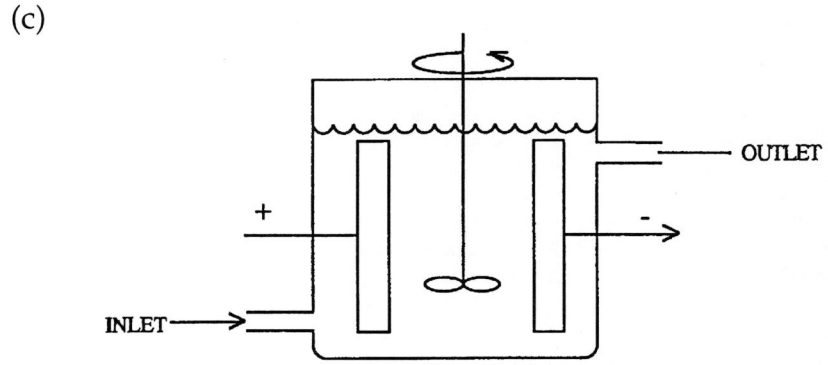

FIGURE 2.26. Schematic diagrams of single batch (a), plug flow (b), and continuous stirred tank (c) electrochemical reactors. (Source: F. Coeuret, "Introduccion a la Ingenieria Electroquimica." Reverte, Barcelona, 1992.)

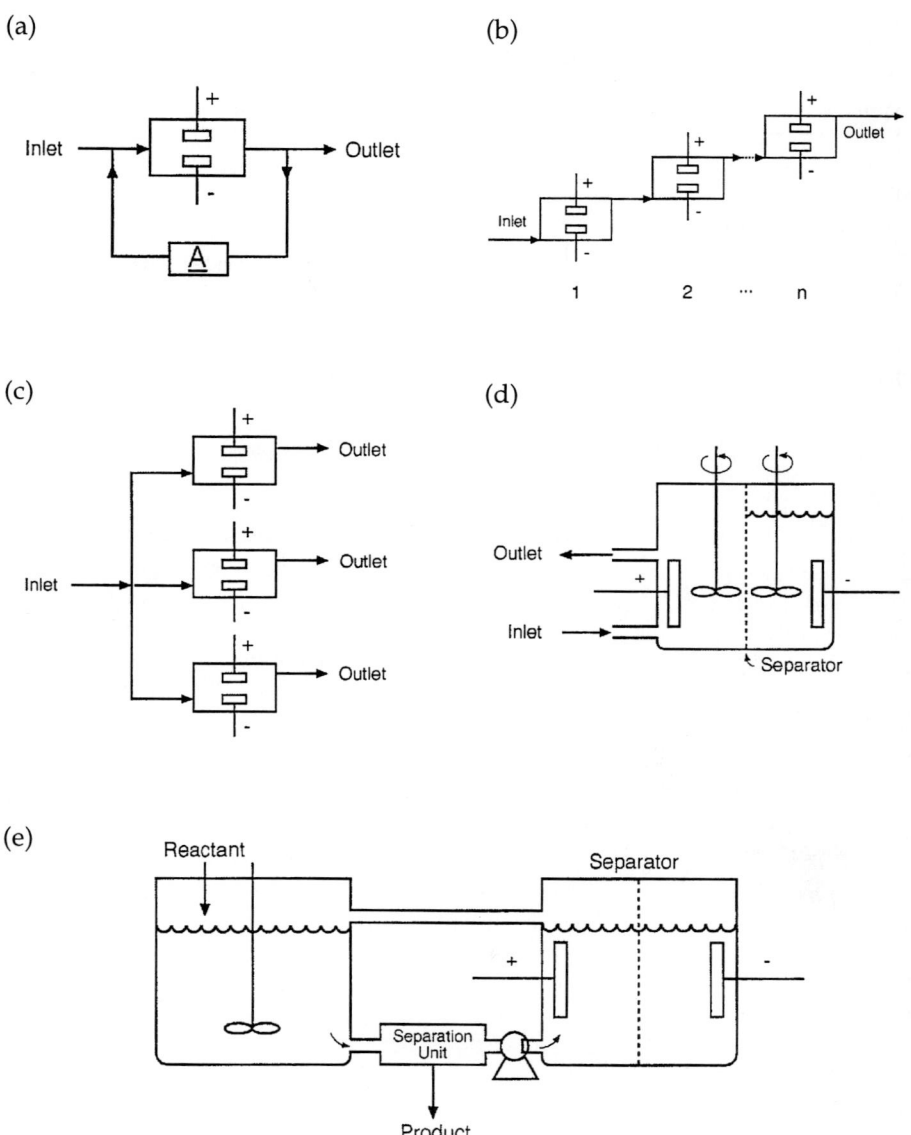

Figure 2.27. Plug flow and continuous stirred tank electrochemical reactors in cascade. Complex designs comprising multiple compartments are also shown (e.g., d and e). The module A in (a) may be just a connector or else a stirred batch-type reactor (SBR) without electrodes. (Adapted from D.J. Pickett, "Electrochemical Reactor Design." Elsevier, Amsterdam, 1977.)

Table 2.5. Expressions for the Fractional Conversion for the Three Major Types of Electrochemical Reactor Designs

Expression	Comments
SBR	
$x^{SBR} = 1 - \exp[(-k_m A / V_R)t]$	Reactor is loaded and then stirred for a given time, during which the reaction occurs to a certain extent
PFR	
$x^{PFR} = 1 - \exp[(-k_m A / V_R)\tau_R]$	Single pass
$x^{PFR,R} = 1 - \exp[-x^{PFR}(t/\tau_t)]$	With batch recirculation
$x^{PFR,C} = 1 - \exp\left(\dfrac{-nk_m A}{\dot{V}}\right)$	PFR in cascade
CSTR	
$x^{CSTR} = 1 - \left[1 + \left(\dfrac{k_m A}{\dot{V}}\right)\right]^{-1}$	Single pass
$x^{CSTR} = 1 - \exp[-x^{CSTR}(t/\tau_t)]$	With batch recirculation
$x^{CSTR,C} = 1 - \left[1 + \left(\dfrac{k_m A}{\dot{V}}\right)^n\right]^{-1}$	CSTR in cascade

Notes: V_R = reactor volume, k_m = mass transfer coefficient, τ_R = average residence time, τ_t = average residence time in the recirculation tank, n = cascade reactor number, $\dot{V}$ = volumetric flow rate.

Source: F. Walsh, "A First Course in Electrochemical Engineering." The Electrochemical Consultancy, Great Britain, 1993.

2.9. Electrochemical Reactor Design

Table 2.4 summarizes the important decisions regarding the design features in electrochemical reactors.[11] A few of these are elaborated here:

(1) Type of electrical connections: Multiple cells can be connected in a monopolar or bipolar configuration, as depicted in Figure 2.28.

Monopolar configuration. Cells are connected in parallel and therefore have the same potential difference across each anode–cathode pair. Each cell consumes its own current; hence, this configuration requires a low voltage source but high current. It has the advantage of low voltages but the disadvantage of high currents, which translates into high power requirements as well as high magnetic fields produced in the workspace.

Bipolar configuration. Cells are externally connected only at the first and last electrodes in such a way that the other electrodes acquire a polarity opposite that of the electrode in front of it and a potential corresponding to a proportional fraction of the overall cell voltage. It has the advantages of low power requirements, lower magnetic fields, and fewer physical connections (in fact, only two are required). This configuration is used whenever possible.

(2) Applied voltage: It must be minimized by decreasing the interelectrode gap whenever possible, increasing the electrical conductivity of the solution, minimizing junction potential drops, selecting appropriate counterelectrode reactions, increasing electrode conductivities, and the like.

(3) Potential and current distribution: These should normally be as homogeneous as possible; otherwise, side reactions, current leakages, local heating, uneven conversions, and such can severely lower the performance of thereactor. Good potential and current distribution can usually be achieved by effective electrode design and cell arrangement.

(4) Heat generation and dissipation: Heat can be produced in an electrochemical reactor due to the reaction itself, because of ohmic drops (IR terms) that produce Joule effects and mechanical friction forces from pumps, agitators, electrode movement, and the rest. It can also be brought to the reactor by means of a heat exchanging fluid, which in some cases may be the same electrolyte. Likewise, heat can be dissipated from the reactor mainly by conduction across its walls, convection at phase boundaries, heat exchanging fluids (e.g., the electrolyte itself), and solvent evaporation. In addition, changes in heat capacities of reactants and products must be taken into consideration for the overall reactor design.

(5) Flow separation: The possibility of undesired reactive mixing of the products of either reaction (cathodic or anodic) with solution components or with the products of the reaction at the other electrode, as well as the possibility of having a product from one electrode reaction undergo the opposite process at the other electrode, warrant the need of a separator. This is also the case where migration effects across a membrane with ion exchange

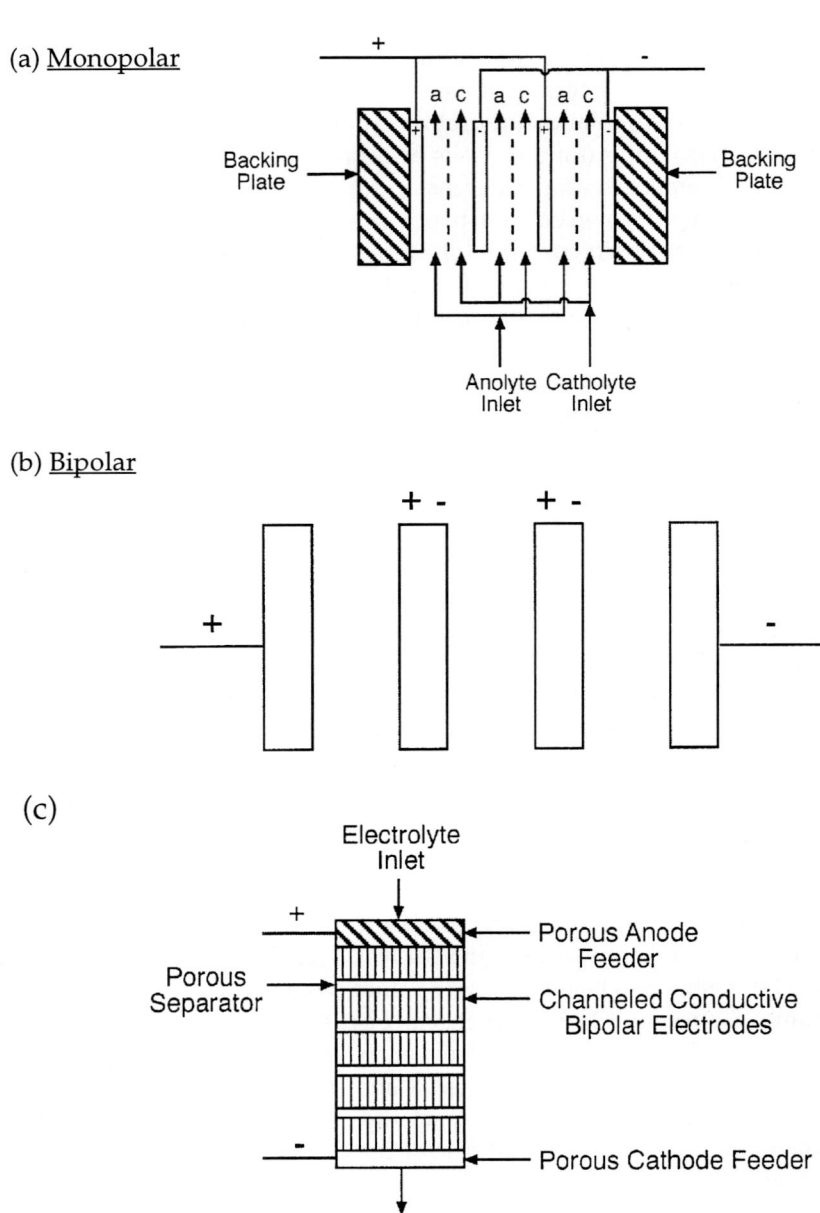

FIGURE 2.28. Electrochemical reactors featuring monopolar (a) or bipolar (b and c) connections. (Adapted from D. Pletcher and F. Walsh, "Industrial Electrochemistry." 2nd ed., Chapman and Hall, London, 1990.)

2.9. Electrochemical Reactor Design

properties are used to split a salt or to desalt a fluid. Physical separators primarily prevent convective and diffusive mixing of the two fluid streams, although if they are coarse (e.g., polymer nets) they act primarily as mechanical supports for ion exchange membranes, as turbulence promoters, and as insulators to prevent electrical short circuits in thin-gap cells. Separators are usually made of porous materials, including porous glass, asbestos, quartz, and polymers such as polytetrafluoroethylene (PTFE) and polyvinyl chloride (PVC).

Electrolytic cell designs are further classifiable as undivided or divided types, static three-dimensional electrodes, and moving electrode configurations. A further discussion of these types of cells is deferred until Chapter 5.

2.9.1. Performance Figures of Merit of an Electrochemical Reactor

A number of quantities may be computed as a concise statement of the performance of a given electrochemical reactor.[11] It has been emphasized[11] that (a) such quantities must be calculated under realistic operating conditions, (b) the importance of each quantity must be weighted in the light of process experience and process objectives, and (c) when comparing reactors, these figures of merit must pertain to similar operating conditions, particularly with regard to electrolyte composition, reactor temperature, and the reactant levels in solution.

Consider a solution phase reactant undergoing convective diffusion to a solid electrode under mass transport control. Then the maximum reaction rate is given by

$$\frac{I_L}{zF} = k_m A C^* \qquad (2.67)$$

where k_m is the mass-transport coefficient (m s^{-1}). For a given reaction and a given reactant level, the product $k_m A$ must be as large as possible. One strategy to accomplish this is via the use of *three-dimensional* electrodes. In this case, the active electrode area per unit volume

$$A_e = \frac{A}{V_e}$$

(where V_e is the electrode volume) will be high, resulting in a compact reactor design. A volumetric mass transport coefficient, $k_m A_e$ (s^{-1}) can be defined (see Eq. (2.67)) as

$$k_m A_e = \frac{I_L}{zFC^*V_R} \qquad (2.68)$$

for a two-dimensional electrode, V_R being the reactor volume, and

$$k_m A_e = \frac{I_L}{zFC^*V_e} = \frac{k_m A}{V_e} \qquad (2.69)$$

for a three-dimensional electrode. Therefore, the product $k_m A_e$ must be maximized in either case to achieve the highest conversion in the reactor.

Other important figures of merit are listed in tabular form in Table 2.6.[11] In concluding this section, we note that electrochemical reactor engineering is of little practical utility unless performed on a reasonable scale. Miniature electrodes and ill-defined flow conditions do not facilitate effective scale-up exercises. In this regard, a number of reactor designs and modular cells (incorporating a variety of electrode materials, separators, and turbulence promoters) have become commercially available. These are discussed in Chapter 8.

2.9.2. Economic Criteria

Economic criteria normally set the target quantity to be optimized, assuming this quantity to be the total cost per unit of production. Strategies have been presented for the optimization of an electrochemical reactor as a function of current density. This analysis assumes that other factors involved are already at their optimum (e.g., thermal energy generation, consumption, conservation and exchange, construction and electrode materials, corrosion prevention, reactor geometry, recycling ratio, and the like), and that personnel, raw materials and utility costs as well as the overall selectivity are independent of current density.

Figure 2.29 contains a representative plot of the cost as a function of the current density. The interested reader is referred to the specialized literature listed at the end of this chapter, for further details.

2.10. ELECTROKINETIC PHENOMENA

Electro-osmotic flow and electrophoresis are important phenomena in membrane-based electrochemical procedures and in electrokinetic remediation of contaminated soils. Similarly, electroanalytical methods such as capillary electrophoresis (CE) are based on these phenomena.

The capacitive property of an electrode/electrolyte junction was discussed in Section 2.6. The *interfacial capacitance* is due to an *electrical double layer*, which exists at the junction between two dissimilar phases regardless of whether they are electrically conductive or not. Consider an insulating wall (e.g., capillary tube, soil) in contact with an electrolyte. The effect of the double layer is magnified when the wall area is large and the solution volume is small. This is

2.10. Electrokinetic Phenomena

Table 2.6. Important Performance Figures of Merit for an Electrochemical Reactor

Figure of merit	Expression	Comments
Fractional conversion, x (dimensionless)	$x = \dfrac{N^o - N^t}{N^o}$ (SBR) $x = \dfrac{N^{in} - N^{out}}{N^{in}}$ (PFR or CSTR)	The ideal scenario is to have a high conversion per pass at high flow rate
Current efficiency or charge Yield, ϕ' (dimensionless)	$\phi' = \dfrac{NF}{Q} = \dfrac{W_R F}{MQ}$	Ideally, should be close to 100%; Values below 100% translate not only to wasted power but also to poor product selectivity
Energy consumption for electrolysis, E_s (J kg^{-1} or J m^{-3})	$E_s = \dfrac{W_{cell}}{W_R}$ or $E_s = \dfrac{W_{cell}}{V_m}$	Must be as small as possible
Space time yield, ρ_{ST} (mol m^{-3} s^{-1})	$\rho_{ST} = \dfrac{1}{V_R} \cdot \dfrac{dN}{dt}$	Defines the amount of product obtained in a unit reactor volume per unit time; obviously must be as high as possible

Notes: N = number of moles of reactant; the superscripts o and t define the initial condition and the condition at any time, t; Q = charge consumed in electrolysis; W_R = mass of reactant; M = its molar mass; W_{cell} = electrolytic power required; V_R = reactor volume; V_m = volume of electrolyte.

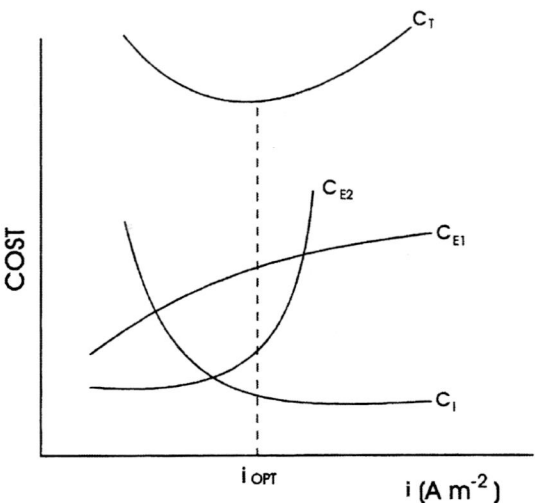

Figure 2.29. The dependence of cost on current density for an electrochemical process. C_T = total cost, C_{EI} = electrolytic power cost, C_{E2} = electric power cost, and C_I = investment cost. (Adapted with permission from D. Pletcher and F.C. Walsh, "Industrial Electrochemistry." Second Edition, Chapman and Hall, London, 1990.)

the situation when electrolyte flows through a capillary tube or soil matrix. Suppose that the wall is negatively charged, and a section of length L is exposed to the electrolyte. The negative charge density on the wall will be balanced by an oppositely charged diffuse layer (in the so-called Gouy–Chapman layer) in the adjacent solution. Thus, there will be an ion cloud whose volumetric charge density will be high close to the wall but will fall off to near zero within a few Debye lengths. The bore of the flow channel will be much larger than the diffuse layer thickness. Hence the distance coordinate, x, directed toward the axis of the flow channel from the wall, will be effectively infinite from a molecular dimension perspective.

Figure 2.30 contains a schematic of an electrokinetic system. The flow channel could be either an interconnected pore in a soil matrix or a silica capillary tube as in a CE experiment. When an electric field, ΔE is applied as shown in Figure 2.30, the solution begins to move. If the solution columns on either side are balanced such that there is no pressure difference, the phenomena is called *electro-osmotic flow*. If this flow is allowed to continue, a pressure will build up on one side to oppose this flow, much like an osmotic pressure builds up on one side of a semi-permeable membrane. The volume flow rate, $\dot{V}$ (m^3 s^{-1}) in electro-osmotic flow is given by

2.10. Electrokinetic Phenomena

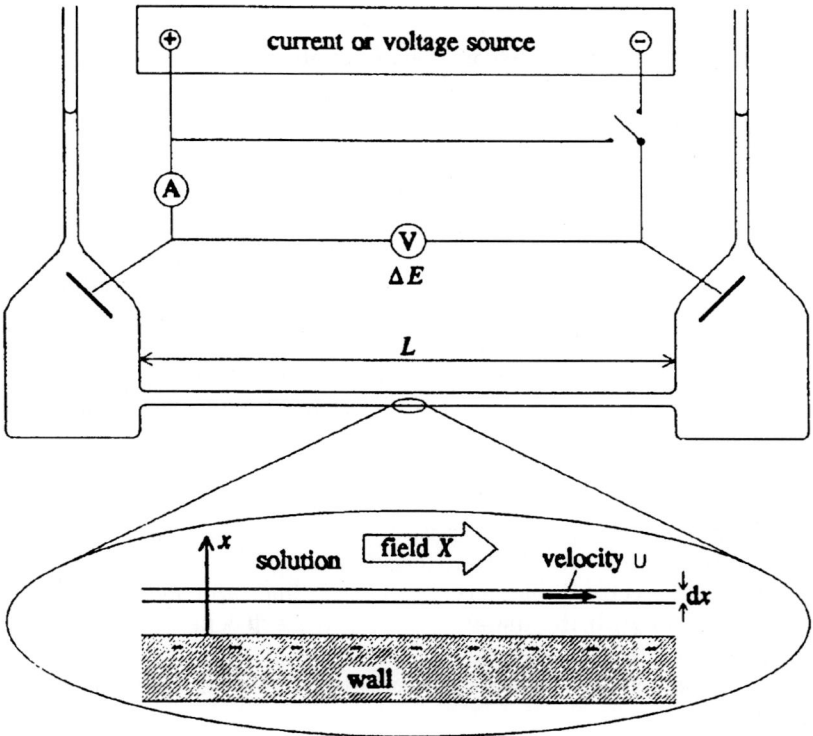

FIGURE 2.30. An electrokinetic system with two reservoirs and a flow channel of length L. (Reproduced with permission from K. B. Oldham and J. C. Myland, "Fundamentals of Electrochemical Science." Academic Press, San Diego, CA, 1994.)

$$\dot{V} = \frac{A \varepsilon \xi}{L \eta} \Delta E \tag{2.70}$$

where ε is the dielectric permitivity (F m^{-1}), η is the fluid viscosity, and ξ is known as the *electrokinetic* or, simply, the *zeta potential*. If we recall that the molecular layer in contact with the wall does not move, ξ defines the potential at the plane beyond which motion occurs (the so-called slip plane).

The zeta potential is dependent on the electrolyte concentration as well as on the wall's charge density. It is also a strong function of pH. In dilute neutral aqueous solution containing only univalent ions at 25°C, ξ is ca. -150 mV at a glass surface, and about half that value at silica.

2.11. SEMICONDUCTOR ELECTROCHEMISTRY

Up until now, we have only considered metal (or metal-like) electrodes. However, the use of a semiconductor as an electrode material brings with it interesting consequences associated mainly with the property of a semiconductor (SC) to absorb light of suitable wavelength:

$$SC \xrightarrow{h\nu > E_g} e^- + h^+ \qquad (2.71)$$

Where $h\nu$ is the photon energy, ν is the frequency of light, and h is the Planck's constant, and E_g is called the *optical band-gap* of the semiconductor. Equation (2.60) suggests that irradiation of the electrode causes the production of a *photocurrent*, since a current flow after all is related to the separation of charge. This brings up an important point; namely, we have to implement a means for *separating* the optically generated carriers to "collect" a current. Before considering this aspect, Eq. (2.71) shows that

- Photoexcitation of a semiconductor is a *quantized* process; that is, only light of energy exceeding the bandgap is absorbed. (Subbandgap excitation does occur in some cases but the transition is weak.)
- Two types of electronic carriers, namely, *electrons* and *holes*, are produced. This contrasts with the usual situation in electrochemistry, where only electrons need be considered.

In an *intrinsic* semiconductor, the electron and hole densities are equal. For example, for silicon at 25°C,

$$n_i = p_i \approx 1.4 \times 10^{10} \text{ cm}^{-3} \qquad (2.72)$$

This situation is akin to the ionic equilibrium that exists in water in the absence of an external acid or base. Thermal generation of carriers can be neglected at temperatures near 25°C because $E_g \gg kT$ (k is the Boltzmann constant). Thus, intrinsic semiconductors are electrical insulators.

Introduction of controlled amounts of impurities into the host material (a process termed *doping*) produces an *extrinsic* semiconductor. These impurities introduce additional electrons or holes because of ionization and thus perturb the thermal equilibrium. Again, the situation is analogous to the alteration of pH because of acid–base ionization. A semiconductor thus doped with a *donor* impurity is termed *n*-type, and that doped with an *acceptor* impurity is called a *p*-type semiconductor. The ionization is almost complete at room temperature (strong "acid–base" behavior in the electrolytic solution analog) such that the electron (or hole) density in an *n*-type (or *p*-type) semiconductor is approximately equal to the donor (or acceptor) density. For example, if the amount of donor dopant is ~1 ppm, the donor density, N_D will be ~5×10^{16} cm^{-3}. The electron density, n will be essentially the same. Thus, the hole density p is much smaller because of the equilibrium conditon:

$$n \simeq n_i^2 \quad (2.73a)$$

and

$$p = \frac{n_i^2}{N_D} \quad (2.73b)$$

For this particular example, p will be $\approx$4,000 cm^{-3} at 25°C. For this reason, electrons are *majority* carriers in an *n*-type semiconductor and holes are termed *minority* carriers. The opposite is true for a *p*-type semiconductor. The doping process thus makes the material electronically conducting via majority carrier flow in the dark under a bias potential.

Upon bandgap illumination, the minority carrier population is greatly enhanced whereas the majority carrier concentration is not significantly perturbed (in the so-called low-level injection regime). Therefore, we ought to expect a significant concentration of holes in an irradiated *n*-type semiconductor. These optically generated electrons and holes will simply recombine with one another if they are not separated. Contact of a semiconductor with an electrolyte fortunately creates a *junction* that separates the carriers much like what happens at a metal/semiconductor contact (or a Schottky barrier). To understand how this occurs, we must consider the *band model* for a semiconductor.

2.11.1. Band Structure of a Semiconductor and Band Bending at a Semiconductor/Liquid Junction

Consider the formation of a solid lattice of, say, Si. When the isolated atoms, comprising filled and vacant orbitals, are assembled into a lattice containing ~5 × 10^{22} atoms/cm^3, new *molecular orbitals* (MO) form. These orbitals are so closely spaced that they merge into bands; the filled bonding orbitals form the *valence band* (VB) and the unfilled antibonding orbitals form the *conduction band* (CB). Figure 2.31 illustrates the situation. The optical bandgap, E_g in the preceding section, separates VB and CB, respectively. In a photoresponsive molecular system, light absorption entails the excitation of an electron from a *highest-occupied* MO (HOMO) to a *lowest-occupied* MO (LUMO). In the solid-state semiconductor counterpart, this process results in electron excitation from the VB to the CB, leaving a hole behind in the VB (see Eq. (2.72)).

The other important concept in semiconductor electrochemistry is the *Fermi level*, E_F. This is defined (strictly at 0 K) as that energy where the probability of a level being occupied is 1/2. For an intrinsic semiconductor (at 25°C), E_F lies approximately midway between the VB and CB. For a doped semiconductor, the position of E_F depends on N_A or N_D. For moderately doped semiconductors

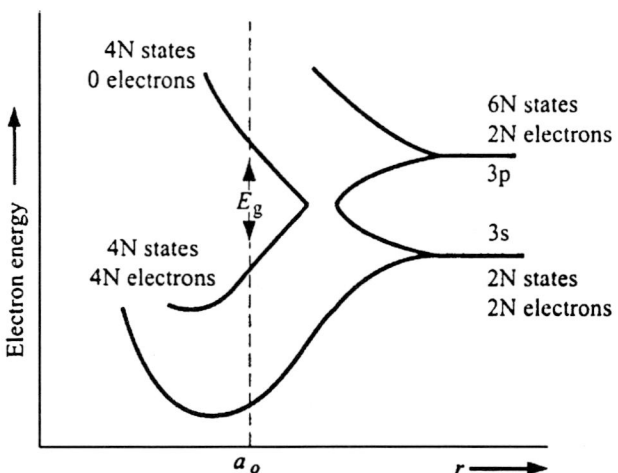

FIGURE 2.31. Formation of energy bands from discrete states for silicon. The electron energy is shown as a function of a distance coordinate, with the equilibrium separation given by a_o. E_g is the energy bandgap of the resultant semiconductor. (Adapted from: W. Shockley, "Electrons and Holes in Semiconductors." Van Nostrand, New York, 1950.)

(N_A, N_D >10^{17} cm^{-3}), E_F lies slightly (within a few kT) below the CB edge or just above the VB edge for an *n*-type or a *p*-type semiconductor respectively.

For the solution phase, the Fermi level is defined as $E^{o\prime}$.[12] Now consider what happens when an *n*-type semiconductor contacts a solution containing a redox couple whose formal potential is $E^{o\prime}$ (Figure 2.32a). Clearly, we have a nonequilibrium situation and electrons will have to flow from the semiconductor to the solution contact until the Fermi levels equalize. We thus have *band bending* at the electrode/solution interface (Figure 2.32b). This band bending serves the useful electrostatic role of separating the optically generated electron–hole pairs at the interface, much like what happens at a metal/semiconductor contact.

Figure 2.32b importantly illustrates that a *space-charge* layer of a few Å width exists within the semiconductor. This contrasts with a metal electrode whose conductivity and carrier concentration are such that no layer can be supported within it, and most of the voltage is dropped across the solution side (the so-called Helmholtz layer) of the metal/electrolyte junction. On the other hand, a *space-charge layer capacitance* can actually be measured for a semiconductor/electrolyte junction.

Figure 2.32c shows that when light of energy, $h\nu > E_g$ is incident on the semiconductor, e$^-$–h$^+$ pairs are created (see Eq. (2.71)). A few of these recombine, but some are driven in opposite directions by the built-in electric field. Thus,

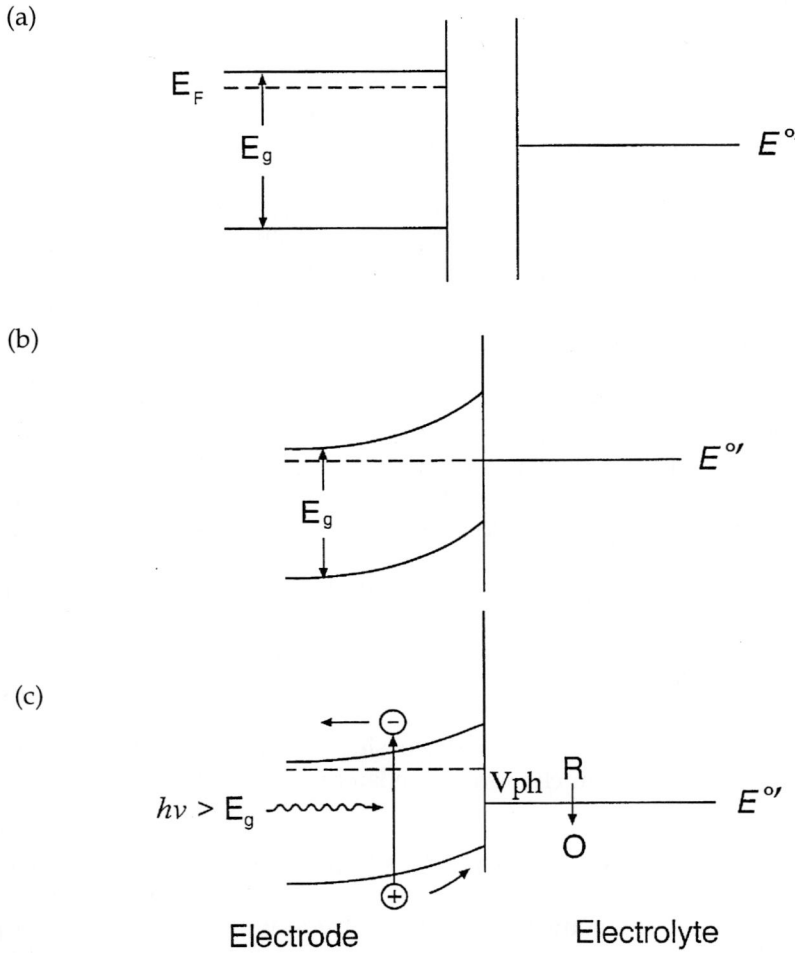

FIGURE 2.32. Contact of an n-type semiconductor with an electrolyte (with redox potential $E^{o\prime}$) (a) gives rise to band bending at the resulting interface (b). Irradiation of this interface with photons of energy greater than E_g gives rise to a photovoltage, V_{ph}, and oxidation of R to O by the photogenerated holes (c).

the electrons can be collected at the rear (ohmic) contact of the semiconductor electrode, and the holes flow to the semiconductor surface, where they are available for oxidation of the reduced half (R) of the solution redox couple. The net result is the flow of a *photocurrent*.

Different types of *photoelectrochemical* cells can be constructed based on these principles[13]:

- Liquid-junction photovoltaic cells: If a reversible redox couple, O/R is used in the electrolyte, a photoelectrochemical cell can be constructed with an n-type semiconductor *photoanode* and a metal cathode (or, alternatively, an n-type photoanode and a p-type photocathode) such that R is oxidized at the irradiated photoanode, and it is reduced back at the cathode (or the irradiated photocathode). Thus, there is no chemical change in the cell, and the device can be used much like a solid-state photovoltaic solar cell to produce electric power at an external load. Figure 2.33a contains a schematic of such a device.
- Photoelectrosynthetic cells: In this case, the reaction at the counterelectrode is different than that at the semiconductor, and the net cell reaction is driven by light in the nonspontaneous direction ($\Delta G > 0$) (Figure 2.33b). In this device, the radiant energy is stored as chemical energy. The photoelectrolysis of water and the photoreduction of CO_2 (an important environmental system) are examples of reactions involving this strategy.
- Photocatalytic cells: These are similar to the photoelectrosynthetic devices except that the relative locations of the potentials of the O/R and O'/R' couples are changed (Figure 2.33c). In this case, the reaction is driven in the spontaneous direction (which is presumably very slow in the dark) ($\Delta G < 0$). The light energy is used here to overcome the *activation energy* for the chemical reaction.

We shall have opportunities for discussing photoelectrosynthetic and photocatalytic reactions in much more detail later in this book.

2.12. PHOTOEMISSION AT METAL ELECTRODES

Under certain circumstances, a metal electrode will emit electrons on suitable irradiation. This arises from the *photoelectric effect*. Thus, when the *work function* of the metal electrode is exceeded by the incident light energy, electrons will be ejected from the metal phase. Unlike the situation in a vacuum, however, the energy requirements for this process are ameliorated by the energy gained via *solvation* of the electron when the metal contacts water. These *solvated* electrons, (e_{aq}^-) are highly reactive, and produce interesting chemistry when scavengers are available to interact with them. For example, N_2O is one such scavenger[14, 15]:

$$e_{aq}^- + H_2O + N_2O \rightarrow N_2 + OH^- + {}^\bullet OH \qquad (2.74)$$

In acidic media, e_{aq}^- also generate $H^\bullet$:

$$H_3O^+ + e_{aq}^- \rightarrow H^\bullet + H_2O \qquad (2.75)$$

2.13. Instumentation   121

(a)

(b)

(c)

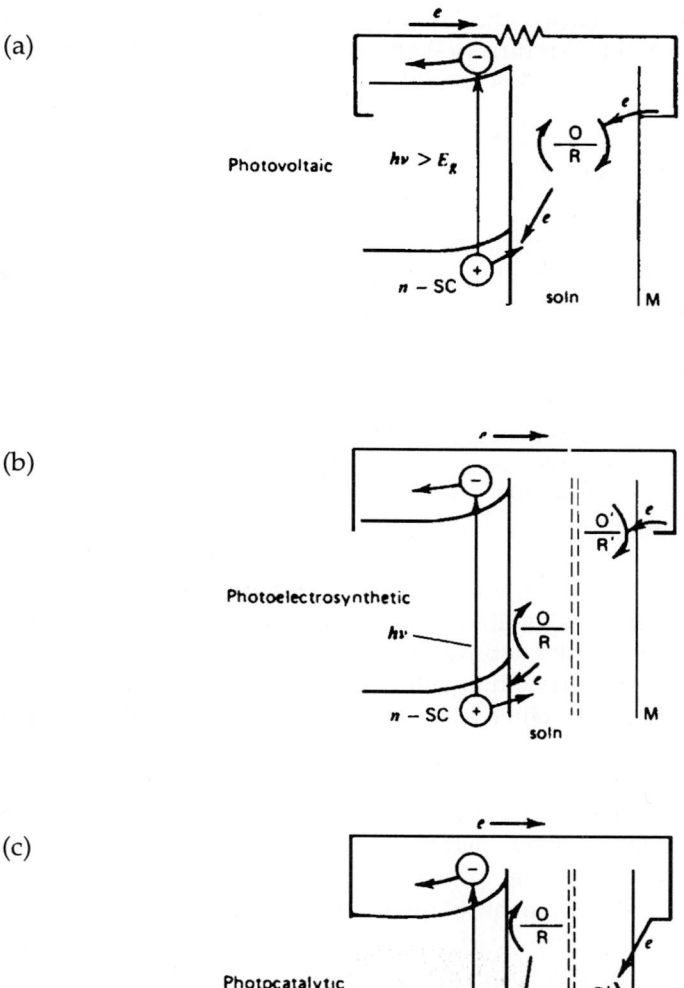

FIGURE 2.33. Schematic representation of different types of photoelectrochemical cells using an n-type semiconductor. (Reproduced with permission from A.J. Bard and L.R. Faulkner, "Electrochemical Methods." Wiley, New York, 1980.)

Species such as e^-_{aq}, $H^\bullet$ and $^\bullet OH$ have interesting redox behavior and, more important in terms of relevance to this book, are very reactive toward a variety of pollutants. Thus, we shall return to their electrochemistry and photoelectrochemistry in Chapters 5–7 of this book.

2.13. INSTRUMENTATION

An important technological advantage with electrochemical pollution abatement is the relative simplicity and low cost of equipment. Electronic circuitry for *controlling* and *measuring* voltage, current, or charge is relatively easy to assemble both in the analog and digital domains. Either constant current or constant potential procedures may be used for bulk electrolytic treatment of pollutants. In either case, provision must be made to ensure low *IR* voltage drops, and to facilitate efficient heat dissipation in the electrochemical system, especially since plant-scale processes may involve currents and voltages of several amperes or volts, respectively.

On a laboratory scale, a *potentiostat/galvanostat* is used for controlling the potential or current in an electrochemical cell. Additionally, a programmer and/or a timing circuit is used in voltammetry for ramping the electrode potential (cyclic and linear sweep voltammetry) or for stepping the potential to the desired limit, where a voltage pulse is superimposed and the current response is sampled at appropriate time intervals (Figure 2.17). The advent of computer-controlled instrumentation has made electroanalysis reliable and user-friendly.

Photoelectrochemical experiments or remediation procedures additionally require suitable light sources and wavelength isolation devices. These include broad-band gas-discharge sources such as Hg or Xe arc lamps (for the UV region), and black-body emission lamps such as the W-halogen system (for the visible range). For isolating a narrow band of wave lengths, either a filter or a monochromator may be employed. Monochromatic sources such as lasers are useful for laboratory experiments, but may be prohibitively expensive for routine, large-scale use. Finally, outdoor experiments may be designed to utilize sunlight as the semiconductor excitation source.

Various forms of radiometers or light meters are commercially available for measuring the incident light intensity (usually in units of W/m^2). A difficulty in quantifying this parameter is that associated with the estimation of the amount of light lost via scattering at the cell wall, at the electrode/photocatalyst surface, and via electrolyte absorption. Usually, the quoted light intensity is not corrected for these spurious effects.

A lock-in amplifier (LIA) is useful in photoelectrochemical experiments, where the photocurrent is small, and the signal/noise ratio is consequently high. This device is used in conjunction with a light chopper, which provides a reference (carrier) signal for the LIA. Alternatively, the LIA can also be used for measurement of the interfacial electrode/solution capacitance. A small-amplitude (nominally 10 mV peak/peak) AC source provides the reference signal in this case, and the LIA measures the in-phase and out-of-phase components of the cell impedance as a function of the AC source frequency and the electrode potential, the latter being controlled via a potentiostat.

2.14. SUMMARY

In this chapter, basic concepts in electrochemistry and photoelectrochemistry have been reviewed. They will serve as building-blocks for the material that appears in later sections of this book, particularily in Chapters 4–6. In the next chapter, we examine the electrochemical behavior of classes of organic and inorganic environmental pollutants.

REFERENCES

1. S. Trasatti, The Absolute Electrode Potential: An Explanatory Note. *J. Electroanal. Chem. Interfacial Electrochem.* **209**, 417 (1986); The Absolute Electrode Potential — The End of the Story. *Electrochim. Acta* **35**, 269 (1990).
2. D. Britz, IR Elimination in Electrochemical Cells. *J. Electroanal. Chem. Interfacial Electrochem.* **88**, 309 (1978); 100% IR Compensation by Damped Positive Feedback. *Electrochim. Acta* **25**, 1449 (1978).
3. G. P. Power and I. M. Ritchie, Mixed Potentials: Experimental Illustrations of an Important Concept in Practical Electrochemistry. *J. Chem. Educ.* **60**, 1022 (1983).
4. J. Heinze, Cyclic Voltammetry — Electrochemical Spectroscopy. *Angew. Chem., Intl. Ed. Engl.* **23**, 831 (1984).
5. R. S. Nicholson and I. Shain, Theory of Stationary Electrode Polarography. *Anal. Chem.* **36**, 706 (1964).
6. M. Rudolph, D. P. Reddy and S. W. Feldberg, A Simulator for Cyclic Voltammetric Responses. *Anal. Chem.* **66**, 589A (1994).
7. D. K. Gosser, Jr., Cyclic Voltammetry: Simulation and Analysis of Reaction Mechanisms. VCH, New York, 1993.
8. D. Jagner and A. Granelli, Potentiometric Stripping Analysis. *Anal. Chim. Acta* **83**, 19 (1976); Instrumental Approach to Potentiometric Stripping Analysis of Some Heavy Metals. *Anal. Chem.* **50**, 1924 (1978).
9. M. Fleischmann and H. R. Thirsk, Metal Deposition and Electrocrystallization. *Adv. Electrochem. Electrochem. Eng.*, **3**, 123 (1963); R. G. Barradas

and E. Bosco, Models of Two-Dimensional Electrocrystallization with Couplings to Diffusion and/or Dissolution. *Electrochim. Acta* **11**, 949 (1986); B. R. Scharifker and J. Mostany, Three-Dimensional Nucleation with Diffusion-Controlled Growth. *J. Electroanal. Chem. Interfacial Electrochem.* **177**, 13 (1984); J. H. O. J. Wijenberg, W. H. Mulder, M. Sluyters-Rehbach, and J. H. Sluyters, On the Rate Equation of Nucleation and the Concept of Active Sites in Electrodeposition. *ibid.* **256**, 1 (1988), and references therein.
10. J. J. Lingane, Interpretation of the Polarographic Waves of Complex Metal Ions. *Chem. Rev.* **29**, 1 (1941).
11. G. Kreysa, Performance Criteria and Nomenclature in Electrochemical Engineering. *J. Appl. Electrochem.* **15**, 175 (1985); F. Walsh and G. Reade, Design and Performance of Electrochemical Reactors for Efficient Synthesis and Environmental Treatment. Part I. Electrode Geometry and Figures of Merit. *Analyst (London)* **119**, 791 (1994).
12. H. Gerischer and W. Ekardt, Fermi Levels in Electrolytes and the Absolute Scale of Redox Potentials. *Appl. Phys. Lett.* **43**, 393 (1983).
13. A. J. Bard, Design of Semiconductor Photoelectrochemical Systems for Solar Energy Conversion. *J. Phys. Chem.* **86**, 172 (1982); A. J. Bard, Photoelectrochemistry. *Science* **207**, 139 (1980).
14. G. C. Barker, D. McKeown, M. J. Williams, G. Bottwea, and V. Concialini, Charge Transfer Reactions Involving Intermediates Formed by Homogeneous Capture of Laser-Produced Photoelectrons. *Faraday Discuss. Chem. Soc.* **56**, 41 (1974).
15. G. C. Barker, Electrochemical Effects Produced by Light Induced Electron Emission. *Ber. Bunsenges. Phys. Chem.* **75**, 728 (1971).

Supplementary Reading

Books and Monographs on Electrochemistry

1. R. D. Adams, "Electrochemistry at Solid Electrodes." Dekker, New York and Basel, 1969.
2. D. T. Sawyer and J. L. Roberts, Jr., "Experimental Electrochemistry for Chemists." Wiley, New York, 1974.
3. D. J. Pickett, "Electrochemical Reactor Design." Elsevier, Amsterdam, 1977.
4. A. M. Bond, "Modern Polarographic Methods in Analytical Chemistry." Dekker, New York and Basel, 1980.
5. A. J. Bard and L. R. Faulkner, "Electrochemical Methods." Wiley, New York, 1980.
6. C. J. H. King, Comments on the Design of Electrochemical Cells. In "Tutorial Lectures in Electrochemical Engineering and Technology" (R. Alkire

and T. Beck, eds.), AIChE Symp. Ser., Vol. 77, No. 204. AIChE, New York, 1981.
7. The Southampton Electrochemistry Group, "Instrumental Methods in Electrochemistry." Ellis Horwood, Chichester, 1985.
8. J. Wang, "Stripping Analysis: Principles, Instrumentation and Applications." VCH Publishers, Deerfield Beach, FL., 1985.
9. B. W. Rossiter and J. F. Hamilton, eds., "Physical Methods of Chemistry;" Vol. 11. Wiley, New York, 1986.
10. E. Heitz and G. Kreysa, "Principles of Electrochemical Engineering: Extended Version of a DECHEMA Experimental Course." VCH, Weinheim, 1986.
11. J. T. Stock and M. V. Orna, eds., "Electrochemistry, Past and Present." ACS Symp. Ser. No. 390. American Chemical Society, Washington, DC, 1989.
12. D. Pletcher and F. C. Walsh, "Industrial Electrochemistry." 2nd ed., Chapman & Hall, London, 1990.
13. J. D. Genders and D. Pletcher, eds., "Electrosynthesis: From Laboratory, To Pilot, To Production." The Electrosynthesis Co., East Amherst, NY, 1990.
14. J. S. Newman, "Electrochemical Systems." Prentice-Hall, Englewood Cliffs, NJ, 1991.
15. J. D. Genders and N. Weinberg, eds., "Electrochemistry for a Cleaner Environment." The Electrosynthesis Co., East Amherst, NY, 1992.
16. J. O'M. Bockris and S. U. M. Khan, "Surface Electrochemistry." Plenum, New York, 1993.
17. P. H. Rieger, "Electrochemistry." Prentice-Hall, Englewood Cliffs, NJ, 1987.
18. P. A. Christensen and A. Hamnett, "Techniques and Mechanisms in Electrochemistry." Chapman & Hall, London, 1993.
19. F. C. Walsh, "A First Course in Electrochemical Engineering." The Electrochemical Consultancy, Great Britain, 1993.
20. J. Koryta, J. Dvorak and L. Kavan, "Principles of Electrochemistry." 2nd ed. Wiley, New York, 1993.
21. K. B. Oldham and J. C. Myland, "Fundamentals of Electrochemical Science." Academic Press, San Diego, CA, 1994.
22. J. Wang, "Analytical Electrochemistry." VCH, New York, 1994.

Books and Monographs on Photoelectrochemistry

1. S. R. Morrison, "The Chemical Physics of Surfaces." Plenum, New York and London, 1977.
2. S. J. Fonash, "Solar Cell Device Physics." Academic Press, San Diego, CA, 1981.
3. S. Chandra, "Photoelectrochemical Solar Cells." Gordon & Breach, New York, 1985.

4. H. O. Finklea, ed., "Semiconductor Electrodes." Elsevier, Amsterdam, 1988.
5. Yu. V. Pleskov, "Solar Energy Conversion." Springer-Verlag, Berlin, 1990.

Journals in Electrochemistry and Photoelectrochemistry

1. *Electrochimica Acta*
2. *Journal of Electroanalytical Chemistry and Interfacial Electrochemistry*
3. *Journal of Applied Electrochemistry*
4. *Journal of the Electrochemical Society*
5. *Electroanalysis*
6. *Solar Energy Materials and Solar Cells*

Electrochemical articles of environmental interest also frequently appear in these journals:

1. *Talanta*
2. *Analytical Chemistry*
3. *The Analyst*
4. *Analytica Chimica Acta*
5. *Environmental Science and Technology*

CHAPTER THREE

3.1. INTRODUCTION

In this chapter, we briefly examine what is known about the electrochemical behavior of organic, inorganic, and organometallic pollutants (Table 1.3). This review will form the framework within the context of which specific pollutant assay and cleanup procedures are discussed in Chapters 4–6 in this book.

The concept of oxidation number or oxidation state is useful in the discussion of redox reactions. It is defined as the charge that a bonded atom would have if the electrons in each bond were given to the more electronegative atom. In the water molecule, for example, the bond electrons are assigned to oxygen, it being the more electronegative element. The oxidation number of oxygen and each hydrogen then becomes -2 and +1, respectively, and the total adds up to zero for the neutral molecule. Note that the oxidation number or state for a given atom is shown with a sign preceding the number, unlike the corresponding charge designation for an ion, which has the opposite sequence; that is, a number followed by the sign (e.g., O^{2-} and H^+ for oxide anions and protons, respectively).

The principal oxidation states are contained in Figure 3.1 for some environmentally relevant elements. Some general trends appear. The oxidation numbers of main group elements do not exceed the group number. Sulfur, a

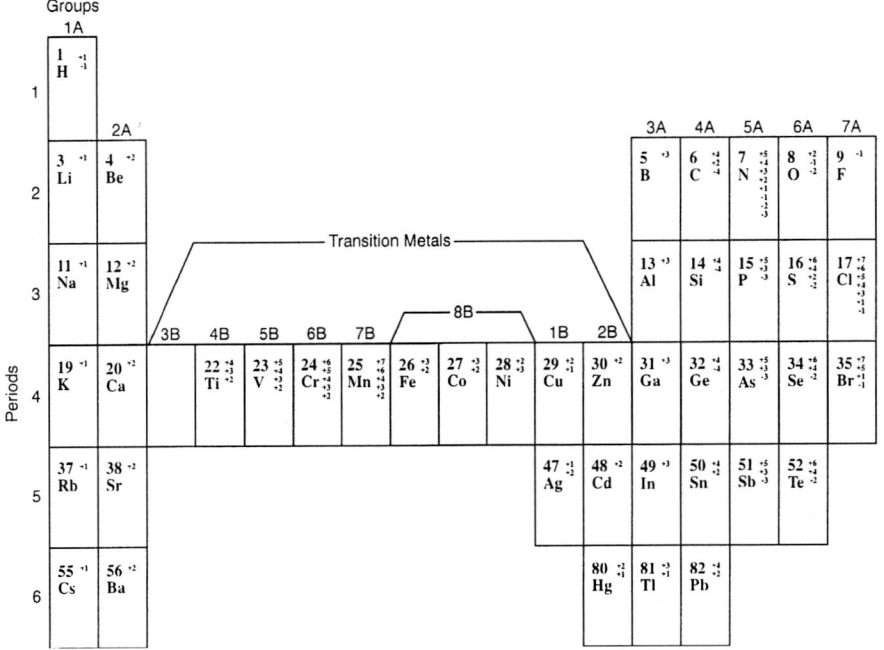

FIGURE 3.1. Periodic Table showing the oxidation states of environmentally important elements. The radioactive elements are not shown.

Group 6 element, has +6 as its highest oxidation state. Metals have only positive oxidation states but nonmetals can have both positive and negative states. The lowest (most negative) oxidation state for a nonmetal is 8 (the octet population) minus the group number. Thus, carbon, a Group 4 element, has -4 as its lowest oxidation state.

An oxidation reaction results in an increase of the oxidation number; a reduction causes a decrease. With the oxidation number concept, the path of a redox reaction may be readily identified even for organic substances whose redox behavior often is not as transparent as for example, metallic systems. The redox transformation of a hydrocarbon to CO_2 illustrates this point. Consider the redox progression of methane:

hydrocarbon	alcohol	aldehyde	acid	
CH_4 →	CH_3OH →	$HCHO$ →	$HCOOH$ →	CO_2
-4	-2	0	+2	+4

The oxidation number of carbon is shown in each case and clearly illustrates the oxidation of each compound, leading finally to CO_2. Contrast this example with the oxidation of iron:

$$Fe(s) \xrightarrow{-2e^-} Fe^{2+}(aq) \xrightarrow{-e^-} Fe^{3+}(aq)$$

To keep this chapter to a manageable size, discretion had to be exercised as to what data could be included for each category of pollutants. The reader interested in any one particular category may wish to delve into the specialized literature provided at the end of the chapter for a fuller discussion. In this regard, the material in this chapter is *not* meant to be a compendium of all the electrochemical information that exists on each of the elements and compounds reviewed in it.

3.2. ELECTROCHEMISTRY OF ORGANIC POLLUTANTS

3.2.1. *General Considerations*

The electrochemistry of organic substances is invariably associated with a chemical reaction preceding or subsequent to the electron transfer step. Consider the generic electrochemical reaction:

$$O + ne^- \rightleftarrows R$$

In many cases, O may not be present initially but may be produced from another, nonelectroactive species. A case in point is the electrochemical reduction of formaldehyde in aqueous media. Formaldehyde exists as a nonreducible hydrated form, $H_2C(OH)_2$, in equilibrium with the reducible $H_2C=O$ form:

$$H_2C(OH)_2 \rightleftarrows H_2C=O + H_2O \tag{3.1}$$

The equilibrium constant favors the hydrated form. Thus Reaction 3.1 precedes the electrochemical step, and under some conditions, the measured current will be governed by the kinetics of this reaction (yielding a so-called kinetic current).

In other cases, R may react with the solvent or supporting electrolyte to yield new species, which may or may not be electroactive. The electrochemical oxidation of organics to yield radical cations (which subsequently undergo ho-

Table 3.1. Mechanistic Schemes for Electrode Processes Coupled With Homogeneous Reactions

Scheme	Reactions	Comments
C_rE_r	$Y \rightleftarrows O$ $O + ne \rightleftarrows R$	Chemical reaction preceding reversible electrode reaction
C_rE_i	$Y \rightleftarrows O$ $O + ne \rightarrow R$	Chemical reaction preceding irreversible electrode reaction
E_rC_r	$O + ne \rightleftarrows R$ $R \rightleftarrows Y$	Reversible chemical reaction follows the electrode process
E_rC_i	$O + ne \rightleftarrows R$ $R \rightarrow Y$	Irreversible chemical reaction follows the electrode process
E_rC_{2i}	$O + ne \rightleftarrows R$ $2R \rightarrow X$	The coupled reaction is a dimerization process
$E_rC'_i$	$O + ne \rightleftarrows R$ $R \rightarrow O$	The coupled reaction regenerates the reactant
$E_rC_iE_r$	$O_1 + n_1e \rightleftarrows R_1$ $R_1 \rightarrow O_2$ $O_2 + n_2e \rightleftarrows R_2$	The chemical reaction is interspersed between two reversible electrode processes

Note: Other variants are possible and are denoted by appropriate subscripts. For example, a quasi-reversible charge transfer is denoted by the symbol, E_q. Other mechanistic subtleties such as the DISP mechanism, where the second electron transfer in an EC scheme *occurs in solution*, are not considered here nor are schemes such as ECE', where the reduction of O_2 takes place at more negative potentials than O_1 (i.e., $E^0{}_1 > E^0{}_2$).

mogeneous chemistry) falls into this category. Table 3.1 contains a summary of mechanistic possibilities where electrode reactions are coupled with homogeneous chemical reactions. A classification of the possible reaction schemes is facilitated by using letters to signify the nature of the step. Thus E stands for a simple electronic transfer step to or from the electrode surface. C stands for a homogeneous chemical reaction. C' denotes a catalytic step. O and R represent electroactive chemical species in the solution. Finally, X and Y represent chemical species in the solution that are nonelectroactive within the potential range of interest and under the experimental conditions.

The effect that a perturbing chemical reaction will have on the measured parameters of the electrode reaction depends on the time window of the experiment or process. This can vary widely, as Table 3.2 shows. The advent of *ultra-*

3.2. Electrochemistry of Organic Pollutants

Table 3.2. Characteristic Time Windows for Electrochemical Measurements and Processes

Technique	Time Window (s)
AC polarography	$2 \times 10^{-4} - 0.1$
Rotating disk electrode voltammetry	$10^{-3} - 0.3$
Chronopotentiometry / Chronoamperometry	$10^{-3} - 50$
Cyclic or linear scan voltammetry [a]	$10^{-4} - 1$
DC polarography	$1 - 5$
Coulometry / Macroscale electrolysis	$100 - 3{,}000$

[a] Will depend on potential scan rate. Scan rates up to 10^6 V/s have been reported for ultramicroelectrodes; the time window in these instances will be in the μs domain. (Adapted from A.J. Bard and L.R. Faulkner, "Electrochemical Methods." Wiley, New York, 1980.)

microelectrodes (i.e., electrodes of μm dimensions, Chapter 4) has shrunk the time window to the millisecond or even microsecond domain. Under these conditions, many electrode processes with coupled chemical reactions of moderate kinetic facility are only mildly perturbed by the latter. On the other hand, in macroscale electrolysis, organic electrochemistry seldom if ever involves "clean" electron transfer with no coupled chemical reactions.

Organic substances can undergo electron transfer in a variety of circumstances and media. The main factors that determine their electrochemical behavior are the nature of the electroactive group in the organic molecule, the nature of the solvent and the supporting electrolyte, the electrode material, the applied potential, and the temperature.

An electron added during reduction usually goes to an antibonding orbital, and the resulting highly reactive anionic species can undergo a variety of reactions, including conformational rearrangement. Likewise, an electron removed from a bonding orbital during oxidation leads to the formation of a cation radical, which will usually undergo dimerization, bond cleavage, nucleophilic attack or rearrangement. Some general trends appear for the electrochemical reduction and oxidation of organic compounds, which are now discussed in turn. The discussion here largely follows the treatment in the book by Lund and Baizer.

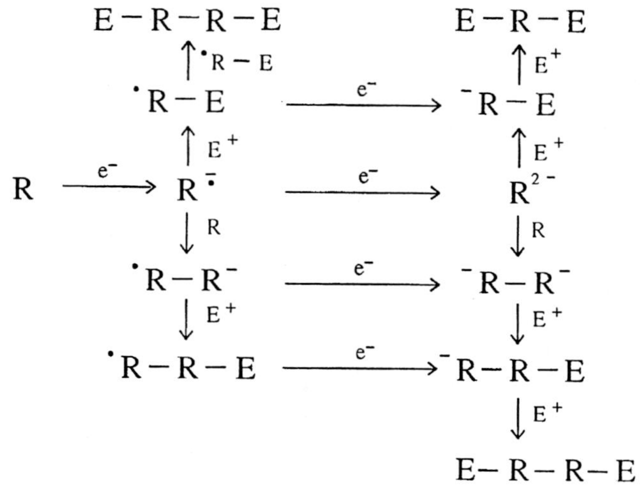

$$
\begin{array}{c}
\text{E}-\text{R}-\text{R}-\text{E} \\
\uparrow \cdot\text{R}-\text{E} \\
\cdot\text{R}-\text{E} \xrightarrow{e^-} {}^-\text{R}-\text{E} \\
\uparrow \text{E}^+ \qquad\qquad \uparrow \text{E}^+ \\
\text{R} \xrightarrow{e^-} \text{R}^{\cdot -} \xrightarrow{e^-} \text{R}^{2-} \\
\downarrow \text{R} \qquad\qquad \downarrow \text{R} \\
\cdot\text{R}-\text{R}^- \xrightarrow{e^-} {}^-\text{R}-\text{R}^- \\
\downarrow \text{E}^+ \qquad\qquad \downarrow \text{E}^+ \\
\cdot\text{R}-\text{R}-\text{E} \xrightarrow{e^-} {}^-\text{R}-\text{R}-\text{E} \\
\downarrow \text{E}^+ \\
\text{E}-\text{R}-\text{R}-\text{E}
\end{array}
$$

FIGURE 3.2. Mechanistic scheme for the electroreduction of an organic compound, denoted by R. E^+ is an electrophile. (Source: J.M.Costa, "Fundamentos de Electrodica." Alhambra Universidad, Madrid, 1981.)

Reduction. Figure 3.2 contains a generic scheme for the reductive conversion exemplified by R going to RE_2. In this scheme, E^+ is an electrophilic species (e.g., H^+). In addition, other homogeneous reactions may also occur; perhaps the most important belong to the so-called DISP(disproportionation)-type mechanism:

$2 R^{\cdot -} = R^{2-} + R$, and $R^{\cdot -} + RH^{\cdot} \rightarrow RH^- + R$

Another pathway for homogeneous reduction involves the reaction:

$R + H^- = RH^-$

where the hydride ion is produced by reduction of H_2 (ads) at a low hydrogen overvoltage electrode (e.g., Pt).

For a two-electron–two-proton electrochemical reduction process from R to RH_2, the possible paths are six (excluding any other possible side chemical or electrochemical reactions). Figure 3.3 illustrates the "square scheme" for this case. (Note that electron transfers are depicted horizontally whereas proton transfers are depicted vertically.) A good example of this type of mechanism involves the electrochemical reduction of a quinone to the corresponding hydroquinone.

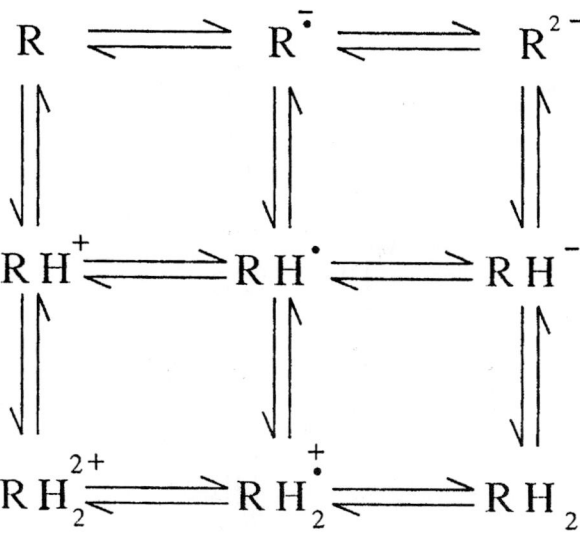

FIGURE 3.3. A square scheme for the electroreduction and protonation of an organic hydrocarbon, R converting ultimately to RH_2. (Source: H. Lund and M.M. Baizer, eds., "Organic Electrochemistry." Chapter 2. Dekker, New York, 1991.)

Halogenated aromatic organic compounds (ArX), including halogenated biphenyls as well as many other organic substances containing "good leaving groups" in addition to the halides (e.g., CN^-, SCN^-, OR^-, SR^-, NR^-, RSO_2^-), may undergo reductive elimination reactions as illustrated in Figure 3.4a. In this scheme, BH is a specie containing an electrophilic group (e.g., a solvent containing hydrogen), B• is the corresponding radical and Nu⁻ is a nucleophile. The ions and radicals produced will usually react further to yield various types of products. It is noteworthy that, for monohalogenated benzenes, there is a linear correlation between the rate constant for the cleavage of the radical anion ArX•⁻ and the negative of the standard potential of the ArX/ArX•⁻ couple.

If the initial electron transfer is slow (as with mercury electrodes), homogeneous electron transfer can occur with another substance that shows more facile heterogeneous kinetics at the electrode surface (e.g., polycyclic aromatic hydrocarbons, PAH, or transition metal complexes, ML). This indirect reduction process is further discussed in Chapter 5. Redox mediators can also be incorporated in a chemically modified electrode; for example, by covalent binding to a species attached to the electrode. This catalytic cycle yields the parent compound, ArH as shown in the scheme in Figure 3.4b.

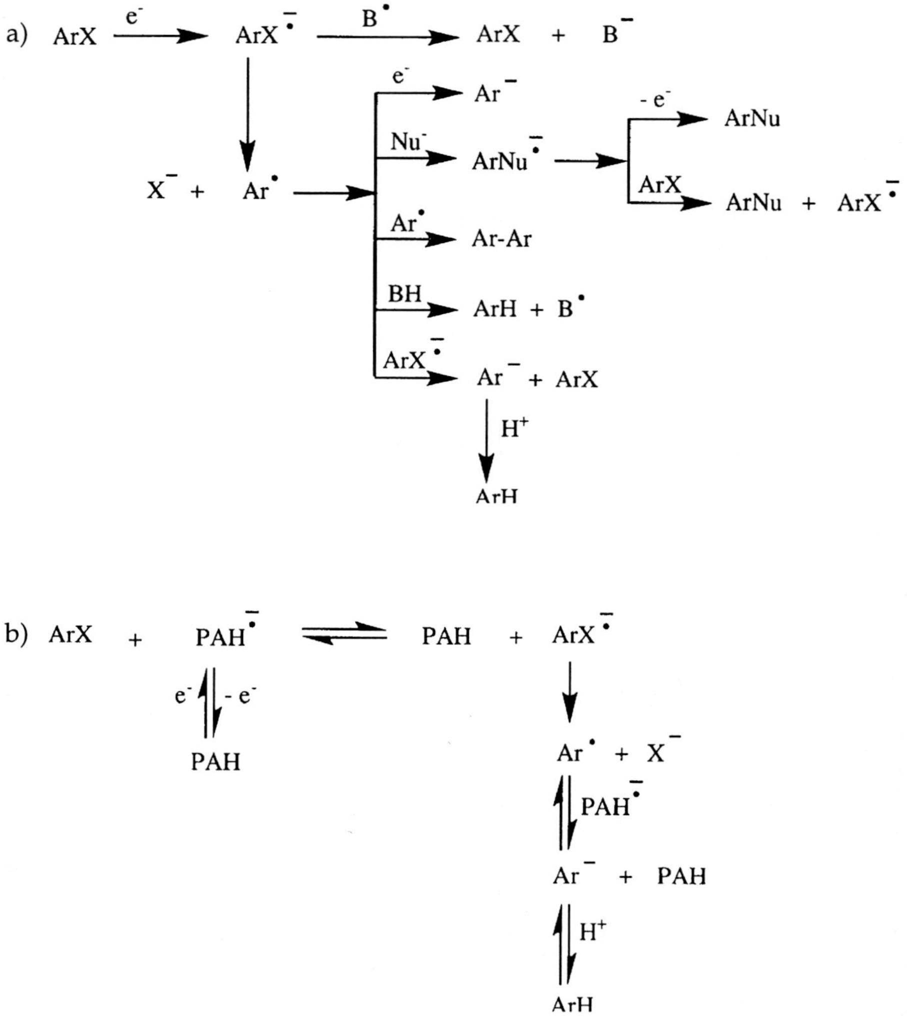

FIGURE 3.4. Electrochemical reduction of a halogenated aromatic hydrocarbon (ArX) either proceeding directly (a) or via a mediator such as a polycyclic aromatic hydrocarbon, PAH (b). BH is a solvent containing hydrogen and Nu⁻ is a nucleophile. (Source: P.H. Rieger, "Electrochemistry." Prentice-Hall, New Jersey, 1987. H. Lund and M. Baizer, eds., "Organic Electrochemistry." Chapter 7. Dekker, New York, 1991.)

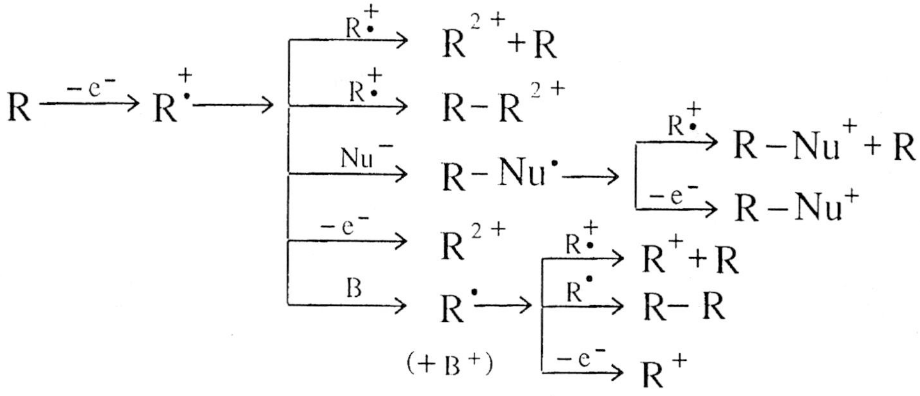

3.2. Electrochemistry of Organic Pollutants

$$R \xrightarrow{-e^-} R^{+\bullet} \longrightarrow \begin{cases} \xrightarrow{R^{+\bullet}} R^{2+} + R \\ \xrightarrow{R^{+\bullet}} R-R^{2+} \\ \xrightarrow{Nu^-} R-Nu^{\bullet} \longrightarrow \begin{cases} \xrightarrow{R^{+\bullet}} R-Nu^+ + R \\ \xrightarrow{-e^-} R-Nu^+ \end{cases} \\ \xrightarrow{-e^-} R^{2+} \\ \xrightarrow{B} R^{\bullet} \longrightarrow \begin{cases} \xrightarrow{R^{+\bullet}} R^+ + R \\ \xrightarrow{R^{\bullet}} R-R \\ \xrightarrow{-e^-} R^+ \end{cases} \\ (+B^+) \end{cases}$$

FIGURE 3.5. Mechanistic scheme for the electrooxidation of an organic compound, B is a base and Nu⁻ is a nucleophile.

<u>Oxidation</u>. The electrochemical oxidation of organic substances usually yields highly reactive intermediates that will normally produce overall reactions of substitution, addition, coupling with elimination, coupling with addition or further electron transfer as exemplified by the generic scheme in Figure 3.5. In this scheme, B denotes a base; that is, $RH^{+\bullet} + B \rightarrow BH^+ + R^{\bullet}$, where $RH^{+\bullet}$ is the initially generated radical cation of the compound. Such deprotonation behavior is exemplified by alkylated aromatic hydrocarbons.

In the case of aromatic substances, the intermediate cations are stable if one or more of the following situations exist: (a) the cation possesses a high degree of charge delocalization that prevents the existence of highly positively charged sites capable of reaction with a nucleophile; (b) the reactive sites are blocked by electroinactive substituents; or (c) the cation is stabilized by functional groups that extend the π system and promote a higher degree of charge delocalization. If the cation is not stable, follow-up reactions will lead to species such as those shown in the previous reaction scheme.

Indirect oxidation can also occur by homogeneous electron transfer with an oxidized specie in solution as exemplified by the scheme in Figure 3.6. The radicals thus produced can undergo the reactions described earlier.

We survey next the electrochemical behavior of classes of organic substances that are environmentally significant.

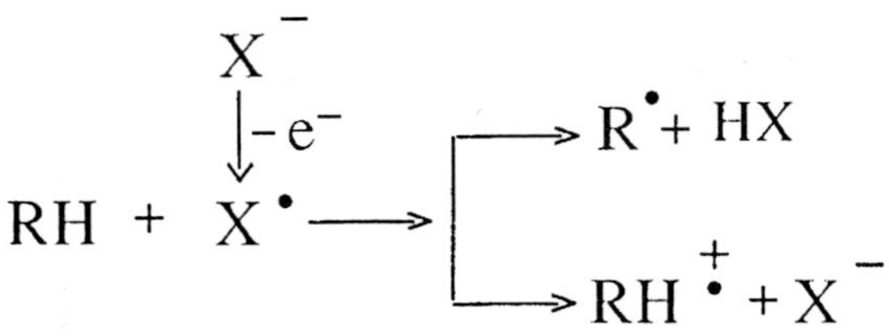

$$\text{RH} + \text{X}^{\bullet} \xrightarrow{\overset{\text{X}^-}{\downarrow -e^-}} \begin{cases} \rightarrow \text{R}^{\bullet} + \text{HX} \\ \rightarrow \text{RH}^{\overset{+}{\bullet}} + \text{X}^- \end{cases}$$

FIGURE 3.6. Indirect oxidation of an organic compound, RH mediated by a solution-confined species, X^-. (Source: H. Lund and M. Baizer, eds., "Organic Electrochemistry." Chapter 13. Dekker, New York, 1991.)

3.2.2. Survey of the Electrochemistry of Environmentally Important Organics

Table 3.3 contains a listing of classes of organic compounds of environmental import. The electrochemical behavior of each of these classes is briefly discussed next. No attempt is made to treat this topic exhaustively, and the reader is instead referred to the monographs and reviews listed at the end of this chapter.

<u>Hydrocarbons</u>. The oxidation and reduction of hydrocarbons normally occurs by initial abstraction or injection of one electron. Saturated aliphatic hydrocarbons have very low electron affinities and thus are mostly electroinactive under normal conditions, but can undergo electrooxidation at high temperatures in strongly acidic or basic media. They can also be oxidized in acetonitrile at potentials of about $1.7 < E_{1/2} < 3.9$ (V vs. Ag/Ag$^+$). The oxidation may involve cleavage of C-H or C-C bonds. In principle, deep oxidation of any hydrocarbon should yield CO_2 although this is seldom achieved.

Although ethylene cannot be electrochemically reduced, the double bond of other olefinic hydrocarbons can be usually reduced at lower potentials than the corresponding saturated compounds. This bond becomes saturated and may also dimerize or intradimerize, as illustrated in Figure 3.7. Olefin oxidation normally produces a cation radical capable of undergoing follow-up reactions including nucleophilic addition, allylic substitution, dimerization, and the like, as schematized in Figure 3.8. The $E_{1/2}$ values for their oxidation in acetonitrile generally fall within the range 2–3 V (vs. Ag/Ag$^+$ reference). In the presence of O_2, the electrogenerated alkene cation radical will form the corresponding dioxyethane. Reduction of alkynes is much more difficult and requires a concerted 4 e$^-$, 4 H$^+$ mechanism to attain saturation.

Aromatic hydrocarbons are much more important from an environmental point of view, primarily due to their hazardous nature and their heavy use as fuels and

3.2. Electrochemistry of Organic Pollutants

Table 3.3. Important Classes of Organic Compounds of Environmental Significance[a]

Compound Category	Examples
1. Hydrocarbons	
Aliphatics	Propane
Aromatics	Benzene
Polycyclic aromatic hydrocarbons	Chrysene
2. Halogenated organics	
Aliphatics	Chloroform
Olefins	Trichloroethylene
Aromatics	Chlorobenzene
Multi-ring systems	PCBs
3. Nitrogen-containing organics	
Nitro derivatives	Nitrobenzene
Aliphatic and aromatic amines	Aniline
Azo compounds	Dyes
Amides	Benzamide
Nitriles	Acetonitrile
4. Alcohols and phenols	
Aliphatics	Methanol
Aromatics	Phenol
5. Aldehydes and ketones	
Aliphatics	Formaldehyde
Aromatics	Benzophenone
6. Acids	
Aliphatics	Formic acid
Aromatics	Salicyclic acid
7. Miscellaneous	
Nitrophenols	
Chlorophenols	
Heterocyclic compounds	

[a] Also refer to Tables 1.5 and 1.6.

in industrial operations. Many of their electrochemical reaction pathways have been summarized already. Most of the electrochemical data, however, originate from studies utilizing nonaqueous media (such as acetonitrile). These solvents are used because of the relatively low solubilities of these compounds in water and because their oxidation–reduction potentials usually fall outside the stability range of water. Further, water is a source of protons and nucleophiles or electrophiles that facilitate reactions coupled to the electron transfer steps, thus complicating the elucidation of reaction mechanisms and the detection or isolation of intermediates.

At low-hydrogen overpotential cathodes, the reduction of aromatic hydro-

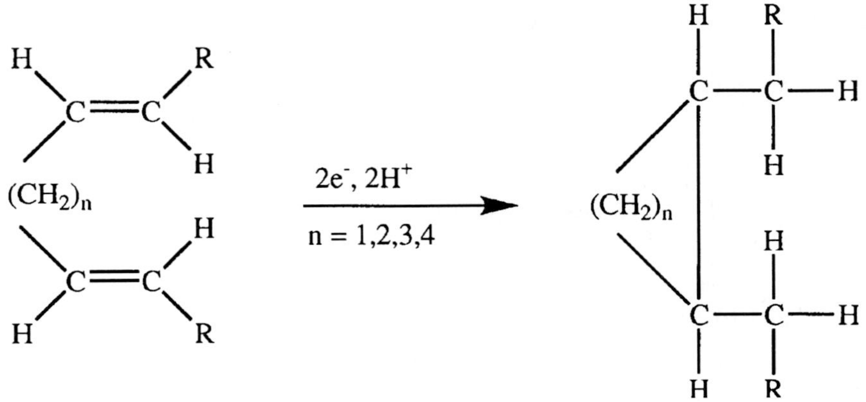

FIGURE 3.7. Electroreduction of olefinic double bonds with intramolecular dimerization. (Source: J.M. Costa, "Fundamentos de Electrodica." Alhambra Universidad, Madrid, 1981.)

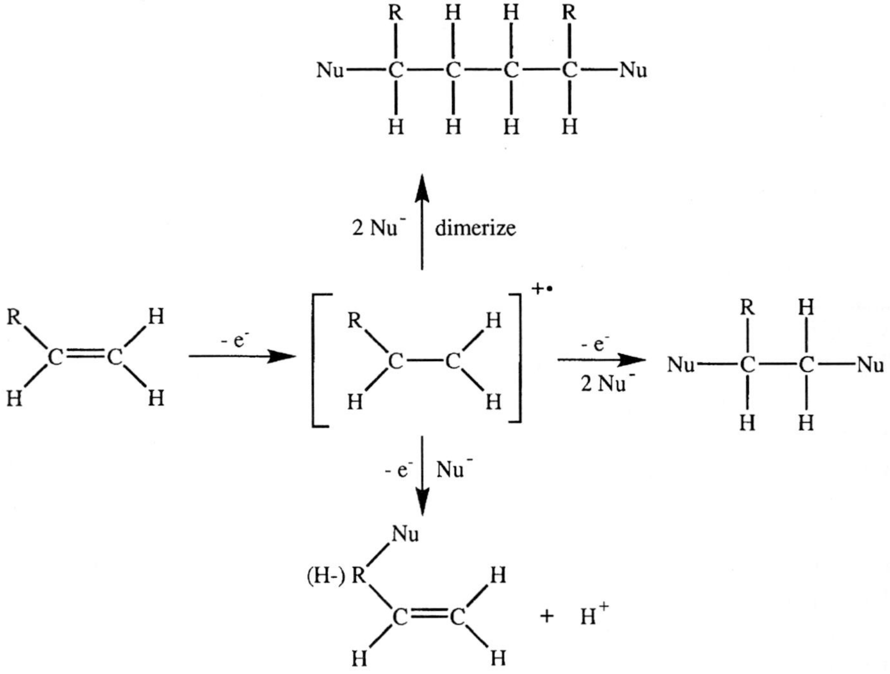

FIGURE 3.8. Various mechanistic possibilities for the oxidation of olefins.

carbons may yield different products, as can be seen in the following example of benzene reduction:

$$C_6H_6 + 6\,e^- + 6\,H^+ \rightarrow \text{cyclohexane}$$
$$C_6H_6 + 4\,e^- + 4\,H^+ \rightarrow \text{cyclohexene}$$
$$C_6H_6 + 2\,e^- + 2\,H^+ \rightarrow \text{cyclohexadiene}$$

In aqueous media, the dominant reaction route is the first one. After initial formation of the radical anion, most aromatics can accept a second electron forming the dianion (notable exceptions are benzene and naphthalene); the potential separation between these two steps has been found to average about 500 mV, except in cases where ion-pairing effects are strong and facilitate the second electron transfer. In highly protic media, cyclic conjugated hydrocarbons normally yield 1,4–dihydro-derivatives. For example, polycyclic aromatic hydrocarbons such as naphthalene and anthracene can be polarographically reduced at potentials between 2 V and 3 V $vs.$ SCE in a two-electron–two-proton reaction to yield 3,4-dihydronaphthalene and 9,10-dihydroanthracene, respectively. Further reduction of naphthalene (with two more e^- and two H^+) yields 1,2,3,4–tetrahydronaphthalene. Biphenyl yields cyclohexa-2,5-dien-1-ylbenzene under similar conditions.

The electrochemical oxidation of aromatic hydrocarbons occurs more easily than that of aliphatic hydrocarbons and, as can be seen in the schemes discussed previously, their intermediates (and their stabilities) as well as products largely depend on the structure of the parent compound and on the experimental conditions. Predictions of the specific site reactivity of PAHs and of the stability of their oxidation products can be made with the aid of the values of the corresponding atomic orbital coefficients from simple Hückel molecular orbital (HMO) calculations. By squaring these coefficients one gets a good estimate of the unpaired electron density at the various sites in the radical after initial removal of an electron. If this density is well distributed throughout the radical, the latter will probably be stable for sometime, whereas if one finds a high unpaired electron density at a particular site, the radical will most likely undergo subsequent chemical reaction. Should these reactive sites be blocked (substituted) with groups that are either electroinactive (e.g., methyl, phenyl) or else extend the charge delocalization (e.g., donor groups such as amino, methoxy), stable cations can be expected to form upon oxidation of the PAH.

Halogenated Organics. Monochlorinated aliphatic hydrocarbons are commonly reduced in the potential interval $-2.7 \leq E_{red} \leq -2$ V $vs.$ SCE, whereas their polychlorinated counterparts undergo reduction in the interval $-2.2 \leq E_{red} \leq -0.7$ V $vs.$ SCE. When comparing the different halogenated hydrocarbons, it is observed that their proclivity for reduction (i.e., less negative potentials required) follows the order I > Br > Cl > F. Monohalogenated aliphatics undergo 1 e^- or 2 e^- reduction, as illustrated in Figure 3.9. The ease of reduction of alkyl

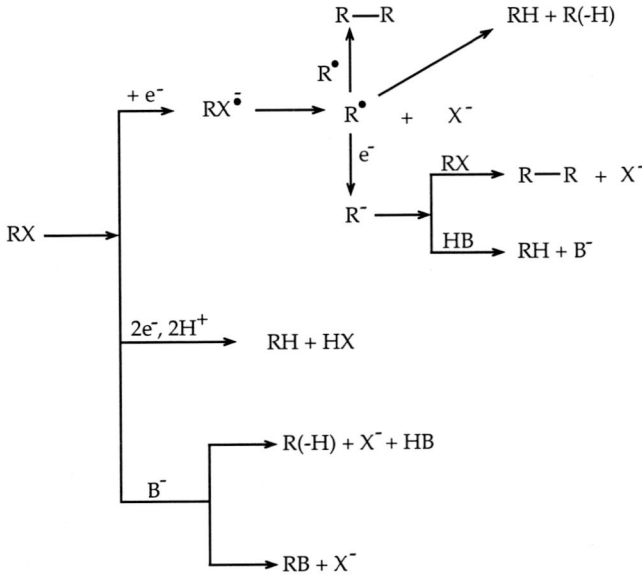

FIGURE 3.9. Electroreduction of monohalogenated aliphatic compounds. HB is an acid in this scheme.

halides increases in the order: primary < secondary < tertiary. When these reductions are performed for example at Hg, Pb, or Sn cathodes, the corresponding organometallic compounds are formed.

Monohalogenated alkenes (vinyl halides) constitute an important category of organic pollutants (Table 1.5). They may be electrochemically reduced, as shown by the scheme in Figure 3.10. Halide ions are generated from these reductions.

The electrochemical behavior of monohalogenated aromatics was already discussed in the preceding section.

Polyhalogenated compounds usually undergo stepwise reduction by successive breaking of the R-X bonds at different potentials and are more easily reduced than their monohalogenated counterparts. For example, the reduction of species containing the trichloromethyl moiety involves the following steps:

$$RCl_4 \xrightarrow[-0.78 \text{ V}]{2\,e^-, H^+} RHCl_3 \xrightarrow[-1.67 \text{ V}]{2\,e^-, H^+} RH_2Cl_2 \xrightarrow[-2.3 \text{ V}]{2\,e^-, H^+} RH_3Cl$$

The potentials in this scheme pertain to CCl_4 solvent and are expressed relative to the SCE reference.

When two halogens are contiguous in an alkane, the pathways illustrated in Figure 3.11 may prevail. Aromatics containing two different halogens can

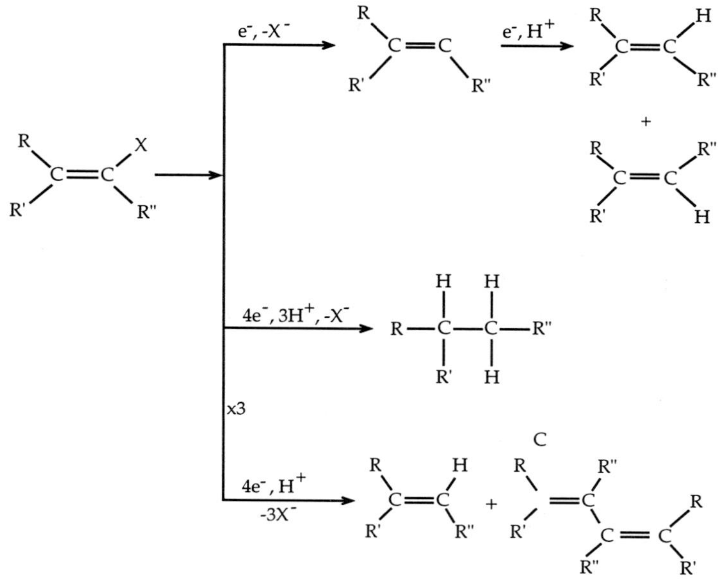

3.2. Electrochemistry of Organic Pollutants

FIGURE 3.10. Electroreduction of monohalogenated alkenes (vinyl halides).

be stepwise reduced in the sequence I > Br > Cl > F. If the halogens are the same, selective dehalogenation at different potentials can often be achieved in 2 e⁻, 1 H⁺ steps:

$$ArX_n \xrightarrow{2 e^-, H^+} ArX_{n-1} + X^-$$

Halogenated hydrocarbons are difficult to oxidize, although iodo compounds can be normally oxidized through the corresponding cations. Chlorinated and brominated derivatives of PAHs can be oxidized on Pt at $1.3 \leq E_{1/2} \leq 2$ V vs. SCE in acetonitrile. The bromo derivatives are normally oxidized at potentials more positive than their parent compounds. Halogenated biphenyls, benzenes, and PAHs have been analyzed in terms of their reduction potentials and electron affinities.[1] All the chlorobenzenes, 37 chloronaphthalenes, and 31 chlorinated biphenyls were considered in this study. The incentive for this study was to explore the feasibility of using $E_{1/2}$ data for determining the electron affinities of compounds that undergo dissociative and nondissociative electron capture. The reversible half-wave reduction potentials ($E_{1/2}$) in aprotic solvents have been related to the electron affinity (EA) in the gas phase and $-\Delta\Delta G°_{sol}$, the difference in solvation energy between the neutral specie and the anion[2]:

$$E_{1/2} = EA - \Delta\Delta G°_{sol} + E_{ref} = EA - C$$

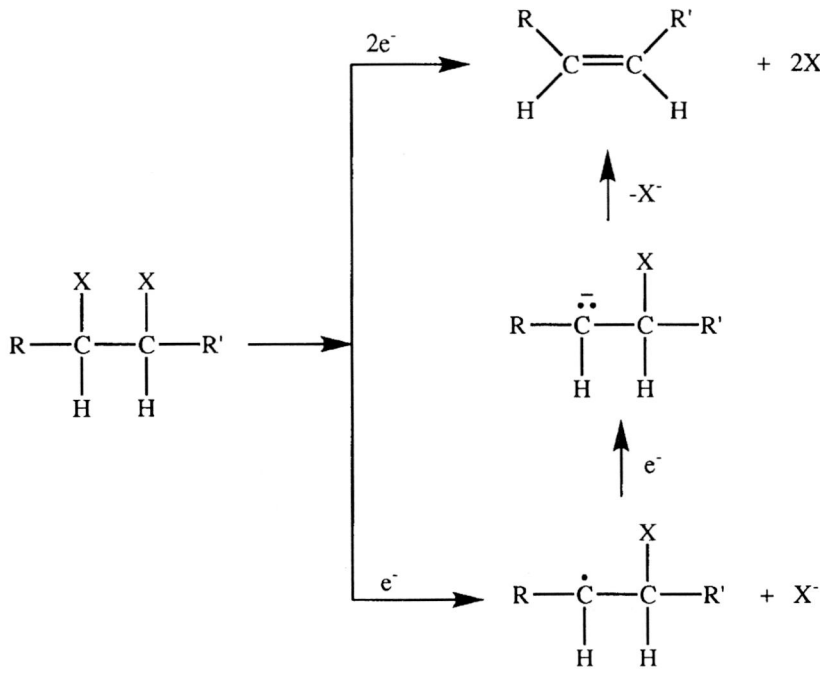

FIGURE 3.11. Electroreduction of a *gem*-dihalide. R and R' are alkyl groups and X is a halogen. (Source: Peters.[17])

The electrochemical data needed for this correlation were taken from earlier work[3,4] using interrupted sweep voltammetry. In the preceding expression, E_{ref} represents a correction factor for the reference electrode used. Table 3.4 contains $E_{1/2}$ and EA data thus generated for the chlorobiphenyls and chlorobenzenes.

A kinetic method has been used[5] to estimate the standard reduction potential ($E^{\circ\prime}$) for a series of PCBs and PBBs in N,N-dimethylformamide. Note that the tabulation in the earlier work[1] uses $E_{1/2}$ (and not $E^{\circ\prime}$) data. *Direct* electrochemical measurement of E° for halogenated compounds is complicated by the follow-up reaction embodied in the scheme:

$$ArX + e^- \rightarrow ArX^{\overline{\cdot}} \qquad (3.2a)$$
$$ArX^{\overline{\cdot}} \rightarrow Ar^{\cdot} + X^- \qquad (3.2b)$$

Because of the chemical reaction (Eq. (3.2b)), equilibrium of the two redox forms, ArX and ArX$^{\overline{\cdot}}$ will not be achieved under practical conditions.

Table 3.4. Reduction Potentials and Calculated Electron Affinities for Chlorobenzenes and Chlorobiphenyls

Compound	Electron Affinity, eV	$-E_{1/2}$, V vs. SCE
Benzene	Not reduced	—
Chlorobenzenes		
1-	-0.14	2.79
1,2-	0.08	2.57
1,3-	0.10	2.55
1,4-	0.10	2.55
1,2,3-	0.34	2.31
1,2,4-	0.30	2.35
1,3,5-	0.32	2.34
1,2,3,4-	0.54	2.11
1,2,4,5-	0.49	2.16
1,2,3,5-	0.51	2.14
1,2,3,4,5-	0.73	1.92
1,2,3,4,5,6-	0.98	1.67
Biphenyl	-0.11	2.76
Chlorobiphenyls		
2-	0.20	2.45
3-	0.19	2.46
4-	0.24	2.61
2,3-	0.34	2.31
2,4-	0.32	2.33
2,5-	0.36	2.29
3,4-	0.43	2.22
3,5-	0.40	2.25
2,3,5-	0.52	2.13
2,3,6-	0.36	2.29
2,4,5-	0.46	2.19
2,4,6-	0.33	2.32
3,4,5-	0.60	2.05
2,3,4,5-	0.62	2.03
2,3,4,6-	0.52	2.13
2,3,5,6-	0.51	2.14
2,3,4,5,6-	0.73	1.92
2,2'-	0.17	2.58
3,3'-	0.27	2.38
4,4'-	0.30	2.35
2,4'-	0.26	2.39
2,6,2',6'-	0.18	2.47
2,5,2',5'-	0.40	2.25
3,4,3',4'-	0.54	2.11
3,5,3',5'-	0.58	2.07
2,5,2',4',5'-	0.53	2.12
2,4,5,2',4',5'-	0.54	2.11
2,4,6,2',4',6'-	0.39	1.76
Decachlorobiphenyl	0.89	1.76

Source: Wiley et al.[1]

However, an electrocatalytic scheme affords a route to the determination of $E^{\circ\prime}$:

$$P + e^- \rightleftharpoons Q^{\bar{\cdot}} \qquad E^\circ_{PQ}$$

$$ArX + Q^{\bar{\cdot}} \underset{k_{-1}}{\overset{k_1}{\rightleftharpoons}} ArX^{\bar{\cdot}} + P$$

Where P and Q are the oxidized and reduced forms of the catalyst. The method involves electrochemical measurement of k_1. Since the reverse rate constant is at the diffusion limit (k_d), k_1/k_d is the equilibrium constant, K for electron transfer. Thus, the Gibbs free energy can be computed from

$$\Delta G^\circ = -RT \ln K$$

which in turn affords a route to $E^{\circ\prime}$:

$$\Delta G^\circ = -nF\,E^{\circ\prime}$$

(see Chapter 2). The rate constant k_1 was calculated from analysis of the catalytic increase in current above that obtained for the catalyst alone. Table 3.5 contains data thus obtained from linear sweep voltammetry measurements[5] for 16 different chlorinated and brominated biphenyl derivatives. The catalysts used and their redox potentials are also contained in the same tabulation.

Nitrogen-Containing Organics. Aromatic organic molecules containing nitro, nitroso, and azo groups are generally easy to reduce. On the other hand, amines are rather difficult to reduce. Heterocyclic nitrogen compounds (as well as most nitrogen-containing organics) catalyze the cathodic production of H_2 except in basic solutions, where reactive free radicals are formed. Figure 3.12 contains a mechanistic scheme for the reduction of a mononitro-aromatic derivative.

A nucleophile may react with cationic intermediates and substitute a hydrogen atom in the aromatic amine product. For example, in the reduction of nitrobenzene in sulfuric acid, water acts as a nucleophile yielding p-aminophenol. In addition, the arylhydroxylamine may undergo a series of reactions including acid-catalyzed rearrangement, condensation, further reduction or oxidation, dehydration, and the like. Dinitroaromatics yield different products according to the relative position of the nitro groups (i.e., o, m, or p). These products can be a diamino compound, a nitrohydroxylamine, or a dihydroxylamine. When mul-

3.2. Electrochemistry of Organic Pollutants   145

Table 3.5. Electron Transfer Rate Constants and Estimated Standard Potentials for PCBs and PBBs

Substrate	Catalyst	N^a	$\log k_1$ L mol^{-1} s^{-1}	$-E^{\circ\prime\,b}$ V vs. SCE
2-CB	1	11	0.75 ± 0.05	2.369
4,4'-PCB	1	18	1.72 ± 0.11	2.317
4,4'-PBB	1	57	3.96 ± 0.19	2.181
4,4'-PBB	2	15	2.83 ± 0.08	2.228
4,4'-PBB	3	18	2.09 ± 0.08	2.175
4,4'-PBB	4	15	2.25 ± 0.16	2.141
2,2',4,4'-PCB	3	12	2.27 ± 0.06	2.161
2,2',4,4'-PCB	5	12	1.18 ± 0.02	2.121
2,2',6,6'-PCB	5	12	0.96 ± 0.09	2.139
3,3',5,5'-PBB	5	21	3.82 ± 0.04	1.970
3,3',4,4',5,5'-PCB	6	15	3.59 ± 0.17	1.899
3,3',4,4',5,5'-PCB	7	18	2.80 ± 0.09	1.820
2,2',4,4',6,6'-PCB	5	15	1.89 ± 0.04	2.024
2,2',4,4',6,6'-PCB	6	20	1.67 ± 0.23	2.009
2,2',3,3',4,4'-PCB	5	16	2.71 ± 0.08	2.034
2,2',3,3',5,5',6,6'-PCB	6	20	3.76 ± 0.04	1.891

No.	Catalyst	$-E^{\circ}_{PQ}$, V vs. SCE
1	9,10-diphenylanthracene	1.845
2	4-methoxybenzophenone	1.825
3	4-cyanopyridine	1.725
4	benzophenone	1.705
5	perylene	1.623
6	tetracene	1.540
7	rubrene	1.410

[a] Number of LSV scans analyzed.
[b] All data pertain to DMF/0.1 M tetrabutylammonium iodide.
Source: Data from Rusling and Miau.[5]

tiple nitro groups are present, stepwise polarographic reduction waves can be observed.

Azo linkages can be easily reduced as shown in Figure 3.13. Similarly, nitriles can be reduced to the corresponding amines at cathodes with low H_2 overvoltage (e.g., Pt). Alternatively, they are dimerized or polymerized in aprotic media at cathodes with high H_2 overvoltage (e.g., Hg) (Figure 3.14).

Nitrogen-containing organics are relatively easy to oxidize as exemplified by aromatic amines. For example, the electrooxidation of aniline is environmentally important because of its extensive use in dyes, and the like. (Chapter 5). Figure 3.15 illustrates the complexity of the oxidation reaction. However,

146 Chapter 3

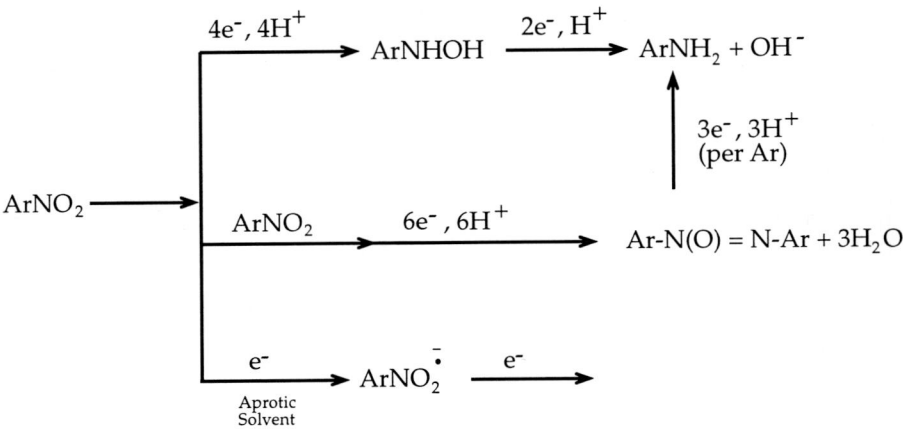

FIGURE 3.12. Electroreduction of a mononitroaromatic compound, $ArNO_2$.

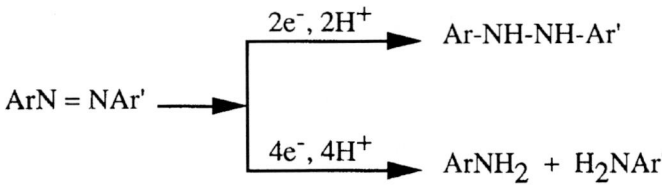

FIGURE 3.13. Electroeduction of an azo linkage. Ar and Ar' are arly groups.

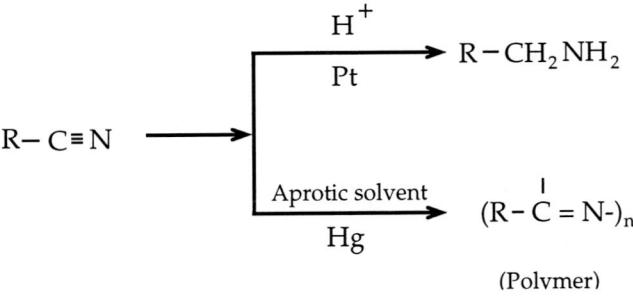

FIGURE 3.14. Electroreduction of an organic nitrile derivative. (Source: J.M. Costa, "Fundamentos de Electrodica." Alhambra Universidad, Madrid, 1981.)

3.2. Electrochemistry of Organic Pollutants

the initial product in all cases is invariably the aniline radical cation. The oxidation of amides usually proceeds through the abstraction of 1 e⁻ from the nitrogen lone pair followed by 1 e⁻, 1 H⁺ removal and reaction with a nucleophile (normally a basic moiety, B⁻ from the solvent, BH):

$$R-Y-\underset{\underset{R'}{|}}{\overset{\overset{H}{|}}{N}}-\underset{\underset{H}{|}}{\overset{}{C}}-R'' \xrightarrow{-e^-} R-Y-\underset{\underset{R'}{|}}{\overset{\overset{H}{|}\,\bullet+}{N}}-\underset{\underset{H}{|}}{\overset{}{C}}-R'' \xrightarrow{-e^-,\,-H^+} R-Y-\underset{\underset{R'}{|}}{\overset{\overset{+}{|}}{N}}-\underset{\underset{H}{|}}{\overset{}{C}}-R''$$

$$\Big\downarrow B^-$$

$$R-Y-\underset{\underset{R'}{|}}{\overset{\overset{B}{|}}{N}}-\underset{\underset{H}{|}}{\overset{}{C}}-R''$$

In this scheme, Y = CO, PO, or SO$_2$. The electrooxidation of amides is normally "clean" and yields only one major product.

Nitrogen-containing compounds derived from pyrazole (a five-membered hetero-ring system containing two N atoms) have been used as herbicides. One such compound is difenzoquat, which is distributed as 1,2-dimethyl-3, 5-diphenyl-pyrazolium methylsulfate:

[pyrazolium structure with Ph groups at 3,5 positions, CH$_3$ groups on both N atoms]⁺ CH$_3$SO$_4^-$

This was introduced under the registered trademarks Avenge and Finaven as a herbicide controlling wild oats (*Avena fatua*, etc.). The lethal dose is quite high, LD$_{50}$ = 470 mg/kg (about twice that of KCN or arsenic).

The electrochemical and adsorption properties of the cation have been described in water, methanol, and acetonitrile.[6,7] The compound was found to undergo 2 e⁻ reduction in the range: -1.27 – -1.52 V (*vs.* Ag/AgCl). Examination of the UV spectra of the electrolysis products allowed the authors to identify a dimerization intermediate and substituted pyrazoline as

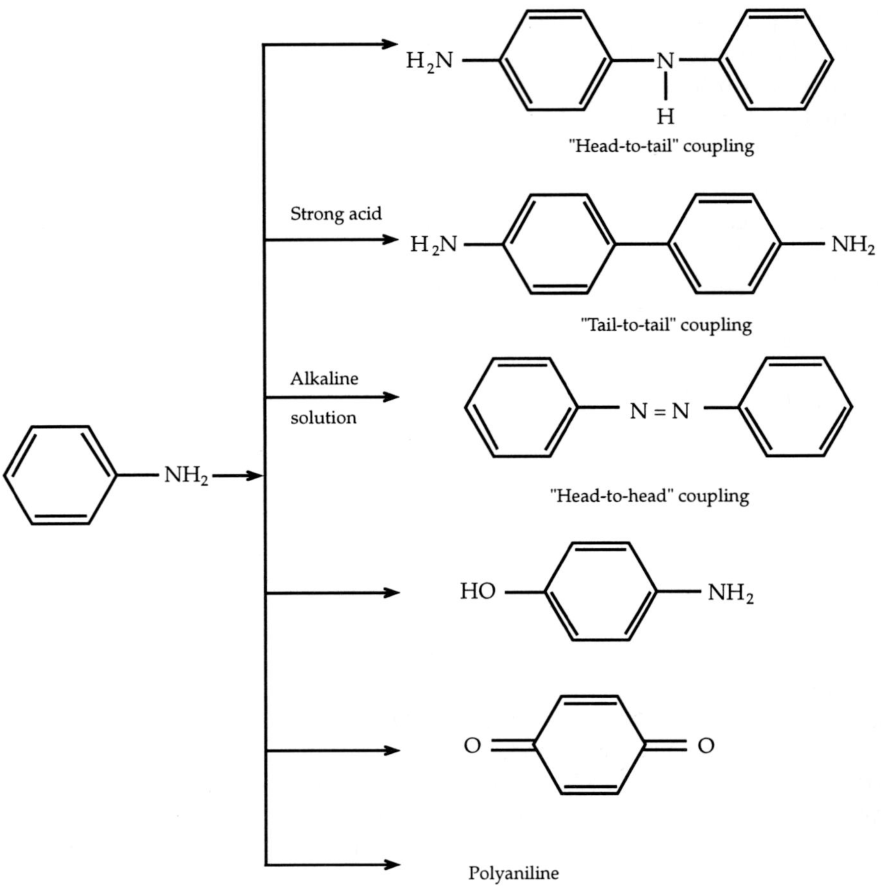

FIGURE 3.15. Electrooxidation of aniline.

the final product.[6] Strong adsorption of this compound was also noted at the mercury electrode at µM concentrations and in a wide potential range.[7] These data suggest the implementation of an adsorptive voltammetry approach as a trace analysis tool for this herbicide (see Chapter 4).

Oxygen-Containing Organics. A generalized pathway for the oxidation of alcohols, aldehydes, and short-chain acids at electrocatalytic electrodes (e.g., Pt) involves the following steps: adsorption, hydrogen abstraction, electron transfer to the electrode, proton abstraction, and oxidation by •OH radicals. Reduction pathways may differ substantially, and a generalization cannot be made.

A summary of the electrochemistry of alcohols, aldehydes, ketones, carboxylic acids, and phenols follows. Follow-up reactions of anions or cations formed by electron transfer or otherwise will depend on the nature and stability of each individual species. All potentials are given vs. SCE unless otherwise specified.

The oxidation of aliphatic alcohols is not easy and the potentials required are rather positive. For example, for aliphatic C_1-C_5 alcohols, $E_{1/2} = 2.6 \pm 0.14$ V (vs. Ag/Ag$^+$ 0.01 M).[8] Short-chain alcohols do not yield substantial amounts of aldehyde during oxidation. In general, primary alcohols and aldehydes can be oxidized as follows:

$$R\text{--}CH_2OH \xrightarrow{-2\,e^-} R\text{--}CHO \xrightarrow{-2\,e^-} R\text{--}COOH$$

The mechanism and products of alcohol oxidation are different in acidic and basic solutions (Figure 3.16).[8] The mechanism in absolute ethanol involves electron removal from the alkoxy ion or the alcohol molecule with formation of the corresponding alkoxy radical, which then disproportionates to the aldehyde (or CO) and the alcohol.[9] A major effort has focused on methanol oxidation because of its application in the fuel cell field.

Secondary alcohols yield the corresponding ketones upon anodic oxidation in acidic media. Mediated, indirect electrochemical oxidation of alcohols can give much higher yields than direct anodic oxidation.

Glycol oxidation involves C-C bond cleavage:

$$\underset{\underset{OH\;OH}{|\;\;\;\;|}}{R\text{--}\overset{\overset{R'\;\;R''}{|\;\;\;|}}{C\text{--}C}\text{--}R'''} \xrightarrow[CH_3OH]{-2\,e^-} \underset{R}{\overset{R'}{\diagdown}}C{=}O \;+\; O{=}C\underset{R'''}{\overset{R''}{\diagup}}$$

In alkaline solution, electroadsorption of ethylene glycol is followed by oxidation of a primary alcohol group to aldehyde. The final oxidation product is oxalic acid.[10]

The oxidation of primary benzyl alcohols (in acetonitrile) yields the corresponding aldehydes. If water is present, the acid can be obtained. Secondary alcohols yield the corresponding ketones if the chain substituent is small. For large substituents, the aldehyde is obtained with simultaneous cleavage of the substituent from the benzyl moiety.[8]

Carbonyls are normally reducible except in those cases involving masking of the carbonyl by hydration or by formation of hemiacetals or acetals. Polyhydroxycarbonyl compounds derived from biomass (e.g., glucose) are ordinarily easy to reduce.

Aldehydes are in general easily reducible to the corresponding alcohols. The reduction rates are a function of pH. The $E_{1/2}$ values of saturated chain aldehydes are normally around -1.9 V (in Hg), except for formaldehyde (-1.5 V).

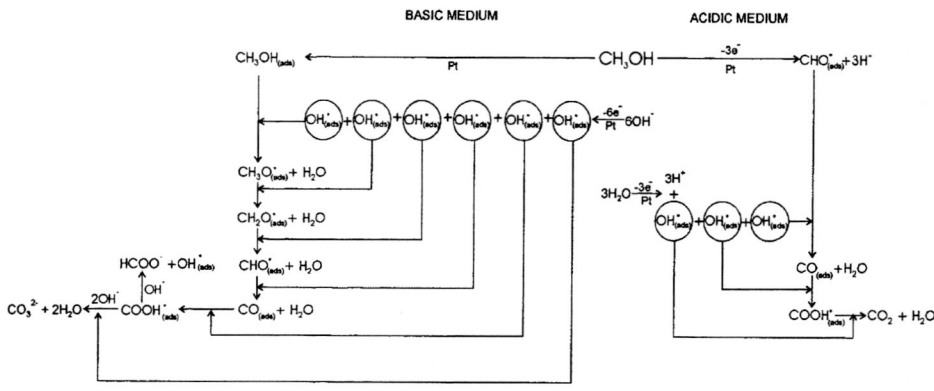

FIGURE 3.16. Oxidation of methanol at Pt electrode in acidic and alkaline aqueous media. (Source: H. Lund and M. Baizer, "Organic Electrochemistry." Chapter 16. Dekker, New York, 1991.)

Unsaturated chain aldehydes may undergo 2 e⁻, 2 H⁺ reduction to achieve saturation and then another 2 e⁻, 2 H⁺ step to yield the corresponding alcohol.

Aromatic aldehydes normally yield the corresponding alcohol in a two-step sequence (one electron and one proton in each step). The aryl groups act as electron withdrawing groups and facilitate reduction of carbonyls. For example, the $E_{1/2}$ for the reduction of benzaldehyde is -1.26 V (pH 7), which is substantially lower than the potentials mentioned previously.

Ketones are more difficult to reduce than aldehydes. For example, the $E_{1/2}$ for acetone is -2.46 V. Double bonds adjacent to the carbonyl group facilitate its reduction; lowering of the $E_{1/2}$ required for reduction by ~1 V has thus been observed. Aromatic ketones behave like unsaturated aliphatic ketones. Their reductive coupling normally yields a pinacol.

A summary of the common routes for carbonyl group reduction is given in Figure 3.17.

Low-hydrogen overvoltage cathodes (e.g., Pt, Ni, Cu) favor the direct aromatic carbonyl reduction to the corresponding alcohol instead of the pinacol formation shown in Figure 3.17. Typical examples of glycol/pinacol formation from aldehyde/ketones in alkaline solutions at high-hydrogen overvoltage cathodes (e.g., Hg, Cd, Zn) are furfural (-1.14 V), acetone (-1.16 V), and acetophenone (-1.6 V).[11] It has been proposed that alcohol formation occurs at the electrode surface, whereas pinacol formation occurs in solution. This last reaction is greatly favored by the presence of certain transition metal ions(e.g., Cr(III)).

An environmentally relevant aldehyde–formaldehyde, behaves rather uniquely.

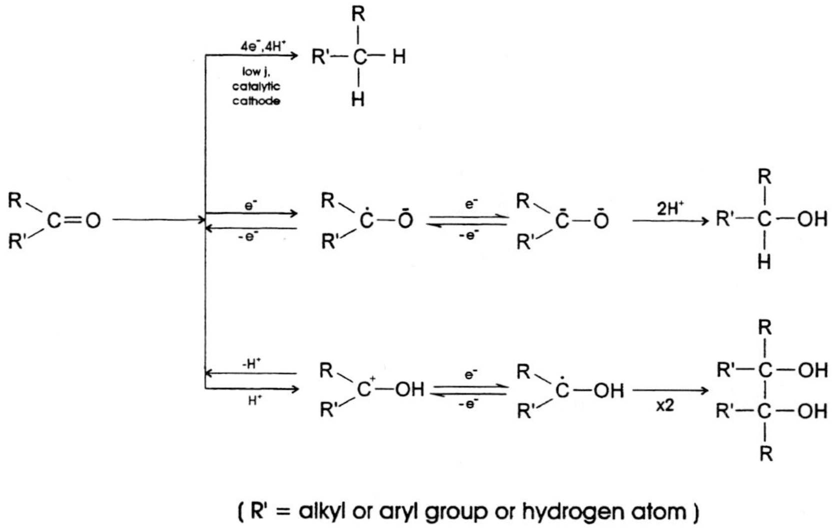

3.2. Electrochemistry of Organic Pollutants

FIGURE 3.17. Summary of the electroreduction routes for the carbonyl group. (R' = alkyl or aryl group or hydrogen atom)

For example,
- It hydrodimerizes to ethylene glycol at graphite cathodes.[12]
- In alkaline solutions, it produces formate ions upon relatively mild oxidation (at potentials ~0.75–1.25 V) or carbonate ions at higher potentials and at catalytic metal oxide electrodes.[13]
- Hydroxymethanolate, $HOCH_2O^-$ (also called methylene glycol or *gem*-diolate) has been invoked as a intermediate during oxidation of formaldehyde in alkaline media at Group 1B metals (e.g., Cu and alloys).[14]
- During electroless metal plating in acidic solution, its oxidation starts with dissociative adsorption that leads to dehydrogenation and CO_2 production. In basic solution, such oxidation may or may not be accompanied by hydrogen evolution, depending on the type of metal surface.[15]

Possible oxidation pathways for aldehydes include (*a*) direct electrochemical oxidation on a nonoxidized surface, and (*b*) reaction of a *gem*-diol or nonhydrated form of the aldehyde with a surface oxide.[16]

Organic carboxylic acids are not easily reduced in acidic or alkaline solutions. In strong acidic media, the protonated acid form can accept electrons more easily. A summary of their cathodic reduction behavior is given in Figure 3.18. As with carbonyls, electron-withdrawing groups facilitate their reduction. An example of this is the negative shift of about half a volt observed

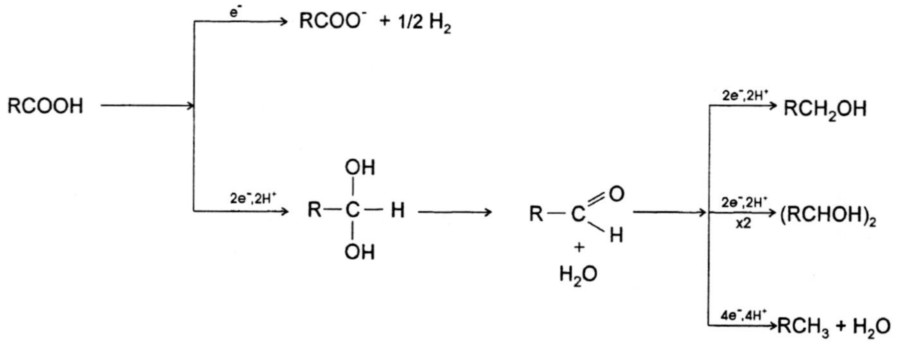

FIGURE 3.18. Electroreduction of a carboxylic group.

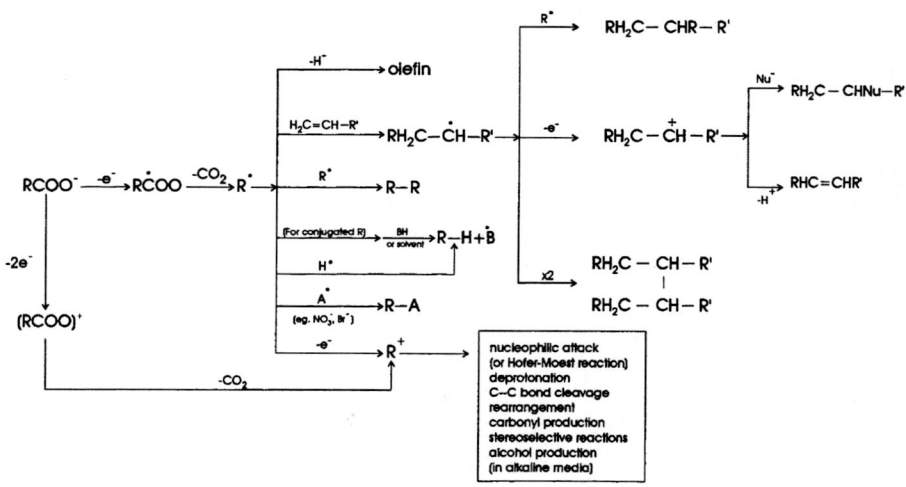

FIGURE 3.19. Kolbe-type oxidative decarboxylation of organic acids.

during the reduction of 1-ethyl-4-carboxypyridinium bromide when going from strongly acidic to strongly basic conditions.

Aromatic acids and esters (e.g., benzoic acid, alkyl benzoates) follow the 4 e-, 4 H+ path shown in Figure 3.18 at very negative potentials on high-hydrogen overvoltage cathodes such as Pb and Hg. The corresponding alcohols are then produced. Reduction of aromatic vicinal diacids (e.g., phthalic acid)

3.2. Electrochemistry of Organic Pollutants

does not yield the expected dialdehydes (e.g., catechol) since ring reduction occurs first.

As mentioned earlier, masked aldehydes (i.e., carbonyl group protected by hydration or by hemiacetal or acetal formation) are difficult to reduce, whereas their nonmasked counterparts are easy to reduce (even easier than their corresponding acids). For this reason, if one desires to reduce an acid to produce the corresponding aldehyde, the latter must be present in the protected form or else it will also be reduced.

Carboxylic acid oxidation usually follows the well-known Kolbe-type mechanism, summarized in Figure 3.19. The short-lived acyloxy radicals undergo fast decarboxylation to yield the neutral radicals R$^\bullet$ that may react in various ways as shown in the scheme. Simple aromatic acids, where the carboxylate group is bonded to the aromatic ring, do not follow the Kolbe mechanism; but when this group is far removed from the ring, normal reaction is observed.[17]

Other cross-coupling reactions can also occur. Examples include

$$RCOO^- + R'COO^- \xrightarrow{-2\,e^-,\,-2\,CO_2} R\text{--}R' + R\text{--}R + R'\text{--}R'$$

$$\begin{array}{c} ROOC\text{--}(CH_2)_x\text{--}COO^- \\ + \\ R'OOC\text{--}(CH_2)_y\text{--}COO^- \end{array} \xrightarrow{-2\,CO_2} ROOC\text{--}(CH_2)_{x+y}\text{--}COOR'$$

(Brown-Walker reaction)

$$RCOOH + R'CHO \xrightarrow{-e^-,\,-CO_2} R'\text{--}C(O)R$$

Phenols are an environmentally important group of oxygen-containing organic compounds. Generally phenols are not cathodically reduced. Their 1 e$^-$ or 2 e$^-$ oxidation can lead to a variety of products including quinones.[18] Film formation due to electropolymerization makes mechanistic studies difficult.[17] Such films have been studied in connection with corrosion protection, redox active film formation, analysis of surface states in semiconductors, and so forth.[19,20]

Anodic peak potentials for phenols (in CH_2Cl_2–$HFSO_3$) commonly occur at 1.3–1.7 V (*vs.* SCE). The oxidation of phenols at different electrode materials is discussed later in this book (Chapter 5). Their electrochemical behavior is summarized in the mechanistic scheme in Figure 3.20.

The intermediate phenolic cation radicals are strongly acidic (their pK_a values are normally in the range -2 to 0) and deprotonate to form neutral phenoxy radicals, ArO$^\bullet$ ($t_{1/2}$ ~10^{-9}–10^{-10} s)[17] which are further oxidized (either heterogeneously or homogeneously) to the phenoxonium ions, ArO$^+$. These

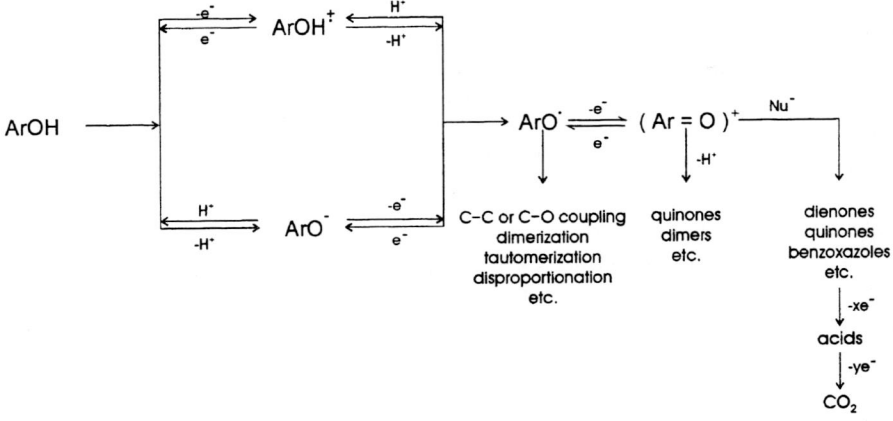

FIGURE 3.20. Electrochemical behavior of phenols. (Adapted from Ref. 8.)

cations may react with a nucleophile and yield dienones, quinones, benzoxazoles, and so on. Upon further oxidation, they can be converted into acids or even CO_2. Phenol can adsorb on oxide electrode surfaces (e.g., PbO_2) and react with •OH radicals and with O_2, forming the peroxydihydroxycyclohexadienyl radical.[18]

The existence of phenoxy radicals has been shown by *in situ* MIRFTIRS (multiple internal reflection Fourier transform infrared spectroscopy).[21] They may undergo reactions such as C–C coupling, C–O coupling, dimerization, tautomerization, disproportionation, and the like (see Figures 3.20 and 3.21). The nature of the substituents frequently determines the type of reaction. For example, if the radicals are substituted with nonbulky groups or are not substituted at all, couplings of the type "head-to-head," "tail-to-tail," and "head-to-tail" are possible. This last type leads to polymer formation. Likewise, if the radicals are substituted with bulky groups, the coupling of the radicals cannot be of the "head-to-tail" type, and a "tail-to-tail" dimer will form instead of a polymer. Other common products include biphenyls, diphenols, diquinones, and peroxides.

Nitro groups in nitrophenols (e.g., 2-amino-4-nitro-phenol) may undergo 6 e^-, 6 H^+ reduction to the corresponding amino group.

Strong similarities between the electrochemical and photochemical oxidation of phenols have been suggested.[17] In addition to the direct electron transfer at an electrode surface, indirect electrochemical processes using a mediator or intermediate have also been used, as discussed in Chapter 5.

3.3. ELECTROCHEMISTRY OF INORGANIC POLLUTANTS

Inorganic pollutants occur in a variety of metallic and nonmetallic forms, including the elements themselves, as cations, anions, or organometallic compounds (see Section 1.4.2). We shall consider each category in turn. The thermodynamic data in this discussion have been culled from the compilations by Pourbaix and by Bard, Parsons, and Jordan.

3.3.1. Metals

Copper. Copper forms compounds in two oxidation states, +1 and +2. The +3 oxidation state is also known, although trivalent copper tends to be unstable in aqueous media. The following standard potentials are available:

$$Cu^{2+} + e^- \rightleftarrows Cu^+ \qquad E° = 0.16 \text{ V}$$

$$Cu^+ + e^- \rightleftarrows Cu^0 \qquad E° = 0.52 \text{ V} \qquad (3.3)$$

$$Cu^{2+} + 2e^- \rightleftarrows Cu^0 \qquad E° = 0.34 \text{ V}$$

$$2Cu^+ \rightleftarrows Cu^{2+} + Cu^0 \qquad \log K = 5.73 \qquad (3.4)$$

In the absence of complexing ligands such as the chloride ion,[22] copper(I) is unstable, disproportionating as just shown.

Figure 3.22 contains Pourbaix diagrams for the copper–water system at 25°C. In basic pH, the hydroxide $Cu(OH)_2$ is less stable than the oxide CuO considered in this diagram.

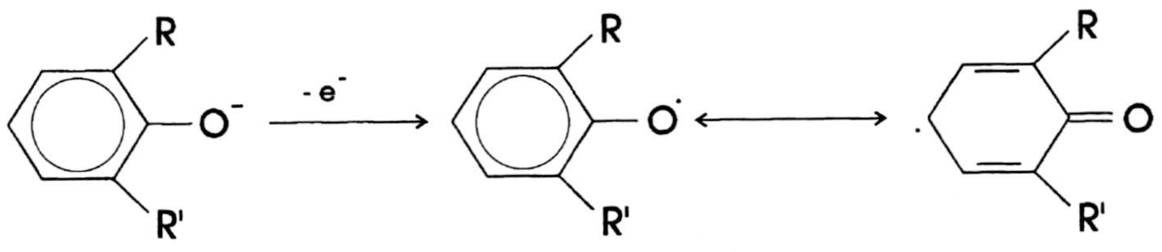

FIGURE 3.21. Electrooxidation of phenolate anion.

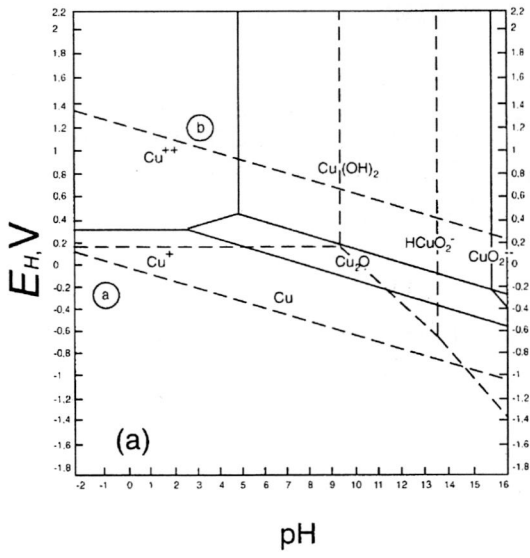

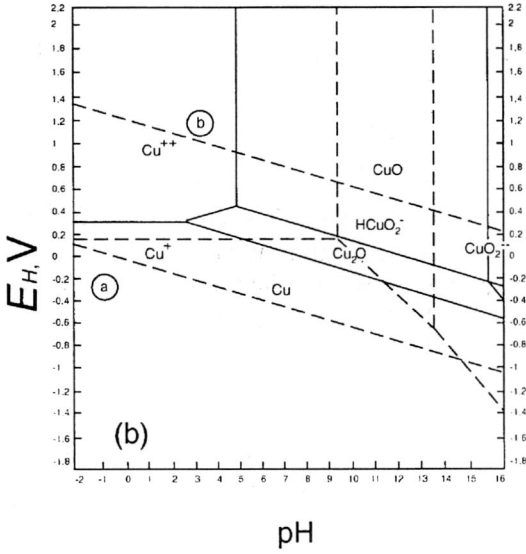

FIGURE 3.22. Pourbaix diagrams of the copper-water system at 25° C. The dashed lines labeled *a* and *b* are the solvent (H$_2$O) decomposition limits. (Adapted with permission from M. Pourbaix, "Atlas of Electrochemical Equilibria." Pergamon, Oxford, 1966.)

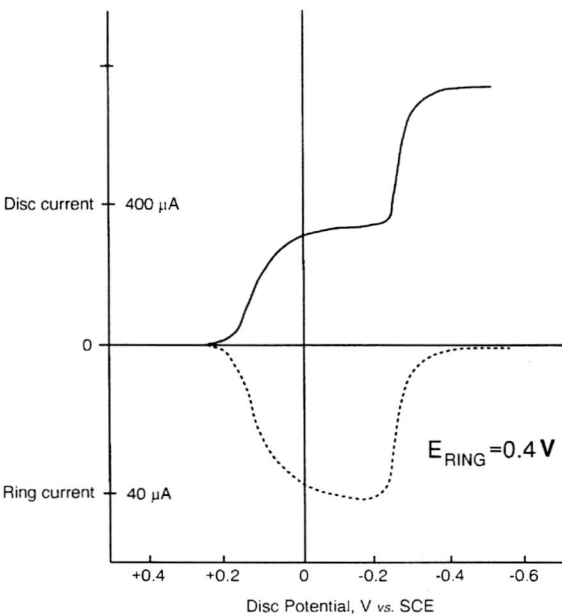

FIGURE 3.23. Disk (solid-line) and ring (dashed-line) voltammograms in 10^{-3} M $CuCl_2$ + 0.5 M KCl. (Reproduced with permission from Napp et al.[23])

The voltammetric behavior of the Cu(II) ion has been studied both in aqueous media[23] as well as in nonaqueous solvents.[24] Figure 3.23 contains rotating ring-disk voltammetry (RRDE) data from early work[23] in 0.5 M KCl. The disk scan (solid line) shows the two-step reduction of Cu(II) to metallic copper. The first wave can be assigned to the $Cu^{2+} \rightarrow Cu^+$ redox process as supported by the ring "collection" scan (dashed line). The ring was held at +0.4 V (vs. SCE reference) to reoxidize the Cu(I) back to Cu(II).

The stabilizing influence of Cl^- on Cu(I) is clearly seen in the CV scans in Figure 3.24.[25] The increase in the amplitude of the anodic wave at +0.23 V (vs. Ag/AgCl) in Figure 3.24a (compare with the set of waves for the $Cu^{2+/+}$ couple in Figure 3.24b) as the potential is swept to encompass the Cu^0 regime is a direct consequence of the coupled chemical reaction (Eq. (3.4)).

We shall further consider the complexation chemistry involving copper ions later in this chapter.

<u>Cadmium.</u> The only stable oxidation state of cadmium in aqueous media is +2. Figure 3.25 contains a Pourbaix diagram for the cadmium–water system at 25°C. In acidic media, the $Cd^{0/2+}$ couple is dominant:

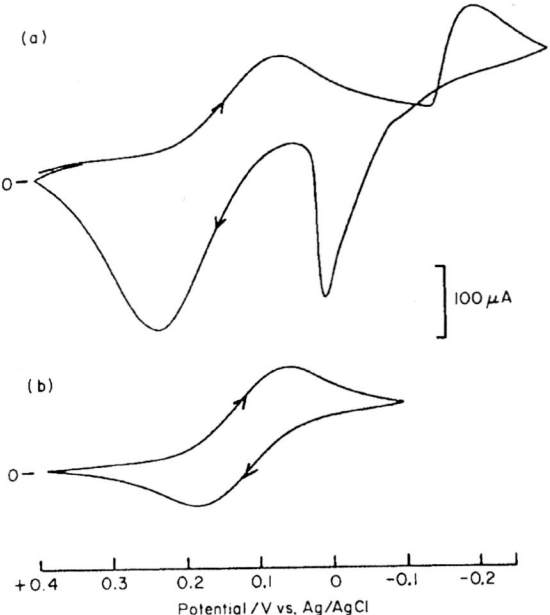

FIGURE 3.24. Cyclic voltammograms (potential scan rate, 0.02 V s^{-1}) at a glassy carbon electrode for a 3 mM solution of CuCl$_2$ in 0.1 M KCl: (a) the negative going scan is carried into the Cu^0 deposition regime; (b) the Cu$^{2+/+}$ redox electrochemistry manifests in a quasi-reversible manner is a more restrictive scan range. (Reproduced with permission from Basak et al.[25])

$$Cd^{2+}(aq) + 2\,e^- \rightleftarrows Cd(s) \qquad E° = -0.402\text{ V} \qquad (3.5)$$

The standard potential for cadmium amalgam formation

$$Cd^{2+}(aq) + Hg + 2\,e^- \rightleftarrows Cd(Hg) \qquad E° = -0.352\text{ V} \qquad (3.6)$$

lies positive of this value. In basic media, as Figure 3.25 illustrates, Cd(OH)$_2$ is the dominant insoluble phase:

$$Cd(OH)_2(s) + 2\,e^- \rightleftarrows Cd(s) + 2\,OH^- \qquad E° = -0.824\text{ V} \qquad (3.7)$$

This hydroxide forms a protective (passivation) layer on the parent metal surface on anodization in alkaline solutions.

The overpotential for hydrogen evolution on cadmium is higher than that for its Group 2 counterpart, Zn. For example, for a current density of 1 A/dm^2, η is -0.746 V for Zn and -1.134 V for Cd.

3.3. Electrochemistry of Inorganic Pollutants 159

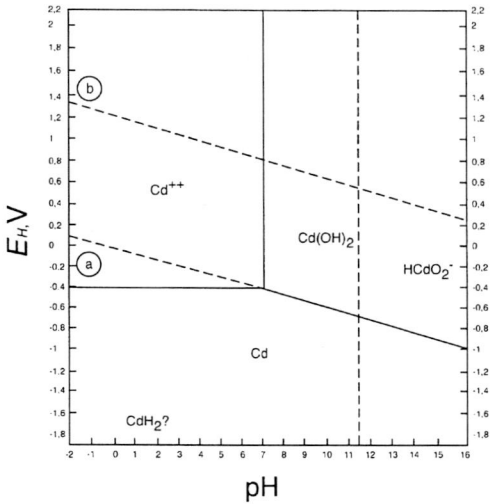

FIGURE 3.25. Pourbaix diagram for the cadmium-H$_2$O system at 25° C. (Adapted with permission from M. Pourbaix, "Atlas of Electrochemical Equilibria." Pergamon, Oxford, 1966.)

The electrochemical reduction of Cd(II) ions at a mercury electrode has frequently served as a model reaction in the study of electrode kinetics. A number of studies have addressed the measurement of $k°'$ and α (see Chapter 2) for this electrode reaction, and the influence of variables such as the supporting electrolyte, solvent, or added inhibitor or catalyst (e.g., thiourea) on the charge-transfer kinetics.[26] While the early studies hinged on the assumption that the two electrons are simultaneously transferred in one reaction, recent evidence shows that the reduction is more complex. Demodulation voltammetry[27] in NaClO$_4$ electrolyte at various water activities shows that the reduction at Hg proceeds via

- Fast loss of 12.5 water molecules in a preceding equilibrium
- A slow "chemical" step that is not a desolvation
- Slow transfer of the first electron
- Fast loss of four water molecules
- Slow transfer of the second electron
- Fast loss of the remaining nine water molecules

Other effects have been reported[28] such as potential oscillations during the electrocrystallization of cadmium from alkaline cyanide solutions under galvanostatic conditions and the influence of solution composition on the UPD of cadmium at single crystal Ag surfaces.[29] Especially relevant from an environmental perspective is a recent study that concerns the influence of

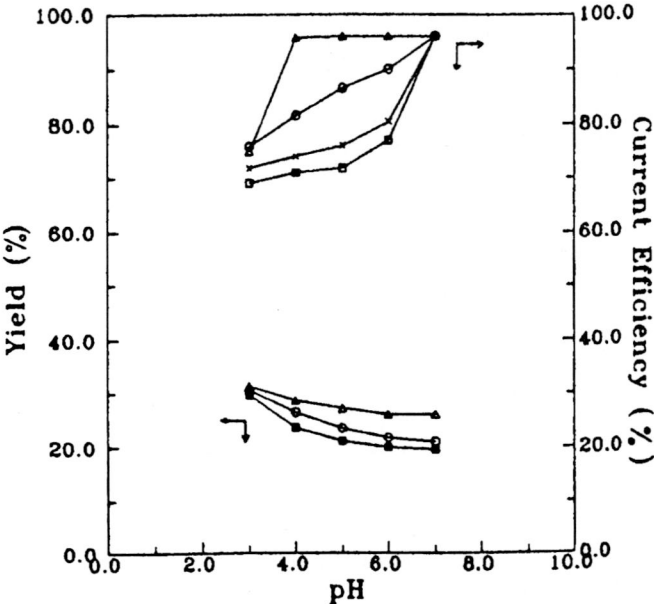

FIGURE 3.26. Cadmium yield and current efficiency as a function of the solution pH at different molar ratio of cadmium and citrate: △, 2:1; ○ 1:1, × ,1:2, and □ 1:3. T=25° C. (Reproduced with permission from Lai and Ku.[30])

complexing agents such as citrate on cadmium deposition.[30] Figure 3.26 contains representative data from this study on the cadmium yield and current efficiency as a function of the solution pH with the cadmium:citrate mole ratio as a parameter. These data have been interpreted[30] in terms of the free Cd(II) ion availability and its dependence on the two variables along with mass transport effects of Cd(II) ion complexation.

Chromium. The most important and stable oxidation states are +3 and +6. All Cr(VI) compounds are oxo compounds in aqueous solutions and are very strong oxidizing agents in acidic media:

$$HCrO_4^-(aq) + 7 H^+ + 3 e^- \rightleftarrows Cr^{3+} + 4 H_2O(l) \qquad E° = 1.38 \text{ V} \qquad (3.8)$$

$$Cr_2O_7^{2-}(aq) + 14 H^+ + 6 e^- \rightleftarrows 4 Cr^{3+}(aq) + 7 H_2O(l) \quad E° = 1.36 \text{ V} \qquad (3.9)$$

3.3. Electrochemistry of Inorganic Pollutants

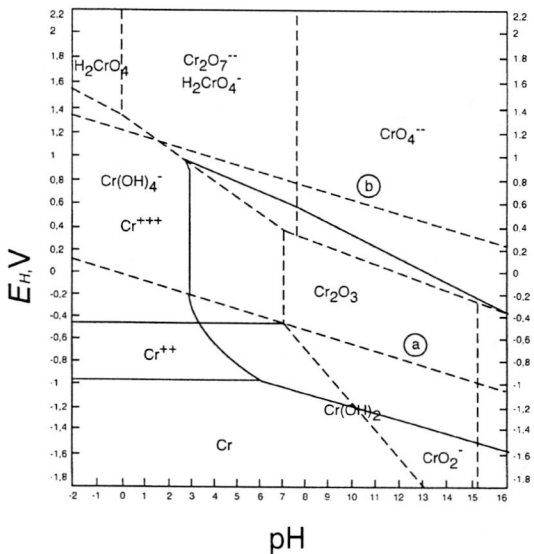

FIGURE 3.27. Simplified Pourbaix diagram for the Cr-H$_2$O system at 25° C. (Adapted with permission from M. Pourbaix, "Atlas of Electrochemical Equilibria." Pergamon, Oxford, 1966.)

The two Cr(VI) species are in equilibrium with one another:

$$2\ HCrO_4^-(aq) \rightleftarrows Cr_2O_7^{2-} + H_2O(l) \qquad K = 33.9 \qquad (3.10)$$

The Pourbaix diagram for chromium is exceedingly complex because of the wide variety of species that can coexist in aqueous media. However, a simplified diagram is contained in Figure 3.27.

Chromium displays the important technological phenomenon of *passivity*. Figure 3.28 illustrates this behavior. A chromium electrode dissolves readily in acidic medium:

$$Cr(s) \rightarrow Cr^{2+}(aq) + 2\ e^- \qquad E° = -0.90\ V \qquad (3.11)$$

However, when the potential reaches ca. -0.2 V, the *Flade potential*, the currently remarkably drops to almost zero. Not until a potential of ~+1.2 V is reached does appreciable current flow again occur in response to the reaction

$$Cr(s) + 4\ H_2O(l) \rightarrow CrO_4^{2-}(aq) + 8\ H^+(aq) + 6\ e^- \qquad (3.12)$$

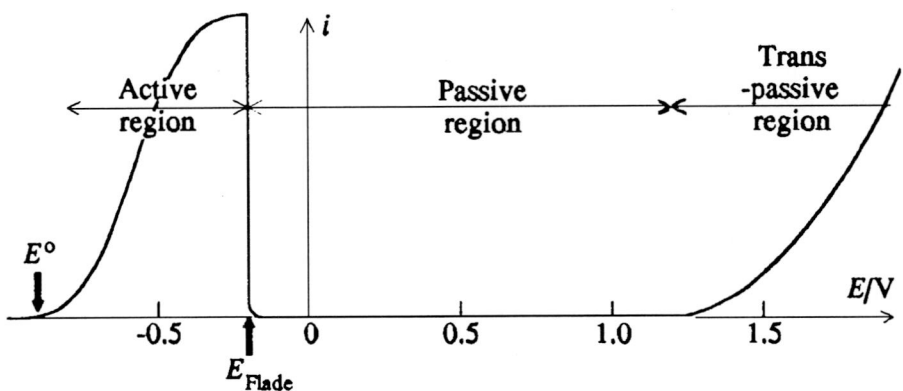

FIGURE 3.28. The electrochemical passivation behavior of chromium. (Reproduced with permission from K.B. Oldham and J.C. Myland, "Fundamentals of Electrochemical Science." Academic Press, San Diego, CA, 1994.)

The passivity is caused by an impermeable oxide layer on the chromium surface (see Figure 3.27), which protects the underlying metal from oxidation. This peculiar behavior has led to the important technological use of chromium for corrosion protection applications. The "plating baths" used for *chrome electroplating* are generally based on Cr(VI) (e.g., CrO_3 or effectively H_2CrO_4). The solubility product of precipitated chromium hydroxide is given by:

$$Cr(OH)_3(s) \rightarrow Cr^{3+} + 3\,OH^- \qquad K_{sp} = 10^{-30} \qquad (3.13)$$

Excess hydroxide ion converts $Cr(OH)_3(s)$ and various Cr(III) species to the soluble chromite ion:

$$Cr(OH)_3(s) + OH^- \rightleftarrows Cr(OH)_4^- \qquad K = 10^{-0.4} \qquad (3.14)$$

Both these species are reduced to metallic chromium as follows:

$$Cr(OH)_3(s) + 3\,e^- \rightleftarrows Cr(s) + 3\,OH^- \qquad E° = -1.33\text{ V} \qquad (3.15)$$

$$Cr(OH)_4^-(aq) + 3\,e^- \rightleftarrows Cr(s) + 4\,OH^- \qquad E° = -1.33\text{ V} \qquad (3.16)$$

The mechanism of the reduction of Cr(VI) to Cr(0) has been the topic of much discussion.[31-33] It is now established that anions such as SO_4^{2-} and Cl^- are an essential requirement for the deposition of good-quality metallic chromium. The role of sulfate appears to be to prevent the precipitation of interme-

3.3. Electrochemistry of Inorganic Pollutants

diate oxidation states. The first step undoubtedly is the reduction of Cr(VI) to Cr(III). This process is diffusion controlled with respect to sulfate.[32] The potential for Cr(VI) → Cr(III) reduction at Cu, Ni, Cr, and C is reported to be determined by the extent of removal of an oxidized film from the metal surface. The presence of a subsequent passivity film has been confirmed by double-layer capacity measurements, current interrupter techniques, X-ray photoelectron spectroscopy, glow discharge–optical emission spectroscopy, and via optical and electron microscopy. This film appears to be a mixed oxidation state chromium oxide, and radiotracer evidence points to the existence of sulfate in it.[32] Chromium metal deposits at potentials negative of the formation of this layer. The reduction of the passivity films to Cr metal is postulated to proceed via a chromous hydroxide intermediate, and HSO_4^- is proposd as a catalyst for this process.[31] Other ions such as Ni(II) have also been shown to improve the current efficiency of chromium deposition.[33] The effect of chloride ions has been discussed within the framework of a CE mechanism; namely,

$$A \rightleftarrows D \qquad \text{slow}$$
$$D + ne^- \to C \qquad \text{fast}$$

where species D has been proposed as the CrO_3Cl^- complex ion.[34,35]

The underpotential deposition of ions such as Bi^{3+} on Pt has been reported to catalyze the reduction of Cr(VI) in alkaline media.[36] The formation of insoluble $Cr(OH)_3$ was found to impede the reduction at the parent electrode surface. The electrocatalytic action of the Bi UPD adatoms at Pt was interpreted in terms of the elimination of CrO_4^{2-} adsorption.

The electrooxidation of Cr(III) has also been fairly well studied. A mechanism involving an adsorbed Cr(IV) intermediate has been proposed[37]:

$$H_2O - e^- \to {}^\bullet OH_{ads} + H^+$$

$$Cr^{3+} + {}^\bullet OH_{ads} \to [CrO]^{2+}{}_{ads} + H^+$$

$$2\,[CrO]^{2+}{}_{ads} + 5\,H_2O - 4\,e^- \to Cr_2O_7^{2-} + 10\,H^+$$

Interesting enough, the catalytic activity of the electrode surface appears to correlate with the bond strength of chemisorbed oxygen-containing species, increasing activity corresponding to decreasing bond strength.[37] Thus, a material such as PbO_2 exhibits excellent characteristics with an ordering of catalytic activity as follows: titanium alloys, steels < Ti/MnO_2 < Pt/PtO_x < Ni/NiO_x < Pb/PbO_2. The Cr(III) → Cr(VI) process is environmentally important from a recycling perspective (see Chapter 5).

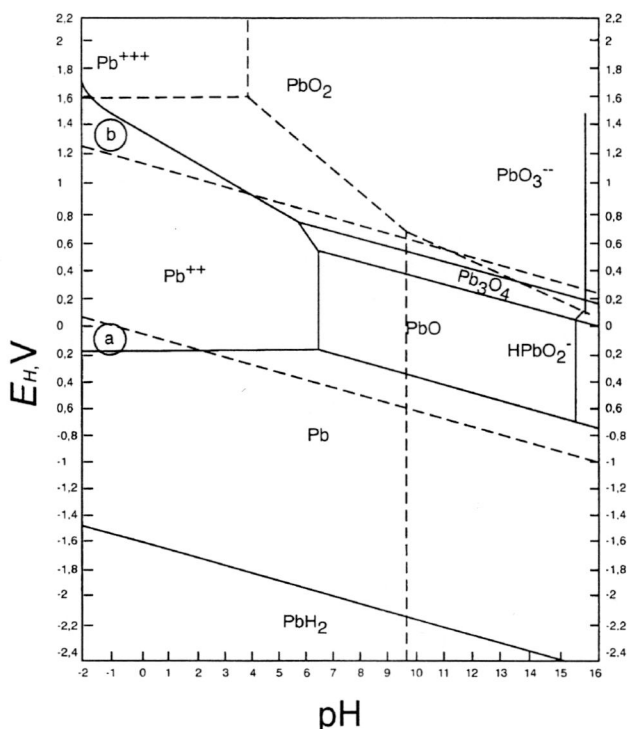

FIGURE 3.29. Pourbaix diagram for the lead–water system at 25 °C. (Adapted with permission from M. Pourbaix, "Atlas of Electrochemical Equilibria." Pergamon, Oxford, 1966.)

The electrochemical properties of Cr(II) ions in aqueous media have not been extensively investigated.[38,39] The formal potential of the Cr(II)/Cr(0)/Hg couple, however, has been determined to be -1.165 V.[38] The Cr(0) formed at the mercury electrode apparently undergoes a fast follow-up reaction, leading to its deactivation. The oxidation of Cr(II) to Cr(III) has also been studied by the same author. The electrode reaction

$$\text{Cr(III)X} + e^- \rightleftarrows \text{Cr(II)X} \tag{3.17}$$

has been the subject of mechanistic studies.[39,40] An inner-sphere electron transfer pathway has been proposed for this system, and for X = OH$^-$, NCS$^-$, Br$^-$ the rate constants are especially high.

<u>Lead</u>. The most important oxidation state is +2. Figure 3.29 shows the Pourbaix diagram for the lead–water system at 25°C. The plumbous ion is

3.3. Electrochemistry of Inorganic Pollutants

stable both in acid and in strongly alkaline solutions and exists as Pb^{2+} and $HPbO_2^-$ ions, respectively. The Pb(IV) oxidation state exists primarily as PbO_2. Anodic oxidation in concentrated (10 M) NaOH generates soluble Pb(IV) as PbO_3^{2-} ions. The oxide Pb_3O_4 may be considered as plumbous plumbate, $2\,PbO \cdot PbO_2$.

The standard potentials in the Pb(IV)-Pb(II)-Pb(0) system are as follows:

$$Pb^{4+} + 2\,e^- \rightleftarrows Pb^{2+}(aq) \qquad E° = 1.69\ V \qquad (3.18)$$

$$Pb^{2+}(aq) + 2\,e^- \rightleftarrows Pb(s) \qquad E° = -0.125\ V \qquad (3.19)$$

Since mercury electrodes are often used for the assay of Pb (see Chapter 4), the lead–lead amalgam potential is relevant, and this is about 5.8 mV, the solid Pb electrode being more negative.

Much of the electrochemical study of Pb has been prompted by its applicability to battery development, and perhaps not surprisingly, the Pb–PbSO$_4$–H$_2$SO$_4$ system has been intensively studied. This topic has been reviewed.[41] On the other hand, as we shall see in later chapters, the Pb/PbO$_2$ anode is also technologically important in many other industrially and environmentally important electrolytic oxidations: chloride to chlorate, bromide to bromate, sulfate to peroxydisulfate, Cr(III) to Cr(VI), and sulfite to sulfate, to mention a few.

Mercury. The important oxidation states are +1 and +2, the corresponding species in acidic media being Hg_2^{2+} and Hg^{2+}, respectively. Even at very low concentrations, the mercurous ion is virtually undissociated[42,43]:

$$Hg_2^{2+}(aq) \rightleftarrows 2\,Hg^+(aq) \qquad K < 10^{-7} \qquad (3.20)$$

Figure 3.30 contains the Pourbaix diagram for the mercury–water system at 25°C. The best value of the standard potential for the $Hg^{+/0}$ couple in acidic media is as follows:

$$Hg_2^{2+}(aq) + 2\,e^- \rightleftarrows 2\,Hg(l) \qquad E° = 0.796\ V \qquad (3.21)$$

This redox couple is important in the construction of the saturated calomel electrode (SCE):

$$Hg_2Cl_2(c) + 2\,e^- \rightleftarrows 2\,Hg(l) + 2\,Cl^-(aq) \qquad E° = 0.268\ V \qquad (3.22)$$

The reaction

$$Hg^{2+}(aq) + 2\,e^- \rightleftarrows Hg(l) \qquad E° = 0.854\ V \qquad (3.23)$$

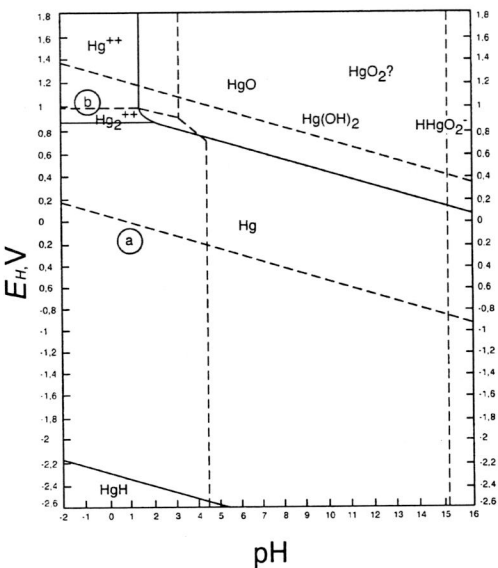

FIGURE 3.30. Pourbaix diagram for the Hg-water system at 25° C. (Adapted with permission from M. Pourbaix, "Atlas of Electrochemical Equilibria." Pergamon, Oxford, 1966.)

underlines the anodic instability of the metal. On the other hand, mercury has a very large hydrogen overpotential, attesting to its popularity in electroanalysis as a working electrode. Combination of $E°$ (3.21) and $E°$ (3.23) leads to the equilibrium constant of the disproportionation reaction:

$$Hg^{2+}(aq) + Hg(l) \rightleftarrows Hg_2^{2+}(aq) \qquad K = 87.9 \qquad (3.24)$$

It follows from this that a fairly powerful oxidizing agent (e.g., Br_2) is required to oxidize the mercurous ion to the mercuric state.

HgO is the predominant insoluble phase at positive potentials:

$$HgO(s) + 2 H^+(aq) + 2 e^- \rightleftarrows Hg(l) + H_2O(l) \qquad E° = 0.926 \text{ V} \qquad (3.25)$$

In basic medium, this redox reaction becomes

$$HgO(s) + H_2O(l) + 2 e^- \rightleftarrows Hg(l) + 2 OH^- \qquad E° = 0.098 \text{ V} \qquad (3.26)$$

This value, in conjunction with the $E°$ of Reaction (3.25), yields

$$Hg(s) + H_2O(l) \rightleftarrows Hg^{2+}(aq) + 2 OH^- \qquad K_{sp} = 2.8 \times 10^{-26} \qquad (3.27)$$

The existence of the Hg(IV) oxidation state is questionable (Figure 3.30).

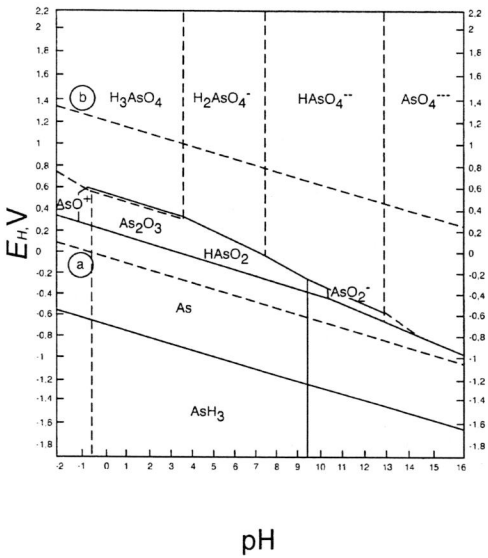

FIGURE 3.31. Pourbaix diagram for the As-water system at 25°C. (Adapted with permission from M. Pourbaix, "Atlas of Electrochemical Equilibria." Pergamon, Oxford, 1966.)

3.3.2. Nonmetals (Metalloids)

Arsenic. The important oxidation states are +5, +3, and -3, and the corresponding species being arsenic acid (H_3AsO_4), arsenious acid ($HAsO_2$ or, alternatively, H_3AsO_3), and arsine (AsH_3). Figure 3.31 contains a Pourbaix diagram for the As–H_2O system at 25°C. The acids undergo stepwise dissociation as the pH is increased as follows:

$$H_3AsO_3 \xrightarrow{K_1} H_2AsO_3^- \xrightarrow{K_2} HAsO_3^{2-}$$

with K_1 and K_2 being 6×10^{-10} and 7.41×10^{-13}, respectively. Similarly for H_3AsO_4:

$$H_3AsO_4 \xrightarrow{K_1} H_2AsO_4^- \xrightarrow{K_2} HAsO_4^{2-} \xrightarrow{K_3} AsO_4^{3-}$$

with K_1, K_2, and K_3 being 5.98×10^{-3}, 1.04×10^{-7}, and 3.98×10^{-12}, respectively.

In most natural waters and aerobic soils, the available data[44] suggest a predominance of As(V) over As(III). An inspection of Figure 3.31 and Figure 3.32, which is a pE–pH diagram of the environmentally significant As–S–H_2O

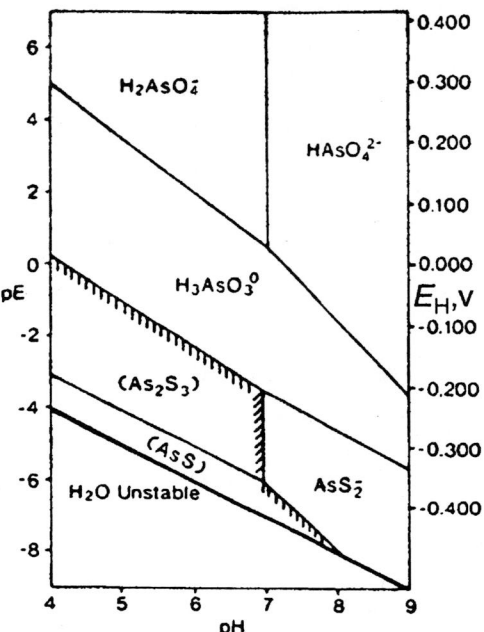

FIGURE 3.32. pE-pH diagram for the As-S-H_2O system at 25°C with total dissolved As and S species set at 50 ppb and 32 ppm, respectively. The area within the hatched lines denotes that the solid phases are predominant (i.e., total dissolved As species <5 ppb.) (Reproduced with permission from Cullen and Reimer.[44])

system,[44] suggests that at these natural redox potentials (0.25–0.40 V), a decrease in pH should increasingly favor the more toxic arsenite over the less toxic arsenate. However, thermodynamically predicted As(V)/As(III) ratios rarely are observed, and a multitude of other factors influence the relative concentrations of these species.[44]

The important redox potentials (standard reduction potentials) in acidic and basic media follow.

Acid

$$\underset{H_3AsO_4}{+5} \xleftrightarrow{0.560} \underset{HAsO_2}{+3} \xleftrightarrow{0.240} \underset{As}{0} \xleftrightarrow{-0.225} \underset{AsH_3}{-3}$$

Base

$$\underset{AsO_4^{3-}}{} \xleftrightarrow{-0.67} \underset{AsO_3^{3-}}{} \xleftrightarrow{-0.68} \underset{As}{} \xleftrightarrow{1.37} \underset{AsH_3}{}$$

3.3. Electrochemistry of Inorganic Pollutants

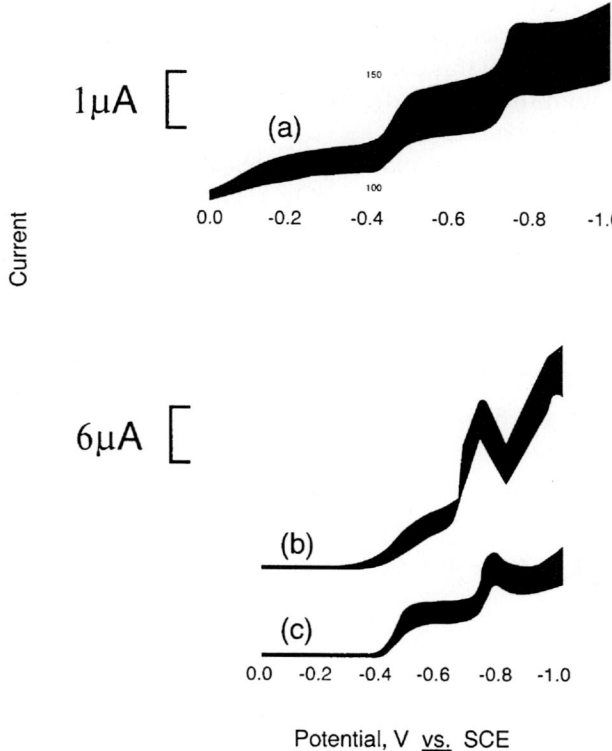

FIGURE 3.33. Polarograms in 2 M $HClO_4$/0.5 pyrogallol for As(V) and As(III). (Reproduced with permission from White and Bard.[49])

The polarographic behavior of As(V) in highly acidic (11.5 M HCl) medium has been discussed.[45] Two waves were observed corresponding to the scheme: As(V) → As(0) → As(-III). However, the wave shape was complex at As(V) levels above 0.4 mM, and in non–complexing media. Similar observations have been made in alkaline media.[46] Arsenic in the +5 oxidation state is also electroactive in phosphoric acid or acidic solutions of polyhydroxy compounds such as catechol,[47,48] pyrogallol,[47,49] and D-mannitol.[50] Figure 3.33 contains representative polarographic data in 2 M $HClO_4$/0.5 M pyrogallol.[49] Three well-defined waves in Figure 3.33a are consistent with the scheme: As(V) → As(III) → As(0) → As(-III). The $E_{1/2}$ (half-wave) potentials for the second and third waves are comparable to those obtained for As(III) (Figure 3.33b), diagnostic of the fact that the reduction of the As(V)–pyrogallol complex, initially formed, proceeds to *uncomplexed* As(III) as the product.

In summary, As(V) is electroinactive in conventional electrolytic media (e.g., 1 M HCl), but is converted to an electroactive complex in the presence of certain agents. The nature of this complexation chemistry remains largely unclear at present although there are indications that a 3:1 complex is formed in the case of As(V) and pyrogallol.

Selenium. The important oxidation states are -2 in selenides (Se^{2-}), +4 in selenites (SeO_3^{2-}), and +6 in selenates (SeO_4^{2-}). Figure 3.34 contains the Pourbaix diagram for the Se–H_2O system at 25°C. As with arsenic, all the preceding species are involved in protonation–deprotonation equilibria:

$$H_2Se \underset{}{\overset{K_1}{\rightleftharpoons}} HSe^- \underset{}{\overset{K_2}{\rightleftharpoons}} Se^{2-}$$

$$K_1 = 1.88 \times 10^{-4}, K_2 = 1 \times 10^{-14}$$

$$H_2SeO_3 \underset{}{\overset{K_1}{\rightleftharpoons}} HSeO_3^- \underset{}{\overset{K_2}{\rightleftharpoons}} SeO_3^{2-}$$

$$K_1 = 2.7 \times 10^{-3}, K_2 = 2.63 \times 10^{-7}$$

$$HSeO_4^- \underset{}{\overset{K_2}{\rightleftharpoons}} SeO_4^{2-}$$

$$K_2 = 8.9 \times 10^{-3}$$

The standard reduction potentials in acidic and basic media are as follows:

Acid

$$\underset{H_2Se}{-2} \xleftarrow{-0.11} \underset{Se}{0} \xleftarrow{0.74} \underset{H_2SeO_3}{+4} \xleftarrow{1.1} \underset{SeO_4^{2-}}{+6}$$

Base

$$\underset{Se^{2-}}{-2} \xleftarrow{-0.67} \underset{Se}{0} \xleftarrow{-0.36} \underset{SeO_3^{2-}}{+4} \xleftarrow{0.03} \underset{SeO_4^{2-}}{+6}$$

The electrochemistry of Se(IV) in aqueous media has been the topic of many studies.[51–64] The behavior is rather complex, and several (interrelated) factors can be attributed to it including, for example:

(1) Sensitivity to the electrode surface: Deposition of a Se(0) layer tends to passivate many electrode surfaces.[51–54] For example, well-defined waves have been observed for Se(IV) reduction at soft graphite electrodes, but poorly defined voltammetry features at glassy carbon.[52]

(2) Underpotential deposition: Some electrode surfaces (e.g., Au) promote the underpotential deposition of Se(0) via strong metal–Se interactions. Thus, two waves have been reported[55] for the reduction of Se(IV) at gold; the one at the more positive potential was assigned to the UPD process.

(3) Compound (selenide) formation involving the electrode material: This complication is evident even in some early studies.[56] For example, changes in the polarographic waves of Se(IV) upon standing[57] were attributed to "HgSe" compound formation. Evidence for the formation of a Au–Se intermetallic has been presented.[55] Other support electrode materials such as Ag,[51] Cd,[51,54] Cu,[51,58-61] and In[58,59] are all known to react chemically either with Se(0) or with the element in the -2 oxidation state.

(4) Coupled chemical reactions: This was also noted very early on in the history of Se(IV) electrochemistry. For example, the following chemical reaction has been postulated to account for certain anomalies in polarographic data[62]:

$$H_2SeO_3 + 2\,H_2Se \rightleftarrows 3\,Se + 3\,H_2O \tag{3.28}$$

In this scheme, the electrogenerated selenide ions react with incoming selenious acid molecules in the diffusion layer, resulting in the precipitation of (colloidal) selenium in the homogeneous solution phase.

(5) Lack of electroactivity of Se(IV) reduction products: It has been reported[53] that electrodeposited Se(0) or the element introduced into graphite paste electrodes is often electroinactive. Evidence has been presented[62] for the formation of at least two forms of Se(0) ("red" and "gray") in the Se(IV) system, and it appears that the red variety is generated via Reaction (3.28). The same group has shown that the selenate ion (SeO_4^{2-}) is electroinactive at the carbon paste electrode.[63] On the other hand, other authors[64] claim that the gray Se(0) formed via the direct electrochemical reduction of Se(IV) is electroinactive.

These complications have prompted a recent re-examination of Se(IV) electrochemistry using a combination of voltammetry and electrochemical quartz crystal microgravimetry (EQCM).[65] Figure 3.35 contains simultaneous linear sweep voltammetry (LSV)-EQCM data on Au without (Figure 3.35a) and with (Figure 3.35b) Se(IV) in the electrolyte. These data are best interpreted in terms of the following scheme. The anodic wave around -0.10 V (vs. Ag/AgCl) (along with the mass gain) is assigned to the UPD of Se(0) on Au:

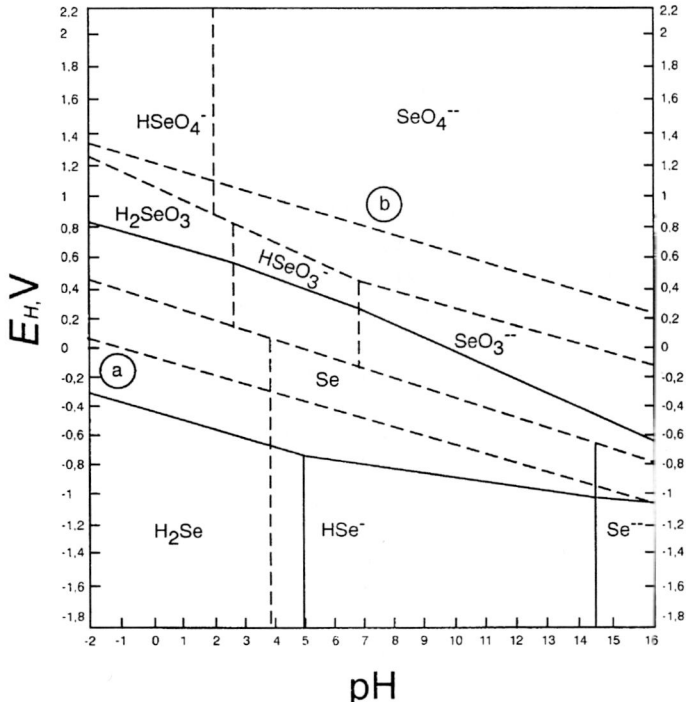

FIGURE 3.34. Pourbaix diagram for the Se-H$_2$O system at 25°C. (Adapted with permission from M. Pourbaix, "Atlas of Electrochemical Equilibria." Pergamon, Oxford, 1966.)

$$H_2SeO_3 + 4\,H^+ + 4\,e^- \rightleftarrows Se(Au) + 3\,H_2O \qquad (3.29)$$

Wave A is assigned to a combination of Reactions (3.30) and (3.31):

$$H_2SeO_3 + 4\,H^+ + 4\,e^- \rightleftarrows Se + 3\,H_2O \qquad (3.30)$$

$$H_2SeO_3 + 6\,H^+ + 6\,e^- \rightleftarrows H_2Se + 3\,H_2O \qquad (3.31)$$

Evidence for the occurrence of Reaction (3.31) at noble metal electrodes (Au and Pt) has been presented.[51] These authors concluded that at high Se(IV) concentrations, the rate of the chemical reaction (Eq. (3.28)) is fast so that Reaction (3.31) couples effectively with Reaction (3.28) leading to an *overall* process for Se(0) generation (i.e., Reaction (3.31) + Reaction (3.28)) that is equivalent to Reaction (3.30). Thus the mass gain observed concomitant with wave A is a manifestation of both the 4 e$^-$ reduction pathway (Reaction (3.30)) and the 6 e$^-$ route coupled with the chemical reaction (Reactions (3.31) + (3.28)).

3.3. Electrochemistry of Inorganic Pollutants 173

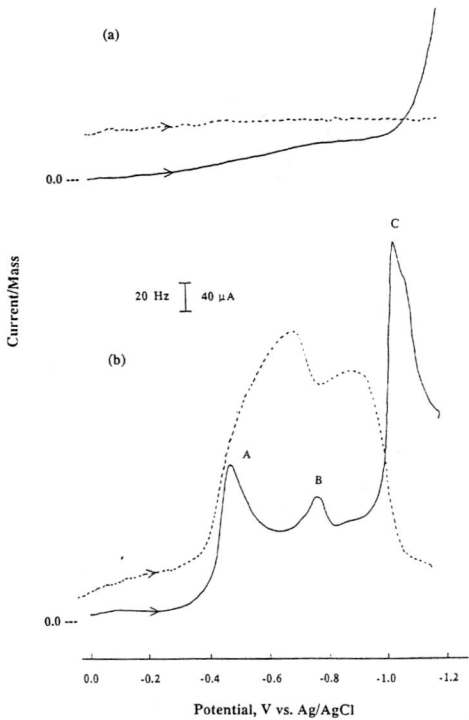

FIGURE 3.35. Combined cathodic linear sweep voltammetry (solid-line) and microgravimetry (dashed-line) scans of Au in 0.05 M Na_2SO_4 with and without 4 x 10^{-3} M H_2SeO_3. (Reproduced with permission from Wei et al.[65])

The initial mass loss associated with wave B (Figures 3.35b and 3.36b) is assigned to Reaction (3.32):

$$Se + 2 H^+ + 2 e^- \rightleftarrows H_2Se \qquad (3.32)$$

However, the mass gain observed at the trailing edge of wave B indicates that stripping is incomplete and *redeposition* of Se(0) occurs via Reaction (3.28). Thus the competition between these two processes appears to be delicately balanced and is dictated by the potential and the concentration of H_2Se in the solution (see later).

Finally, wave C is assigned to the complete stripping of Se(0) from the Au surface, including the UPD layer. The mass loss in this regime is monotonic, although note that the frequency (mass) does not return to its initial value (at 0 V *vs.* Ag/AgCl) until potentials more positive than -0.8 V are reached on the return cycle. This can be attributed to the interference from the proton reduction reaction.

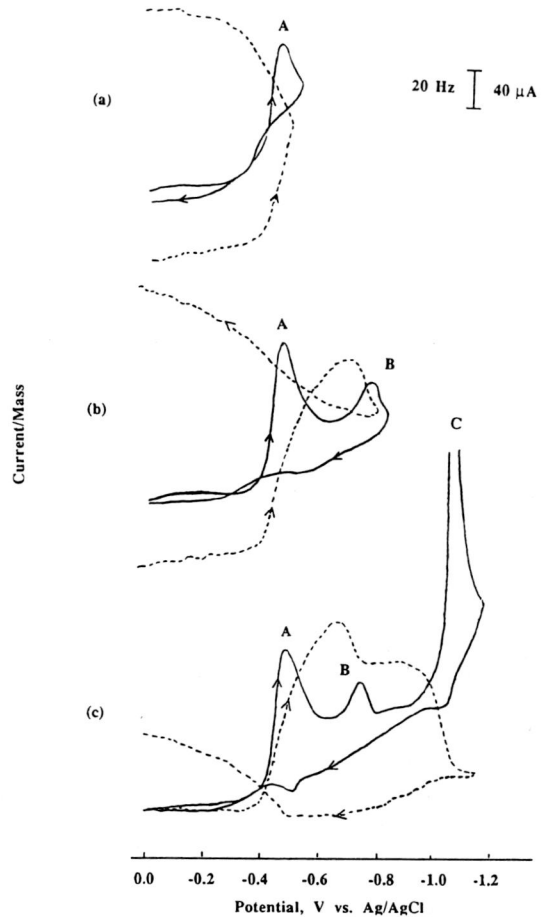

Figure 3.36. Combined cyclic voltammograms (solid-line) and microgravimetry (dashed-line) scans of Au in the same solution as in Figure 3.35. Switching potentials for the traces were as follows: (a) -0.55 V; (b) -0.85 V; (c) -1.20 V. (All potentials vs. the Ag/AgCl reference. (Reproduced with permission from Wei et al.[65])

The importance of the 6 e⁻ reduction pathway at relatively positive potentials (e.g., corresponding to the wave A regime in this study) appears not to have been recognized in the majority of the earlier studies, at least in neutral and acidic media. The early polarographic data cited earlier were interpreted in terms of a Se(IV) → Se(0) → Se(-II) scheme. This is the case even in more recent studies.[52,63] It now appears more likely that the direct 6 e⁻ Se(IV) → Se(-II) reduction pathway competes with the initial 4 e⁻ process.[65] Coupling with a subsequent (fast) chemical reaction between Se(-II) and Se(IV) results in the further deposition of Se(0) at the electrode surface. The electrochemical behavior at more negative potentials reflects a complex interplay of the 4 e⁻, 6 e⁻ reduction processes and the chemical reaction along with the effect of Se(-II) ions stripped off from the initial selenium layer. Thus, the delicate balance between these is influenced by two variables; namely, the potential and the Se(-II) concentration in the electrolyte. Further data in support of this mechanistic scheme have been presented,[65] using a Au disk electrode oriented "face down" in the cell as well as by hydrodynamic voltammetry measurements.

3.3.3. Inorganic Anions and Gaseous Species

We next examine the electrochemistry of anionic species of sulfur, carbon, and nitrogen such as sulfite, cyanide, and nitrate as well as gaseous compounds such as SO_2, NO, CO, and CO_2. All these species have environmental import, see Section 1.4.2.

<u>Sulfur Dioxide, Sulfite, and Bisulfite.</u> The electrochemistry of sulfur–oxygen compounds has been recently reviewed.[66] The sulfur compounds are presented in order of the oxidation state, and Pourbaix diagrams are given in each case. We focus here mainly on the electrochemisry of SO_2 and the sulfite and bisulfite ions. Thermodynamic data on aqueous systems reveal that SO_2 exists mostly as the dissociated species, HSO_3^- and SO_3^{2-}:

$$H_2SO_3 \underset{}{\overset{K_1}{\rightleftharpoons}} HSO_3^- \underset{}{\overset{K_2}{\rightleftharpoons}} SO_3^{2-}$$

$$K_1 = 1.7 \times 10^{-2}, K_2 = 5 \times 10^{-6}$$

Oxidation of SO_3^{2-} or HSO_3^- yields the sulfate ion:

$$SO_3^{2-} + 2\,OH^- \rightleftharpoons H_2O + SO_4^{2-} + 2\,e^- \qquad E° = -0.930\ V \qquad (3.33)$$

$$HSO_3^- + H_2O \rightleftharpoons 3\,H^+ + 2\,e^- + SO_4^{2-} \qquad E° = 0.158\ V \qquad (3.34)$$

The electrooxidation of SO_2, aside from environmental importance, also has an important side benefit in terms of hydrogen production by water electrolysis at low cell voltage levels. Thus, this electrochemical scheme allows the removal of SO_2 from flue gas and the production of sulfuric acid.[67–69] The difficulty is the sluggish kinetics and the high overpotential required for SO_2 oxidation. The electrochemical oxidation of SO_3^{2-} and HSO_3^- in sodium sulfate electrolyte has been studied at graphite anodes.[69] The i–E curves were modeled by the Butler–Volmer formalism (see Chapter 2). The overall oxidation has been modeled as first order in sulfite and bisulfite over a wide concentration range. Molecular SO_2 was observed to be oxidized with a Tafel slope of 280 mV/decade either on graphite or glassy carbon,[68,70] and the oxidation kinetics was found to decrease with increasing pH.[71] Other studies on the oxidation of sulfite or molecular SO_2 have been reviewed.[69]

The reduction of sulfite has been studied by several authors using polarography on Hg.[66] The proposed reduction product in acidic medium is dithionite. On the other hand, decomposition of dithionite has been implicated in the production of the sulphoxylate ion, which subsequently dimerizes to thiosulfate.[72] Other authors conclude that the ultimate reduction product on Hg is sulfur and sulfur dioxide.[73] For Pt and gold electrodes, both sulfide and sulfur have been reported as the reduction products.[66,74,75] A recent voltammetric study proposes the following scheme for reduction of Pt[75a]:

$$Pt + H_2O + e^- \rightleftarrows Pt \cdot H + OH^-$$

$$Pt \cdot HSO_3^- + e^- \rightleftarrows Pt \cdot SO_3^{2-} + 1/2\, H_2$$

$$Pt \cdot H_2SO_3 + 2\, e^- \rightleftarrows PtSO_3^{2-} + H_2$$

$$Pt \cdot H_3O^+ + e^- \rightleftarrows H_2O + 1/2\, H_2 + Pt$$

This scheme features differential adsorption of HSO_3^- and SO_3^{2-} on Pt as a function of pH and concomitant hydrogen evolution.

The reduction of SO_2 in nonaqueous media such as dimethylformamide is electrochemically reversible and generates the radical anion, $SO_2^{\bullet -}$.[76] Dimerization of these radicals yields the dithionite anion, $S_2O_4^{2-}$ and $S_3O_6^{2-}$. This process has been studied by UV–VIS spectroelectrochemistry.[77] Much of the interest in the nonaqueous electrochemistry of SO_2 stems from its applicability in Li/SO_2 battery development.

Cyanide. Carbon forms with nitrogen, compounds such as cyanogen, $(CN)_2$ with properties similar to halogens. These "pseudohalogens" may be reduced:

$$(CN)_2(g) + 2\, H^+ + 2\, e^- \rightleftarrows 2\, HCN(aq) \qquad E° = 0.373\ V \qquad (3.35)$$

$$(CN)_2(g) + 2\, e^- \rightleftarrows 2\, CN^- \qquad E° = -0.176\ V \qquad (3.36)$$

These reactions, as well as the oxidation of cyanide to cyanate;

$$OCN^- + 2\, H^+ + 2\, e^- \rightleftarrows CN^- + H_2O \qquad E° = -0.14\ V \qquad (3.37)$$

are electrochemically irreversible.

3.3. Electrochemistry of Inorganic Pollutants

<u>Nitrate and NO$_x$</u>. The oxidation state of nitrogen in nitrate (NO$_3^-$) is +5. The reduction of NO$_3^-$:

$$NO_3^- + 2\,H^+ + 2\,e^- \rightarrow NO_2^- + H_2O \tag{3.38}$$

proceeds very slowly at pH 3.0 because of proton depletion at the electrode surface. On the other hand, aqueous nitric acid is a strong oxidizing agent because of the following redox equilibria:

$$NO_3^-(aq) + 4\,H^+(aq) + 3\,e^- \rightleftarrows NO(g) + 2\,H_2O(l) \qquad E° = 0.964 \tag{3.39}$$

$$NO_3^-(aq) + 3\,H^+(aq) + 2\,e^- \rightleftarrows HNO_2(aq) + H_2O(l) \qquad E° = 0.928 \tag{3.40}$$

The addition of a metal ion (e.g., Ni^{2+}) which can be precipitated as a hydroxide increases the rate of NO$_3^-$ reduction because of the concomitant lowering of the interfacial pH. Further reduction of this species proceeds as follows:

$$\underset{HNO_2}{+3} \xleftarrow{0.983} \underset{NO}{+2} \xleftarrow{1.678} \underset{N_2}{0} \xleftarrow{0.275} \underset{NH_4^+}{-3}$$

Hydrazine (N$_2$H$_4$) is an intermediate reduction product of nitrogen that also has environmental significance; the oxidation state of nitrogen in this compound is -2. The reduction of hydrazine proceeds as follows in acidic and basic media, respectively:

$$N_2H_5^+(aq) + 3\,H^+(aq) + 2\,e^- \rightleftarrows 2\,NH_4^+(aq) \qquad E° = 1.275\ V \tag{3.41}$$

$$N_2H_4(aq) + 2\,H_2O(l) + 2\,e^- \rightleftarrows 2\,NH_3(aq) + 2\,OH^-(l) \quad E° = 0.10\ V \tag{3.42}$$

It must be stressed that the redox behavior of the compounds of nitrogen are all characterized by electrochemical irreversibility. Therefore, nitrogen fixation via

$$N_2(g) + 3\,H_2(g) \rightarrow 2\,NH_3(g) \tag{3.43}$$

$$2\,N_2(g) + 5\,O_2(g) + 2\,H_2O \rightarrow 4\,NO_3^- + 4\,H^+ \tag{3.44}$$

while thermodynamically feasible does not occur in practice because of the kinetic irreversibility.

Spectroscopic identification of the intermediates in the electroreduction of NO_3^- on Ag has been carried out *in situ* using Raman spectroscopy[78] and mass spectrometry.[79] Interestingly enough, the UPD of Cd on single crystal Ag(111) electrodes has been observed to catalyze the reduction of NO_3^- ions.[80] Thus, the half-wave potential for the electroreduction of nitrate is most positive for Ag and Cd followed by Cu and Zn. Interesting effects have also been observed of nitrate reduction during lead UPD on gold.[81] On the other hand, Tl inhibits the reduction on the low-index single crystal face of Ag.[82]

<u>Carbon Dioxide and Carbon Monoxide.</u> The electrolytic conversion of CO_2 to useful fuels has both fundamental and practical significance. The relevant standard reduction potentials in acidic and basic media are as follows:

<u>Acid</u>

$$CH_4(g) \xleftarrow{0.59} CH_3OH(aq) \xleftarrow{0.232} HCHO(aq) \xleftarrow{0.034}$$

$$HCOOH(aq) \xleftarrow{0.034} CO_2(g)$$

<u>Base</u>

$$CH_4(g) \xleftarrow{-0.2} CH_3OH(aq) \xleftarrow{-0.59} HCHO(aq) \xleftarrow{-1.07}$$

$$HCOO^-(aq) \xleftarrow{0.034} CO_3^{2-}(aq)$$

As with many of the reactions presented in this chapter, the preceding potentials have only thermodynamic significance, and all these electrode reactions proceed with a large overpotential. The electrode reaction of the system CO_2/CO:

$$CO_2(g) + 2\,H^+(aq) + 2\,e^- \rightleftarrows CO(g) + H_2O \quad E° = -0.106\text{ V} \quad (3.49)$$

is also kinetically irreversible. In theory, CO_2 or CO_3^{2-} may be reduced to elemental C:

$$CO_2(g) + 4\,H^+ + 4\,e^- \rightleftarrows C(s) + 2\,H_2O \quad E° = 0.206\text{ V} \quad (3.50)$$

$$CO_3^{2-}(aq) + 6\,H^+ + 4\,e^- \rightleftarrows C(s) + 3\,H_2O \quad E° = 0.475\text{ V} \quad (3.51)$$

Again, these reactions have very large overpotentials and are not observed in aqueous media.

3.3. Electrochemistry of Inorganic Pollutants

The following standard potentials are available for the electroreduction of CO:

$$CO(g) + 6\,H^+ + 6\,e^- \rightarrow CH_4(g) + H_2O \qquad E° = 0.260\text{ V} \qquad (3.52)$$

$$CO(g) + 2\,H^+ + 2\,e^- \rightarrow C(s) + H_2O \qquad E° = 0.517\text{ V} \qquad (3.53)$$

However, the hydrogeneration of CO is only slight[83] and CO conversion yields either squarate anions or dimeric species in nonaqueous solvents[84] and in media such as liquid ammonia.[85]

Ozone. An interesting aspect of the oxidizing action of ozone (O_3) is that only one of the oxygen atoms is reduced while the other two atoms form dioxygen. The following standard reduction potentials are available for the ozone-water couple in acid and base, respectively. In an acid,

$$O_3(g) + 2\,H^+(aq) + 2\,e^- \rightleftarrows O_2(g) + H_2O(l) \qquad E° = 2.075\text{ V} \qquad (3.54)$$

In a base,

$$O_3(g) + H_2O(l) + 2\,e^- \rightleftarrows O_2(g) + 2\,OH^-(aq) \qquad E° = 1.246\text{ V} \qquad (3.55)$$

The high standard reduction potentials indicate that ozone ranks alongside fluorine in the category of powerful oxidizing agents. In solution, ozone is unstable with respect to decomposition into dioxygen:

$$O_3(g) + H_2O(l) \rightarrow 2\,O_2(g) + 2\,H^+(aq) + 2\,e^- \qquad E° = -0.383\text{ V} \qquad (3.56a)$$

Ozone exhibits Tafel slopes ranging from 0.10 V to 0.12 V in alkaline media on a bright Pt cathode.[86] These data are consistent with a rate-determining electron-transfer step;

$$O_3(g) + e^- \rightarrow O_3^- \qquad (3.56b)$$

followed by the reaction of the ozonide ion with water:

$$O_3^- + H_2O \rightarrow O_2 + {}^\bullet OH + OH^- \qquad (3.56c)$$

$$O_3^- + {}^\bullet OH \rightarrow OH^- + O_3 \qquad (3.56d)$$

3.3.4. Organometallic Compounds

The severe curtailment in the use of leaded gasoline reduces the significance of organolead compounds in industrial effluents, although in many countries, the use of tetramethyl and tetraethyl lead as gasoline antiknock additives still continues. The effluents from industries manufacturing these compounds contain soluble trialkyl and dialkyl lead compounds, which present a serious disposal problem.

The electrolytic formation of radicals opens up a route for the treatment of these effluents. For example:

$$R_3Pb^- \rightarrow R_3Pb^\bullet + e^-$$

These radicals then dimerize, disproportionate, or decompose to give hydrocarbons, tetraalkyl lead, or metallic lead, respectively. The dimers and trimers are usually water soluble but higher oligomers form insoluble products, which may be removed by filtration or sedimentation.

Early studies on tetraalkyl lead compounds found them to be electroinactive.[87] More recently, polarography has been performed on tetraphenyl-[88] and tetraethyl- and tetramethyl lead[89] in dichloromethane. The oxidation of tetraphenyl lead is believed to proceed by the following sequence,[89]

$$2\ Ph_4Pb + Hg \rightarrow 2\ Ph_3Pb^+ + Ph_2Hg + 2\ e^-$$

$$2\ Ph_3Pb^+ + Hg \rightarrow 2\ Ph_2Pb^{2+} + Ph_2Hg + 2\ e^-$$

followed by

$$Ph_2Hg + Hg \rightarrow 2\ PhHg^+ + 2\ e^-$$

On the other hand, the oxidation of tetraalkyl lead compounds involves an one-electron process; for example,[90]

$$Et_4Pb + Hg \rightarrow Et_3Pb^\bullet + EtHg^+ + e^-$$

$$2\ Et_3Pb^\bullet \rightarrow Et_6Pb_2$$

Longer time domain measurements indicate further reactions of the kind

$$Me_6Pb_2 + MeHg^+ \rightarrow Me_5Pb_2^+ + Me_2Hg$$

3.3. Electrochemistry of Inorganic Pollutants

The major difference between the two groups of organolead compounds therefore is that the triphenyl lead radical dimerizes more slowly and remains available for follow-up chemistry.

In the preceding scheme, diphenyl mercury was polarographically detected by its oxidation wave at +0.68 V (vs. Ag/AgCl/satd. LiCl, CH_2Cl_2 reference).[89] On the other hand, $PhHg^+$ is electroinactive at mercury electrodes in the potential window from 0 to ~+0.80 V (vs. Ag/AgCl/ satd. LiCl, CH_2Cl_2 reference). Phenyl mercury compounds are generally reduced in two steps, both steps being irreversible. Ethyl mercury chloride also yields two polarographic waves, and the first wave is electrochemically reversible.[91] In the presence of sulfite at alkaline pH, this wave is shifted to a more negative potential indicating complexation of the sulfite with the compound. The second reduction wave is drawn out and indicates an irreversible electron transfer process.

3.3.5. Organo Derivatives of Arsenic, Phosphorus, and Sulfur

Organo arsenicals are generally reduced at potentials negative of -1.1 V, although the reduction mechanism is complex and not very well characterized. The electrolytic reduction of organo arsenic acids such as methylarsonic acid (MMA, 1) and dimethyl arsinic acid (DMA, 2) (also see Table 1.8) has been studied.[92]

$$CH_3As(O)(OH)_2 \qquad\qquad (CH_3)_2As(O)(OH)$$
$$1 \qquad\qquad\qquad\qquad\qquad 2$$

In buffers between pH 1.9 and 5.4, DMA displays a single polarographic reduction wave that is well separated from the hydrogen evolution reaction (HER) regime. On the other hand, MMA yields a wave very close to the HER background. Electron stoichiometry values of 1.1 and 0.8 were obtained for the bulk electrolysis of DMA and MMA, respectively.[92] The electrochemical data for DMA suggest a mechanism based on the reaction

$$(CH_3)_2As(O)OH + H^+ + e^- \rightarrow (CH_3)_2As(O)OH_2$$

Production of dimethyl arsine from DMA involves the reaction

$$(CH_3)_2As(O)(OH) + 4\,e^- + 4\,H^+ \rightarrow (CH_3)_2AsH + 2\,H_2O$$

Mass spectrometric analysis of the trapped gaseous effluent from the bulk reduction indicated that volatile arsines were the major products.[92] In this regard, parallels have been drawn between the electrolytic route and natural mechanisms for arsine generation from the acid in moist soils.[92]

Other polarographic studies on DMA and MMA have reached essentially similar conclusions.[93] Cyclic voltammetry has been performed on these compounds and on 3-acetamido-4-hydroxyphenyl arsonic acid (**3**, acetarsone)[94]:

AsO(OH)$_2$

NHCOCH$_3$

OH

3

Only acetarsone gave a reduction wave in the cathodic scan (see Figure 3.37). The reduction wave at ca. -0.8 V (*vs.* SCE) was assigned to the initial reduction of **3** to the corresponding arsine.[94] The anodic waves on the return scan were attributed to the adsorption–oxidation of the reduction product.

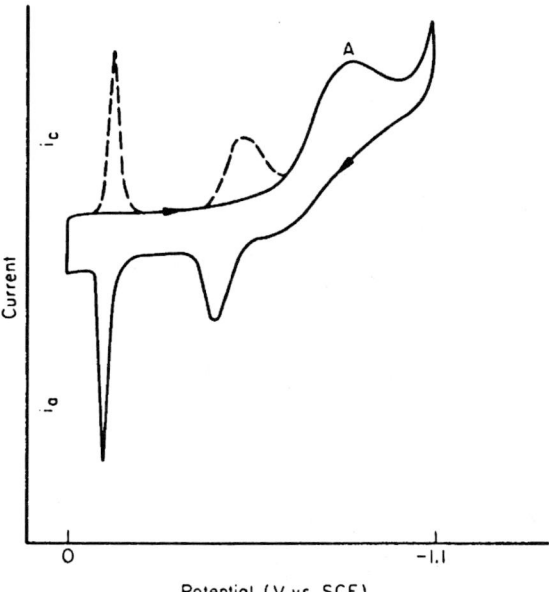

FIGURE 3.37. Cyclic voltammetry of 10^{-5} M acetarsone in 0.1 M sulfuric acid (scan rate 100 mV s^{-1}, delay 6 s): (solid-line) anodic scan from 0 V; (dashed-line) cathodic scan from -1.1 V. (Reproduced with permission from Spini *et al.*[94])

3.3. Electrochemistry of Inorganic Pollutants

Clearly more work is needed to elucidate the mechanistic details associated with the electrochemistry of organo arsenicals in general.

Organo phosphorus compounds are important in many pesticide formulations. The polarographic behavior of O,O-diethyl O-p-nitrophenyl thiophosphate, **4** (parathion)

$$(CH_3O)_2\overset{\overset{\displaystyle S}{\|}}{P}-O-\underset{}{\bigcirc}-NO_2$$

4

was investigated with a view to implementation as an assay procedure.[95] The electroactivity of this compound was associated with the reduction of the nitro group to the amino group (see Section 3.2) although the reduction mechanism was not investigated. A "decomposition potential" of -0.30 V (vs. SCE) and a half-wave potential of -0.39 V (vs. SCE) was quoted in this study. p-Nitrophenyl, a major contaminant of technical grade parathion, was not found to interfere although diethyl p-nitrophenyl phosphate, the oxygen analog of parathion, was found to exert an interference.

As with the arsenic derivatives, mechanistic details are largely lacking and the studies have been motivated mainly by their analytical utility (see Chapter 4).

Mechanistic information on the reduction reaction is, however, available for model compounds such as tris (p-nitrophenyl) phosphate (TNP), **5**, triphenyl phosphine (TPP), **6**, and triphenyl phosphine oxide (TPPO), **7**.

$$(NO_2C_6H_4O)_3P=O \qquad (C_6H_5)_3P \qquad (C_6H_5)_3P=O$$
$$\mathbf{5} \qquad\qquad\qquad \mathbf{6} \qquad\qquad\quad \mathbf{7}$$

The polarographic and voltammetric studies[96,97] were conducted in a nonaqueous medium (dimethyl formamide). The electrochemical reduction of **6** is complex and appears to proceed via an initial 2 e⁻ reaction:

$$(NO_2C_6H_4O)_3PO + 2\,e^- \rightarrow (NO_2C_6H_4O)_3PO^{2-}$$

followed by hydrogen abstraction from the solvent, leading to

$$(NO_2C_6H_4O)P\overset{OH}{\underset{OH}{-}}O^-$$

as a possible intermediate. 4,4'-Dinitrobiphenyl was detected as an end product, indicating that further reduction occurs by cleavage of the C–O bond while the phosphorus–oxygen bond remains intact.

Reduction of TPP and TPPO appears to proceed via the formation of anion radicals[96]:

$$(C_6H_5)_3P + e^- \rightarrow (C_6H_5)_3P^{\bullet -}$$

$$(C_6H_5)_3PO + e^- \rightarrow (C_6H_5)_3PO^{\bullet -}$$

These radicals undergo follow-up chemical reactions, leading in the TPP case to diphenyl phosphioic acid [$(C_6H_5)_2PH$] and, ultimately, biphenyl, a sequence reminiscent of the case of **6**.

The electrochemistry of organic derivatives of P and As, particularly of the type Ph_2MX has been reported.[97]

Sulfur forms a variety of organic derivatives including sulfones, sulfoxides, sulfides, disulfides and mercaptans or thiols. Some of these compounds are notoriously malodorous. The following redox chemistry is representative:

sulfone sulfoxide
$$(CH_3)_2SO_2(g) + 2\,e^- + 2\,H^+(aq) \rightleftarrows (CH_3)_2SO(g) + H_2O(l) \qquad E° = 0.238\ V \qquad (3.57)$$

sulfoxide sulfide
$$(CH_3)_2SO(g) + 2\,e^- + 2\,H^+(aq) \rightleftarrows (CH_3)_2S(g) + H_2O(l) \qquad E° = 0.769\ V \qquad (3.58)$$

disulfide thiol
$$(CH_3)_2S_2(g) + 2\,e^- + 2\,H^+(aq) \rightleftarrows 2\,CH_3SH(g) \qquad E° = 0.176\ V \qquad (3.59)$$

These redox reactions are largely irreversible in terms of charge transfer kinetics at the electrode-solution interface.

Aromatic and aliphatic compounds containing sulfur have been extensively studied using a variety of electroanalytical techniques. Reviews of this body of work are available.[98–100] These studies include those of diaryl sulfides and sulfoxides,[101–104] dimethyl sulfides,[105] thioethers,[106] phenyl disulfides,[107] and compounds of the sort shown in structures **8–14**:

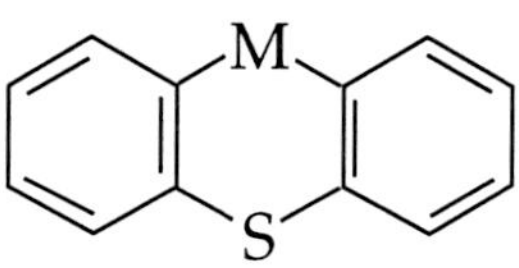

8: $M = CH_2$

9: $M = CO$

10: $M = S$

11: $M = NH$

3.3. Electrochemistry of Inorganic Pollutants

[Structures: 12, 13, 14]

Thus the electrochemical oxidation of **8, 10, 12,** and **14** has been described.[108–113] Similarly, the electrolytic reduction of **8, 9, 10, 12,** and **14** has been studied.[90,110] In all these cases, initial radical cation (in the case of oxidation) and radical anion (in the case of reduction) formation is rapidly followed by chemical reactions involving these radicals. In some instances (e.g., thianthrene), the radical cation and even the dication are exceptionally stable in nonaqueous solvents.

Severe loss of electrode activity is associated with the anodic oxidation of many sulfur compounds. Reproducible I–E behavior is often obtained only when the electrode is repeatedly cycled between the positive and negative limits of the solvent window prior to a recorded scan. A method to circumvent this electrode passivation has been developed, based on pulsed amperometry.[114]

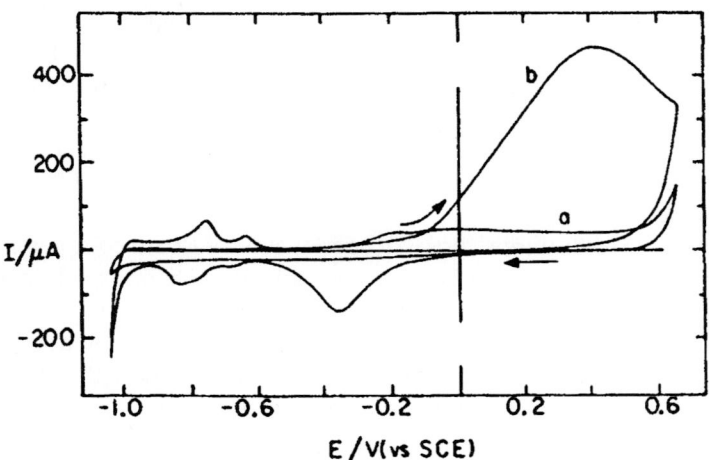

Figure 3.38. Current-potential curves for thiourea on a Pt RDE by cyclic voltammetry in 0.25 M NaOH. Conditions: potential scan rate 6.0 V min^{-1}; electrode rotation speed; 94 rad s^{-1}; concentrations; (a) 0.00 mM; (b) 1.00 mM. (Reproduced with permission from Polta and Johnson.[114])

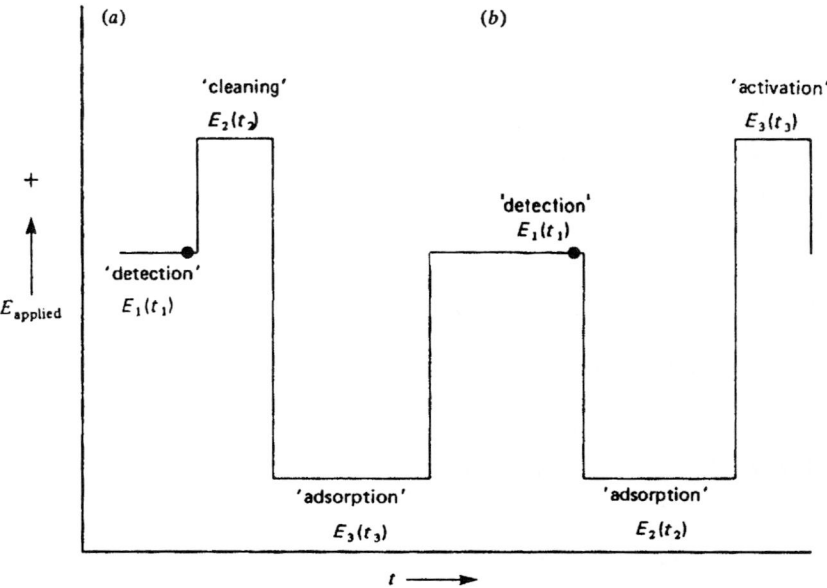

Figure 3.39. Triple-step potential waveforms for pulsed amperometric detection (PAD). Solid dot indicates measurement of current. (Reproduced with permission from Polta and Johnson.[114])

The case study of thiourea is representative of this approach although the model substrate is not toxic.[114] A typical cyclic voltammogram for thiourea is shown in Figure 3.38 for a Pt rotating disc electrode (RDE). The set of waves negative of ~-0.6 V (vs. SCE) correspond to the hydrogen evolution reaction. The anodic wave at $E > -0.3$ V (vs. SCE) and the cathodic peak at -0.3 V (vs. SCE) are caused by surface oxide formation and reduction, respectively, on Pt (in the absence of thiourea). Thus, the O_2 evolution reaction manifests at $E > 0.6$ V (vs. SCE) in this case.

In the presence of thiourea, the hydrogen waves of Pt are completely suppressed indicating that thiourea is adsorbed onto Pt, and no free Pt sites are available for hydrogen adsorption. Oxidation of thiourea produces the large anodic wave and is severely inhibited by surface oxide formation as noted by the dramatic current decrease on scan reversal. However, note the absence of the surface oxide *reduction wave* in the presence of thiourea.

A pulsed amperometric detection (PAD) waveform was used, as illustrated in Figure 3.39,[114] and anodic peaks were detected for thiourea. A mechanism consistent with these data is the 2 e⁻ reduction of thiourea at E_3 in the region of platinum oxide reduction ($E < -0.4$ V vs. SCE):

3.3. Electrochemistry of Inorganic Pollutants

$$2\ H_2N-\underset{\underset{NH_2}{|}}{C}=S \longrightarrow \left[H_2N-\underset{\underset{NH}{\|}}{C}-S\right]_2 + 2\ H^+ + 2\ e^-$$

The adsorbed formamidine disulfide was concluded as the culprit for loss of electrode activity.[114]

We shall return to the analytical implications of PAD in Chapter 4.

The determination of trace amounts of organic compounds using voltammetric stripping techniques is a growing discipline (see Chapter 4). An important strategy for preconcentration of the analyte at the electrode (usually a Hg film) surface is adsorption. A variety of sulfur compounds and several non sulfur-containing organics (e.g., pterins, flavins, purines) have been determined in this manner; in many instances surface complex formation is involved, as for example, in the mercury–thiol system.[115] In fact, alkyl and phenyl mercury compounds have long been used for the amperometric titration of sulfhydryl and disulfide groups.[91]

The polarographic behavior of the phenyl mercury and alkyl mercury compounds has also been discussed in this connection,[91] and the effect of added sulfur derivation such as sulfite, cysteine, cystine, and glutathione has been noted.[91,116] For example, the first wave for the reduction of ethyl mercury chloride (EMC) is reduced to half its value at a mole ratio of cystine to EMC of 2:1, and a new wave at more negative potential appears corresponding to the reduction of ethyl mercury cysteinate, C_2H_5HgSR.[91] Similarly, mercury(I) complexes of glutathione (GSH) are known to adsorb on the Hg drop and are reduced:

$$GSHg + H^+ + e^- \rightleftarrows GSH + Hg$$

The thiol group in biologically important sulfur compounds such as cysteine and glutathione can also complex with metal ions such as Cu(I). For example, the following electrode mechanism has been proposed[116] for the Cu(I)–cysteine (RSH) system:

$$Cu(II) + e^- \rightleftarrows Cu(I)$$

$$Cu(I) + RSH \rightarrow RSCu(I) + H^+$$

$$RSCu(I) + Hg + 2e^- + H^+ \rightarrow RSH + Cu(Hg)$$

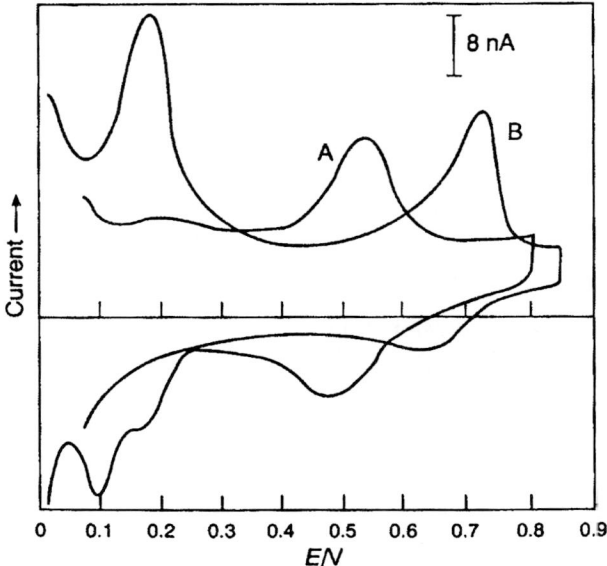

FIGURE 3.40. Cyclic voltammogram of 5 μM glutathione, A, before and, B, after addition of 1 μM copper in pH 8.5 seawater. Scan rate, 50 mV s^{-1}; adsorption time, 60 s; adsorption potentials, A - 0.05 and B 0 V. (Reproduced with permission from LeGall and Vanden Berg.[118])

In the case of a disulfide such as cystine (RSSR), the complexing mechanism is more complicated, as this compound must be reduced first,

$$RSSR + 2H^+ + 2e^- \rightarrow 2RSH$$

before complexation with Cu(I) can take place.

The degree of complexation is often quite sensitive to the oxidation state of the metal ion. For example, the formation constants of Cu(I) with cysteine and glutathione are quite high: log K = 20 and 13.8, respectively.[117,118] On the other hand, Cu(II) complexes are not very stable (e.g., log K for the Cu(II) complex with cystine is only 7.0[119]).

The cyclic voltammetry data in Figure 3.40 are representative of these points.[118] The voltammogram A (in the absence of added Cu(I)) contains a set of reversible waves centered at ~-0.50 V corresponding to the reduction of the Hg(I)–glutathione complex. Note the small difference in the peak potentials for the forward and reverse waves, consistent with a surface confined complex. Addition of Cu(I) to the system yields the sets of waves at ~-0.15 V (cathodic) and -0.10 V (anodic) associated with *uncomplexed* Cu(I). Note that the Cu(I) oxidation state is stabilized in the chloride matrix (see Section 3.3.1) as diagnosed by the fact that two 1 e⁻ waves are seen for the reduction of Cu(II) in this medium.[118] The reduction wave at -0.65 V then corresponds to the Cu(I)–glutathione complex. Both the 1:1 complex and its 1:2 (i.e., $Cu(GS)_2$) counterpart have been identified by voltammetric experiments in this medium.[118] As is to be expected, the 1:1 complex is dominant at glutathione concentrations below ~0.3 mM whereas the 1:2 complex is predominant at higher concentrations.[118] Perhaps not surprising, such complexation effects are very metal specific. For example, in a related study of complexes of Cu(I) with purines, no complexation was noted when 10^{-7} M of several metal ions including Pb(II), Cd(II), Mn(IV), Ni(II), was added to the medium.[120]

We shall defer further discussion of the electroanalytical implications of these findings until Chapter 4.

In closing this section, we must note that organosulfur and organonitrogen compounds have long been used as models to probe adsorption phenomena and their kinetic consequences at electrode/electrolyte interfaces.[121] Apart from their importance in biochemical reactions, amino acids such as cysteine and cystine have also been employed as model substrates to study the electrocatalytic activity of chemically modified electrode surfaces.[122] The electrochemistry of cystine and cysteine has been recently reviewed.[123]

3.4. SUMMARY

In this chapter, the electrochemical behavior of the three major (chemical) categories of environmental pollutants, has been reviewed. This knowledge base is very dispersed indeed, and an attempt has been made to consolidate it, in this chapter. However, because of space constraints, many important details undoubtedly have been omitted, and the reader is referred to the primary sources listed as References 3, 5, 6, 9, and 10 in the Supplementary Reading list that follows below.

REFERENCES

1. J. R. Wiley, E. C. M. Chen, E. S. D. Chen, P. Richardson, W. R. Reed, and W. E. Wentworth, The Determination of Absolute Electron Affinities of Chlorobenzenes, Chloronaphthalenes and Chlorinated Biphenyls from Reduction Potentials. *J. Electroanal. Chem. Interfacial Electrochem.* **307**, 169 (1991).
2. H. Shalev and D. H. Evans, Solvation of Anion Radicals: Gas Phase vs. Solution. *J. Am. Chem. Soc.* **111**, 2667 (1989), and references therein.
3. S.O. Farwell, F. A. Beland, and R. D. Geer, Reduction Pathways of

Organohalogen Compounds. Part II. Polychlorinated Biphenyls. *J. Electroanal. Chem. Interfacial Electrochem.* **61**, 315 (1975).
4. S.O. Farwell, F.A. Beland, and R.D. Geer, Interrupted-Sweep Voltammetry for the Identification of Polychlorinated Biphenyls and Naphthalenes. *Anal. Chem.* **47**, 895 (1975).
5. J. F. Rusling and C. L. Miau, Kinetic Estimation of Standard Reduction Potentials of Polyhalogenated Biphenyls. *Environ. Sci. Technol.* **23**, 476 (1989).
6. L. Pospisil, M. P. Columbini, R. Fuoco, and V. V. Strelets, Electrochemical Properties of Difenzoquat Herbicide (1,2-dimethyl-3,5-diphenyl-pyrazolium). *J. Electroanal. Chem. Interfacial Electrochem.* **310**, 169 (1991).
7. L. Pospisil, J. Hanzlik, R. Fuoco, and N. Fanelli, Growth of Compact Layers at the Interface. Part VI. Adsorption Properties of Difenzoquat Herbicide (1,2-dimethyl-3,5-diphenyl-pyrazolium). *J. Electroanal. Chem. Interfacial Electrochem.* **334**, 309 (1992).
8. O. Hammerich and B. Svensmark, Anodic Oxidation of Oxygen-Containing Compounds. In "Organic Electrochemistry" (H. Lund and M. M. Baizer, eds.), 3rd ed., Chapter 16. Dekker, New York, 1991.
9. Y. B. Vassiliev and B. M. Lotvin, Specific Features of Electrooxidation of Alcohols on Platinum in Absolute Alcohol Solutions. Role of Water in Chemisorption and Dehydrogenation Processes. *Electrochim. Acta* **30**, 1345 (1985).
10. E. Santos and M. C. Giordano, Electrocatalytic Oxidation of Organic Molecules in Alkaline Solutions. II. Electroadsorption and Electrooxidation in Ethylene Glycol at Platinum. *Electrochim. Acta* **30**, 871 (1985).
11. M. M. Baizer, Electrolytic Reductive Coupling. *In* "Organic Electrochemistry" (H. Lund and M. M. Baizer, eds.), 3rd ed., Chapter 22. Dekker, New York, 1991.
12. M. M. Baizer, Carbonyl Compounds. *In* "Organic Electrochemistry" (H. Lund and M. M. Baizer, eds.), 3rd ed., Chapter 10. Dekker, New York, 1991.
13. E. J. M. O'Sullivan and J. R. White, Electrooxidation of Formaldehyde on Thermally Prepared RuO_2 and Other Noble Metal Oxides. *J. Electrochem. Soc.* **136**, 2576 (1989).
14. K.-I. Machida, K. Nishimura, and M. Enyo, Amorphous Copper-Palladium-Zirconium Ternary Alloys for Formaldehyde Oxidation Anode Materials. *J. Electrochem. Soc.* **133**, 2522 (1986).
15. P. Bindra and J. Roldan, Mechanisms of Electroless Plating. II. Formaldehyde Oxidation. *J. Electrochem. Soc.* **132**, 2581 (1985).
16. M. Beltowska-Brzezinska, Electrochemical Oxidation of Formaldehyde on Gold and Silver. *Electrochim. Acta* **30**, 1193 (1985).

17. D. G. Peters, Halogenated Organic Compounds. In "Organic Electrochemistry" (H. Lund and M. M. Baizer, eds.), 3rd ed., Chapter 8. Dekker, New York, 1991.
18. B. Fleszar and J. Ploszynska, An Attempt to Define Benzene and Phenol Electrochemical Oxidation Mechanism. *Electrochim. Acta* **30**, 31 (1985).
19. R. L. McCarley, Characterization of Molecular Films on Electrode Surfaces Formed by the Electropolymerization of Phenols. *J. Electrochem. Soc.* **137**, 218C (1990).
20. G. Mengoli and M. M. Musiani, Protective Coatings on Iron by Anodic Oxidation of Phenols in Oxalic Acid Medium. *Electrochim. Acta* **31**, 201 (1986).
21. M. C. Pham, F. Adami, and P. C. Lacaze, In Situ Study by Multiple Internal Reflection Fourier Transform Infrared Spectroscopy (MIRFTIRS) of the Phenoxy Radical During Anodic Oxidation of Phenol Derivatives on Iron. *J. Electrochem. Soc.* **136**, 677 (1989).
22. G. Gunawardena, G. Hills, and I. Montenegro, Electrochemical Nucleation. Part IV. Electrodeposition of Copper onto Vitreous Carbon. *J. Electroanal. Chem. Interfacial Electrochem.* **184**, 357 (1985).
23. D. T. Napp, D. C. Johnson, and S. Bruckenstein, Simultaneous and Independent Potentiostatic Control of Two Indicator Electrodes: Application to the Copper(II)/Copper(I)/Copper System in 0.5 M KCl at the Rotating Ring-Disc Electrode. *Anal. Chem.* **39**, 481 (1967).
24. M. D. Benari and G. T. Hefter, Electrochemical Characteristics of the Copper(II)/Copper(I) Redox Couple in Dimethyl Sulfoxide Solutions. *Aust. J. Chem.* **43**, 1791 (1990).
25. S. Basak, P. S. Zacharias, and K. Rajeshwar, Binding and Surface Coordination Chemistry of Copper(II) Macrocycles at Nafion-modified Glassy Carbon Electrodes. *J. Electroanal. Chem. Interfacial Electrochem.* **319**, 111 (1991).
26. R. M. Sonto, M. Sluyters-Rehbach, and J. H. Sluyters, On the Catalytic Effect of Thiourea on the Electrochemical Reduction of Cadmium(II) Ions at the DME from Aqueous 1 M KF Solutions. *J. Electroanal. Chem. Interfacial Electrochem.* **201**, 33 (1986).
27. M. Saakes, M. Sluyters-Rehbach, and J. H. Sluyters, The Mechanism of the Reduction of Cadmium(II) Ions at a Dropping Mercury Electrode from $NaClO_4$ Base Electrolyte Solutions with Varied Water Activity. *J. Electroanal. Chem. Interfacial Electrochem.* **259**, 265 (1989).
28. N. Kaneko, H. Nezu, and N. Shinohara, Potential Oscillations During the Electrocrystallization of Cadmium from Alkaline Cyanide Solutions Under Galvanostatic Conditions. *J. Electroanal. Chem. Interfacial Electrochem.* **252**, 371 (1988).
29. V. D. Jovic, B. M. Jovic, and A. R. Despic, The Influence of Solution Com-

30. C.-C. Lai and Y. Ku, Effect of Chelate Formation on the Kinetics of Cadmium Electrodeposition in Citrate Solution. *Electrochim. Acta* **37**, 631 (1992).
31. J. P. Hoare, On the Mechanisms of Chromium Electrodeposition. *J. Electrochem. Soc.* **126**, 190 (1979).
32. J. Lin-Cai and D. Pletcher, The Electrochemical Study of a Chromium Plating Bath. II. Chromium Metal and Surface Film Formation. *J. Appl. Electrochem.* **13**, 245 (1983), and references therein.
33. K. Nishimura, H. Fukushima, T. Akiyama, and K. Higashi, Effect of Sulfate and Nickel on Chromium Electrodeposition. *Met. Finish.*, March, p.45 (1987), and references therein.
34. A. Bennouna, B. Durand, and O. Vittori, Influence des Ions Chlorure sur la Reduction Electrochimique du Chrome(VI). *Electrochim. Acta* **31**, 831 (1986).
35. S. B. Faldini, S. M. L. Agostinho, and H. C. Chagas, The Effect of Chloride Ions on the Electrochemical Reduction of Cr(VI) to Cr(II) at a Rotating Disk Electrode. *J. Electroanal. Chem. Interfacial Electrochem.* **284**, 173 (1990).
36. D. Sazou and G. Kokkinidis, Electrocatalytic Reduction of Chromate(VI) Ions in Alkaline Media on Pt Modified by Underpotential Deposition of Heavy Metals. *J. Electroanal. Chem. Interfacial Electrochem.* **271**, 221 (1989).
37. F. I. Danilor and A. B. Velichenko, Electrocatalytic Activity of Anodes in Reference to Cr(III) Oxidation Reaction. *Electrochim. Acta* **38**, 437 (1993).
38. P. K. Wrona, The Oxidation of Cr(0) at the Mercury Electrode. *J. Electroanal. Chem. Interfacial Electrochem.* **197**, 395 (1986).
39. P. K. Wrona, Electrochemical Behavior of Cr(II) and Cr(III) Ions in Weakly Acidic Solutions. *J. Electroanal. Chem. Interfacial Electrochem.* **322**, 119 (1992).
40. M. J. Weaver and F. C. Anson, Potential Dependence of the Electrochemical Transfer Coefficient. Further Studies of the Reduction of Chromium(III) at Mercury Electrodes. *J. Phys. Chem.* **80**, 1861 (1976).
41. A. T. Kuhn, ed., "The Electrochemistry of Lead." Academic Press, London, 1979.
42. H. C. Moser and A. F. Voigt, Dismutation of the Mercurous Dimer in Dilute Solutions. *J. Am. Chem. Soc.* **79**, 1837 (1957).
43. S. Fujita, H. Horii, T. Mori, and S. Taniguchi, Pulse Radiolysis of Mercuric Oxide in Neutral Aqueous Solutions. *J. Phys. Chem.* **79**, 960 (1975).
44. W. R. Cullen and K. J. Reimer, Arsenic Speciation in the Environment. *Chem. Rev.* **89**, 713 (1989).

45. L. Meites, Polarographic Characteristics of +3 and +5 Arsenic in Hydrochloride Acid Solutions. *J. Am. Chem. Soc.* **76**, 5927 (1954).
46. J. P. Arnold and R. M. Johnson, Polarography of Arsenic. *Talanta* **16**, 1191 (1969).
47. T. Ferri, R. Morabito, B. M. Petronio, and E. Pitti, Differential Pulse Polarographic Determination of Arsenic, Selenium and Tellurium at μg Levels. *Talanta* **36**, 1259 (1989).
48. C. McCrory-Joy and J. M. Rosamilia, Differential Pulse Polarography of Germanium(IV), Tin(IV), Arsenic(V), Antimony(V), Selenium(IV) and Tellurium(VI) at the Static Mercury Drop Electrode in Catechol-Perchlorate Media. *Anal. Chim. Acta* **142**, 231 (1982).
49. M. G. White and A. J. Bard, Polarography of Metal-Pyrogallol Complexes. *Anal. Chem.* **38**, 61 (1966).
50. D. Chakraborti, R. L. Nichols, and K. J. Irgolic, Determination of Arsenite and Arsenate by Differential Pulse Polarography. *Z. Anal. Chem.* **319**, 248 (1984).
51. M. S. Kazacos and B. Miller, Studies in Selenious Acid Reduction and CdSe Film Deposition. *J. Electrochem. Soc.* **127**, 869 (1980).
52. G. Jarzabek and Z. Kublik, Cyclic and Stripping Voltammetry of Se(+4) and Se(-2) at Carbon Electrodes in Acid Solutions. *J. Electroanal. Chem. Interfacial Electrochem.* **114**, 165 (1980).
53. E. O. Portnyagina, A. G. Stromberg, and A. A. Kaplin, Electrochemical Accumulation of Selenium in Stripping Voltammetry. *Zh. Anal. Khim.* **39**, 493 (1984).
54. K. K. Mishra and K. Rajeshwar, A Re-examination of the Mechanisms of Electrodeposition of CdX and ZnX (X = Se, Te) Semiconductors by the Cyclic Photovoltammetric Technique. *J. Electroanal. Chem. Interfacial Electrochem.* **273**, 169 (1989).
55. R. W. Andrews and D. C. Johnson, Voltammetric Deposition and Stripping of Selenium(IV) at a Rotating Gold Disk Electrode. *Anal. Chem.* **47**, 294 (1975).
56. G. D. Christian, E. C. Knoblock, and W. C. Purdy, Polarography of Selenium(IV). *Anal. Chem.* **35**, 1228 (1963).
57. G. D. Christian, E. C. Knoblock, and W. C. Purdy, Use of Highly Acid Supporting Electrolytes in Polarography-Observed Changes in Polarographic Waves of Selenium(IV) Upon Standing. *Anal. Chem.* **37**, 425 (1965).
58. P. Carbonnelle and L. Lamberts, A Voltammetric Study of the Electrodeposition Chemistry of the Cu + Se System. *J. Electroanal. Chem. Interfacial Electrochem.* **340**, 53 (1992).
59. K. K. Mishra and K. Rajeshwar, A Voltammetric Study of the Electrodeposition Chemistry in the Cu + In + Se System. *J. Electroanal. Chem. Interfacial Electrochem.* **271**, 279 (1989).

60. Y. Ueno, H. Kawai, T. Sugiura, and H. Minoura, Electrodeposition of CuInSe$_2$ Films from a Sulphate Bath. *Thin Solid Films* **157**, 159 (1988).
61. G. Mattsson, L. Nyholm, and Å. Olin, Cathodic Stripping Voltammetry of Cu$_2$Se at Mercury Electrodes. *J. Electroanal. Chem. Interfacial Electrochem.*, (in press).
62. J. J. Lingane and L. W. Niedrach, Polarography of Selenium and Tellurium. II. The +4 State. *J. Am. Chem. Soc.* **71**, 196 (1949).
63. A. M. Espinosa, M. L. Toscon, M. D. Vasquez, and P. S. Batanero, Electroanalytical Study of Selenium(+4) at a Carbon Paste Electrode with Electrolytic Binder and Electroactive Compound Incorporated. *Electrochim. Acta* **37**, 1165 (1992).
64. D. Liu, Y. Zhang, and S. Zhou, Studies on Electrochemical Reduction Mechanism of H$_2$SeO$_3$ by Rotating Ring-Disc Electrode. *J. Xiamen Univ., Nat. Sci.* **28**, 495 (1989).
65. C. Wei, N. Myung, and K. Rajeshwar, A Combined Voltammetry and Electrochemical Quartz Crystal Microgravimetry Study of the Reduction of Aqueous Se(IV) at Gold. *J. Electroanal. Chem. Interfacial Electrochem.* **375**, 109 (1994).
66. T. Hemmingsen, The Electrochemical Reaction of Sulphur-Oxygen Compounds. Part I. A Review of Literature on the Electrochemical Properties of Sulphur/Sulphur-Oxygen Compounds. *Electrochim. Acta* **37**, 2775 (1992).
67. P. W. T. Lu and R. L. Ammon, An Investigation of Electrode Materials for the Anodic Oxidation of Sulfur Dioxide in Concentrated Sulfuric Acid. *J. Electrochem. Soc.* **127**, 2610 (1980), and references therein.
68. G. Kreysa, J. M. Bisang, W. Kochanek, and G. Linzback, Fundamental Studies on a New Concept of Desulphurization, *J. Appl. Electrochem.* **15**, 639 (1985).
69. T. Hunger and F. Lapicque, Electrochemistry of the Oxidation of Sulfite and Bisulfite Ions at a Graphite Surface: An Overall Approach. *Electrochim. Acta* **36**, 1073 (1991).
70. I. P. Voroshilov, N. N. Nechiporenko, and E. P. Voroshilova, *Electrokhimiya* **10**, 1378 (1974).
71. M. R. Tarasevich and E. I. Khruscheva, Electrocatalytic Properties of Carbon Materials, *Mod. Aspects Electrochem.* **19**, pp. 295-358 (1989).
72. A. Benayada and J. Bessiere, Comportement Chimique et Electrochimique de SO$_2$ dans les Milieux Acides Phosphoriques Concentres, *Electrochim. Acta* **30**, 593 (1985).
73. E. Jacobsen and D. T. Sawyer, Electrochemical Reduction of Sulfur Dioxide at a Mercury Electrode, *J. Electroanal. Chem. Interfacial Electrochem.* **15**, 181 (1976).
74. Z. Samec and J. Weber, Study of the Oxidation of SO$_2$ Dissolved in 0.5 M

H_2SO_4 on a Gold Electrode - I. Stationary Electrode, *Electrochim. Acta* **20**, 403 (1975).
75. M. J. Foral and S. H. Langer, Characterization of Sulfur Layers from Reduced Sulfur Dioxide on Porous Platinum Black/Teflon Electrodes. *J. Electroanal. Chem. Interfacial Electrochem.* **246**, 193 (1988)
75a. T. Hemmingsen, The Electrochemical Reaction of Sulfur–Oxygen Compounds - Part II. Voltammetric Investigation Performed on Platinum. *Electrochim. Acta* **37**, 2785 (1992).
76. F. Magno, G. A. Mazzocchin, and G. Bontempelli, Voltammetric Behavior of Sulphur Dioxide at a Platinum Electrode in Dimethylformamide. *J. Electroanal. Chem. Interfacial Electrochem.* **57**, 89 (1974), and references therein.
77. D. Knittel, Electrolytically Generated Sulfur Dioxide Anion Radical $S_2O_4^-$, its Absorption Coefficient and Some of its Decay Reactions. *J. Electroanal. Chem. Interfacial Electrochem.* **195**, 345 (1985).
78. J. C. G. Thanos, An In-Situ Raman Spectroscopic Study of the Reduction of HNO_3 at a Rotating Silver Electrode. *J. Electroanal. Chem. Interfacial Electrochem.* **200**, 231 (1986).
79. K. Nishimura, K. Machida, and M. Enyo, On-Line Mass Spectroscopy Applied to Electroreduction of Nitrite and Nitrate Ions at Porous Pt Electrode in Sulfuric Acid Solutions. *Electrochim. Acta* **36**, 877 (1991).
80. X.-K. Xing and D. A. Scherson, Electrocatalytic Reduction of Nitrate Ions Induced by the Underpotential Deposition of Cadmium. *J. Electroanal. Chem. Interfacial Electrochem.* **199**, 485 (1986).
81. J. Garcia-Domenech, M. A. Climent, A. Aldaz, J. L. Vazquez, and J. Clavilier, Adsorption and Desorption of Lead at Polycrystalline Gold Electrode and its Effect on the Superimposed Reduction of Nitrate Anion. *J. Electroanal. Chem. Interfacial Electrochem.* **159**, 223 (1983).
82. C. Mayer, K. Jüttmer, and W. J. Lorenz, Influence of Lead and Thallium Underpotential Adsorbates at Silver Single Crystal Surfaces on Different Redox Reactions. *J. Appl. Electrochem.* **9**, 161 (1979).
83. K. Ogura and M. Takagi, Electrocatalytic Reduction of Carbon Monoxide with a Photocell. *J. Electroanal. Chem. Interfacial Electrochem.* **195**, 357 (1985).
84. G. Silvestri, S. Gambino, G. Filardo, G. Spardo, and L. Palmisano, The Electrochemistry of Carbon Monoxide Reductive Cyclotetramerization to Squarate Anion. *Electrochim. Acta* **23**, 413 (1978).
85. F. A. Uribe, P. R. Sharp, and A. J. Bard, Electrochemistry in Liquid Ammonia. Part VI. Reduction of Carbon Monoxide. *J. Electroanal. Chem. Interfacial Electrochem.* **152**, 173 (1983).
86. C. Fabyan, Die Kathodische Reduktion von Ozon in Alkalischen Electrolyten. *Monatsh. Chem.* **106**, 513 (1975).

87. R. E. Dessy, W. Kitching, and T. Chivers, Organometallic Electrochemistry: I. Derivatives of Group IV B Elements. *J. Am. Chem. Soc.* **88**, 453 (1965).
88. A. M. Bond and N. M. McLachlan, Oxidation Processes at Mercury Electrode for Tetraphenyllead and Related Compounds in Dichloromethane. *J. Electroanal. Chem. Interfacial Electrochem.* **182**, 367 (1985).
89. A. M. Bond and N. M. McLachlan, Oxidation Processes for Tetraethyl Lead and Tetramethyl Lead in Dichloromethane at Mercury Electrodes. *J. Electroanal. Chem. Interfacial Electrochem.* **194**, 37 (1985).
90. R. Gerdil and E. A. C. Lucken, A Polarographic and Spectroscopic Study of Dibenzothiophene and Some of its Isologs. *J. Am. Chem. Soc.* **88**, 733 (1966).
91. W. Stricks and S. K. Chakravarti, Amperometric Titration of Sulfhydryl and Disulfide Groups with Organic Mercury Compounds at the Rotated Dropping Mercury Indicator Electrode. *Anal. Chem.* **33**, 194 (1961).
92. R. K. Elton and W. E. Geiger, Jr., Analytical and Mechanistic Studies of the Electrochemical Reduction of Biologically Active Organoarsenic Acids. *Anal. Chem.* **50**, 712 (1978).
93. A. Watson and G. Svehla, Polarographic Studies on Some Organic Compounds of Arsenic. Part 1. Substituent Effects and the Arsonic Acids. *Analyst (London)* **100**, 489 (1975).
94. G. Spini, A. Profumo, and T. Soldi, Voltammetric Determination of Some Organic Compounds of Arsenic Reduced at the Mercury Electrode. *Anal. Chim. Acta* **176**, 291 (1985).
95. C. V. Bowen and F. I. Edwards, Jr., Polarographic Determination of O,O-Diethyl O-p-Nitrophenyl Thiophosphate (Parathion). *Anal. Chem.* **22**, 706 (1950).
96. K. S. V. Santhanam, L. O. Wheeler, and A. J. Bard, Electrochemistry of Organophosphorus Compounds. I. Electroreduction of Tris (*p*-nitrophenyl) Phosphate. *J. Am. Chem. Soc.* **89**, 3386 (1967); K. S. V. Santhanam and A. J. Bard, Electrochemistry of Organophosphorus Compounds. II. Electroreduction of Triphenylphosphine and Triphenylphosphine Oxide. *ibid.* **90**, 1118 (1968).
97. R. E. Dessy, T. Chivers, and W. Kitching, Organometallic Electrochemistry. III. Organometallic Anions Derived from Group V Elements. *J. Am. Chem. Soc.* **88**, 467 (1966).
98. J. Q. Chambers, *in* "Encyclopaedia of Electrochemistry of the Elements" (A. J. Bard and H. Lund, eds.), Vol. 12, p. 329. Dekker, New York, 1978.
99. J. Grimshaw, Electrochemistry of the Sulphonium Group. *in* "The Chemistry of the Sulfonium Group" (C. J. M. Stirling, ed.), Chapter 7, p. 142. Wiley, Chichester, 1981.

100. B. Svensmark, Anodic Oxidation of Sulphur-Containing Compounds, in "Organic Electrochemistry" (M. M. Baizer and H. Lund, eds.), Chapter 17, p. 519. Dekker, New York, 1984.
101. H. E. Imberger and A. A. Humffray, Anodic Oxidation of Diaryl Sulphides. I. Diphenyl Sulphide in Sulphate and Perchlorate Media. *Electrochim. Acta* **17**, 1421 (1972).
102. H. E. Imberger and A. A. Humffray, Anodic Oxidation of Diaryl Sulphides. II. Diphenyl Sulphide in Halide Media. *Electrochim. Acta* **17**, 1435 (1972).
103. F. Magno and G. Bontempelli, Electrochemical Behavior of Diphenyl Sulfide in Acetonitrite Medium at a Platinum Electrode. *J. Electroanal. Chem. Interfacial Electrochem.* **36**, 389 (1972).
104. G. Bontempelli, F. Magno, G.-A. Mazzocchin, and R. Seeber, Anodic Oxidation of Diphenylsulphoxide in Aprotic Solvent. *J. Electroanal. Chem. Interfacial Electrochem.* **55**, 109 (1974), and references therein.
105. P. T. Cottrell and C. K. Mann, Electrochemical Oxidation of Aliphatic Sulfides under Nonaqueous Conditions. *J. Electrochem. Soc.* **116**, 1499 (1969).
106. R. Gerdil, The Polarographic Reduction of Organic Sulphides in N,N-Dimethylformamide. *J. Chem. Soc. B*, p.1071 (1966).
107. G. Bontempelli, F. Magno, and G. A. Mazzocchin, Electrochemical Oxidation of Phenyldisulfide in Acetonitrile Medium. *J. Electroanal. Chem. Interfacial Electrochem.* **42**, 57 (1973).
108. P. T. Kissinger, P. T. Holt, and C. N. Reilley, Electrochemical Studies of Thioxanthene, Thioxanthone and Related Species in Nonaqueous Media. *J. Electroanal. Chem. Interfacial Electrochem.* **33**, 1 (1971).
109. H. E. Imberger and A. A. Humffray, Anodic Oxidation of Aryl Sulphides. IV. Thianthrene. *Electrochim. Acta* **18**, 373 (1973).
110. E. W. Tsai, L. Throckmorton, R. McKellar, M. Baar, M. Kluba, D. S. Marynick, K. Rajeshwar, and A. L. Ternay, Jr., Electrochemistry of Thioxanthene, Thioxanthone and Related Compounds in Acetonitrile. *J. Electroanal. Chem. Interfacial Electrochem.* **210**, 45 (1986).
111. D. S. Houghton and A. A. Humffray, Anodic Oxidation of Diaryl Sulphides. III. Dibenzothiophen in Sulphate, Perchlorate and Halide Media. *Electrochim. Acta* **17**, 2145 (1972).
112. G. Bontempelli, F. Magno, G.-A. Mazzocchin, and S. Zecchin, Cyclic and A. C. Voltammetry Study on Dibenzothiophene in Acetonitrile Medium. *J. Electroanal. Chem. Interfacial Electrochem.* **43**, 377 (1973).
113. J. Sroyl, M. Janda, I. Sliter, J. Kos, and V. Vyskoyil, Electrochemical Oxidation of Benzothiophenes. *Collect. Czech. Chem. Commun.* **43**, 2015 (1978).
114. T. Z. Polta and D. C. Johnson, Pulsed Amperometric Detection of Sulfur Compounds. I. Initial Studies at Platinum Electrodes in Alkaline Solutions. *J. Electroanal. Chem. Interfacial Electrochem.* **209**, 159 (1986).

115. T. M. Florence, Cathodic Stripping Voltammetry. Part I. Determination of Organic Sulfur Compounds, Flavins and Porphyrins at the Sub-Micromolar Level. *J. Electroanal. Chem. Interfacial Electrochem.* **97**, 219 (1979).
116. C. M. G. van den Berg, B. C. Househam, and J. P. Riley, Determination of Cystine and Cysteine in Seawater Using Cathodic Stripping Voltammetry in the Presence of Cu(II). *J. Electroanal. Chem. Interfacial Electrochem.* **239**, 137 (1988).
117. I. M. Koltoff and W. Stricks, Polarographic Investigation of Reactions in Aqueous Solutions Containing Copper and Cystine (Cystine). II. Reactions in Ammoniacal Medium in the Presence and Absence of Sulfite. *J. Am. Chem. Soc.* **73**, 1728 (1951).
118. A.-C. Le Gall and C. M. G. van den Berg, Cathodic Stripping Voltammetry of Glutathione in Natural Waters. *Analyst (London)* **118**, 1411 (1993).
119. A. E. Martell and R. M. Smith, "Critical Stability Constants," Vol. I. Plenum, New York, 1984.
120. B. C. Househam, C. M. G. van den Berg, and J. P. Riley, The Determination of Purine in Fresh and Sea Water by Cathodic Stripping Voltammetry After Complexation with Copper(I). *Anal. Chim. Acta* **200**, 291 (1987).
121. R. Srinivasan and R. De Levie, Condensed Thymine Films at the Mercury/Water Interface. Part II. Effects on Electrode Kinetics. *J. Electroanal. Chem. Interfacial Electrochem.* **201**, 145 (1986).
122. X. Chen, B. Xia, and P. He, Dynamics of the Electrocatalytic Oxidation of Cysteine at Nafion Film Coated Electrodes. *J. Electroanal. Chem. Interfacial Electrochem.* **281**, 185 (1990).
123. T. R. Ralph, M. L. Hitchman, J. P. Millington, and F. C. Walsh, The Electrochemistry of L-Cystine and L-Cysteine. Part I. Thermodynamic and Kinetic Studies. *J. Electroanal. Chem. Interfacial Electrochem.* **375**, 1 (1994).

SUPPLEMENTARY READING

1. J. Koryta, J. Dvorak, and L. Kawn, "Principles of Electrochemistry." Wiley, New York, 1993.
2. T. Shono, "Electroorganic Synthesis." Academic Press, London, 1991.
3. H. Lund and M. M. Baizer, eds., "Organic Electrochemistry." Dekker, New York, 1991.
4. J. A. Plambeck, "Electroanalytical Chemistry: Basic Principles and Applications." Wiley(Interscience), New York, 1982.
5. A. J. Bard, R. Parsons, and J. Jordan, eds., "Standard Potentials in Aqueous Solution." Dekker, New York and Basel, 1985.
6. M. Pourbaix, "Atlas of Electrochemical Equilibria." Pergamon, Oxford, 1966.

7. J.M. Costa, "Fundamentos de Electrodica." Alhambra Universidad, Madrid, 1981.
8. P.H. Rieger, " Electrochemistry." Prentice-Hall, New Jersey, 1987.
9. D.K. Kiriacov, "Basics of Electroorganic Synthesis." Wiley, New York, 1981.
10. R.D. Little and N.L. Weinberg, eds., "Electroorganic Syntesis: Fetschrift for Manuel M. Baizer." Marcel Dekker, New Tork, 1991.
11. P.T. Kissinger and W.R. Heineman, eds., "Laboratory Techniques in Electroanalytical Chemistry." Marcel Dekker, New York, 1984.
12. A.J. Bard and L.R. Faulkner, "Electrochemical Methods. Fundamentals and Applications." Wiley, New York. 1980.

CHAPTER FOUR

4.1. INTRODUCTION

Electrochemical methods for sensing pollutants may be categorized into those based on (a) potentiometric, (b) amperometric—coulometric, (c) voltammetric, and (d) conductometric approaches. Each of these will be discussed in turn. A further point of distinction pertains to whether the pollutant is gaseous (e.g., ozone, H_2S, CO, O_2, NO_x, SO_x) or whether it is confined to the solution phase. An electrochemical detector is also often a component of a chromatography system for separating the constituents of a complex mixture. Two such classes of composite instrumental systems, capillary electrophoresis and ion chromatography, have made spectacular advances in recent years. Alternatively, in scenarios wherein large numbers of similar samples are to be analyzed, a flow-injection analysis (FIA) system can be used in conjunction with electrochemical detection. Finally, this chapter explores situations in which electrochemical principles can be utilized not for detection purposes but in a supportive role, as for example, for speciation studies using furnace atomic absorption spectrometry.

Chromatography and atomic spectroscopy are the traditional workhorses in environmental analysis laboratories against which electroanalytical approaches will have to compete for routine adoption. In this regard, comparative studies are available. For example, HPLC with amperometric de-

Table 4.1. Relative Sensitivity of Some Electrochemical Techniques

Electrochemical Technique (Electrode)[a]	Limit of Detection for Lead/M
DC polarography (DME)	2×10^{-6}
DC polarography (SMDE)	1×10^{-7}
DP polarography (DME)	8×10^{-8}
DP polarography (SMDE)	1×10^{-7}
DP anodic stripping voltammetry (HMDE)	2×10^{-10}
SW anodic stripping voltammetry (HMDE)	1×10^{-10}
DC anodic stripping voltammetry (MTFE)	5×10^{-11}
DP anodic stripping voltammetry (MTFE)	1×10^{-11}
SW anodic stripping voltammetry (MTFE)	5×10^{-12}

[a] DC = direct current; DP = differential pulse; SW = square wave; DME = dropping mercury electrode; SMDE = static mercury drop electrode; HMDE = hanging mercury drop electrode; MTFE = mercury thin film electrode.

Source: From T. M. Florence, *Analyst (London)* 111, 489 (1986).

tection was used for the trace analysis of 3,4-dichloroaniline and found to outperform GC and HPLC/UV techniques.[1] Similarly, electrochemical and UV detection were compared for microcolumn HPLC determination of trace quantities of phenylurea herbicides in water.[2] Electrochemical detection was found to offer the advantage of higher selectivity when heavily contaminated water was used by detecting only the redox active components. Other studies report a 10 fold improvement in detection limit with electrochemical detection as opposed to UV absorbance and fluorescence methods.[3] Another pulsed amperometric detection (see Section 3.3) study of amino acid separation by anion-exchange chromatography reports greater sensitivity than ninhydrin derivatization with UV absorbance detection.[4] Obviously, it is difficult to make generalizations when comparing variant approaches; nonetheless two positive features with electroanalysis of pollutants can hardly be refuted: (a) they are more amenable to use at field sites because of their portability, and (b) they

4.1. Introduction

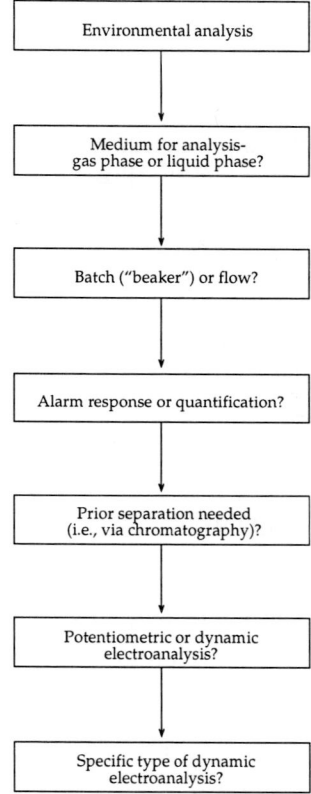

FIGURE 4.1. Decision tree in the adoption of electrochemical techniques for environmental analyses.

are likely to be more cost effective. Electroanalytical instrumentation is significantly less expensive than its spectroscopic and chromatography counterparts, and it is unlikely that this trend will change in the near future. The scales are tilted even more in favor of electrochemical methods when the current trend toward the use of hyphenated methods (e.g., ICP-MS) is considered.

Table 4.1 provides a taste of what can be achieved in terms of detection limits using electroanalyses.[5] However, these methods do suffer from several drawbacks. Perhaps the most vexing are problems associated with fouling of the sensing electrode surface. This has disastrous consequences for on-line continuous monitoring applications in terms of stable, long-term performance. While a variety of cleansing and surface regeneration strategies are being refined,[6] the solution may well lie with the use of low-cost, disposable electrochemical sensors. This certainly appears to be the trend in the related biosensor field.

Electrochemical methods appear to hold much promise for trace metal speciation applications.[5] Nonetheless, electrochemical methods in general do not afford knowledge on *individual* ion activities. Ion-selective electrode (ISE) potentiometry does have this virtue but suffers from rather poor sensitivity for trace analyses. Further, dynamic electroanalytical methods such as voltammetry and polarography (see Section 2.6.2) utilize current signals that may perturb ionic equilibria in the environmental medium to be sampled. However, this difficulty is not unique to electroanalyses.

Figure 4.1 contains a decision tree for the implementation of electroanalyses in environmental applications. The performance criteria widely vary. For example, monitoring of the performance of waste treatment equipment requires only an on–off alarm response from the sensor. On the other hand, "end-of-the-pipe" discharge monitoring and monitoring of drinking water quality require stringent performance from the sensor at ppm and ppb levels.

4.2. FLOW INJECTION ANALYSIS

The flow injection method is a relatively new concept for continuous flow analysis of discrete samples.[7] This method can be combined with any of the four types of electroanalytical techniques to be reviewed next. Flow injection analysis is based on the injection of small (usually a few μL) aliquots of a liquid sample into a moving, nonsegmented continuous carrier stream of a suitable liquid. The injected sample forms a zone (see Figure 4.2), which is then trans-

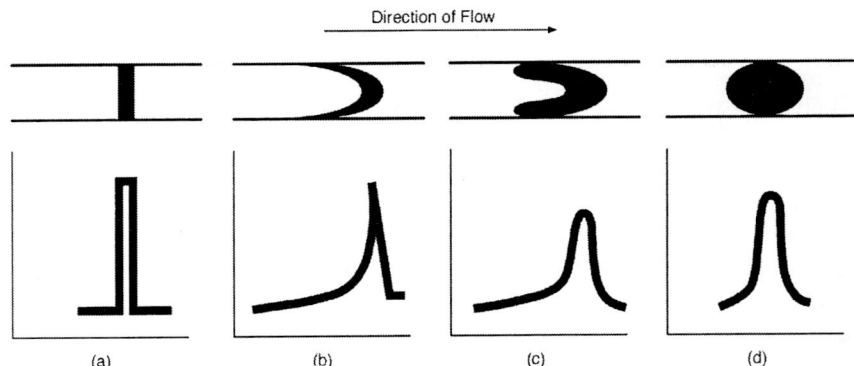

FIGURE 4.2. Diagrammatic representation of effects of convection and radial diffusion on concentration profiles of samples monitored at a suitable distance downstream from injection: (a) no dispersion; (b) dispersion predominantly by convection; (c) dispersion by convection; and (d) dispersion predominantly by diffusion. [Reproduced with permission from *Anal. Chem.* 50, 836A (1978)]

ported toward a detector. A peak-shaped response is usually obtained at the detector as the sample "plug" flows past it (see Figure 4.3). The simplest FIA system consists of a pump, an injection port, and a reaction coil in which the sample zone disperses (and often reacts with a derivatizing agent) in the carrier stream (see Figure 4.3). The requirements for the pump and the injection port are not unlike those used in a HPLC system. The tube dimensions (nominally a few mm internal diameter) and the volumetric flow rates employed (a few mL/min) are also comparable in the two cases. A major difference between FIA and HPLC is that the former operates around ambient pressures whereas substantially higher pressures (well over 70 atm) are employed in HPLC because the liquid has to be forced through a column in this case. The basic operating principle is also different in the two cases. Column separation is based on equilibrium partitioning and elution phenomena. In FIA, on the other hand, the sample zone travels with the same speed as the carrier stream through an open narrow tube usually at conditions approaching laminar flow. The key process in FIA is sample *dispersion*.

Figure 4.4 illustrates the concept of dispersion. The dispersion parameter D' is given by

$$D' = \frac{C^0}{C^{max}} = \frac{H^0}{H} \frac{const'}{const''} \tag{4.1}$$

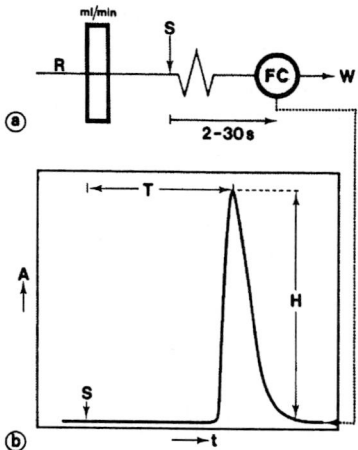

FIGURE 4.3. Single-line FIA manifold (a) with typical recorder output (b) as obtained with a spectrophotometric flow-through cell: R, carrier stream of a reagent; S, sample injection; FC, flow-through cell; W, waste; H, peak height; and T, residence time. (Reproduced with permission from J. Ruzicka and E.H. Hansen, "Flow Injection Analysis." Wiley, New York, 1981.)

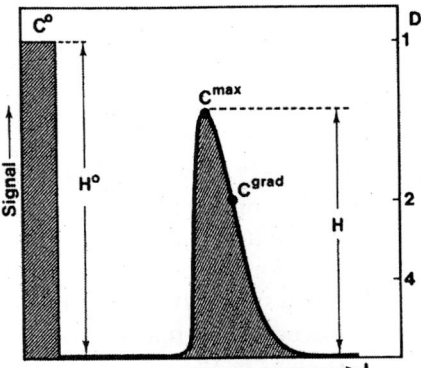

FIGURE 4.4. Dispersion D' in the FIA system defined as the ratio between the original concentration $C°$ and the concentration of the dispersed species (C^{max} or C^{grad}). (Reproduced with permission from J. Ruzicka and E.H. Hansen, "Flow Injection Analysis." Wiley, New York, 1981.)

Thus a value of $D' = 2$ corresponds to a 1:1 dilution of the sample with the carrier stream. Given that dispersion dictates the time between sample injection, considerable effort has gone into modeling dispersion in cylindrical tubes.[8-10] These models show that a bolus introduced into a flowing stream will be distorted from its original shape as a result of convective and dispersive forces. The Vanderslice model[10] numerically integrates the convection–diffusion equation for laminar flow, and affords estimates for t_i (the initial appearance time) and Δt (the base width of a peak after the injection of sample).

These expressions are as follows:

$$t_i = \frac{109\, a^2 D^{0.025}}{f} \left(\frac{L}{u}\right)^{1.025} \tag{4.2}$$

$$\Delta t = \frac{35.4\, a^2 f}{D^{0.36}} \left(\frac{L}{u}\right)^{0.64} \tag{4.3}$$

In Eqs. (4.2) and (4.3), a is the tube radius (cm), L is the tubing length (cm), u is the flow rate (mL/min), D is the diffusion coefficient of the sample (cm²/s), and f is a correction factor to take into account the variation of detector sensitivity to afford equal peak height under different experimental conditions (e.g., different flow rates).

The range of validity of Eqs. (4.2) and (4.3) and the underlying model assumptions have been discussed.[11] Experimentally, the simplest way for characterizing the dispersion in a given FIA system is to inject a well-defined volume of dye solution into a colorless buffer stream and continuously monitor the absorbance. The height of the signal (H) obtained under these conditions can then be compared with the signal recorded (H^0) when the flow cell is filled with undiluted dye (see Figure 4.4). An interesting *electrochemical* approach to this measurement using ultramicroelectrodes (e.g., electrodes of μm dimensions) has been described.[12]

The requirements for electrochemical detectors in FIA systems center mainly around the fact that if the analyte is not effectively transferred from the bulk of the solution to the diffusion layer and across the diffusion layer to the sensor surface, it simply will not be sensed. The detector geometries shown in Figure 4.5 illustrate this; the straight measuring channel in Figure 4.5a has the difficulty that is exacerbated if the sensor surface (S), because of faulty construction, is recessed in the wall. Wall-jet detector cells are more reliable and yield faster response and higher sensitivity (see later). These are commonly used for voltammetric detectors as we shall soon see. A cascade-type flow cell (Figure 4.5b) is a further extension of this principle.[13] A wire-type detector (Figure 4.5c)[14,15] samples the central, most rapidly moving section of the dispersion zone and has the advantage that the sensing surface is rapidly washed when the (distorted) sample bolus has moved downstream.

We next review the capabilities of each of the four classes of electroanalytical methods along with combined approaches for the monitoring of environmental pollutants.

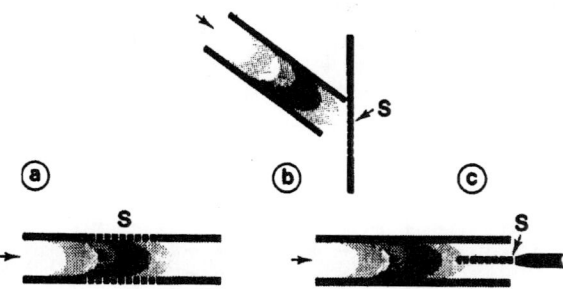

FIGURE 4.5. Electrochemical detectors (S, sensitive surface): (a) annular sensor; (b) cascade-type sensor; (c) wire-type sensor. (Reproduced with permission from J. Ruzicka and E.H. Hansen, "Flow Injection Analysis." Wiley, New York, 1981.)

4.3. POTENTIOMETRIC SENSORS

These types of sensors rely on the relationship between the voltage in an electrochemical cell and the concentration (strictly, the activity) of the chemical species in the sample (Section 2.6.1). In the broadest sense, the voltage measurement can be performed at any cell current; however, the vast majority of these sensors operate in the "zero current potentiometry" mode. A further point of subdivision is whether the sensor is a *symmetrical* device (e.g., ISE) or has an *asymmetrical* configuration of the selective membrane with respect to the sample. The latter has the advantage over the symmetrical ISEs of ruggedness. Potentiometric sensors can also be categorized into those utilizing an internal reference solution and solid-state sensors requiring no internal solution. Much work has been devoted in recent years to the development of the latter due to the ease with which these sensors can be miniaturized. Solid-state gas sensors are invariably asymmetrical. Table 4.2 contains a classification of potentiometric sensors based on the nature of the sensor membrane material. Many of these are now commercially available. Figure 4.6 schematizes the underlying principle of an ISE.

The most critical component of a potentiometric sensor is the nature of the membrane itself. The membrane electrode should equilibrate rapidly, should change only in response to an ion of interest, and preferably change linearly with the concentration of this species. The origins of the membrane potential are not fully understood but are believed to be related to Donnan equilibria (ion partitioning) on both sides of the membrane as well as diffusion through the membrane (see Section 2.6.1). The understanding of transport phenomena in polymer–membrane systems is an area of active research.

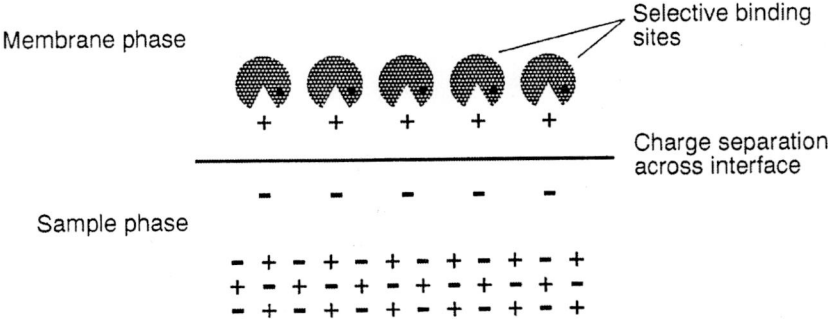

FIGURE 4.6. The ISE membrane extracts target ions leaving behind the co-ions. The resulting charge separation established across the sample–membrane interface generates a potential that is related to the concentration of the target ions in the sample. (Pac Man is a licensed TM, ©1980, Bally Midway Mfg. Co., all rights reserved. Reproduced with permission from Czaban[21].)

Table 4.2. Types of ISEs Based on the Nature of the Sensor Membrane Material

Analyte Ion	Membrane	Description
H^+, alkali metal cations, Ag^+	Glass	Usually formed from lithia, aluminosilicate or multicomponent glasses. Replacement of silica with alumina affords ISEs with increased selectivity towards alkali and silver ions
Various cations or anions (refer to text)	Liquid ion exchanger	An organic (hydrophobic) liquid phase incorporating mobile ions such as hydrophobic acids, bases, or salts is employed
Cations	Neutral carrier liquid	An organic solution of an electrically neutral ionophore is held in an inert polymer matrix
Anions and cations	Solid state	The membrane contains an insoluble salt of one of the species sensed or has sites capable of interacting with the analyte ion.

The selectivity of glass membranes made of Na_2O–Al_2O_3–SiO_2 mixtures depends on the composition of the glass and are used for the determination of hydrogen ion activity (pH). Some compositions are more selective to the alkali metal ions than toward the hydrogen ion and vice versa. At low pH, the electrodes primarily are sensitive only to the hydrogen ion, while at higher pH values responses of the alkali metal ions, sodium, lithium, and potassium become more influential. The pH glass electrode was the first commercial ISE and remains one of the most important standard laboratory devices.

Liquid membrane electrodes with electrically charged ion exchange sites generally show *permselectivity* to oppositely charged counterions. The selectivity between different counterions of the same charge is dictated mainly by the extraction behavior of the solvating membrane medium. The partition coefficients are small for counterions that are strongly hydrated in the aqueous phase. Therefore, the following selectivity ordering applies for cations and anions, respectively:

$$R^+ > Cs^+ > Rb^+ > K^+ > Na^+ > Li^+$$
$$R^- > ClO_4^- > I^- > NO_3^- \sim Br^- > Cl^- > F^-$$

Table 4.3. Representative Examples of Studies of Potentiometric Ion Sensors Based on Conducting Polymers and Chemically Modified Electrode Surfaces

Membrane/Coating	Analyte Ion(s)
Graphite modified with poly(pyrenamine)	H^+
Metal (Ir, Pb, Pd) oxides	H^+
Polyaniline or polyphenol	H^+
Poly(vinyl chloride) modified with a calixarene ionophore and a polypyrrole solid contact	Na^+
PVC modified for improved adhesion	NH_4^+, K^+
PVC modified with Sn(IV) and Mn(III) porphyrins	ClO_4^-, IO_4^-, SCN^-, I^-, etc.
Glassy carbon modified with electropolymerized Co(III) porphyrin	SCN^-, ClO_4^-, I^-, NO_3^-, etc.
Cyclodextrin	Anionic surfactants

In the presence of complexation between the ionic sites and the counterions, the potentiometric selectivity trends become more complex.

Neutral carrier membrane electrodes exploit the intrinsically outstanding specificity of certain natural and synthetic ionophores. The ion binding selectivity of such *electrically neutral* complexing agents can be fully exploited, unlike in the liquid ion–exchanger counterparts. For example, the selectivity of neutral carrier membranes among different cations of the same charge is virtually determined by the stability constants of the ion–carrier complexes involved. The following selectivity applies to the ionophore valinomycin in both biological and artificial membrane systems:

$$Rb^+ \gtrsim K^+ \gtrsim Cs^+ \gg Na^+ > Li^+$$

Solid-state membrane electrodes are used primarily as sensors for those kinds of ions that are constituents of the insoluble salt forming the membrane. In addition, they enable the detection of other species interacting with the ionic sites of the membrane material. Perhaps the best-known examples of this category are the silver halide membranes. These serve as sensors for silver ions, halide ions, and sulfide ions as well as for ligands that form stable complexes with the silver ion (e.g., SCN^-, CN^-). The selectivity sequence is as follows:

$$S^{2-} \gg I^- > Br^- \sim SCN^- > Cl^-$$

4.3. Potentiometric Sensors

Table 4.4. Examples of Non-glass Potentiometric Membrane Electrodes of Environmental Import

Analyte(s)	Active Material	Matrix/Membrane[a]
Al^{3+}, PO_4^{3-}	Manganese(III) phosphate, aluminum oxine	Silicone rubber
Tl^+	Tl salts of molybdo (or tungsto) phosphoric acid	Epoxy
Cd^{2+}	CdS + Ag_2S (or CdS)	—
Pb^{2+}	PbS + Ag_2S PbS	Polyethylene Silicone rubber
Hg^{2+}	HgS + Ag_2S (coated on Ag wire)	—
Cu^{2+}	CuS + Ag_2S $Cu_{1.8}Se$ CuS	— — —
F^-	LaF_3 crystal	Silicone rubber
Selenium	3,3'-diamino benzidine	Liquid membrane
NO_3^-	Nickel phenanthroline	Carbon paste
CN^-	AgI + Ag_2S AgI	— Silicone rubber
Cr(VI)	Iron phenanthroline	Liquid membrane
Surfactant	Detergent + quaternary ammonium chloride	PVC-coated Pt wire
	Ag_2S	Silicone rubber

[a] The matrix is shown only for the heterogeneous membrane cases. In the other cases, the membrane is homogeneous in composition as given by the second column.

Another advanced solid-state membrane is the lanthanum fluoride crystal doped with EuF_2 for the determination of fluoride. These electrodes exhibit a stable and reproducible response selective for fluoride, provided the solution pH remains neutral or lower.

The scope of solid-state membrane sensors has considerably broadened with the advent of new materials (e.g., electronically conductive polymers) and electrode surface modification schemes.[6,16] Table 4.3 contains some examples of the new possibilities. Finally, Table 4.4 contains further examples of potentiometric sensor designs, especially those of environmental import. The contents of Tables 4.2 – 4.4 show that the range and versatility of potentiometric sensor designs ultimately may be limited only by the ingenuity with which new chemical schemes may be devised as a basis for sensor operation.

An active area of research has centered on the question of transport rates across membranes containing ionophores and membrane morphological issues such as microviscosity. For example, studies of PVC membranes by nuclear magnetic resonance spectrometry indicate the existence of a water-rich surface region.[17] The ratio of such ionic sites to carriers appears to be an important factor in membrane selectivity.[18]

4.3.1. Speciation Studies

The virtues of ISEs for speciation analyses were alluded to in an introductory section. Copper serves as a good model system for illustrating this point. Copper(II) is present in natural water in a variety of forms: Cu^{2+}, $CuCO_3$, $Cu(CO_3)_2^{2-}$, $CuHCO_3^+$, $CuOH^+$, $CuCl^+$, and so forth. On the other hand, in seawater only 0.02 – 2% of dissolved Cu(II) is accounted for by inorganic species; the remainder is present as organic species.[19] Importantly, it appears that free Cu(II) and not the total Cu(II) is responsible for this metal's toxicity. Thus, environmental monitoring of marine water quality requires a *free* Cu(II) ion sensor. A Cu(II) ISE has the potential to fulfill this requirement although chloride ion intereference is a major problem with its implementation.

Most Cu(II) ISEs are based on the use of a chalcogenide layer (e.g. CuS, $Cu_{1.8}Se$ or copper/silver sulfide). The potential of this ISE is then dictated by the amount of Cu^+ released by a surface exchange reaction, for example,

$$CuS_{(s)} + Cu^{2+}_{(aq)} \rightleftharpoons S_{(s)} + 2\,Cu^+_{(aq)} \tag{4.4}$$

In the presence of chloride, however, stable copper(I)–chloro complexes are formed decreasing the level of Cu^+ and the electrode potential. A Nafion membrane coating has been found to alleviate the chloride interference in Cu(II) ISEs.[20] On the other hand, kinetic limitations of Reaction (4.4) have been claimed to also ameliorate the chloride interference.[19] These limitations arise from the very low concentration of free Cu(II) ($10^{-13} - 10^{-11}$ M) in seawater. Thus $Cu_{1.8}Se$ and CuS/Ag_2S electrodes have been reported to yield Nernstian response in the range $10^{-16} - 10^{-9}$ M free Cu(II) in Cu(II)–ethylenediamine buffers also containing 0.6 M NaCl.[19] Figure 4.7 contains representative data from this study.

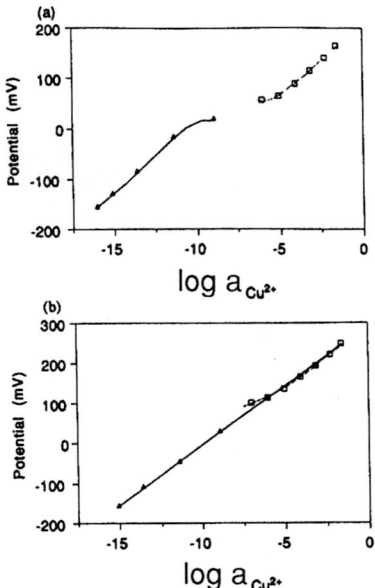

FIGURE 4.7. Potentiometric response curves of Cu(II) electrodes in Cu(II)–ethylenediamine (en) buffers: (a) $Cu_{1.8}Se$; (b) $Ag_{1.5}Cu_{0.5}S$. The graphical marks indicate the following: (□) bare electrode in chloride-free standards above 10^{-7} M Cu^{2+}; (Δ) bare electrode in Cu(II)-en buffers containing 0.6 M NaCl. (Reproduced with permission from DeMarco[19].)

4.3.2. Solid State and Gas Sensors

Potentiometric gas sensors have been a topic of intense activity in recent years. A major technological incentive for these studies is in the area of automotive engine combustion and exhaust emission monitors. In one type of sensor design, gas permeable membranes (e.g., Nafion and poly(ethylene oxide)) are used to separate the analyte solution from an internal reference solution. The gas diffuses through the membrane into the internal reference solution, which is in contact with an ISE. The electrolyte and ISE are chosen to introduce selectivity into the assay. A thin, semi-permeable membrane along with a thin layer of the internal reference solution ensures fast equilibration times. Figure 4.8 illustrates the operating principle of a potentiometric gas electrode with CO_2 shown as the test analyte.[21] In this design, CO_2 diffuses through the membrane, and then is hydrolyzed to carbonic acid. The dissociation of the latter generates protons and thus alters the pH of the buffer layer. This pH change is sensed by a pH-reference electrode combination.

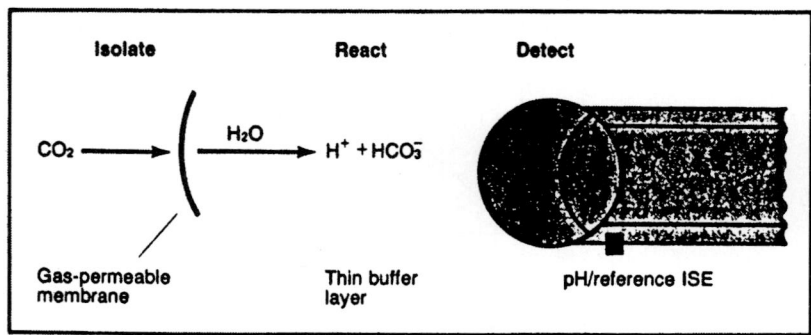

FIGURE 4.8. Schematic diagram of a potentiometric gas electrode. In the CO_2 electrode shown here, CO_2 diffuses through the membrane and is hydrolyzed to carbonic acid. The carbonic acid dissociates, the pH of the buffer solution is changed, and the pH change is monitored by the pH-reference electrode combination. (Reproduced with permission from Czaban[21].)

Solid-state gas potentiometric sensors have the advantage that they contain no water or liquid electrolyte and are thus amenable to miniaturization. The technological developments in this area have been made possible by the advent of solid electrolytes such as fast or super-ion conductors (e.g., NASICON). The selectivity in these all-solid-state devices is built in, using a suitable sensing layer composed of, for example, carbonates (for CO_2), sulfates (for SO_x), nitrates (for NO_x), or oxides (for O_2). Problems with suitable reference electrodes in these systems have been recognized and discussed.[22] Sensors with β-alumina as the solid electrolyte have been shown to exhibit Nernstian response to SO_x and NO_x at temperatures in the range 500–700°C and 150–220°C, respectively.[23,24] The temperature dependence of the response of similar arsine sensors has been analyzed.[25] Other electrolytes include the use of zirconia and lanthanide fluorides.

The rapid developments in microelectronics have spurred exploration of the utility of semiconductor-based devices for environmental sensing. Such ISFETs or CHEMFETs (the acronym, FET denotes field-effect transistor) can be used in place of ISEs. A major advantage with their use is their small size and their low-impedance output. Most ISEs have very high output impedance (ca. $10^{-5} – 10^9$ Ω), placing stringent requirements on the input impedance of the measuring instrument. Measurements on high-impedance devices are also prone to capacitive and electrostatic noise interference, necessitating special shielding and/or data treatment procedures. On the other hand, a major difficulty with the use of ISFETs is the preparation of ion-selective membranes on the FET surfaces. The devices also require frequent calibration and the signal is often not very stable. Nonetheless, complex bilayers and calix(4)arene/PVC membranes applied to the ISFET gate are claimed to yield a Nernstian response to Na^+ and anions, respectively.[26] Polyaniline–mercury composite forms a charge-transfer complex with gaseous HCN that can be measured potentiometrically either with a Kelvin probe or using a FET.[27]

Gas adsorption often alters the barrier height at semiconductor–metal Schottky diode contacts as a result of surface dipole effects.[28] This modulation can be exploited for sensing purposes in a potentiometric device.

4.3.3. *Flow Systems*

Flow-through systems employing ISEs can be used for continuous monitoring of environmental processes especially in the analysis of waters. They have often displaced the more complicated flame photometric and absorbance measurement-based detection systems, especially in FIA applications. Unlike spectrometric methods, measurements with ISEs are not affected by the original sample color or medium turbidity. The measurement is also nondestructive unlike flame photometry. The ISE-based FIA systems also feature a simpler manifold.

Flow cells with ISEs also suffer from several drawbacks. First is the response time. Under optimal conditions in sufficiently concentrated solutions of the analyte, where mass transport limitations are minimized, solid-state potentiometric sensors have a response time of a few ms. Liquid membrane-based ISEs have a corresponding time constant of a few seconds. However, at low analyte concentrations, the response time increases to tens of seconds or even several minutes. This situation generally can be ameliorated by enhancement of the analyte mass transfer (by employing high flow rates, turbulence, etc.), but the improvements are modest. Hence, ISEs are rarely used for HPLC detection. Another difficulty in this context is that related to the selectivity of the ISE electrode. Sensor selectivity is a virtue in continuous monitoring, but it is a drawback for HPLC detection where the detector must respond to *all* the eluents from the column!

A second difficulty with ISEs is associated with the reference electrode. Reference electrodes (especially those of the "second kind") are difficult to calibrate under flow conditions. They also exhibit poorly reproducible liquid junction potentials and are often prone to blockage at the bridge or frit. These problems are tackled in three ways. (1) The measurement is made without a reference electrode by employing two ISEs. This approach then yields the ratio of the activities of two ions, such as Na^+ and K^+. (2) A reference electrode with slow electrolyte leakage through a capillary junction can be used. (3) A reference electrode with the electrolyte immobilized in a gel or polymeric matrix is used. It is always good practice to place the reference electrode downstream from the ISE, to avoid intereference from the electrolyte leaking from the bridge.

A third difficulty with flow potentiometry is parasitic potentials. For example, streaming potentials (see Section 2.10) that arise from electrokinetic phenomena may amount to several hundred mV in worst-case scenarios. The effect of streaming potentials decreases with increasing concentration of the supporting electrolyte and is usually negligible at concentrations higher than ~0.1 M, provided that the flow rate and the liquid composition are constant. Peristaltic pumps (often used in FIA systems) cause considerable fluctuation of the potentials. This problem can be solved by the use of pulseless pumps or alleviated to an extent by electrical isolation of the pump from the detector cell or by the placement of a grounding electrode in the liquid stream.

Flow cells for use with ISEs were discussed earlier and can vary from the simplest flow-through designs to more complicated versions. Representative designs are contained in Figure 4.5. The wall-jet design is widely used for electrochemical flow detectors (see earlier). Note that, in the cascade-type cell (Figure 4.5b), the reference electrode is placed in the electrolyte and the ISE is *not* in contact with the latter.[13] Electrical contact is provided by a thin film of liquid flowing along the ISE membrane. Coated-wire ISEs, in which the internal reference solution is eliminated, have the advantage for flow measurements of better selectivity and amenability to miniaturization.[29] However, their response often is unstable. Other designs of low-volume, flow-through potentiometric electrodes include an ISE fitted with a plastic cap with an inlet and outlet for the flow stream.[30,31] Multi-ion detectors based on ISFETs have also been combined with a miniaturized FIA system that was assembled by micromachining.[32]

Potentiometric detection has been combined with ion chromatography,[33] open-tubular column liquid chromatography,[34] and capillary zone electrophoresis.[35] Along with miniaturization, another important development has been the advent of ISE arrays.[36] This multiple detector approach partially circumvents the difficulty with the use of ISE detectors in liquid chromatography of selectivity (see earlier). For example, an array ISE detector can be designed for response to a targeted set of analytes along with a sparingly selective electrode as a "universal" detector.[37,38] An array of five ISEs and one reference electrode was used to eliminate Cl⁻ and Br⁻ interferences in the environmental determination of Hg in water.[39] Chemometrics can be a useful tool in the interpretation of data from array electrodes.[40–42] A review of electrochemical sensor arrays is available.[42]

4.3.4. Examples of Environmental Applications of Potentiometric Sensors

Bromide, chloride, and fluoride ISEs were used for the analysis of snow and rain accumulated at different locations in and around industrial Hamilton, Ontario.[43] Seasonal variations in the aerosol composition were registered and

4.3. Potentiometric Sensors

attributed to variables such as the origin and physical character of storms and continental versus maritime influences in Canada. Bromide and chloride values were reasonably constant, while high fluorides correlated with localized industrial pollution.[43]

Ion-selective electrodes have been widely employed for measuring fluoride, nitrate, cyanide, sulfide, carbonates, and ammonia/ammonium in potable waters, rivers, estuaries, effluents, boilerfeeds, stream condensates, and seawater. These analyses are relatively inexpensive and, more important, unlike optical assays are unaffected by sample color and turbidity. The hydrochemical conditions in natural waters can be conveniently characterized via the master variables pH and pe (pe = $-\log_{10} a_e$, where a_e is the electron activity). Commercial pH and pe probes have problems in certain aquatic environments associated with the temperature and pressure sensitivity of reference electrode potentials, and the susceptibility of the latter to hydrogen sulfide interference. (Sulphide bearing waters, for example, occur in the hypolimnion of stratified lakes, anoxic fjords, and the Black Sea.) A further operational difficulty is the transmission of the very weak glass electrode output signals. These difficulties have been recently addressed via the development of a new liquid–junction potentiometric probe for the *in situ* monitoring of pH, pH_2S, and redox potential.[44] Field data from Lake Kinneret, Israel are reported in this study.

Fluoride is an environmental double-edged sword that needs monitoring, since it is harmful at levels >1.2 – 1.5 ppm in potable waters, while too little promotes dental cavities. Industrial fallout rapidly increases the fluoride levels in local vegetation and surface waters. An early study[45] on different types of waters (city, well, sea, and waste) evaluated several techniques for fluoride assay and concluded that the potentiometric approach surpassed the others in accuracy, convenience and recovery capacity.

Total inorganic nitrogen as ammonia/ammonium, nitrate, or nitrite is of considerable analytical importance in pollution and process control. High nitrate and low ammonia in river waters, for example, are diagnostic of an oxidative environment. An analytical problem often encountered is that of nitrate reduction (to nitrite) upon storage. Thus, water analyses could give acceptable nitrate levels that are grossly misleading in that toxic nitrites may be present due to delayed analysis after sampling.

Chloride in boilerfeeds and steam condensates causes corrosive acid conditions and must be kept below specified limits. Thus, chloride ISEs can be profitably employed in these continuous analysis scenarios. A related application is monitoring of sodium water levels in power stations. The detection of sodium in boilerfeeds and condensates is important for diagnosing the ingress of cooling water from condenser leaks. This may be accomplished via a sodium ISE.

At one stage in Kraft pulping, highly alkaline spent liquor is oxidized to offset sulfur losses and reduce the release of volatile odors. Potentiometry has been used for monitoring the species evolving during oxygen or air sparging of black liquors.[46] These authors concluded that simple inorganic sulfides (which are essentially uncomplexed) are rapidly oxidized first while the organosulfur compounds (mercaptans, alkyl sulfides, etc.) are oxidized much more slowly.

Plating baths must be maintained within strict specification limits to assume trouble-free and high-quality performance. The ISE approach can be used for monitoring practically all ions in industrial plating baths. These include cyanide, chromate, fluoride, copper, chloride, and sulfate.

Potentiometric detection can be combined with a FIA system for handling large numbers of samples. This combined approach has been used for the determination of nitrate and total nitrogen in environmental samples,[47–49] calcium and sulfide in potable and natural waters,[50] and for the indirect assay of sulfate with a lead ISE.[51] Figure 4.9 contains a schematic of a FIA system for the determination of free cyanide in flow streams based on diffusion of gaseous HCN through a gas-permeable membrane and subsequent indirect determination at a silver ISE.[52] A detection limit of ~4 ppb is quoted for this technique. The heart of the system is the gas-permeable membrane that sequesters the HCN from the acidified sample solution into the detector cell. A microporous polytetrafluoroethylene membrane was found to afford the best permeability and sensitivity of the various membranes that were tested in this study.[52]

Other examples include a flow-through potentiometric sensor based on a PVC-membrane for the continuous flow monitoring of nitrate in ground water,[52a] a chalcogenide glass sensor for the determination of Tl^+ in natural and wastewater,[52b] and an ISFET devise based on a PVC–sebacate membrane that is selective to anionic detergents.[52c]

Potentiometric gas sensors of the type illustrated in Figure 4.8 are simple and reliable, but they tend to have rather slow response and recovery time (a few seconds to even minutes). The two most advanced sensors in this category are the ammonia/ammonium and the CO_2 gas electrodes. The ammonia gas electrode works by a principle similar to its CO_2 counterpart; that is, the NH_3 molecules diffuse through a membrane and alter the pH of the internal filling solution via hydrolysis. Amines will cause an interference as will a nonaqueous sample or one containing surfactants, which "wet" the (hydrophobic) membrane and allow liquid diffusion.

The determination of ammonium ion in airborne particulates is significant since atmospheric ammonium sulfate is related to health, corrosion, and visibility problems. The colorimetric assay for ammoniacal nitrogen based on indophenol dye generation is fraught with interference problems from sample components and reagents invariably containing nitrogen contaminants. Thus ammonium particulates and ammonia in cigarette smoke have been measured

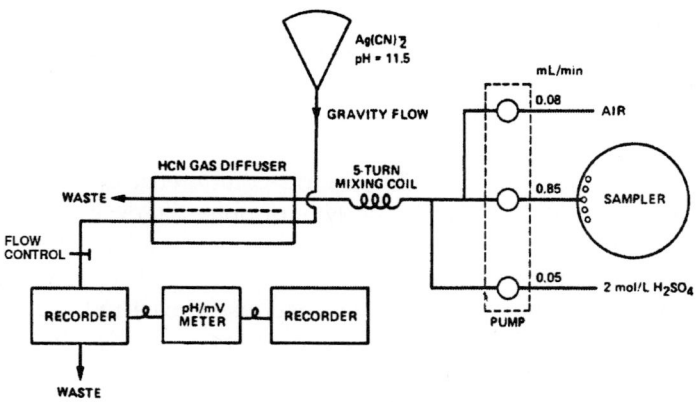

FIGURE 4.9. Schematic diagram of the cyanide flow analyzer. (Reproduced with permission from Durst.[52])

with gas sensor electrodes.[53,54] Ammonia in cigarette smoke exists as NH_4^+ species since the pH of such medium varies from 5.4 to 6.4.

Oxides of nitrogen (NO_x) are singularly important species in the nitrogen cycle (see Section 1.3.1). Over 60% of the gaseous NO_x, largely as nitric oxide, emitted in the United States is from stationary combustion sources, although automobile exhaust also contributes to atmospheric NO_x pollution. These gaseous nitrogen species can be converted to nitrate and then measured with a nitrate ISE. Alternately, nitrate can be in turn reduced to nitrite or ammonia and then measured with the appropriate potentiometric sensor. The use of sensors employing solid electrolytes for NO_x determination was referred to earlier in Section 4.3.2. A ceramic sensor employing NASICON as the solid electrolyte and $NaNO_2$ solid electrode as the active material has been reported for the low-temperature sensing of NO_x.[55]

Sulfur dioxide levels in flue gases can often exceed 5,000 ppm. The content of several flue gases has been indirectly measured after adsorption and oxidation to sulfate by potentiometric titration with lead perchlorate.[56] Alternatively, sulfate could be determined with barium perchlorate to a potentiometric endpoint assay with a barium ISE.[57] This brings up an important point: potentiometric titration often yields more reliable results than direct potentiometry since the assay is based on many potential readings and standards are not needed.

Examples of other gaseous pollutants determined by potentiometry include arsine and H_2S. Thus an arsine sensor has been developed based on silver β-alumina as solid electrolyte; the estimated detection limit is 0.05 ppm.[58]

4.4. AMPEROMETRIC–COULOMETRIC AND VOLTAMMETRIC–POLAROGRAPHIC DETECTION IN FLOW SYSTEMS

4.4.1. Cell Design: General Considerations

As reviewed in Chapter 2, amperometric–coulometric and voltammetric–polarographic approaches differ in the type of potential excitation applied to the cell. In the former case, a constant potential is employed; in the latter case, either a potential ramp or various pulse routines are applied (see Figure 2.17). In amperometry and voltammetry, the response measured is the current, whereas in coulometry, charge is the measured parameter. Another distinction between amperometry or voltammetry, on the one hand, and coulometry, on the other, relates to the extent of electrolyte conversion of the substrate at the working (indicator) electrode surface. In the former, the degree of conversion is low (a few percent) whereas 100% conversion is targeted in coulometry. In this regard, coulometry is advantageous in that, if complete conversion can be achieved and side reactions eliminated, the measurement can be *absolute* (i.e., no calibration is required). Unfortunately, this ideal is seldom achieved in practice.

All four techniques are employed most often in conjunction with a flow stream, and usually as a detector for liquid chromatography column or FIA system. Therefore, we shall dicuss aspects related to hydrodynamics and cell design as one unit in this section. Some general criteria can be identified at the outset; the flow cell detector must
- Afford a high signal-to-noise ratio
- Have a low dead volume
- Exhibit well-defined hydrodynamics
- Have small ohmic drop
- Enable easy dismantling, electrode cleaning, polishing, and the like.

As with the potentiometric detectors discussed earlier, the reference (and counterelectrodes) must be located on the downstream side of the indicator electrode to avoid interference from reference electrode solution leakage and from a new complication not present in the potentiomery case; namely, the influence of products generated at the counterelectrode.

An important factor in the design of a coulometric flow cell relates to the requirement of complete electrolysis. This necessitates a large working electrode area/cell volume ratio. The mobile phase flow rate must also be very low if LC detection is the task at hand. These requirements are not compatible with optimal separation in the LC column. On the other hand, the advent of microbore and capillary LC columns and microporous working electrode materials (e.g., reticulated vitreous carbon) has revived interest in coulometric flow detection. Other important advantages with coulometric detection include those

4.4. Amperomatric-Coulometric and Voltammetric-Polarographic Detection

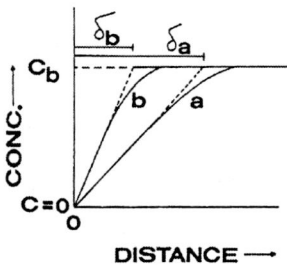

FIGURE 4.10. Concentration–distance profiles at a working-electrode/electrolyte interface under two different hydrodynamic conditions: slow (a) and fast (b) rates of convective transport. Refer to Figure 2.25. (Reproduced with permission from J. Wang, "Analytical Electrochemistry." VCH, New York, 1994.)

related to the flow-rate independence of the measured signal and electroactive derivative generation for the determination of electroinactive solutes in a flow stream. Unlike in amperometry where the current is often very sensitive to variations in the mobile phase flow rate, a time-integrated signal is accumulated in coulometric detection.

In Chapter 2, the influence of substrate depletion at the working electrode surface on the shape of the current–potential curve was discussed. We again ignore the influence of electron-transfer kinetics at the electrode/electrolyte interface (i.e., we assume a Nernstian regime) and examine the effect of convective diffusion imposed by motion of the fluid relative to the working electrode surface. The relevant expression for the mass-transfer limited regime of a hydrodynamic voltammogram is

$$I_L = \frac{nFAD_O C_O^*}{\delta} \tag{4.5}$$

Again, the reduction of an electroactive species O is assumed and I_L is the corresponding limiting current. The parameter δ is the diffusion layer thickness (see Figure 4.10).

Suppose that the electrolyte is now stirred at a rate U. The bulk concentration (C_o^*) is maintained at a *fixed* distance from the electrode surface by the stirring rate, that is, the diffusion layer is "pinned" by the value of U. An increase in U shrinks the diffusion layer thickness according to the expression

$$\delta = \frac{B}{U^\alpha} \tag{4.6}$$

where B and α are constants for a given system of electrode geometry and

Table 4.5. Limiting Current Response of Various Working Electrode Configurations

Electrode Geometry	Limiting Current Expression[a]
Rotating disk	$I = 0.62\, nFC^* A\, D^{2/3} \omega^{1/2} \nu^{-1/6}$
Tubular	$I = 1.61\, nFC^* (DA/r)^{2/3} U^{1/3}$
Planar (parallel flow)	$I = 0.68\, nFC^* D^{2/3} \nu^{-1/6} (A/b)^{1/2} U^{1/2}$
Planar (perpendicular flow)	$I = 0.903\, nFC^* D^{2/3} \nu^{-1/6} A^{3/4} U^{1/2}$
Thin layer	$I = 1.47\, nFC^* (DA/b)^{2/3} u^{1/3}$
Wall-jet	$I = 0.898\, nFC^* D^{2/3} \nu^{-5/12} a^{-1/2} A^{3/8} U^{3/4}$

Source: Adapted from H. B. Hanekamp, P. Bos and R. W. Frei, *Trends Anal. Chem.* **1**, 135 (1982); and J. Wang, "Analytical Electrochemistry." VCH, New York, 1994.

[a] a = diameter of inlet; A = electrode area; b = channel height; C^* = bulk concentration (mM); D = diffusion coefficient; ν = kinematic viscosity; r = electrode radius; U = average volume flow rate; u = flow velocity; ω = electrode rotation rate (rad/s).

electrolyte hydrodynamics (see later). Consequently, the concentration gradient becomes steeper (compare curves *a* and *b* in Figure 4.10) and the limiting current increases proportionately.

The fluid velocity is not zero within the diffusion layer but increases from a value of zero at the electrode surface ($x = 0$) to the value in the solution bulk. This hydrodynamic boundary layer, δ_H, termed the *Prandtl* layer (see Section 2.8) is given by

$$\delta \approx (D_o/\nu)^{1/3} \delta_H \qquad (4.7)$$

where ν is the kinematic viscosity. For water, $\nu = 10^{-2}$ cm²/s and if we assume a nominal D_o value for substrates in aqueous media (~10^{-5} cm²/s), δ_H is seen to be ca. 10-fold larger than δ, indicating negligible convection within the diffusion layer (Figure 2.25).

Substitution of Eq. (4.6) into (4.5) yields a general current response expression for flow systems:

$$I_L = nFA\, k_m\, C_o^*\, U \qquad (4.7a)$$

where k_m is the mass-transport coefficient. A more rigorous approach utilizes

4.4. Amperometric-Coulometric and Voltammetric-Polarographic Detection

solution of the three-dimensional convective diffusion equation with appropriate initial and boundary conditions. The resultant equations for a variety of electrode geometries (discussed next) are summarized in Table 4.5.

4.4.2. Specific Cell Designs and Electrode Geometries

Thin-layer (channel) and wall-jet configurations perhaps represent the two most popular designs for electrochemical flow detection. Representative cells of this type are illustrated in Figure 4.11 along with the tubular geometry (see Table 4.5) and a flow-through cell incorporating a Hg-coated rotating disk working electrode.

The thin-layer cell[59,60] relies on a thin layer of solution that flows parallel to the (planar) electrode surface, which is imbedded in the flow channel (Figure 4.11a). The flow channel is defined by two plastic blocks pressing against a Teflon gasket affording a small dead volume (~1 μL). The placement of the counter- and reference electrodes relative to the indicator electrode is crucial to maintaining a low ohmic drop, and cell designs to optimize this factor have been described.[61]

The term wall-jet was first introduced to describe the flow due to a jet of fluid that spreads out over a planar surface, the fluid outside the jet being at rest (Figure 4.11b).[62,63] Application to electrochemical detection in flow streams soon followed.[64] The liquid is often restricted from flowing in all directions along the electrode surface (as in the classical design in Figure 4.11b) in order to obtain the shortest possible distances (and hence the smallest ohmic drop) between the three electrodes. The modified wall-jet design in Figure 4.12 exemplifies this.[64] An ingenious detector design, which makes it possible to rapidly switch from a wall-jet configuration to thin-layer geometry by simply turning the bottom part of the cell 180°, has also been described.[65] Large-volume cells have also been described using the wall-jet geometry to avoid problems associated with the precise machining and polishing of small flow channels.[66,67]

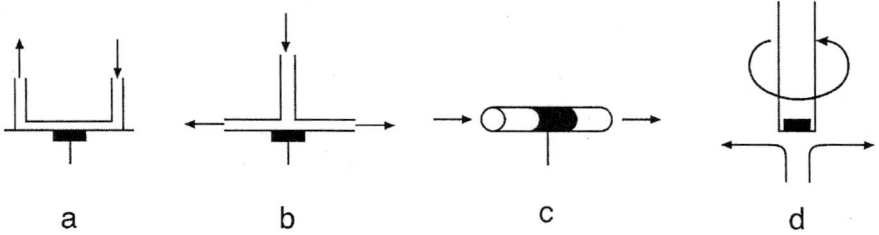

FIGURE 4.11. Schematic diagrams of thin-layer (a), wall-jet (b) tubular (c), and rotating (d) working electrode configurations.

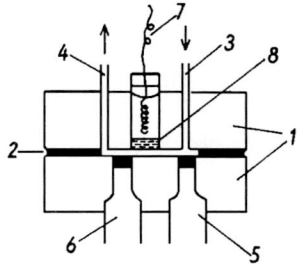

FIGURE 4.12. Modified wall-jet cell assembly: 1, PTFE blocks; 2, PTFE spacer; 3, inlet; 4, outlet; 5, working electrode; 6, counter electrode; 7, reference electrode; 8, porous glass. (Reproduced with permission from Fleet and Little.[64])

However, these cells present rather large dead volumes (several µL), limiting the range of liquid flow rates that can be used.

Tubular electrode designs (Figure 4.11c) have not been used as often as the thin-layer and wall-jet geometries for HPLC detection. This is possibly because of difficulties in making them small enough for routine and reproducible use. Tubular electrodes are also more difficult to clean and maintain. Nonetheless, tubular cell geometry is particularly compatible with coulometric detection. A variety of tubular cell designs have been discussed in the literature.[68-72]

The rotating disk electrode configuration in Figure 4.11d offers a powerful approach for diagnostic studies on electron-transfer kinetics at bare and chemically modified electrode surfaces. The *Levich* equation (Table 4.5) provides the underlying theoretical framework for analysis of RDE voltammetry data. The application of RDE geometry to the flow-through cell was prompted by consideration of the dependence of the analytical signal on the flow rate.[73] The rationale here is that, since mass transport to the electrode surface is governed primarily by the electrode rotation speed, the analytical sensitivity is not severely compromised even at low fluid flow-rates.

Polarographic detectors employing the dropping mercury electrode (DME) have been constructed, ranging from rather large cells for process stream monitoring to smaller ones for HPLC detection. In spite of the key advantage of the periodically renewed electrode surface, these detectors are less often used in practice than their counterparts employing solid electrodes. This is because of their rather complicated construction and problems with mechanical failure (e.g., capillary blockage) during routine use. Nonetheless, a variety of designs have been described,[74-76] and a review is available.[77] Detector cells with stationary mercury electrodes have much simpler construction, and both stationary Hg drop (Kemula design)[78,79] and Hg film electrodes can be employed[80-82] for flow detection. Of course, these do have the disadvantage of a narrowed posi-

4.4. Amperomatric-Coulometric and Voltammetric-Polarographic Detection

tive potential window and are best suited for electroreducible solutes.

Finally, the cylindrical electrode geometry based on the use of wire (or fiber) working electrodes has been claimed to eliminate most of the problems associated with the manufacture of low-volume detector cells. The effective detector volume can be varied by varying the fiber length. Other attractive features include simplicity in detector geometry and very high analytical sensitivity (pg detection limits are easily attained) coupled with flow-rate independence and low noise levels. The drawbacks include those associated with the fabrication and polishing of these *ultramicroelectrodes*. Figure 4.13 shows an example of a carbon-fiber based amperometric detector for use with a capillary zone electrophoresis system.[83] Electrochemical detection in microcolumn separations has been recently reviewed,[84] and a variety of microdetector cell geometries (with nL volume) have been described.[85-92]

The performance of several voltammetric detectors, namely a tubular detector with a Pt working electrode and thin-layer and wall-jet geometries with

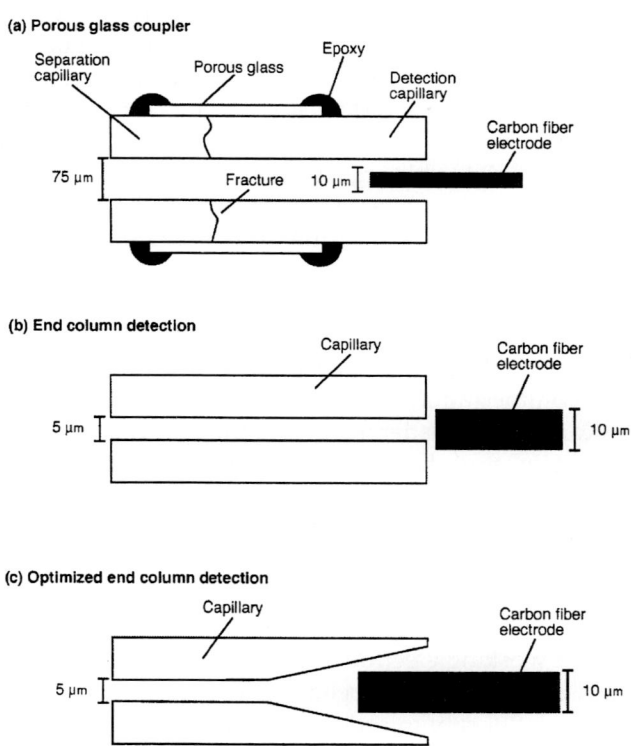

FIGURE 4.13. Electrochemical detection schemes for capillary zone electrophoresis. (Reproduced with permission from Ewing.[84])

glassy carbon and graphite paste has been compared, using DC and differential pulse excitation.[93] The parameters studied were the residual (background) current, accessible potential windows, the detection limit, sensitivity, reproducibility, the time constant, the response volume, and the dependence of the signal on the flow rate. The theoretical response (see Table 4.5) was also calculated for a test substrate (adrenaline) and compared with the experimental values. The detectors exhibited somewhat lower sensitivities than theoretical predictions. This was attributed to errors in the latter caused by uncertainty in the magnitude of parameters in the theoretical expressions (e.g., D), the effect of analyte adsorption, and deviations from laminar flow in the thin-layer cell geometry.

Strategies for improving the sensitivity and detection limits in flow voltammetric detection and sources of noise in LC-EC detection systems have been discussed.[94,95] A recent review also describes the principles, requirements and applications of LC-EC systems.[96]

We close this section with a discussion of coulometric detectors and predetector reactors. Classical designs employing large-area *planar* electrodes lead to unacceptably large dead volumes and analyte zone broadening in FIA and HPLC applications. However, such designs are still useful as predetector reactors. For example, the signal-to-noise (S/N) ratio of a polarographic detector can be improved by the insertion of an upstream coulometric cell, where dioxygen, metals, and reducible organics are reduced.[97] Similarly, an upstream coulometric cell has been combined with voltammetric detector cells for stripping analysis applications.[98,99] Other dual coulometric-amperometric cell combinations have been described.[100,101]

The advent of porous flow-through electrodes[102–104] and high surface area materials (e.g., RVC) has changed the situation regarding the scope and applicability of coulometric detectors. For example, the use of RVC provides a large surface area-to-volume ratio, which facilitates rapid coulometric conversion without significant band broadening in FIA and HPLC detection.[105] The sensitivity was found to approach the theoretical value for this cell design at flow rates less than 1.0 mL/min, and detection limits in the picomole range were reported.[105]

4.4.3. Dual-Electrode and Array Detectors

The first dual working electrode cell appears to have been developed in 1976,[106] and this design has since been applied to thin-layer and wall-jet cell geometries.[107] The latter configuration has many features common to the rotating ring-disk electrode (RRDE) geometry[108] and is therefore termed *wall-jet ring-disk electrode* (WJRDE) or simply *wall-jet electrode with collection*.[109] There are three possible arrangements of two working electrodes in a flow stream: par-

4.4. Amperometric-Coulometric and Voltammetric-Polarographic Detection

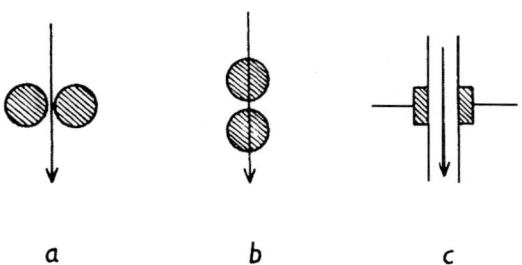

FIGURE 4.14. The parallel–adjacent (a), series (b), and parallel–opposed (c) arrangement of two working electrodes in a thin-layer cell.

allel–adjacent, series, and parallel–opposed (Figure 4.14). All three arrangements are built around the thin-layer cell geometry.[110] On the other hand, the WJRDE affords only a series configuration of the disk and the ring electrodes.

The parallel–adjacent arrangement (Figure 4.14a) enables the detection of analytes that are not well separated in the column, provided that their electrode reactions occur at disparate potentials. The two electrodes can be then held at these potentials to enable their differentiation. Other possibilities have been discussed, including analyte identification by comparison of peak height ratios at the two indicator electrodes with those of standards[111] and enhancement of the S/N ratio by passing the test fluid over one electrode and the (blank) carrier fluid over the other.[112]

The series configuration (Figure 4.14b) can be advantageously utilized in at least three different ways; (1) Analytes that undergo reversible (or quasi-reversible) electrochemical reactions at high potentials (with attendant background noise problems) can be converted at the upstream electrode at the high potential and then reconverted in the measurement mode at the downstream electrode at a substantially lower potential. The result is an enhanced S/N ratio.[113] (2) The upstream electrode can be used as a scavenger electrode to remove or "filter out" interferents. (3) The two electrodes are held at identical applied potentials with analyte conversion at the upstream electrode and residual current measurement at the downstream electrode. A ratio mode can then be employed to enhance the S/N ratio. Alternatively, this strategy is useful in gradient elution HPLC as the baseline shift can be suppressed to a degree by difference detection at the two electrodes.[114]

A series dual-electrode arrangement requires a high degree of electrolytic conversion at the upstream electrode in all these scenarios. Further, the downstream electrode is designed for maximal collection efficiency. This can be achieved either by using the configurations shown in Figure 4.15 or, better still, by the WJRDE geometry. Interestingly, the analyte need not be electroactive in these instances. For example, an electroactive titrant (e.g., Br_2) can be gener-

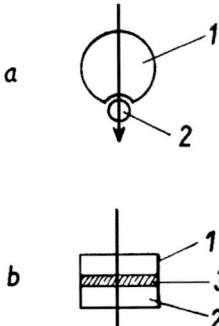

FIGURE 4.15. A series dual electrode cell: (a) generating (1) and detecting (2) electrode in the "horseshoe" arrangement; (b) generating (1) and detecting (2) electrodes in the "sandwich" arrangement (3-polyethylene spacer). (Reproduced with permission from MacCrehan and Durst.[113])

ated at the disk and collected at the ring. In the presence of an analyte that reacts with the titrant in the diffusion layer, the ring current is reduced in proportion to the analyte concentration.[109]

The parallel–opposed electrode arrangement (Figure 4.14c) enables amplification of the detector signal for reversible (or quasi-reversible) systems through multiple redox of the analyte on passage through the dual electrodes. For the amplification effect to be appreciable, the working electrodes must be large, the cell thin, and the liquid flow rate very low. Under these conditions, the amplification factor has been found to range from ~2 to 20 for flow rates in the µL/min range.[115] While these flow rates are compatible with microcolumn HPLC, the requirement of large electrodes is problematic. Theory for these types of cell designs is available,[116,117] and experimental verification has been reported.[115]

An electrochemical detector for FIA has been designed, based on biamperometric end-point detection.[118] This detector consists of two opposing planar platinum electrodes in contact with the carrier flow stream. A small potential difference (10–500 mV) is imposed between the two electrodes and the resulting current is measured. In this design, the current will flow only when both the reduced and oxidized forms of a reversible (or quasi-reversible) couple are present in the flow cell. Thus for an interferent to generate a signal, it must satisfy the (rather stringent) dual requirements that it must be of the opposite form as that of the redox reagent, and it must undergo electrolysis at a potential near that of the reagent couple. As with the WJRDE case discussed earlier, the analyte itself need not be electroactive as long as it is capable of undergoing a homogeneous chemical reaction with one-half of the redox reagent.

4.4. Amperometric-Coulometric and Voltammetric-Polarographic Detection 229

A new type of dual working electrode cell, namely, scanning electrochemical microscopy (SECM),[119] has been inspired by developments in scanning tunneling microscopy. The sensor applications of this approach are slowly emerging, and very high analytical sensitivity (i.e., the detection of a few molecules!) appears to be feasible.[120] Figure 4.16 contains a schematic of the SECM setup, and depending on the nature of the "substrate" electrode (i.e., whether it is conductive or insulating), the tip signal experiences either positive or negative feedback respectively.

The trend toward electrochemical array detectors (ECADs) mirrors corresponding developments in spectroscopy. The advent of ultramicroelectrodes with at least one electrode dimension in the µm domain,[121] new composite materials (e.g., Kelgraf),[122] and the application of microelectronic fabrication technology (e.g., lithography)[123-127] have combined to accelerate the development of ECADs. These arrays offer a degree of versatility far surpassing that attainable with their dual-electrode counterparts discussed earlier. For example, tailored arrays of voltammetric electrodes can be designed based on different electrode materials,[128] surface-modification agents (with different voltammetric characteristics),[129,130] or on the use of different potentials or surface pretreatments.[131,132] In this manner, the array detector can be made to respond to a range of analytes simultaneously. This obviously opens up a new dimension

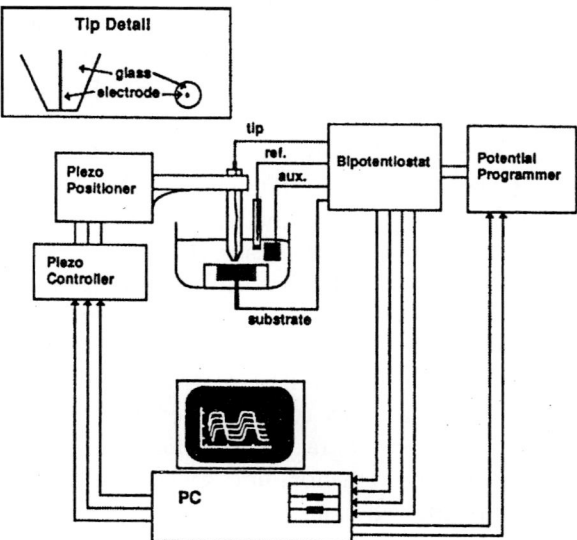

FIGURE 4.16. Block diagram of the SECM apparatus. (Reproduced with permission from Bard et al.[119])

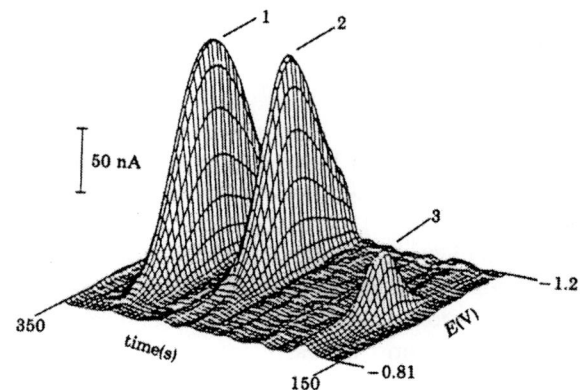

FIGURE 4.17. Three-dimensional "chromatovoltammogram" for two nitrosamines (peaks 1 and 2) and an unknown impurity (peak 3). (Reproduced with permission from Samuelesson et al.[139])

to LC-EC detection in much the same manner that UV–VIS diode–array detectors revolutionized optical absorbance or fluorescence detection. In an operational sense, single-electrode and array LC-EC detection bear the same juxtaposition to their fixed wavelength and diode array UV–VIS detector counterparts, respectively.

Another important advantage with array detection pertains to the S/N ratio. It has been found that improved S/N ratios are obtained for carbonaceous electrodes immobilized in an inert binder.[122] This phenomenon was investigated in detail by varying the ratio of carbon to binder in a Kelgraf electrode matrix.[122] A maximum in the S/N ratio was obtained for electrodes containing a higher percentage of the binder such that the graphite particles existed as individual islands (~100 μm diameter) on the electrode surface. This theme has been further explored by the use of an electrode comprising 100 disks (each of 5 μm radius) that was constructed from carbon fibers.[133] These individual disks were spaced such that diffusional cross talk did not occur. This ultramicroelectrode array was found to exhibit a reduced dependence of the amperometric signal on the flow rate. This behavior is rationalizable on the basis of the time independence of the faradaic current flow at ultramicroelectrodes in the absence of convection.[121,134]

Three-dimensional chromatographic detection has been described, based on the use of multielectrode detectors and gradient elution.[135] The range of compounds that can be identified and the confidence with which they can be matched with known standards are both shown to be enhanced in this manner. A 16-channel ultramicroelectrode array detector has been described for use with FIA and HPLC.[136] The 16-microband electrodes were held at different

4.4. Amperometric-Coulometric and Voltammetric-Polarographic Detection 231

potentials in a stepwise manner for 16-channel detection. An alternative scheme for 80-channel detection was based on the use of a five-step potential staircase on the 16 electrodes.[136] Figure 4.17 contains an example of a three-dimensional chromatogram based on multichannel amperometric detection.[137]

Array detection is not the only route to the acquisition of three-dimensional chromatograms. The use of ultramicroelectrodes along with fast-scan(several V/s) techniques enables complete voltammograms to be recorded as the analytes exit the separation column. Examples of this approach include the use of rapid-scan swept-potential techniques for detection based on square-wave voltammetry/polarography,[138–140] differential pulse polarography at a DME,[141] phase-sensitive AC polarography,[142] normal-pulse voltammetry,[143] adsorptive stripping voltammetry,[144] and coulostatic measurements.[145,146]

4.4.4. Electrode Materials and Electrode Surface Modification

In Section 2.6, the positive and negative features associated with the use of mercury as a working electrode were discussed. Notwithstanding its toxicity and limited positive potential window, mercury is still the electrode material of choice, especially in stripping analyses. Aside from its high overpotential for hydrogen discharge, mercury-based electrodes generally have very low and reproducible residual (background) currents because of their surface homogeneity. Mercury thin-film electrodes (MTFE), obtained by electrodepositing a mercury film onto a suitable support (carbon, noble metals), have become popular in recent years. It must be noted, however, that MTFEs lack the main advantage of the DME configuration; namely, that of periodic renewal of the electrode surface. Further, deposition of mercury onto a noble metal surface (e.g., Pt, Au) produces an *amalgam* and not a pure mercury phase. Amalgam electrodes do suffer from a somewhat reduced hydrogen overvoltage relative to mercury and consequently a narrower potential window in the cathodic regime. There is also the potential danger of film damage at high fluid flow rates. Generally, mercury films with the best properties are obtained on a gold support surface. The construction and operation of detectors based on mercury drop electrodes have been reviewed.[147]

Solid electrodes for voltammetry (and amperometry) include the noble metals, glassy carbon, pyrolytic graphite, and carbon paste. The criteria most often employed for the choice of material candidates may be grouped according to those related to performance, ease of maintenance, and cost. In the first category could be included factors such as good kinetics for a wide range of redox couples, wide accessible potential windows, low residual currents and background noise, and a low susceptibility to surface fouling and passivation. Of course, not all these ideals are *simultaneously* achieved with a given electrode material in practice. For example, carbon paste electrodes have a high

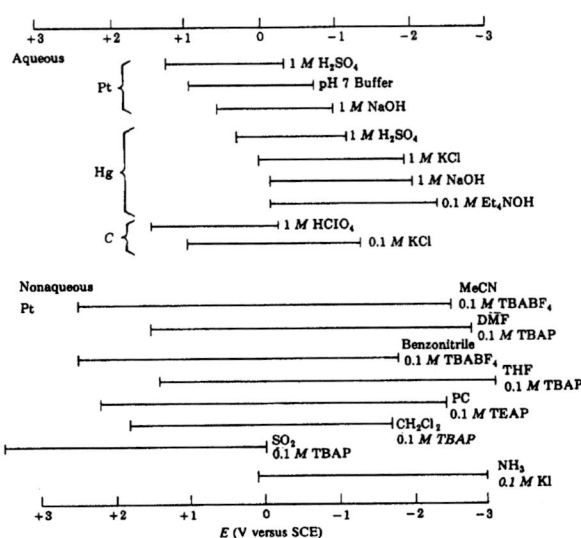

FIGURE 4.18. Estimated potential ranges of various combinations of working electrode and solvent-supporting electrolyte systems. (Reproduced with permission from A. J. Bard and L. R. Faulkner, "Electrochemical Methods." Wiley, New York, 1980.)

S/N ratio in aqueous media leading to very low amperometric detection limits even though electrochemical reactions are generally slower at these surfaces than on well-polished noble metals or glassy carbon.

Figure 4.18 contains a comparison of the potential windows of several electrode materials in aqueous media. In using this information, it must be borne in mind that the limits shown are sensitive to several factors, such as electrolyte composition, pH and dissolved O_2 and impurity content. The positive potential limit is governed by electrolytic breakdown of the base electrolyte, the solvent, or by anodic oxidation of the electrode surface. Hence, noble metals such as gold and platinum are oxidized at positive potentials to yield surface oxides with well-defined redox behavior (see Figure 4.19).[148–151] The cathodic limit is usually imposed by hydrogen discharge and is lowest for active metals such as Pt and highest for materials such as carbon paste. Often the cathodic residual current is high because of adsorbed O_2, and this problem is particularly severe for materials such as carbon paste. In these cases the O_2 is trapped in the matrix during electrode preparation. Finally, carbonaceous materials also have various functional groups on the surface that have both redox and acid–base properties. The surface concentration of these groups is very sensitive to the electrode pretreatment. It is clear from the preceding discussion that the electrochemical performance of a solid electrode material is

4.4. Amperometric-Coulometric and Voltammetric-Polarographic Detection 233

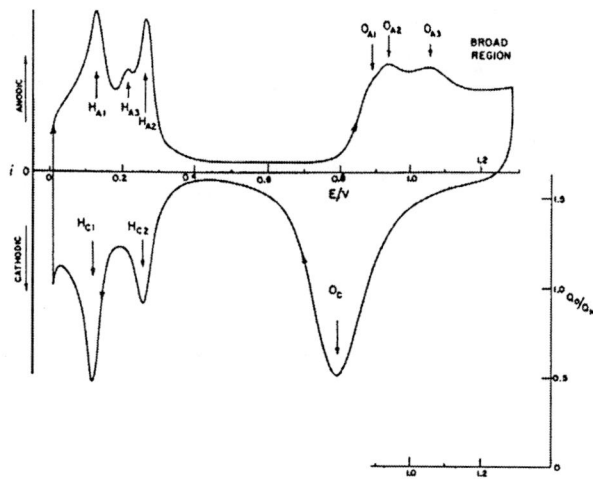

FIGURE 4.19. Cyclic voltammogram illustrating platinum surface oxide formation and reduction in 0.5 M H_2SO_4. (Reproduced with permission from Ross.[151])

crucially dependent on its surface condition and prior history. Unfortunately, much of our current knowledge of solid electrode surfaces (especially at an atomic level) is fragmentary and optimization of a solid electrode surface still remains an art rather than an exact science. Of course, this situation is not unique to electrochemistry but pervades all surface-oriented phenomena (e.g., catalysis) in general, especially if we omit scenarios where the system is operated in pristine conditions (i.e., ultrahigh vacuum). We shall briefly examine each type of electrode material next.

<u>Mercury Thin Films and Drops</u>. The mercury films are either preplated onto a suitable support or generated *in situ*.[152] In general, the average thickness, d of a MTFE may be calculated from the expression:[153]

$$d = 2.43 \, It/r^2 \tag{4.8}$$

I is the mercuric ion deposition current, t is the deposition time, and r is the support disk radius. The film thickness ranges from 10 to 1000 nm. In the *in situ* case, 1-5 x 10^{-5} M mercuric nitrate is directly added to the sample solution; the mercury film and the metals are thus simultaneously deposited.[152] The bare disk surface must be polished prior to each analysis to ensure reproducible behavior (see later). A rotating glassy carbon electrode has been coupled with an *in situ* plated mercury film for stripping analyses.[152,154]

Other supports have been used for the MTFE including RVC,[155] wax-impregnated graphite,[156] epoxy-bonded graphite,[157] pyrolytic carbon,[158] carbon fi-

FIGURE 4.20. Kemula-type hanging mercury drop electrode. (Photograph courtesy of Brinkmann Instruments Co., Westbury, New York.)

ber,[159] Kelgraf (fabricated from compression molding of Kel–F resin and powdered graphite),[122,160] and a graphite spray support made by spraying carbon powder on a conducting surface to create a thin film of uniform particles.[161]

A hanging mercury drop electrode is often used for stripping analyses. In the Kemula design, mercury is displaced from a reservoir through a glass capillary by a screw-driven plunger (Figure 4.20). Reproducible drops are thus generated at the capillary tip by adjustment of a calibrated micrometer. The static mercury drop electrode (SMDE) configuration allows use in both the HMDE or DME modes by simple flip of a panel switch. The drop size is claimed to be more reproducible and stable than in the Kemula design.[162] Importantly, the SMDE has a constant area when the I–E curve is recorded, thus essentially eliminating changing current due to drop growth.

A comparative assessment of MTFE and HMDE is difficult because the electrode requirements are often unique to a given analytical problem. Nonetheless, the MTFE has a surface-area-to-volume ratio several orders of magnitude greater than the HMDE. The small area in the latter configuration reduces the metal plating efficiency for stripping analyses, and the large volume translates to a low concentration of metal in the amalgam. In general, the MTFE offers greater resolution and sensitivity whereas the HMDE is less susceptible to surface fouling and changes in electrolyte composition.

Carbon-Based Electrodes. Glassy carbon is a popular choice for the electrode material, and a review of its physical and electrochemical properties is available.[163] With well-polished surfaces, fast electron-transfer kinetics can be achieved for a variety of redox couples. The residual currents at this surface are also reasonably low, and glassy carbon has a wide potential window for use (Figure 4.18).

Carbon paste electrodes[164–167] are prepared by mixing finely powdered graphite or other carbonaceous materials with a liquid such as Nujol, paraffin

4.4. Amperometric-Coulometric and Voltammetric-Polarographic Detection

oil, silicone grease, or bromonaphthalene. These electrodes have the virtues of easy preparation, low cost, surface renewability, amenability to chemical modification, and very low background currents. A disadvantage of carbon paste is the tendency of the organic binder to dissolve in solutions containing an appreciable fraction of organic solvent. Thus, the advantage with the use of these solvent mixtures as the mobile phase for HPLC (e.g., lower residual current and noise) is mitigated when carbon pastes are used. In general, carbon paste electrodes work best with aqueous solutions whereas glassy carbon, pyrolytic graphite, and noble metals are suitable for use with mixed media.

Composite carbon electrodes have a solid polymer matrix in place of the liquid diluent. Such electrodes are mechanically rugged, can be polished, and are more resistant to organic media. A number of such electrodes have been described, based on the dispersion of carbon or graphite in waxes,[168] Kel–F resin, [122,160] polyethylene,[169] and polypropylene.[116] In many instances, the electrochemical behavior of carbon pastes or composites can be effectively modeled in terms of arrays of individual ultramicroelectrodes that are embedded in an inert (i.e., electroinactive) matrix.

Carbon fiber electrodes are increasingly being used for electroanalysis.[159,170,171] These fibers are produced by high-temperature pyrolysis of polymers or via catalytic chemical vapor deposition. The high-modulus type of fiber microstructure is most suitable for electroanalysis because of its well-ordered graphitic structure and low porosity. The fibers are generally of 5–20 μm diameter and are typically mounted at the tip of a capillary with epoxy adhesive.

Carbon aerogel composite electrodes have been prepared by sol–gel condensation of resorcinol and formaldehyde in a basic medium followed by drying and pyrolysis.[172] The electrodes possess high porosity, surface area between 400–1000 m^2/g, a nominal pore size of less than 50 nm, and a solid matrix composed of interconnected colloidal-like particles or fibrous chains with characteristic diameters of 10 nm. The porosity and particle size can be varied over a wide range given the flexibility in the fabrication procedure.

Composite metal–carbon electrodes have been recently described.[173,174] Stainless steel fibers (2 μm diameter) and carbon fiber bundles (20 μm diameter) are combined with cellulose fibers, used as a binder, into a continuous interwoven paper preform. The composite paper preform is then cut into electrodes with desired dimensions and geometry and sintered to a stainless steel foil substrate. These electrodes possess surface areas, as measured by the BET technique, of ca. 750 m^2/g. The resultant high surface area allows high accessibility to gases and electrolytes while providing adjustable porosities and void volumes. The electrodes can be prepared with other metal fibers such as Ni.

Pyrolytic carbon films offer rates of electron transfer comparable to or even better than those attainable at polished glassy carbon without electrode pretreatment.[175-177] New types of substrates such as Macor are being used for these films,[178] and these have the important advantage of machinability not available with quartz and glassy carbon supports. Thus Macor-supported pyrolytic carbon film electrodes can be machined to unusual sizes and shapes. The issues of long-term stability and film reactivation have been addressed in a recent study.[179] Electrochemical anodization was found to be most effective for restoring film activity degraded by exposure to atmospheric O_2.

Other novel carbon-based electrodes beginning to emerge include carbon foam composites,[180] epoxy-impregnated reticulated vitreous carbon,[181] doped carbon thin films and doped glassy carbon.[182,183] Finally, diamond is an intriguing electrode material for electroanalysis. Normally, the insulating character of virgin diamond (>10^{12} ohm-cm) would preclude the use of this material as an electrode. However, the resistivity of chemical vapor deposited diamond thin films can be made as low as 0.01 ohm-cm by doping with boron.[184] Chemical inertness and low electrode capacitance are positive attributes of this material. Its electrocatalytic activity toward a variety of redox probes remains to be tested.

<u>Noble Metals</u>. Platinum and gold are the most widely used electrode materials in this category. They have fast electron kinetics toward a wide range of redox couples and a wide positive potential range. However, these metals are also catalytic toward the HER, which limits the negative potential window (unlike mercury). Another problem is the high background current associated with the redox activity of the surface oxide and hydride films[150,151] at these metals (see Figure 4.19). Of course, this problem is less severe in nonaqueous media. As with carbon, ultramicroelectrodes and array electrodes can also be constructed from platinum and gold, using capillary-based and microelectronics fabrication procedures, respectively.

<u>Chemically Modified Working Electrode Surfaces</u>. Two major problems with electroanalysis are fouling of the electrode surface by unwanted precipitation or adsorption processes and the slow electrochemical reaction rates of some analytes, which consequently require considerable overpotential. Both of these problems are solved in principle by *deliberately* modifying the electrode surface. In fact, this concept was in part borne out of the frustrated electrochemist's desire to exert direct control of the chemistry at the electrode surface.[185] The underlying rationale is simple in that, by deliberately attaching chemical reagents to an electrode surface, one hopes that the resultant chemically modified electrode (CME) will take on the chemical properties of the attached reagent. We have already encountered this idea in connection with imparting potentiometric selectivity to a parent electrode surface (see Section 4.3). Such reagent-based control of the CME surface therefore includes desirable features such as fast electron-transfer rates, selectivity, immunity to ad-

4.4. Amperometric-Coulometric and Voltammetric-Polarographic Detection 237

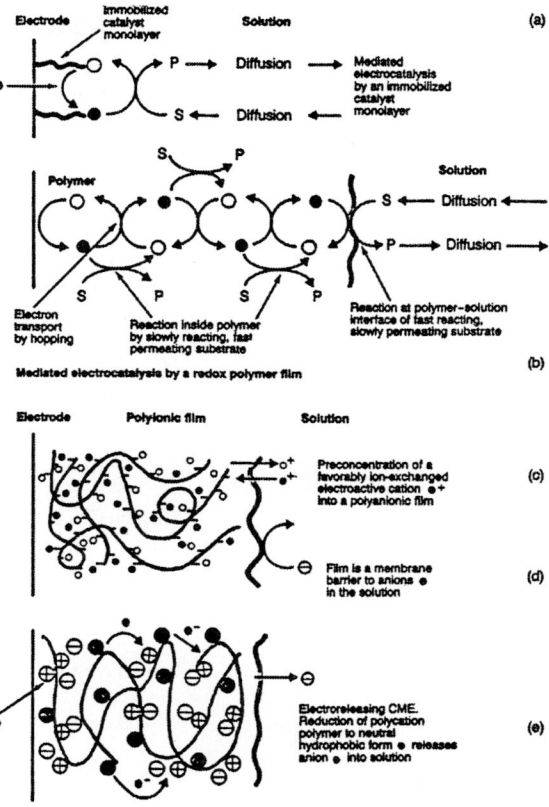

FIGURE 4.21. Schematic illustrations of modified electrodes used in electrocatalysis: (a) monolayers and (b) mulitmolecular polymer layers of catalyst sites ○ and ●; (c) preconcentration of electroactive cations with polyanionic polymer films; (d) membrane barriers to undesired anions using polyanionic films; and (e) electroreleasing of counterions into solution by reduction by polyanionic electroactive polymer. (Reproduced with permission from Murray et al.[185])

sorption and electrode fouling, and access to new properties such as optically excited electronic states. Figure 4.21 and Table 4.6 outline some of the possibilities with CMEs in terms of sensor applications. Other aspects of these systems of relevance to electrocatalysis, photoelectrochemistry, and clinical assays have been reviewed.[186-193]

As Figure 4.21 and Table 4.6 indicate, polymers have played a particularly important role in the selection of chemical reagents for surface chemical modification. Thus, these reagent moieties are either anchored to the polymer backbone (in a premodification step) or are simply partitioned into the electrode

Table 4.6. Commonly Used Coating Materials for CMEs and the Underlying Analyte Binding Mechanism

Coating Material	Mechanism
Cellulose acetate	Size exclusion
Polyphenol	Size exclusion
Poly(1,2-diaminobenzene)	Size exclusion
Phospholipid	Hydrophobic interactions
Self-assembled thiols	Hydrophobic interactions
Nafion	Charge (Donnan) exclusion
Polystyrenesulfonate	Charge (Donnan) exclusion
Polyvinylpyridine	Charge (Donnan) exclusion
Polyaniline	Mixed control
Polypyrrole	Mixed control
Cellulose acetate/Nafion	Mixed control
Nafion/α-cyclodextrin	Mixed control
Clays and zeolites	Size exclusion

matrix by exposing the (initially) polymer-coated electrode to a solution containing the reagent. As we shall see in a later section, the latter strategy has opened a new route to stripping analyses based on adsorptive preconcentration of the target analyte molecule or ion. Figure 4.22 contains a summary of the variant ways in which the polymer film and the reagent can be attached to a parent electrode surface. A simple and elegant route involves the generation of hydroxyl groups at the support electrode surface (e.g., Pt, Au, carbon, tin-doped indium oxide) followed by hydrolytic coupling of an organosilane reagent to which a redox moiety (e.g., ferrocene) can be preattached (Figure 4.22).

Physical adsorption and spontaneous chemisorption of macromolecular assemblies constitute yet another powerful way to modify a given electrode surface. Thus, spontaneously adsorbed Langmuir–Blodgett films of n-alkane thiols $[X(CH_2)_nSH, n \geq 10]$ at (111) oriented gold surfaces are particularly well

4.4. Amperometric-Coulometric and Voltammetric-Polarographic Detection

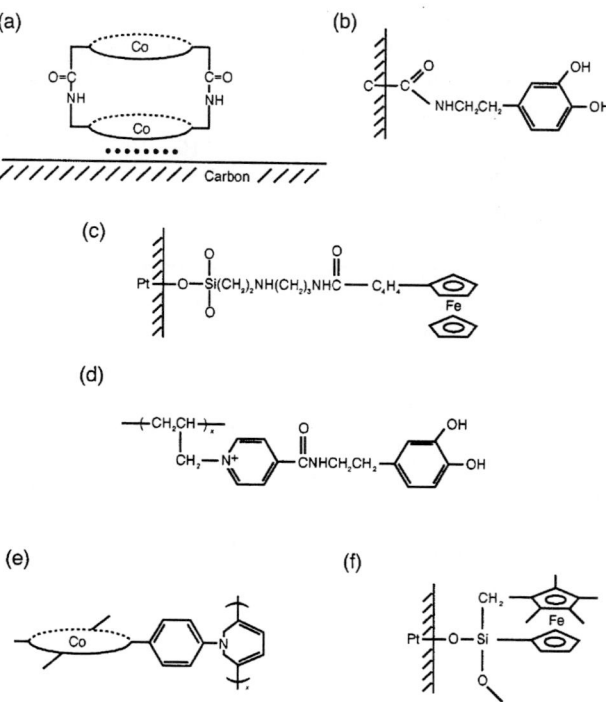

FIGURE 4.22. Summary of strategies for attaching a polymer film and electroactive reagent to an electrode surface. (Reproduced with permission from Murray et al.[185])

suited for controlling the reactivity at the interface. This can be done through the use of different head group (X) chemistry. Descriptions of such self-assembled monolayers (SAMs) are available.[194,195]

Other (physical) methods for attaching a polymer film (with or without a chemical reagent) to an electrode surface include spin casting. In this method, a few drops of the solution of the polymer are dropped onto the electrode surface. The solvent is then allowed to evaporate, often by spinning the electrode to accelerate the process. Nafion and polyvinylchloride-modified electrodes are fabricated in this manner.

Finally, electropolymerization can be profitably employed for generating a chemically modified surface layer. Films of electronically conductive polymers such as polypyrrole and polyaniline (Figure 4.23) are prepared in this manner (see also Section 5.9). It has been claimed that the size-exclusion selectivity of such films can be controlled by the use of electropolymerization conditions such as polymerization time and monomer concentration.[196] Polymer films containing bipyridyl reagent moieties are also synthesized using electropolymerization.[189] This approach is also amenable to the preparation of *three-dimensional* surface assemblies by *in situ* polymerization of a targeted monomer (e.g., pyrrole) in a dispersion containing nanosized particles of a catalyst (e.g., Pt)[197] or matrix modifier (e.g., carbon black).[198,199] Polypyrrole films containing a variety of chemical reagents including metal and metal oxide catalyst,[200,201] electrochromics,[202] or semiconductors[203,204] have been prepared in this manner.

In principle, all of the electrode materials discussed earlier could be used as supports for the chemically modified surface layers. Chemical modification of carbon pastes is particularly simple – the carbon and the modifying agent (including metal particles, see Ref. 205) are simply mixed together in a mortar and pestle. Similarly, *chemically* polymerized polypyrrole (or polyaniline) can be mixed in powder form with a variety of reagents[206] and subsequently used in a packed electrode configuration. Even mercury films have been used[207] as a support for electropolymerized polypyrrole – the oxidized mercury being rapidly re-reduced (and redeposited) at negative potentials before it escapes from the electrode surface (as Hg_2^{2+} and Hg^{2+} ions). The resultant "polymer dispersed–mercury modified electrode" was used in a flow-through electrochemical cell in the reversed-pulse amperometric detection mode.[207]

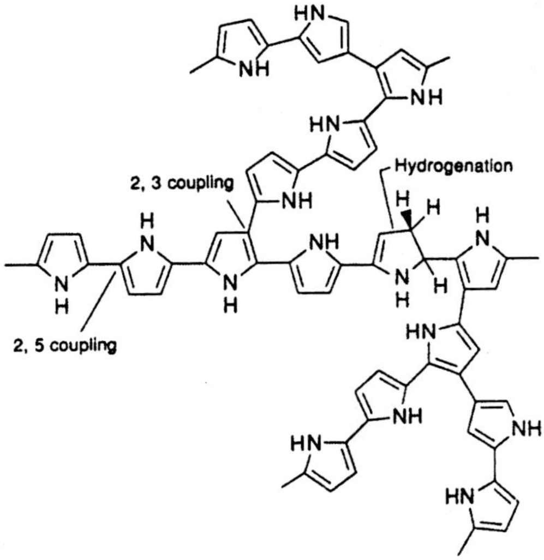

FIGURE 4.23. Electrochemical growth of a polypyrrole film. (Reproduced with permission from M. Kanatzidis, *Chem. Eng. News*, December 3, p. 44 (1990).)

4.4. Amperometric-Coulometric and Voltammetric-Polarographic Detection

Our knowledge of the morphology of the chemically modified surface layer continues to expand with the availability of scanning probe microscopies. The polymer microstructure has a profound impact on the dynamics of electron transfer within the matrix, and this aspect continues to be addressed.[208,209] Another problem is the estimation of the electroactive reagent concentration within the surface layer arising from the uncertainty in the knowledge of the actual (wetted) polymer thickness. Use of the *dry* film thickness clearly is unacceptable in most instances as the polymer swells (often by factors of 100 or more) on uptake of the solvent.[210–212] Again, the use of new *in situ* and real time probes such as ellipsometry[213] and electrochemical quartz crystal microgravimetry[212,214–217] promises to afford new insights and thus solutions to these analytical problems.

We shall next examine some representative scenarios where CMEs enhance the scope of the electroanalysis scheme relative to the parent electrode (unmodified) configuration.

Often the desired electrode reaction has high overpotential; that is, it occurs at an appreciable rate only at potentials substantially higher than the thermodynamic redox potential. Such reactions can be catalyzed by attaching to the surface a fast electron transfer mediator. Knowledge of the corresponding homogeneous solution kinetics of this (reversible) redox couple is useful for selecting a catalyst candidate. The mediator functions by shuttling the charge between the analyte and the electrode support for the chemically modified surface layer:

$$M_{ox} + ne^- \rightarrow M_{red} \tag{4.8}$$

$$M_{red} + A_{ox} \rightarrow M_{ox} + A_{red} \tag{4.9}$$

Where M is the mediator and A is the analyte (Figure 4.24). Note that the electron transfer now takes place between the electrode and the mediator and *not* directly between the electrode and the analyte. Reaction (4.9) is the heterogeneous analog of the (homogeneous) redox reaction between M_{red} and A_{ox}. Models for *heterogeneous redox catalysis* are available.[187,218] In many cases, a one-electron redox couple is used as the mediator, and an additional multielectron storage catalyst (e.g., Pt, RuO_2) is used as an additional means of shuttling charge between the solution and the underlying electrode.[219] Clearly, CMEs offer a powerful route to assembling, *at a molecular level*, multicomponent systems with complementary functions.

Preconcentrating surface layers designed for binding target analytes onto

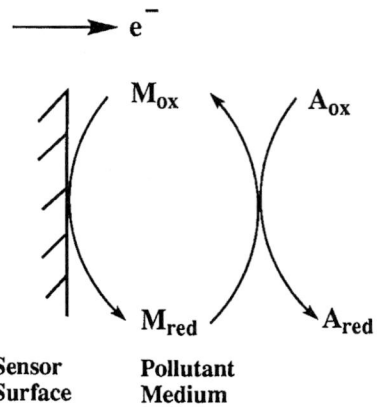

FIGURE 4.24. Redox mediator scheme for electroanalysis: M is the mediator and A is the analyte.

the sensor surface (Figure 4.21) are reminiscent of the strategy discussed earlier for imparting selectivity to potentiometric ISEs. In this scheme, the target analyte is first partitioned from the dilute sample into the preconcentrating surface layer and then quantified via either amperometry or voltammetry. The mechanisms involved for such partitioning schemes are variant and include surface complexation, size exclusion, electrostatic binding, and schemes utilizing the hydrophobicity or lipophilicity of a surface layer (e.g., a lipid coating). These are illustrated in "cartoon" fashion in Figure 4.21. Table 4.7 contains examples of analytical strategies utilizing these binding mechanisms. Permselective membranes obviously play a crucial role in the chemical tailoring of sensor surfaces, and candidate membranes include ionomeric systems such as Nafion, polystyrene sulfonate, poly(1,2-diaminobenzene), and conducting polymers such as polypyrrole and polyaniline (see Table 4.6). In all these instances, the CME surfaces not only provide enhanced sensitivity and selectivity but also facilitate the presence of built-in membrane barriers that serve the (important) additional function of excluding potential interferents and agents responsible for electrode passivation and fouling.

Agents for chemical modification of electrode surfaces are not limited to organic, inorganic, or organometallic systems. Microorganisms and other biological components can also be profitably employed. In fact, enzyme-based sensors have been traditionally used for glucose assay, and CME designs incorporating glucose oxidase in a polymeric matrix are now commercially available. Table 4.7 contains additional examples of CME sensors based on biochemical components.

4.4. Amperometric-Coulometric and Voltammetric-Polarographic Detection

Table 4.7. Examples of Analytical Schemes Employing CMEs

CME System	Comments	Reference(s)
Carbon paste modified with 2,9-dimethyl-1,10-phenanthroline	Copper binding and assay via surface complexation	220
Carbon paste modified with dimethyl glyoxime or porphyrin	Same as preceding but for Ni(II) trace analysis	221,222
Platinum modified with vinyl ferrocene/vinyl pyridine copolymer incorporating complexation reagents (e.g., sulfonated batho phenanthroline)	Used for copper and iron binding and assay	223,224
Glassy carbon modified with quaternized polyvinylpyridine (QPVP) loaded with cyano metallate complexes (e.g., $Fe(CN)_6^{4-}$)	Silver ions preconcentrated in an open circuit followed by voltammetry and stripping	225
Carbon paste modified with a chelating resin with a polythioether backbone	Detection limit after 5 min. preconcentration of Ag(I) is 3×10^{-10} M	226
Glassy carbon modified with polyvinylpyridine and $M(bpy)_2$ complex (M = Os, Ru; bpy = and Fe(III) 2,2'-bipyridyl)	A dual-electrode configuration employed for speciation of Fe(II)	227
Pt wire modified with p-mercuribenzoate using silylation chemistry	The CME is designed for specificity to sulphydryl compounds	228
Carbon paste modified with crown ethers and cryptands	Used for trace measurements of lead	229
Carbon paste modified with phenylenediamine	Used for redox catalysis of target analytes	230
Carbon paste modified with cobalt phthalocyanine	Used for the assay of inorganic anions containing sulfur, selenium and nitrogen	231
Platinum coated with cellulose acetate (CA) and loaded with H_2O_2	The CA film eliminates protein fouling and ascorbate interference; used for blood serum and urine analyses	232
Glassy carbon modified with Nafion and 18-crown-6	Silver preconcentrated and determined by stripping analysis with a limit of detection of 2×10^{-12} M	233
Carbon paste modified with metal particles (e.g., Ru, Pt)	Used for redox catalysis of organoperoxides and hydrazine compounds	234
A composite coating of phosphatidylcholine, cholestrol and stearic acid on a silver-band electrode	Used for CN$^-$ assay in a FIA configuration	235
Gold electrodes modified with a self-assembled monolayer of 4-aminothiophenol	Used for binding of a variety of solution species	236
Gold electrodes modified with SAMs of thiols and thioctic acid	Selectivity and sensitivity toward solution species (e.g., $Ru(NH_3)_6^{3+}$ and $Fe(CN)_6^{3-}$) probed	237

Cont. Table 4.7. Examples of Analytical Schemes Employing CMEs

CME System	Comments	Reference(s)
Carbon paste modified with a zeolite	Used for preconcentration of Cu(II) ions	238
A SnO_2 glass electrode coated with a montmorillonite clay film, containing Δ-Ru(phen)$_3^{2+}$ (phen = 1,10-phenanthroline)	Stereoselectivity demonstrated for the surface modified film	239
A MTFE modified with ω-mercapto carboxylic acid-based SAM	Used for trace determination of Cd(II) with a limit of detection of 0.45 ppt	240
A hanging mercury drop electrode modified with alizarin complexone and fluoride ion	Used for preconcentration and assay of La(III)	241
Glassy carbon disk modified with QPVP	Used for preconcentration and determination of Cu(II) ions	242
Glassy carbon modified with chitosan, β-1,4-poly-D-glucosamine	Used for trace determination of Pb(II) in water samples	243
MTFE modified with a cellulose triacetate dialysis membrane	Lead determined in a variety of matrices including tap water, river water, seawater, and whole blood samples	244
A carbon electrode modified with Nafion and 2,2'-biquinoline	Used for preconcentration of Cu(I) ions	245
Platinum wire coated with polypyrrole	Used for amperometric sensing of chloride, nitrate, nitrite, perchlorate, bromide, carbonate, sulfate and phosphate with a detection limit in the 0.1-1.0 ppm range	246
A gold electrode coated with overoxidized polypyrrole	Used for Cr(VI) assay	247
Chlorella vulgaris immobilized in a porous polycarbonate membrane	Assembly used as a photomicrobial amperometric sensor for phosphate	248
A membrane filter O_2 electrode loaded with *S. typhimurium*	Used for mutagenic screening of carcinogens	249
A glassy carbon electrode coated with cellulose acetate and loaded with algae (e.g., *Eisenia* bicyclis)	Used for ion exchange preconcentration of redox analytes	250
Carbon paste modified with sulfite oxidase enzyme	Used for assay of sulfite ion in aqueous media and $SO_2(g)$	251
Glassy carbon disk modified with polypyrrolic viologen and loaded with a nitrate reductase enzyme	Used for amperometric detection of nitrate	252
Glassy carbon modified with a phosphatidylcholine layer	Redox active amphiphiles extracted into the cast lipid layer	253

4.4 Amperometric-Coulometric and Voltammetric-Polarographic Detection

Cont. Table 4.7. Examples of Analytical Schemes Employing CMEs

CME System	Comments	Reference(s)
Carbon paste modified with Amberlite XAD-2 resin	Used for determination of paraquat by cathodic stripping voltammetry	254
Glassy carbon spin-coated with 4,7,13,16,21, 24 -hexa-oxa-1,10-diazabicyclo [8.8.8]hexa-cosane (Kryptofix-222)	Used for trace determination of mercury by square-wave voltammetry	254a

<u>Electrode Activation</u>. The most vexing problem with the use of solid electrode materials for electroanalysis is electrode fouling. Even in less severe instances, there is a gradual loss of electrochemical activity at the surface. This is manifested in the form of drawn-out voltammetric waves for fast redox probes such as $[Fe(CN)_6]^{3-/4-}$. Electrochemical activation can be employed in such instances to restore the original activity of the surface. This "activation" technique often involves polarization of the working electrode in the supporting electrolyte at suitable potentials. One method comprises fast cycling of the electrode potential between the O_2 and H_2 evolution potential limits. Anodic activation has been also employed, and the mechanistic aspects of electrode activation have been addressed both for metal oxide (e.g. Sn-doped indium oxide)[255] and glassy carbon[256,257] electrodes. Activation procedures for Pt electrodes also have been reviewed.[258]

The idea of pulse amperometric detection [259,260] was mentioned elsewhere in this book in a different context (Section 3.3.5). This concept is based on the premise that the Faradaic signal for oxidative desorption of organic compounds and free radicals is applicable to the quantitative detection of all organic compounds that can adsorb on the electrode surface. For example, many anodic processes involve O-atom transfer from H_2O in the solvent phase to the oxidation product(s). However, this process is kinetically inhibited because of its complexity and occurs only at substantial overpotential, such that the electrode surface is concomitantly passivated[261]:

$$H_2O + Pt \rightarrow PtOH + H^+ + e^-$$

$$As(OH)_3 + PtOH \rightarrow Pt + OAs(OH)_3 + H^+ + e^-$$

A normal (i.e., pulse-free) voltammetric determination at these high positive potentials would also be precluded by the large background signal due to the formation of the inert PtO:

$$PtOH \rightarrow PtO + H^+ + e^-$$

The PAD waveform regulates the electrocatalytic properties of noble-metal electrode surfaces by inhibiting this reaction. In fact, many compounds that were previously considered to be electroinactive at a constant applied potential can now be detected with enhanced sensitivity by using the PAD procedure.

This procedure is particularly useful for anodic detection at Pt electrodes, where the loss of electroactivity is attributed to strong adsorption of the analyte and/or free-radical reaction products. This method uses a triple-step potential waveform of the sort shown earlier in Figure 3.19. Loss of response is avoided in PAD by measurement of the anodic current signal a short time (e.g., 50–250 ms) after application of the detection potential (E_{det}) followed by sequential application of large positive (E_{ox}) and negative (E_{red}) potentials for oxidative and reductive cleaning, respectively, of the electrode surface prior to the next de-

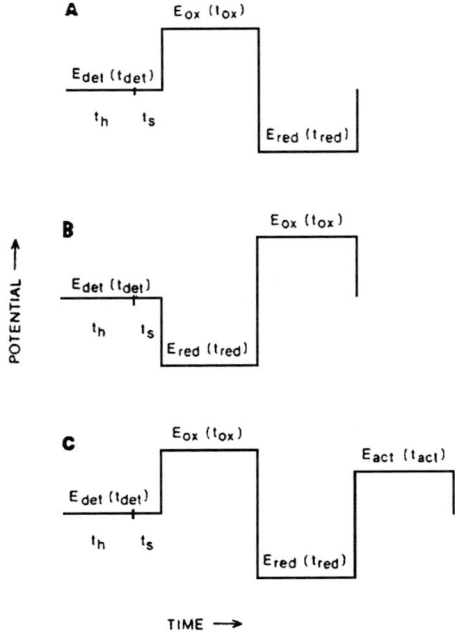

FIGURE 4.25. Waveforms used in pulsed amperometric detection (PAD): (A) normal PAD; (B) reverse PAD; (C) activated PAD. (Reproduced with permission from Williams and Johnson.[261])

4.4 Amperometric-Coulometric and Voltammetric-Polarographic Detection

tection cycle. The frequency of the waveform can be sufficiently high (ca. 1–2 Hz) to facilitate application in chromatographic and flow-injection systems.

Recent innovations of PAD include the use of modified pulse routines aimed at minimizing the background signal associated with PtO formation (Figure 4.25). Thus the reversed pulse waveform (RPAD) has been reported to result in a significant decrease in the background signal for several sulfur compounds at Au electrodes.[262] In the activated PAD mode (APAD) (Figure 4.25c), a brief activation pulse (E_{act}) is applied prior to the detection pulse. Thus, in the APAD waveform, the oxidative processes of surface activation and cleaning are managed separately. The key is to ensure that the PtOH needed for activation is not substantially converted to the electroinactive PtO.

The PAD procedure has been applied successfully to the anodic determination of alcohols, polyalcohols, and carbohydrates on Pt and Au flow-through electrodes[263] and amino acids,[264] organic sulfur compounds,[262,265] As(III)[261] and Cl⁻ and CN⁻[263] on Pt electrodes. Reviews of PAD are available.[260,261,266]

4.4.5. Dioxygen Removal, Derivatization and Other Practical Considerations

Dioxygen has a molar solubility in aqueous solutions of ~10^{-3} M at room temperature and pressure. Dissolved O_2 interferes in amperometry based on reductive electrolysis of the analyte solution. As in the case of classical polarography, dissolved O_2 is also a problem in stripping analysis. Depending on the solution pH, O_2 is reduced in two steps.

Step 1, $O_2 + 2H^+ + 2e^- \rightleftarrows H_2O_2$ (acid)
            $O_2 + 2H_2O + 2e^- \rightleftarrows H_2O_2 + 2OH^-$ (base)

Step 2, $H_2O_2 + 2H^+ + 2e^- \rightleftarrows 2H_2O$ (acid)
            $H_2O_2 + 2e^- \rightleftarrows 2OH^-$ (base)

The $E_{1/2}$ values for these steps are ~-0.05 V and -0.9 V, respectively. These reductions result in an increased background current in amperometry. In stripping voltammetry, the analytical waves of interest are often obscured by these processes (see Figure 4.26). Other complications, especially in stripping voltammetry, include the following: (a) Dioxygen may oxidize the metals preconcentrated as the amalgam; (b) The OH⁻ ions formed during the reduction of O_2 can precipitate the metal ions (as the hydroxide) in the vicinity of the working electrode. For these reasons, O_2 must be rigorously excluded from the cell and the electrolyte prior to amperometric or voltammetric analyses.

The most common tactic consists of bubbling an electroinactive gas (N_2, Ar, etc.) through the solution. Prepurified N_2 can be used for this purpose; it is

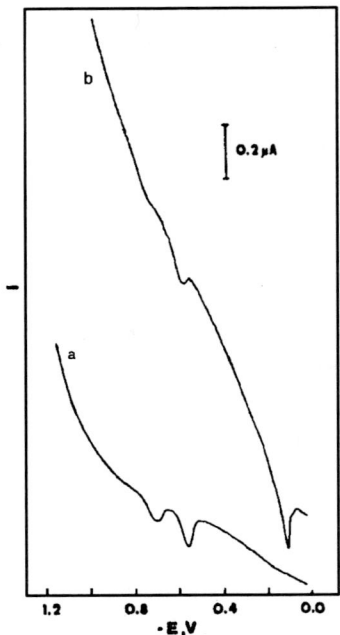

FIGURE 4.26. Comparison of stripping voltammograms recorded at the "downstream" detector utilizing potentials of -0.50 V (a) and 0.0 V (b) at the "upstream" cell 5×10^{-8} M lead and cadmium in 0.1 M KNO_3; 3-min depositions at -1.3 V; flow rate, 0.36 mL/min; linear scan of 1 V/min. (Reproduced with permission from Wang and Dewald.[98])

commercially available and inexpensive. The solution is thoroughly flushed with N_2 for a few minutes prior to analyses. The "purge gas" tube is then lifted from within the solution and poised a little distance away from the solution surface. The gas flow rate is turned down, and a continuous stream is maintained so as to ensure a positive pressure of N_2 above the solution phase. Other methods include the use of the N_2 stream as a convenient stirring mode during the metal preconcentration in stripping voltammetry[267] and a clever strategy that uses a CO_2–N_2 mixture to maintain both solution pH control and deoxygenation.[268]

Chemical methods of O_2 removal (for example, via additions such as sodium sulfite or ascorbic acid[269]) are less preferable because of the potential for electrochemical interference and solution contamination.

Dissolved O_2 is particularly a problem in reductive LC-EC detection because of the variant level of O_2 in the flow stream and the consequent degradation of the S/N ratio. Scrubber columns have been used (see Ref. 270) to reduce and eliminate the dissolved O_2 in an acidic mobile phase in LC. The scrubber column is made by packing a column with zinc amalgam particles and

4.4 Amperometric-Coulometric and Voltammetric-Polarographic Detection 249

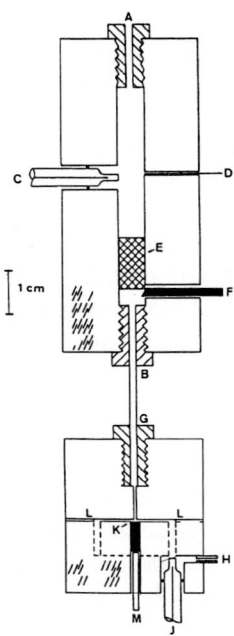

FIGURE 4.27. Dual flow cell assembly: (A, G) solution inlets; (B, H) solution outlets; (C, J) reference electrodes; (D) auxiliary electrode; (E) RVC cylinder; (F) lead to RVC; (K) glassy carbon disk; (L) Teflon spacers; (M) lead to glassy carbon disk. (Reproduced with permission from Wang and Dewald.[99])

mounting it between the pump and the injection device. The zinc consumed in the redox reaction with O_2 must be periodically refurbished.

Other strategies for dissolved O_2 removal include either dual-cell or dual-electrode configurations (see Section 4.3.3). An example of the former approach is contained in Figure 4.27.[99] The upstream working electrode is held at a scrubbing potential of -0.50 V. A flow-through cell with RVC as the working electrode is used for the upstream cell to obtain high electrolytic yields (close to 100%). Thus the deoxygenated solution enters the downstream detector cell. Figure 4.26 contains an example of the sort of improvement that is possible in stripping voltammetry detection of lead and cadmium.[98] Of course, this approach is not suitable for the determination of metals with positive reduction potentials (e.g., Cu, Bi, Sb). On the other hand, the majority of metal ions determined by anodic stripping voltammetry (e.g., Cd, Pb, Tl, Zn) are not reduced at the potential used for deoxygenation.

In the dual-electrode approach,[107,113,271,272] the (upstream) generator electrode is held at a potential more negative than the half-wave potential of the analyte to be determined via reductive amperometry. The detector electrode is poised at a potential suitable for the oxidation of the product generated by the first electrode. Because the actual detection occurs in the oxidative mode, there is no interference from O_2. Obviously, this approach is applicable only to those analytes that undergo reversible electrochemical reactions.

Capillary electrophoresis has been recently shown to be superior to reductive LC-EC detection in terms of very short deoxygenation times.[273] The feasibility of reductive CE-EC detection has been evaluated for the determination of organic compounds, including nitroaromatics and quinones.[273]

Square-wave and staircase voltammetry have the advantage of use at high frequencies. Thus, dissolved O_2 contributes only very little to the Faradaic current, as its reduction is electrochemically irreversible, especially at high frequencies (1–2 kHz). Hence, the use of high-frequency modulation waveforms for electroanalysis has the crucial advantage that stringent deaeration is not necessary as a prelude to analyses.

Turning next to postcolumn derivatization, many electroinactive or difficult-to-oxidize (or reduce) analytes can be converted to electroactive solutes using a variety of schemes. Electrochemical, chemical, photolytic, and enzymatic derivatization schemes have been developed for this.[274,275] Such methods of generation of detectable species are also accomplished with negligible band broadening. The series dual-electrode configuration is particularly amenable to on-line electrochemical derivatization. Chemical derivatization, with *in situ* generation of dithiocarbamate complexes, has been used for the detection of various metal ions.[276-278] Postcolumn generation of Br_2 has been used for the determination of phenolic ethers.[279] Enzyme derivatization[280] is advantageous in that the very good selectivity of enzyme reactions can be profitably employed for the selective determination of target analytes from a "cocktail." We shall discuss photolytic approaches along with other photoelectrochemical approaches later in this chapter.

4.5. AMPEROMETRIC SENSORS FOR ENVIRONMENTAL POLLUTANTS: SOME EXAMPLES

The common types of organic pollutants that can be detected by oxidative amperometry are contained in Table 4.8. The limits for reductive detection are usually less favorable than for oxidative detection owing to the background resulting from the reduction of traces of O_2, hydrogen ions, and trace metals. However, many metal pollutants, organometallics, and pesticides have been successfully determined via reductive amperometry (see later), and organics containing reducible groups (e.g., nitrocompounds, see Table 4.9) can be reductively determined at a sensitivity comparable or better in some cases, to ultraviolet photometry.

4.5. Amperometric Sensors for Environmental Pollutants

Table 4.8. Compounds of Environmental Importance Suitable for Oxidative Amperometric Detection

	Type of Compound	Formula
1.	*Aromatic hydroxy compounds*	
	Antioxidants	R—C₆H₃(C(CH₃)₃)—OH
	Catechols	(HO)₂C₆H₃—R
	Halogenated phenols	Hal—C₆H₃—OH
	Halogenated hydroxybiphenyls	Hal—C₆H₃—OH—C₆H₄—OH
	Phenols	R—C₆H₄—OH
	Methoxyphenols	HO—C₆H₃(OCH₃)—R
2.	*Aromatic amines*	
	Chlorinated anilines	Cl—C₆H₄—NH₂
	2,3-dichloroaniline	2,3-Cl₂C₆H₃—NH₂
	Benzidines	H₂N—C₆H₄—C₆H₄—NH₂
	Anilines	R—C₆H₄—NH₂

Cont. Table 4.8. Compounds of Environmental Importance

	Type of Compound	Formula
3.	Aliphatic amines	
	N-nitrosoamines (e.g. N-nitrosodimethylamine)	$(CH_3)_2NNO$
4.	Sulfur compounds	
	Mercaptans	R–SH
	Disulfides	R–S–S–R
5.	Purine derivatives	(structure with OH or NH_2)

Source: J. Frank, *Chimia* **35**, 24 (1981).

Table 4.9. Types of Organic Groups Amenable to Reductive Amperometric Detection

Bond Grouping	Example
NO_2	$O_2N-\text{C}_6\text{H}_4-R \xrightarrow{4\,e^-,\,4\,H^+} HONH-\text{C}_6\text{H}_4-R + H_2O$
C–Hal	$C_6H_6Cl_6 \xrightarrow{6\,e^-} C_6H_6 + 6\,Cl^-$
S–S	$R-S-S-R \xrightarrow{2\,e^-,\,2\,H^+} 2\,RSH$
C=O	$Ar-CO-Alk \xrightarrow{e^-,\,H^+} Ar-COH^\bullet-Alk$
N=N	$Ar-N=N-Ar \xrightarrow{2\,e^-,\,2\,H^+} Ar-NH-NH-Ar$

Source: K. Stulik and V. Pacakova, "Electroanalytical Measurements in Flowing Liquids." Ellis Horwood, Chicester, 1987.

4.5. Amperometric Sensors for Environmental Pollutants

Table 4.10 contains further examples of specific substances of environmental interest that are detectable by oxidative and reductive amperometry. In the vast majority of these instances, amperometric detection has been coupled with HPLC separation. This is because the matrices of environmental interest are often quite complex, and an efficient separation step is necessary prior to analyses. Organic analytes such as pesticides traditionally have been determined by gas chromatography (GC) coupled with electron capture detection (ECD). However, derivatization, which is often needed for GC detection, introduces problems with sample contamination and loss of selectivity. For example, cationic organometals are derivatized with a halogen prior to GC–ECD. However, extensive sample "clean up" is required prior to analysis. In other cases (e.g., sulfur compounds), GC analysis is precluded by the thermal instability of the analytes. Other pollutants such as benzidine are base-extractable aromatic amines that are very difficult to extract, concentrate, and quantify by GC/MS. In view of these difficulties, HPLC provides an attractive alternative or supplement to GC analyses.

There are many detector candidates for use with HPLC. The traditional workhorse is UV photometry. In many cases, the analytes (e.g., phenols) are only weakly chromophoric. Fluorescence detection is another possibility, and in many instances, nonfluorescent analytes may be converted to fluorophores by on-line derivatization. However, as with the GC case, this requires additional sample handling, treatment and work up and provides additional scope for error. Graphite furnace atomic absorption spectrometry has been used as a detection system with automated sampling and atomization of the chromatographic eluent. However, many analytes (e.g., organometals) have poor atomization efficiency, and thus the detection limits are rather poor for elements such as Hg and As with this approach.

We shall now discuss LC (or FIA)-EC studies of environmental pollutants in terms of groupings of analytes into specific categories.

Phenols. The important substances in this group include chlorinated phenols and naphthols, alkyl phenols that are used as antioxidants, and hydroxyl derivatives of biphenyls and PCB metabolites (Tables 4.8 and 4.10). They are present in wastes from the manufacture of formaldehyde resins, lacquers and binders, pharmaceuticals and pesticides, from coking and coal distillation plants, and in the soil and vegetable residues. Chlorophenols and nitrophenols, which are used in industry and agriculture and as wood preservatives can also be present in raw water as a result of spillage or accidents. Chlorophenols can be formed during water chlorination, causing taste and odor problems even at very low (ppb) levels. Because of these problems, the U.S. EPA has created a list of the 11 most important phenols as high priority pollutants.[281]

Table 4.10. Environmental Pollutants That Have Been Determined by Oxidative and Reductive Amperometry Coupled with HPLC

Oxidation

PCB Metabolites
    2-chloro-4-biphenylol
    3-chloro-2-biphenylol
    3-chloro-4-biphenylol
    4'-chloro-4-biphenylol
    5-chloro-2-biphenylol
    2-chloro-5-biphenylol
    2',5'-dichloro-2-biphenylol
    2',5'-dichloro-4-biphenylol
    2',5'-dichloro-3-biphenylol
    3,5-dichloro-2-biphenylol
    4,4'-dichloro-3-biphenylol
    3,4'-dichloro-4-biphenylol
    4,4'-dichloro-3,3'-biphenyldiol
    2,2',5'-trichloro-5-biphenylol
    2,2',5'-trichloro-4-biphenylol
    2',3,5-trichloro-4-biphenylol
    2',4',6'-trichloro-4-biphenylol
    2',5,5'-trichloro-2-biphenylol
    3,4',5-trichloro-4-biphenylol
    2',3',4',5'-tetrachloro-4-biphenylol
    2',3',4',5'-tetrachloro-3-biphenyl
    3,3',5,5'-tetrachloro-4,4'-biphenyldiol
    2',3',4',5,5'-pentachloro-2-biphenylol
    2',3,3',4',5'-pentachloro-2-biphenylol

Hydroxylated Biphenyls
    2-biphenylol
    3-biphenylol
    4-biphenylol
    3,3'-biphenyldiol
    4,4'-biphenyldiol
    2,2'-biphenyldiol
    2,5-biphenyldiol
    3,4-biphenyldiol

Amines
    N-nitrosodimethylamine
    N-nitrosodi-n-propylamine
    N-nitrosodiphenylamine
    4-nitroso-N,N-diethylaniline
    benzidine
    3,3'-dichlorobenzidine
    1,2-diphenylhydrazine

Chlorinated Phenols
    2-chlorophenol
    3-chlorophenol
    4-chlorophenol
    2,3-dichlorophenol
    2,4-dichlorophenol
    2,5-dichlorophenol
    2,6-dichlorophenol
    3,4-dichlorophenol
    3,5-dichlorophenol
    2,3,4-trichlorophenol
    2,3,5-trichlorophenol
    2,3,6-trichlorophenol
    2,4,5-trichlorophenol
    2,4,6-trichlorophenol
    3,4,5-trichlorophenol
    2,3,4,5-tetrachlorophenol
    2,3,4,6-tetrachlorophenol
    2,3,5,6-tetrachlorophenol
    pentachlorophenol
    3,5-dimethylphenol
    2,3,6-trimethylphenol
    2,3,5-trimethylphenol

Chlorinated Naphthols
    4-chloronaphthol
    2,4-dichloronaphthol

Chlorinated Anilines
    2-chloroaniline
    3-chloroaniline
    4-chloroaniline
    2,3-dichloroaniline

4.5. Amperometric Sensors for Environmental Pollutants

Cont. Table 4.10. Environmental Pollutants That Have Been Determined by Oxidative and Reductive Amperometry Coupled with HPLC

Reduction

Organocompounds
Methylmercury
Ethylmercury
Phenylmercury
Diphenylmercury
Diphenylthalium
Triethyllead
Triphenyllead
Trimethyllead

Pesticides
Carbamates (see Table 4.11)

Source: Reproduced with permission from Bioanalytical Systems, Inc.

The standard EPA method (Method 604) for the determination of phenols is liquid–liquid extraction and derivatization followed by GC-ECD. Limits of detection between 0.58 – 2.2 ppb have been obtained for the 11 high priority phenolic pollutants.[282] As mentioned earlier, HPLC with UV detection (at 254 nm) is nonselective and is problematic because of the weak adsorption of these chromophores. High-speed LC with electrochemical detection is a method suitable for screening large numbers of samples quickly.

In view of the polar nature of phenols, reversed-phase LC with chemically bonded stationary phases is usually used with weakly acidic mobile phases containing ion-pairing agents. Traces of phenols have been determined in such variant matrices as potable water,[169,282,283] soil,[284] river water,[282] commercial beverages,[285,286] orange rind,[287] pimento,[285] and spices.[288]

Dual-electrode detection is particularly useful in the identification of phenols in complex matrices. Figure 4.28 contains simultaneous oxidative and reductive LC-EC chromatograms of the 11 higher priority phenolic pollutants at parallel–adjacent glassy carbon electrodes at -0.75 V to +0.98 V (vs. Ag/AgCl).[107] The upper trace corresponds to four phenols containing reducible nitro groups; the lower recording contains the peaks for all the phenols that can be oxidized.

Pesticides. Pesticides of the carbamate class and amine-derived compounds (Table 4.11) can be oxidatively determined. Detection limits as low as 40 pg have been obtained after separation by a C_{18} reverse-phase LC column.[289,290] These limits are superior to other competitive methods for the direct determination of carbamate pesticides without preconcentration, including direct-in-

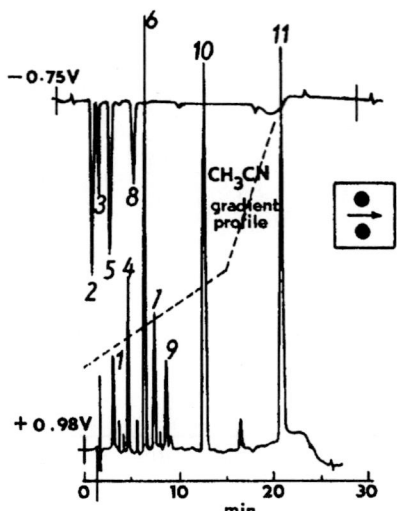

FIGURE 4.28. Simultaneous oxidation and reduction chromatograms of priority pollutant phenols with parallel-adjacent glassy carbon electrodes at -0.75 and +0.98 V (vs. Ag/AgCl). Linear gradient from 30% v/v acetonitrile, 70% 0.05 M NaClO$_4$/0.005 M citrate, pH 3.8 to a 50/50% mixture within 15 min, and then to an 80/20% mixture within 20 min. Flow rate, 2.0 mL/min. Column, Biophase ODS, 5 μm, 250 × 4.6 mm. 1, phenol; 2, 2,4-dinitrophenol; 3, p-nitrophenol; 4, o-chlorophenol; 5, 4,6-dinitro-o-cresol; 6, 2,4-dimethylphenol; 7, 4-chloro-3-methylphenol; 8, o-nitrophenol; 9, 2,4-dichlorophenol; 10, 2,4,6-trichlorophenol; 11, pentachlorophenol. (Reproduced with permission from Roston et al.[107])

jection GC (≥10 ng), LC with variable-wavelength UV detection (1–10 ng) and fluorescence detection after derivatization (1–10 ng). Microarray Kelgraf electrodes are also reported to yield lower detection limits (50–430 pg range) than other electrode materials including glassy carbon.[290] Representative chromatographic traces for a river water sample are shown in Figure 4.29.[290]

Derivatives of thiourea (TU) such as alkylthiourea, ethylenethiourea, N,N'-diphenylthiourea as well as other pesticides can also be oxidized at modest potentials and thus determined directly or via complexation of the thiocarbonyl group with mercury ions[291,292]:

$$Hg + 2\,TU \rightarrow Hg(TU)_2^{2+} + 2\,e^-$$

The limiting current for this oxidation is proportional to the ligand concentration. The detection at +0.19 V is not very sensitive (the detection limit is ~10 ng), but the analysis has good selectivity. Therefore ethylthiourea can be determined in urine without sample pretreatment.

4.5. Amperometric Sensors for Environmental Pollutants

Table 4.11. Carbamate and Amine Pesticides

Trade Name	Chemical Name	Chemical Structure
Aminocarb	4-(dimethylamino)-*m*-tolyl methylcarbamate	
Asulam	methyl [(4-aminobenzene)-sulfonyl]carbamate	
BPMC	2-*sec*-butylphenyl *N*-methylcarbamate	
Carbaryl	1-naphthyl *N*-methyl-carbamate	
Carbendazim (MBC)	methyl 2-benzimidazole-carbamate	
Chloramben	3-amino-2,5-dichloro-benzoic acid	

Cont. Table 4.11. Carbamate and Amine Pesticides

Trade Name	Chemical Name	Chemical Structure
Chlorpropham	isopropyl N-(3-chlorophenyl)carbamate	
Desmedipham	3-ethoxycarbonylaminophenyl N-phenylcarbamate	
Dichloran	2,6-dichloro-4-nitroaniline	
Phenmedipham	3-methoxycarbonylaminophenyl N-(3'-methylphenyl)carbamate	
Picloram	4-amino-3,5,6-trichloropicolinic acid	

Source: Anderson and Chesney.[289]

4.5. Amperometric Sensors for Environmental Pollutants   259

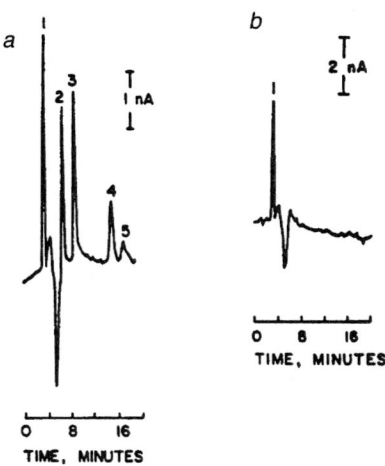

FIGURE 4.29. Chromatographic separation of pesticides in river water. Eluant methanol/aqueous acetate-phosphate buffer (58:42); flow rate, 0.90 mL/min; background current offset, 129 nA; applied potential +1.1 V vs. Ag/AgCl; working electrode, Kelgraf (15% graphite by weight), sheathed in Kel–F. All samples injected in 20 µL volume. (a) River water spiked with 7.50×10^{-7} M concentrations of all pesticides: (1) solvent front, (2) carbendazim (2.8 ng), (3) aminocarb (3.1 ng), (4) desmedipham (4.5 ng), (5) dichloran (3.1 ng). (b) Unspiked river water: (1) solvent front. (Reproduced with permission from Anderson et al.[290])

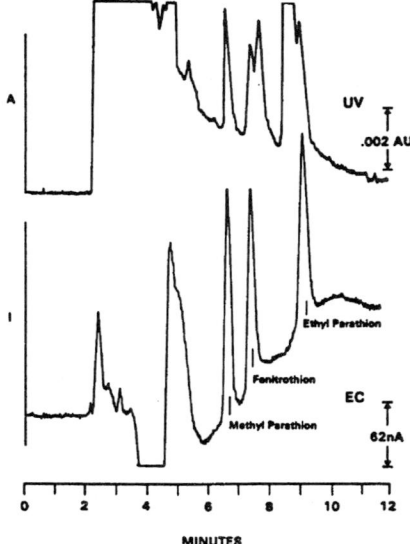

FIGURE 4.30. Chromatographs of a turnip green sample using series ultraviolet and electrochemical detection: 40.04 ng of methyl parathion and 44.52 ng of ethyl parathion injected; mobile phase, 64% acetonitrile, 36% 0.05 M ammonium acetate, pH 5.0; flow rate, 1 mL/min; UV detector at 270 nm; EC detector at -0.97 V vs. Ag/AgCl. (Reproduced with permission from Clark et al.[294])

Table 4.12. A Listing of Agricultural Chemicals Suitable for HPLC-hυ-EC

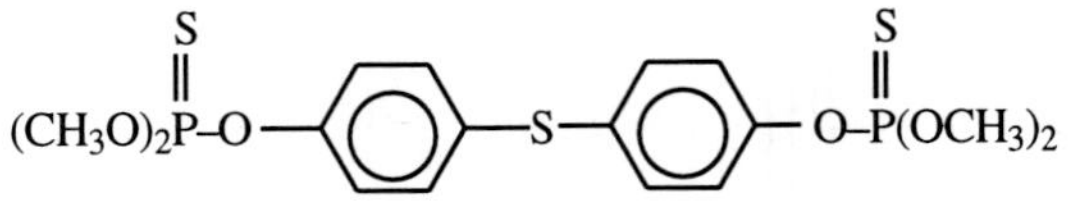

$(CH_3O)_2\overset{S}{\overset{\|}{P}}-O-\underset{}{\bigcirc}-S-\underset{}{\bigcirc}-O-\overset{S}{\overset{\|}{P}}(OCH_3)_2$

Abate

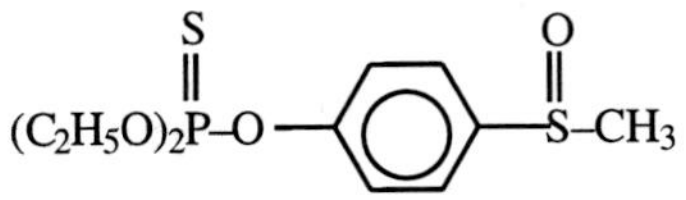

$(C_2H_5O)_2\overset{S}{\overset{\|}{P}}-O-\underset{}{\bigcirc}-\overset{O}{\overset{\|}{S}}-CH_3$

Dasanit

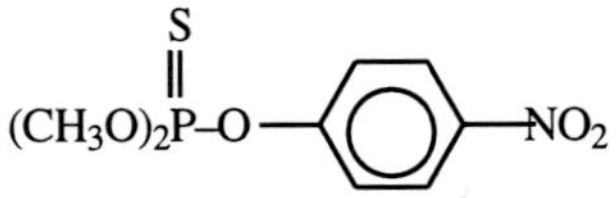

$(CH_3O)_2\overset{S}{\overset{\|}{P}}-O-\underset{}{\bigcirc}-NO_2$

Parathion

Guthion: benzotriazinone with N–CH$_2$SP(OCH$_3$)$_2$ (P=S)

$(CH_3O)_2\overset{S}{\overset{\|}{P}}-S-\underset{CH_2COOC_2H_5}{\overset{H}{\overset{|}{C}}}-COOC_2H_5$

Malathion

Dioxathion: dioxane ring with two S–P(=S)(OC$_2$H$_5$)$_2$ substituents

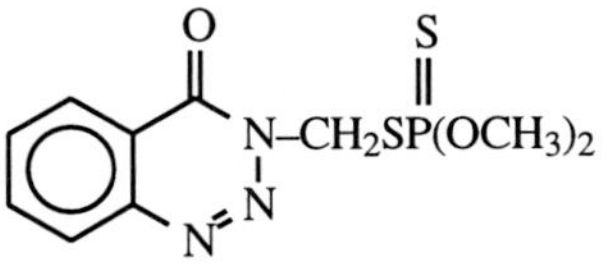

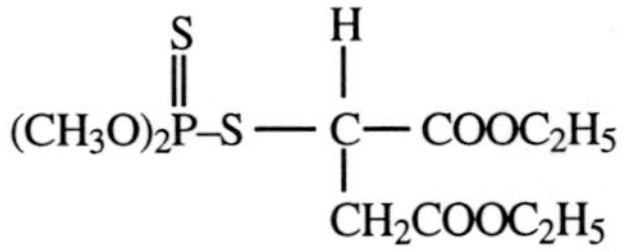

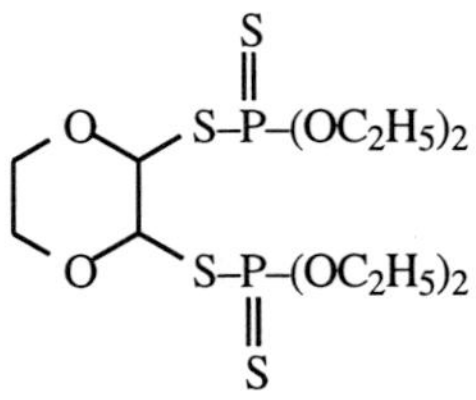

$(C_2H_5O)_2\overset{S}{\overset{\|}{P}}-CH_2-S-C_2H_5$

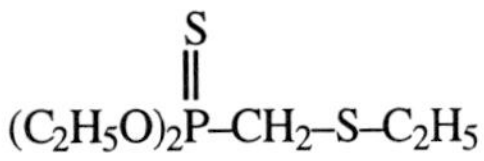

Thimet

$(C_2H_5O)_2-\overset{S}{\overset{\|}{P}}-S-CH_2-S-\overset{S}{\overset{\|}{P}}-(OC_2H_5)_2$

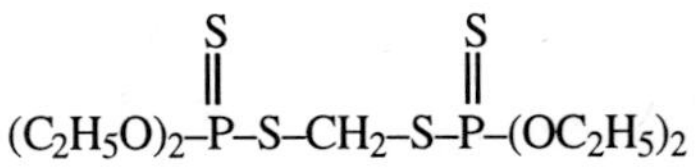

Ethion

4.5. Amperometric Sensors for Environmental Pollutants

Cont. Table 4.12. A Listing of Agricultural Chemicals Suitable for HPLC-hυ-EC

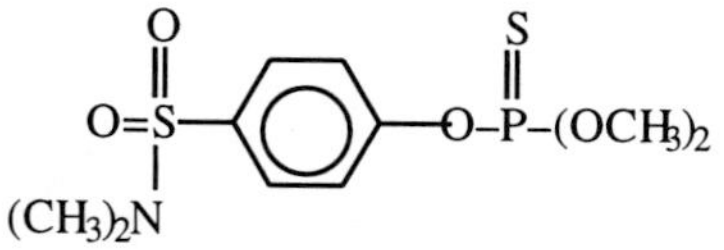

Famphur

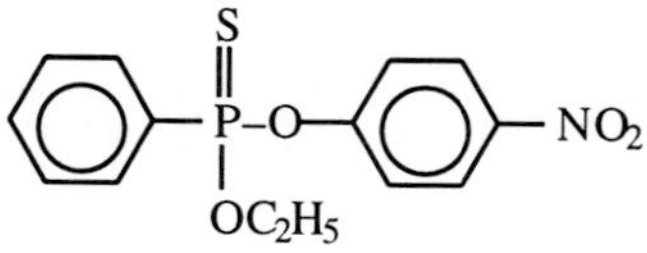

EPN

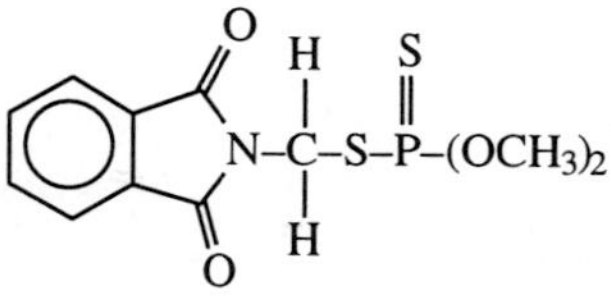

Imidan

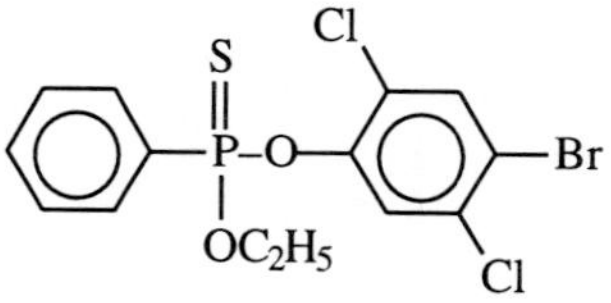

Leptophos

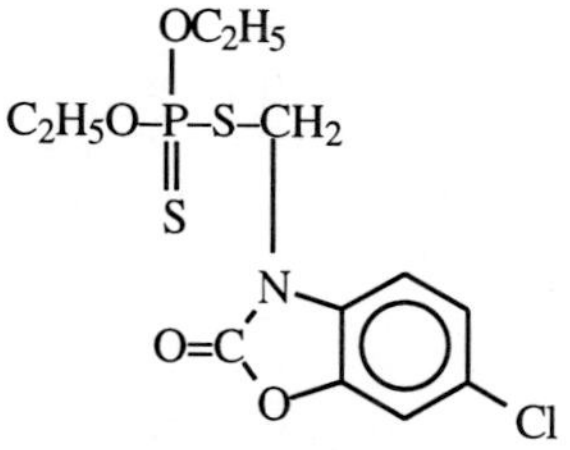

Phosalone

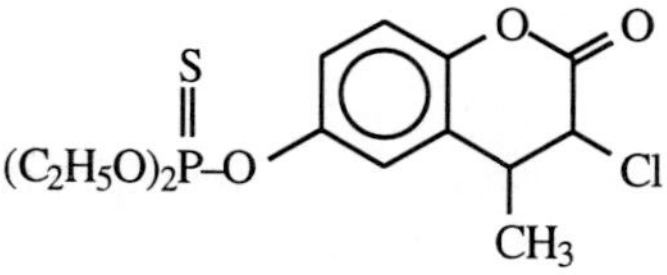

Coumphos

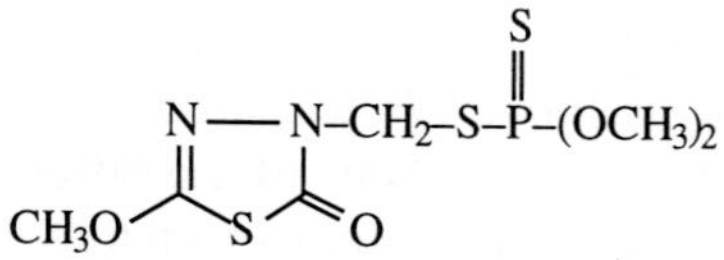

Supracide

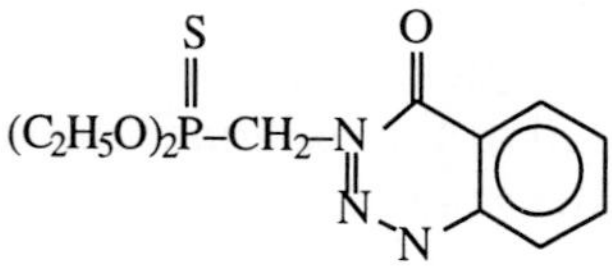

Ethyl Guthion

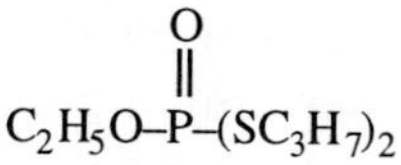

Mocap

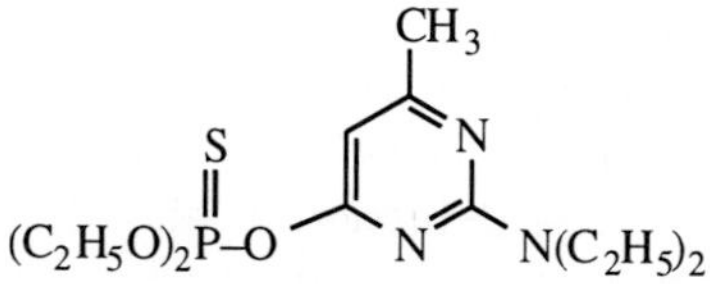

Pipimiphos Ethyl

Source: Dung and Krull.[295]

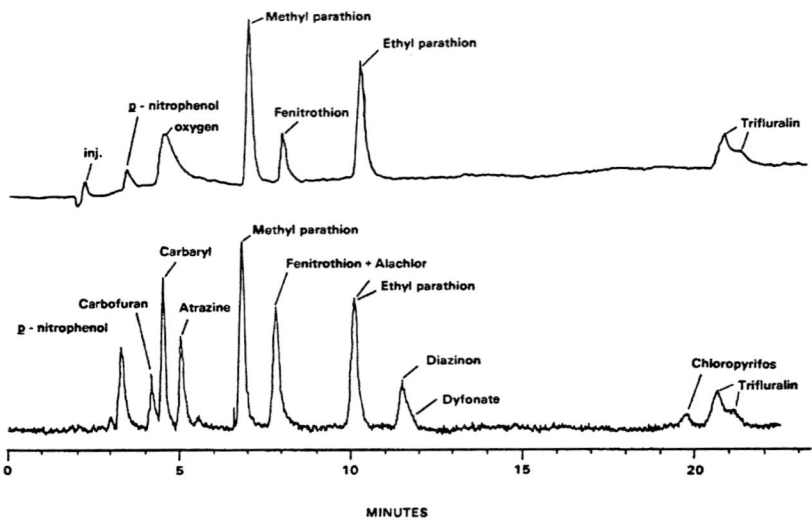

FIGURE 4.31. Comparison of selectivity of electrochemical (upper trace) and ultraviolet (lower trace) detectors for a group of pesticides possibly present in runoff waters from middle Tennessee. Approximately 35 ng of each pesticide was injected. Chromatographic conditions were the same as those in Figure 4.30. (Reproduced with permission from Clark et al.[294])

Organophosphorous compounds such as ethyl and methyl parathion (Table 4.12) used to be popular pesticides and have been applied to a variety of crops including green vegetables. These applications leave residue on the produce and also in surface waters that drain from the cropland. Reductive amperometric detection has been employed for the analysis of these pesticides in runoff water[293,294] as well as in green vegetables.[294] In the latter case, a series UV/EC detection scheme was employed for the analysis of the effluent from an octadecyl silane column. The UV detector responded to a greater number of compounds, on the one hand, while the greater selectivity of the electrochemical detector afforded quantitation capability without the necessity of good chromatographic separation from the interfering plant matter. Figure 4.30 contains both types of chromatograms from a turnip green sample.[294] The identification of the pesticide peaks is more reliable in the electrochemical case. Figure 4.31 and Table 4.13 contain data from a further suite of 14 pesticides.[294]

Oxidative detection of malathion, parathion, and other similar pesticides has been reported using on-line UV photolysis coupled with either FIA or HPLC[295] and dual-electrode oxidative detection. Table 4.12 lists the agricultural chemicals that have been successfully analyzed by this HPLC-hυ-EC

4.5. Amperometric Sensors for Environmental Pollutants

Table 4.13. Electrochemical and Ultraviolet Response Characteristics of Some Common Pesticides

Compound	Relative Retention Time	Absorbance max, nm	Electroactive (at -0.85 V)[a]
Alcohol	1.20	235	No
Atrazine	0.67	265	No
Carbaryl	0.64	280	No
Carbofuran	0.62	270	No
Chloropyrifos	3.86	290	No
Diazinon	1.73	245	No
Dyfonate	1.80	240	No
Ethyl parathion	1.63	270	Yes
Fenitrothion	1.16	265	Yes
Lannate	0.47	233	No
Methyl parathion	1.00	270	Yes
p-Nitrophenol	0.48	310	Yes
Orthene	0.43	220	No
Trifluralin	4.57	270	Yes

[a] Potential versus Ag/AgCl reference.

Source: G. J. Clark, R. R. Goodin, and J. W. Smiley, *Anal. Chem.* **57**, 2223 (1985); D. C. Paschal, R. Bicknell, and D. Dresbach, ibid. **49**, 1551 (1977).

method. Figure 4.32 illustrates typical chromatograms for a mixture of six standard thiophosphates at the 0.4 - 2.0 ppm levels.[295] These dual traces were obtained from a single injection with the two electrodes in the parallel mode. No response was obtained with the UV lamp off, suggesting that (as yet unidentified) electroactive species are photolytically generated.

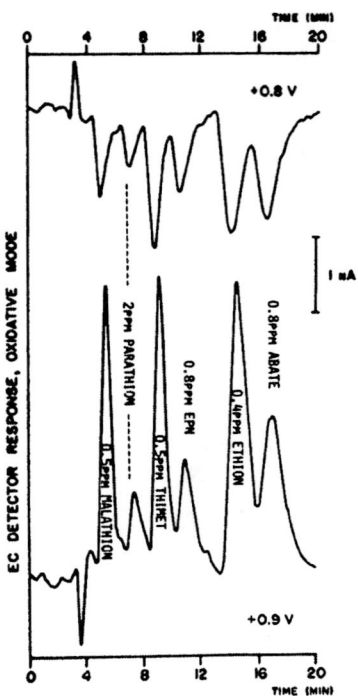

FIGURE 4.32. HPCL-*hυ*-EC dual detector chromatograms for a mixture of standard thiophosphate agricultural chemicals. HPCL used an RP C-18 10 μm column with MeOH–0.2 M NaCl (70:30) mobile phase at a 1.2 mL/min flow rate and BAS GC dual electrodes. (Reproduced with permission from Dung and Krull.[295])

Organophosphorous pesticides can also produce phenolic compounds through ester cleavage. For example, Fenitrothion (**I**)

produces 3-methyl-4-nitrophenol upon base hydrolysis.[296] These phenols can be oxidatively determined as discussed earlier. Detection limits in the 0.1–2 ng range were quoted for these compounds using a bonded octadecyl silane column fitted with a carbon fiber-based electrochemical flow detector.[296]

4.5. Amperometric Sensors for Environmental Pollutants 265

Amines. These compounds enter the environment either directly as industrial wastes or indirectly as degradation products of pesticides and azo dyes. Many of these substances are highly toxic or mutagenic or both. Because of their polar nature, GC analysis requires derivatization to prevent problems associated with adsorption and peak tailing. Detection strategies for use with GC or HPLC include flame ionization, flame thermionic emission, mass spectrometry, thermal energy analysis, conductivity, UV-visible absorption, polarography, amperometry, and pulsed photoelectrochemistry.[297]

Aromatic amines are readily oxidized at modest potentials (Chapter 3). The half-wave potentials for important carcinogenic aromatic amines are compiled in Table 4.14. Aliphatic amines (e.g., N-nitrosamines) require higher potentials, although reductive detection can also be employed in this instance. However, a difficulty is dissolved O_2 interference – the deaeration step can be problematic because of the volatility of the N-nitrosamines. A chemically modified glassy carbon electrode has been used for oxidative detection of these compounds, where ruthenium oxide stabilized by cyano crosslinkages acts as an electron mediating catalyst atop the glassy carbon electrode. A detection limit of ~10 nM was quoted by the authors for these compounds.[297] For the determination of aromatic amines, a UV detector was combined with LC detection to provide specificity and an alternative to the latter for those cases where UV interferences at 280 nm are negligible or can be removed from the water by an acidic extraction cleanup.[298] As with the cases shown in Table 4.15, the UV detection was about 50 times less sensitive than oxidative amperometry. In all these cases, the LC was operated in the reverse-phase mode. Aromatic amines and their derivatives (e.g., chloranilines) have been determined in body fluids,[299,300] atmosphere,[301] soil,[302] and water.[298] For the air analyses, the compounds were first adsorbed on fiber filters and then extracted with methanol prior to analyses.[301]

Sulfur Compounds. As mentioned earlier, the use of GC for the analyses of these compounds is often limited by the lack of volatility and the thermal instability in these compounds. An LC approach not requiring efficient separation is attractive in this regard, and this is facilitated by selective detection. Sulfur-containing pesticides (e.g., parathion, malathion) have been discussed earlier. A variety of other sulfur compounds (e.g., thiols, sulfides, disulfides, isothiocyanates, sulfones, thioureas) are amenable to amperometric detection of both types, via oxidative and reductive electrolyses. Table 4.15 compares the detection limits for UV photometric and amperometric detection.[303] Aryl thiols and sulfides have lower detection limits (relative to the alkyl and cycloalkyl) counterparts because of the stray UV absorption of the benzene chromophore. For thiols, amperometric detection is advantageous, since in addition to better sensitivity, it offers a degree of selectivity.

Table 4.14. Half-Wave (Peak) Potential Values of Some Aromatic Amines and the Detection Limits Obtained in Ultraviolet Photometric and Amperometric Detection

Substance	$E_{1/2}$ (E_p) (V)	Detection Limit (ng) Photometric	Amperometric
Benzidine (4,4'-diaminobiphenyl)	+0.36	3	0.003
o-Dianisidine (3,3'-dimethoxybenzidine)	+0.29 (+1.23)	4	0.05
3,3'-Dichlorobenzidine	(+0.51)	16	0.45
3,3'-Diaminobenzidine	+0.17	3	0.05
o-Tolidine (3,3'-dimethylbenzidine)	+0.33	4	0.03
4-Aminobiphenyl	(+0.58)	10	1.1
4-Nitrobiphenyl	-0.73 (DME)	12	—
1-Naphthylamine	(+0.51)	11.4	1.4
2-Naphthylamine	(+0.58)	11.4	9.0
2,5-Diaminotoluene	+0.45	12.2	0.06
4,4'-Methylenebis-(o-chloroaniline)	+0.29 (+0.63)	5	4.6

Source: J. Barek, V. Pacakova, K. Stulik and J. Zima, *Talanta* 32, 279 (1985).

Metabolites. The majority of toxic compounds are in themselves relatively inert and require metabolic activation to form species that initiate chemical damage within an organism. *In vitro* techniques are useful because of their ability to focus on a specific metabolic action. For example, the liver microsomal fraction has become popular for studying the initial metabolism of xenobiotics. The reactive intermediates that are formed enzymatically are formed in extremely low concentrations, and new analytical methodology to determine the mechanism by which they form, has become necessary. This is because conventional procedures require either lengthy preconcentration steps to detect metabolites or the use of radiolabels followed by chromatography. Some compounds (e.g., hydroxylamine) may decompose during these work-up procedures. In these instances, LC-EC detection is effective and has been

4.5. Amperometric Sensors for Environmental Pollutants

Table 4.15. Detection Limits in HPLC Analysis of Sulfur Compounds with UV and Amperometric Detection

Compounds	UV Detector $D_{ng}{}^a$	UV Detector $D_{ppm}{}^b$	Amperometric Detector $D_{ng}{}^a$	Amperometric Detector $D_{ppm}{}^b$
Thiols				
Alkyl	380	4.8	42^c	0.54
Aryl	0.8	0.01	0.24^c	0.003
Cycloalkyl	1.30	1.7	17^c	0.21
Sulfides				
Alkyl	700	10		
Aryl	0.7	0.01		
Cycloalkyl	200	2.5		
Disulfides				
Alkyl	50	0.6		
Aryl	10	0.1		
Isothiocyanates				
Aryl	0.6	0.009	9^d	0.12
Thioamides				
Alkyl	0.5	0.007	3^d	0.03
Aryl	0.2	0.003	5^d	0.06
Sulfones				
Aryl	25	0.3		
Thioureas				
Alkyl	1	0.01	0.25^c	0.003
Aryl	0.4	0.005		

[a] Nanograms on-column required to produce a signal/noise ratio of 2.
[b] Injection volume assumed to be 100 μL.
[c] Detector potential E = +0.5 V vs. Ag/AgCl.
[d] Detector potential E = -1.0 V vs. Ag/AgCl.

Source: J. A. Cox and A. Przyjazny, *Anal. Lett.* 10, 869 (1977).

used for studying the oxidative metabolism of such compounds as aromatic amines,[304,305] benzene,[306] and azo dye intermediates.[271] For example, a dual-electrode LC-EC detector in the series mode permitted determination of 4-nitroaniline and its metabolites (2-amino-5-nitrophenol, N-hydroxyl-4-nitroaniline) from the commonly used textile dye, Disperse Orange 3, in incubation media.[271] Detection limits for these compounds were in the subpicomole range. Similarly, preconcentration of microsomal metabolites of benzene via solvent extraction of the incubation mixture permitted LC-EC detection of hydroquinone, catechol, and dihydroxy derivatives in addition to the major metabolite, phenol.[306]

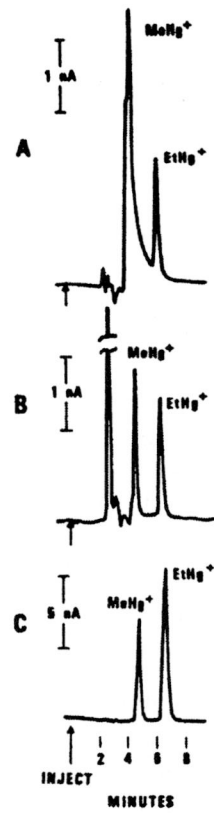

FIGURE 4.33. Eliminating the interference of heavy metals on the organomercury measurement: (A) simple amperometric detection at -0.83 V of a 5×10^{-6} M solution of MeHg$^+$, Cu^{2+}, Cd^{2+} and Pb^{2+}; (B) same conditions except 10^{-4} M EDTA has been added to the sample; (C) chromatogram of the same sample as A except the detection is in the differential pulse mode of detection at -0.74 V. (Reproduced with permission from MacCrehan and Durst.[307])

4.5. Amperometric Sensors for Environmental Pollutants 269

Organometals. The difficulties with analysis of "real-world" samples for organometals were presented in an earlier paragraph. In general, GC separation requires thermally stable, strong complexes of cationic organometals (e.g., CH_3Hg^+) to be made by derivatization before analysis. Reductive amperometric detection has been coupled with reverse-phase LC for the determination of a variety of organometals including methyl-, ethyl-, and phenylmercury and trimethyl- and triethyllead.[82,307] Figure 4.33 contains representative data.[307] Simultaneous collection of other reducible species (e.g., Cd^{2+}, Pb^{2+}, and Cu^{2+}) is a problem that can be circumvented by altering either the separation chemistry (see Figure 4.33B) or the selectivity of the detector by employing a differential pulse waveform (see Figure 4.33C). In the former instance, ethylenediamine tetraacetate (EDTA) was added to the samples prior to injection. The EDTA apparently forms stronger complexes with the divalent ions than 2-mercaptoethanol (which was added as a complexing agent to neutralize the positive charge on the organometal cations),[82] but methyl- and ethylmercury do not. In the latter instance, the detector response is limited to a small potential window (see Figure 4.34), thus providing discrimination from other coeluting species. This approach has the advantage of minimizing sample manipulation and handling prior to analysis. The detection limit for methylmercury was 40 pg, and this substance was determined in tuna fish and shark meat.[307] A recent study also combines HPLC and PAD for the separation and determination of alkyl lead compounds.[308]

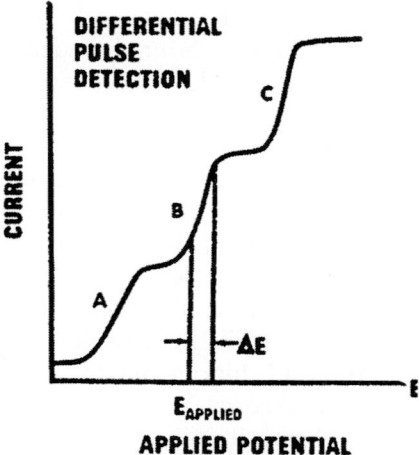

FIGURE 4.34. Hydrodynamic voltammetry of three reducible, coeluting species. The potential is pulsed in a small window, ΔE to provide analyte discrimination. (Reproduced with permission from MacCrehan and Durst.[307])

270   Chapter 4

<u>Metals and Inorganic Anions</u>. Atomic absorption spectrometry is most commonly used for the determination of "heavy" metals such as Pb, Cd, Hg, Co, Ni, Cr, and Cu. However, this method is time consuming if more than one element is to be determined. Dithiocarbamates (II and III)

$$\underset{S^-}{\overset{S}{\underset{\|}{\|}}}\!\!C\!-\!N\underset{CH_2CH_3}{\overset{CH_2CH_3}{\diagup}}\qquad\qquad \underset{S^-}{\overset{S}{\underset{\|}{\|}}}\!\!C\!-\!N\!\!\diagup\!\!\Box$$

II   III

(dtc⁻) have been used for complexing a variety of metal ions and determining them by a combination of HPLC and either UV photometric or amperometric detection.[276-288,309,310] Many of these complexes undergo ligand-based oxidation, and this irreversible reaction can be used as a basis for oxidative amperometry. This contrasts with the reductive electrochemical approach commonly used for the analysis of these metal ions (see later). One problem with the amperometric detection was the interference from the excess ligand. This could be partially circumvented via the use of a suppressor (guard) column to bind the dtc⁻ ligand via anion exchange. Limits of detection "substantially less than 1 ng" have been quoted by the authors.[277]

Inorganic anions such as nitrite, phosphate, and bromide have been determined either by FIA[311] or ion chromatography[312] coupled with amperometric detection. In the former case, acidic bromide and molybdate were added to the flow stream and the resultant nitrosyl bromide and molybdophosphate were determined amperometrically.[311] The electrochemistry underlying the detection strategy was not described in detail by these authors. Bromide in snow samples was determined with a detection limit of less than 1 ppb (without preconcentration) by using a silver electrode coated with a AgI layer.[312] The usual method of ion chromatography coupled with conductivity detection fails in this case because of the large excess (~100-fold) of other anions (e.g., NO_3^-) normally present in environmental samples. Other interesting applications of amperometric sensors for inorganics include an in-line method for monitoring pulp bleaching with chlorine dioxide.[313] The reduction of chlorine dioxide yields chlorite, and the concentrations of chlorine dioxide and chlorite measured by the in-line amperometric method correlates well with the standard titrimetric method.[313] The purpose of chemical pulp bleaching is primarily to remove lignin, and in the final (bleaching) stage to remove colored compounds from the pulp. These results suggest that the amperometric approach may also be used for monitoring chlorine dioxide in water treatment plants. Chlorine dioxide is now being considered a chlorine substitute for water disinfection because of the problems associated with the latter, of disinfection by-products (Chapter 7).

4.6. AMPEROMETRIC GAS SENSORS

The earliest successful amperometric gas sensor is the Clark electrode used for O_2 determination.[314] This sensor design uses a membrane-covered planar electrode through which the analyte gas diffuses into a thin (5–15 µm) electrolyte layer. Once in the layer, the dissolved O_2 molecules are electrochemically reduced, and the magnitude of the resultant (cathodic) current provides a measure of the concentration of O_2 in the sampled gas. Figure 4.35 illustrates the principle. Interestingly, the same electrode can be used for detecting peroxide in the oxidative mode. Polypropylene or polyethylene are the membrane materials of choice; the membrane thickness is typically 20–25 µm.[315] Both these polymers are good electrical insulators and their permeability to O_2 is practically unchanged in the presence of water vapor. This facilitates their use in matrices where the relative humidity may vary from sample to sample. Two-electrode configurations are usually used for the Clark sensor and the reference/counter electrode is usually silver–silver chloride.

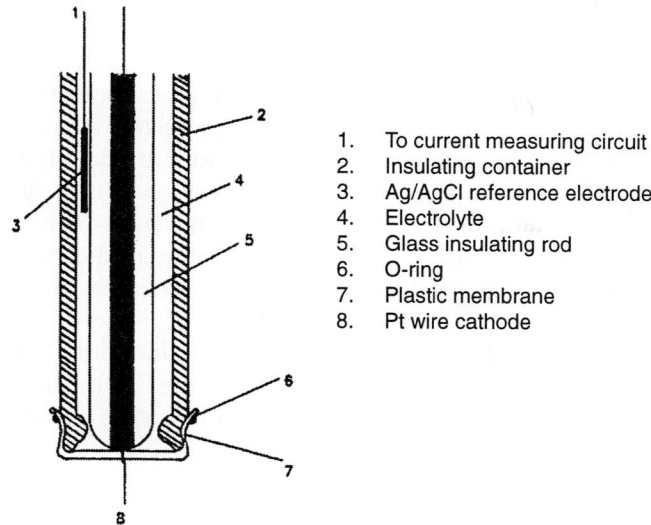

1. To current measuring circuit
2. Insulating container
3. Ag/AgCl reference electrode
4. Electrolyte
5. Glass insulating rod
6. O-ring
7. Plastic membrane
8. Pt wire cathode

FIGURE 4.35. Schematic outline of a practical Clark amperometric O_2 cell. (Reproduced with permission from Chang et al.[315])

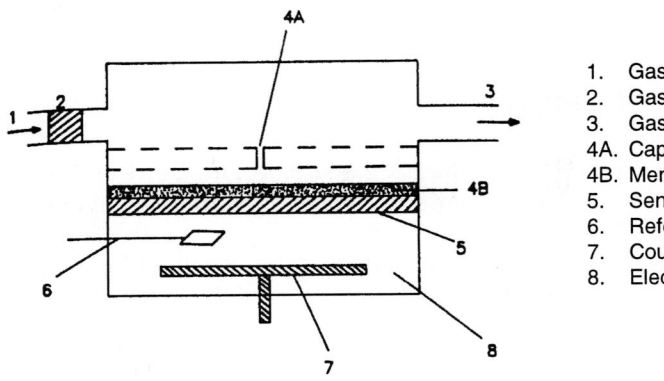

FIGURE 4.36. Schematic diagram of the amperometric gas sensor system. (Reproduced with permission from Chang et al.[315])

Figure 4.36 contains a generic structure of an amperometric gas sensor system comprising six major parts: filter, membrane (or capillary), working (sensor) electrode, electrolyte, counter, and reference electrodes. Developments in fuel cell technology have had a major impact on new design configurations for gas sensing electrodes. For example, Teflon-bonded gas-diffusion electrodes and the use of Nafion as the membrane material largely have their origin in fuel cells. In fabricating a gas sensor with fast response time, it is useful to analyze the time constants associated with the discrete steps that form the sensor strategy.[316] The following steps have been identified for the gaseous analyte (see Figure 4.37).[316] (a) transport or permeation to the surface of the gas-permeable membrane; (b) diffusion through the membrane; (c) dissolution in the internal electrolyte; (d) further reaction–dissociation in the electrolyte; (e) diffusion to the sensing electrode surface; and (f) electrochemical reaction at the sensor surface.

Progress in mathematical modeling of these individual steps has been reviewed for important sensor designs including the Clark electrode.[315] The overall system response can be viewed as a "resistance" to current passage as follows:

$$1/R(s) = 1/R(d) + 1/R(m) + 1/R(k) \tag{4.10}$$

4.6. Amperometric Gas Sensors

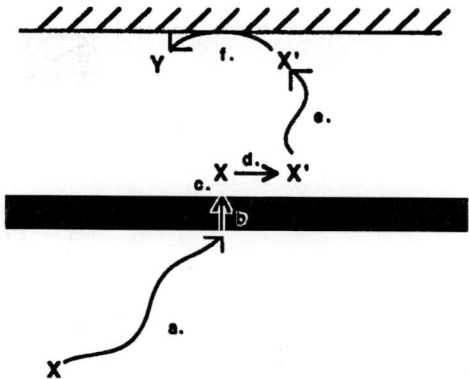

FIGURE 4.37. Schematic diagram of the process of electrochemical gas sensing. The shaded region comprises the membrane. Processes *a* through *e* are discussed in the text. (Reproduced with permission from Tierney and Kim.[316])

where $R(s)$ is the system resistance, $R(d)$ is the diffusion resistance, $R(m)$ is the membrane (electrolyte solubility and mass transport) resistance, and $R(k)$ is the "resistance" of the electrocatalytic reaction. In most sensor designs, the second and third terms on the right hand side of (4.10) are the rate-controlling steps. This is because membrane diffusion (D: ~$10^{-8} - 10^{-10}$ cm^2/s) and mass transport in the internal electrolyte are slow relative to mass transport in the gas phase. Further, steps *d* and *f* are usually fast, and the latter can be further optimized by suitable electrocatalytic modification of the sensor surface. The time for step *e* can be reduced by pressing the sensor electrode against the gas-permeable membrane so that the electrolyte layer is thin. Usually step *b* is the rate-limiting step.

Recent innovations in ensuring fast sensor response have included the use of an extremely thin (2 µm) Teflon film as the gas-permeable membrane,[317] and the elimination of a separate gas-permeable membrane altogether.[316] In the former instance, a response time of 40 ms was attained. In the latter case, a porous substrate with thin-film electrodes deposited on it such that the electrodes were porous as well, was utilized. A solid layer of polymer electrolyte (Nafion gel) covers the electrodes but does not protrude into the pores of the substrate. The electrochemical reduction (or oxidation) of the analyte gas begins when the gas reaches the "triple points," sites where the electrode, the electrolyte, and the gas phase meet (Figure 4.38A).[316] A response time of 150 ms was attained with this design for O_2 detection (Figure 4.38B). The same sensor design was also used for the detection of NO and CO_2, the latter in the potentiometric mode (see Figure 4.7).

(A)

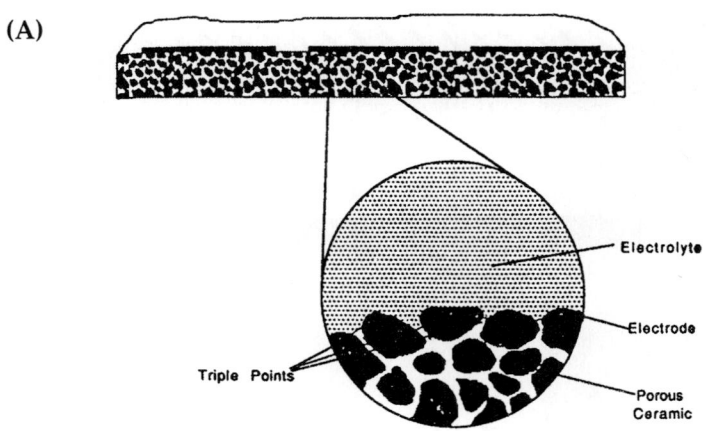

(B)

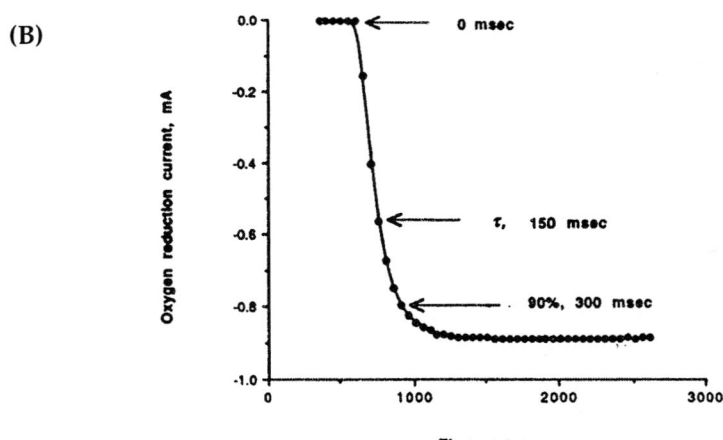

FIGURE 4.38. Structure of a solid polymer electrolyte membrane/electrolyte interface and response of the O_2 sensor to a step change from 0 to 100% O_2. Applied potential -0.6 V vs. Ag/Ag$_2$O. (Reproduced with permission from Tierney and Kim.[316])

A variety of analytes have been determined using amperometric gas sensors, including hydrocarbons and carbon–oxygen compounds, oxides of nitrogen and sulfur, other sulfur compounds (e.g., thiols, H_2S), reduced nitrogen species (e.g., NH_3, N_2H_4, amines), and miscellaneous species such as O_3, Cl_2, HCl, HCN, PH_3, AsH_3, and CFCs. Table 4.16 provides a representative listing. Detection limits for most of these analytes are in the ppm range, although in

4.6. Amperometric Gas Sensors

Table 4.16. Examples of Amperometric Gas Sensor Reactions

Analyte	Sensor Electrochemistry
NO_2	$NO_2 + 2\,H^+ + 2\,e^- \rightarrow NO + H_2O$
NO	$NO + 2\,H_2O \rightarrow NO_3^{2-} + 4\,H^+ + 2\,e^-$
N_2O	$N_2O + H_2O + 2\,e^- \rightarrow N_2 + 2\,OH^-$
CO	$CO + H_2O \rightarrow CO_2 + 2\,H^+ + 2\,e^-$
CO_2[a]	$CO_2 + H_2O + 2\,e^- \rightarrow HCOO^- + OH^-$
O_3	$O_3 + 2\,e^- + 2\,H^+ \rightarrow O_2 + H_2O$

[a] Indirect sensing strategies have been developed; refer to text.

some cases, they have been extended to the ppb range. A variety of electrode materials have been used including Au, Ag, Pt,[315] and porous carbon felt.[318] Some of these materials afford selectivity in detection. For example, the oxidation of CO at Pt is facile whereas it is very slow at Au. In contrast, H_2S is very reactive to the Au surface. Therefore a mixture of H_2S and CO can be selectively analyzed for the former at a Au sensor surface. Frequently, the potential can also be controlled for selective detection. For example, NO oxidation does not occur at potentials below 1.0 V but NO_2 can be reduced at this surface at ~0.8 V. Hence NO_2 can be detected in the presence of NO by poising the gold at 0.8 V.

An interesting example where the analyte is not directly electrolyzed is contained in recent reports on a new CO_2 sensor strategy.[319,320] The sensor response arises in this case from changes in the electrochemistry of a solution of Cu(II) (bis-1,3-propanediamine) ion in aqueous potassium chloride when it is contacted by an atmosphere of CO_2. The addition of CO_2 to the electrolyte leads to a decrease in its pH and the conversion of the diamine complex to uncomplexed Cu^{2+}; the equilibrium concentration of Cu^{2+} is detected amperometrically in this sensor.

Applications of amperomeric gas sensors have been recently reviewed.[315,321] Apart from environmental applications related to process (flue gas) monitoring, industrial hygiene/safety, and indoor air pollution, another major realm of applicability is the clinical area. The latter application has been reviewed.[21] For example, there has been commercialization of O_2 sensors for transcutaneous monitoring in infants. Measurements of the partial pressure of O_2 in blood constitute a major clinical application of amperometric gas sensors.

Finally, an important application for gas-phase amperometry is as a water or humidity sensor. Perhaps the most advanced sensor design in this category is that based on the use of P_2O_5.[322] This material is well known for its affinity to water. The electrolysis of syrupy H_3PO_4 as a film between two electrodes yields P_2O_5 as a paste; uptake of H_2O by P_2O_5 yields a current on imposition of sufficient potential across the two electrodes for the electrolysis of water:

$$H_2O(g) \rightleftarrows H_2O \text{ (film)} \rightarrow H_2 \text{ (cathode)} + \frac{1}{2}O_2 \text{ (anode)}$$

The relative fragility of the P_2O_5 film and its decomposition on exposure to high humidity levels continue to be problems. To this end, amperometric microsensors for water have been recently described that incorporate perfluorosulfonate ionomer (PFSI) membranes (e.g., Nafion, Dow PFS1) into the sensor design.[323,324] These sensors are reported to have fast response times, no significant hysteresis or flow-rate dependence, and feature amenability to operation from sub-ppm to saturation levels of water.

4.7. STRIPPING ANALYSIS: SPECIALIZED ASPECTS

Stripping voltammetry was introduced in Section 2.6.2. It was seen to be a two-step technique incorporating a preconcentration (analyte accumulation) step followed by a "probe" scan. We shall explore aspects of this versatile trace-analysis approach in more detail in this section. Anodic stripping voltammetry (ASV) is the forerunner of this category of techniques. Interestingly enough, it has been pointed out[325] that much of the progress since the 1960s has been due to an improved understanding of the chemistry underlying the preconcentration step rather than because of instrumentation or computation advances. This trend is distinct from other areas of analytical chemistry, such as spectroscopy and chromatography, where many of the technological advancements are computationally driven (e.g., multiplex spectroscopy, array detection). Much of the innovation in the area of stripping analyses is related to expanding the scope of the technique and enhancing the sensitivity or the limit of detection. Therefore, the advent of adsorptive stripping voltammetry (discussed next) has expanded the range of measurable analytes from about 25 trace metals originally to about 45 elements.[326]

4.7.1. Adsorptive Stripping Voltammetry

The preconcentration step in classical ASV consists of deposition of the analyte (metal ions) onto a mercury electrode. Thus, amalgam-forming elements are collected efficiently, whereas elements that do not form an amalgam

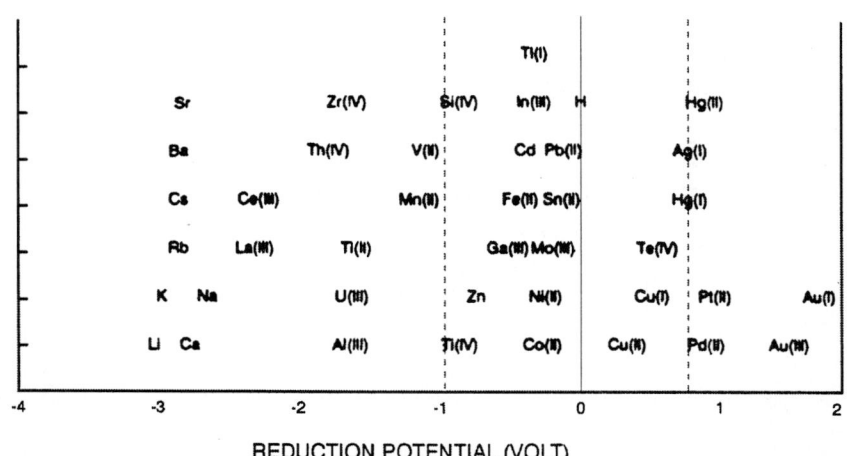

FIGURE 4.39. Standard reduction potentials to the metallic state of a number of metals in noncomplexing and acidic conditions. (Reproduced with permission from van den Berg. [325])

with mercury or are reduced to the metallic state at very negative potentials cannot be determined.

Figure 4.39 lists the environmentally important metals that can be determined by ASV at mercury-based electrodes, along with many others that are precluded. The dotted lines in the figure show the boundaries that encompass the analyzable metals – their boundaries are defined by the overpotential of hydrogen discharge (at the negative limit) and the susceptibility of mercury to oxidation beyond ~0.8 V. Unfortunately, however, the positive limit is shifted toward more negative potentials (by ca. 0.4 V) by chloride complexation of Hg(I)[325] such that measurements in "real-world" matrices (e.g., seawater) are limited to a rather smaller potential window than that indicated by Figure 4.39. The reduction potentials in Figure 4.39 are also shown for noncomplexing and acidic media.

The other practical difficulty with conventional ASV is that many of the metals are clustered together in Figure 4.39 (e.g., Sn, Pb, Tl, and Cd). The concentrations of Pb and Cd, though low, tend to be much higher than Sn and Tl so that only the former two metals are determinable in natural water. These and Cu and Zn are the only species present in uncontaminated waters at detectable levels so that ASV (without further preconcentration steps such as solvent extraction or ion exchange and masking or complexation techniques) of natural water is limited to these four metals.[325]

Adsorptive stripping voltammetry exploits the unique selectivity of interfacial adsorption for the preconcentration of analytes at the electrode surface. In that the initial deposition step does not utilize conversion to the metallic state, *any* oxidation state can be collected with two concomitant advantages:[325]

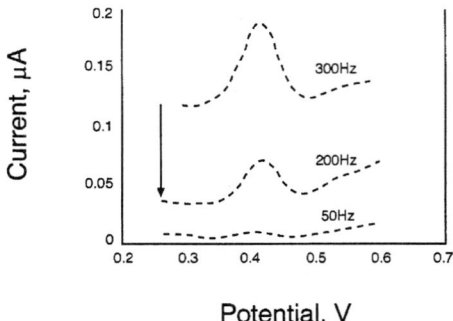

FIGURE 4.40. Effect of increasing the square-wave frequency on the peak current for 2 nM Cu in seawater. The measurements were performed on board a ship in the North Atlantic. The conditions were 25 µM quinolin-8-ol, 0.01 M borate buffer (pH 8.3), 90 s adsorption at -0.25 V. (Reproduced with permission from van den Berg. [325])

- Any element with a reduction potential (any reduction potential, not just that to the metallic state) falling within the stability range of mercury and hydrogen, is accessible.
- The material is collected as a monolayer on the surface, so that *all* the material is instantaneously accessible to assay. Thus, there are no diffusion limitations (as in conventional ASV) and fast potential scan techniques (e.g., square-wave or staircase voltammetry) can be employed, producing larger signals.

The analyte material that has been collected via adsorption is usually quantified by reduction of either the metal or the ligand in the complex. Unlike in ASV, the oxidative analysis is less frequently employed. Occasionally, the catalytically accelerated reduction of dissolved reactants can be utilized to quantify the adsorbed material.[325] Table 4.17 lists environmentally important species that have been successfully determined by adsorptive stripping voltammetry.[325] The sensitivity of adsorptive cathodic stripping voltammetry (ACSV) is very high, with typical limits of detection being at the $pM - nM$ level. Importantly, the analysis time can be much reduced (relative to ASV) by using square-wave modulation at high frequencies by reducing the width of the potential step. Figure 4.40 provides an example of the signal enhancement obtained with high-frequency modulation for the specific instance of Cu assay of seawater.[325] These data were acquired on site on a measurement vessel in the North Atlantic.

One difficulty with ACSV is interference from competitive adsorption by surface–active organic compounds. Thus the natural surfactants present in fresh or estuarine waters diminish significantly the sensitivity of ACSV. This is

Table 4.17. Examples of the Application of Adsorptive Cathodic Stripping Voltammetry to Environmentally Important Species

Analyte	Reagent[a]	Buffer (pH)[b]	Limit of Detection[c]	Scan Type[d]
Reduction of the Element				
As	Copper	1 M HCl	3	DP
Cd	Quinolin-8-ol	HEPES (7.8)	0.1	DP
Co	DMG	NH_4^+ (9.2)	0.1	DP
Cu	Quinolin-8-ol	Borate (8.5)	0.2	DP
Cu	Catechol	Borate (8.5)	0.3	DP
Cu	Tropolone	Borate (8.5)	0.4	DP
Fe	Catechol	PIPES (6.8)	2	DP
I	Copper	Acetate	3	DP
I	Mercury(I)	Natural pH	0.6	SW
Pb	Quinolin-8-ol	HEPES (7.8)	0.3	DP
Se	Copper	$(NH_4)_2SO_4$	3	DP
U	Quinolin-8-ol	PIPES (6.8)	0.2	DP
Reduction of the Ligand				
Mn	Erichrome Black T	PIPES	5	
U	Mordant Blue	Acetate	1	
Catalytic Effects				
Cr	DTPA[e]	Acetate	0.5	

a DMG = dimethylglyoxime; DTPA = diethylenetriaminepentaacetic acid.
b HEPES = N-hydroxyethylpiperazine-N'-2-ethane-sulfonic acid; PIPES = piperazine-N,N'-bis-2-ethanesulfonic acid.
c Quoted in ppm for 60 s accumulation time.
d DP = differential pulse; SW = square wave.
e Nitrate is used as the solution reactant, and the catalytic oxidation of Cr(II) to Cr(III) by NO_3^- is used as a measure of the chromium concentration.

Source: C. M. G. van den Berg, Anal. Chim. Acta 250, 265 (1991).

less of a problem with seawater analyses especially for samples of deep-sea origin. Another problem is the strong complexation of trace metals by the organic matter present in natural waters. Of course, this problem is not unique to ACSV. Intraelemental interferences can occur in ACSV if chelates with several elements adsorb on the electrode and produce overlapping peaks. Approaches aimed at solving these problems have been discussed in a recent review.[325]

280   Chapter 4

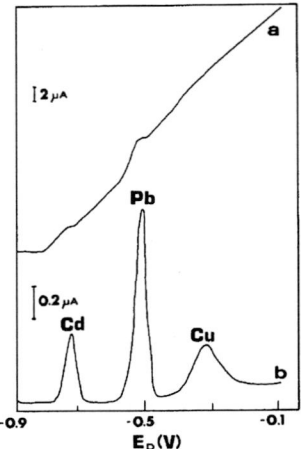

FIGURE 4.41. Voltammogram for Cd, Pb, and Cu in seawater by anodic stripping with collection: (a) disk, (b) ring. $[Hg^{2+}] = 1.63 \times 10^{-5}$ M; rotation speed 1500 rpm; deposition time 5 min; scan rate 3 V/min. The ordinate contains the current response. (Reproduced with permission from Brihaye and Duyekaerts.[328])

4.7.2. Anodic Stripping with Collection

This approach[327] eliminates the background charging (i.e., capacitive) current that is a nuisance with most types of electroanalyses (Section 2.6.2). Specifically, an RRDE is used, and the signal is measured at the ring. The ring potential is maintained constant so that the capacitive contribution is minimized. Figure 4.41 compares stripping (disk) and collection (ring) voltammograms for Cd, Pb, and Cu in seawater.[328] It has been demonstrated[329] that the ring collection signal can be enhanced by increasing the potential scan rate at the disk, but without paying the price of a concomitant increase in the capacitive component. The collection approach has been compared with other stripping modes.[330] Anodic stripping with collection has been described for two mercury-coated glassy carbon tubular electrodes connected in series.[331] The effect of various parameters on the collection signal has also been investigated.[328] In general, ASV with collection has not enjoyed widespread use thus far in trace metal analyses.

4.7.3. Subtractive Sripping Voltammetry

In this approach, the background current is subtracted from the total current to yield only the stripping component. This is done using either a "two-channel" approach, employing dual working electrodes, or via a single-channel computerized configuration. In the former case, the same deposition po-

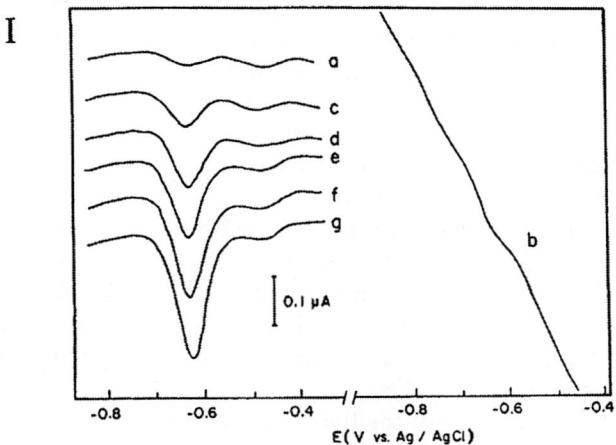

FIGURE 4.42. Standard addition of Cd(II) to seawater: (a) and (b), subtractive and conventional stripping voltammograms, respectively, recorded with the original seawater sample; (c)–(g), subtractive voltammograms obtained after successive standard additions of Cd(II), each increasing the sample concentration by steps equal to 5.7×10^{-9} M. Original concentration, 3.8×10^{-9} M Cd(II), deposition time, 4 min at -1.1 V; scan rate, 200 mV/s, solution flow rate, 95 mL/min (potentials vs. Ag/AgCl reference). The ordinate contains the current response. (Reproduced with permission from Wang and Ariel.[332])

tential is imposed on both electrodes but for different times. During the stripping step, the difference between the two stripping currents is measured. In this manner, all background current components independent of the deposition time, are canceled out. Interestingly enough, this approach corrects for both the nonfaradaic (i.e., capacitance) and Faradaic components of the background current. Most pulse waveforms (see Section 2.6.2) correct only for the former. However, major sources of interference in stripping analyses include redox reactions of soluble species, mercury oxidation, and electrolyte decomposition. For example, copper near the mercury oxidation regime and zinc near the HER limit are easily quantified using the subtractive mode. An added advantage is that fast scan rates can be used since the charging current is compensated. Figure 4.42 contrasts the stripping voltammogram obtained for Cd(II) in the subtractive mode (a) with that from the conventional mode (b).[332]

The major difficulty in two-channel subtractive stripping analysis is the preparation of two *symmetrical* electrodes with identical geometry, morphology, solution hydrodynamics, and the like. Further, the two electrodes are initially held at disparate potentials (e.g., the "background" electrode at 0 V and the "analytical" electrode at the deposition potential), thus potentially destroying the electrochemical identity of the two electrode surfaces. Ingenious attempts have been made to attack this problem. One involves using the same (deposition) potential at the two electrodes, followed by a short positive potential pulse at the "background" electrode to dislodge the deposited analyte prior to the stripping step.[333] The second involves modulation of the solution hydrodynamics (e.g., the "analytical" electrode is rotated while the "background" electrode is held steady) at the two electrodes held at the same deposition potential.[334] Nonetheless, the background current is not perfectly compensated for in any of these innovations thus far.

A variety of electrode configurations have been proposed for the subtractive mode including two HMDEs,[335] two rotating glassy-carbon based MTFEs,[336] a rotating split disk electrode,[337] and two stationary MTFEs.[332] Another approach is based on the use of two *cells*, one containing the blank solution and the other, the sample.[338] Most of these studies have utilized a linear scan for the stripping step.

The second approach to subtractive stripping voltammetry relies on computerized data acquisition. Thus, the "background" voltammogram for zero deposition time is recorded after the normal stripping scan. Subtraction of the computer-stored background current data from the total current yields the current of interest. Here again (as with the dual-electrode case discussed previously), a point of concern is that the electrode surface may have changed in the time elapsed between the two "runs." Another route to obtaining a subtractive response with a single electrode is possible with FIA. Thus, the "analytical" and "background" stripping voltammograms can be recorded while the sample and carrier solutions flow alternately through the detector.[339]

An interesting approach combines the background correction capability of subtractive stripping voltammetry with the signal enhancement capability of multiple (co-added) scans.[340] Figure 4.43 illustrates representative data for bismuth assay of a seawater sample.[340] In this approach, a fast linear scan follows the deposition step, and the data are stored in the on-board computer. At the end of the "probe" scan, the deposition potential is reapplied, and thus a fraction of the ions is replated. After a 1-s delay, a new scan is initiated and the data coadded with the previous data. This process is repeated for multiple (10 to 30) consecutive scans. At the end of the sequence, the background current is recorded after the analyte is completely stripped off from the electrode surface and after sufficient time has elapsed for the resultant ions to have diffused away from the electrode. The result, after background subtraction, is enhanced S/N because of the "multiplex" data acquisition and processing.

4.7.4. Potentiometric Stripping Analysis (PSA)

This approach[341-344] is similar to conventional (voltammetric) stripping analysis in that the metals are preconcentrated either as amalgams (in a thin mer-

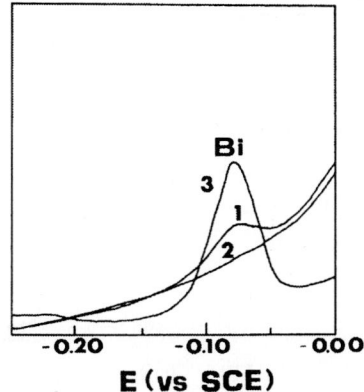

FIGURE 4.43. Multiple scanning subtractive stripping measurements. The sum of 30 analytical scans (curve 1), 30 background scans (curve 2), and the difference between these sums (curve 3) for the analysis of a standard seawater sample spiked with 0.1 ppb bismuth at pH 0.8. The ordinate contains the current response. (Reproduced with permission from Kryger and Jagner.[340])

cury film), electrodeposited (as a thin film) on a solid electrode, or accumulated at the interface via adsorption. However, the stripping mode differs in PSA, and either an oxidant (e.g., Hg^{2+}, O_2, or Cr^{6+}) or a constant current is used to remove the analyte, and the potential is monitored as a function of time. A potential–time curve similar to Figure 4.44 is obtained in the former case.[341] A sharp potential change accompanies the depletion of each metal from the electrode surface. The stripping step therefore can be regarded as a redox titration with continuous delivery of titrant, that is diffusion of oxidant to the electrode surface. The time lapse between two consecutive equivalence points (the "transition time") is proportional to the analyte concentration:[344]

$$t_{M,strip} \alpha \frac{[M^{n+}]t_d \delta_2}{\delta_1 D_{ox}[A_{ox}]} \tag{4.11}$$

Where δ_1 and δ_2 are the diffusion layer thicknesses during the preconcentration and stripping steps, respectively; t_d is the deposition time; and D_{ox} and $[A_{ox}]$ are the diffusion coefficient and concentration of the oxidant. For a given concentration of the oxidant and solution hydrodynamics, Eq. (4.11) reduces to

$$t_{M,strip} \alpha [M^{n+}]t_d \tag{4.12}$$

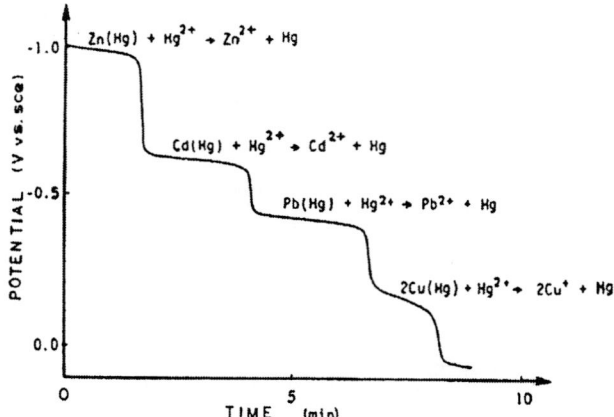

FIGURE 4.44. Time–potential curve for a mercury drop electrode after preelectrolysis for 3 min at -1.25 V vs. SCE in a 0.5 M sodium chloride solution containing 1.5 µM of Zn(II), Cd(II), Pb(II), and Cu(II) and 500 µM of Hg(II). (Reproduced with permission from Jagner and Graneli.[341])

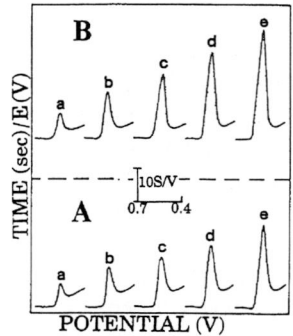

FIGURE 4.45. Stripping potentiograms obtained at the gold-coated glassy carbon disk (A) and carbon-strip (B) electrodes, for increasing mercury concentration in 10 µg L^{-1} steps (a–e). Preconcentration for 2 min at -0.10 V, from a stirred, non-deaerated solution. Constant stripping current, +2 µA; electrolyte, 0.05 M HCl. (Reproduced with permission from Wang and Tian.[345])

Thus, the transition time can be used as a measure of the analyte concentration for quantitative purposes.

Alternatively, a constant current can be applied and the E–t profile recorded. Representative stripping "potentiograms" are shown in Figure 4.45 for mercury for two types of electrode surfaces.[345] These calibration curves were generated by increasing the mercury concentration in 10 ppb steps from a to e, with the preconcentration and stripping conditions as shown. Figure 4.46 shows data from the same study[345] for a 25 ppb mercury solution with different

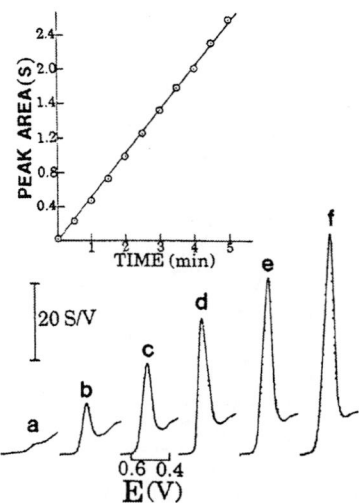

FIGURE 4.46. Potentiometric stripping response to 25 µg L^{-1} mercury after different preconcentration times: (a) 0, (b) 1, (c) 2, (d) 3, (e) 4, and (f) 5 min. Other conditions as in Figure 4.45. Also shown (as insert) is the resulting plot of peak area vs. time. (Reproduced with permission from Wang and Tian.[345])

preconcentration periods (0–5 min., a–f). As would be expected from the chemical oxidant counterpart (see Eq. (4.12)), the stripping signal (e.g., peak area) is proportional to t_d. This stripping mode has also been termed *stripping chronopotentiometry*. It must be noted that stripping chronopotentiometry is not a new technique. For example, the theoretical principles for "voltammetry at constant current" were reviewed in 1955,[346] and experimental evaluation soon followed.[347] Other early examples are available.[348] However, the routine application of this technique to the analysis of natural water does not appear to have occurred until ca. 1987.[349,350] As has been pointed out,[351] a possible reason for this is the advent of microprocessor-controlled instrumentation. The signal shape in constant current PSA (see Figures 4.45 and 4.46) is similar to that obtained in stripping voltammetry, except that, in the latter case, the current is plotted as a function of potential.

Qualitative identification of the analyte is based either on the location of the reoxidation potential (i.e., the plateau in Figure 4.44) or on the peak potential in the constant current mode. Alternatively, the derivative signal can be employed to locate the inflection in the transition for each analyte in the chemical oxidant route.[342] This strategy is particularly useful for background correction as illustrated in Figure 4.47; the time lapse between two consecutive points on the background ("blank") trace can be taken as a measure of the background contribution for the particular element oxidized in this potential range. Thus,

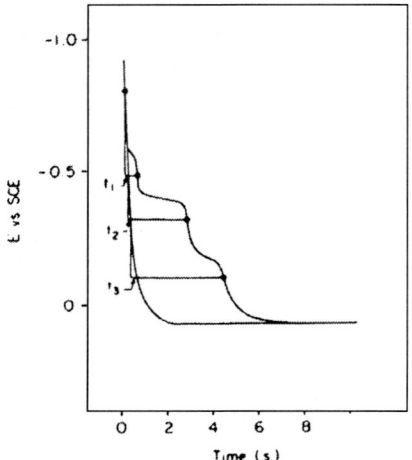

FIGURE 4.47. PSA and background curves for 1 ppb cadmium, 5.8 ppb lead, and 2.8 ppb copper. Plateaus t_1–t_3 represent the background corrections for cadmium, lead, and copper, respectively. (Reproduced with permission from Jagner and Aren.[342])

the times t_1, t_2 and t_3 are the background contributions to cadmium, lead, and copper, respectively.[342]

Several advantages have been claimed for PSA, particularly in the chemical oxidant mode. These include insensitivity to redox-active dissolved species including O_2 and the double-layer charging process. The interference from dissolved O_2 and the double-layer charging current to stripping voltammetric measurements has been noted earlier. Further, PSA has been claimed to be better suited[341] to field experiments; for example, for measurements on board a ship. Frequency and voltage fluctuations in the main power supply in such remote locations would be potential sources of noise in a DPASV or a SWASV experiment. The instrumentation for PSA (in either mode) tends to be rather simpler than for the voltammetry counterpart. Finally, PSA appears to be affected less by adsorbed organics for metal determinations than ASV.[344,352]

For PSA at ppb or sub-ppb range, the stripping is much too rapid unless very long deposition periods are employed. An on–board microcomputer is particularly useful in such cases, and signal enhancement strategies such as background subtraction,[353] differentiation,[354] and multiple stripping and re-reduction of the metal analyte[355] can be used in a manner similar to that discussed in earlier sections in a stripping voltammetry context.

Historically, "chemical" stripping analysis[356] is a forerunner of PSA. In this technique, chemical stripping of potentiostatically deposited metals was used for trace determination of strong oxidants. Thus, the time required to remove a known amount of plated silver was found to be inversely proportional to the

Table 4.18. Historical Development of Working Electrodes for Stripping Analysis

Period	Type of Electrode
1950s	Hanging mercury drop (Kemula type)
1960s	Pre-plated mercury thin film
1970s	*In situ* plated mercury thin film
1980s	Ultramicroelectrodes
1990s	Screen-printed electrodes Electrode arrays

Source: Adapted from Wang.[370]

concentration of the oxidant (see Eqn. (4.11)); for example, Ce(IV), permanganate or Fe(III). In the vast majority of cases involving metal analyte ions, PSA has been employed in the oxidative stripping mode. However, interfacial adsorptive accumulation of the analyte can also be coupled with *cathodic* stripping, much like the adsorptive cathodic stripping *voltammetry* (ACSV) case. A variety of matrices have been examined by PSA including sea water,[357,358] urine,[359] beverages,[360] fly-ash,[361] lake or river water,[362] tap water,[362] groundwater,[363] sediments and sludge[364] and blood serum.[351] Thus, elements such as Pb, Tl, Cd, Bi, Cu, Sn, Zn, and Hg have been determined. The PSA technique has been extended to alkali and alkaline earth metals (see Figure 4.39) by using organic solvents and water–organic solvent mixture.[358] PSA has also been combined with FIA for automated determination of metals in groundwater.[363] Finally, the influence of electrochemical parameters on the efficacy of "film" PSA, that is, the direct concentration of the test ion on a solid electrode (without a mercury film),[345,366,367] has also been discussed.[365]

We shall discuss further applications of PSA for speciation studies later on in this chapter.

4.7.5. *Miscellaneous Aspects of Stripping Analysis*

The electrode materials for stripping analyses were discussed in Section 4.4.4. Table 4.18 provides a further historical perspective of electrode materials within the particular context of stripping analysis. While MTFEs provide unsurpassed sensitivity for trace analyses and speciation studies (Table 4.1), there are many advantages in switching to the use of solid electrode surfaces especially for on-site use in remote locations. In particular, the choice of solid elec-

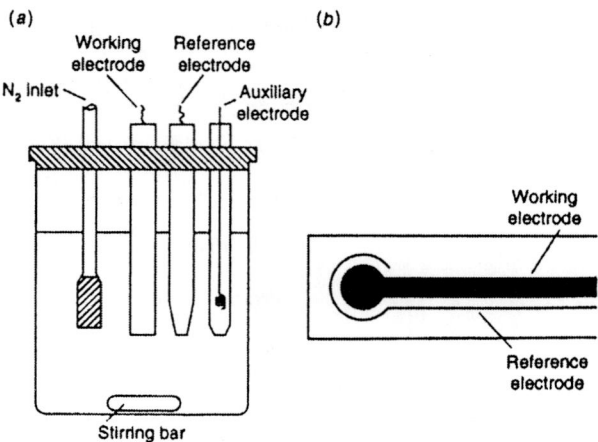

FIGURE 4.48. (a) Conventional and (b) screen-printed cells for stripping analysis. (Reproduced with permission from Wang.[370])

trode materials is compatible with the recent trend toward miniaturization. This is exemplified both by the recent development of ultramicroelectrodes (UMEs) and screen-printed electrodes. We have already discussed the former. Screen-printed electrodes (SPEs) are a relatively recent development.[345,359,366–369] Figure 4.48 contrasts the conventional stripping analysis cells and the compact approach based on the use of SPEs.[370] The conventional cells (Figure 4.48a) obviously are cumbersome, and their operation involves careful cleaning, prolonged deaeration time, solution stirring during deposition, and other steps associated with working electrode pretreatment, conditioning, and the rest. The SPEs comprise planar carbon or gold working electrode strips and a silver quasi-reference which are printed on an inexpensive plastic or ceramic support (Figure 4.48b). The entire assembly is disposable and the sample droplet to be analyzed is placed on it. Because of the efficient nonlinear diffusion at the individual microdiscs (each of ~μm dimension), SPEs eliminate the need for solution stirring during analyte preconcentration. By coupling the sensor array to a square–wave modulation waveform or PSA, the need for time-consuming solution deaeration is also eliminated. Recent applications of SPEs have been reviewed.[370] Interestingly enough, this recent trend toward the use of solid electrodes goes against "conventional wisdom," where materials such as gold were used for the stripping analyses of only elements (e.g., Se, As, Te) more electropositive than mercury.[371,372] Indeed, the many positive features associated with the use of Hg (see Chapter 2, Section 2.6) as an electrode material have largely outweighed concerns about its toxicity in the past.

4.7. Stripping Analyses

The use of UMEs in stripping analysis is also becoming more widespread, particularly for applications related to on-site use (e.g., portable blood [Pb] analyzers). This approach is especially attractive because of the requirement of small sample volumes. Such portable electrochemical analyzers would be compatible with scenarios where, for example, large numbers of infants need to be screened for blood [Pb] levels. As with the screen-printed electrode technology discussed previously, UMEs also eliminate the need for stirring because of hemispherical diffusion. Use of array detection also results in an improvement in the S/N ratio proportional to the square root of the number of elements in the array.[373] Interestingly enough, this last advantage is unavailable with PSA, where the stripping time, t_d is independent of electrode area.[343] Carbon-based UMEs have been described for blood [Pb] analyses by a number of authors.[351,374] Both SWASV[374] and PSA[351] have been used in these studies.

Finally, a new method has been described[375] for transporting sample to the surface of a glassy carbon MTFE for stripping analysis. In this technique, instead of rotating the electrode or stirring the solution, a flat disk having a conically shaped hole is positioned below the glassy carbon electrode. The disk is then vibrated at high frequency in the vertical plane, forcing solution onto the electrode surface in a jet stream. It has been shown that, with proper optimization of the geometries of the vibrating disk and the conical hole, higher sensitivity than rotation or electrode stirring can be achieved.[376]

Stripping analysis techniques are eminently suited for use with flow systems for environmental surveillance or industrial quality control. Flow systems for automated stripping analysis of discrete samples based on either FIA or the autoanalyzer concepts have been described.[363,377-379] The autoanalyzer is based on sequential sampling and washing on gas-segmented flow streams. One of the earliest automated stripping analysis systems appears to have been developed in 1966, wherein the sequence of preconcentration and stripping at a HMDE was computer programmed.[380] The advent of operational amplifier modules saw the development of a fully automated apparatus for stripping voltammetry soon thereafter.[381] This setup was used for the determination of sub–pg quantities of triphenylation compounds. Various flow-through or "flow-onto" stripping voltammetry systems have been described and applied for the determination of Se,[382] Tl,[383] As(III)[384] and Cd.[377,378,385] Flow-through configurations have also been used for on-line adsorption stripping voltammetry of U[386] and anodic stripping *coulometry* with collection for absolute trace analysis of Pb.[387] The latter approach is interesting in that, if *complete* electrochemical deposition of the trace metals is achieved, then *absolute* analysis could be achieved and calibration with standard solutions or standard additions would not be required.[387]

An automated on-line stripping analyzer has been developed for on-ship use for trace metal analyses.[388] An open tubular electrode geometry was em-

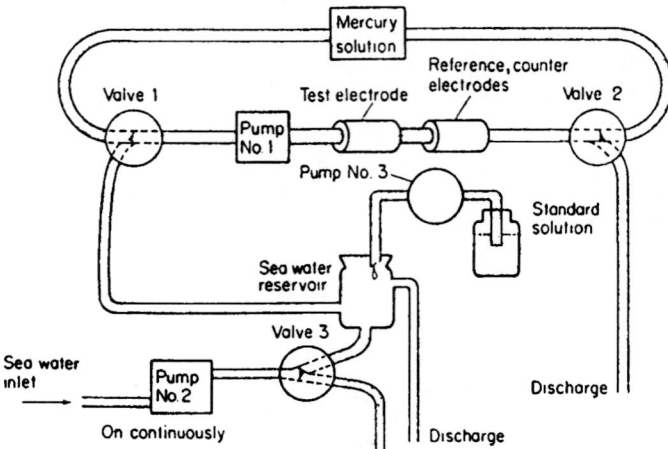

FIGURE 4.49. Schematic diagram of electrolysis cell for automated anodic stripping voltammetry. (Reproduced with permission from Zirino et al.[388])

ployed through which the sample and mercury solution flowed alternately. A programmer controls the potentiostat, valves, pumps and data acquisition; an early version of this system is contained in Figure 4.49.[388] The design and applications of on-line stripping systems have been reviewed.[389]

4.7.6. Trace Element Speciation in Water by Stripping Analyses

As pointed out earlier in this book (Section 1.4.2), measurement of the *total* concentration of a trace element in a natural water sample is insufficient as a means of assessing the toxicity of the water to aquatic organisms and human beings. Depending on the *chemical* and *redox* forms of the element, a water with a high total concentration of the pollutant may in fact be *less* toxic than another sample with lower concentration. For example, it is believed that ionic copper is far more toxic to aquatic organisms than complexed copper, and the more stable the copper complex, the lower is its toxicity. Similarly, Cr(VI) is much more toxic than Cr(III). The reverse is true for the As case, that is the lower redox state, As(III), is more toxic than the oxidized form, As(V). *Speciation analysis* of an element in a water sample may be defined as the determination of the concentrations of the different physicochemical forms of the element that together make up the total level in the sample.[5] The individual physicochemical forms may include particulate matter and dissolved forms such as simple inorganic species, organic complexes, and the element adsorbed on a variety of colloidal particles (Table 1.5). All these species can coexist and may or may not be in thermodynamic equilibrium with one another.[5]

4.7. Stripping Analyses

Aside from biological (or toxicological) implications, speciation analysis also affords information on the geochemical cycling of the elements. Variation in the speciation of an element, for example, can affect the degree of its adsorption on suspended matter (a phenomenon well studied and understood in colloid science), its rate of transfer to the sediment, and its overall mass transport in a water system. Thus, speciation analysis can assist in the predictive modeling of distances over which a river will be affected by an effluent discharged from a point source.

Many sensitive analytical techniques such as atomic absorption spectrometry (AAS) and neutron activation analysis are not directly applicable to speciation studies because they measure only the total analyte concentration. Electroanalysis is a powerful technique in this regard because the electrochemical behavior of a given element is very sensitive to its physicochemical state. Many comprehensive speciation schemes have been developed and reviewed. One representative example[5,390-393] is contained in Table 4.19 for the procedure based on ASV.[5] Some major speciation steps are as follows:

(1) Filtration: This step removes the particulate matter. In more refined procedures, a size distribution analysis can be carried out based on the separation of species according to molecular size. In general, in the case of metal speciation, the smaller the metal complex, the higher is its biological activity.
(2) Chelating resin separation: Metal that cannot be removed from a water sample by a chelating resin column (e.g., Bio-Rad Chelex-100) represents metal bound in highly stable or inert complexes or metal associated with colloidal particles (which are *not* removed by filtration, see earlier).
(3) UV irradiation: This can be done either at natural pH of the water sample (see Aliquot 3, Table 4.19) or after acidification (Aliquot 1, Table 4.19). Hydrogen peroxide is added to assist the photodegradation of the dissolved organic matter (see Chapter 6). In the former case, only metal associated with organic matter (e.g., metal–humic acid, metal–fulvic acid) will be liberated. On the other hand, UV irradiation coupled with acidification releases all forms of metal, including inorganic colloids.
(4) Solvent extraction: Lipid–soluble metal complexes (e.g., alkylmercury compounds) are particularly toxic forms of heavy metals because they can diffuse rapidly through a biomembrane and carry both metal and ligand into the cell interior (see Figure 1.7). These can be removed via solvent extraction or by passage of the water through a column of Bio-Rad SM2 resin.[5]

The importance of reporting *all* analytical details associated with the speciation procedure has been stressed,[5,390-393] especially given that the speciation analysis trends are often dependent on operational parameters. For example, the lability of a metal complex depends not only on its dissociation kinetics, but also on the *effective measurement time*.[5,394,395] In a stripping analysis context,

Table 4.19. Recommended Speciation Scheme for Copper, Lead, Cadmium, and Zinc in Waters

Sample (unacidified)

Filter through a 0.45 μm membrane filter. Reject particulates and store filtrate unacidified at 4°C.

Filtrate analysis

Aliquot No.	Volume/mL	Operation	Interpretation
1	20	Acidify to 0.05 M HNO_3, add 0.1% H_2O_2 and UV irradiate for 8 h, then ASV[a]	Total metal
2	10	ASV at natural pH for sea water; add 0.025 M acetate buffer (pH 4.7) for fresh waters.	ASV-labile metal
3	20	UV irradiate with 0.1% H_2O_2 at natural pH, then ASV[b]	(3)–(2) = organically bound labile metal
4	20	Pass through small column of Chelex 100 resin; ASV on effluent[c]	Very strongly bound metal
5	20	Extract with 5 mL of hexane – 20% butan-1-ol; ASV on acidified, UV-irradiated aqueous phase[d]	(1)–(5) = lipid-soluble metal

[a] Adjust to pH 4.7 with acetate buffer.
[b] Not valid if [Fe] > 100 μg L^{-1}.
[c] Optional step.
[d] Dissolved solvent in aqueous phase must be removed first.

Source: From Florence.[5]

labile metal refers to the free (i.e., hydrated) metal ion and metal that can dissociate in the double layer from complexes or colloidal particles. With the constant preconcentration time used for stripping analysis, the extent of metal complex dissociation depends on its residence time in the diffusion layer, δ, which in turn is governed by the solution hydrodynamics (see Section 4.4.1).

The kinetic criterion for labile or inert discrimination is based on the pa-

4.7. Stripping Analyses

Table 4.20. Criteria for Electrochemical Lability of Lead Complexes

Description	Lability Criterion[a]	Lead Complexes Concerned
Labile	$i_k/i_d > 0.99$ $\log(\beta_1[L]^{1/2}) < 2$	$PbCl^+$, $PbSO_4$
Quasi-labile	$i_k/i_d < 0.99$ $i_k/i_d > (1+\sigma)^{-1}$	$PbCO_3$, $PbOH^+$ Pb–humic acid
Inert	$i_k/i_d = (1+\sigma)^{-1}$	Pb–EDTA

$$ML \underset{k_d}{\overset{k_f}{\rightleftharpoons}} M^{2+} + L^{2-}$$

$\beta_1 = [ML]/[M][L] = k_f/k_d$

[a] β_1 = stability constant for 1:1 complex; $[L]$ = concentration of ligand; i_k = kinetic current; i_d = diffusion current; $\sigma = \beta_1[L]$.

Source: From Turner and Whitfield.[395]

rameter I_k/I_d as shown in Table 4.20. The diffusion current, I_d, is the current observed for the *same* concentration of the metal ion but in the absence of the ligand, L. In the absence of kinetic control (i.e., total dissociation), $I_k/I_d = 1$. Attempts have been made[396] to couple the influence of the electroanalytical observation process and the intrinsic dissociation kinetics using reaction layer theory to arrive at a classification of complex lability as follows:

labile: $k_d k_f^{-1/2} t_e^{1/2} \gg 1$
quasi labile: $k_d k_f^{-1/2} t_e^{1/2} \approx 1$
inert: $k_d k_f^{-1/2} t_e^{1/2} \ll 1$

A large excess of ligand was assumed in this classification scheme.[396] The parameter t_e is the effective measurement time, which is technique dependent.[5] It is obvious that a stripping analysis definition of complex lability depends not only on the kinetic parameters of the metal complex dissociation but also on other factors such as ligand concentration and diffusion layer thickness.

There has been discussion[397] on whether the process of speciation analysis itself perturbs the existing or natural state of the water sample. For example, the sample pH is adjusted by buffer addition (see Table 4.19) and Hg^{2+} ions are added to the water sample for *in situ* generation of the Florence electrode for stripping analyses. However, it has been pointed out[5,39] that stripping analyses are *dynamic* techniques (unlike, for example, potentiometry); so the very act of measurement disturbs the equilibrium. For these reasons, it is again important to stress that the speciation data are *operationally defined*, and for the user to have a proper understanding of their significance, all aspects of the speciation protocol must be carefully specified.

<u>Pseudo-Polarography and Pseudo-Voltammetry.</u> An implicit assumption in the preceding discussion is that the metal complex, ML, is not *directly* reducible; that is only the dissociated metal ions are assumed to be electroactive. However, direct electrochemical reduction of some complexes can occur, as has been discussed earlier within the context of cathodic stripping voltammetry. The presence of such complexes can be detected from the effect of the deposition potential, E_d, on the peak current. The peak current will increase continuously with increasing E_d (Figure 4.50) instead of increasing from zero to a limiting value over a small range of E_d. Thus a pseudo-polarogram is a plot of the stripping peak current versus deposition potential. The half-wave potential (E_c) of a pseudo-polarogram of a metal is related to (but not identical with) the polarographic half-wave potential ($E_{1/2}$).[398] Pseudo-polarography has been utilized for identifying lead–carbonato complexes in seawater,[399,400] the complexation of cadmium in marine samples,[401] and the speciation of As, Cd, and Pb in various natural water, including geothermal reservoirs.[398] The theory for pseudo-voltammetry of both reversible and irreversible systems has been given.[402,403]

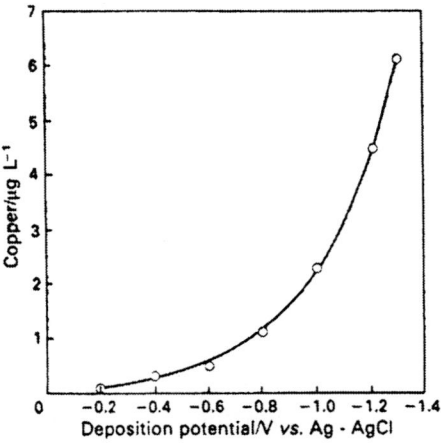

FIGURE 4.50. Pseudo-polarogram of copper in sewage plant effluent water. (Reproduced with permission from Florence.[5])

4.7. Stripping Analyses

<u>Complexing Capacity.</u> Complexing capacity is an important water quality parameter because it is a measure of the "safe" discharge level of a heavy metal into an effluent stream before release of the free metal ion by the water. This parameter is determined by titration of the water sample with a heavy metal ion. Copper(II) is usually chosen as the titrant because it is commonly found in a variety of waters and the free metal ion is highly toxic to aquatic organisms. Complexing capacity is then defined as the concentration of Cu(II) ion (in mol liter^{-1}) that must be added to a water sample before free Cu^{2+} appears. Near-shore surface seawater has a Cu-complexing capacity of ~2 x $10^{-8} M$, whereas river water ranges from 1 x 10^{-8} M to 50 x 10^{-8} M.[5]

Conventional methods of measuring the complexing capacity include bioassays, ion-exchange chromatography, chemical exchange, and copper salt solubilization. Electrochemical methods such as ISE potentiometry, amperometry, and voltammetry have all been used,[5] although stripping analyses such as ASV or adsorptive CSV are eminently more suitable for this analysis. In either case, aliquots of a standard Cu solution are added to the sample, and the stripping signal is measured and plotted against either the concentration of added copper or the total dissolved copper. Both plotting formats are contained in Figure 4.51.[5,325] Both the complexing capacity and the conditional stability constant of the metal–organic complex can be estimated from analyses of such plots.

The ASV approach suffers, however, from several difficulties.[5] (a) Some Cu complexes (e.g., Cu–NTA, NTA = nitrilotriacetic acid), although thermodynamically *stable*, are kinetically *labile* and dissociate extensively in the diffusion layer. These kinetic currents (see earlier) can yield erroneous results for the complexing capacity. (b) Organic matter adsorbed on the electrode may cause a depression of the metal ASV peak although no actual complex formation may have occurred. (c) Formation of the Cu complex may be exceedingly slow. While these problems can be corrected to varying degrees,[5] the technique of competitive ligand exchange[325,404] is ideally suited to the adsorptive CSV procedure. Thus, competition between the natural ligands and those intentionally added, such as catechol or tropolene, can be set up, followed by cathodic stripping by the adsorbed Cu–ligand complex. This procedure has the advantage that double-layer dissociation of the complex is eliminated, and organic adsorption may be inhibited by the preferential adsorption of the Cu–ligand complex. These points are illustrated in Figure 4.51b.[325] The "plating" approach (similar to the ASV procedure) affords a lower complexing capacity, suggesting that the neutral complexes may have dissociated during deposition at negative potentials.

<u>Matrix (or Medium) Exchange.</u> In the ideal case, the relative heights of the ASV stripping peaks for labile and total metal in the sample, and hence the calculated fraction of labile metal, is controlled solely by the parameters asso-

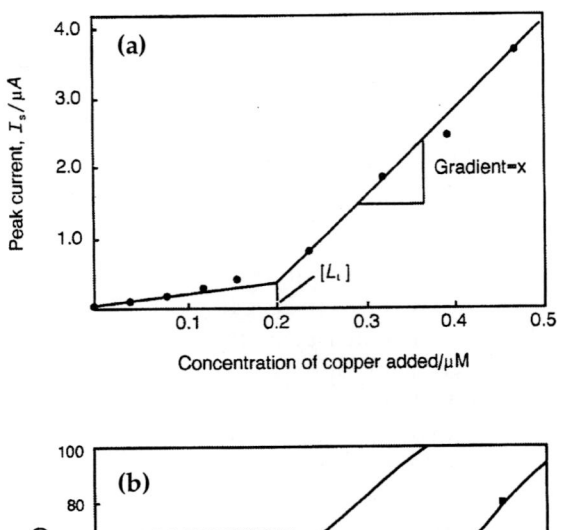

FIGURE 4.51. Complexing capacity titration for copper in a natural fresh water (top) and the North Sea (bottom). (Reproduced with permission from Florence[5] (Figure 4.51a) and Van den Berg[325] (Figure 4.51b).)

ciated with the preconcentration step, such as deposition time. However, under certain circumstances, the kinetics of the stripping step may have an influence on the measured signal. This can occur if a complexing agent that is present in the sample (but not in the standards) affects the stripping chemistry or kinetics. A common example of this is the chloride ion effect in seawater. The chloride ion stabilizes the intermediate valency state of Cu [Cu(I)], leading to a smaller number of electrons associated with the stripping step[5]:

Standard: $Cu^0 \rightarrow Cu^{2+} + 2e^-$
Sample: $Cu^0 + 2Cl^- \rightarrow CuCl_2^- + e^-$

The result is a diminished stripping signal for the sample. Other complications with the effects associated with surface–active substances[405] or precipita-

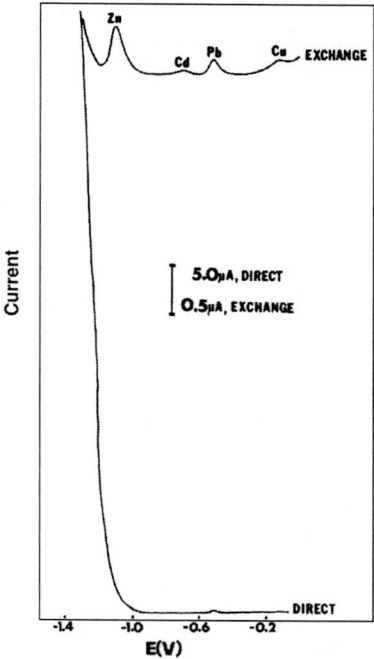

FIGURE 4.52. The effect of applying the mediun-exchange method on the analysis of river water (pH 1.2): lower curve, direct method; upper curve, medium-exchange method (exchange solution used, 0.1 M ammonium acetate, pH 6.4). Deposition: 3 min at -1.4 V; flow rate 4.8 mL/min. Stripping: differential pulse ramp with 5mV/s rate. The ordinate contains the current response. (Reproduced with permission from Wang and Greene.[410])

tion of the oxidized metal ion onto the electrode surface[406] on the stripping peak current have been noted.

An effective way to circumvent these difficulties is medium exchange,[407,408] where the sample solution, after preconcentration, is replaced with a new supporting electrolyte in which the stripping step is performed.[409] The new electrolyte can be chosen to yield reversible, reproducible stripping peaks for the element under study. Interestingly enough, the medium-exchange strategy is not new, but its widespread use was hindered by a myriad of practical difficulties, such as exposing the electrode to the atmosphere, break of the electrical circuit, or incomplete exchange of the sample solution with the stripping medium. Consequently, considerable loss of the deposited metals was reported.[407] The advent of flow systems for stripping analyses (see Section 4.7.5), and system automation, has altered the situation. Figure 4.52 provides a rather dramatic illustration of the efficacy of medium exchange for stripping analysis experiments.[410]

Redox Speciation. The redox speciation of eluents over different oxidation states can be determined if one of the oxidation states is stabilized. For example, As(V) is not electroactive. Therefore As(III) can be determined separately, and As(V) converted to As(III) by chemical reduction. The difference between the total As assay and As(III) previously determined, yields the As(V) concentration. Similarly, Se(VI) is not electroactive. Therefore Se(IV) can be specifically determined (in the presence of Se(VI)) and Se(VI) subsequently after a separate reduction step under UV irradiation at pH 8.[411] In a sense, redox state speciation may be regarded as a special case of labile/inert determination (see earlier). The redox speciation cannot be determined if both oxidation states are electrochemically reversible. An example is the copper system wherein CSV is unsuccessful for determining the Cu(I)/Cu(II) ratio.[325]

Polarography and/or ASV have been used for distinguishing between Fe(III)/(II),[412] Cr(VI)/(III),[413] Tl(III)/(I),[414,415] Mn(IV)/(II),[5] Se(VI)/(IV)[416] and As(V)/(III).[417] Similarly, redox speciation of chromium has been accomplished by either adsorptive[418] or catalytic CSV.[419] A CSV procedure based on added copper has been used for redox speciation of selenium.[411]

Arsenic and selenium are two elements that are being increasingly regulated – their maximum allowable concentrations in the environment are likely to be lowered from present standards (Chapter 1). Early polarographic and stripping voltammetry work[420,421] on these elements was already reviewed in Chapter 3. Since As(V) is electroinactive, for total arsenic assay, it must be converted to the (electroactive) As(III) state. This has been done by the use of reducing agents such as Na_2SO_3,[422] or Cu_2Cl_2.[423] Interestingly enough, Se(IV) was found to have a useful "catalytic" effect on the electrochemical reduction of As(III) which was exploited for the determination of the latter by CSV.[424] Similarly, As(III) is not usually amenable to determination by stripping voltammetry at a MTFE or HMDE because of the insolubility of arsenic in mercury. However, in the presence of copper, arsenic forms an "inter-metallic" compound that is soluble in mercury.[425] Thus, arsenic was determined in drinking water contaminated with Cu(II) by CSV at a HMDE.[425]

Correlation Between ASV Lability and Toxicity. Although there is accumulating evidence that a free metal ion is the most toxic metal form, the situation is not completely clear.[5] It has been noted that lipid-soluble Cu and Hg complexes are extremely toxic, and there is evidence[426] that the hydroxy, citrate, and ethylenediamine complexes of Cu are also toxic. Attempts to correlate ASV data of metal lability with toxicity have met with varying degrees of success, as reviewed elsewhere.[5] When the marine diatom *Nitzschia closterium* was used as a bioprobe, good correlation was obtained for the copper system[427] when natural complexing agents (e.g., fulvic, humic, tannic, and alginic acids) were present in the growth medium. However, when synthetic ligands such as NTA were added, there was no sensible correlation.[5] Obviously, more research is needed in this area.

4.7.7. Electrochemical Speciation Data on Environmentally Important Elements

Table 4.21 provides a representative listing of stripping analysis and speciation studies that have been conducted on various water samples using ASV, CSV, and PSA.[428-445] This listing is not comprehensive and the interested reader may want to peruse the material given at the end of the chapter for further details and entry into the literature. We close this section with a brief summary of the electrochemical speciation data on some environmentally important elements. For further details, recent reviews[5,390-393,446] may be consulted.

Copper. Computer models predict that inorganic Cu exists in seawater mainly as the carbonato and hydroxy complexes. These species are ASV labile.[5] The percentage of inert organic complexes varies between 40% and 60% for coastal surface seawater. The Cu-binding ligands are siderophorous, metallothioneins, or porphyrins, although some of the copper is also adsorbed on inorganic colloids (e.g., hydrated iron oxide). Most fresh water streams have little ASV-labile Cu and the fraction of organically bound copper is high. Industrially polluted waters feature electroactive copper complexes.

Lead. Computer modeling of fresh water suggests the predominant inorganic Pb species to be the carbonato complexes. In seawater, Pb speciation is divided between carbonato complexes and chloro complexes. The available electrochemical data[5] indicate that these are only partially ASV labile.

Unlike copper, a significant fraction of Pb is bound to hydrated Fe_2O_3, and the affinity for organic ligands appears to be low. Most natural water has little ASV-labile Pb.[5]

Cadmium. In seawater, Cd exists mainly as the chloro complexes, $CdCl^+$ and $CdCl_2$. In river water, the dominant inorganic forms are Cd^{2+} and $CdCO_3$ depending on the pH. A high proportion of this element is ASV labile in both sea and fresh water. Very little Cd is present as pseudo-colloids even at relatively high pH.[5]

Manganese. As with arsenic, the natural water chemistry of Mn is dominated by nonequilibrium behavior. Oxidation of Mn(II) to Mn(IV), that is MnO_2, is thermodynamically favored in seawater and high pH fresh waters, but the oxidation is extremely sluggish. Colloidal MnO_2, is troublesome in water treatment plants because it clogs filters.

Chromium. The principal oxidation states are Cr(VI) and Cr(III) in natural waters. At the natural pH of seawater, the predominant Cr(VI) specie is CrO_4^{2-}, and the major Cr(III) forms are $Cr(OH)_3$ and $Cr(OH)_2^+$. The oxidation state of chromium is readily changed as Cr(III) can be oxidized by natural oxidants such as MnO_2 and H_2O_2.

Selenium and Arsenic. The speciation of these elements has been discussed in Chapter 1. Inorganic selenium exists in seawater mainly as Se(VI), but the presence of Se(IV) suggests that the ratio of the two species can be used as an indicator of the redox potential of natural waters.[411] Electrochemical data appear to be sparse on the organic speciation of these elements, either in natural or polluted waters.

Table 4.21. Stripping Analyses and Speciation Studies in Various Water Samples

Entry No.	Analyte/Matrix	Stripping Mode	Comments	Reference(s)
1.	As/polluted water	DP	—	428
2.	As/natural water	DP	—	429
3.	Cd (and Cu, Pb, Zn)/ river water	DP	Medium exchange used; influence of ligand and surfactant (Triton X-100) explored in the estimation of ASV lability	409
4.	Cd (and Cu, Pb)/ seawater	DP	—	430
5.	Cd/lake water	Linear scan	—	431
6.	Cd (and Pb, Cu)/ synthetic water	PSA	Effect of pH buffers studied	432, 433
7.	Cu (and Pb, Cd, Zn)/ synthetic water	DP	Stability constants determined using the Lingane method for carbanato complexes	434
8.	Cu (and Zn, Pb, Cd)/ fresh water	DP	An in-flow automated system developed for speciation studies	435
9.	Cu (and Cd, Zn)/ seawater	DP	Metal–organic (marine fulvic acids) interactions explored	436
10.	Cu (and Zn)/ estuarine water	Constant current–cathodic PSA	Adsorption collection used to determine copper lability	437

4.7. Stripping Analyses

Cont. Table 4.21. Stripping Analyses and Speciation Studies in Various Water Samples

Entry No.	Analyte/Matrix	Stripping Mode	Comments	Reference(s)
11.	Cu (and Cd, Pb)/ seawater	DP	Cathodic stripping voltammetry used after collection with 8-hydroxy quinoline	438
12.	Cu (and Zn, Cd, Pb)/ lake water	DP	—	439
13.	Cr/seawater	Catalytic CSV	Adsorption-collection with DTPA used	419
14.	Cr/seawater	CSV	Cr(III) adsorbed on silica	418
15.	Hg/seawater	Linear scan	Medium exchange used	440
16.	Hg/river water	DP	—	441
17.	Pb/various waters	—	See entry nos. 3, 4, 6, 7, 8, 11	430, 432–435, 438
18.	Se/seawater	CSV	Preconcentrated as a Cu_2Se complex	411
19.	Tl/seawater	DP	—	415
20.	Tl/natural water	DP	—	442
21.	Glutathione/natural water	CSV	Adsorption of complexes of Hg(I) and Cu(I) with glutathione used	443
22.	Purines/fresh (and sea) water	CSV	Determined as the Cu(I) complex of the purine	444
23.	Cystine and cysteine/ seawater	CSV	Determined as preceding as the copper complexes	445

Table 4.22. Examples of the Types of Environmental Matrices That Have Been Analyzed by Stripping Voltammetry (or PSA)

Matrix	Technique(s)	Comments	Reference(s)
Particulate Matter			
Suspended particulates in water and plankton	ASV	Six metals analyzed	447
Atmospheric precipitation (e.g., rain, snow)	ASV	Toxic metal content determined	448–450
Airborne particulates	ASV, CSV, PSA	Pb, Cd, Zn, Cu and Tl measured in air samples collected from urban locales as well as in particulates such as coal fly ash	361, 451–459
Biological Matrices			
Urine, fish muscle	PSA	Sample analyzed for Hg content with medium exchange	359
Blood	ASV, PSA	Extra precaution needed to preclude interference from protein adsorption, etc.	358, 351, 374, 460
Urine	ASV, PSA	Samples analyzed for heavy metal content	359, 461
Hair, teeth, nail, eye tissue	ASV	Analyzed for various trace metals; hair is a known depository for trace metals in the body; similarly, teeth may be useful indicators of past exposure to trace metals	462–465
Other miscellaneous samples including feces, bone, etc.	ASV	Used for medical management and post-mortem studies	466
Bovine liver, fish	ASV	Selenium, copper, lead, and cadmium determined	467

Cont. Table 4.22. Examples of the Types of Environmental Matrices That Have Been Analyzed by Stripping Voltammetry (or PSA)

Matrix	Technique(s)	Comments	Reference(s)
Geological Samples			
Soils and plants	ASV	EDTA extracts of soils have been used because they correlate well with plant uptake of metals	468, 469
Sediments and sludge	PSA	—	364
Rocks and minerals	ASV, AC stripping voltammetry, linear scan	In some cases, the electrochemical method found to be superior to flameless (furnace) AAS	470–474
Foods and Wines			
Fruit juices and commercial beverages	ASV, PSA	Various heavy metals including lead and tin determined	475–479
Milk	ASV	Lead and cadmium determined	480, 481
Wines	PSA	Lead in red wine determined	482
Hamburger meat, butter, and rice	ASV	Fusion pretreatment applied for destroying the organic matter; heavy metals determined	483
Tuna fish, eggs, and lobster	CSV	Selenium determined following wet digestion with HNO_3 and magnesium nitrate	484
Cereals	Staircase stripping voltammetry	—	485
Fruits	ASV	—	486
Soybean and oyster samples	ASV	—	487

4.7.8 Other Studies on Environmental Pollutants by Stripping Analyses

Stripping analyses (particularly stripping voltammetry) have been used for the determination of pollutants in a wide variety of matrices *other than water samples*. The compilation in Table 4.22 provides an example of the very wide range of sample matrices that can be analyzed. The challenge in the analyses of these complex matrices is to circumvent interference from the matrix components; for example, organic matter in foods and geological samples, and proteins in biological samples.

As the last few examples in Table 4.21 show, a relatively recent capability of stripping analyses is the determination of organic compounds. This has been rendered possible largely by the advent of adsorptive interfacial accumulation. The mercury–thiol complex has had a long history in this respect although the range of organics has now extended beyond sulfur compounds.[488] Attempts have been made recently[489] to systematize the trends and changes in the accumulation behavior of organic compounds at carbon paste electrodes.[489] Partition experiments and solvent extraction studies were correlated with voltammetric data for this purpose. The vast bank of knowledge on adsorption trends at other electrode surfaces (including mercury) can be profitably tapped for the development of new protocol for adsorptive stripping analyses of organic analytes. An important class of compounds to be determined in this manner are pesticides. Thus, azinphos-methyl (guthion) was determined by adsorptive stripping voltammetry at a hanging mercury drop electrode.[490] Organochlorine pesticides such as endosulfan 6,7,8,9,10,10a-hexachloro-1,5,5a,6,9,9a-hexahydro-6,9-methano-2,4,3-benzo[e]-di-oxathiepin-3-oxide) was similarly determined[491] in a soil sample using initial voltammetric reduction followed by square-wave stripping at a stationary glassy carbon electrode in an aqueous acetonitrile medium.

We close the discussion on stripping analyses by referring the reader to several excellent reviews for further details.[325,371,389,428,492,493]

4.8. DIRECT VOLTAMMETRIC (OR POLAROGRAPHIC) DETERMINATION OF POLLUTANTS

In the stripping voltammetry procedure, the difficulty (with linear sweep or cyclic voltammetry) of inadequate sensitivity is overcome by coupling the preconcentration step with the use of various pulse routines (e.g., differential pulse, square wave) in the stripping step. However, in many cases, the adsorptive accumulation strategy affords adequate analyte levels at the interface to perform a *direct* voltammetric assay even with a linear scan. The term *adsorption voltammetry* can be employed to describe this approach. Of course, in many

4.8. Direct Voltammetric Determination of Pollutants

respects, adsorptive CSV (described in an earlier section) is a close analytical cousin. In this section, we describe several instances where pollutants have been analyzed by various voltammetric (or polarographic) techniques without the use of a separate stripping step. That is, even in those cases where a metal complex is formed in the accumulation step, the complex is directly reduced in the analytical step. In other words, Step 1 rather than Step 2 occurs:

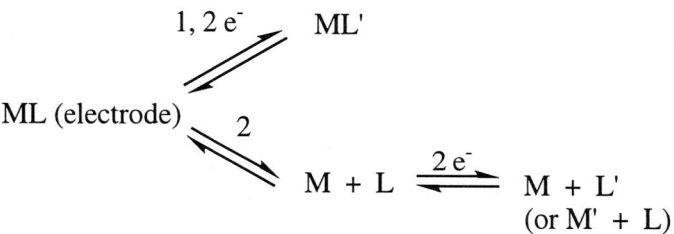

Where, ML', M, and L' are the reduced complex, metal, and the ligand, respectively.

To review the new advances, it is perhaps instructive to take a step back and consider the history of polarography (and voltammetry) in trace analysis. As has been pointed out,[494] these techniques have had a checkered career. Polarography gained early acceptance in the 1940s as an instrumental method for trace analysis. However, this technique almost disappeared from analytical use in the 1950s and 1960s, the reasons for which are multivarious but are undoubtedly coupled with the spectacular developments in atomic spectroscopies, mass spectrometry, and chromatography. The advent of the operational amplifier saw a rebirth of polarography and voltammetry in the 1970s. This trend has continued into the 1980s and 1990s thanks to the advent of computers and user-friendly electrochemical instrumentation.[495]

We must examine the early advances in the electrochemical analysis of pollutants. Hence, procedures for the normal polarographic analysis of pesticides such as parathion and malathion[496,497] were followed by the use of pulse polarography for the analysis of organohalide pesticides.[498] Pulse polarography (mainly differential pulse polarography) was also extensively used for the analysis (at the ppb level) of elements such as As and Se in the 1970s and 1980s.[494] This contrasts with the corresponding trend for the trace analysis of heavy metals during this era, which was almost exclusively done via ASV.

The advent of *selective* adsorption of the analyte as a preconcentration strategy in the 1980s,[499–502] has resulted in a substantial enhancement of the differential pulse voltammetry (or polarography) measurement. The resulting peak currents depend on the length of the adsorption time and the rate of solution mass transport to the electrode surface. It must be recognized, however, that the adsorptive accumulation step enhances mostly the analytical peak current. The background current, especially the double-layer component, may or may not be affected. The surface redox reaction contribution, adventitious impurity (or O_2) currents and solvent decomposition are definitely not influenced by adsorption. The overall result nonetheless is an enhancement of the S/N ratio of the measured current signal relative to the "normal" procedure involving no adsorptive accumulation. The detectability is even further improved by using the subtractive procedure outlined earlier. Subtractive differential pulse voltammetry following adsorptive accumulation has been described for various organic compounds on carbon paste and Pt working electrodes.[503] A detection limit of ~1×10^{-9} M was obtained for chlorpromazine with a 10 min. preconcentration period. Adsorptive preconcentration was also coupled with FIA followed by differential pulse quantitation.[504] The determination of chlorpromazine in a 10^2-fold excess of nonadsorbable solution species with similar redox potentials (e.g., ferrocyanide ion and ascorbic acid) was demonstrated. More impressive, reproducible determination of chlorpromazine in urine was claimed to be possible with no sample treatment. Related work by the same group has shown that poly-3-methylthiophene-coated glassy carbon electrodes are not passivated in the presence of several phenolic compounds.[505] Figure 4.53 contains representative data for p-chlorophenol. The bare electrode (B) shows rapid fouling by the phenol oxidation product(s) and consequent loss of signal. The polymer-coated electrode on the other hand retains the electroactivity over successive cycles.

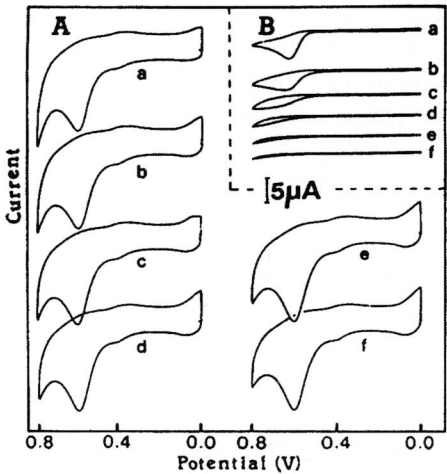

FIGURE 4.53. Successive (a–f) cyclic voltammograms for 2×10^{-4} M chlorophenol at poly(thiophene)-coated (A) and bare (B) glassy carbon electrodes. (Reproduced with permission from Wang and Li.[505])

4.9. Electrochemistry as Auxiliary Tool to Atomic Spectroscopies

Examples of the use of adsorptive voltammetry for the analysis of trace pollutants include the use of 2-(5'-bromo-2'-pyridylazo)-5-diethylamino-phenol for the determination of Cr(III) on mercury,[506] o-phenylenediamine for the reductive determination of the Se(IV) complex by a HMDE[507], and diphenylcarbazone for the adsorptive voltammetry determination of trace amounts of Cr(VI).[508]

We have discussed mainly organic analytes and elements such as As, Se, and Cr in this section. Finally, we note that techniques such as cyclic voltammetry can even be used for the detection and determination of microbial cells. Thus, an anodic current signal was obtained for a cell suspension of *Saccharomyces cervesiae* at a basal plane pyrolytic graphite electrode.[509] The peak current was linearly proportional to the cell concentration in the range (0.1 – 1.9) x 10^8 cells mL^{-1} illustrating the analytical utility of this assay procedure.

4.9. ELECTROCHEMISTRY AS AN AUXILIARY TOOL TO ATOMIC SPECTROSCOPIES

Electrolytic preconcentration followed by analyses using electrothermal (furnace) atomic absorption spectrometry (FAAS) or ICP-MS is an effective strategy for the determination of elements such as Ni, Co, Cr and Mn that are difficult to determine by direct ASV. Electrolysis can also serve to separate the analyte from matrix species that are problematic in FAAS or ICP-MS such as chloride. Many matrix components that are especially troublesome in ICP-MS including alkali metals and alkaline earths have reduction potentials that differ from those of transition metals by 1 V or more, affording easy separation by controlled potential deposition. Interestingly enough, alkali metal halides cause serious interferences in FAAS, yet they are ideal media for electrochemical studies!

A number of studies have therefore advantageously combined the capabilities of both types of techniques.[510–527] Pre-concentration of the analyte has been electrolytically accomplished on Hg-coated graphite furnace,[512–514] tungsten wire,[517,518] porous carbon rod,[519] or Hg-coated pyrolytic graphite platform,[516,520,521] followed by insertion of these heating elements into a FAAS system. Incorporation of a voltammetry cell into a flow system for ICP-MS has also been described.[522–526] This arrangement affords convenient medium exchange and on-line operation with the ICP-MS (or ICP-AES) detector placed downstream from the electrochemical cell. A wide range of analytes has been determined with the combined ASV-ICP-MS (or ASV-ICP-AES) system including Cd and Cu,[523] As(III) and Se(IV),[524] Cr(VI) and V(V),[526] and Pb and Tl.[525] Polyatomic ions that have the same nominal mass as the analyte (e.g., ArC$^+$, ArCl$^+$) are problematic with ICP-MS analyses performed in the conventional manner; these are efficiently eliminated using the ASV pretreatment protocol.

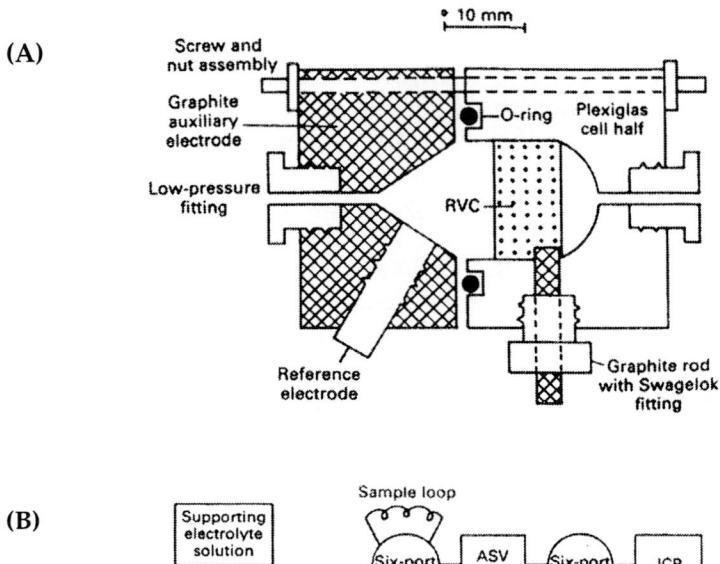

FIGURE 4.54. Schematic diagrams of ASV cell with RVC working electrode, and an ASV-ICP system. (Reproduced with permission from Pretty et al.[522])

In the early development of the combined approach, the recovery of the analyte from the sample was generally poor and also slow. If quantitation deposition is desired, prohibitively long electrolysis times are required. On the other hand, the use of short deposition times (a few seconds) results in the recovery of only a small fraction of the analyte. In some instances, this inefficient recovery has been tied to the presence in the sample matrix of an oxidant, such as nitric acid.[527] Thus the electrolysis step involves a competition between the (desired) deposition and dissolution of the metal. High voltages have been used to sustain high electrolysis currents and the effect of convection created by vigorous gas evolution at both electrodes has been used to enhance the deposition rate.[521] Schematic diagrams of the ASV cell with RVC working electrode and the combined ASV-ICP system are contained in Figure 4.54.[522]

4.9. Electrochemistry as Auxiliary Tool to Atomic Spectroscopies 309

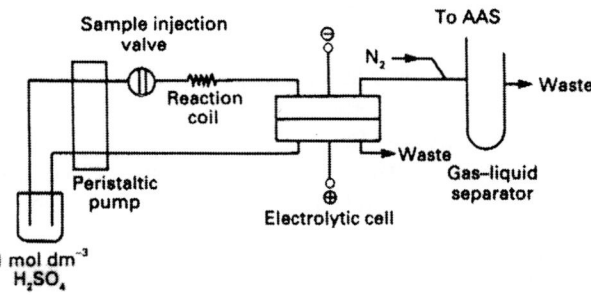

FIGURE 4.55. Schematic diagram of the flow injection–electrochemical hydride generation system. (Reproduced with permission from Lin et al.[528])

Another interesting application of electrochemistry in atomic absorption spectrometry relates to electrochemical hydride generation (EC-HG).[528] Elements such as As and Sn are determined by AAS using the hydride generation technique. The $NaBH_4$–acid reduction technique widely used for hydride generation suffers from several problems. The reagent can contaminate the sample and is expensive. Further, the aqueous solution is unstable and has to be prepared fresh each working day. Excessive H_2 gas is also evolved that, although not a problem with AAS, would alter the impedance of the ICP torch. Then, extra power is needed to keep the plasma stable, and in extreme instances, the plasma can also be extinguished. In the electrochemical approach, the hydride is produced electrolytically and is swept by an inert gas stream into the (downstream) U–tube gas-liquid separator (Figure 4.55).[528] The hydride species and other gases are subsequently transported into the AAS assembly for detection and quantitation. As with the combined preconcentration approach described earlier, the EC-HG strategy has also seen an evolution from batch treatment to on-line flow systems.

The use of electrolysis is not limited to AAS and ICP-MS (or ICP-AES) systems. Recent reports espouse the merits of electrolytic eluent production for ion chromatography (IC).[529-531] In this approach, electrolysis was coupled with the use of a cation-exchange membrane (e.g., Nafion) to yield ultrapure NaOH. The use of NaOH as an IC eluent has several attractive features:[529] (a) under nominal conditions, the suppressed product conductance is much lower with OH^-, proportionately lowering the limits of reduction; (b) unlike CO_3^{2-}/HCO_3^-, which is widely used as an IC eluent, OH^- is not expected to show analyte response nonlinearity; and (c) OH^- is superior for gradient IC applications. Nonetheless, the widespread use of OH^- eluent has been hindered by purity considerations.

4.10. CONDUCTIVITY DETECTORS

Conductometric (sometimes called *conductimetric*) sensors depend on some form of modulation of the electronic or ionic conductivity of the sensing layer by introduction of the analyte. Most conductometric gas sensors rely on alterations in the electronic conductivity of a solid (i.e., the active) layer on the sensor surface. On the other hand, in liquid-based sensors, the change in the ionic conductivity of the electrolyte is monitored. Yet another form of conductometric sensor relies on the modulation of the electronic conductivity of a gate (or a channel) by interaction with the analyte. All three types of sensor approaches are briefly considered in this section.

Gas sensors of the Taguchi type rely on analyte interaction at the active surface while the modulation and measurement of the conductance are done along the surface of the device. The most advanced sensor designs in this category utilize SnO_2 or ZnO thin films as the active layer. Selectivity in these layers is achieved via doping.[532] Thus, doping with Al enhances the selectivity to H_2 and isobutane, Sb doping affords methane detection, and V/In doping yields layers sensitive to NO_2 with no reported cross-sensitivity to CO, CO_2, H_2 or CH_4. Other sensing possibilities with judicious dopant selection are listed in Table 4.23. Thin films of SnO_2 or ZnO on sapphire substrates are also reported to be sensitive to PH_3 (0.02 – 0.42 ppm) and AsH_3 (0.015 – 0.3 ppm).[533] A film temperature of 350°C was used to monitor the conductivity changes in these cases. Other active materials have been used; for example, mixed Cu/Ba/Sn oxide, zeolites, NASICON, ZrO_2/MoO_3, and In_2O_3/CaO for CO_2 detection.[532] Gold-doped WO_3 is reported to have a broad dynamic range for NH_3 detection at 450°C.[534] Conductivity changes in carbon black–PVC composite membranes have been exploited for the determination of various chlorinated hydrocarbons.[535] Similarly, the conductance change of a polystyrene film has been used for NO_x detection.[536]

Phthalocyanines are organic semiconductors that have been widely used for the detection of a range of gaseous analytes including organic nitro compounds (e.g., trinitrotoluene and ethylene glycol dinitrate),[537] NH_3, NO_2, H_2, and dimethyl methyl phosphonate.[532] The underlying mechanism for conductivity modulation in most of these cases, however, is not yet well understood. Finally, conducting polymer films such as polypyrrole and polyaniline have been used as active materials in conductometric sensors for NH_3 and humidity.[538,539]

The possibility of exploiting for sensor purposes other property changes in a semiconductor on exposure to an analyte (e.g., capacitance) has been discussed,[540] although the *selectivity* of such property changes is questionable.

In liquid-based conductometric sensors, the modulation and measurement of the conductance take place in a direction perpendicular to the fluid surface.

4.10. Conductivity Detectors

Table 4.23. Examples of Enhancement of Selectivity of SnO_2-Based Conductometric Detectors by Doping

Dopant	Targeted Analyte
Al	H_2 and isobutane
Sb	Methane
V/In	NO_2
La	Alcohols and CO_2
Ti	Hydrocarbons
W	Formaldehyde
Cu	Hydrogen sulfide
Pd	Hydrogen

Source: Based on data in Janata *et al.*[532]

Flow-through conductometric detectors have enjoyed a long history of use as industrial monitors. They are also most often used in conjunction with an IC or CZE system for detection of anions and cations. Classically, the electrical (ionic) conductivity of quiescent solutions has been measured with an AC bridge at frequencies ranging from a few kHz to even MHz. High values of the frequency are preferable for two reasons: first, polarization at the electrode/electrolyte interface is avoided, and second, high frequencies permit the use of low capacitance cells with volumes in the μL range without loss of measurement sensitivity.

A bipolar pulse technique has been based on the application of two equal successive constant voltage pulses of opposite polarity to the cell and measurement of the current at the end of the second pulse.[541-543] In this manner, capacitance effects are largely eliminated in the conductance measurement. Applications of this technique have been described[541-543] and commercial detectors are available based on this principle. Conductometric detectors for flow-through detection are generally simple and easy to construct with a small internal volume. The linear dynamic range of this type of detector is also high; for example a dynamic range of six decades in concentration has been reported for one particular cell design.[544]

The obstacles to DC measurement of conductance are largely absent in flow-through cell configurations. This is because continuous flow reduces concentration polarization at the interfaces. Thus simple cell designs can be devised, as exemplified by a recent study describing the measurement of the current in response to a DC potential that is imposed across two narrow gauge hypodermic needles in a polytetrafluoroethylene tube.[545] This affords an inexpensive, low volume detector for IC, with effective cell volumes as low as 50 nL being claimed.[545] However, it must be noted that the voltage (6 V) applied is such that electrolysis is actually involved here, and thus this detector is not appropriately classified as of the conductometric type.

4.11. PHOTOASSISTED DETECTION OF POLLUTANTS

The vast majority of the electroanalytical methods discussed in this chapter have been performed without the use of photoexcitation. Photolysis *was* used in isolated instances, as for example, for degrading the organic complexing ligands prior to trace metal speciation analyses (see Section 4.7.6) and for amperometric detection of some pesticides (see Section 4.5). In this section, strategies for coupling photoexcitation with electrochemical detection will be described. Either the electrode itself or the solution may be subjected to photoexcitation. For example, irradiation of a metal with light of sufficient energy causes photoemission of electrons from the metal into the solution phase (see Section 2.12). Alternatively, electron-hole pairs may be created in a semiconductor electrode by irradiating it with light of energy exceeding the optical band-gap of the semiconductor (i.e., $h\upsilon > E_g$). Such "photoelectrochemical" (PEC) processes were also discussed in Chapter 2. Finally, the analyte molecules (or ions) in the solution may be photoexcited using light of a specific wavelength. The resultant modulation of the current signal may be used to advantage for its selective detection. Unfortunately, this latter strategy has also been termed *photoelectrochemical*.[94] To avoid confusion with the present definition of PEC processes (as those involving photoexcitation of the electrode), we suggest *photogalvanic* as a possible alternative terminology to describe photoexcitation of species in the homogeneous solution. Such a definition would be consistent with the accepted use of the term *photogalvanic cells*.[546]

Finally, spectroscopic detection can be coupled with electrochemistry in a "spectroelectrochemical cell." Here, the species are detected *optically* (rather than electrochemically) and electrochemistry is used solely to generate the optically responsive specie(s) of interest. Figure 4.56 compares and contrasts the photoelectrochemical and spectroelectrochemical approaches. The interesting symmetry in the two analytical philosophies – namely, light is used in the former for (optical) excitation and current is measured, whereas current (or potential) is used for (electrical) excitation and light is the measured signal – is worthy of

4.11. Photoassisted Detection of Pollutants

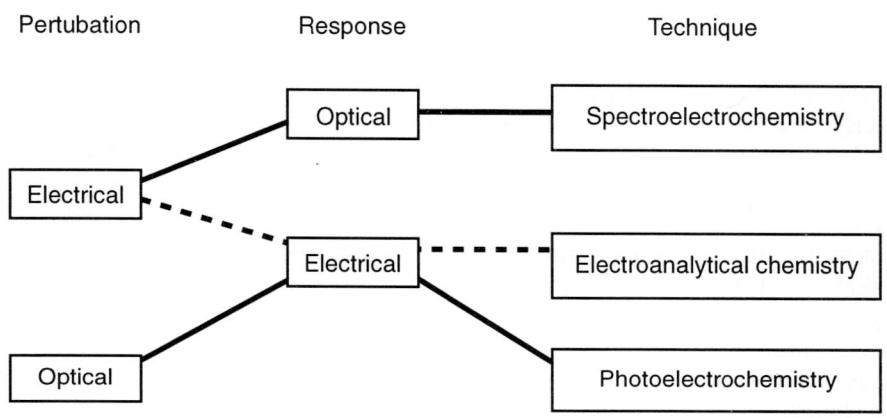

FIGURE 4.56. Comparison of photoelectrochemical and spectroelectrochemical methods of analysis.

note.[547] Almost the entire gamut of spectroscopic tools available to the chemist can be brought to bear in this powerful approach. This includes the entire range of the electromagnetic spectrum from X-rays (picometer or 10^{-12} m wavelength) to radio-frequency (meter) waves. Thus, electrochemistry has been combined with X-ray spectroscopy, UV/VIS spectrophotometry, Fourier transform IR spectroscopy, electron paramagnetic resonance spectroscopy, and nuclear magnetic resonance spectrometry. Even combination of electrochemistry with mass spectrometric detection of electrolysis products has been described.[548]

The use of spectroscopic detection probes is not limited to absorption-based techniques. Photoexcitation of an electrochemically generated product often results in *luminescent light emission* that furnishes an exceedingly sensitive and selective probe of the compound under study. Under certain circumstances, electrochemical excitation directly furnishes an optical emission signal (i.e., without photoexcitation) as exemplified by the phenomenon of *electrogenerated chemiluminescence*. In the case of a semiconductor/electrolyte interface, a similar strategy is possible, where modulation (e.g., quenching) of the *electroluminescence* signal can provide very sensitive detection of an adsorbed analyte.

Obviously, the range of analytical possibilities is very broad indeed, and space will afford us an opportunity to only briefly explore a few of these. In general, it can be stated that the interesting possibilities with photoassisted detection have not been tapped to a signficant extent in the environmental analysis field. The possible lone exception is the HPLC-*hv*-amperometry detection scheme[295] discussed in Section 4.5.

4.11.1. Photoelectrochemical Detection

A photoelectrochemical detector has been used[549,550] in conjunction with a HPLC system, and both studies have used TiO_2 (see Chapter 6) as the semiconductor electrode. The sensitivity of many organic functional groups to PEC oxidation on irradiated semiconductor surfaces (a topic covered in more detail in Chapter 6) prompted the evaluation of a PEC detector for the analysis of organic mixtures.[549] The detector design can be similar to that in the LC-EC case and, in this particular case, constitutes a two-electrode PEC cell fashioned inside a polyethylene block through which a hole was cut to permit the passage of excitation light.[549] Detection limits ranging from 1.6 to 210 µg were obtained for aniline and benzyl alcohol respectively. In principle, the PEC detection scheme allows for the sensitive detection of analytes with low UV absorptivity. In fact, the analytes need not be chromophoric at all and should show better response when they are optically transparent and hence non–filtering. However, the advantage of PEC detection over the LC-EC detection approach in this regard is not clear-cut. Also problematic are the facts that the magnitude of the photocurrent is dependent on the eluent flow rate[549] and the photoresponse appears to be significantly influenced by the adsorption of the analyte on the semiconductor surface.

More intriguing is the possibility, with the PEC approach, of light-addressable detection. This is particularly appealing in connection with whole column detection chromatography. The whole column detection concept[551] has evolved from the success of *on-column* detection.[552] On-column detection has potential advantages over *post-column* detection that include elimination of (a) band broadening associated with the detector dead volume, (b) sensitivity loss associated with post-column dilution, and (c) the need to make difficult postcolumn connections. Interestingly enough, classical chromatography featured the use of vertical glass tubes affording continual visual information on the elution process. Whole-column detection could be achieved in principle by scanning a light beam along the length of a transparent HPLC column containing a semiconductor electrode in the form of a wire or a coating on the wall. This concept has been recently tested[550] using quartz tubing into which a TiO_2 working electrode was inserted along its length. The latter was synthesized by thermal oxidation of a titanium wire.

The system was operated in the FIA mode by injecting aliquots of the analyte into an acetonitrile mobile phase. Some 26 organic analytes were tested, and the photocurrent response was linear over two orders of magnitude with detection limits of 40 and 140 pmol for *p*-aminoacetanilide and diethylamine respectively.[550] One problem with this detector for use with HPLC could be the high background current that is observed for water as the mobile phase. The authors of this pilot study[550] attempted to circumvent it by chemically modifying the TiO$_2$ surface with a non-polar silane reagent. The hope was to remove the detector response to water, to increase the rate of oxidation of the solution redox species, enhance sensitivity, decrease analyte adsorption (see earlier), and decrease the potential of the photogenerated holes. However, the results were disappointing.[550] Nonetheless, the PEC detection approach merits further scrutiny, perhaps with other semiconductor electrode materials (with lower proclivity to photooxidize water) as the sensor candidate.

4.11.2. Photogalvanic and Luminescent Detection

Electronically excited states are better oxidizing and reducing agents than the corresponding ground-state species. Consequently it should be possible to selectively increase the reactivity of certain classes of analytes. To use the light to analytical advantage, chemistry must occur in the excited state between the (photosensitive) analyte and a molecule in solution. Then, in principle, either of the two species involved in the reaction can be selectively detected.

Carbonyl compounds or carbonyl derivatives have been detected in this manner using HPLC coupled with a photoelectrochemical detector (PED).[553,554] The PED used a conventional thin-layer amperometric detection system, modified to permit continuous irradiation of the working electrode surface with high-intensity UV light. As the analyte band passes over the working electrode surface, the analyte molecules are optically transformed to the excited state. The high degree of selectivity with these compounds (without chemical derivatization) was attributed to the long-lived triplet state of the carbonyl chromophore and near-unity quantum efficiency of triplet generation.[555] As a demonstration of the HPLC-PED approach, benzaldehyde was determined in a variety of foodstuffs including food flavoring, liqueur, and tea.[555] Since benzaldehyde is used in many food and pharmaceutical products, an accurate and sensitive assay procedure is needed, especially since this compound is a skin irritant and, at high levels, a central nervous system depressant. There is no simple, direct HPLC method available for the determination of ng quantities of benzaldehyde in complex matrices.

An alternative approach is to use a photochemically active specie as a specific label or derivatizing agent in chromatography or in binding assays, such as immunoassay. One such well-studied complex, Ru(bpy)$_3^{2+}$ (bpy = 2,2'-bipyridine) has been used to develop an understanding of the capabilities and limitations of this detection approach.[556] Hence, Ru(bpy)$_3^{2+}$ is first excited followed by electron-transfer quenching to an intentionally added quencher. The photooxidized Ru(bpy)$_3^{3+}$ is then cathodically reduced at the working electrode to generate the photocurrent signal:

$$D \xrightarrow{h\upsilon} D^*$$
$$D^* + Q \rightarrow D^+ + Q^-$$
$$D^+ \xrightarrow{e^-} D$$

where D is Ru(bpy)$_3^{2+}$, D^* represents the triplet state, and Q is the quencher molecule. The efficacy of this scheme relies crucially on the kinetics of the (thermal) back reaction between Q^- and D^+:

$$Q^- + D^+ \rightarrow D + Q$$

If this happens, the original reactants are regenerated and no signal ensues. Another criterion is that the quencher must not yield any photobackground signal; that is it must not be photochemically active. Several compounds have been tested as electron-transfer quenchers, and the dependence of the photocurrent on variables including the light intensity, quencher concentration and Ru(bpy)$_3^{2+}$ concentration has been quantified.[556]

While this approach builds upon the wealth of information available on Ru(bpy)$_3^{2+}$ photochemistry,[557,558] the advantages of this detection approach have not been realized yet. As the authors themselves mention, the challenge is to apply the photochemistry to analytes of interest. Currently, this would require labeling the analyte in a precolumn reaction with attendant problems of contamination and the like as noted in an earlier section.

There are several other examples of selective photochemical (electronic) excitation of species in solution and measurement of photocurrents.[559–561] However, all these studies have been mechanistically motivated and do not appear to have an analytical significance. Fluorescence measurements have also been reported on electrogenerated intermediates, and these measurements have revealed the participation of previously unsuspected species in the overall electrode reaction.[562] Again, the analytical implications of these studies are not clear at present.

Electrogenerated chemiluminescence (ECL)[563,564] arises from the electrochemical generation of radical species that subsequently undergo high energy electron transfer reactions in solution to generate excited state molecules:

$$C \rightarrow C^{\cdot+} + e^-$$
$$A + e^- \rightarrow A^{\cdot-}$$
$$A^{\cdot-} + C^{\cdot+} \rightarrow A^* + C$$
$$A^* \rightarrow A + h\upsilon$$

Where, $C^{\ddot{+}}$ and $A^{\dot{-}}$ are the cation and anion radicals generated by the electrochemical oxidation and reduction of donor and acceptor molecules in solution. Typical reactions producing ECL involve the radical ions of rubrene, N,N,N',N'-tetramethyl-p-phenyldiamine, p-benzoquinone and also metal complexes such as $Ru(bpy)_3^{2+}$. The early work on ECL was oriented mainly toward mechanistic aspects, and its potential as a detection technique in analytical chemistry has only been recently explored.[555] Recent studies have utilized flow configurations with a view to exploring the applicability for LC and FIA detection. Thus metals such as cobalt and copper have been determined at trace levels by using luminol as an ECL reagent in alkaline medium.[565-567] Peroxide ion concentrations have been similarly determined at a limit of detection of 66 pmol.[568]

An ECL detector has been used in conjunction with reversed-phase HPLC for the detection of polycyclic aromatic hydrocarbons such as naphthalene, anthracene, fluoranthene and chrysene.[568] Both AC and DC methods have been used[569-571] for the electrogeneration of the ECL reactants in reversed-phase HPLC detection applications. However, electrode fouling remains a major problem. The ECL approach has the advantage over conventional chemiluminescent derivatization in that some of the reagents needed for the reaction are produced *in situ*. Thus, in principle, the reaction is tunable via the electrode potential since the emission is concentrated near the electrode surface. However, this advantage must be balanced against the need for more complex instrumentation and the aforementioned electrode stability problems. At present, it can be stated that this method of detection has not attained its full potential.

4.11.3. Spectroelectrochemical Detection

The spectroelectrochemical approach has been used in few studies for detection purposes, either in batch processes or in flow streams. In one such study,[572] a flow-through spectroelectrochemical detector was described that facilitates the simultaneous (real-time) monitoring of both the redox and the spectral properties of analytes eluting from a chromatographic column. For example, added selectivity is achieved (much like dual-electrode detection, see Section 4.4.3) in cases where only one specie is electroactive and the other absorbs, both are electroactive but only one absorbs the light, or both absorb but only one is electroactive. Mixtures of nitro- and chlorophenols were used to illustrate the spectroelectrochemical approach. Both these groups of compounds are oxidized at the same potential, such as +1.6 V (*vs.* Ag/AgCl). However, the optical absorption maxima of the two groups differ and lie at 280 nm and 350 nm, respectively for chloro- and nitrophenols. Therefore, amperometric detection yields a combined response, whereas it is possible to differentiate photometrically between the two groups.

Spectroelectrochemical dector designs are generally more complex, and necessitate the use of optically transparent working electrodes (e.g., Sn-doped indium oxide, RVC, or gold minigrid) and thin-layer cell geometry. The positive features associated with this approach of selectivity enhancement and the like are rather more easily realized with dual-electrode detection. This is probably the reason why spectroelectrochemical detection has not yet gained widespread popularity. Nonetheless, optical spectroscopy has much to offer in environmental and process monitoring, especially given the availability now of inexpensive and versatile detector technology coupled with automated data processing. Further, IR and long path Fourier transform IR spectroscopies have a long history in monitoring of air emissions from hazardous waste sites and automobile exhaust.

Infrared detection technology does suffer from severe interferences from H_2O and CO_2, and in many instances is incompatible for use with aqueous media. In this regard, Raman spectroscopy offers several advantages, not the least of which is the possibility of use with aqueous media and quartz/glass optics. New analytical possibilities have also emerged with the discovery of the surface-enhanced Raman scattering (SERS) phenomenon.[547,573,574] The SERS effect potentially offers both the necessary sensitivity for detecting trace organics and the exceptional selectivity inherent with Raman spectroscopy. However, several challenges must be overcome before SERS-based approaches become viable for routine adoption in environmental analysis applications: (a) Real–world environmental matrices poison the electrode surface; (b) The variability in the electrolyte content and sample matrix may affect the SERS response; and (c) The technique, at least in its current state of refinement, suffers from a limited dynamic range. However, we have already seen that these problems are common to most of the electroanalytical techniques reviewed in this chapter. As in these earlier instances, clever manipulation of the surface chemistry can offer effective solutions as exemplified by the electrode surface modification strategy (Section 4.4.4). Similarly, the use of coated hybrid SERS substrates appears to be a promising approach.[575] As reviewed elsewhere,[575] this circumvents the problem with electrode poisoning, and equally important, the analyte is not required to undergo electrochemical transformation.

4.12. SUMMARY

Electrochemical methods for sensing pollutants are applicable to a wide range of analytical scenarios from the laboratory to the field. These methods may be implemented either in the batch mode or in flow streams. They can also be used as the detector as part of a composite system as in chromatography or in capillary electrophoresis. Finally, the use of light in conjunction with electrochemistry (photoelectrochemical detection) brings an added dimension to the versatility of electroanalytical methods for pollutant sensing and quantification.

REFERENCES

1. M. Maruyama, Determination of Trace Amounts of 3,4-dichloroaniline by High-Performance Liquid Chromatography with Amperometric Detection and its Application to Pesticide Residue Analysis. *Fresenius J. Anal. Chem.* **343**, 890 (1992).
2. R. Boussenadji, P. Dufek, and M. Porthault, *LC-GC* **11**, 450, 452, 454 (1993).
3. M. Rose and G. Shearer, Determination of Tranquilizers and Carazolol Residues in Animal Tissue Using High-Performance Liquid Chromatography with Electrochemical Detection, *J. Chromatogr.* **624**, 471 (1992).
4. D. A. Martens and W. T. Frankenberger, Jr., Pulsed Amperometric Detection of Amino Acids Separated by Anion Exchange Chromatography. *J. Liq. Chromatogr.* **15**, 423 (1992).
5. T. M. Florence, Electrochemical Approaches to Trace Element Speciation in Waters. A Review. *Analyst (London)* **111**, 489 (1986).
6. B. Fleet and H. Gunasingham, Electrochemical Sensors for Monitoring Environmental Pollutants. *Talanta* **39**, 1449 (1992).
7. J. Ruzicka and E. H. Hansen, "Flow Injection Analysis." Wiley, New York, 1981.
8. M. J. E. Golay and J. G. Atwood, Early Phases of the Dispersion of a Sample Injected in Poiseuille Flow. *J. Chromatogr.* **186**, 353 (1979).
9. J. G. Atwood and M. J. E. Golay, Dispersion of Peaks by Short Straight Open Tubes in Liquid Chromatography Systems. *J. Chromatogr.* **218**, 97 (1981).
10. J. T. Vanderslice, K. K. Stewart, A. G. Rosenfeld, and D. J. Higgs, Laminar Dispersion in Flow-Injection Analysis. *Talanta* **28**, 11 (1981).
11. J. T. Vanderslice, G. R. Beecher, and A. G. Rosenfeld, Dispersion and Diffusion Coefficients in Flow Injection Analysis. *Anal. Chem.* **56**, 292 (1984).
12. E. W. Kristensen, R. L. Wilkins, and R. M. Wightman, Dispersion in Flow Injection Analysis Measured with Microvoltammetric Electrodes. *Anal. Chem.* **58**, 986 (1986).
13. E. H. Hansen, J. Ruzicka, and A. K. Ghose, Flow Injection Analysis for Calcium in Serum, Water and Waste Waters by Spectrophotometry and by Ion-Selective Electrode. *Anal. Chim. Acta* **100**, 151 (1978).
14. B. Karlberg and S. Thelander, Determination of Readily Oxidized Compounds by Flow Injection Analysis and Redox Potential Detection. *Analyst (London)* **103**, 1154 (1978).
15. C.-M. Wolff and H. A. Mottola, Enzymic Substrate Determination in Closed Flow-Through Systems by Sample Injection and Amperometric Monitoring of Dissolved Oxygen Levels. *Anal. Chem.* **50**, 94 (1978).
16. K. Rajeshwar, J. G. Ibanez, and G. M. Swain, Electrochemistry and the Environment. *J. Appl. Electrochem.* **24**, 1077 (1994).

17. A. D. C. Chan, X. Li, and D. J. Harrison, NMR Study of the State of Water in Ion-Selective Electrode Membranes. *Anal. Chem.* **65**, 32 (1993).
18. R. Eujster, V. E. Spichlger, and W. Simon, Membrane Model for Neutral Carrier based Membrane Electrodes Containing Ionic Sites. *Anal. Chem.* **65**, 689 (1993).
19. R. DeMarco, Response of Copper(II) Ion-Selective Electrodes in Seawater. *Anal. Chem.* **66**, 3202 (1994).
20. B. Hoyer, Calibration of a Solid-State Copper Ion-Selective Electrode in Cupric Ion Buffers Containing Chloride. *Talanta* **38**, 115 (1991).
21. J. D. Czaban, Electrochemical Sensors in Clinical Chemistry: Yesterday, Today and Tomorrow. *Anal. Chem.* **57**, 345A (1985).
22. H. Schettler, J. Jui, W. Weppner, and R. A. Huggins, Investigation of Solid Sodium Reference Electrodes for Solid-State Electrochemical Gas Sensors. *Appl. Phys.* **A57**, 31 (1993).
23. N. Rao, O. T. Sorensen, J. Schoonman, and C. M. Van der Bleek, *Key Eng. Mater.* **59/60**, 367 (1991).
24. N. Rao, G. M. Van der Bleek, and J. Schoonman, Potentiometric NO_x (x = 1,2) Sensors with Ag^+-β''-Alumina as Solid Electrolyte and Ag Metal as Solid Reference. *Solid State Ionics* **52**, 339 (1992).
25. J. Kirchrova and C. W. Bale, Solid-State Na-β-Alumina Potentiometric Sensor for Measuring Gaseous Arsenic Oxides. *Solid State Ionics.* **59**, 109 (1993).
26. J. A. J. Brunink, J. R. Haak, J. G. Boner, D. N. Reinhondt, M. A. McKervey, and S. J. Harris, Chemically-Modified Field Effect Transistors: A Sodium Ion Selective Sensor Based on Calix[4]arene Receptor Molecules. *Anal. Chim. Acta* **254**, 75 (1991).
27. J. Langmaier and J. Janata, Sensitive Layers for Electrochemical Detection of Hydrogen Cyanide. *Anal. Chem.* **64**, 523 (1992).
28. D. E. Aspnes and A. Heller, Photoelectrochemical Hydrogen Evolution and Water-Photolyzing Semiconductor Suspensions: Properties of Platinum Group Metal Catalyst-Semiconductor Contacts in Air and in Hydrogen. *J. Phys. Chem.* **87**, 4919 (1983).
29. C. R. Martin and H. Freiser, Coated-Wire Ion Selective Electrodes and Their Application to the Teaching Laboratory *J. Chem. Educ.* **57**, 512 (1980).
30. M. E. Meyerhoff and P. M. Kovach, An Ion-Selective Electrode/Flow Injection Analysis Experiment. *J. Chem. Educ.* **60**, 766 (1983).
31. R. Llenado and G. A. Rechnitz, Ion Electrode Based Automatic Glucose Analysis System. *Anal. Chem.* **45**, 2165 (1973).
32. B. van der Schoot, S. Jeanneret, A. van den Berg, and N. de Rooji, Microsystems for Flow Injection Analysis. *Anal. Methods Instrum.* **1**, 38 (1993).
33. I. Isildak and A. Covington, Ion Selective Electrode Potentiometric Detection in Ion Chromatography. *Electroanalysis* **5**, 815 (1993).

34. A. Manz and W. Simon, Picoliter Cell Volume Potentiometric Detector for Open-Tubular Column LC. *J. Chromatogr. Sci.* **21**, 326 (1983).
35. A. Nann, I. Silverstri, and W. Simon, Quantitative Analysis in Capillary Zone Electrophoresis Using Ion-Selective Microelectrodes as On-Column Detectors. *Anal. Chem.* **65**, 1662 (1993).
36. M. Otto and J. D. R. Thomas, Model Studies on Multiple Channel Analysis of Free Magnesium, Calcium, Sodium, and Potassium at Physiological Concentration Levels with Ion Selective Electrodes. *Anal. Chem.* **57**, 2647 (1985).
37. R. J. Forster, F. Regan, and D. Diamond, Modeling of Potentiometric Electrode Assays for Multicomponent Analysis. *Anal. Chem.* **63**, 876 (1991).
38. D. Diamond, J. Lu, Q. Chen, and J. Wang, Multicomponent Batch-Injection Analysis Using an Array of Ion-Selective Electrodes. *Anal. Chim. Acta* **281**, 629 (1993).
39. K. Beebe, D. Verz, J. Sandifer, and B. Kowalski, Sparingly Selective Ion-Selective Array for Multicomponent Analysis. *Anal. Chem.* **60**, 66 (1988).
40. T. Pearce, J. Gardner, S. Freil, P. Bartlett, and N. Blair, Electronic Nose for Monitoring the Flavour of Beers. *Analyst (London)* **118**, 371 (1993).
41. A. Newman, Electronic Noses. *Anal. Chem.* **63**, 585A (1991).
42. D. Diamond, Progress in Sensor Array Research. *Electroanalysis* **5**, 795 (1993).
43. R. C. Harris and H. H. Williams, Specific-Ion Electrode Measurements on Br, Cl and F in Atmospheric Precipitation. *J. Appl. Metereol.* **8**, 299 (1969).
44. W. Eckert, T. Frevert, and H. G. Trüper, A New Liquid-Junction Free Probe for the In Situ Determinaton of pH, pH_2S and Redox Values. *Water Res.* **24**, 1341 (1990).
45. N. T. Crosby, A. L. Denns, and J. G. Stevens, An Evaluation of Some Methods for the Determination of Fluoride in Potable Waters and Other Aqueous Solutions. *Analyst (London)* **93**, 643 (1968).
46. J. L. Swartz and T. S. Wright, Analysis of Alkaline Pulping Liquor with Sulfide Ion-Selective Electrode. *Tappi* **53**, 90 (1970).
47. E. H. Hansen, A. K. Ghose, and J. Ruzicka, Flow Injection Analysis of Environmental Samples for Nitrate Using an Ion-Selective Electrode. *Analyst (London)* **102**, 705 (1977).
48. E. H. Hansen, F. J. Krey, A. K. Ghose, and J. Ruzicka, Rapid Determination of Nitrogen, Phosphorus and Potassium in Fertilizers by Flow Injection Analysis. *Analyst (London)* **102**, 714 (1977).
49. J. Ruzicka, E. H. Hansen, and E. A. Zagatto, Flow Injection Analysis. Part VII. Use of Ion-Selective Electrodes for Rapid Analysis of Soil Extracts and Blood Serum. Determination of Potassium, Sodium and Nitrate. *Anal. Chim. Acta* **88**, 1 (1977).

50. J. D. R. Thomas, in "Electrochemical Detection in Flow Analysis: Modern Trends in Analytical Chemistry" (E. Pungor and I. Buzás, ed.), Part A, p. 314. Elsevier, Amsterdam, 1984.
51. M. Trojanowicz, Continuous Potentiometric Determination of Sulphate in a Differential Flow System. *Anal. Chim. Acta* **114**, 293 (1980).
52. R. A. Durst, Continuous Monitoring of Free Cyanide by Means of Membrane Diffusion of Gaseous HCN and an Electrode Indicator Technique. *Anal. Lett.* **10**, 961 (1977).
52a. O. Wassmus and K. Cammann, A Nitrate Sensor System for Continuous Flow Monitoring. *Sens. Actuators* **B18-B19**, 362 (1994).
52b. M. S. Miloshova, B. L. Seleznev, and E. A. Bychkev, Chalcogenide Glass Chemical Sensors for Determination of Thallium in Natural and Waste Water. *Sens. Actuators* **B18-B19**, 373 (1994).
52c. L. Campanella, M. Battilotti, A. Borraccino, C. Colapicchioni, M. Tomassetti, and G. Vicco, A New ISFET Device Responsive to Anionic Detergents. *Sens. Actuators* **B18-B19**, 321 (1994).
53. M. L. Eagan and L. DuBois, The Determination of Ammonium Ion in Airborne Particulates with Selective Electrodes. *Anal. Chim. Acta* **70**, 157 (1974).
54. C. H. Sloan and G. P. Morie, Determination of Ammonia in Tobacco and Tobacco Smoke with an Ammonia Electrode. *Anal. Chim. Acta* **69**, 243 (1974).
55. S. Yao, Y. Shimizu, N. Miura, and N. Yamazoe, Use of Sodium Nitrite Auxiliary Electrode for Solid Electrolyte Sensor to Detect Nitrogen Oxides. *Chem. Lett.*, p. 587 (1992).
56. J. N. Driscoll, K. Mahoney, and M. Young, Potentiometric Determination of Sulfur Dioxide in Flue Gases with an Ion-Selective Lead Electrode. *Anal. Chem.* **45**, 2283 (1973).
57. A. M. Y. Jaber, G. J. Moody, and J. D. R. Thomas, Solvent Mediator Studies on Barium Ion-Selective Electrodes Based on a Sensor of the Tetraphenylborate Salt of the Barium Complex of a Nonylphenoxypoly(ethyleneoxy)ethanol. *Analyst (London)* **101**, 179 (1976).
58. J. Kirchnerova, C. W. Bale, and J. M. Skeaff, Ag-β-Al$_2$O$_3$ Solid Electrolyte for Sensing Arsine, AsH$_3$. *Sens. Actuators* **B2**, 7 (1990).
59. P. T. Kissinger, C. Refshauge, R. Dreiling, and R. N. Adams, An Electrochemical Detector for Liquid Chromatography with Picogram Sensitivity. *Anal. Lett.* **6**, 465 (1973).
60. K. Stulík and V. Pacáková, Comparison of Several Voltammetric Detectors for High Performance Liquid Chromatography. *J. Chromatogr.* **208**, 269 (1981).
61. K. Stulík, V. Pacáková, and B. Stárková, Carbon Pastes for Voltammetric Detectors in High-Performance Liquid Chromatography. *J. Chromatogr.* **213**, 41 (1981).

62. H. Matsuda, Zur Theorie Der Stationären Strom-Spannungs-Kurven Von Redox-Elektrodenreaktionen In Hydrodynamischer Voltammetric. I. Laminare Staupunktströmungen. *J. Electroanal. Chem. Interfacial Electrochem.* **15**, 109 (1967).
63. J. Yamada and H. Matsuda, Limiting Diffusion Currents in Hydrodynamic Voltammetry. III. Wall Jet Electrodes. *J. Electroanal. Chem. Interfacial Electrochem.* **44**, 189 (1973).
64. B. Fleet and C. J. Little, Design and Evaluation of Electrochemical Detectors. *J. Chromatogr. Sci.* **12**, 747 (1974).
65. M. Podolák, K. Stulík, and V. Pacáková, Construction and the Checking of the Functional Sample of a Voltammetric Detector for High-Performance Liquid Chromatography. *Chem. Listy* **76**, 1106 (1982).
66. H. Gunasingham, Large-Volume Well Jet Cells as Electrochemical Detectors for High-Performance Liquid Chromatography. *Anal. Chim. Acta* **159**, 139 (1967).
67. H. Gunasingham and B. Fleet, Wall-Jet Electrode in Continuous Monitoring Voltammetry. *Anal. Chem.* **55**, 1409 (1983).
68. W. J. Blaedel and Z. Yim, Flow-Through Electrochemical Cell with Open Liquid Junction. *Anal. Chem.* **50**, 1722 (1978).
69. W. J. Blaedel and Z. Yim, Rapid Pulsed Flow Voltammetry. *Anal. Chem.* **52**, 564 (1980).
70. P. L. Meshi and D. C. Johnson, Flow-Rate Independent Component of the Steady-State Current in Tubular Electrodes. *Anal. Chem.* **52**, 1304 (1980).
71. P. L. Meshi and D. C. Johnson, The Amperometric Response of Tubular Electrodes Applied to Flow-Injection Determinations. *Anal. Chim. Acta* **124**, 303 (1981).
72. P. L. Meshi and D. C. Johnson, The Coulometric Response of Tubular Electrodes Applied to Flow-Injection Determinations. *Anal. Chim. Acta* **124**, 315 (1981).
73. J. Wang and M. Ariel, The Rotating Disc Electrode in Flowing Systems. Part I. An Anodic Stripping Monitoring System for Trace Metals in Natural Waters. *Anal. Chim. Acta* **99**, 89 (1978).
74. E. Scarano, M. G. Bonicelli, and M. Forina, A Cell for Continuous Analysis in Flowing Solutions with the Rapidly Dropping Mercury Electrode. *Anal. Chem.* **42**, 1470 (1970).
75. K. Stulík, V. Pacáková, and M. Podolák, Effect of Various Measuring Techniques on the Response of a Polarographic High-Performance Liquid Chromatographic Detector. *J. Chromatogr.* **262**, 85 (1983).
76. S. J. Lyle and M. I. Saleh, Observations on a Dropping-Mercury Electrochemical Detector for Flow Injection Analysis and HPLC. *Talanta* **28**, 251 (1981).
77. K. Stulík and V. Pacáková, Electrochemical Detection in High Performance Liquid Chromatography. *CRC Crit. Rev. Anal. Chem.* **14**, 297 (1984).

78. D. L. Rabenstein and R. Saetre, Mercury-Based Electrochemical Detector for Liquid Chromatography for the Detection of Glutathione and other Sulfur-Containing Compounds. *Anal. Chem.* **49**, 1036 (1977).
79. J. Janata and J. Ruzicka, Combination of Flow Injection Analysis and Voltammetry. *Anal. Chim. Acta* **139**, 105 (1982).
80. W. A. MacCrehan, Differential Detection in Liquid Chromatography and Application to the Measurement of Organometal Cations. *Anal. Chem.* **53**, 74 (1981).
81. R. C. Buchta and L. J. Papa, Electrochemical Detector for Liquid Chromatography. *J. Chromatogr. Sci.* **14**, 713 (1976).
82. W. A. MacCrehan, R. A. Durst and J. M. Bellama, Electrochemical Detection in Liquid Chromatography: Application to Organometallic Speciation. *Anal. Lett.* **10**, 1175 (1977).
83. P. Curry, C. Engström, and A. G. Ewing, Electrochemical Detection for Capillary Electrophoresis. *Electroanalysis* **3**, 587 (1991).
84. A. G. Ewing, J. M. Mesaros, and P. F. Gavin, Electrochemical Detection in Microcolumn Separations. *Anal. Chem.* **66**, 527A (1994).
85. J. A. Loun, R. Koile, and D. C. Johnson, Amperometric Flow-Through Wire Detector. A Practical Design with High Sensitivity. *Anal. Chim. Acta* **116**, 33 (1980).
86. K. A. Rubinson, T. W. Gilbert, and H. B. Mark, Jr., Liquid Chromatography Electrochemical Detector with a Porous Membrane Separator. *Anal. Chem.* **52**, 1549 (1980).
87. K. Slais and D. Kourilová, Electrochemical Detector with a 20 nL Volume for Micro-Column Liquid Chromatography. *Chromatographia* **16**, 265 (1982).
88. L. A. Knecht, E. J. Guthrie, and J. W. Jorgenson, On-Column Electrochemical Detector with a Single Graphite Fiber Electrode for Open-Tubular Liquid Chromatography. *Anal. Chem.* **56**, 479 (1984).
89. R. L. St. Claire, III and J. W. Jorgenson, Characterization of an On-Column Electrochemical Detector for Open-Tubular Liquid Chromatography. *J. Chromatogr. Sci.* **23**, 186 (1985).
90. V. F. Ruban, Electrochemical Detection for Trace Analysis by Capillary HPLC. *J. High Resolution Chromatogr., Chromatogr. Commun.* **16**, 663 (1993).
91. J. Mattusch, T. Welsch, and G. Werner, HPLC-Electrochemical Detector with a Carbon Fiber Working Electrode. *J. Prakt. Chem./Chem.-Ztg.* **334**, 49 (1992).
92. L. Loub, F. Opekav, V. Pacáková, and K. Stulík, Solid Polymer Electrolyte Amperometric Detector for FIA and HPLC with Mobile Phases of Low Conductivity. *Electroanalysis* **4**, 447 (1992).
93. K. Stulík and V. Pacáková, Comparison of Several Voltammetric Detectors for High-Performance Liquid Chromatography. *J. Chromatogr.* **208**, 269 (1981).

94. S. E. Weber and J. T. Long, Detection Limits and Selectivity in Electrochemical Detectors. *Anal. Chem.* **60**, 903A (1988).
95. J. T. Long and S. G. Weber, Voltammetry in Static and Flowing Solutions with a Large Amplitude Sine Wave Potential. *Electroanalysis* **4**, 429 (1992).
96. J. Wang, in "HPLC Detection" (G. Patonay, ed.), Chapter 5, pp. 91-109. VCH, New York, 1992.
97. H. B. Hanekamp, W. H. Voogt, P. Bos, and R. W. Frei, An Electrochemical Scrubber for the Elimination of Eluent Background Effects in Polarographic Flow-Through Detection. *Anal. Chim. Acta* **118**, 81 (1980).
98. J. Wang and H. D. Dewald, Dual Coulometric-Voltammetric Cells for On-Line Stripping Analysis. *Anal. Chem.* **55**, 933 (1983).
99. J. Wang and H. D. Dewald, Deposition of Metals at a Flow-Through Reticulated Vitreous Carbon Electrode Coupled with On-Line Monitoring of the Effluent. *J. Electrochem. Soc.* **130**, 1814 (1983).
100. G. W. Shieffer, Dual Coulometric-Amperometric Cells for Increasing the Selectivity of Electrochemical Detection in High-Performance Liquid Chromatography. *Anal. Chem.* **52**, 1994 (1980).
101. R. Eggli and R. Asper, Electrochemical Flow-Through Detector for the Determination of Cystine and Related Compounds. *Anal. Chim. Acta* **101**, 253 (1978).
102. J. A. Trainham and J. Newman, A Flow-Through Porous Electrode Model: Application to Metal-Ion Removal from Dilute Streams. *J. Electrochem. Soc.* **124**, 1528 (1977).
103. R. E. Sioda, Flow-Through Electrodes Composed of Parallel Screens. *Electrochim. Acta* **22**, 439 (1977).
104. T. Fujinaga and S. Kihara, Electrolytic Chromatography and Coulopotentiography. A Rapid Electrolysis at the Column Electrode Used for the Preparation, Separation, Concentration and Estimation of Trace and/or Unstable Substances. *CRC Crit. Rev. Anal. Chem.* **6**, 223 (1977).
105. D. J. Curran and T. P. Toryas, Electrochemical Detector Based on a Reticulated Vitreous Carbon Working Electrode for Liquid Chromatography and Flow Injection Analysis. *Anal. Chem.* **56**, 672 (1984).
106. C. L. Blank, Dual Electrochemical Detector for Liquid Chromatography. *J. Chromatogr.* **117**, 35 (1976).
107. D. A. Roston, R. E. Shoup, and P. T. Kissinger, Liquid Chromatography/Electrochemistry: Thin-Layer Multiple Electrode Detection. *Anal. Chem.* **54**, 1417A (1982).
108. W. J. Albery and S. Bruckenstein, Ring-Disc Electrodes. Part 2. Theoretical and Experimental Collection Efficiencies. *Trans. Faraday Soc.* **62**, 1920 (1966).
109. W. J. Albery, L. R. Swanberg, and P. Wood, *J. Electroanal. Chem. Interfacial Electrochem.* **162**, 29, 45 (1982).

110. D. A. Roston and P. T. Kissinger, Series Dual-Electrode Detector for Liquid Chromatography/Electrochemistry. *Anal. Chem.* **54**, 429 (1982).
111. G. S. Mayer and R. E. Shoup, Simultaneous Multiple Electrode Liquid Chromatographic-Electrochemical Assay for Catecholamines, Indole-Amines and Metabolites in Brain Tissue. *J. Chromatogr.* **255**, 533 (1983).
112. K. Brunt and C. H. P. Bruins, New Electrochemical Detector for High-Performance Liquid Chromatography. The Differential Amperometric Detector. *J. Chromatogr.* **161**, 310 (1978).
113. W. A. MacCrehan and R. A. Durst, Dual-Electrode, Liquid Chromatographic Detector for Determination of Analytes with High Redox Potentials. *Anal. Chem.* **53**, 1700 (1981).
114. U. R. Tjaden, J. de Jong, and C. F. M. Van Valkenburg, Gradient Elution of Biogenic Amines and Derivatives in Reversed Phase Ion-Pair Partition Chromatography with Electrochemical and Fluorometric Detection. *J. Liq. Chromatogr.* **6**, 2255 (1983).
115. M. Goto, G. Zou, and D. Ishii, Current Amplification in Dual Electrochemical Detection for Micro High-Performance Liquid Chromatography. *J. Chromatogr.* **268**, 157 (1983).
116. S. G. Weber and W. C. Purdy, The Behavior of an Electrochemical Detector Used in Liquid Chromatography and Continuous Flow Voltammetry. Part 1. Mass Transport-Limited Current. *Anal. Chim. Acta* **100**, 531 (1978).
117. S. G. Weber and W. C. Purdy, Electrochemical Detection with a Regenerative Flow Cell in Liquid Chromatography. *Anal. Chem.* **54**, 1757 (1982).
118. T. P. Tougas, J. M. Jannetti, and W. G. Collier, Theoretical and Experimental Response of a Biamperometric Detector for Flow Injection Analysis. *Anal. Chem.* **57**, 1377 (1985).
119. A. J. Bard, G. Dennault, C. Lee, D. Mandler, and D. O. Wipf, Scanning Electrochemical Microscopy: A New Technique for the Characterization and Modification of Surfaces. *Acc. Chem. Res.* **23**, 357 (1990).
120. A. J. Bard, F. F. Fan, D. T. Pierce, P. R. Unwin, D. O. Wipf, and F. Zhou, Chemical Imaging of Surfaces with the Scanning Electrochemical Microscope. *Science* **254**, 68 (1991).
121. J. Heinze, Ultramicroelectrodes—A New Dimension in Electrochemistry? *Angew. Chem., Int. Ed. Engl.* **30**, 170 (1991); J. F. Cassidy and B. F. Foley, Microelectrodes—Potential Invaders. *Chem. Br.* September, p. 764 (1993).
122. D. E. Weisshaar, D. E. Tallman, and J. L. Anderson, Kel-F-Graphite Composite Electrode as an Electrochemical Detector for Liquid Chromatography and Application to Phenolic Compounds. *Anal. Chem.* **53**, 1809 (1981).
123. D. Bélanger and M. S. Wrighton, Microelectrochemical Transistors Based on Electrostatic Binding of Electroactive Metal Complexes in Protonated

Poly(4-vinylpyridine): Devices That Respond to Two Chemical Stimulii. *Anal. Chem.* **59**, 1426 (1987).
124. G. Kittlesen, H. White, and M. Wrighton, Chemical Derivatization of Microelectrode Assays by Oxidation of Pyrrole and N-Methylpyrrole: Fabrication of Molecule-Based Electronic Devices. *J. Am. Chem. Soc.* **106**, 7389 (1984).
125. J. J. Hickman, D. Ofer, P. E. Laibinis, G. M. Whitesides, and M. S. Wrighton, Molecular Self-Assembly of Two-Terminal Voltammetric Microsensors with Internal References. *Science* **252**, 688 (1991).
126. P. N. Bartlett and P. Birkin, Enzyme Switch Responsive to Glucose. *Anal. Chem.* **65**, 1118 (1993).
127. M. Nichizawa, T. Matsue, and I. Uchida, Penicillin Sensor Based on a Microarray Electrode Coated with pH-Responsive Polypyrrole. *Anal. Chem.* **64**, 2642 (1992).
128. R. S. Glass, S. P. Perone, and D. R. Ciarlo, Application of Information Theory to Electroanalytical Measurements Using a Multielement, Microelectrode Array. *Anal. Chem.* **62**, 1914 (1990).
129. Q. Chen, J. Wang, G. D. Rayson, B. Tian, and Y. Lin, Sensor Array for Carbohydrates and Amino Acids Based on Electrocatalytic Modified Electrodes. *Anal. Chem.* **65**, 251 (1993).
130. J. Wang, G. D. Rayson, Z. Lu, and H. Wu, Coated Amperometric Electrode Arrays for Multicomponent Analysis. *Anal. Chem.* **62**, 1924 (1990).
131. J. Stetter, P. C. Jurs, and S. L. Rose, Detection of Hazardous Gases and Vapors: Pattern Recognition Analysis of Data from an Electrochemical Sensor Array. *Anal. Chem.* **58**, 860 (1986).
132. P. Fielden and T. McCreedy, Voltammetric Information from Arrays of Individually Controlled Electrodes: Their Potential for Industrial Process Measurements. *Anal. Chim. Acta* **273**, 111 (1993).
133. W. L. Caudell, J. O. Howell, and R. M. Wightman, Flow Rate Independent Amperometric Cell. *Anal. Chem.* **54**, 2532 (1982).
134. R. M. Wightman, Microvoltammetric Electrodes. *Anal. Chem.* **53**, 1125A (1981).
135. C. N. Svendsen, Multi-Electrode Array Detectors in High-Performance Liquid Chromatography: A New Dimension in Electrochemical Analysis. *Analyst (London)* **118**, 123 (1993).
136. A. Aoki, T. Matsue, and I. Uchida, Multichannel Electrochemical Detection with a Microelectrode Array in Flowing Streams. *Anal. Chem.* **64**, 44 (1992).
137. J. Hoogvliet, J. Reijn, and W. van Bennekom, Multichannel Amperometric Detection System for Liquid Chromatography and Flow Injection Analysis. *Anal. Chem.* **63**, 2418 (1991).

138. J. Wang, F. Ouziel, C. Yarnitzky, and M. Ariel, A Flow Detector Based on Square-Wave Polarography at the Dropping Mercury Electrode. *Anal. Chim. Acta* **102**, 99 (1978).
139. R. Samuelesson, J. O'Dea, and J. Osteryoung, Rapid Scan Square Wave Voltammetric Detector for High-Performance Liquid Chromatography. *Anal. Chem.* **52**, 2215 (1980).
140. P. A. Reardon, G. E. O'Brien, and P. E. Sturrock, A Swept-Potential Electrochemical Detector for Flow Streams. *Anal. Chim. Acta* **162**, 175 (1984).
141. J. J. Scanlon, P. A. Flaquer, G. W. Robinson, G. E. O'Brien, and P. E. Sturrock, High-Performance Liquid Chromatography of Nitrophenols with a Swept Potential Electrochemical Detector. *Anal. Chim. Acta* **158**, 169 (1984).
142. A. Trojanek and H. G. de Jong, Fast-Scan AC Voltammetry for Better Resolution of Chromatographically Overlapping Peaks. *Anal. Chim. Acta* **141**, 115 (1982).
143. W. Caudill, A. G. Ewing, S. Jones, and R. M. Wightman, Liquid Chromatography with Rapid Scanning Electrochemical Detection at Carbon Electrodes. *Anal. Chem.* **55**, 1877 (1983).
144. L. Zhang and P. E. Sturrock, HPLC Separation and Square-Wave Adsorptive Stripping Detection of Human Immunoglobulin G and A. *Electroanalysis* **2**, 289 (1990).
145. T. A. Last, Coulostatic Voltammetric Liquid Chromatography Detector. *Anal. Chem.* **55**, 1509 (1983).
146. R. K. Trubey and T. A. Nieman, Rapid Scanning Coulostatic Liquid Chromatography/Electrochemistry in Dual Electrode Cells. *Anal. Chem.* **58**, 2549 (1986).
147. W. W. Kubiak, Use of Mercury Drop Electrodes for Polarographic Measurements in Flowing Systems. *Electroanalysis* **1**, 379 (1989).
148. H. Angerstein-Kozlowska, B. E. Conway, and W. Sharp, The Real Condition of Electrochemically Oxidized Platinum Surfaces. Part 1. Resolution of Component Processes. *J. Electroanal. Chem. Interfacial Electrochem.* **43**, 9 (1973).
149. S. Hadzi-Jordanov, H. Angerstein-Kozlowska, M. Vukovic, and B. E. Conway, The State of Electrodeposited Hydrogen at Ruthenium Electrodes. *J. Phys. Chem.* **81**, 2271 (1977).
150. L. D. Burke and M. E. G. Lyons, Electrochemistry of Hydrous Oxide Films. *In* "Modern Aspects of Electrochemistry" (R. E. White, J. O'M. Bockris, and B. E. Conway, eds.), Chapter 4, pp. 169-248. Plenum, New York and Canada, 1990.
151. P. N. Ross, Jr., Structure Sensitivity in the Electrocatalytic Properties of Pt. I. Hydrogen Adsorption on Low Index Single Crystals and the Role of Steps. *J. Electrochem. Soc.* **126**, 67 (1979).

152. T. M. Florence, Anodic Stripping Voltammetry with a Glassy Carbon Electrode Mercury-Plated In Situ. *J. Electroanal. Chem. Interfacial Electrochem.* **27**, 273 (1970).
153. R. A. Osteryoung and J. H. Christie, Theoretical Treatment of Pulsed Voltammetric Stripping at the Thin Film Mercury Electrode. *Anal. Chem.* **46**, 351 (1974).
154. W. Lund and M. Salberg, Anodic Stripping Voltammetry with the Florence Mercury Film Electrode. Determination of Copper, Lead and Cadmium in Sea Water. *Anal. Chim. Acta* **76**, 131 (1975).
155. J. Wang, Reticulated Vitreous Carbon—A New Versatile Electrode Material. *Electrochim. Acta* **26**, 1721 (1981).
156. W. R. Matson, D. K. Roe, and D. E. Carrit, Composite Graphite-Mercury Electrode for Anodic Stripping Voltammetry. *Anal. Chem.* **37**, 1594 (1965).
157. J. Wang, Anodic Stripping Voltammetry at Graphite-Epoxy Microelectrodes for In Vitro and In Vivo Measurements of Trace Metals. *Anal. Chem.* **54**, 221 (1982).
158. I. Gustavsson and K. Lundström, A Pyrolytic Carbon Film Electrode for Voltammetry. III. Application to Anodic Stripping Voltammetry. *Talanta* **30**, 959 (1983).
159. M. R. Cushman, B. G. Bennet, and C. W. Anderson, Electrochemistry at Carbon Fibers. Part 1. Characteristics of the Mercury Film Carbon Fiber Electrode in Differential Pulse Anodic Stripping Voltammetry. *Anal. Chim. Acta* **130**, 323 (1981).
160. J. E. Anderson, D. E. Tallman, D. J. Chesney, and J. L. Anderson, Fabrication and Characterization of a Kel-F-Graphite Composition Electrode for General Voltammetric Applications. *Anal. Chem.* **50**, 1051 (1978).
161. J. M. Kaufman, A. Laudet, G. J. Patriarche, and G. D. Christian, Preparation and Characterization of Graphite-Coated Metallic Electrodes: The Graphite-Sprayed Electrode. *Talanta* **29**, 1077 (1982).
162. W. R. Peterson, The Static Mercury Drop Electrode. *Am. Lab.*, p. 11, December (1979).
163. W. E. van der Linden and J. W. Dieker, Glassy Carbon as Electrode Material in Electroanalytical Chemistry. *Anal. Chim. Acta* **119**, 1 (1980).
164. C. Olson and R. N. Adams, Carbon Paste Electrodes. Application to Anodic Voltammetry. *Anal. Chim. Acta* **22**, 582 (1960).
165. C. Urbaniczky and K. Lundström, Voltammetric Studies on Carbon Paste Electrodes. The Influence of Paste Composition on Electrode Capacity and Kinetics. *J. Electroanal. Chem. Interfacial Electrochem.* **176**, 169 (1984).
166. M. Rice, Z. Galus, and R. N. Adams, Graphite Paste Electrodes. Effects of Paste Composition and Surface States on Electron Transfer Rates. *J. Electroanal. Chem. Interfacial Electrochem.* **143**, 89 (1983).

167. R. N. Adams, Carbon Paste Electrodes. *Anal. Chem.* **30**, 1586 (1958).
168. R. J. Fenn, S. Siggia, and D. J. Curran, Liquid Chromatography Detector Based on Single and Twin Electrode Thin-Layer Electrochemistry: Application to the Determination of Catecholamines in Blood Plasma. *Anal. Chem.* **50**, 1067 (1978).
169. D. N. Armentrout, J. D. McLean, and M. W. Long, Trace Determination of Phenolic Compounds in Water by Reversed Phase Liquid Chromatography with Electrochemical Detection Using a Carbon-Polyethylene Tubular Anode. *Anal. Chem.* **51**, 1039 (1979).
170. E. Csoregi, L. Gorton, and G. Marho-Vega, Carbon Fibres as Electrode Materials for the Construction of Peroxidase-Modified Amperometric Biosensors. *Anal. Chim. Acta* **273**, 59 (1993).
171. T. E. Edmonds and J. Guoliang, Carbon Fibre Microelectrodes in the Differential Pulse Voltammetry of Copper Ions. *Anal. Chim. Acta* **151**, 99 (1983).
172. J. Wang, L. Angnes, H. Tobias, R. A. Roesner, K. C. Hong, R. S. Glass, F. Kong, and R. W. Pekala, Carbon Aerogel Composite Electrodes. *Anal. Chim. Acta* **65**, 2300 (1993).
173. D. Kohler, J. Zabasajja, A. Krishnagopalan, and B. J. Tatarchuk, Metal-Carbon Composite Materials from Fiber Precursors. I. Preparation of Stainless Steel-Carbon Composite Electrodes. *J. Electrochem. Soc.* **137**, 136 (1990).
174. D. Kohler, J. Zabasajja, F. Rose, and B. J. Tatarchuk, Metal-Carbon Composite Materials from Fiber Precursors. II. Electrochemical Characterization of Stainless Steel-Carbon Structures *J. Electrochem. Soc.* **137**, 1750 (1990).
175. A. Beilby and A. Carlsson, A Pyrolytic Carbon Film Electrode for Voltammetry. Part V. Characterization and Comparison with the Glassy Carbon Electrode by Electrochemical Pretreatment in Basic Solution. *J. Electroanal. Chem. Interfacial Electrochem.* **248**, 283 (1988).
176. A. Eriksson, A. Norekrans, and J. Carlsson, Surface Structural and Electrochemical Investigations of Pyrolytic Carbon Film Electrodes Prepared by Chemical Vapor Deposition Using Ethene as Carbon Source. *J. Electroanal. Chem. Interfacial Electrochem.* **324**, 291 (1992).
177. C. F. McFadden, P. R. Melaragno, and J. A. Davis, Fabrication of Pyrolytic Carbon Film Electrodes by Pyrolysis of Methane on a Machinable Glass Ceramic. *Anal. Chem.* **62**, 742 (1990).
178. C. F. McFadden, L. L. Russell, P. R. Melaragno, and J. A. Davis, Low-Temperature Pyrolytic Carbon Films. Electrochemical Performance and Surface Morphology as a Function of Pyrolysis Time, Temperature and Substrate. *Anal. Chem.* **64**, 1521 (1992).

179. J. K. Clark, W. A. Schilling, C. A. Wijayawardhana, and P. R. Melaragno, Reactivation and Stability of Pyrolytic Carbon Film Electrodes on Macor Substrates. *Anal. Chem.* **66**, 3528 (1994).
180. J. Wang, A. Brennsteiner, and A. Sylwester, Composite Electrodes Based on Carbonized Poly(acrylonitrile) Foams. *Anal. Chem.* **62**, 1102 (1990).
181. N. Sleszynski, J. Osteryoung, and M. Carter, Arrays of Very Small Voltammetric Electrodes Based on Reticulated Vitreous Carbon. *Anal. Chem.* **56**, 130 (1984).
182. M. R. Callström, T. X. Neenan, R. L. McCreery, and D. C. Alsmeyer, Doped Glassy Carbon Electrode (DGC): Low-Temperature Synthesis, Structure and Catalytic Behavior. *J. Am. Chem. Soc.* **112**, 4954 (1990).
183. D. Ingersoll and D. H. Huskisson, Preparation of Doped Carbon Films. *J. Electroanal. Chem. Interfacial Electrochem.* **307**, 281 (1991).
184. G. M. Swain and R. Ramesham, The Electrochemical Activity of Boron-Doped Polycrystalline Diamond Thin Film Electrodes. *Anal. Chem.* **65**, 345 (1993).
185. R. W. Murray, A. G. Ewing, and R. A. Durst, Chemically Modified Electrodes. Molecular Design for Electroanalysis. *Anal. Chem.* **59**, 379A (1987).
186. L. R. Faulkner, Chemical Microstructures on Electrodes. *Chem. Eng. News*, February 27, p. 28 (1984).
187. R. W. Murray, Chemically Modified Electrodes. *Electroanal. Chem.* **13**, 191-367 (1984).
188. "Charge Transfer in Polymeric Systems," *Faraday Discussion Chemical Society*, No. 88. Royal Society of Chemistry, London, 1989.
189. H. D. Abruna, Coordination Chemistry in Two Dimensions: Chemically Modified Electrodes. *Coord. Chem. Rev.* **86**, 135 (1988).
190. K. D. Snell and A. G. Keenan, Surface Modified Electrodes. *Chem. Rev.* **8**, 259 (1979).
191. M. Kaneko and A. Yamada, Solar Energy Conversion by Functional Polymers. *Adv. Polym. Sci.* **55**, 1-47 (1984).
192. M. Kaneko and D. Wöhrle, Polymer-Coated Electrodes: New Materials for Science and Industry. *Adv. Polym. Sci.* **84**, 141-228 (1988).
193. G. Inzelt, Role of Polymeric Properties in the Electrochemical Behavior of Redox Polymer-Modified Electrodes. *Electrochim. Acta* **34**, 83 (1989).
194. P. E. Laibinis, G. Whitesides, D. L. Allara, Y.-T. Tao, A. N. Parikh, and R. G. Nuzzo, Comparison of the Structure and Wetting Properties of Self-Assembled Monolayers of n-Alkanethiols on the Coinage Metal Surfaces, Cu, Ag, Au. *J. Am. Chem. Soc.* **113**, 7152 (1991).
195. C. E. D. Chidsey, Free Energy and Temperature Dependence of Electron Transfer at the Metal-Electrolyte Interface. *Science* **251**, 919 (1991).
196. J. Wang, S.-P. Chen, and M. S. Lin, Use of Different Electropolymerization Conditions for Controlling the Size-Exclusion Selectivity at Polyaniline,

Polypyrrole and Polyphenol Films. *J. Electroanal. Chem. Interfacial Electrochem.* **273**, 231 (1989).
197. C. S. C. Bose and K. Rajeshwar, Efficient Electrocatalyst Assemblies for Proton and Oxygen Reduction: The Electrosynthesis and Characterization of Polypyrrole Films Containing Nanodispersed Platinum Particles. *J. Electroanal. Chem. Interfacial Electrochem.* **333**, 235 (1992).
198. W. A. Wampler, C. Wei, and K. Rajeshwar, Electrocomposites of Polypyrrole and Carbon Black. *J. Electrochem. Soc.* **141**, L13 (1994).
199. W. A. Wampler, C. Wei, and K. Rajeshwar, Composites of Polypyrrole and Carbon Black. 2. Electrosynthesis, Characterization and Influence of Carbon Black Characteristics. *Chem. Mater.* **7**, 585 (1995).
200. G. K. Chandler and D. Pletcher, The Electrodeposition of Metals onto Polypyrrole Films from Aqueous Solution. *J. Appl. Electrochem.* **16**, 62 (1986).
201. R. Noufi, The Incorporation of Ruthenium Oxide in Polypyrrole Films and the Subsequent Photooxidation of Water at n-GaP Photoelectrode. *J. Electrochem. Soc.* **130**, 2126 (1983).
202. H. Yoneyama and Y. Shoji, Incorporation of WO_3 into Polypyrrole and Electrochemical Properties of the Resulting Polymer Films. *J. Electrochem. Soc.* **137**, 3826 (1990).
203. K. Kawai, N. Mihara, S. Kuwabata, and H. Yoneyama, Electrochemical Synthesis of Polypyrrole Films Containing TiO_2 Powder Particles. *J. Electrochem. Soc.* **137**, 1793 (1990).
204. F. Beck, M. Dahlhaus, and N. Zahedi, Anodic Codeposition of Polypyrrole and Dispersed TiO_2. *Electrochim. Acta* **37**, 1265 (1992).
205. J. Wang, N. Naser, L. Angnes, H. Wei, and L. Chen, Metal-Dispersed Carbon Paste Electrodes. *Anal. Chem.* **64**, 1285 (1992).
206. C. S. C. Bose, S. Basak, and K. Rajeshwar, Preparation, Voltammetric Characterization and Use of a Composite Containing Chemically Synthesized Polypyrrole and a Carrier Polymer. *J. Electrochem. Soc.* **139**, L75 (1992).
207. H. Ge, H. Zhao, and G. G. Wallace, Development of a Polymer Dispersed-Mercury Modified Electrode. *Anal. Chim. Acta* **238**, 345 (1990).
208. H. Zhang and R. W. Murray, Monomer and Polymer Solvent Dynamic Control of an Electron-Transfer Cross-Reaction Rate at a Redox Polymer/Solution Interface. *J. Am. Chem. Soc.* **113**, 5183 (1991).
209. T. T. Worster, M. L. Longmire, M. Watanabe, and R. W. Murray, Diffusion and Heterogeneous Electron-Transfer Rates in Acetonitrile and in Polyether Polymer Melts by Alternating Current Voltammetry at Microdisk Electrodes. *J. Phys. Chem.* **95**, 5315 (1991).
210. A. H. Schroeder, F. B. Kaufman, V. Patel, and E. M. Engler, Comparative Behavior of Electrodes Coated with Thin Films of Structurally Related

Electroactive Polymers. *J. Electroanal. Chem. Interfacial Electrochem.* **113**, 193 (1980).
211. A. H. Schroeder and F. B. Kaufman, The Influence of Polymer Morphology on Polymer Film Electrochemistry. *J. Electroanal. Chem. Interfacial Electrochem.* **113**, 209 (1980).
212. G. Inzelt and J. Bácskai, Electrochemical Quartz Crystal Microbalance Study of the Swelling of Poly(vinylferrocene) Films. *Electrochim. Acta* **37**, 647 (1992).
213. Y. T. Kim, R. W. Collins, K. Vedam, and D. L. Allara, Real Time Spectroscopic Ellipsometry: In Situ Characterization of Pyrrole Electropolymerization. *J. Electrochem. Soc.* **138**, 3266 (1991).
214. S. Bruckenstein and A. R. Hillman, Consequences of Thermodynamic Restraints on Solvent and Ion Transfer During Redox Switching of Electroactive Polymers. *J. Phys. Chem.* **92**, 4837 (1988).
215. A. R. Hillman, N. A. Hughes, and S. Bruckenstein, Solvation Phenomena in Polyvinylferrocene Films: Effect of History and Redox State. *J. Electrochem. Soc.* **139**, 74 (1992).
216. A. Glidle, A. R. Hillman, and S. Bruckenstein, Dynamic Film Rigidity Observations During Electrochemical Deposition of Polybithiophene Films. *J. Electroanal. Chem. Interfacial Electrochem.* **318**, 411 (1991).
217. A. P. Clarke, J. G. Vos, and A. R. Hillman, The Influence of Prior Electrode History on Mobile Species Transfer Characteristics of Polymer Modified Electrodes. *J. Electroanal. Chem. Interfacial Electrochem.* **356**, 287 (1993).
218. M. E. G. Lyons, Electrocatalysis Using Electroactive Polymers, Electroactive Composites and Microheterogeneous Systems. *Analyst (London)* **119**, 805 (1994).
219. H.-Y. Liu and F. C. Anson, Redox Mediation of Dioxygen Reduction Within Nafion Electrode Coatings Containing Colloidal Platinum as Catalyst. *J. Electroanal. Chem. Interfacial Electrochem.* **158**, 181 (1983).
220. S. Prabhu, R. Baldwin, and L. Kryger, Chemical Preconcentration and Determination of Copper at a Chemically Modified Carbon Paste Electrode Containing 2,9-Dimethyl-1,10-phenanthroline. *Anal Chem.* **59**, 1074 (1987).
221. R. P. Baldwin, J. K. Christensen, and L. Kryger, Voltammetric Determination of Traces of Nickel(II) at a Chemically Modified Electrode Based on Dimethylglyoxime-Containing Carbon Paste. *Anal Chem.* **58**, 1790 (1986).
222. T. Malinski, A. Ciszewski, J. Fish, and L. Czuchajowski, Conductive Polymeric Tetrakis(3-methoxy-4-hydroxyphenyl)porphyrin Film Electrode for Trace Determination of Nickel. *Anal Chem.* **62**, 909 (1990).
223. A. R. Guadalupe and H. D. Abruna, Electroanalysis with Chemically Modified Electrodes. *Anal Chem.* **57**, 142 (1985).

224. L. M. Wier, A. R. Guadalupe, and H. D. Abruna, Multiple-Use Polymer-Modified Electrodes for Electroanalysis of Metal Ions in Solution. *Anal Chem.* **57**, 2009 (1985).
225. E. Lorenzo and H. D. Abruna, Determination of Silver with Polymer-Modified Electrodes. *J. Electroanal. Chem. Interfacial Electrochem.* **328**, 111 (1992).
226. P. Li, Z. Gao, Y. Xu, G. Wang, and Z. Zhao, Determination of Trace Amounts of Silver with a Chemically Modified Carbon Paste Electrode. *Anal. Chim. Acta* **229**, 213 (1990).
227. A. P. Doherty, R. J. Forster, M. R. Smyth, and J. G. Vos, Speciation of Iron(II) and Iron(III) Using a Dual Electrode Modified with Electrocatalytic Polymers. *Anal. Chem.* **64**, 572 (1992).
228. E. Yu. Katz and A. A. Solovév, Chemically Modified Electrodes with Affinity to Sulphydryl Compounds. *J. Electroanal. Chem. Interfacial Electrochem.* **261**, 217 (1989).
229. S. Prabhu, R. P. Baldwin, and L. Kryger, Preconcentration and Determination of Lead(II) at Crown Ether and Cryptanol Containing Chemically Modified Electrodes. *Electroanalysis* **1**, 13 (1989).
230. K. Ravichandran and R. P. Baldwin, Phenylenediamine-Containing Chemically Modified Carbon Paste Electrodes as Catalytic Voltammetric Sensors. *Anal. Chem.* **55**, 1586 (1983).
231. E. G. Cookeas and C. E. Efstathiou, Flow Injection Amperometric Determination of Thiocyanate and Selenocyanate at a Cobalt Phthalocyanine-Modified Carbon Paste Electrode. *Analyst (London)* **119**, 1607 (1994).
232. G. Sittampalam and G. S. Wilson, Surface-Modified Electrochemical Detector for Liquid Chromatography. *Anal. Chem.* **54**, 1608 (1983).
233. S. Dong and Y. Wang, Anodic Stripping Voltammetric Determination of Trace Lead with a Nafion/Crown Ether Film Electrode. *Talanta* **35**, 1891 (1988).
234. J. Wang, N. Naser, L. Angnes, H. Wu, and L. Chen, Metal-Dispersed Carbon Paste Electrodes. *Anal. Chem.* **64**, 1285 (1992).
235. S. D. Nikolic, E. B. Milosavljevic, J. L. Hendrix, and J. H. Nelson, Flow Injection Amperometric Determination of Cyanide on a Modified Silver Electrode. *Analyst (London)* **117**, 47 (1992).
236. L. Sun, B. Johnson, T. Wade, and R. M. Crooks, Selective Electrostatic Binding of Ions by Monolayers of Mercaptan Derivatives Adsorbed to Gold Substrates. *J. Phys. Chem.* **94**, 8869 (1990).
237. Q. Cheng and A. Brajter-Toth, Selectivity and Sensitivity of Self-Assembled Thioctic Acid Electrodes. *Anal. Chem.* **64**, 1998 (1992).
238. N. El-Murr, M. Kerkeni, A. Sellami, and Y. Ben Taarit, The Zeolite-Modified Carbon Paste Electrode. *J. Electroanal. Chem. Interfacial Electrochem.* **246**, 461 (1988).

239. A. Yamagishi, and A. Aramata, A Clay-Modified Electrode with Stereoselectivity. *J. Chem. Soc., Chem. Commun.*, p. 452 (1984).
240. I. Turyan and D. Mandler, Self-Assembled Monolayers in Electroanalytical Chemistry: Application of ω-Mercaptocarboxylic Acid Monolayers for Electrochemical Determination of Ultralow Levels of Cadmium. *Anal. Chem.* **66**, 58 (1994).
241. W. Jin, J. Wang, X. Zang, and S. Wang, On the Adsorption Voltammetry of the Lanthanum(III)-Alizarin Complexone-Fluoride System. *J. Electroanal. Chem. Interfacial Electrochem.* **281**, 221 (1990).
242. J. F. Cassidy and K. Tokuda, Preconcentration and Voltammetric Determination of Copper Ions in Aqueous Chloride Solutions at a Crosslinked Poly(4-vinylpyridine)-Coated Electrode. *J. Electroanal. Chem. Interfacial Electrochem.* **285**, 287 (1990).
243. X. Jinrui and L. Bin, Preconcentration and Determination of Lead Ions at a Chitosan-Modified Glassy Carbon Electrode. *Analyst (London)* **119**, 1599 (1994).
244. J. H. Aldstadt, D. F. King, and H. D. Dewald, Flow Injection Potentiometric and Voltammetric Stripping Analysis Using a Dialysis Membrane Covered Mercury Film Electrode. *Analyst (London)* **119**, 1813 (1994).
245. Z. Gao, A. Ivaska, and P. Li, *Anal. Sci.* **8**, 337 (1992); *Chemical Abstr.* **117**, 204245c (1992).
246. P. Ward and M. R. Smyth, Development of a Polypyrrole-Based Amperometric Detector for the Determination of Certain Anions in Water Samples. *Talanta* **40**, 1131 (1993).
247. H. Ge, J. Zhang, and G. G. Wallace, Use of Overoxidized Polypyrrole as a Chromium(VI) Sensor. *Anal. Lett.* **25**, 429 (1992).
248. T. Matsunaga, T. Suzuki, and R. Tomola, Photomicrobial Sensors for Selective Determination of Phosphate. *Enzyme Microb. Technol.* **6**, 355 (1984).
249. I. Karube, T. Nakahara, T. Matsunaga, and S. Suzuki, *Salmonella* Electrode for Screening Mutagens. *Anal. Chem.* **54**, 1725 (1982).
250. J. Wang, T. Martinez, and D. Darnall, Uptake of Redox Couples by Algae Immobilized on Porous Cellulosic Films. *Electrochim. Acta* **35**, 1377 (1990).
251. P. Abu Nader, S. S. Vives, and H. A. Mottola, Studies with a Sulfite Oxidase-Modified Carbon Paste Electrode for Detection/Determination of Sulfite Ion and $SO_2(g)$ in Continuous-Flow Systems. *J. Electroanal. Chem. Interfacial Electrochem.* **284**, 323 (1990).
252. S. Cosnier, C. Innount, and Y. Jouanneau, Amperometric Detection of Nitrate via a Nitrate Reductase Immobilized and Electrically Wired at the Electrode Surface. *Anal. Chem.* **66**, 3198 (1994).
253. O. J. Garcia, P. A. Quintela, and A. E. Kaifer, Electrodes Modified with a Film of Phosphatidylcholine: Electrochemistry Inside a Lipid Layer. *Anal. Chem* **61**, 979 (1989).

254. E. Alvarez, M. Teresa Sevilla, J. M. Pinilla and L. Hernandez, Cathodic Stripping Voltammetry of Paraquat on a Carbon Paste Electrode Modified with Amberlite XAD-2 Resin. *Anal. Chim. Acta* **260**, 19 (1992).
254a. I. Turyan and D. Mandler, Electrochemical Mercury Detection. *Nature (London)* **362**, 703 (1993).
255. H. H. Thorp, Electrochemistry of Proton-Coupled Redox Reactions. Role of the Electrode Surface. *J. Chem. Educ.* **69**, 250 (1992).
256. G. E. Cabaniss, A. A. Diamantis, W. R. Murphy, Jr., R. W. Linton, and T. J. Meyer, Electrocatalysis of Proton-Coupled Electron Transfer Reactions at Glassy-Carbon Electrodes. *J. Am. Chem. Soc.* **107**, 1845 (1985).
257. R. J. Rice, N. M. Pontikos and R. L. McGreery, Quantitative Correlations of Heterogeneous Electron-Transfer Kinetics with Surface Properties of Glassy Carbon Electrodes. *J. Am. Chem. Soc* **112**, 4617 (1990).
258. D. C. Johnson, J. A. Polta, T. Z. Polta, G. G. Neubürger, J. Johnson, A. P-C. Tang, I.-H. Yeo, and J. Baur, Anodic Detection in Flow-Through Cells. *J. Chem. Soc., Faraday Trans. I* **82**, 1081 (1986).
259. D. C. Johnson and W. R. LaCourse, Pulsed Electrochemical Detection at Noble Metal Electrodes in Liquid Chromatography. *Electroanalysis* **4**, 367 (1992).
260. W. R. LaCourse, Pulsed Electrochemical Detection at Noble Metal Electrodes in High Performance Liquid Chromatography. *Analusis* **21**, 181 (1993).
261. D. G. Williams and D. C. Johnson, Pulsed Voltammetric Detection of Arsenic(III) at Platinum Electrodes in Acidic Media. *Anal. Chem.* **64**, 1785 (1992).
262. T. Z. Polta and D. C. Johnson, Pulsed Amperometric Detection of Sulfur Compounds. Part I. Initial Studies at Platinum Electrodes in Alkaline Solutions. *J. Electroanal. Chem. Interfacial Electrochem.* **209**, 159 (1986).
263. J. A. Polta and D. C. Johnson, Pulsed Amperometric Detection of Electroinactive Adsorbates at Platinum Electrodes in a Flow Injection System. *Anal. Chem.* **57**, 1373 (1985).
264. J. A. Polta and D. C. Johnson, The Direct Electrochemical Detection of Amino Acids at a Platinum Electrode in an Alkaline Chromatographic Effluent. *J. Liq. Chromatogr.* **6**, 1727 (1983).
265. T. Z. Polta, D. C. Johnson, and G. R. Lweke, Pulsed Amperometric Detection of Organic Compounds. Part II. Dependence of Response on Adsorption Time. *J. Electroanal. Chem. Interfacial Electrochem.* **209**, 171 (1986).
266. D. C. Johnson and W. R. LaCourse, Liquid Chromatography with Pulsed Electrochemical Detection at Gold and Platinum Electrodes. *Anal. Chem.* **62**, 589A (1989).

267. A. Zirino and M. L. Healy, pH-Controlled Differential Voltammetry of Certain Trace Transition Elements in Natural Waters. *Environ. Sci. Technol.* **6**, 243 (1972).
268. I. Cukrowski, E. Cukrowska, and K. Sykut, Subtractive Anodic Stripping Voltammetry at a Blocked Set of Electrodes. *J. Electroanal. Chem. Interfacial Electrochem.* **125**, 53 (1981).
269. T. M. Florence and Y. J. Farrar, Removal of Oxygen from Polarographic Solutions with Ascorbic Acid. *J. Electroanal. Chem. Interfacial Electrochem.* **41**, 127 (1973).
270. A. Bergens, Reductive Electrochemical Detection in Liquid Chromatography with a Zinc Amalgam Scrubber Colum. *J. Chromatogr.* **598**, 195 (1992).
271. D. M. Radzik, J. S. Brodbelt and P. T. Kissinger, Determination of Toxic Azo Dye Metabolites In Vitro by Liquid Chromatography/Electrochemistry with a Dual-Electrode Detector. *Anal. Chem.* **56**, 2927 (1984).
272. A. Bergens, Determination of Nitrodiphenylamines by Liquid Chromatography and Dual-Electrode Amperometric Detection. *J. Chromatogr.* **410**, 437 (1987).
273. M. A. Malone, P. L. Weber, M. R. Smyth, and S. M. Lunte, Reductive Electrochemical Detection for Capillary Electrophoresis. *Anal. Chem.* **66**, 3782 (1994).
274. P. T. Kissinger, K. Bratin, G. C. Bratin, and L. A. Pachla, The Potential Utility of Pre- and Post-Column Chemical Reactions with Electrochemical Detection in Liquid Chromatography. *J. Chromatogr. Sci.* **17**, 137 (1979).
275. I. S. Krull, X. D. Ding, C. Selavka, and R. Nelson, *LC* **2**, 214 (1984).
276. A. M. Bond and G. G. Wallace, Automated Determination of Nickel and Copper by Liquid Chromatography with Electrochemical and Spectrophotometric Detection. *Anal. Chem.* **55**, 718 (1983).
277. A. M. Bond and G. G. Wallace, Simultaneous Determination of Copper, Nickel, Cobalt, Chromium(VI) and Chromium(III) by Liquid Chromatography with Electrochemical Detection. *Anal. Chem.* **54**, 1706 (1982).
278. A. M. Bond and G. G. Wallace, Liquid Chromatography with Electrochemical and/or Spectrophotometric Detection for Automated Determinaton of Lead, Cadmium, Mercury, Cobalt, Nickel and Copper. *Anal. Chem.* **56**, 2085 (1984).
279. W. Kok, U. A. Brinkman, and R. W. Frei, On-Line Electrochemical Reagent Production for Detection in Liquid Chromatography and Continuous Flow Systems. *Anal. Chim. Acta* **162**, 19 (1984).
280. L. Dalgard, Immobilized Enzymes as Post-Column Reactors in High Performance Liquid Chromatography. *Trends Anal. Chem.* **7**, 185 (1986).
281. L. H. Keith and W. A. Telliard, Priority Pollutants I — A Perspective View. *Environ. Sci. Technol.* **13**, 416 (1979).

282. J. Ruana, I. Urbe, and F. Borrull, Determination of Phenols at the ng/L Level in Drinking and River Waters by Liquid Chromatography with UV and Electrochemical Detection. *J. Chromatogr.* **A655**, 217 (1993).
283. R. E. Shoup and G. S. Mayer, Determination of Environmental Phenols by Liquid Chromatography/Electrochemistry. *Anal. Chem.* **54**, 1164 (1982).
284. C. Webster, M. Smith, P. Wilson, and M. Cooke, Determination of the Total Phenol Content of Soils by High Speed Liquid Chromatography with Electrochemical Detection. *J. High Resolution Chromatogr., Chromatogr. Commun.* **16**, 549 (1993).
285. R. M. Smith and S. Beck, High-Performance Liquid Chromatographic Analysis of Eugenol in Pimento Using Ultraviolet and Electrochemical Detection. *J. Chromatogr.* **291**, 424 (1984).
286. D. A. Roston and P. T. Kissinger, Identification of Phenolic Constituents in Commercial Beverages by Liquid Chromatography. *Anal. Chem.* **53**, 1695 (1981).
287. D. E. Ott, Evaluation of Amperometric Detector for Liquid Chromatographic Determination of 2-Phenylphenol Residues in Orange Rind. *J. Assoc. Off. Anal. Chem.* **61**, 1465 (1978).
288. R. M. Smith, Analysis of the Pungent Properties of Ginger and Grains of Paradise by High-Performance Liquid Chromatography with Electrochemical Detection. *Chromatographia* **16**, 155 (1982).
289. J. L. Anderson and D. J. Chesney, Liquid Chromatographic Determination of Selected Carbamate Pesticides in Water with Electrochemical Detection. *Anal. Chem.* **52**, 2156 (1980).
290. J. L. Anderson, K. K. Whiten, J. D. Brewster, T.-Y. Ou, and W. K. Nonidez, Microarray Electrochemical Flow Detection at High Applied Potentials and Liquid Chromatography with Electrochemical Detection of Carbamate Pesticides in River Water. *Anal. Chem.* **57**, 1366 (1985).
291. H. B. Hanekamp, P. Bos, and R. W. Frei, Design and Selective Application of a Dropping Mercury Electrode Amperometric Detector in Column Liquid Chromatography. *J. Chromatogr.* **186**, 489 (1979).
292. J. W. Lawrence, F. Iverson, H. B. Hanekamp, P. Bos, and R. W. Frei, Liquid Chromatography with UV Absorbance and Polarographic Detection of Ethylenethiourea and Related Sulfur Compounds. Application to Rat Urine Analysis. *J. Chromatogr.* **212**, 245 (1981).
293. K. Bratin, P. T. Kissinger, and C. J. Bruntlett, Reductive Mode Thin-Layer Amperometric Detector for Liquid Chromatography. *J. Liq. Chromatogr.* **4**, 1777 (1981).
294. G. J. Clark, R. R. Goodin, and J. W. Smiley, Comparison of Ultraviolet and Reductive Amperometric Detection for Determination of Ethyl and Methyl Parathion in Green Vegetables and Surface Water Using High-Performance Liquid Chromatography. *Anal. Chem.* **57**, 2223 (1985).

295. X-D. Ding and I. S. Krull, Trace Analysis for Organothiophosphate Agricultural Chemicals by High-Performance Liquid Chromatography-Photolysis-Electrochemical Detection. *J. Agric. Food Chem.* **32**, 622 (1984).
296. H. B. Wan, H. Chi, M. K. Wong, C. Y. Mok, and A. K. Hsieh, Determination of the Ester- Cleavage Products of Some Organophosphorus Pesticides by Liquid Chromatography with Electrochemical Detection. *J. Liq. Chromatogr.* **16**, 4049 (1993).
297. W. Gorski and J. A. Cox, Amperometric Determination of N-nitrosamines in Aqueous Solution at an Electrode Control with a Ruthenium-Based Inorganic Polymer. *Anal. Chem.* **66**, 2771 (1994), and references therein.
298. D. N. Armentrout and S. S. Cutie, Determination of Benzidine and 3,3'-Dichlorobenzidine in Wastewater by Liquid Chromatography with UV and Electrochemical Detection. *J. Chromatogr. Sci.* **18**, 370 (1980).
299. E. M. Lores, F. C. Meakins, and R. F. Moseman, Determination of Halogenated Anilines in Urine by High-Performance Liquid Chromatography with an Electrochemical Detector. *J. Chromatogr.* **188**, 412 (1982).
300. J. R. Rice and P. T. Kissinger, Determination of Benzidine and its Acetylated Metabolites in Urine by Liquid Chromatography. *J. Anal. Toxicol.* **3**, 64 (1979).
301. C. J. Purnell and C. J. Warwick, Application of Electrochemical Detection to the Measurement of 4,4'-Methylenebis(2-chloroaniline) and 2-Chloroaniline Concentrations in Air by High-Performance Liquid Chromatography. *Analyst (London)* **105**, 861 (1980).
302. J. R. Rice and P. T. Kissinger, Liquid Chromatography with Precolumn Sample Preconcentration and Electrochemical Detection: Determination of Aromatic Amines in Environmental Samples. *Environ. Sci. Technol.* **16**, 263 (1980).
303. J. A. Cox and A. Przyjazny, High Pressure Liquid Chromatography of Selected Sulfur Compounds. *Anal. Lett.* **10**, 869 (1977).
304. D. M. Radzik and P. T. Kissinger, Determination of Aniline and Metabolites Produced in Vitro by Liquid Chromatography/Electrochemistry. *Anal. Biochem.* **140**, 74 (1984).
305. J. R. Rice and P. T. Kissinger, Cooxidation of Benzidine by Horseradish Peroxidase and Subsequent Formation of Possible Thioether Conjugates of Benzidine. *Biochem. Biophys. Res. Commun.* **104**, 1312 (1982).
306. D. A. Roston and P. T. Kissinger, Isolation and Identification of Benzene Metabolites In Vitro with Liquid Chromatography/Electrochemistry. *Anal. Chem.* **54**, 1798 (1982).
307. W. A. MacCrehan and R. A. Durst, Measurement of Organomercury Species in Biological Samples by Liquid Chromatography with Differential Pulse Electrochemical Detection. *Anal. Chem*, 2108 (1978).

308. M. Robecke and K. Cammann, Determination of Tetraethyllead and Tetramethyllead Using High Performance Liquid Chromatography and Electrochemical Detection. *Fresenius J. Anal. Chem.* **341**, 555 (1991).
309. A. M. Bond and G. G. Wallace, Determination of Copper as a Dithiocarbamate Complex by Reversed-Phase Liquid Chromatography with Electrochemical Detection. *Anal. Chem.* **53**, 1209 (1981).
310. R. M. Smith and L. E. Yankey, Determination of Metal Ions by Liquid Chromatography Incorporating Dithiocarbamates in the Eluent. *Analyst (London)* **107**, 744 (1982).
311. A. G. Fogg and N. K. Bsebsu, Sequential Flow Injection Voltammetric Determination of Phosphate and Nitrite by Injection of Reagents into a Sample Stream. *Analyst (London)* **109**, 19 (1984).
312. S. Seefeld and U. Battensperger, Determination of Bromide in Snow Samples by Ion Chromatography with Electrochemical Detection. *Anal. Chim. Acta* **283**, 246 (1993).
313. A. Ivaska, P. Forsberg, and R. Heikka, Application of an Amperometric Sensor to In-Line Monitoring of Pulp Bleaching with Chlorine Dioxide. *Anal. Chim. Acta.* **238**, 223 (1990).
314. L. C. Clark, R. Wolf, D. Granger, and Z. Taylor, Continuous Recording of Blood Oxygen Tensions by Polarography. *J. Appl. Physiol.* **6**, 189 (1953).
315. S. C. Chang, J. R. Stetter, and C. S. Cha, Amperometric Gas Sensors. *Talanta* **40**, 461 (1993).
316. M. J. Tierney and H.-O. L. Kim, Electrochemical Gas Sensor with Extremely Fast Response Times. *Anal. Chem.* **65**, 3435 (1993).
317. W. O. Friesen and M. B. McIlroy, Rapidly Responding Oxygen Electrode for Respiratory Gas Sampling. *J. Appl. Physiol.* **29**, 258 (1970).
318. S. Uchiyama, T. Ikarugi, M. Mori, K. Kasama, Y. Ishikawa, M. Kaneko, and A. Umezawa, Highly Sensitive Electrochemical Sensor for Ozone in Water Using Porous Carbon Felt Electrode. *Electroanalysis* **5**, 121 (1993).
319. J. Evans, D. Pletcher, P. R. G. Warburton, and T. K. Gibbs, Amperometric Sensor for Carbon Dioxide: Design, Characteristics and Performance. *Anal. Chem.* **61**, 577 (1989).
320. J. Evans, D. Pletcher, P. R. G. Warburton and T. K. Gibbs, A New Electrochemical Sensor for Carbon Dioxide. Part 2. Study of the Sensor Chemistry. *J. Electroanal. Chem. Interfacial Electrochem.* **262**, 119 (1989).
321. J. M. Skeaff and A. A. Dubreuil, Electrochemical Measurement of SO_3-SO_2 in Process Gas Streams. *Sens. Actuators* **B10**, 161 (1993).
322. F. A. Keidel, Determination of Water by Direct Amperometric Measurement. *Anal. Chem.* **31**, 2043 (1959).
323. H. Huang and P. K. Dasgupta, Amperometric Microsensor for Water. *Anal. Chem.* **62**, 1935 (1990).

324. H. Huang, P. K. Dasgupta, and S. Ronchinsky, Perfluorosufonate Ionomer – Phosphorus Pentoxide Composite Thin Films as Amperometric Sensors for Water. *Anal. Chem.* **63**, 1570 (1991).
325. C. M. G. van den Berg, Potentials and Potentialities of Cathodic Stripping Voltammetry of Trace Elements in Natural Waters. *Anal. Chim. Acta* **250**, 265 (1991).
326. J. Wang, Advances in Adsorptive Stripping Voltammetry. *Anal. Proc.* **24**, 325 (1987).
327. D. C. Johnson and R. E. Allen, Stripping Voltammetry with Collection at a Rotating Ring-Disk Electrode. *Talanta* **20**, 305 (1973).
328. C. Brihaye and G. Duyckaerts, Determination of Traces of Metals by Anodic Stripping Voltammetry at a Rotating Glassy Carbon Carbon Ring Disc Electrode. Part 1. Method and Instrumentation with Evaluation of Some Parameters. *Anal. Chim. Acta* **143**, 111 (1982).
329. D. Laser and M. Ariel, Anodic Stripping with Collection, Using Thin Mercury Films. *J. Electroanal. Chem. Interfacial Electrochem.* **49**, 123 (1974).
330. C. Brihaye and G. Duyckaerts, Determination of Traces of Metals by Anodic Stripping Voltammetry at a Rotating Glassy Carbon Carbon Ring-Disc Electrode. Part 2. Comparison Between Linear Anodic Stripping Voltammetry with Ring Collection and Various Other Stripping Techniques. *Anal. Chim. Acta* **146**, 37 (1983).
331. G. W. Schieffer and W. J. Blaedel, Anodic Stripping Voltammetry with Collection at Tubular Electrodes for the Analysis of Tap Water. *Anal. Chem.* **50**, 99 (1978), and references therein.
332. J. Wang and M. Ariel, Anodic Stripping Voltammetry in a Flow-Through Cell with Fixed Mercury Film Glassy Carbon Disc Electrodes. Part II. The Differential Mode (DASV). *J. Electroanal. Chem. Interfacial Electrochem.* **85**, 289 (1977).
333. L. Sipos, S. Kozar, I. Kontusic, and M. Branica, Subtractive Anodic Stripping Voltammetry with Rotating Mercury Coated Glassy Carbon Electrode. *J. Electroanal. Chem. Interfacial Electrochem.* **87**, 347 (1978).
334. J. Wang and M. Ariel, Subtractive Anodic Stripping Voltammetry with Twin Identical Mercury Film Electrodes Differing in Their Convection Transport During Deposition. *Anal. Chim. Acta* **128**, 147 (1981).
335. W. Kemula, Polarographic Methods of Analysis. *Pure Appl. Chem.* **15**, 283 (1967).
336. M. Ariel and J. Wang, *in* "Improved Anodic Stripping Voltammetry," NBS Spec. Publ. No. 422, p. 881. U. S. Gov. Printing Office, Washington, DC 1976.
337. L. Sipos, P. Valenta, H. W. Nürnberg, and M. Branica, Applications of Polarography and Voltammetry to Marine and Aquatic Chemistry. IV. A New Voltammetric Method for the Study of Mercury Traces in Sea Water

and Inland Waters. *J. Electroanal. Chem. Interfacial Electrochem.* **77**, 263 (1977).

338. E. Steeman, E. Temmerman, and R. Verbinnen, Subtractive Anodic Stripping Voltammetry at Twin Mercury Film Electrodes. *Anal. Chim. Acta* **96**, 177 (1978).
339. J. Wang and H. D. Dewald, Subtractive Anodic Stripping Voltammetry with Flow Injection Analysis. *Anal. Chem.* **56**, 156 (1984).
340. L. Kryger and D. Jagner, Computerized Electroanalysis. Part III. Multiple Scanning Anodic Stripping and its Application to Sea Water. *Anal. Chim. Acta* **80**, 255 (1975).
341. D. Jagner and A. Graneli, Potentiometric Stripping Analysis. *Anal. Chim. Acta.* **83**, 19 (1976).
342. D. Jagner and K. Aren, Derivative Potentiometric Stripping Analysis with a Thin Film of Mercury on a Glassy Carbon Electrode. *Anal. Chim. Acta.* **100**, 375 (1978).
343. D. Jagner, Instrumental Approach to Potentiometric Stripping Analysis of Some Heavy Metals. *Anal. Chem.* **50**, 1924 (1978).
344. D. Jagner, Potentiomeric Stripping Analysis. *Analyst (London)* **107**, 593 (1982).
345. J. Wang and B. Tian, Screen-Printed Electrodes for Stripping Measurements of Trace Mercury. *Anal. Chim. Acta* **274**, 1 (1993).
346. P. Delahay and G. Mamantov, Voltammetry at Constant Current. Review of Theoretical Principles. *Anal. Chem.* **27**, 478 (1955).
347. C. N. Reilly, G. W. Everett, and R. H. Johns, Voltammetry at Constant Current. Experimental Evaluaton. *Anal. Chim. Acta* **27**, 483 (1955).
348. W. Kemula and J. W. Strojek, Controlled Chronopotentiometric Stripping of Metals Deposited on the Hanging Mercury Drop Electrode. *J. Electroanal. Chem. Interfacial Electrochem.* **12**, 1 (1966).
349. C. Hua, D. Jagner, and L. Renman, Automated Determination of Molybdenum(VI) in Seawater by Means of Constant-Current Reduction of the Adsorbed 8-Quinolinol Complex in a Computerized Flow Potentiometric Stripping Analyzer. *Anal. Chim. Acta* **192**, 103 (1987).
350. H. Huiliang, D. Jagner, and L. Renman, Flow Constant Current Stripping Analysis for Antimony(III) and Antimony(V) with Gold Fibre Working Electrodes. Application to Natural Waters. *Anal. Chim. Acta.* **202**, 123 (1987).
351. A. Almestrond, D. Jagner, and L. Renman, Automated Determination of Cadmium and Lead in Whole Blood by Computerized Flow Potentiometric Stripping with Carbon Fiber Electrodes. *Anal. Chim. Acta.* **193**, 71 (1987).

352. D. Jagner, M. Josefson, and S. Westerlund, Simultaneous Determination of Cadmium and Lead in Urine by Means of Computerized Potentiometric Stripping Analysis. *Anal. Chim. Acta.* **128**, 155 (1981).
353. A. Graneli, D. Jagner, and M. Josefson, Microcomputer System for Potentiometric Stripping Analysis. *Anal. Chem.* **52**, 2220 (1980).
354. J. L. Christensen, K. Keiding, L. Kryger, J. Rasmussen, and H. J. Skov, Analog Instrument for Oxidative and Reductive Potentiometric Stripping Analysis. *Anal. Chem.* **53**, 1847 (1981).
355. L. Kryger, Differential Potentiometric Stripping Analysis. *Anal. Chim. Acta* **120**, 19 (1980).
356. S. Bruckenstein and J. W. Bixler, Chemical Stripping Analysis. *Anal. Chem.* **37**, 786 (1965).
357. D. Jagner and K. Aren, Potentiometric Stripping Analysis for Zinc, Cadmium, Lead and Copper in Sea Water. *Anal. Chim. Acta* **107**, 29 (1979).
358. J. F. Coetzee, A. Hussam and T. R. Petrick, Extension of Potentiometric Stripping Analysis to Electropositive Elements by Solvent Optimization. *Anal. Chem.* **55**, 120 (1983).
359. D. Jagner and K. Aren, Flow Potentiometric Stripping Analysis for Mercury(II) in Urine, Sediment and Acid Digest of Biological Material. *Anal. Chim. Acta* **141**, 157 (1982).
360. D. Jagner and S. Westerlund, Determination of Lead, Copper and Cadmium in Wine and Beer by Potentiometric Stripping Analysis. *Anal. Chim. Acta* **117**, 159 (1980).
361. J. K. Christensen, L. Kryger, and N. Pind, The Determination of Traces of Cadmium, Lead and Thallium in Fly Ash by Potentiometric Stripping Analysis. *Anal. Chim. Acta* **141**, 131 (1982).
362. E. Beinrohr, P. Csemi, F. J. Rojas, and H. Hofbauerova, Determination of Manganese in Water Samples by Galvanostatic Stripping Chronopotentiometry in a Flow-Through Cell. *Analyst (London)* **119**, 1355 (1994).
363. A. Hu, R. E. Dessy, and A. Graneli, Potentiometric Stripping with Matrix Exchange Techniques in Flow Injection Analysis of Heavy Metals in Groundwaters. *Anal. Chem.* **55**, 320 (1983).
364. P. Pheiffer Madsen, I. Draback, and J. Sorensen, The Determination of Copper and Lead in Sediments by Potentiometric Stripping Analysis. *Anal. Chim. Acta* **151**, 479 (1983).
365. C. Labar, R. Muller, and L. Lamberts, Studies on Film Potentiometric Stripping Analysis: Effects of Electrochemical Parameters. *Electrochim. Acta* **36**, 2103 (1991).
366. J. Wang and B. Tian, Screen-Printed Stripping Voltammetric/Potentiometric Electrodes for Decentralized Testing of Trace Lead. *Anal. Chem.* **64**, 1708 (1992).

367. J. Wang and B. Tian, Mercury-Free Disposable Lead Sensors Based on Potentiometric Stripping Analysis at Gold-Coated Screen-Printed Electrodes, *Anal. Chem.* **65**, 1529 (1993).
368. S. Wring and J. P. Hart, Chemically-Modified Screen-Printed Carbon Electrodes. *Analyst (London)* **117**, 1281 (1992).
369. S. Wring, J. Hart, L. Bracey, and B. Birch, Development of Screen-Printed Carbon Electrodes, Chemically Modified with Cobalt Phthalocyanine for Electrochemical Sensor Applications. *Anal. Chim. Acta* **231**, 203 (1990).
370. J. Wang, Decentralized Electrochemical Monitoring of Trace Metals: From Disposable Strips to Remote Electrodes. *Analyst (London)* **119**, 763 (1994).
371. T. R. Copeland and R. K. Skogerboe, Anodic Stripping Voltammetry. *Anal. Chem.* **46**, 1257A (1974).
372. R. Posey and R. Andrew, Stripping Voltammetry of Tellurium(IV) in 0.1 M Perchloric Acid at Rotating Gold Disk Electrodes. *Anal. Chim. Acta* **119**, 55 (1980).
373. H. Reller, E. Kirowa-Eisner, and E. Gileadi, Ensembles of Microelectrodes. A Digital Simulation. *J. Electroanal. Chem. Interfacial Electrochem.* **138**, 65 (1982).
374. B. J. Feldman, J. D. Osterloh, B. H. Hata, and A. D'Alessandro, Determination of Lead in Blood by Square-Wave Anodic Stripping Voltammetry at Carbon Disk Ultramicroelectrode. *Anal. Chem.* **66**, 1983 (1994).
375. T. Magjer and M. Branica, A New Electrode System with Efficient Mixing of Electrolyte. *Croat. Chem. Acta* **49**, L1 (1977).
376. C. Kramer, Y. Guo-Hui, and J. C. Duinker, Optimization and Comparison of Four Mercury Working Electrodes in Speciation Studies by Differential-Pulse Anodic Stripping Voltammetry. *Anal. Chim. Acta* **164**, 163 (1984).
377. J. Wang and M. Ariel, The Rotating Disk Electrode in Flowing Systems. *Anal. Chim. Acta* **101**, 1 (1978).
378. J. Wang, H. D. Dewald, and B. Greene, Anodic Stripping Voltammetry of Heavy Metals with a Flow Injection System. *Anal. Chim. Acta* **146**, 45 (1983).
379. J. A. Wise, W. R. Heineman, and P. T. Kissinger, Flow Injection System for Stripping Voltammetry. *Anal. Chim. Acta* **172**, 1 (1985).
380. R. Neeb and D. Saur, Eine programmgesteuerte Anordnung zur inversen Voltammetrie. *Z. Anal. Chem.* **222**, 200 (1966).
381. M. D. Booth, M. J. D. Brand, and B. Fleet, A Fully Automated Apparatus for Stripping Voltammetry. *Talanta* **17**, 1059 (1970).
382. R. W. Andrews and D. C. Johnson, Determination of Se(IV) by Anodic Stripping Voltammetry in Flow Systems with Ion Exchange Separation. *Anal. Chem.* **48**, 1056 (1976).

383. W. R. Seitz, R. Jones, L. N. Klatt, and W. D. Mason, Anodic Stripping Analysis at a Tubular Mercury-Covered Graphite Electrode. *Anal. Chem.* **45**, 841 (1973).
384. J. Wang and B. Greene, Characteristics of a Flow Cell for the Determination of Arsenic(III) by Stripping Voltammetry. *J. Electroanal. Chem. Interfacial Electrochem.* **154**, 261 (1983).
385. E. B. Buchanan, Jr. and D. D. Soleta, Automated Square-Wave Anodic Stripping Voltammetry with a Flow-Through Cell and Medium Exchange. *Talanta* **29**, 207 (1982).
386. J. Wang, R. Setiadji, L. Chen, J. Lu, and S. G. Morton, Automated System for On-Line Adsorptive Stripping Voltammetric Monitoring of Trace Levels of Uranium. *Electroanalysis* **4**, 161 (1992).
387. E. Beinrohr, M. Nemeth, P. Tschopel, and G. Tolg, Anodic Stripping Coulometry with Collection for Absolute Trace Analysis of Lead, Using Flow-Through Electrochemical Electrodes. *Fresenius J. Anal. Chem.* **344**, 93 (1992).
388. A. Zirino, S. H. Lieberman, and C. Clavell, Measurement of Cu and Zn in San Diego Bay by Automated Stripping Voltammetry. *Environ. Sci. Technol.* **12**, 73 (1978).
389. J. Wang, On-Line Sensors for Trace Metals Based on Stripping Analysis. *Am. Lab.* p. 14, July (1983).
390. H. W. Nürnberg, The Voltammetric Approach in Trace Metal Chemistry of Natural Waters and Atmospheric Precipitation. *Anal. Chim. Acta* **164**, 1 (1984).
391. T. M. Florence, Trace Metal Species in Fresh Waters. *Water Res.* **11**, 681 (1977).
392. T. M. Florence, The Speciation of Trace Elements in Waters. *Talanta* **29**, 345 (1982).
393. T. M. Florence, Development of Physico-Chemical Speciation Procedures to Investigate the Toxicity of Copper, Lead, Cadmium and Zinc Towards Aquatic Biota. *Anal. Chim. Acta* **141**, 73 (1982).
394. D. R. Turner and M. Whitfield, The Reversible Electrodeposition of Trace Metal Ions from Multi-Ligand Systems. Part 1. Theory. *J. Electroanal. Chem. Interfacial Electrochem.* **103**, 43 (1979).
395. D. R. Turner and M. Whitfield, The Reversible Electrodeposition of Trace Metal Ions from Multi-Ligand Systems. Part II. Calculations on the Electrochemical Availability of Lead at Trace Levels in Seawater. *J. Electroanal. Chem. Interfacial Electrochem.* **103**, 61 (1979).
396. H. P. Van Leeuwen, Kinetic Classification of Metal Complexes in Electroanalytical Speciation. *J. Electroanal. Chem. Interfacial Electrochem.* **99**, 93 (1979).

397. R. K. Skogerboe, S. A. Wilson, and J. G. Osteryoung, Exchange of Comments on Scheme for Classification of Heavy Metal Species in Natural Waters. *Anal. Chem.* **52**, 1960 (1980); T. M. Florence and G. E. Batley, *ibid.*, p. 1962.
398. S. P. Brown and B. R. Kowalski, Pseudopolarographic Determination of Metal Complex Stability Constants in Dilute Solutions by Rapid Scan Anodic Stripping Voltammetry. *Anal. Chem.* **51**, 2133 (1979).
399. P. Valenta, *in* "Trace Element Speciation in Surface Waters" (G. G. Leppard, ed.), p. 49. Plenum New York, 1983.
400. H. W. Nürnberg, P. Valenta, L. Mart, B. Raspor, and L. Sipos, Applications of Polarography and Voltammetry to Marine and Aquatic Chemistry. II. The Polarographic Approach to the Determination and Speciation of Toxic Trace Metals in the Marine Environment. *Z. Anal. Chem.* **282**, 357 (1976).
401. M. Branica, D. M. Novak, and S. Budic, Application of Anodic Stripping Voltammetry to Determination of the State of Complexation of Traces of Metal Ions at Low Concentration Levels. *Croat. Chem. Acta* **49**, 56 (1977).
402. A. Zirino and S. P. Kounaves, Anodic Stripping Peak Currents: Electrolysis Potential Relationships for Reversible Systems. *Anal. Chem.* **49**, 56 (1977).
403. M. S. Shaman and J. L. Cromer, Pseudopolarograms: Applied Potential-Anodic Stripping Peak Current Relationships. *Anal. Chem.* **51**, 1546 (1979).
404. C. M. G. van den Berg, Determination of the Complexing Capacity and Conditional Stability Constants of Complexes of Copper(II) with Natural Organic Ligands in Seawater by Cathodic Stripping Voltammetry of Copper-Catechol Complex Ions. *Mar. Chem.* **15**, 1 (1984).
405. G. A. Bhat, R. A. Saar, R. B. Smart, and J. H. Weber, Titration of Soil-Derived Fulvic Acid by Copper(II) and Measurement of Free Copper(II) by Anodic Stripping Voltammetry and Copper(II) Selective Electrode. *Anal. Chem.* **53**, 2275 (1981).
406. J. Buffle, Calculation of the Surface Concentration of the Oxidized Metal During the Stripping Step in the Anodic Stripping Techniques and its Influence on Speciation Measurements in Natural Waters. *J. Electroanal. Chem. Interfacial Electrochem.* **125**, 273 (1981).
407. S. L. Phillips and I. Shain, Application of Stripping Analysis to the Trace Determination of Tin. *Anal. Chem.* **34**, 262 (1962).
408. M. Ariel, U. Eisner, and S. Gottesfeld, Trace Analysis by Anodic Stripping Voltammetry. II. The Method of Medium Exchange. *J. Electroanal. Chem. Interfacial Electrochem.* **7**, 307 (1964).
409. T. M. Florence and K. J. Mann, Anodic Stripping Voltammetry with Medium Exchange in Trace Element Speciation. *Anal. Chim. Acta* **200**, 305 (1987).

410. J. Wang and B. Greene, Stripping Analysis of Zinc in Natural Waters Utilizing the Medium Exchange Method. *Water Res.* **17**, 1635 (1983).
411. C. M. G. van den Berg and S. H. Khan, Determination of Selenium in Sea Water by Adsorptive Cathodic Stripping Voltammetry. *Anal. Chim. Acta* **231**, 221 (1990).
412. L. E. Leon and D. T. Sawyer, Simultaneous Determination of Iron(II) and Iron(III) at Micromolar Concentrations by Differential Pulse Polarography. *Anal. Chem.* **53**, 706 (1981).
413. G. E. Batley and J. P. Matousek, Determination of Chromium Speciation in Natural Waters by Electrodeposition on Graphite Tubes for Electrothermal Atomization. *Anal. Chem.* **52**, 1570 (1980).
414. G. E. Batley and T. M. Florence, An Evaluation and Comparison of Some Techniques of Anodic Stripping Voltammetry. *J. Electroanal. Chem. Interfacial Electrochem.* **55**, 23 (1974).
415. G. E. Batley and T. M. Florence, Determination of Thallium in Natural Waters by Anodic Stripping Voltammetry. *J. Electroanal. Chem. Interfacial Electrochem.* **61**, 205 (1975).
416. T. W. Hamilton, J. Ellis, and T. M. Florence, Determination of Selenium and Tellurium in Electrolytic Copper by Anodic Stripping Voltammetry at a Gold Film Electrode. *Anal. Chim. Acta* **110**, 87 (1979).
417. F. T. Henry, T. O. Kirch, and T. M. Thorpe, Determination of Trace Level Arsenic(III), Arsenic(V) and Total Inorganic Arsenic by Differential Pulse Polarography. *Anal. Chem.* **51**, 215 (1979).
418. M. Boussemart and C. M. G. van den Berg, Preconcentraton of Chromium(III) from Sea-Water by Adsorption on Silica and Voltammetric Determination. *Analyst (London)* **119**, 1349 (1994).
419. M. Boussemart, C. M. G. van den Berg, and M. Ghaddaf, The Determinaton of the Chromium Speciation in Sea Water Using Catalytic Cathodic Stripping Voltammetry. *Anal. Chim. Acta* **262**, 103 (1992).
420. J. P. Arnold and R. M. Johnson, Polarography of Arsenic. *Talanta* **16**, 1191 (1969).
421. D. J. Ayers and J. Osteryoung, Determination of Arsenic(III) at the Parts per Billion Level by Differential Pulse Polarography. *Anal. Chem.* **45**, 267 (1973).
422. G. Forsberg, J. W. O'Laughlin, R. G. Megargle, and S. R. Koirtyohann, Determination of Arsenic by Anodic Stripping Voltammetry and Differential Pulse Anodic Stripping Voltammetry. *Anal. Chem.* **47**, 1586 (1975).
423. P. H. Davis, G. R. Dulude, R. M. Griffin, W. R. Matson, and E. W. Zink, Determination of Total Arsenic at the Nanogram Level by High-Speed Anodic Stripping Voltammetry. *Anal. Chem.* **50**, 137 (1978).
424. W. Holak, Determination of Arsenic by Cathodic Stripping Voltammetry with a Hanging Mercury Drop Electrode. *Anal. Chem.* **52**, 2189 (1980).

425. R. S. Sadana, Determination of Arsenic in the Presence of Copper by Differential Pulse Cathodic Stripping Voltammetry at a Hanging Mercury Drop Electrode. *Anal. Chem.* **55**, 304 (1983).
426. R. D. Guy and A. R. Kean, Algae as a Chemical Speciation Monitor - I. A Comparison of Algal Growth and Computer Calculated Speciation. *Water Res.* **141**, 891 (1980).
427. T. M. Florence, B. G. Lumsden, and J. J. Fardy, Evaluation of Some Physico-Chemical Techniques for the Fraction of Dissolved Copper Toxic to the Marine Diatom *Nitzschia Closterium*. *Anal. Chim. Acta* **151**, 281 (1983).
428. P. C. Leung, K. S. Subramanian, and J. C. Meranger, Determination of Arsenic in Polluted Waters by Differential Pulse Anodic Stripping Voltammetry. *Talanta* **29**, 515 (1982).
429. F. G. Bodewig, P. Valenta, and H. W. Nürnberg, Trace Determination of As(III) and As(V) in Natural Waters by Differential Pulse Anodic Stripping Voltammetry. *Z. Anal. Chem.* **311**, 187 (1982).
430. W. Lund and M. Salberg, Anodic Stripping Voltammetry with the Florence Mercury Film Electrode. Determination of Copper, Lead and Cadmium in Sea Water. *Anal. Chim. Acta* **76**, 131 (1975).
431. J. E. Poldshi and G. E. Glass, Anodic Stripping Voltammetry at a Mercury Film Electrode: Baseline Concentrations of Cadmium, Lead, and Copper in Selected Natural Waters. *Anal. Chim. Acta* **101**, 79 (1978).
432. C. Labar and L. Lamberts, pH Buffer Solutions in Potentiometric Stripping Analysis of Lead, Cadmium and Copper. *Electrochim. Acta* **33**, 1405 (1988).
433. C. Labar, Potentiometric Stripping Analysis and the Speciation of Heavy Metals in Environmental Studies. *Electrochim. Acta* **38**, 807 (1993).
434. R. Ernst, H. E. Allen, and K. H. Mancy, Characterization of Trace Metal Species and Measurement of Trace Metal Stability Constants by Electrochemical Techniques. *Water Res.* **9**, 969 (1975).
435. W. Martinetti, G. Querirazza, F. Realine, and G. Ciceri, In-Flow Speciation of Copper, Zinc, Lead and Cadmium in Fresh Waters by Differential Pulse Anodic Stripping Voltammetry. *Anal. Chim. Acta* **261**, 323 (1992).
436. S. R. Plotrowicz, M. Springer-Young, J. A. Puig, and M. J. Spencer, Anodic Stripping Voltammetry for Evaluation of Organic-Metal Interactions in Seawater. *Anal. Chem.* **54**, 1367 (1982).
437. C. M. G. van den Berg, Monitoring of Labile Copper and Zinc in Estuarine Waters Using Cathodic Stripping Chronopotentiometry. *Mar. Chem.* **34**, 211 (1991).
438. C. M. G. van den Berg, Determination of Copper, Cadmium and Lead in Seawater by Cathodic Stripping Voltammetry of Complexes with 8-Hydroxyquinoline. *J. Electroanal. Chem. Interfacial Electrochem.* **215**, 111 (1986).

439. Y. K. Chau and K. L. Shue-Chan, Determination of Labile and Strongly Bound Metals in Lake Water. *Water Res.* **8**, 383 (1974).
440. R. Fukai and L. Huynh-Ngoc, Direct Determination of Mercury in Sea-Water by Anodic Stripping Voltammetry with a Graphite Electrode. *Anal. Chim. Acta* **83**, 375 (1976).
441. K. Kristsotakis, N. Laskowski, and H. J. Tobschall, *Int. J. Environ. Anal. Chem.* **6**, 203 (1979).
442. J. E. Bonneli, H. E. Taylor, and R. K. Skogerboe, A Direct Differential Pulse Anodic Stripping Voltammetric Method for the Determination of Thallium in Natural Waters. *Anal. Chim. Acta* **118**, 243 (1980).
443. A-C. LeGall and C. M. G. van den Berg, Cathodic Stripping Voltammetry of Glutathione in Natural Waters. *Analyst (London)* **118**, 1411 (1993).
444. B. C. Househam, C. M. G. van den Berg and J. P. Riley, The Determination of Purines in Fresh and Seawater by Cathodic Stripping Voltammetry After Complexation with Copper(I). *Anal. Chim. Acta* **200**, 291 (1987).
445. C. M. G. van den Berg, B. C. Househam, and J. P. Riley, Determination of Cystine and Cysteine in Seawater Using Cathodic Stripping Voltammetry in the Presence of Cu(II). *J. Electroanal. Chem. Interfacial Electrochem.* **239**, 137 (1988).
446. J. Wang, Anodic Stripping Voltammetry as an Analytical Tool. *Environ. Sci. Technol.* **16**, 104A (1982).
447. G. Gillian, Studies of Pretreatments in the Determination of Zn, Cd, Pb, Cu, Sb and Si in Suspended Particulate Matter and Plankton by Differential-Pulse Anodic Stripping Voltammetry with a Hanging Mercury Drop Electrode. *Talanta* **29**, 651 (1982).
448. V. D. Nguyen, P. Valenta, and H. W. Nürnberg, Voltammetry in the Analysis of Atmospheric Pollutants. The Determination of Toxic Trace Metals in Rain Water and Snow by Differential Pulse Stripping Voltammetry. *Sci. Total Environ.* **12**, 151 (1979).
449. M. P. Landy, An Evaluation of Differential Pulse Anodic Stripping Voltammetry at a Rotating Glassy Carbon Electrode for the Determination of Cadmium, Copper, Lead and Zinc in Antarctic Snow Samples. *Anal. Chim. Acta* **121**, 39 (1980).
450. U. Eisner and H. B. Mark, Jr., The Anodic Stripping Voltammetry of Trace Silver Solutions Employing Graphite Electrodes. Application to Silver Analysis of Rain and Snow Samples from Silver Iodide Seeded Clouds. *J. Electroanal. Chem. Interfacial Electrochem.* **24**, 345 (1970).
451. R. G. Dhaneshwar and L. R. Zarapkar, Simultaneous Determination of Thallium and Lead at Trace Levels by Anodic Stripping Voltammetry. *Analyst (London)* **105**, 386 (1980).

452. G. Colvos, G. S. Wilson, and J. Moyers, The Determination of Trace Amounts of Zinc, Cadmium, Lead and Copper in Airborne Particulate Matter by Anodic Stripping Voltammetry. *Anal. Chim. Acta* **64**, 457 (1973).
453. B. L. Dennis, G. S. Wilson, and J. Moyers, The Determination of Bromide, Chloride and Lead in Airborne Particulate Matter by Stripping Voltammetry. *Anal. Chim. Acta* **86**, 27 (1976).
454. R. N. Khandekar, R. G. Dhaneshwar, M. M. Palrecha, and L. R. Zarapkar, Simultaneous Determination of Lead, Cadmium and Zinc in Aerosols by Anodic Stripping Voltammetry. *Z. Anal. Chem.* **307**, 365 (1981).
455. K. E. MacLeod and R. E. Lee, Selected Trace Metal Determination of Spot Tape Samples by Anodic Stripping Voltammetry. *Anal. Chem.* **45**, 2380 (1973).
456. R. P. Harrison and J. W. Winchester, Area-Wide Distribution of Lead, Copper and Cadmium in Air Particulates from Chicago and Northwest Indiana. *Atmos. Environ.* **5**, 863 (1971).
457. M. Taddia, Anodic Stripping Determination of Mercury in Air with a Glassy Carbon Electrode. *Microchem. J.* **23**, 64 (1978).
458. L. K. Hoeflich, R. J. Gale, and M. L. Good, Differential Pulse Polarography and Differential Pulse Anodic Stripping Voltammetry for Determination of Trace Levels of Thallium. *Anal. Chem.* **54**, 1591 (1983).
459. E. Casassas, A. M. Perez-Vendrell, and L. Puignon, An Improved Voltammetric Procedure for the Determination of Zn, Pb, Cd and Cu in Atmospheric Aerosols. *Int. J. Environ. Anal. Chem.* **45**, 55 (1991).
460. M. Oehme and W. Lund, Comparison of Digestion Procedures for the Determination of Heavy Metals (Cd, Cu, Pb) in Blood by Anodic Stripping Voltammetry. *Z. Anal. Chem.* **298**, 260 (1979).
461. W. Lund and R. Eriksen, The Determination of Cadmium, Lead and Copper in Urine by Differential Pulse Anodic Stripping Voltammetry. *Anal. Chim. Acta* **107**, 37 (1979).
462. G. Chittiborough and B. J. Steel, The Determination of Zinc, Cadmium, Lead and Copper in Human Hair by Differential Pulse Anodic Stripping Voltammetry at a Hanging Mercury Drop Electrode After Nitrate Fusion. *Anal. Chim. Acta* **119**, 235 (1980).
463. M. Oehme, W. Lund, and J. J. Jonsen, The Determination of Copper, Lead, Cadmium and Zinc in Human Teeth by Anodic Stripping Voltammetry. *Anal. Chim. Acta* **100**, 389 (1978).
464. I. M. Shapiro, H. L. Needleman, and O. C. Tuncay, The Lead Content of Human Deciduous and Permanent Teeth. *Environ. Res.* **5**, 467 (1972).
465. T. R. Williams, O. P. Foy, and C. Benson, The Determination of Zinc in Human Eye Tissues by Anodic Stripping Voltammetry. *Anal. Chim. Acta* **75**, 250 (1975).

466. J. T. Kinard, Diagnosis of Metal Poisoning and Evaluation of Chelation Therapy by Differential Pulse Anodic Stripping Voltammetry Coupled to a Novel Digestion Procedure. *Anal. Lett.* **10**, 1147 (1977).
467. S. B. Adeloju, A. M. Bond, and H. C. Hughes, Determination of Selenium, Copper, Lead and Cadmium in Biological Materials by Differential Pulse Stripping Voltammetry. *Anal. Chim. Acta* **148**, 59 (1983).
468. T. E. Edwards, P. Guogong and T. S. West, The Differential Pulse Anodic Stripping Voltammetry of Copper and Lead and Their Determination in EDTA Extracts of Soils with the Mercury Film Glassy Carbon Electrode. *Anal. Chim. Acta* **120**, 41 (1980).
469. S. Forbes, G. P. Bounds, and T. S. West, Determination of Selenium in Soils and Plants by Differential-Pulse Cathodic Stripping Voltammetry. *Talanta* **26**, 473 (1979).
470. G. Calderoni and T. Ferri, Determination of Thallium at Subtrace Level in Rocks and Minerals by Coupling Differential Pulse Anodic Stripping Voltammetry with Suitable Enrichment Methods. *Talanta* **29**, 371 (1982).
471. I. Liem, G. Kaiser, M. Sager, and G. Tölg, The Determination of Thallium in Rocks and Biological Materials at ng g^{-1} Levels by Differential Pulse Anodic Stripping Voltammetry and Electrothermal Atomic Absorption Spectrometry. *Anal. Chim. Acta* **158**, 179 (1984).
472. A. M. Bond, T. A. O'Donnell, A. B. Waugh, and R. J. W. McLaughlin, Use of Polarographic Methods for the Determination of Tin on Geological Samples. *Anal. Chem.* **42**, 1168 (1970).
473. K. Camman, A Critical Comparison Between Flameless Atomic Absorption Spectroscopy and an Improved Electrochemical Anodic Stripping Technique in the Case of a Rapid Trace Determination of Lead in Geological Samples. *Z. Anal. Chem.* **293**, 97 (1978).
474. K. A. Khasgiwale, R. Parthasarathy, and M. Sankar Das, Determination of Lead in Geological Samples by Anodic Stripping Analysis. *Anal. Chim. Acta* **59**, 485 (1972).
475. E. W. Zink, W. R. Matson, S. L. Pfeiffer, and A. F. Pietrzyk, Direct Analysis of Lead in Fruit Juice by Anodic Stripping Voltammetry with No Prior Sample Preparation. *J. Assoc. Off. Anal. Chem.* **61**, 653 (1978).
476. P. Debacker, J. L. Vandenbalck, G. J. Partriarche, and G. D. Christian, Simultaneous Determination of Tin and Lead by Anodic Stripping Voltammetry in Aqueous-Alcoholic Medium. Application to the Direct Determination of These Elements in Canned Foods. *Microchem. J.* **26**, 192 (1981).
477. A. Andruzzi, A. Trazza, and G. Marrosu, The Use of an Automated Device in the Direct Simultaneous Determination of Heavy Metals in Commercial Coca Cola by DPASV. *Anal. Lett.* **16**, 853 (1983).

478. S. Mannino, Simultaneous Determination of Lead and Tin in Fruit Juices and Soft Drinks by Potentiometric Stripping Analysis. *Analyst (London)* **108**, 1257 (1983).
479. S. Mannino, Potentiometric Stripping Analysis of Lead and Tin with a Continuous Flow System. *Analyst (London)* **109**, 905 (1984).
480. D. G. Cornell and M. J. Pallansch, Cadmium Analysis of Dried Milk by Pulse Polarographic Techniques. *J. Dairy Sci.* **56**, 1479 (1973).
481. A. M. Sulek, E. R. Elkins, and E. W. Zink, Lead in Evaporated Milk by Anodic Stripping Voltammetry and Atomic Absorption Spectrophotometry: Cooperative Interlaboratory Study. *J. Assoc. Off. Anal. Chem.* **61**, 931 (1978).
482. L. Anderson, D. Jagner, and M. Josefson, Potentiometric Stripping Analysis in Flow Cells. *Anal. Chem.* **54**, 1371 (1983).
483. W. Holak, Determination of Heavy Metals in Foods by Anodic Stripping Voltammetry After Sample Decomposition with Sodium and Potassium Nitrate Fusion. *J. Assoc. Off. Anal. Chem.* **58**, 777 (1975).
484. W. Holak, Determination of Arsenic and Selenium in Foods by Electroanalytical Techniques. *J. Assoc. Off. Anal. Chem.* **59**, 650 (1976).
485. F. Szydlowski, K. L. Monti, S. P. Michalak, and D. L. Dunmire, A Practical Voltammetric Method for the Analysis of Lead in Cereal and Feed Products Using a Tubular Mercury Thin Film Electrode. *Anal. Lett.* **13**, 529 (1980).
486. M. D. Booth and B. Fleet, Electrochemical Behavior of Triphenyltin Compounds and Their Determination at Submicrogram Levels by Anodic Stripping Voltammetry. *Anal. Chem.* **42**, 825 (1970).
487. T. P. DeAngelis, R. E. Bond, E. E. Brooks and W. R. Heineman, Thin Layer Differential Pulse Voltammetry. *Anal. Chem.* **49**, 1792 (1977).
488. T. M. Florence, Cathodic Stripping Voltammetry. Part I. Determination of Organic Sulfur Compounds, Flavins and Porphyrins at the Sub-Micromolar Level. *J. Electroanal. Chem. Interfacial Electrochem.* **97**, 219 (1979).
489. J. Wang, B. K. Deshmukh, and M. Bonakdar, Solvent Extraction Studies with Carbon Paste Electrodes. *J. Electroanal. Chem. Interfacial Electrochem.* **194**, 339 (1985).
490. P. Hernandez, E. Lorenzo, and L. Hernandez, Determination of Azinophos-Methyl by Adsorptive Stripping Voltammetry. *Anal. Chim. Acta* **238**, 383 (1990).
491. H. G. Prabu and P. Manisankar, Determination of Endosulfan by Stripping Voltammetry. *Analyst (London)* **119**, 1867 (1994).
492. H. Siegerman and G. O'Dom, Differential Pulse Anodic Stripping of Trace Metals. *Am. Lab.* p. 59, June (1972).
493. W. M. Peterson and R. V. Wong, Fundamentals of Stripping Voltammetry. *Am. Lab.*, November (1981).

494. A. M. Bond, Developments in Polarographic (Voltammetric) Analysis in the 1980's. *J. Electroanal. Chem. Interfacial Electrochem.* **118**, 381 (1981).
495. J. Osteryoung, Voltammetry for the Future. *Acc. Chem. Res.* **26**, 77 (1993).
496. C. V. Bowen and F. I. Edwards, Jr., Polarographic Determination of O,O-Diethyl O-*p*-Nitrophenyl Triphosphate (Parathion). *Anal. Chem.* **22**, 706 (1950).
497. W. H. Jura, Polarographic Determination of S-(1,2-Dicarbethoxyethyl)-O,O-Dimethyl Dithiophosphate (Malathion). *Anal. Chem.* **27**, 525 (1955).
498. W. H. Chan, A. W. M. Lee, and P. X. Cai, Differential Pulse Polarographic Microdetermination of Reactive Organohalides is a In Situ Generation of S-Alkylisothiouranium Salts. *Analyst (London)* **117**, 185 (1992).
499. T. B. Jarbawi and W. R. Heineman, Pre-concentration of Chlorpromazine at a Wax-Impregnated Graphite Electrode. *Anal. Chim. Acta* **135**, 359 (1982).
500. R. Kalvoda, Adsorptive Accumulation in Stripping Voltammetry. *Anal. Chim. Acta* **138**, 11 (1982).
501. J. F. Price and R. P. Baldwin, Preconcentration and Determination of Ferrocenecarboxaldehyde at a Chemically Modified Platinum Electrode. *Anal. Chem.* **52**, 1940 (1980).
502. E. N. Chaney and R. P. Baldwin, Electrochemical Determination of Adriamycin Compounds in Urine by Preconcentration at Carbon Paste Electrodes. *Anal. Chem.* **54**, 2556 (1982).
503. J. Wang and B. A. Freiha, Subtractive Differential Pulse Voltammetry Following Adsorptive Accumulation of Organic Compounds. *Talanta* **30**, 837 (1983).
504. J. Wang and B. A. Freiha, Selective Voltammetric Detection Based on Adsorptive Preconcentration for Flow Injection Analysis. *Anal. Chem.* **55**, 1285 (1983).
505. J. Wang and R. Li, Highly Stable Voltammetric Measurements of Phenolic Compounds at Poly(3-methylthiophene)-Coated Glassy Carbon Electrodes. *Anal. Chem.* **61**, 2809 (1989).
506. J. Lu, W. Jin, S. Wang and T. Sun, Determination of Trace Chromium(III) by Linear Sweep Adsorption Voltammetry. *J. Electroanal. Chem. Interfacial Electrochem.* **291**, 49 (1990).
507. J. Wang, C. Sun, and W. Jin, Adsorption Voltammetry of Selenium in the Presence of Phenylenediamine (*o*-PDA). *J. Electroanal. Chem. Interfacial Electrochem.* **291**, 59 (1990).
508. C. Elleouet, F. Quentel, and C. Madec, Determination of Trace Amounts of Chromium(VI) in Water by Electrochemical Methods. *Anal. Chim. Acta* **257**, 301 (1992).
509. T. Matsunaga and Y. Namba, Detection of Microbial Cells by Cyclic Voltammetry. *Anal. Chem.* **56**, 798 (1984).

510. K. Josstad, B. Salba, and A. C. Pappas, Multielement Analysis of Human Blood Serum by Neutron Activation and Controlled Potential Electrolysis. *Anal. Chem.* **53**, 1398 (1981).
511. J. Muhlbaier, C. Stevens, D. Graczyk, and T. Tisue, Determination of Cadmium in Lake Michigan by Mass Spectromeric Isotopic Dilution Analysis or Atomic Absorption Spectrometry Following Electrodeposition. *Anal. Chem.* **54**, 496 (1982).
512. G. E. Batley and J. P. Matsousek, Determination of Heavy Metals in Seawater by Atomic Absorption Spectrometry after Electrodeposition on Pyrolytic Graphite-Coated Tubes. *Anal. Chem.* **49**, 2031 (1977).
513. G. E. Batley and J. P. Matsousek, Determination of Chromium Speciation in Natural Waters by Electrodeposition on Graphite Tubes for Electrothermal Atomization. *Anal. Chem.* **52**, 1570 (1980).
514. G. Volland, P. Tschöpel, and G. Tölg, Elektrolytische Abscheidung Ion Hydrodynamischen System Von ng-Mengen Eisen, Kobalt, Zink und Wismut in Graphitrohr. *Anal. Chim. Acta* **90**, 15 (1977).
515. Y. Thomassen, B. V. Larsen, F. J. Langougher, and W. Lund, The Application of Electrodeposition Techniques to Flameless Atomic Absorption Spectrometry. Part IV. Separation and Preconcentration on Graphite. *Anal. Chim. Acta* **83**, 103 (1976).
516. G. E. Batley, In Situ Electrodeposition for the Determination of Lead and Cadmium in Sea Water. *Anal. Chim. Acta* **124**, 121 (1981).
517. E. J. Czobik and J. P. Matsousek, The Application of Electrodeposition on a Tungsten Wire to Furnace Atomic Absorption Spectrometry. *Spectrochim. Acta Part B.* **35B**, 741 (1980).
518. Y. Hoshino, T. Utsunomiya and K. Fukui, Graphite Furnace Atomic Absorption Spectrometry Utilizing Selective Adsorption of Metal Ions Onto Tungsten Wire in Aqueous Solutions. *Chem. Lett.*, p.947 (1976).
519. C. Fairless and A. J. Bard, Electrodeposition Techniques for Carbon Rod Flameless Atomic Absorption Analysis. *Anal. Lett.* **5**, 433 (1972).
520. J. Shirwatana and J. P. Matsousek, Electrodeposition on Pyrolytic Graphite Platforms for Electrothermal Atomic Absorption Spectroscopic Determination of Labile Lead in Saline Water. *Talanta* **38**, 375 (1991).
521. J. P. Matsousek and H. K. J. Powell, Analyte Preconcentration and Separation for Small Volumes by Electrodeposition for Electrothermal Atomic Absorption Spectroscopy. *Talanta* **40**, 1829 (1993).
522. J. R. Pretty, E. H. Evans, E. A. Blubaugh, W.-L. Chen, J. A. Caruso, and T. M. Davidson, Minimization of Sample Matrix Effects and Signal Enhancement for Trace Analytes Using Anodic Stripping Voltammetry with Detection by Inductively Coupled Plasma Atomic Emission Spectrometry and Inductively Coupled Plasma Mass Spectrometry. *J. Anal. At. Spectrom.* **5**, 437 (1990).

523. J. R. Pretty, E. A. Blubaugh, E. H. Evans, J. A. Caruso, and T. M. Davidson, Determination of Copper and Cadmium Using an On-Line Anodic Stripping Voltammetry Flow Cell with Detection by Inductively Coupled Plasma Mass Spectrometry. *J. Anal. At. Spectrom.* **7**, 1131 (1992).
524. J. R. Pretty, E. A. Blubaugh, and J. A. Caruso, Determination of Arsenic(III) and Selenium(IV) Using an On-Line Anodic Stripping Voltammetry Flow Cell with Detection by Inductively Coupled Plasma Atomic Emission Spectrometry and Inductively Coupled Plasma Mass Spectrometry. *Anal. Chem.* **65**, 3396 (1993).
526. J. R. Pretty and J. A. Caruso, Signal Enhancement of Lead and Thallium in Inductively Coupled Plasma Atomic Emission Spectrometry Using On-Line Atomic Stripping Voltammetry. *J. Anal. At. Spectrom.* **8**, 545 (1993).
526. J. R. Pretty, E. A. Blubaugh, J. A. Caruso, and T. M. Davidson, Determination of Chromium(VI) and Vanadium(V) Using an On-Line Anodic Stripping Voltammetry Flow Cell with Detection by Inductively Coupled Plasma Mass Spectrometry. *Anal. Chem.* **66**, 1540 (1994).
527. J. L. Anderson and R. E. Sioda, Electrodeposition as a Preconcentration Step in Analysis of Multicomponent Solutions of Metallic Ions. *Talanta* **30**, 627 (1983).
528. Y. Lin, X. Wang, D. Yuan, P. Yang, B. Huang, and Z. Zhuang, Flow-Injection - Electrochemical Hydride Generation Technique for Atomic Absorption Spectrometry. *J. Anal. At. Spectrom.* **7**, 287 (1992).
529. D. L. Strong, P. K. Dasgupta, K. Friedman, and J. R. Stillian, Electrodialytic Eluent Production and Gradient Generation in Ion Chromatography. *Anal. Chem.* **63**, 480 (1991).
530. D. L. Strong and P. K. Dasgupta, Electrodialytic Production of Gas-Free Sodium Hydroxide Based on Donnan Breakdown. *J. Membr. Sci.* **57**, 321 (1991).
531. D. L. Strong, C. U. Joyng, and P. K. Dasgupta, Electrodialytic Eluent Generation and Suppression. Ultralow Background Conductance Suppressed Anion Chromatography. *J. Chromatogr.* **546**, 159 (1991).
532. J. Janata, M. Josowicz, and D. M. DeVaney, Chemical Sensors. *Anal. Chem.* **66**, 207R (1994).
533. A. E. Varfolomeev, A. I. Volkov, A. V. Eryshlin, V. V. Malyshev, A. S. Rasumov, and S. S. Yakimov, Detection of Phosphine and Arsine in Air by Sensors Based on SnO_2 and ZnO. *Sens. Actuators* **B7**, 727 (1992).
534. T. Maekawa, J. Tamaki, N. Miura, and N. Yamazoe, Gold-Loaded Tungsten Oxide Sensor for Detection of Ammonia in Air. *Chem. Lett.* p.639 (1992).
535. P. Talik, M. Zabkowska-Waclawek, and W. Waclawek, Sensing Properties of the CB-PVC Composites for Chlorinated Hydrocarbon Vapours. *J. Mater. Sci.* **27**, 6807 (1992).

536. W. H. Christensen, D. N. Sinha, and S. F. Agnew, Conductivity of Polystyrene Film Upon Exposure to Nitrogen Dioxide: A Novel NO_2 Sensor. *Sens. Actuators* **B10**, 149 (1993).
537. A. Wilson, M. Tamizi, and J. P. Wright, Detection of Nitro Compounds by Organic Semiconductor Sensors. *Sens. Actuators* **B18-B19**, 511 (1994).
538. J. J. Miasik, A. Hooper, and B. C. Tofield, Conducting Polymer Gas Sensors. *J. Chem. Soc., Faraday Trans. 1* **82**, 1117 (1986).
539. J. G. Rabe, G. Bischoff, and W. F. Schmidt, Electrical Conductivity by Polypyrrole Films as Affected by Adsorption of Vapors. *Jpn. J. Appl. Phys.* **28**, 518 (1989).
540. I. K. Kovacs and G. Horvai, Possibilities of Chemical Sensing at the Semiconductor/Electrolyte Interface. *Sens. Actuators* **B18-B19**, 315 (1994).
541. J. M. Keller, Bipolar-Pulse Conductivity Detector for Ion Chromatography. *Anal. Chem.* **53**, 344 (1981).
542. K. J. Casuta, F. J. Holler, S. R. Cronch, and C. G. Enke, Computer-Controlled Bipolar Pulse Conductivity System for Applications in Chemical Rate Determinations. *Anal. Chem.* **50**, 1534 (1978).
543. C. R. Powley, R. F. Geiger, Jr., and T. A. Neiman, Bipolar Pulse Conductance Measurement with a Calcium Ion-Selective Electrode. *Anal. Chem.* **52**, 705 (1980).
544. V. Svoboda and J. Marsal, A Conductimetric Detector with a Wide Dynamic Range for Liquid Chromatography. *J. Chromatogr.* **148**, 111 (1978).
545. D. Qi, T. Okada and P. K. Dasgupta, Direct Current Conductivity Detection in Ion Chromatography. *Anal. Chem.* **61**, 1383 (1989).
546. For example, W. J. Albery, Development of Photogalvanic Cells for Solar Energy Conversion. *Acc. Chem. Res.* **15**, 142 (1982).
547. K. Rajeshwar, R. O. Lezna, and N. R. de Tacconi, Light in an Electrochemical Tunnel? Solving Analytical Problems in Electrochemistry via Spectroscopy. *Anal. Chem.* **64**, 429A (1992).
548. B. Bittins-Cattaneo, E. Cattaneo, P. Königshoven, and W. Vielstich, New Developments in Electrochemical Mass Spectroscopy. *Electroanal. Chem.*, **17**, 181 (1991).
549. M. A. Fox and T. Tien, A Photoelectrochemical Detector for High-Pressure Liquid Chromatography. *Anal. Chem.* **60**, 2278 (1988).
550. G. N. Brown, J. W. Birks, and C. A. Koval, Development and Characterization of a Titanium Dioxide-Based Semiconductor Photoelectrochemical Detector. *Anal. Chem.* **64**, 427 (1992).
551. D. G. Gelderloos, K. L. Rowlen, J. W. Birks, J. P. Avery, and C. G. Enke, Whole Column Detection Chromatography. Computer Simulations. *Anal. Chem.* **58**, 900 (1986).
552. E. J. Guthrie and J. Jorgenson, On-Column Fluorescence Detector for Open-Tubular Capillary Liquid Chromatography. *Anal. Chem.* **56**, 483 (1984).

553. W. R. LaCourse and I. S. Krull, Photoelectrochemical Detection in Analytical Chemistry. *Trends Anal. Chem.* **4**, 118 (1985).
554. W. R. LaCourse, I. S. Krull, and K. Bratin, Photoelectrochemical Detector for High-Performance Liquid Chromatography and Flow Injection Analysis. *Anal. Chem.* **57**, 1810 (1985).
555. W. R. LaCourse and I. S. Krull, Photoelectrochemical Detection of Benzaldehyde in Foodstuffs. *Anal. Chem.* **59**, 50 (1987).
556. J. M. Elbicki, D. M. Morgan, and S. G. Weber, Photoelectroanalytical Chemistry: Electrochemical Detection of a Photochemically Active Species, Tris(2,2'-bipyridine)ruthenium(II). *Anal. Chem.* **57**, 1746 (1985).
557. T. J. Meyer, Chemical Approaches to Artificial Photosynthesis. *Acc. Chem. Res.* **22**, 163 (1989).
558. E. Krausz and J. Ferguson, The Spectroscopy of the [Ru(bpy)$_3$]$^{2+}$ System. *Prog. Inorg. Chem.* **37**, 293 (1989).
559. T. Nagaoka, D. Griller, and D. D. M. Wagner, Digital Simulation of Photomodulation Voltammograms: Reactivity of the Diphenylmethyl Carbanion and Carbocation in Acetonitrile. *J. Phys. Chem.* **95**, 6264 (1991).
560. R. G. Compton, R. A. W. Dryfe, and A. C. Fisher, Photoelectrochemical Dehalogenation of *p*-halo-nitrobenzenes. *J. Electroanal. Chem. Interfacial Electrochem.* **361**, 275 (1993).
561. A. M. Waller and R. G. Compton, The Photoelectrochemical Reduction of Pyrene in Acetonitrile Solution. *Electrochim. Acta* **35**, 1837 (1990).
562. R. G. Compton, A. C. Fisher, R. G. Wellington, and J. Winkler, Spectrofluorimetric Hydrodynamic Voltammetry: Theory and Practice. *J. Phys. Chem.* **96**, 8153 (1992).
563. T. Kuwana, Photoelectrochemistry and Electroluminescence. *Electroanal. Chem.* **1**, 197 (1966).
564. D. M. Hercules, Chemiluminescence from Electron-Transfer Reactions. *Acc. Chem. Res.* **2**, 301 (1969).
565. K. E. Haapakka and J. J. Kankare, The Mechanism of the Electrogenerated Chemiluminescence of Luminol in Aqueous Alkaline Solution. *Anal. Chim. Acta* **138**, 263 (1982).
566. K. E. Haapakka, Application of Electrogenerated Chemiluminescence of Luminol to Determination of Traces of Cobalt(II) In Aqueous Alkaline Solution. *Anal. Chim. Acta* **139**, 229 (1982).
569. K. E. Haapakka and J. J. Kankare, Application of the Electrochemiluminescence of Luminol to the Determination of Copper. *Anal. Chim. Acta* **118**, 333 (1980).
568. G. M. Greenway, Analytical Applications of Electrogenerated Chemiluminescence. *Trends Anal. Chem.* **9**, 200 (1990).

569. C. Blatchford and D. J. Malcome-Lawes, Electrochemiluminescence as a Detection Technique for Reversed Phase High-Performance Liquid Chromatography. I. Preliminary Experiments. *J. Chromatogr.* **321**, 227 (1985).
570. C. Blatchford, E. Humphreys, and D. J. Malcome-Lawes, Electrochemiluminescence as a Detection Technique for Reversed-Phase High-Performance Liquid Chromatography. II. Low-Frequency A.C. Electrochemiluminescence. *J. Chromatogr.* **329**, 281 (1985).
571. E. Hill, E. Humphreys, and D. J. Malcome-Lawes, Electrochemiluminescence as a Detection Technique for Reversed-Phase High-Performance Liquid Chromatography. IV. Detection of Fluorescent Derivatives. *J. Chromatogr.* **441**, 394 (1988).
572. H. D. Dewald and J. Wang, Spectroelectrochemical Detector for Flow-Injection System and Liquid Chromatography. *Anal. Chim. Acta* **166**, 163 (1984).
573. M. Fleischmann, P. J. Hendra, and A. J. McQuillan, Raman Spectra of Pyridine Adsorbed at a Silver Electrode. *Chem. Phys. Lett.* **26**, 163 (1974).
574. R. L. Garrell, Surface-Enhanced Raman Spectroscopy. *Anal. Chem.* **61**, 401A (1989).
575. J. M. E. Storey, T. E. Barber, R. D. Shelton, E. A. Wachter, K. T. Carron, and Y. Jiang, Applications of Surface-Enhanced Raman Scattering (SERS) to Chemical Detection. *Spectroscopy* **10**, 20 (1995).

SUPPLEMENTARY READING

Potentiometry and Ion-Selective Electrodes

1. H. Freiser, ed., "Ion-Selective Electrodes in Analytical Chemistry." Vols. 1 and 2. Plenum, New York, 1978 and 1980.
2. A. K. Covington, "Ion-Selective Electrode Methodology." CRC Press, Boca Raton, FL, 1979.
3. W. E. Morf, "The Principles of Ion-Selective Electrodes and of Membrane Transport." Elsevier, Amsterdam, 1981.
4. J. Koryta, "Ions, Electrodes and Membranes." Wiley, New York, 1982.
5. N. Lakshminarayanaiah, "Membrane Electrodes." Academic Press, New York, 1976.

Stripping Analysis

6. J. Wang, "Stripping Analysis." VCH Publishers, Deerfield Beach, FL, 1985.
7. F. Vydra, K. Stulík, and E. Julakwa, "Electrochemical Stripping Analysis." Ellis Horwood, Chichester, 1976.

8. Kh. Brainina and E. Neyman, "Electrochemical Stripping Methods." Wiley, New York, 1994.

Flow Detectors

9. K. Stulík and V. Pacáková, "Electroanalytical Measurements in Flowing Liquids." Ellis Horwood, Chichester, 1987.
10. J. Ruzicka and E. H. Hansen, "Flow Injection Analysis." Wiley, New York, 1981.
11. W. B. Furman, "Continuous Flow Analysis: Theory and Practice." Dekker, New York, 1976.

Voltammetry/Electroanalysis

12. R. N. Adams, "Electrochemistry at Solid Electrodes." Dekker, New York, 1969.
13. Z. Galus, "Fundamentals of Electrochemical Analysis." Ellis Horwood, Chichester, 1976.
14. A. M. Bond, "Modern Polarographic Methods in Analytical Chemistry." Dekker, New York and Basel, 1980.
15. J. Wang, "Analytical Electrochemistry." VCH, New York, 1994.

Miscellaneous

Periodic reviews of various aspects of electroanalysis appear in the series of volumes titled "Electroanalytical Chemistry," edited by A. J. Bard and published by Dekker, New York.

CHAPTER FIVE

5.1. INTRODUCTION

Many electrochemical processes for environmental pollution control involve the direct reaction of species at electrode surfaces, while others involve production of active species at the electrode and further reaction with the targeted pollutants. The term *direct electrolysis* shall be used here in connection with those processes in which an electron transfer reaction to or from the undesired pollutant occurs at the surface of an electrode. Likewise, the term *indirect electrolysis* refers here to processes in which a dissolved redox reagent either exists in or is generated from the electrolyte or from the electrode phase in order to participate in a targeted reaction. The two types of approaches are compared and contrasted in Figure 5.1.

The direct and indirect electrolytic processes discussed in this chapter include the oxidation and reduction of organic and inorganic pollutants. Indirect processes also include processes of electrocoagulation, electroflotation, and electroflocculation. Separate sections in this chapter are devoted to the electrochemical treatment of gases, soils, and membrane–ion exchange based approaches. Due to its special importance and scope, electrochemical disinfection of water will be discussed in a separate chapter (Chapter 7). Likewise, commercial prospects for these electrochemical technologies will be presented in the last chapter (Chapter 8) of this book.

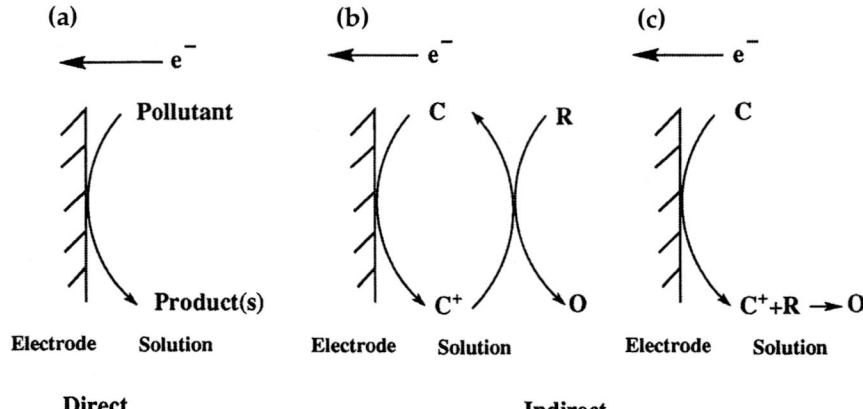

FIGURE 5.1. Schemes for direct (a) and indirect electrolytic treatment of pollutants. The latter can be done both with reversibly (b) and irreversibly (c) electrogenerated reagents. In the last two examples, R is a pollutant and C is a reagent. An oxidative scheme is assumed.

5.2. POSITIVE FEATURES OF ELECTROCHEMICAL APPROACHES TO POLLUTION CONTROL

Electrochemical techniques offer many distinctive advantages relative to the other technologies discussed in Chapter 1:
- Environmental compatibility: The main reagent used is the electron, which is a clean reagent, and usually there is no need for adding extra chemicals.
- Versatility: Electrochemical processes involving direct or indirect oxidation and reduction can generate neutral, positively, or negatively charged inorganic, organic, or biochemical species. They can also deal with solid, liquid, or gaseous pollutants and can induce the production of precipitates, gaseous species, pH changes, or charge neutralization. The products from the electrolysis of pollutants often are even useful. In addition, a plethora of reactor and electrode materials, shapes and configurations can be utilized. Frequently, the same reactor can be used for different electrochemical reactions with only minor changes. In addition, point-of-use production of chemicals is facilitated by electrochemical techniques (e.g., for water disinfection). Last, volumes of fluid from microliters to millions of liters can be treated.
- Energy efficiency: Electrochemical processes often have lower temperature and pressure requirements than those of equivalent nonelectrochemical counterparts (e.g., incineration, supercritical oxidation). The applied potentials can be controlled and electrodes and cells can be designed to minimize power losses due to poor current distribution and voltage drops.

- Safety: Electrochemical processes are safe because of the mild conditions usually employed, and the small quantity and innocuous nature of the added chemicals.
- Selectivity: The applied potential in many cases can be controlled to selectively attack specific bonds and thus avoid production of by-products.
- Amenability to automation: The electrical variables used in electrochemical processes (I, E) are particularly suited for facilitating data acquisition, process automation, and control.
- Cost effectiveness: The required equipment and operations are normally simple and, if properly designed, can also be made relatively inexpensive.

5.3. DIRECT ELECTROLYSIS OF POLLUTANTS

Pollutants capable of undergoing direct electrochemical oxidation or reduction at an electrode can in principle be removed from water streams or reservoirs by the application of appropriate potentials in electrochemical reactors. Here, oxidation or reduction processes occur directly on inert electrodes without the involvement of other substances (e.g., electron mediators, biocidal species). Unfortunately, rather than the removal of unwanted material being the dominant electrode process, side reactions (particularly solvent breakdown) almost always occur; for example,

$$2\,H_2O \rightarrow O_2 + 4H^+ + 4\,e^- \tag{5.1}$$

$$2\,H_2O + 2\,e^- \rightarrow H_2 + 2\,OH^- \tag{5.2}$$

While these reactions do have a deleterious effect on the overall process efficiency, clever schemes can be devised to exploit the production of gases as well as the pH changes associated with Reactions (5.1) and (5.2) to bring about useful reaction chemistry. For example, the production of $H_2(g)$ has been used to facilitate the flotation and removal of unwanted substances in the electrocoagulation–electroflotation process (see later). Likewise, the generation of hydroxyl ions at the cathode (Reaction (5.2)) serves to immobilize the electrolytically generated Cr(III) as the insoluble hydroxide, in the electrochemical reduction of the (very toxic) Cr(IV) to Cr(III).[1,2] Finally, the production of H_3O^+ has been used in the field of electrokinetic or electro-osmotic remediation of soil to increase acidity in soils, solubilize pollutant species, and so forth.

5.3.1. Anodic Oxidation of Sample Organic Pollutants

<u>Figures of Merit</u>. The potentials required for the oxidation of organic compounds are usually high and thus the production of O_2 from the electrochemi-

cal oxidation of H_2O is normally the main parasitic reaction. By measuring the dioxygen flow rate during electrolysis in the presence ($\dot{V}_{O_2,org}$) and in the absence ($\dot{V}_{O_2}$) of an organic species in aqueous media, an instantaneous current efficiency (ICE) at time t for its oxidation can be defined[3-6]:

$$\text{ICE} = \frac{\dot{V}_{O_2} - \dot{V}_{O_2,org}}{\dot{V}_{O_2}} \tag{5.3}$$

In other words, if all the current during electrolysis were used for the oxidation of the organic, then $\dot{V}_{O_2,org} = 0$ and ICE = 1. When the electrolysis products are soluble in the electrolyte, the ICE can also be calculated from the (instantaneous) difference between two values of the *chemical oxygen demand* (COD) of the solution, given in g_{O_2} liter^{-1}, as follows[4]:

$$\text{ICE} = \frac{\left[(\text{COD})_t - (\text{COD})_{t+\Delta t}\right]}{8 I \Delta t} \bullet F \bullet V_s \tag{5.4}$$

where COD_t and $\text{COD}_{t+\Delta t}$ are the values of the chemical oxygen demand at times t and Δt, respectively; V_s is the volume of the electrolyte in liters, I is the current (A), and F is the Faraday constant. If the parameter ICE is plotted as a function of time, the area under the curve integrated, and the result divided by the total time (τ) elapsed up to the point when ICE = 0, an "average current efficiency" is obtained that is called the *electrochemical oxidability index* (EOI):

$$\text{EOI} = \frac{\int_0^\tau \text{ICE} \, dt}{\tau} \tag{5.5}$$

An example of the ICE vs. t curve obtained during the electrochemical oxidation of p-aminotoluenesulfonate on a Pt anode, is shown in Figure 5.2.[3]

The parameter, EOI gives a quantitative estimate of the ease of electrochemical oxidation of organic species; that is, the larger this value, the more easily the species can be oxidized. Sample EOI values for selected aromatic compounds are shown in Table 5.1.[3] It can be seen that electron-withdrawing groups (e.g., -SO$_3$H, -COOH) produce low EOI values indicative of a low electron density available for oxidation, whereas electron-donating groups (e.g., -NH$_2$) produce high EOI values due to the increased electron density available. When both types of groups are present, as in p-aminotoluenesulfonic acid, the inductive effect of the electron-donating group dominates. In fact, an excellent correlation has been found between EOI and the σ (Hammett) constant of a

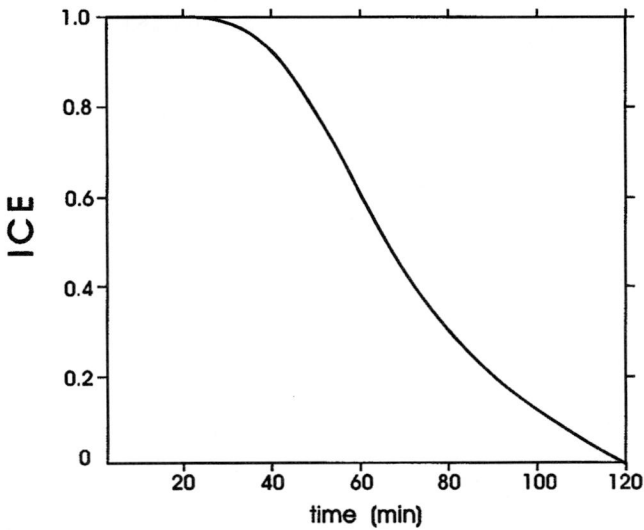

FIGURE 5.2. Instantaneous current efficiency (ICE) during the oxidation of *p*-aminotoluene sulfonic acid on Pt. (Reproduced with permission from Comninellis and Plattner.[3])

substituent (which depends on the nature and location of the substituent) and an electrodic parameter ρ as follows[7]:

$$\log \text{EOI} = \rho\sigma + c \qquad (5.6)$$

where c is a dimensionless constant. For monosubstituted benzenes, this equation becomes

$$\log \text{EOI} = -2\sigma - 1.3 \qquad (5.7)$$

Since $\sigma_{NH2} << \sigma_{NO2}$, the fact that the slope is negative suggests that the limiting step in these anodic oxidations is of an electrophilic nature.

By calculating the fraction of current that oxidizes the organic species (as opposed to the fraction that oxidizes water) and converting it to its equivalent in grams of O_2 per gram of the organic species, the following rationale has been used to define the *electrochemical oxygen demand* (EOD).[3] If all the current in an anodic oxidation were used for O_2 production,

$$2\,H_2O \rightarrow O_2 + 4\,H^+ + 4\,e^- \qquad (5.1)$$

Table 5.1. The Electrochemical Oxidazability Index of Representative Aromatic Substrates

Substrate	EOI
Aniline (NH$_2$–C$_6$H$_5$)	0.56
Benzenesulfonic acid (SO$_3$H–C$_6$H$_5$)	<0.05
4-amino-2-methylbenzenesulfonic acid (NH$_2$, CH$_3$, SO$_3$H substituted benzene)	0.58
4-nitro-2-methylbenzenesulfonic acid (NO$_2$, CH$_3$, SO$_3$H substituted benzene)	0.1
Benzoic acid (COOH–C$_6$H$_5$)	<0.05
Anthranilic acid (COOH, NH$_2$ substituted benzene)	0.55
Phenol (OH–C$_6$H$_5$)	0.2

Source: From Comninellis et al. [3–7]

5.3. Direct Electrolysis of Pollutants

then one would theoretically obtain the following quantity of O_2 after time t:

$$\text{mol } O_{2(theor)} = \left(\frac{1 \text{ mol} O_2}{4 \text{ mol e}^-}\right)\left(\frac{1 \text{ mol e}^-}{F}\right) It = \frac{It}{4F} \tag{5.8}$$

In the presence of a species (Org) contained in the solution, a fraction of the total current will be used to oxidize Org:

$$\text{Org} \xrightarrow{-e^-} \text{Oxidation products} \tag{5.9}$$

The equivalent of this fraction of the current in moles of O_2 is then

$$\text{mol } O_2 = \text{mol } O_{2(theor)} \bullet \text{EOI} = \frac{It}{4F} \bullet \text{EOI} \tag{5.10}$$

If these moles of O_2 were used to chemically oxidize g grams of the species Org, one would obtain the following ratio:

$$\frac{\text{mol } O_2}{g_{org}} = \frac{\left(\frac{It}{4F}\right)}{g_{org}} \bullet \text{EOI} \tag{5.11}$$

This can be expressed as a weight ratio by converting the moles of O_2 to grams of O_2 (g_{O_2}):

$$\frac{g_{O_2}}{g_{org}} = \frac{\left(\frac{It}{4F} \bullet \text{EOI}\right) \text{mol } O_2}{g_{org}} \bullet \left(\frac{32 g_{O_2}}{1 \text{ mol} O_2}\right) \tag{5.12}$$

Since τ is the time elapsed until electrolysis is essentially complete (i.e., ICE = 0), one can replace τ for t; hence, the electrochemical oxygen demand is defined as

$$\text{EOD} = \frac{g_{O_2}}{g_{org}} = 8 \bullet \left(\frac{I\tau}{F \bullet g_{org}}\right) \bullet \text{EOI} \tag{5.13}$$

The degree of oxidation (x) of the organic pollutant (Org) can be defined as the ratio of the (equivalent of the) amount of dioxygen required for the oxidation of one gram of Org to its oxidation products (this amount is the EOD) to the amount of dioxygen initially required for complete oxidation to CO_2. Here,

$$x = \frac{EOD}{COD^*} \quad (5.14)$$

and

$$COD° \cdot \left[\left(\frac{g_{org}}{L}\right)^0\right]^{-1} = COD^* \quad (5.15)$$

where COD^* is expressed in g_{O2}/g_{org}, whereas $COD°$ and $[g_{org}/L]°$ are the initial COD and the initial concentration of the pollutant, respectively. It can be seen that as x approaches unity, the anodic oxidation to CO_2 is more complete. However, it must be borne in mind that the oxidation need not go all the way to CO_2 for the process to render the pollutant unharmful. Intermediates such as oxalic and maleic acids[8] are biodegradable and as such are acceptable as end oxidation products. This translates into high electricity savings, since fewer electrons are anodically removed from the pollutant. A biological or a chemical oxidation post-treatment can then be coupled to the electrochemical treatment, as schematized in Figure 5.3.[9]

The criteria for definitions may vary and be subject to interpretation; however, a degradation of 20% of the initial COD is called *primary degradation* (since it involves a modification of the initial substance), whereas a degradation of 70% is called *final degradation* or *complete mineralization* and the substance is referred to as *intrinsically biodegradable*.[9] Note that the kinetics of these processes are not taken into account for such definitions.

If the organic substance is refractory (i.e., nonbiodegradable) or toxic, then the initial treatment (chemical or electrochemical) has to be designed as to modify it and make it amenable for biological degradation, which is generally

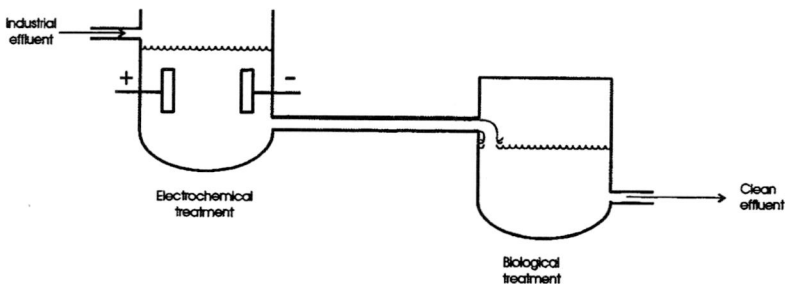

FIGURE 5.3. Coupled biological–electrochemical system for water treatment. (Adapted from Seignez et al.[9])

5.3. Direct Electrolysis of Pollutants

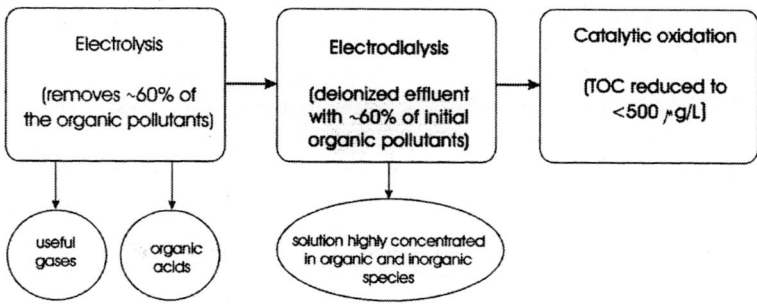

FIGURE 5.4. A combined water treatment system for a space-habitat application.

the process of choice due primarily to its low cost. These hybrid processes may play a key role in waste management for industrial effluents with the characteristics just described. For example, for a space-habitat environment, a system has been designed to produce potable water from a waste stream of laundry, shower, handwash, and urine waste water, as shown in Figure 5.4.[10] This combined process is reported to reduce the total organic carbon of the waste stream by >99.9%. Further, it does not require components that must be replenished and uses only electricity. Most important, the system produces nontoxic gases such as N_2, H_2, and CO_2.

Electrode Materials. The selection of electrodes for the direct anodic oxidation of organic and inorganic pollutants has to take into account the composition and nature of the solution to be treated as well as the stability of the electrode material, its cost, selectivity, and environmental compatibility. Since many organic and inorganic substances require rather high potentials for their oxidation (often higher than that for the oxidation of water), the nature of the electrode must be such that it will not corrode under the application of such potentials. This can be predicted to a certain degree in aqueous solutions with the aid of the corresponding Pourbaix diagram (see Chapter 2). Such predictions, however, may be complicated by the presence of reactive components in the solution other than the pollutant (e.g., complexing agents, dissolved gases, salts). Generally, oxidized noble metal surfaces (e.g., Pt, Ir, Ru) will be suitable for the oxidation of organic substances, although their cost poses a major restriction for their widespread use. On the other hand, cheaper substitutes such as oxidized nickel and lead can be used in aqueous media. Three-dimensional electrodes are known to offer a high surface area per unit volume. In addition, the passage of the solution through these electrodes produces local turbulence; this salutary effect facilitates mass-transfer processes. Commercially available high surface area anodes include graphite, reticulated vitreous carbon, titanium, stainless steel, nickel, and Ebonex (a Ti-based ceramic, see Section 5.9). Care has to be taken in proper selection, since most electrodes (e.g., nickel and stainless steel) are useful in only a limited range of potential and pH.

Another consideration in choosing the electrode materials is the influence of the cathodic reaction on the overall efficiency. For example, during the oxidation of phenol, benzoquinone is produced as an intermediate that can be reduced at the cathode, producing the corresponding hydroquinone that can, in turn, be reoxidized at the anode, thereby decreasing the efficiency of the process of interest and requiring a separator between anode and cathode. On the other hand, the high overpotential used at the SnO_2 anode yields essentially irreversible oxidations, where the products cannot be normally reduced at the cathode; therefore, higher yields are to be expected.

Carbonaceous materials (carbon felt, reticulated vitreous carbon, glassy carbon) are known to have surface oxygenated functional groups that facilitate electron exchange with organic substances and are safe from an environmental point of view. However, noble metals and oxide-covered Pb and Ti substrates are more commonly used for these applications, due in part to their high resistance to severe conditions and the high O_2 evolution overpotentials of the oxide-covered materials. An interesting approach involves the modification of metal electrodes with organic ligands to facilitate the effect just discussed.

The potential at the auxiliary electrode is also a key factor in achieving high current efficiencies. Unfortunately, the reactions normally occur in a rather unselective manner and most of the energy from the power source is spent on side reactions and heat generation. An approach that diminishes this problem consists of the use of electrocatalytic materials on the auxiliary electrode surface. This approach has been reported to afford high selectivity.[11] Furthermore, application of an alternating current facilitates electrochemical reactions at the average electrochemical potentials set by the equilibrium between surface groups and the electrolytic solution. In addition, a self-cleaning effect occurs due to the periodic reversal of current, which changes the nature of the substances produced at each electrode, thus preventing deposits and other undesired cumulative effects.[11]

We shall now proceed to discuss individual classes of organic pollutants.

<u>Phenols</u>. These are aromatic compounds containing one or more hydroxyl groups attached to the aromatic ring. Phenols are produced as wastes in a variety of industries, including dyes, plastics, pharmaceutical, oil refineries, and coke plants. The toxicity and bad smell of phenols warrant further treatment before discharge. Treatment technologies include activated carbon adsorption, solvent extraction, and biological, chemical, or electrochemical oxidation. Normally, biodegradation is used; however, this is not a viable option when the concentrations are rather high or when the composition of the efflu-

5.3. Direct Electrolysis of Pollutants

ent is variable. In these scenarios, chemical oxidation processes are preferred. These include the use of hydrogen peroxide in the presence of Fe^{2+} (Fenton's reagent), ozone, and chlorine (see Section 5.4). Nonetheless, the high cost of chemical oxidizers has led to renewed interest in the study and development of direct electrochemical alternatives.

The electrochemical route involves oxidation at anodes that show high overpotentials for the production of dioxygen, such as PbO_2, graphite, and Ti covered with Ti or Sn oxides. Platinum electrodes have also been used. Depending on the nature of the anode and pH conditions used, different pathways, intermediates, and side products have been identified in the oxidation of phenol. On a graphite fluidized bed, at least 48 compounds were shown to be produced during phenol oxidation.[12] This may be due to the presence of surface oxygenated functional groups at the carbon electrode as discussed previously. Graphite or activated carbon particles can also be mixed with oxidation catalysts (metal oxides from the groups IVA, VA, VIB, and VIIB of the periodic table, particularly MnO_2, Cr_2O_3, Bi_2O_3, and PbO_2) for the treatment of phenolic compounds.

Phenol oxidation on lead dioxide packed bed anodes is initially kinetically controlled until the concentration decreases and the process becomes mass-transfer controlled.[8] Oxidation is faster under acidic conditions, whereas the removal of oxidation products (as measured by the total organic carbon content, TOC) is higher in basic media. The major intermediates in a cell with a Nafion cationic separator are benzoquinone and maleic acid, and the main final product is CO_2.[13] The oxidation can be made essentially complete. It is important to note that at higher phenol concentrations, the production of CO_2 is somewhat inhibited. This can be attributed to the formation of a polymeric film at the anode (also noted during the oxidation of phenol on Pt, see later). When the CO_2 produced is used as a measure of phenol oxidation, the current efficiencies are on the order of 20% (at 1 A), and 11-16% (2 and 3 A). These can be improved by running the reaction potentiostatically instead of galvanostatically. Greater conversion to CO_2 has been achieved by using higher acid concentrations (which promote breakdown of the benzoquinone ring), higher temperatures, and dissolved dioxygen (which increases the rate of oxidation of benzoquinone).[13] Creation of defect sites by incorporation of Bi(V) into PbO_2 films produces an electrocatalytic effect for the oxidation of phenol with only a slight increase in background current due to O_2 evolution.[14] Such efficient oxidation has the additional benefit of preventing the formation of polymeric films that hinder the electron transfer process at the interface. Interestingly enough, these films are deliberately formed for metal coating applications due to their strong affinity for substrate adsorption and pinhole-free nature.[15]

Anodic oxidation of aromatic hydroxy compounds generally leads to the formation of a polymeric film that slows down electron transfer and complicates mechanistic studies.[16] In fact, during the anodic oxidation of phenol a yellow-brown, electrically conducting polymeric film forms at the surface of a Pt cylinder. Its formation is favored by high pH, high temperature, high phenol concentration, and low current density. These findings have led to the proposal of a reaction mechanism based on the phenolate anion ($C_6H_5O^-$) and the formation of a polyoxyphenylene film.[4] The EOI is independent of the current, and so the process is not mass-transfer limited. For this reason, it has been proposed that the reaction occurs by an electrophilic attack of the ·OH radical on the aromatic nucleus, which also explains the observed increase of EOI with pH, since the phenolate ion is more reactive than phenol itself to electrophilic attack. This proposed mechanism is supported by the fact that the intermediates (hydroquinone, catechol, and benzoquinone) are initially formed in large amounts and then further oxidized into aliphatic acids (maleic, fumaric, and oxalic), which are stable to further oxidation. These products are very similar to those observed during the oxidation of phenol with Fenton's reagent, which is known to generate the highly oxidizing ·OH radicals in aqueous solutions (see Chapter 6). Since this reaction mechanism involves ·OH radicals, it is more appropriately discussed within the framework of indirect electrolytic processes later in this chapter. Since the TOC attained via the chemical route is on the order of 30%, whereas that obtained electrochemically is 60%, it is believed that direct oxidation of phenol or its intermediates (but not that of the aliphatic acids, which are only very slowly oxidized) also occurs at the anode.[7]

It has been hypothesized that most of the initial current flow during phenol oxidation on Pt is due to simple, fast reactions in the outer Helmholtz layer.[7] Further oxidation requires that the compound become part of the inner Helmholtz layer, whereby the electron transfer is slowed down due to the presence of a metal oxide layer. Temperature increases or periodic reduction of the metal oxide layer result in higher pollutant oxidation rates.[15] When oxidation of bisphenol-A [2,2-bis(4'-hydroxyphenyl)propane] is performed in the presence of chloride ions, toxic chlorinated aromatic intermediates are formed. Fortunately, under controlled conditions, these intermediates are not found in the reaction mixture, and the final products are short-chain aliphatic acids.[16]

Aromatic Amines. These compounds are known to be quite poisonous due to their reaction with hemoglobin and the consequent suppression of dioxygen uptake. Effects from chronic exposure include anemia, anorexia, weight loss, and cutaneous lesions. These compounds are commonly produced as by-products or wastes in the dye, petroleum, rubber, perfume, coal, and pulp and paper industries. The parent compound of the series, aniline, can be oxidized in an one-electron oxidation process:

$$C_6H_5NH_2 \rightarrow C_6H_5NH_2^+ + e^-$$

5.3. Direct Electrolysis of Pollutants

This cation radical can polymerize in acidic media, forming a dark-green precipitate at the anode.[17] The degradation of aniline in sulfuric acid on PbO_2 electrodes in a fixed bed reactor with a Nafion cationic exchange membrane as a separator between anolyte and catholyte is believed to proceed as follows:

$$C_6H_5NH_2 + 2\,H_2O \rightarrow C_6H_4O_2 + 3\,H^+ + 4\,e^- + NH_4^+$$
$$C_6H_4O_2 + 6\,H_2O \rightarrow C_4H_4O_4 + 12\,H^+ + 12\,e^- + 2\,CO_2$$
$$C_4H_4O_4 + 4\,H_2O \rightarrow 12\,H^+ + 12\,e^- + 4\,CO_2$$

It is noteworthy that the intermediates are similar to those found during the degradation of phenolic compounds and benzoquinone, thus indicative of a similar mechanism. In fact, these intermediates or by-products, together with any unreacted aniline, account for ca. 90% of the initial aniline after 5 hr of reaction time, whereas at short reaction times (0.5 hr) they only account for ~50%. This means that some other intermediates are produced and later transformed into these main compounds. The oxidation of aniline under these conditions is rather fast, since more than 80% is oxidized within 0.5 hr. After 5 hr, 97.5% is degraded and 72.5% completely mineralized to CO_2. The current efficiency for a 5 hr experiment drops from 12% (1 A) to 9% (3 A) due to dioxygen evolution, which may also produce oxidation of the by-products. From these equations, it can be predicted that higher pH values should shift the reactions to the right, since the OH^- ions remove the H^+ present. In fact, complete oxidation is favored at high pH values. The best current efficiencies (~40%) are obtained at short times (0.5 hr) and at the highest initial pH (pH = 11).

The degradation in alkaline solutions shows first-order kinetics on an anodized Pb foil, where 99% of the pollutant molecules (aniline or 4-chloroaniline) are degraded within 2.5–3 hr. In addition, complete disappearance of all organic intermediates is observed.[17,18] The overall reactions are

$$C_6H_5NH_2 + 28\,OH^- \rightarrow 6\,CO_2 + NH_3 + 28\,e^- + 16\,H_2O$$
$$C_6H_5NHCl + 27\,OH^- \rightarrow 6\,CO_2 + NH_3 + 26\,e^- + 15\,H_2O + Cl^-$$

<u>Halogenated and Nitro Derivatives</u>. As mentioned in Chapter 1, many polyhalogenated compounds (e.g., pesticides) have been identified as toxic, and their disposal is expensive. In addition, they usually must be shipped long distances or be incinerated or stored in appropriate sites. Conventional methods used for their treatment include concentration (membrane separation, adsorption, special resins), chemical oxidation (wet air, O_3, UV light), or reduction (e.g., high temperature catalytic hydrogenolysis) and biological methods. Polyhalogenated compounds are in general not amenable to destruction by incineration due to high cost, high temperatures, incomplete combustion, and the possibility of the reaction of Cl_2 with the surroundings. Waste materi-

als containing these compounds as well as nitrated ones (to the level of $1-10^5$ ppm) can be detoxified by adding small amounts of cationic, anionic, on non–ionic micelle-forming compounds. In passing the waste material along the anode (made of Ti, PbO_2 or carbon fibers) of an electrochemical cell, the toxicity and mutagenity are substantially reduced. In addition, a significant decrease in the consumption of energy (~45%) (as compared to the non-micelle process) can be achieved.[19] Anodic dehalogenation of 1,2-dichloroethane on Pt electrodes has been demonstrated in aqueous solutions.[20] The reaction path involves stepwise oxidation to the corresponding alcohol, aldehyde, acid, and CO_2 as the final product together with Cl_2 and $HClO_4$.

<u>Waste Biomass.</u> Waste biomass can be oxidized on Pt or PbO_2 anodes in H_2SO_4 and urine electrolytes; this process has been termed *electrochemical incineration*.[21] High TOC removal is observed (up to ~95%) without coproduction of undesirable gases (CO, NO, NO_2, CH_4, NH_3). When urine is present as the supporting electrolyte, the oxidation occurs through an indirect mechanism due to the anodic production of ClO^- ions. Redox mediators (e.g., $Ce^{4+/3+}$) can also accelerate the rate of oxidation.[21]

<u>Other Organic Compounds.</u> Table 5.2 contains a representative listing of other organics that have been successfully treated using the anodic electrolysis approach.[22-25] In many of the cases shown, however, the mechanistic aspects have not been completely elucidated. As with the compounds discussed previously, the direct oxidation often competes with other electrochemical (and even chemical) routes. For example, the production of radicals from the oxidation of water and nitric acid as well as *chemical* attack on the organics by nitric acid are believed to be responsible for the high rates observed for benzene, methanol, propanol, and cyclohexane.[22]

5.3.2. Anodic Oxidation of Sample Inorganic Pollutants

<u>Cyanide.</u> Chlorine and hypochlorite have been used for oxidizing low concentrations of cyanides in waste water. However, the chemical treatment is rather complex and expensive when dealing with concentrated solutions[26]; furthermore, the possibility of forming potentially toxic halogenated organic compounds warrants the evaluation of other techniques. The electrochemical route has been under investigation, since no extra chemicals are required, no sludge is produced, and high current efficiencies (up to nearly 100%) can be obtained in concentrated alkaline solutions ($[CN^-] > 0.2\ M$) when using a $RuO_2 + TiO_2$ titanium anode covered with an electrodeposited PbO_2 film. In addition, the corrosion rate under these conditions (at 50°C) is negligible. The overall reaction proposed is

5.3. Direct Electrolysis of Pollutants

Table 5.2. Organic Substrates Successfully Treated with the Direct Oxidation Approach

Substrate(s)	Comments	Reference
Formaldehyde, trioxane, and urotropin	Biodegradability of these three compounds is low; electrochemical degradation investigated on Sb-doped SnO_2 film on Ti substrates	22
Benzene, methanol, propanol, and cyclohexane	Rates of destruction at 3 V: ~0.2–0.3 mL $(A-hr)^{-1}$	22
Thymol blue	A fluidized bed of carbon particles is used for the oxidation of this dye; low yields, below the Cl_2 evolution regime in KCl-containing electrolye; at potentials in the Cl_2 evolution regime, the degradation rate increased	22a
Meat extract	Graphite electrode used for anodic oxidation of the organics present in meat extracts	23
Tributylphosphate	Directly oxidized in aqueous HNO_3 at carbon anodes to yield CO_2, H_2O, inorganic phosphates, and a small amount of CO	24
Carboxylic acid anions	Liquid nuclear wastes contain transition metal complexing agents (e.g., EDTA, citrate, formate, acetate) that can be destroyed using Ni or Pt–Ti anodes	25

Notes: In many cases, the direct pathway may be in competition with other routes (refer to text).

$$CN^- + 2\,OH^- = CNO^- + H_2O + 2\,e^-$$

The possible mechanisms include the following.

Mechanism a: $CN^- = CN^{\cdot}_{(ad)} + e^-$
$CN^{\cdot}_{(ad)} + OH^- = HCNO + e^-$
$HCNO + OH^- = CNO^- + H_2O$

Mechanism b: $CN^- = CN^{\cdot}_{(ad)} + e^-$
$CN^{\cdot}_{(ad)} + CN^- = NCCN + e^-$
$NCCN + 2\,OH^- = CN^- + CNO^- + H_2O$

Mechanism c: $2\,CN^- = 2\,CN^{\cdot} + 2\,e^-$
$2\,CN^{\cdot} = NCCN$

Mechanism d:
$$NCCN + 2\,OH^- = CN^- + CNO^- + H_2O$$
$$CN^- = CN^\bullet + e^-$$
$$CN^\bullet + H_2O = CNO^- + 2\,H^+ + e^-$$

The cyanate ions thus produced, CNO^- are approximately 1,000 times less toxic than CN^- ions.

When NaCl is added to the electrolyte to increase its conductivity, the mechanism switches to an indirect one (see later), since Cl_2 is produced and it oxidizes cyanide. Removal rates of practically 100% have been achieved in a bipolar trickle tower reactor.[27] Energy consumption is on the order of 0.78 kW hr m^{-3} when no chemicals are added. To take advantage of the high surface area and the creation of local turbulence by porous electrodes, PbO_2 has been deposited on a reticulated vitreous carbon matrix for the oxidation of cyanide, with excellent results for pollutant destruction and electrode stability and reasonably good faradaic yield (~50%).[28]

Simultaneous oxidation of cyanide ions and reduction of copper ions to metallic Cu (at a cathode separated from the anode by a diaphragm) has been achieved with carbon fiber electrodes modified with copper oxides. The rate of this oxidation is affected by the degree of complexation of the CN^- ions. For example, ML_2 complexes are relatively easy to oxidize whereas the corresponding ML_3 complexes are the most difficult ones.[29]

<u>Thiocyanate</u>. These ions can be effectively oxidized on platinized-titanium electrodes in a batch recirculation reactor, and their concentration can be lowered practically by 100% (at 30 A m^{-2}).[30]

5.3.3. Cathodic Reduction of Pollutants: Electrode Materials

Since H_2 evolution is commonly the side reaction in aqueous solutions, the cathode material should have a large overpotential for this reaction. Carbon electrodes may undergo deterioration for the following reasons: (a) attack by radical species either from the electrolyte or from the solvent, such as peroxide, formed during the cathodic reduction of dioxygen, or (b) in the case of graphite, deterioration due to intercalation of ions or molecules in solution that migrate between its basal planes causing fractures. Three-dimensional carbonaceous cathodes made of partially graphitized (at least 5% of the carbon is graphitic) amorphous carbon and graphite felts have proven to be immune from such difficulties. The graphitization ensures sufficient reactivity enhancement, whereas the amorphous structure provides cross-linking between planes. Other cathode materials include nickel (metallic, alloyed, or in composites), platinized Pt, Pb, Hg, metal hydrides, and the like.

5.3. Direct Electrolysis of Pollutants

Reductions with low valency (e.g. +3) titanium compounds are well known in organic chemistry. However, reagents such as $TiCl_3$ are not easy to handle. Titanium or titanium dioxide cathodes have been suggested[31] as an alternative to these reagents and applied for the reduction of compounds such as nitrobenzene. The idea is to utilize the Ti(IV/III) and Ti(III/II) redox processes in *immobilized layers* for the heterogeneous redox catalysis of targeted reduction reactions. Reaction schemes such as

$$Ti(OH)_4 + e^- + H^+ \rightarrow Ti(OH)_3 + H_2O$$

$$6\,Ti(OH)_3 + C_6H_5NO_2 + 4\,H_2O \rightarrow 6\,Ti(OH)_4 + C_6H_5NH_2$$

(with the product $Ti(OH)_4$ recycled to the first reaction)

have been proposed.[31]

5.3.4. Cathodic Reduction of Sample Organics

<u>Chlorinated Organics</u>. As discussed in Chapter 1, the toxicity of these compounds is directly related to their chlorine content. Reductive dehalogenation is an especially attractive route to the treatment of these compounds. As will be seen in Chapter 6, this is often the method of choice for pollutants such as chloroalkanes (e.g., CCl_4), since the reactivity of species such as $^{\cdot}OH$ to them is rather low. The dehalogenation reaction can be generically represented as

$$R\text{-}Cl + 2\,H^+ + 2\,e^- \rightarrow R\text{-}H + HCl$$

These reactions occur at potentials in the range, -1 to -3 V vs. SCE. Advantages of this method include the following: treatment occurs at ambient temperature, no additional chemicals are required, and the removal of chlorine atoms is selective thus leaving the organic skeleton to be degraded by cheaper methods (i.e., biological).

<u>Polychlorinated Biphenyls</u>. Their detoxification has been successfully achieved by a reduction process whereby PCBs are extracted from a liquid carrier (usually electrically insulating mineral and silicone oils), and then reduced at a mercury cathode, where they lose their Cl content.[32] The proposed reaction is[12]

$$C_{12}H_{10-n}Cl_n + 2n\,e^- + n\,H^+ \rightarrow C_{12}H_{10} + n\,Cl^-$$

To avoid the loss of toxic mercury to the liquid stream, to improve current densities, and to increase surface areas for scale-up, other cathode materials have been studied for this application.[33] The most suitable include partially graphitized amorphous carbon (discussed earlier) and graphite felt; the latter provides up to 99.5% destruction in a propylene carbonate–mineral oil emulsion, with a current efficiency of 53% at 15 mA cm^{-2} in a flow-through configuration. In laboratory cells, the dechlorination of PCBs has been shown to be facilitated by ultrasonication.

<u>o-Chlorobenzoic Acid</u>. This refractory pollutant is resistant to anodic oxidation but has been shown to be cathodically converted to the corresponding alcohol and aldehyde, which are readily oxidizable to aliphatic acids.[34] Dechlorination during reduction is indicated by a positive test with silver nitrate, whereby silver chloride precipitate is formed if there are chloride ions in solution.

<u>Chlorophenols</u>. Several chlorophenols have been dechlorinated at cathodes made of carbon fiber bundles with current efficiencies ranging from 0.4% (in the case of 4-Cl-C$_6$H$_4$NO$_2$) to 75% (in the case of dichlorvos).[35] A commonly used chlorophenol, pentachlorophenol, can be dechlorinated stepwise all the way to phenol from an initial concentration of 50 ppm with a current efficiency of 2% and an energy consumption of 36 kW hr m^{-3}.

5.3.5. Cathodic Reduction of Sample Inorganics

<u>Oxynitrogen Ions</u>. High-level nuclear waste containing nitrate and nitrite can be treated electrolytically to reduce these anions by 99% by conversion to nitrogen-containing gases.[36] Alternatively, electrochemical separation can be achieved by using ion selective membranes (see later). The reduction process has been claimed to decrease the concentration of the pollutant, to recycle chemicals, to minimize the volume of wastes, and to transform corrosive nitrates to less corrosive (or even corrosion-inhibiting) substances.

In this electrolytic reduction process, nitrates are reduced to nitrites[36]:

$$NO_3^- + H_2O + 2\,e^- = NO_2^- + 2\,OH^-$$

Likewise, nitrites are reduced to yield N$_2$O, N$_2$, or NH$_3$. These three substances are generically called A in the following equation:

$$a\,NO_2^- + b\,H_2O + c\,e^- = A + d\,OH^-$$

where a, b, c, and d are the stoichiometric coefficients necessary for the production of 1 mole of A. The hydroxide solution thus produced can be recovered by evaporation and recycled back into fuel processing. Nickel and lead are used as cathode materials and nickel and platinum as anode materials. As an added benefit, pertechnate and ionic ruthenium, Cr(VI), and Hg(II) species are deposited or precipitated on the cathode thus removing undesired non-radioactive species as well as Tc-99 and Ru-106 radioisotopes.[25]

Many industrial chemical operations yield waste solutions containing alkali nitrates and alkali hydroxides.[37] For example, these wastes are generated in the production of nickel–cadmium batteries, in the precipitation of metal hydroxide catalysts, and in the regeneration of anion-exchange resins. It is possible to recover the hydroxide by an electrochemical process that involves first the reduction of nitrate to nitrite and then the reduction of nitrite to nitrogen gas[37]:

$$2\,NO_3^- + 4\,e^- + 2\,H_2O \rightarrow 2\,NO_2^- + 4\,OH^-$$
$$2\,NO_2^- + 6\,e^- + 4\,H_2O \rightarrow N_2 + 8\,OH^-$$

The corresponding anodic reaction is the oxidation of OH^- ions to O_2. Since the intermediate nitrites (and some ammonia produced during their reduction) can be reoxidized at the anode, a diaphragm is used to prevent the back reaction and thus to improve the efficiency to high values.

As discussed before, nitrates and nitrites are also pollutants in drinking water. Stringent legislation has been set due in part to their implication in methemoglobinemia in infants (Chapter 1). Reduction of these oxynitrogen species has been successfully carried out in commercial flow cells on a copper felt cathode, with pollutant reduction of up to ~94% in a single pass.[38] The main final product is claimed to be $N_2(g)$, and neither NH_3 nor NO_x is produced in the process.

<u>Oxychloride Species.</u> Oxyhalide water disinfection by-products (DBPs) such as chlorite (ClO_2^-) and chlorate (ClO_3^-) are either present or formed during the use of ClO_2, Cl_2, or NaOCl (Chapter 7). Concentration ranges from ppm to several weight percent can be reduced up to 100% in a flow cell with a stainless steel fiber felt cathode; the final product is predominantly the environmentally safe Cl^- ion.[38]

The electrochemical treatment of metal ions warrants a separate section in itself because of the huge knowledge bank that exists on it. We shall discuss this technology in the next section.

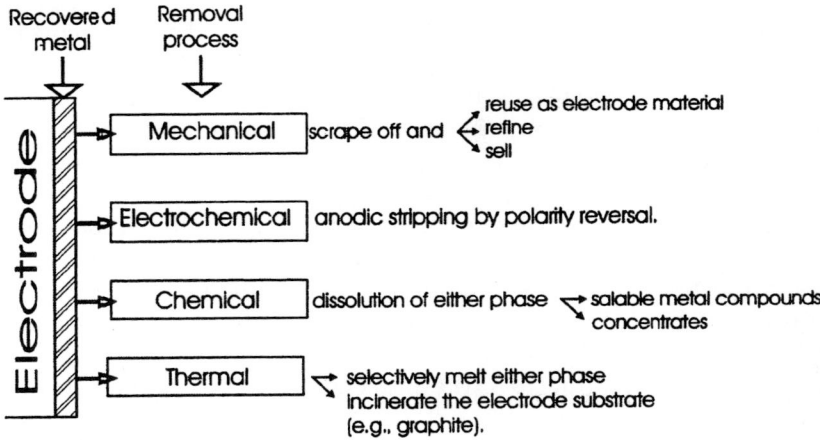

FIGURE 5.5. Processes for the removal of electrodeposited metals.

5.3.6. Cathodic Reduction of Metal Ions

The most commonly used treatment methods for metal ion removal (other than landfill) include
- Precipitation as hydroxides, sulfides, or oxalates (with lime, magnesium oxide, sulfides, oxalate)
- Filtration (mixed media, activated carbon, sand, ultrafiltration)
- Chemical or electrochemical ion exchange (zeolites, chelating ion exchange resins)
- Reverse osmosis
- Chemical or physical adsorption (e.g., on china clay, activated carbon, vermiculite, alumina-silica membranes, biopolymers, chemically or electrolytically produced iron or aluminum hydroxides; see Section 5.4)
- Stabilization or solidification
- Chemical reduction (including cementation)
- Biochemical remediation
- Electrochemical deposition

The last technique is the topic of this section. The electrochemical treatment of metal ion pollutants has several advantages, including the following:
(1) The metal is normally produced in its most valuable form, that is the metallic form, and then reused or recycled (see Figure 5.5).
(2) No extra reagents are added, and so the treated water or solution can often be recycled.

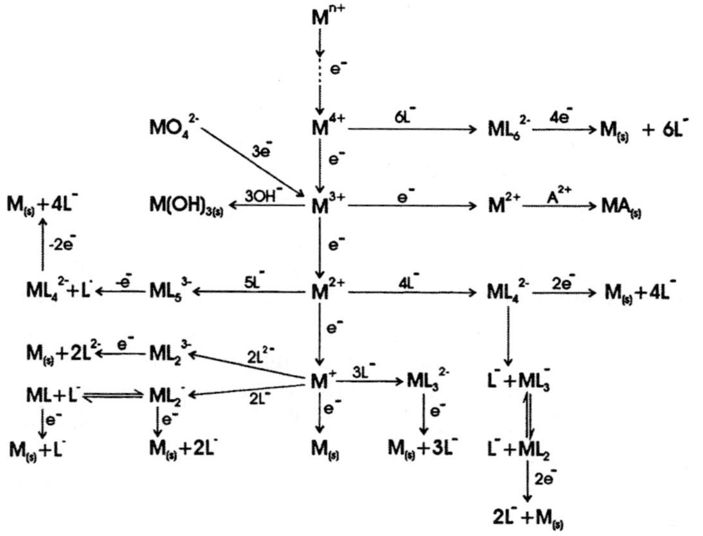

5.3. Direct Electrolysis of Pollutants

FIGURE 5.6. Pathways for the electrochemical removal of metal ions and metal ion complexes from aqueous media.

$M^{m+,n+}$ = metal ion (e.g., Ag^+, Pd^{2+}, Eu^{3+}, Pt^{4+}, Cr^{6+})
L = ligand (e.g., CN^-, $S_2O_3^{2-}$). They can in principle be recycled to form more complex species with the metal ions after metal deposition
A = precipitating anion (e.g., SO_4^{2-})

(3) Control of stream pH can be electrochemically achieved; H^+ ions are produced during water oxidation and OH^- ions during water reduction. Such control can avoid some side reactions or promote the production of desired products (for example, the production of $Cr(OH)_3$ during Cr(VI) reduction).
(4) Sludge production is minimized.
(5) Selective deposition of one metal in mixtures of metal ions may be achieved in some cases (see Figure 5.6) by careful control of the deposition conditions or by the use of a series of electrochemical reactors, each fitted to remove a particular metal ion.
(6) Alloy deposition can be achieved in some cases.
(7) Operating costs are low.
(8) The corresponding anodic reaction can be advantageously used; for example, undesirable complexing agents such as cyanide and various chelating agents and other organics may be destroyed at the anode of the same cell. High-purity dioxygen gas may be produced in a single electrochemical reactor.

(9) Use is simple and compact, as exemplified by the use of electrolytic metal removal-recovery units in many small individual jewelry and watch plating shops in Europe.

On the other hand, the electrochemical approach faces several technological challenges to be overcome, including the following:
(1) Concentrations can be rather low and get even lower during removal. The process then becomes mass-transport limited. For this reason, mass transport enhancement has to be achieved to make the process efficient (e.g., with three-dimensional electrodes, turbulence promoters, flow directional baffles, particle bed design, gas sparging, moving electrodes), and a great deal of effort has focused on the study of such alternatives.
(2) The metal ion depletion with time induces a lowering of the current efficiency. This effect can be minimized by the use of a clever scheme based on the programmed exponential decrease of the operating current with time.[39] Furthermore, chattering control of potential as well as continuous time-varying potential techniques have been shown to be rather effective (i.e., a 25% increase in the yield and 900% in the selectivity) in some electrochemical processes.[40,41]
(3) A supporting electrolyte may have to be added due to low ion concentrations.
(4) Interference from the hydrogen evolution reaction or from dioxygen reduction has to be prevented or minimized. For example, dioxygen reduction becomes very important as the concentration of the metal ion decreases, since its solubility at room temperature is on the order of 8 ppm. Nonetheless, the high-purity hydrogen gas by-product has an important commercial value; in addition, pH control can be used to shift the thermodynamically preferred reaction. For example, removal of cobalt ions from solution competes with hydrogen evolution at low pH values; however, at pH > 4 it can be efficiently removed. Last, hydrogen gas production has been shown in certain cases to improve mass transport near the electrode and yield metal deposits that are easier to remove by mechanical means.
(5) The surface of the cathode, where the metal is deposited, changes in properties with time, and this may necessitate additional process control. On the other hand, in some cases, it may even facilitate the deposition process due to higher exchange currents on the modified surface.
(6) Three-dimensional electrodes may suffer plugging or agglomeration as the metal is deposited, especially in solutions with high metal ion concentration; this effect is common to many chemical engineering processes. Periodic metal removal has been used to avoid this problem (see Figure 5.5).
(7) The deposition rate and the composition of the solution in some cases may favor the production of dendrites or else of loose or spongy deposits.

5.3. Direct Electrolysis of Pollutants

(8) High flow rates favor high limiting currents but also give small residence times, which may lead to smaller removal rates. A competition between higher limiting currents and long residence times may lead to either result, depending on the system. For example, in the case of Cu^{2+}, higher flow rates yield lower removal times and higher efficiencies, whereas in the Hg^{2+} case, higher removal was observed at lower flow rates and higher residence times.[42,43]

(9) Potentiostatic control, which can be used to maximize current efficiencies and minimize the occurrence of undesired side reactions, is difficult to achieve in large cells due to shorting of the reference electrode and the inherent difficulty in measuring the electrode potential across the entire working electrode surface. In this case, the potential can be controlled by a judicious combination of current and mass-transfer control. Potentiostatic control has been shown to lead to faster metal ion depletion rates than the corresponding galvanostatic control.[39]

(10) Other substances present in the solution may interfere with the deposition process.

Thermodynamic, kinetic, thermal, and chemical factors (other than those already mentioned) that are fundamental in metal ion recovery include

- The formal reduction potential: It is a function of the nature of the metal; that is, more negative potentials are required for the reduction of an ion of an active metal (e.g., zinc) than that of a noble metal ion (e.g., silver). The change in reduction potential of certain species with pH, concentration, or complexing agents (see later) can be used to advantage. For example, the electrodeposition of cobalt from highly acidic solutions is hindered by hydrogen evolution, which is thermodynamically favored. However, at pH > 4 electrodeposition is possible due to the lowering of the reduction potential of hydrogen while that of the cobalt couple remains constant.[44]
- The concentration of the metal ion: This is understood by noting that the Nernst equation for calculating equilibrium potentials (see Chapter 2) involves the concentration of the ion to be reduced.
- Interaction of the metal ions and the metallic species with the substrate: For example, extra energy is required to form stable nuclei of many metals on the substrate surface at the beginning of the metal deposition process; this necessitates the application of more negative potentials (i.e., an *overpotential*, see Chapter 2). Likewise, the interaction with the substrate may be so strong that the ions may start depositing at a potential more positive than the corresponding thermodynamic value (i.e., *underpotential deposition*, UPD).
- Coupled chemical reactions: They may affect the deposition potential as well as the rate of the deposition process; for example, the preceding chemi-

cal reactions are known to slow down the deposition process for some metal ions. In the case of chromate ion reduction, Cr(III) intermediates undergo a hydration reaction that impedes its further reduction at reasonable potentials.[45]

- Complexation: As mentioned earlier, complexation is known to affect the formal potential of metal ions (usually toward more negative values). The new formal potential depends on the size of the equilibrium complexation constant(s) involved. This may or may not interfere with the metal ion removal process. For example, in a system for the removal of metal ions for the production of high purity water, it was found that the addition of EDTA impeded the removal of Cd and Pb ions.[46] On the other hand, addition of EDTA facilitates the removal of Au from a Cu-containing solution due to the formation of a strong Cu–EDTA complex that in turn permits the selective deposition of Au.[47] Even strong complexes of some noble metals have been electrodeposited under appropriate conditions.[48]
- Solution temperature.
- Solution composition.

The presence of these complexing agents and other species may not only affect the formal potentials but also the mechanism of the reduction–deposition process (see later). For example, the reduction of Cu^{2+} ions in the presence of Cl^- is well known to involve a Cu (I) moiety due to its stabilization in the presence of Cl^-. Also, chromium electrodeposition is strongly affected by the presence of SO_4^{2-} anions, which may form bridges between the chromium moieties and the substrate, thus facilitating deposition.[49,50] Likewise, F^- and Cl^- ions form a bridged complex with Cr(III) that facilitates electron transfer.[45] The pH of the solution can shift the metallic ion in terms of its location in the corresponding Pourbaix diagram, where its deposition is either facilitated or impeded. In addition, pH changes may alter the relative positions of the reduction of the metal ion and hydrogen evolution under a given set of conditions and also promote dissolution or precipitation of species such as hydroxides or hydroxo complexes. The presence of dissolved gases such as O_2 may promote competitive pathways for the electrons furnished to the system, thus lowering some of the process figures of merit (e.g., current efficiency).

<u>Metal Ion Deposition Mechanisms</u>. Many mechanisms have been proposed for the electrochemical reduction of metal ions and metal ion complexes. Common mechanisms are summarized in Figure 5.6. At potentials near the metal ion formal reduction potential, the deposits are normally compact; at more negative potentials they successively become rough and powdery.[51,52] These characteristics have to be chosen according to the future use of the metal and its desired mode of removal. For example, dendritic metallic deposits are pur-

5.3. Direct Electrolysis of Pollutants

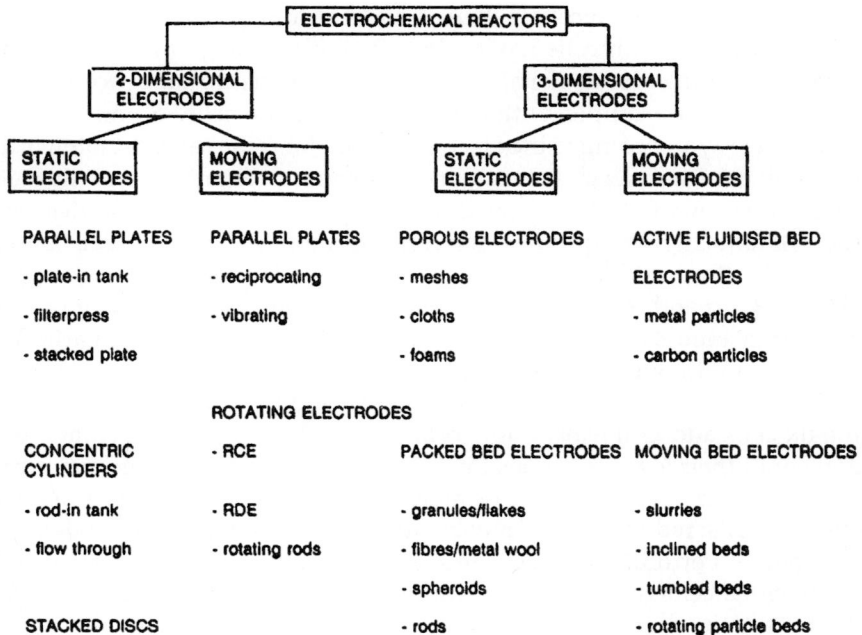

FIGURE 5.7. Classification of electrochemical reactors in terms of electrode geometry and motion. (Reproduced with permission from D. Genders and N. Weinberg, eds., "Electrochemistry for a Cleaner Environment." The Electrosynthesis Co., East Amherst, 1992.)

posely favored for the production of filtering media. The application of a magnetic field during deposition favors porous deposit formation, which is then kept in place by the application of a galvanic coating.[53]

Electrode and Reactor Configurations: Specific Examples. Figure 5.7 contains a classification of electrodes and cells that have been employed for metal ion removal.[54] Assuming that charge transfer at the electrode/electrolyte interface does not play a limiting role and that the metal ion concentration at the electrode surface is negligible, the expression for the limiting current (see Chapter 2) is

$$I_L = k_m A n F C^* \tag{5.16}$$

where I_L is the limiting current, k_m is the local mass transfer coefficient, A is the electrode area, and C^* is the bulk metal ion concentration. Metal ion concentrations greater than ~10^2 ppm are normally treated for recovery with plate-in tank or air sparged plate cells, whereas concentrations lower than ~10 ppm need to be treated with cells having high $k_m A$ values (e.g., rotating cylinder electrode cells, three-dimensional electrode cells). Intermediate concentrations can be treated, for example, with enhanced flow plate cells. Some examples are given below, according to the electrodes used for removal. It is noteworthy that combinations of electrochemical reactors can significantly lower the metal ion concentration of waste streams (for example, two reactors in series, each capable of reducing 90% the metal ion concentration, afford an overall concentration reduction of 99%). The choice of the anode material is not always straightforward and depends on the solution composition (including impurities), desired anodic reaction, electrode durability, and the like (see earlier).

Static two-dimensional electrodes (Figure 5.7) include tank, plate-and-frame, concentric cylinder, perforated cathode, and tubular electrode designs. The tank cells and plate-and-frame cells are used for high concentrations of metals, since their low $k_m A$ values can be compensated by the large concentrations. Concentric cylinders (one or more) have been used for the recovery of precious metals; redox shuttles may be prevented by the use of a divided configuration.[52,55] Perforated cathodes (e.g., perforated carbon plates) can be used for flow enhancement. Their long durability, fabrication simplicity, and low cost are attractive features. Tubular electrodes are simple and their behavior is fairly well understood even for simultaneous cathodic reactions.[56]

As mentioned earlier, the low concentrations of metal ions normally encountered in waste streams make their electrochemical recovery to be mass-transport limited. Transport enhancement can be achieved by the use of moving electrodes in which turbulence promotion leads to increased values of the mass-transport coefficient due to a decrease in the diffusion layer thickness. Some of the popular electrode designs are discussed next.

<u>Rotating Cylinder Electrodes</u>. Rotating cylinder electrodes (RCE) have been found to be simple, versatile, and efficient, producing high space-time yield values. They are the most commonly used moving two-dimensional electrodes.[54,57–60] Here, a metal cathode cylinder (e.g., stainless steel) is rotated in the solution containing the metal ion(s) to be removed (Figure 5.8). A nonuniform metal deposit is obtained; its roughness plays an important role in that it promotes localized turbulence near the surface as well as increased recovery current values due to an increase in surface area. The roughness factor can be obtained from the ratio of the current on the smooth surface to that on the rough surface. The deposited metal can then be removed in different ways (see Figure 5.5), the most common being by scraping. Mass-transport equations for the RCE can be written in terms of the $k_m A$ parameter (see Chapter 2), which is related to the electrode geometry and the electrolyte conditions and composition. This latter parameter can be determined from either limiting

5.3. Direct Electrolysis of Pollutants

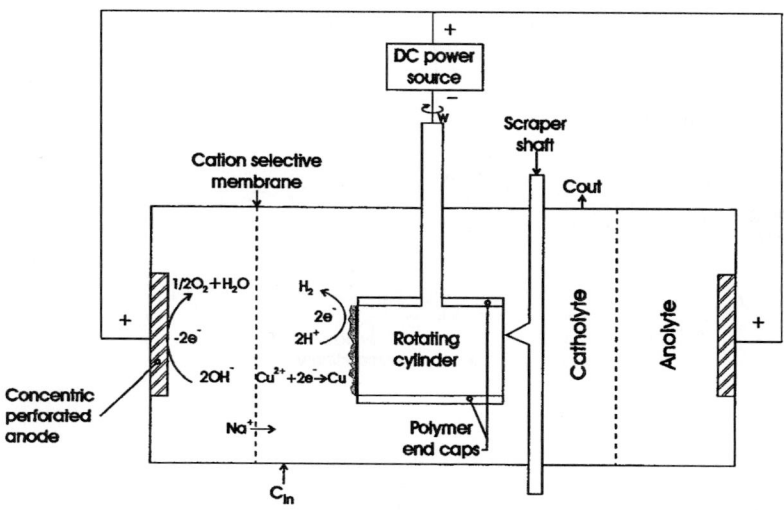

FIGURE 5.8. Schematic diagram of a rotating cylinder electrode reactor. (Adapted from Walsh.[54])

current measurements or from mass transport relationships.[52,58] In this manner, the following equation is obtained:

$$k_m A = 0.079 \, \pi \, d^{0.92} \, l \, v^{-0.59} \, D^{0.67} \, u^{0.92} \tag{5.17}$$

where d and l are the diameter and height of the cylinder, respectively; v is the kinematic viscosity of the liquid; D is the diffusion coefficient; and u is the flow velocity.

For rotating cylinder electrodes the current is usually described by the equation

$$I = K C V^x \tag{5.18}$$

where K is a constant, C is the concentration of the metal ion (in ppm), V is the peripheral velocity (in cm/s), and x is usually near 2/3. However, a value of x = 0.92 has been achieved with a RCE. By virtue of the induced stirring action, these electrodes also prevent surface crystallization of insoluble substances and decrease the local pH increase due to hydrogen production, which normally yields insoluble hydroxides that impede further deposition. This effect has also been minimized by the use of a Pb-RCE, which is known to have a large overpotential for hydrogen production.[61]

Successful simultaneous removal of copper on a plating barrel cathode and cyanide destruction on a packed bed anode can be achieved from waste cop-

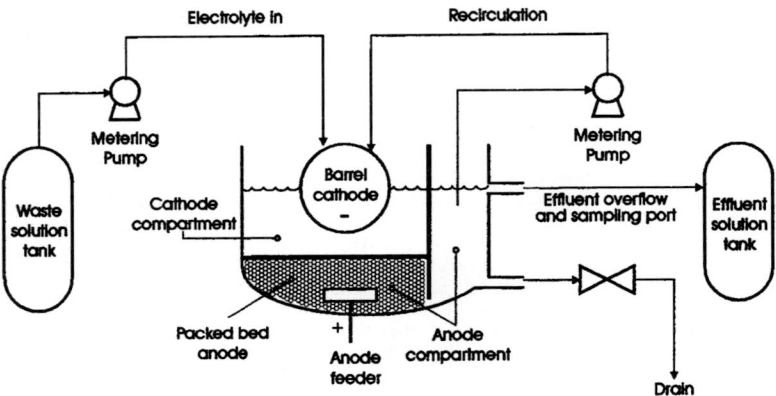

FIGURE 5.9. Simultaneous cathodic metal removal and anodic destruction of pollutants. (Adapted from Zhou and Ching.[63])

per, silver, and cadmium cyanide solutions (see Figure 5.9). This has been reported to be more cost effective than the conventional alkaline chlorinated oxidation for cyanide destruction as well as the electrochemical oxidation on a carbon fiber anode. An increase in temperature leads to an increase in the rates of the simultaneous processes as well as an improvement in current efficiency and a reduction in energy consumption; final concentrations of metal ions and total cyanide can be made smaller than 1 and 10 ppm, respectively by a judicious combination of process variables. Addition of NaCl leads to the formation of hypochlorite at the anode, which also acts as an effective oxidant to cyanide. Coordination chemistry plays an important role here since, under the same operating conditions, the effluent total cyanide concentration as well as the energy consumption follow the same order as would be expected from their complexation constants (i.e., the relative magnitudes of the complexation constants are Cu < Ag < Cd, and the effluent cyanide concentrations in the corresponding solutions are Cu > Ag > Cd). Likewise, the effluent metal ion concentration follows the same ordering as the corresponding reduction potential (i.e., the reduction potentials of the cyanide complexes are Ag > Cu > Cd, and the effluent metal concentrations are Cd > Cu > Ag). Copper can also be recovered from a waste acid copper sulfate stream. Current efficiencies are in the range 30–40%.[62,63]

Hollow perforated cylinders containing metallic pieces of high metal holding capacity can also be used in the recycling mode for the removal or recovery of metals (e.g., Ni, Cu, Pb, Au, Ag, Sn, and Zn) from concentrated as well as dilute solutions, from 10 to 20,000 ppm. If the metal removal medium is made of steel wool, the capacity for metal removal is greatly increased and the final metal content of the solution is substantially reduced.[64]

5.3. Direct Electrolysis of Pollutants

Another variation consists of the use of rotating, perforated concentric non-conducting cylinders filled with conducting particles. Cost effective, high removal ratios of copper ions can be obtained (i.e., from 2400 to 1 ppm, and from 1800 to 4 ppm in the presence of cyanide).[65] Other advantages of the RCE include modular and scaleable geometry and versatility of mode of operation (e.g., batch, batch recirculation, plug flow, continuous stirred tank, single or multiple pass, cascade). High single-pass conversions (50–90%) can sometimes be obtained, although more typical values are on the order of 2–30%. High current efficiencies are also observed in certain cases (up to ~95%). Still another advantage is the possibility of obtaining charge transfer control by using a large inert anode surface and appropriate flow conditions. Uniform current densities can be obtained by appropriate (normally concentric) placement of the anode(s). A CSTR model (see Chapter 2) has been successfully used[66] to describe the behavior of a RCE during metal removal because of the effective electrolyte stirring caused by rotation. The RCE performance is a function of solution, electrode, and flow parameters as well as the operation mode.

Challenges include the following. The current density must be reduced when the metal ion concentration decreases. Scale-up is usually not linear, due to geometric effects (e.g., edge and corner effects). Chemical side effects may occur in some cases (e.g., copper metal redissolution in acidic electrolytes, silver sulfide precipitation during silver recovery from fixing solutions). Determination of $k_m A$ values by the use of limiting currents can be problematic, since too high a potential would not set the system in the full mass-transfer limit region and too low a potential would yield hydrogen gas. Sharp current decreases may be observed after some time due to a decrease in metal ion concentration, which also displaces the reduction potential to more negative values, as per the Nernst equation.

As mentioned already, most applications have been devoted to the recovery of copper from acid or alkaline solutions (from a few to several thousand parts per million), including the simultaneous anodic destruction of cyanide, the recovery of silver from spent fixing photographic solutions, of zinc from viscose rayon plant effluents, of metals from mining liquors, and the treatment of chromium, nickel, and tin wastes. Comprehensive reviews of the characteristics and applications of RCE are available.[54,66]

Static Three-Dimensional Electrodes. Home-made porous electrodes have been used since the last decade of the 19th century[67]; however, their use was by no means widespread. A variety of three-dimensional, porous materials (including reticulated vitreous carbon, carbon and metal felts, metal wool, foams, stacked mesh and perforated plates, and particulate electrode beds) are now commercially available and have been used for many electrochemical processes, including water treatment, organic and inorganic synthesis, energy generation and storage (fuel cells, batteries), and analytical determinations. Versatile home-made reactors with interchangeable porous electrodes can be easily made.[68] Three-dimensional electrode materials can be advantageously used in either flow-through or flow-by modes (see Chapter 2)[69] to increase the electrode surface area and mass transfer of metal ions to the electrode surface to be reduced and removed from solution. Indeed, the flow-by configuration is known to usually allow longer residence times, higher conversions per pass, less sensitivity to clogging and more uniform potential–current distributions than its flow-through counterpart. Scaling up is relatively simple.[70] The area and mass-transfer enhancement permits the continuous (vs. batch) use of processes and more compact systems. Thus, Eq. (5.16) becomes[43,71]:

$$I_L = k_m A_e V_e n F C^* \tag{5.19}$$

where V_e is the volume occupied by the porous electrode and A_e is the specific surface area (active area/unit of volume). Estimation of A_e by other techniques (e.g., pressure drop measurements) permits the separation of the effect of the area from that of mass transfer when enhancement of the limiting current obtained by the use of porous electrodes is evaluated. For example, in the case of Cu^{2+} removal with reticulated vitreous carbon (discussion follows), a current enhancement of 100–250-fold was obtained due to the increase in area, whereas only a 2-fold enhancement was due to increased turbulence.[43] Special attention must be given to the design of the current feeders or current collectors to maximize the possibility for particle contact; fin-type current feeders (similar in shape to many aluminum heat dissipators in electronic circuits) are particularly effective in this regard. Likewise, potential drops through the solution phase must be taken into account and minimized, since they can cause an uneven current–potential distribution with associated problems. Other experimental and modeling studies on porous electrode-based electrochemical reactors are available, although not all of these are oriented toward pollution abatement.[72–81]

<u>Reticulated Vitreous Carbon.</u> This relatively new form of glasslike, open pore carbon is produced by the pyrolysis of a polymeric resin.[82] It combines properties of glass and regular carbon and shows several advantages for its use in water treatment processes. These include high surface to volume ratio (up to some 66 cm^2/cm^3), a high mass-transfer coefficient, relatively low resistance to liquid flow as well as low pressure drop (depending on porosity), high chemical inertness to a variety of pollutants, high thermal and electrochemical stability (usable potential range, 1.2 to -1.0 V vs. SCE at pH 7), machinability, low weight (densities are typically in the range of 0.045–0.062 g cm^{-3}), and commercial availability at a reasonable cost. In addition, RVC materials feature a self-supporting, honeycomb-type structure; electrical contacts can be easily made by either pressing metal sheets or conductive rods on an RVC plate or

else by application of conducting adhesives or paints. One can select from several porosities (on the order of 92–98%) usually denoted by the number of pores per inch (ppi). Typical choices are from 10 to 100 ppi, but compressed RVC of up to 500 ppi is commercially available. Limiting currents are normally much greater than those with solid electrode materials (e.g., in Cu^{2+} removal, a 250-fold increase is observed with a 100 ppi RVC electrode as compared to a stainless steel plate). In addition, the normalized space velocity (see Chapter 2) when using RVC has been shown to be greater than that with other three-dimensional electrodes.[71] Once a metal has been plated on RVC it can be selectively dissolved from it or, alternatively, the carbon substrate may be incinerated to leave the pure metal. The main drawbacks of RVC electrodes include the following: lower conductivity than other electrodes (e.g., metals, solid vitreous carbon), possibility of plugging, exchange current densities lower than with solid glassy carbon, aging effects, fragility, and O_2 entrapment.

Several reactor types and electrode geometries have been used for metal ion removal with RVC in divided or undivided cells, including parallel plate, fixed or moving beds, bipolar trickle tower, CSTR in series, and PFR in series. An arrangement of electrodes in series permits the analysis of individual segments for a better understanding of the entire phenomenon. With respect to the position of the counterelectrode, an upstream arrangement (i.e., the counterelectrode located in such a way that the fluid is closer to it before it passes through the porous working electrode) is usually more effective than its downstream counterpart due to a smaller ohmic potential drop.

Metals that have been removed with RVC include Cu, Cd, Cr, Pb, U, Hg, Ag, and Zn, with removal yields up to virtually 100% in a single pass in optimum cases.[83–85] In other cases,[85] the conversion has been less impressive. Analytical uses for RVC have been considered in Chapter 4.

<u>Porous Graphite Electrodes</u>. Porous graphite electrodes can be advantageously used for metal ion removal and other processes. For example, the removal of Eu^{3+} ions from a lanthanide mixture can be accomplished by reducing them to Eu^{2+} at a flow-through porous graphite cathode; in the presence of sulfate ions, Eu^{2+} forms an insoluble sulfate, which can then be easily removed.[86] This reaction sequence can be observed in Figure 5.6. Enhancement of electrochemical performance can, in some cases, be achieved by anodic oxidation of these materials under acidic conditions, probably due to the disappearance of regions containing graphite debris.

<u>Graphite or Carbon Felt, Cloth, or Fiber Electrodes</u>. The properties of graphite or carbon when combined with a three-dimensional flexible structure provide for a very large area, low-cost, rugged, highly chemically resistant and versatile material for different electrochemical applications.[87–89] These materi-

als have long been known in the fuel cell area. Carbon fiber electrodes[2,90] typically have fiber diameters of 10 μm and a surface area of 1000 m^2/g. These electrodes can be used as monopolar or bipolar electrodes. In addition, scale-up is normally simple. Graphite cloth is available in the range from fine double weave to coarse weave and graphite felt[91] as woven or pressed felt. Modification of these electrodes by impregnation with a surface active substance (anionic, cationic, non-ionic, or amphoteric surfactants) has proven to facilitate metal ion deposition and removal. Possible mechanisms include the promotion of gas bubble release from the electrode surface and the attainment of better surface contact.[88]

The efficiency of these electrodes for removal of metal ions has been found to be a function of several factors.[92–96] For example, at high supporting electrolyte concentrations, the potential distribution throughout the pores is uniform due to the low resistivity of the solution, whereas at low concentrations the distribution becomes uneven and thus the potential necessary for deposition may not be available at all sites, and side processes may occur.[95] Limiting currents of carbon felt, flow-through electrodes (in zinc–bromine batteries) are proportional to the electrolyte velocity to the n power (where $0.61 < n < 0.72$).[93,97] Mathematical models have been developed to account for homogeneous chemical reactions coupled to the main electrode process in graphite felts and to incorporate kinetic data for the evaluation of different reaction conditions.[94] Current efficiencies of ~100% have been obtained in removing silver from highly concentrated solutions in photographic spent fixing baths with graphite felt electrodes.[87] However, typical efficiencies in commercial systems for water purification purposes are on the order of 20–60%.

In spite of the advantages of bipolar *vs.* monopolar electrodes mentioned in Chapter 2, monopolar electrodes perform better for metal ion removal in a wider range of supporting electrolyte concentrations because of their simpler potential distribution requirement.[95] Felt or cloth electrodes can be stacked to facilitate the subsequent analysis of the different sections for a better understanding or evaluation of specific applications. Alternatively, an electrode cartridge can be used and replaced whenever necessary. Highly efficient removal of metal ions from solutions with high or low metal concentrations can be achieved (to <1 ppm). The limiting currents attainable are similar to those in a rotating cylinder system.

As mentioned earlier, the metal deposit can be recovered by several means including the burning of the carbon substrate, which may also act as a reducing agent for metal oxides formed after deposition. Graphite felt is a more efficient choice than graphite granules, graphite cloth, copper screens, or nickel particles for the removal of Hg^{2+} metal ions.[91] A possible explanation is that the felt shows a lower degree of channeling.[96]

5.3. Direct Electrolysis of Pollutants

Another application of graphite felt or fiber electrodes for the removal of metal ions consists of the reduction of Cr(VI) to Cr(III) ions, which form the insoluble hydroxide with the OH⁻ ions electrochemically generated and are then separated from the solution.[1,2] Electrochemical pH control is used here. Hydroxide particles adhere to the cathode due to a positive zeta potential; electrode regeneration is then achieved by potential inversion which electrochemically dissolves the hydroxide. Final Cr(VI) concentrations as low as a few ppb can be obtained.[1,2]

Other metal ions that have been successfully removed from a variety of effluents include Au, Ir, Pt, Ag, Pd, Cd, Pb, Ni, and Hg. Drawbacks include the relatively low current efficiencies sometimes obtained, the high degree of sensitivity to minor limiting current variations at a relatively small Reynolds number, and the difficulties in reproducing results.[97]

Packed Bed Electrodes. Regularly or irregularly shaped conducting particles can be put together and act as packed bed anodes or cathodes for electrochemical processes.[98] Both electrodes in a cell can be packed beds; this has been termed a *double packed bed cell*.[99] Examples of bed packings include graphite and metal powder, rods, flakes, granules, spheres, shot, turnings, wirelets, Raschig rings, and metal wool. The simplicity of design, high mass-transfer rates, versatility, and low cost of the conducting particles are key advantages. Efficiencies and conversions can be high (conversions up to 99.9% in a single pass are possible) even in low conductivity solutions.[48,100] Multibed stacks can be easily arranged. Current collectors can be of many materials and shapes.

These electrodes can be used in the flow-by or flow-through modes (see Chapter 2). For similar conditions, higher flow rates more frequently are possible in the flow-by mode than in the flow-through mode. The mass-transfer coefficient, k_m, is known to be a function of the Reynolds number.[99] For low metal ion concentrations, pulsed current has been shown to increase mass-transfer rates from one to three times under appropriate conditions of frequency and duty cycle. Similar or better removal ratios can be obtained with pulsed average currents equivalent to one-half of those required with direct current, and consequently the energy consumption is much lower. A possible explanation for this behavior is the decrease in the Nernst diffusion layer thickness. In addition, the hydrogen bubbles produced during the "on" time can be dislodged during the "off" time, thus rendering a higher available surface for deposition.[101,102]

To increase the electrochemically accessible specific surface area of granular electrodes, smaller particles can be used. However, a compromise has to be made, since ultrafine particles may drastically decrease the permeability of the fluid. If the bed particles are porous, intraparticle convection must be taken into account for an accurate description of reactant transport.[103] Another improvement consists of the addition of ionic conducting particles to the aforementioned electronically conducting particles. This mixture is intended to increase porosity and ionic conductivity. One such complex porous electrode design, consisting of a mixture of granular graphite and ion exchange resin beads, has been successfully used for the recovery of silver from a silver plating rinse water.[104] Recovery percentages much higher than those obtained with the electronically conducting particles only (i.e., >90%) were obtained and the power consumption was small.[104]

A popular cell design for the use of packed bed electrodes is the filter press type. Cylindrical packed bed electrodes have also been analyzed and their performance shown to be better in some cases than their rectangular counterparts.[105] Bipolar trickle towers, consisting of a tower packed with physically and electrically separated layers of conducting particles (which become bipolar), have been used for scavenging Cu, Ni, Zn, Ag, Pb, and Cd ions from dilute solutions.[106,107] A key factor here is the *fractional active length*; that is, in this case, the fraction of length of the particle that becomes cathodically polarized. Thus, each particle becomes a zoned reactor. If very low final concentrations are desired, these cells can be run in cascade. Simultaneous deposition of metal ions and anodic destruction of cyanide ions can also be achieved with initial high current efficiencies (~ 80%) that decrease with decreasing concentration. Conversions per pass can be as high as 50%; the deposition of the metal ion continues until the CN⁻ concentration decreases to the stoichiometric ratio of the least stable complex (see Figure 5.6). A challenge to be faced in the trickle tower is a possibly substantial variation in the stoichiometry of the involved reactions as a function of the position in the reactor (since the potential field may be different at each point). Likewise, the anodic reaction products may affect the cathodic reaction. For example, the Pb deposition rate in a Cl⁻ solution changes substantially when the pH shifts due to a change in the nature of the anodic reaction (from the production of OCl⁻ to the production of Cl_2).

Yet another innovation uses metal wool as the packing material.[108,109] Advantages include comparable performance to the normal particle packed beds, lower hydraulic pressure and ohmic potential drops, high flexibility, high porosity, low density, and much reduced material amount requirements (on the order of 10%). Under gas production conditions, metal wools show a lower pore electrolyte resistance than that in particle packed bed electrodes, due to the smaller volume void fraction of gas in the wool. In addition, they can be fabricated in a wide range of specific surface areas. Examples include the use of Cu and Ag metal wool for the removal and recovery of copper metal ions.

A combined carbon particle packed bed–gas diffusion electrode scheme has been developed to take advantage of the favorable thermodynamics for the reduction of chromate ions with hydrogen.[110] Both processes occur simultaneously, spontaneously, and rapidly (hydrogen oxidation on the gas diffusion electrode side, chromate reduction on the packed bed side) in this thermodynamically downhill reaction that does not require an external power supply. Another innovation is the use of hydrodynamic modulation of the electrolyte flow in a conducting particle bed.[111]

An important parameter of packed beds, the bed voidage, can be determined easily by calculating the weight difference between a known-volume bed filled with water and that same bed after draining and contacting the particles with damp clothes.[48] Mathematical models based on kinetic parameters have been successfully developed for special flow regimes and different packed bed geometries.[112] The drawbacks of packed beds include the possibility of channeling, the uneven distribution of the electric fields, the need for limiting currents to achieve large conversions and yields, and the lowering of efficiency with increasing bed thickness.[100]

<u>Metal Screens, Nets, or Grids</u>. Stacked metal screens, nets, or grids may act as three-dimensional electrodes for various purposes. For example, silver removal from photographic fixing baths with a cell containing stacked stainless steel grids as electrodes has been achieved, yielding a final silver ion concentration of about 1 ppm. Current responses up to two orders of magnitude higher than those of flat plates have been observed for stacked nets. The flow-through configuration can give higher space–time yields than the flow-by configuration under appropiate conditions.[113] Some examples of metals and alloys commercially available in the form of grids include Al, Cu, Ni, Zn, brass, monel, steel, and stainless steel.

<u>Metal Felts or Foams</u>. Felts and foams made of diverse metals are commercially available (e.g., Ni, Al, Cu, precious metals, alloys).[69,70,114–116] Very high areas (up to some 5,000 cm^2/cm^3 for the case of Ni felt) have been measured by the BET method, although not the entire area is available for electron transfer; they are much less brittle than their RVC counterparts and can be easily cut, shaped, and electrically connected. Interestingly, the tortuosity of these electrodes is fairly constant for different porosities.[69] They have been successfully used as replacements for turbulence promoters. The surface imbalance between a metal foam electrode and a flat plate electrode covered with turbulence promoters as its counterelectrode facilitates a cell design without the need for membrane separators. In addition, flat plate electrodes can often be simply replaced by felts or foams to increase reactor compactness.

An elegant way to compare the performance of metal foam electrodes with other electrodes consists of the analysis of the limiting currents for a mass-transfer controlled system (e.g., ferricyanide reduction in a basic solution) and a similar analysis for a kinetically controlled system (e.g., alcohol oxidation, also in a basic solution).[117] A problem that needs to be addressed is that these materials often trap gas bubbles in their matrix, which lowers the available area and greatly increases the ohmic drop.

Moving bed three-dimensional electrodes (Figure 5.7) are considered next. Here, a conducting or non-conducting particulate bed is set into motion by the passage of a stream of the solution to be treated. The two main types of moving bed three-dimensional electrodes are the fluidized bed and the circulating bed electrodes. They have many common features as is discussed next.

<u>Fluidized Bed Electrodes</u>. Well-designed fluidized bed reactors normally show more uniform potential and current distributions and higher mass transfer coefficients (due in part to the decrease of the diffusion layer) than those found in static porous beds.[117] They can have several geometries: rectangular, annular, or cylindrical. Their scale-up by an order of magnitude yields similar characteristics as long as the characteristic dimension in the direction of current flow is not changed (see Chapter 2). In the case of metal deposition, deposits are also more uniformly distributed and continuous particle removal is possible. Due to the instantaneous contact between conducting particles, the effective conductivity is frequently similar to that of the electrolyte. Particle chains formed during this process may touch either other chains or else the feeder electrode closing the electric circuit. Monopolar and bipolar particles arise from this phenomenon, since some of them will not be electrically connected in the manner described at a given moment. Current efficiencies usually increase with current density.[118]

Continuous metal recovery can be achieved by adding small particles at the top of the bed while removing grown particles at the bottom.[118] Cylindrical, fluidized bed electrodes with different flow modes and counter electrode positions have been shown to be more effective in some cases than their rectangular counterparts.[105] Other important factors that affect the performance of fluidized bed electrodes include bed expansion and bed thickness. High current efficiencies (80–90%) have been reported in some cases as well as the removal of metal ions down to the sub-ppm level.

Applications of fluidized bed electrodes in metal ion removal include the removal of cobalt from 200 to 0.2 ppm,[44] copper from a concentrated copper nitrate waste stream,[117] and cadmium (from a solution containing cadmium and ferric ions) down to 1–5 ppm.[119] A pilot-plant scale fluidized bed cell has also been described.[120] The drawbacks of fluidized bed electrodes include the possibility of formation of inactive zones, preferential growth near the separating membrane (where the cathodic potential is highest), higher applied voltages required for a given applied current density, current efficiencies usually decreasing with time; lowering of the intergranular electric conductivity on fluidization, particle–particle and particle–electrode agglomeration, bipolar particles possibly offer sites for anodic dissolution of deposited metal, and metal

5.3. Direct Electrolysis of Pollutants

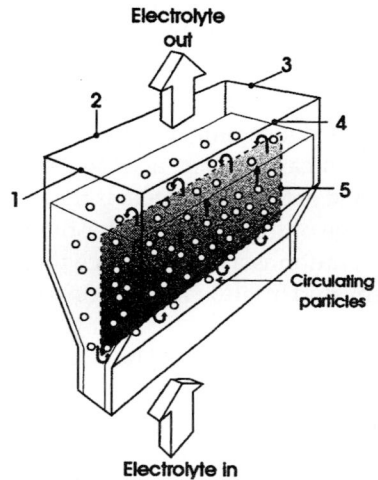

Circulating bed electrodes:

	Spouted bed	Vortex bed	Moving bed
1	wall	diaphragm	wall
2	feeder	wall	diaphragm
3	wall	feeder	wall
4	diaphragm	wall	wall
5	wall	wall	feeder

FIGURE 5.10. Schematic diagram of three types of circulating bed electrodes. (Adapted from Scott.[122])

deposition on the feeder electrode. Last, concentrations below 100 ppm are not efficiently treated.[44,118]

A review of fluidized bed electrodes for metal recovery is available.[118]

<u>Circulating Bed Electrodes</u>. These electrodes differ from the previous ones in that a circulating bed electrode is employed, although they have many features similar to those found in the fluidized electrode systems.

Three common systems that share the same type of circulating particle motion are the spouted bed electrode (SBE), the vortex bed electrode (VBE), and the moving bed electrode (MBE).[39,121,122] Figure 5.10 illustrates the three configurations. There are two regions inside the electrode: (a) a moving particle region, where the particles and the liquid stream move in the same direction, and (b) a packed bed–like region, where they move in the opposite direction. Particle vortex motion can be achieved, for example, by introducing the electrolyte into the bottom of the bed compartment through a slot that has an upward increasing width.[39]

Since electrolyte flow and current path are perpendicular, the flow rate and reactor size (and thus residence time) can be set independent of the potential distribution. These cells can contain a wide range of particle sizes and distributions. One advantage of the SBE system is that it can handle gas bubbles (which usually produce large ohmic drops) better than many other configurations; in addition, agglomeration is avoided, since there is no grain immobilization on the diaphragm and metal deposition on the current feeder is avoided.[118]

Similar efficiencies for metal ion removal have been reported for these three types of moving bed electrodes. In optimum cases, current efficiencies of nearly 100% have been obtained. Examples of metal ions removed include Cu, Sn, Zn, Co, and Mn.

Challenges to be faced include the following. Pressure losses are somewhat higher than those observed in fluidized bed electrodes. A diaphragm is needed in the VBE and MBE configurations. On scale-up, the VBE has regions where the bed may agglomerate. The falling phase has essentially zero electrochemical activity. Last, part of the electrolyte will not be electrolyzed due to its function as a particle carrier and thus a single pass will not achieve total metal ion removal.

<u>Tumbled Bed Electrodes</u>. Commercially available barrel platers can be loaded with conducting particles (e.g., metal spheres) to effect metal ion removal and obtain smooth, adherent, and uniformly distributed metal deposits even under unfavorable conditions, such as simultaneous hydrogen evolution. Tumbled beds offer simple, low maintenance, and reliable means of metal ion removal and deposition.[123]

As indicated at the outset of this section, electrochemical treatment of toxic metals is a rather mature technology. Further details may be found in the specialized literature listed at the end of this chapter.

5.4. INDIRECT ELECTROLYSIS OF POLLUTANTS

As illustrated in Figure 5.1, the idea here is to use an electrochemically generated redox reagent as a chemical reactant (or catalyst) to convert pollutants to less harmful products. The redox reagent acts then as an intermediary for shuttling electrons between the pollutant substrate and the electrode. These redox reagents can be electrochemically generated either in a reversible (Fig-

5.4. Indirect Electrolysis of Pollutants

ure 5.1b) or irreversible (Figure 5.1c) manner. These two strategies are discussed in turn next.

5.4.1. Reversible Processes

The general scheme is

$$C \rightarrow C^+ + e^- \tag{5.16}$$
$$C^+ + R \rightarrow O + C \tag{5.17}$$

for the oxidation case and

$$C + e^- \rightarrow C^- \tag{5.18}$$
$$C^- + O \rightarrow R + C \tag{5.19}$$

for the reduction analog.

Indirect electrolytic processes can be viewed as chemical on–off switches, since they stop when the current supply is discontinued.[124] Among the principal requirements for obtaining high efficiencies in the indirect electrolytic processes are the following[125]:
(1) The potential at which the intermediate species, C, is produced must n o t be near the potential for O_2 or H_2 evolution, since then a large portion of the current will not be employed in the desired reaction.
(2) The rate of generation of C must be large.
(3) The rate of reaction of C with the pollutant R (or O) must be higher than the rates of any competing reactions (for example, $C^+ + H_2O \rightarrow O_2$ + products).
(4) Adsorption of the pollutant (or any other species) must be minimized since electron exchange of C at the electrode often becomes sluggish as a result.

Table 5.3 contains a listing of the reversible redox reagents that have been successfully used for pollutant treatment and the corresponding redox potentials. Selected systems are discussed next.

Ag(I)/Ag(II). Silver in the +2 oxidation state is a powerful oxidizing agent. This specie can be produced by the anodic oxidation of Ag(I) ions:

$$Ag^+ \rightarrow Ag^{2+} + e^-$$

In nitric acid, the following (dark brown) complex is formed:

$$Ag^{2+} + NO_3^- \rightarrow AgNO_3^+$$

Table 5.3. Redox Mediators for the Indirect Electrolysis of Pollutants

Mediator Couple	Standard Reduction Potential V vs. SHE
Ag(I/II)	1.98
Co(II/III)	1.82
Ce(IV/III)	1.44
Fe(II/III)	0.77

which then reacts with water or with an oxidizable pollutant (R) as follows:

$$4\,AgNO_3^+ + 2H_2O \rightarrow 4\,Ag^+ + O_2 + 4\,HNO_3$$
$$a\,AgNO_3^+ + R + b\,H_2O \rightarrow a\,Ag^+ + n\,CO_2 + a\,HNO_3$$

At the end of the reaction the dark brown color fades away, signaling the reduction of Ag(II) to Ag(I). In this equation, the values of a and b depend on the degree of oxidation of the carbon atoms in R. This process is also known to destroy organics containing (hetero) atoms other than C, H, and O (for example, organophosphorous, organosulphur, and chlorinated aliphatic and aromatic compounds).[126]

The corresponding cathodic reaction in HNO_3 medium is

$$HNO_3 + 2\,H^+ + 2e^- \rightarrow HNO_2 + H_2O$$

In the presence of O_2, nitrous acid is oxidized back to nitric acid:

$$2\,HNO_2 + O_2 \rightarrow 2\,HNO_3$$

The net result is then the oxidation of the organic pollutant and/or of water; Ag^+ and HNO_3 are recycled into the system. Aside from a direct attack of Ag(II), chemical reactions generating highly reactive radicals have been implicated[24,126]:

$$Ag(II) + H_2O \rightarrow Ag(I) + H^+ + \cdot OH$$
$$2\,Ag(II) + H_2O \rightarrow 2\,Ag(I) + O\cdot + 2\,H^+$$

Table 5.4 contains a listing of the pollutants that have been successfully treated using this redox reagent.[127–130] The main challenges for further develop-

Table 5.4. Representative Listing of Pollutants Successfully Treated by the Ag(I/II) Mediator Approach

Pollutant	References
Ethylene glycol	127
Isopropanol	128,129
Acetone	128
Organic acids	130
Benzene	127
Tributyl phosphate/kerosene	126

ment of this process include the following: (a) silver ions are considered as hazardous waste themselves (provisions should be made for total recycling or regeneration of the catalyst); (b) chloride ions precipitate Ag^+ and thus reduce the overall destruction efficiency; and (c) silver is relatively expensive.

Fe(II)/Fe(III). Relative to the Ag(I/II) couple, the Fe (II/III) couple has a much less positive $E°$ (0.77 V), which should substantially lower the electrical cost that accounts for up to 90% of the total operating cost.[131] Oxidation efficiencies up to 100% have been reported.[131] Cellulosic materials, fats, urea, cattle manure, sewage sludge, meat packing wastes, ethylene glycol (as surrogate waste), and the like have been treated.[131] Furthermore, the reaction rate of Fe (III) with organic compounds has been shown to increase substantially (one order of magnitude) by the addition of some transition metal ions as co-catalysts to the solution.[131,132] This process has been proven to be amenable to scale-up.[132] Electrolytic grade H_2 gas is produced at the cathode with an efficiency of up to 99%; this can be used in a fuel cell to generate electricity.

Mediated electrochemical oxidation with Fe (III) can also be used to oxidize carbonaceous material (e.g., coal slurry) in acidic media at the anode of an electrochemical cell to partially oxidized carbonaceous materials (e.g., humic acid) or else to carbon oxides.[133] The complementary reaction is hydrogen evolution or metal ion deposition.

Differences in solubility and in formation constants of N- and S-containing organic pollutants (e.g., carbon disulfide, thianaphthene, isoquinoline) in coal liquids with different oxidation states of a metal in a metal ion complex have been exploited in a pollutant removal–concentration cycle.[134] Here, reversible

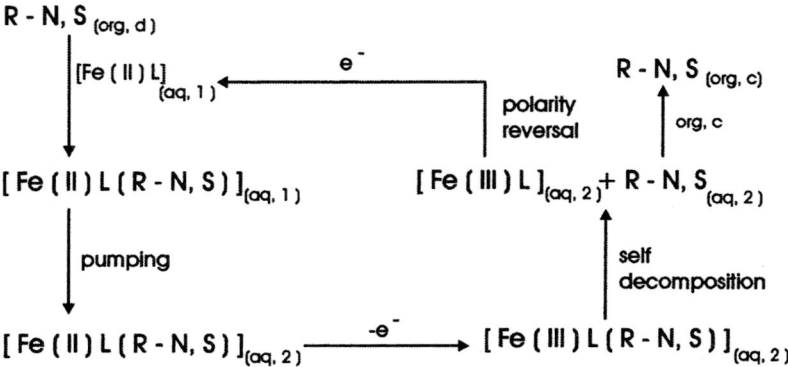

FIGURE 5.11. Electrolytic cycle for the concentration of heteroatom-containing pollutants.

complexation and electrolytic regeneration are utilized for the removal and concentration of these compounds from a polluted hydrocarbon phase (diluted organic phase, symbolized as org,d) to a waste hydrocarbon phase (concentrated organic phase, symbolized as org,c) as shown in Figure 5.11.[134] A Fe(II)L complex is selected in such a way that its formation constant with a nitrogen or sulfur containing organic compound (symbolized as R-N,S) is very large, whereas that for the corresponding Fe(III)L complex with R-N,S is very small. Here, L = Na^+ salt of tetra(4-sulfonato-phenyl) porphyrin. The catholyte side (the number 1 assigned to it) initially contains $[Fe(II)L]_{(aq,1)}$; after equilibration with the dilute organic phase, it will contain $[Fe(II)L(R\text{-}N,S)]_{(aq,1)}$, which is then pumped to the anolyte compartment (number 2) to be oxidized to $[Fe(III)L(R\text{-}N,S)]_{(aq,2)}$. This compound decomposes due to its small equilibrium constant and allows R-N,S to partition into the concentrated hydrocarbon waste. Fe(III)L is then reduced to Fe(II)L to reinitiate the cycle.[134]

Electrolytically generated ferrous iron has been used for the chemical reduction of toxic hexavalent chromium,[135] and the ferric form for the conversion of gaseous H_2S to sulfur and hydrogen.[136] This last process is discussed in more detail later in this chapter (Section 5.6).

The difficulties with Fe(II/III) based processes, are two-fold: low current densities and relatively high temperatures (~100°C) for efficient operation. The Mn(II/III) redox system has been investigated as an alternative in this regard.[133]

Co(II/III). This redox couple has a high standard potential ($E° = 1.82$ V), which makes Co(III) a powerful oxidizing agent. $CoNO_3$ in HNO_3 has been studied for the destruction of different organic substances. However, since metallic Co can be electrodeposited from Co^{3+} at the cathode, a separator be-

5.4. Indirect Electrolysis of Pollutants

tween electrodes is required.[137] This complicates the process somewhat, since the introduction of a membrane commonly involves a higher voltage drop and the possibility of leakage, rupture, fouling, or chemical attack. Furthermore, nitric acid reduction is known to generate some undesirable NO_x at the cathode, although these gases can be treated to prevent their release and produce useful chemicals (see Section 5.6). An application in the destruction of organic radioactive waste materials has been successfully tested with cellulose. The oxidation of one unit of glucose follows[138]:

$$C_6H_{12}O_6 + 24\,Co(III) + 6\,H_2O \rightarrow 6\,CO_2 + 24\,Co(II) + 24\,H^+$$

A process that avoids most of the limitations described previously involves the use of a $CoSO_4$ solution in a sulfuric acid medium; here, the reaction at the cathode cannot be the reduction of the anion but the reduction of H^+, which impedes the deposition of cobalt.[137]

This system has been shown to successfully destroy chlorinated (e.g., 1,3-dichloro-2-propanol) and nonchlorinated organics (e.g., ethylene glycol). In the first case, conversion (90%) has been obtained (chlorine enters the solution as Cl^-), whereas in the second conversion up to 100% can be achieved. This difference has been explained as due to volatilization from the electrolyte. Coulombic efficiencies of 43 and 76% were obtained, respectively.[137,139] One would expect that organic compounds similar to these with lower vapor pressures and equal solubilities should warrant high conversion efficiencies; unfortunately, this is not necessarily the case, as demonstrated with trimsol cutting oil (a halogenated organic) that showed a very low conversion yield (20%) after many hours of reaction.[140] The mechanism for the oxidation of ethylene glycol to CO_2 probably involves the generation of two molecules of formaldehyde and subsequent conversion to formic acid, which is then oxidized to CO_2.

The use of metal ion redox couples in conjunction with reagents such as H_2O_2 (Fenton reagent) will be discussed in a separate section next.

<u>Metal Oxide Electrodes and Soluble Metal Oxide Catalysts</u>. Many metal oxide (e.g., SnO_2) surfaces are hydroxylated and can undergo the following surface chemistry on oxidation:

$$M\text{-}OH \rightarrow M\text{-}O^\bullet + H^+ + e^-$$
$$M\text{-}O^\bullet + OH^- \rightarrow M\text{-}O^- + {}^\bullet OH$$
$$M\text{-}O^- + H_2O \rightarrow M\text{-}OH + OH^-$$

As will be discussed in Chapter 6, the hydroxyl radicals ($^\bullet OH$) are extremely reactive to organic substrates. The bimolecular rate constants involving these species and many organics (especially those containing abstractable hydrogen) are very high and close to the diffusion controlled reaction limit.

Another advantage with metal oxide electrodes such as SnO_2 is that they possess high overvoltage for O_2 evolution.[3-7,140-144] Thus, they are ideally suited for the anodic treatment of pollutants. For example, when doped SnO_2 films (doped for example with Sb) are grown on Ti anodes by the spray hydrolysis method, electrodes with very high overvoltages for O_2 evolution are obtained, which are capable of oxidizing a wide variety of organic substances with average efficiencies (in terms of EOI) typically several times higher than those with Pt or platinized Ti anodes.[145-147] In the case of phenol, the EOI value is 0.25 at SnO_2 as compared to 0.10 at Pt; the corresponding conversions are 0.90 and 0.40. TOC removal was 90% with SnO_2 and 38% with Pt. This can be understood since the aliphatic acid intermediates (see earlier) are rapidly oxidized at the SnO_2 anode but are essentially inactive at Pt.

An alternative approach involves the activation of the metal oxide surface for O-transfer reactions with a minimal increase in the *net* rate of O_2 evolution.[148-151] In this approach, catalytic sites are implanted in the oxide matrix to assist the oxidative discharge of water to produce ·OH radicals and impede massive production of O_2 at the surface. Diverse doping agents, both anionic (e.g., Cl⁻) and cationic (e.g., Fe^{3+}, Bi^{5+}, As^{5+}), have been tested with PbO_2 electrodes for the oxidation of organic and inorganic pollutants with promising results. For example, electrodes prepared with a coating containing Bi(V) have been reported to be useful in the oxidation of phenol, cyanide, and ammonia as well as in the selective oxidation of organic compounds, in the on-site generation of disinfecting species and strong oxidizers, and in the oxidation and extraction of various minerals from ores.[152]

Many inorganic compounds that are normally insoluble and resistant to oxidation have been shown to be electrolyzed in aqueous cationic surfactant systems.[153] The surfactant solutions apparently stabilize these compounds and form hydrophobic films on the electrode surface, thus raising the oxidation potential of water by 0.9–1.7 V. The net result is that higher positive potentials can be applied without interference from the electrolysis of water. In early versions of this approach, barium peroxide was utilized for the destruction of halogenated hydrocarbons.[154] The formation of barium superoxide intermediate was implicated in the reaction where the intermediate displaces the halogen and produces a less toxic product. The barium peroxide can be regenerated via the use of an O_2 cathode.[155] In a later version of this overall approach, copper(II) oxide and manganese(II) oxide have been used as catalysts for the electro-oxidation of organosulfur and halogenated organic compounds.[156] The added oxides form higher-valent derivatives (e.g., copper(III) oxide and Mn(III) and Mn(IV) oxides) to mediate the charge transfer between the anode and the organic substrate.

5.4.2. Irreversible Processes

This category of processes relies on the (irreversible) electrolytic generation of "killer" agents such as H_2O_2, ozone, hypochlorite, or chlorine, which can then be used to attack the pollutant. The same approach can also be used as a disinfection tool, as will be discussed in Chapter 7. Interestingly, reactions involving species such as Cl_2 may also have contributed to pollutant destruction in many of the scenarios discussed previously that purportedly involved *direct* charge transfer between the electrode and the substrate.

Electrolytic generation of Cl_2 has been used for the indirect destruction of CN^- as per the following scheme[157]:

$$2\,Cl^- \rightarrow Cl_2 + 2\,e^-$$
$$2\,H_2O + 2\,e^- \rightarrow H_2 + 2\,OH^-$$
$$Cl_2 + H_2O \rightarrow HCl + HOCl$$

Thus the overall reaction can be written as

$$2\,NaCN + 5\,HOCl \rightarrow 2\,CO_2 + N_2 + H_2O + 3\,HCl + 2\,NaCl$$

The chloride ions thus produced can be recycled back into the anode chamber for further production of Cl_2. In this regard, the overall process is cyclical although the anode reaction is irreversible in an electrochemical sense.

Table 5.5 contains a listing of reagents that can be electrolytically generated for pollutant destruction, including the hydroxyl radical discussed in a preceding section. The corresponding standard reduction potentials of these reagents are also shown. These values are all fairly positive attesting to the ability of these reagents for oxidizing a wide variety of organic substrates.

An important advantage to the electrolytic generation of these reagents is that it can be done *in situ*. Thus, these rather dangerous chemicals need not be transported over long distances. Further, many of the reagents listed in Table 5.5 (e.g., H_2O_2) are unstable on long-term storage. Finally, electrolytic generation of reagents in a pollutant treatment reactor has the virtue of precise process controllability (because of its electrical nature), and the extent of reagent generation can be profitably coupled to the demand imposed by the degree of pollution of the process stream.

It is thus interesting to note that the electrogeneration of H_2O_2 has witnessed a rebirth thanks to the application of this chemical in environmental pollution abatement. The original process for the electrosynthesis of H_2O_2 (which dates back to 1853) was based on the oxidation of sulfate to persulfate on an anode. Hydrogen peroxide is produced via the subsequent hydrolysis of

Table 5.5. Reagents That Can Be Electrolytically Generated for the Anodic Treatment of Pollutants and the Corresponding Standard Potentials

Reagent	Standard Reduction Potential V vs. SHE
Hydroxyl radical	2.80
Atomic oxygen	2.42
Ozone	2.07
Hydrogen peroxide	1.78
Perhydroxyl radical	1.70
Chlorine	1.36

Solution of pH 0 assumed.

the persulfate. This process was developed to the extent that it was the dominant commercial process, in spite of high power consumption. However, virtually all the H_2O_2 produced today uses the anthraquinone process (cyclic hydrogen reduction of an anthraquinone and air oxidation of the anthrahydroquinone), and the last electrochemical plant in the United States closed several years ago.

The renewed interest in the electrochemical production of H_2O_2 can be traced to several factors: (a) only dilute solutions are needed and they must be generated on site (see preceding); and (b) the alkaline *cathodic* process offers the potential of significantly reduced power consumption versus the anodic process. This process, however, generates 1 mole of OH⁻ for every mole of hydroperoxide:

$$O_2 + 2 H_2O + 2 e^- \rightarrow HO_2^- + OH^- \qquad E° = -0.065 \text{ V}$$

Since sodium hydroperoxide is a strong base, the net product is a very highly alkaline solution of peroxide. Yet, peroxide is least stable in alkaline media,

$$2 HO_2^- \rightarrow 2 OH^- + O_2$$

which makes the separation of the two species (HO_2^- and OH⁻) difficult and economically uncompetitive relative to the anthraquinone process. This is where

5.4. Indirect Electrolysis of Pollutants

the environmental application niche plays a crucial role. For example, the demand for peroxide as a Cl_2 substitute in the paper and pulp industry actually requires alkaline solutions.

The third factor in the rebirth of H_2O_2 electrosynthesis technology is the availability of new electrode materials (see Section 5.8), where the first 2 e⁻ reduction step is reasonably facile (leading to the peroxide) but the overpotential for the further 2 e⁻ reduction (to water) is high. Interestingly, this property runs counter to the requirements of fuel cell technology where electrocatalysts that assist the direct 4 e⁻ reduction of O_2 are sought.

Nonetheless, the main technical hurdles in H_2O_2 electrosynthesis technology are the relative slowness of the O_2 reduction reaction, and the mass-transport limitations that result from the low solubility of O_2 in the electrolyte solution. The commercialization efforts in this area will be reviewed later in Chapter 8.

The cathodic generation of H_2O_2 on graphite has been used to study the *in situ* oxidative destruction of formaldehyde.[158,159] The maximum current efficiency for the cathodic reduction of O_2 to H_2O_2 was found to be ~93.5%. The oxidation of formaldehyde by the electrogenerated H_2O_2 was governed primarily by the current density, the solution pH, and the process temperature. In a subsequent study,[159] the kinetics of this process was studied and found to be second order in formaldehyde and first order in H_2O_2, with an activation energy of 36 kJ mol⁻¹.

A proton exchange membrane (PEM)-based flow electrochemical reactor has been used for the electrogeneration of H_2O_2.[160] Commercially available membrane electrode assemblies, which are used in fuel cells, were utilized for this study. The same group subsequently has reported[161] the simultaneous synthesis of ozone and H_2O_2 within the same reactor. Thus, deionized water was oxidized to form ozone at the anode while O_2 was reduced (as earlier) on the cathode. The authors report that the low current efficiencies (~4.5%) and high operating potential (4.5 V) for ozone evolution render the economics of this process currently rather unfavorable. Somewhat higher current efficiencies (14.6%) have been reported for Fe(III)-doped PbO_2 film electrodes in phosphate buffer by another group.[162] Interestingly enough, lower current efficiencies were observed for *undoped* PbO_2. This result was tentatively explained on the basis of a mechanism involving the transfer of oxygen from hydroxyl radicals adsorbed on Pb(IV) sites [adjacent to Fe(III) sites] to O_2 adsorbed at the Fe(III) dopant sites on the PbO_2 surface.[162]

In general, the anodic generation of O_2 is thermodynamically favored over ozone generation:

$$O_3 + 6\,H^+ + 6\,e^- \rightleftharpoons 3\,H_2O \qquad E° = 1.51 \text{ V}$$
$$O_2 + 4\,H^+ + 4\,e^- \rightleftharpoons 2\,H_2O \qquad E° = 1.23 \text{ V}$$

Strategies for increasing the current efficiency for O_3 generation include one or more of the following: (a) choice of anode materials with large overpotential for the O_2 evolution reaction; (b) application of a large anodic current density to achieve a large positive electrode potential; and (c) addition of an adsorbate, such as F^-, BF_4^-, or PF_6^-, to block the O_2 evolution mechanism. Thus, efficiencies as high as 21% for β–PbO_2 in 2 M HPF_6 (750 mA cm^{-2}),[163] 35% for GC in 7.3 M HBF_4 (600 mA cm^{-2}),[164] and 45% for GC in 62 wt.% HBF_4 (200 mA cm^{-2})[165] have been reported.

Interestingly enough, the electrogeneration of H_2O_2 via the 2 e$^-$ reduction of O_2 warrants cell and process design considerations that run counter to the requirements for fuel cell technology. In the latter, 4 e$^-$ (deep) reduction of O_2 (to H_2O) is desired to achieve higher power delivery by the cell. Therefore, unlike in the O_3 case, a *lower* current density is desirable for H_2O_2 generation. Further, rather expensive catalysts such as Pt, which promote the 4 e$^-$ reduction pathway, are not required rendering the overall process more attractive from an economical perspective.

5.4.3. Fenton Reaction Chemistry

In 1894, Fenton reported that ferrous ion strongly promotes the oxidation of maleic acid by hydrogen peroxide. Subsequent work has shown that the combination of H_2O_2 and a ferrous salt, "Fenton's reagent," is an effective oxidant for a wide variety of organic substrates.[166] The Fenton reaction involves the following sequence of major steps:

$$Fe^{2+} + H_2O_2 \rightarrow Fe^{3+} + OH^- + {}^{\bullet}OH \quad (5.20)$$

$$Fe^{2+} + {}^{\bullet}OH \rightarrow Fe^{3+} + OH^- \quad (5.21)$$

$$H_2O_2 + {}^{\bullet}OH \rightarrow HO_2^{\bullet} + H_2O \quad (5.22)$$

$$Fe^{2+} + HO_2^{\bullet} \rightarrow Fe^{3+} + HO_2^- \quad (5.23)$$

Of these, only Reaction (5.20) is desirable from the point of view of ${}^{\bullet}OH$ production. On the other hand, reactions such as (5.21)–(5.23) are H_2O_2 consuming, and thus a substantial portion of the theoretical oxidation equivalent of the H_2O_2 is unavailable for oxidation of the organic substrate.[167]

A further difficulty is that Fe^{3+} accumulates in the system via Reactions (5.20), (5.21), and (5.23) and the Fe^{2+} ion catalyst therefore is not efficiently regenerated. This is also because of the relatively sluggish nature of Fe^{2+}-regenerating reactions, including

$$Fe^{3+} + H_2O_2 \rightarrow Fe^{2+} + HO_2^{\bullet} + H^+ \quad (5.24)$$

$$Fe^{3+} + HO_2^{\bullet} \rightleftharpoons Fe^{2+} + H^+ + O_2 \quad (5.23a)$$

Reactions such as (5.24) also consume H_2O_2.

Photochemical[168] or electrochemical[169] regeneration of Fe^{2+} offers a solution to this problem. Thus, photolysis of iron(III) complexes is known to occur via the following scheme:

$$Fe(III)L_n + light \rightarrow Fe(II) + nL$$

where L is a ligand. Ligands such as citrate and oxalate have been tested.[168,170] Equilibrium speciation computations have shown[168] that oxalate is a good choice as ligand in acidic solutions and that citrate is preferable in neutral aqueous media. The photochemical approach has been termed the *photo-Fenton reaction*.[168]

Most of the early investigations on the Fenton reaction were motivated by possibilities in preparative organic chemistry.[166] Application of this reaction to water treatment is a relatively later development.[171] Recent interest has focused on implications in natural and biological media, since both Fe(II) species and H_2O_2 are ubiquitously present in these environments. For example, many human diseases are traceable to radical generation reactions in the tissue, and so forth. Abundant evidence exists that both Fe(II) and H_2O_2 are formed in sunlit regions of surface waters as well as in atmospheric water drops. Under the conditions typical of most natural water, ferric ions exist mostly as hydroxy complexes and insoluble oxide-hydroxide phases. The photo-Fenton route may be a major pathway for the generation and subsequent oxidation of Fe(II) and thus may be an important route for organic oxidations mediated by ·OH radicals in the environment.

Instead of ferrous ions, ferric salts have also been used *in the dark* in conjunction with H_2O_2 for the oxidation of formaldehyde.[172] A wide variety of iron salts (and even bovine hemin in spite of its low solubility in water) resulted in formaldehyde oxidation regardless of the iron oxidation state. The term *Fenton-like reaction* has been applied in this case,[172] and the underlying mechanism is unclear at the time of this writing. The original Haber–Weiss mechanism (Reactions (5.20), (5.21), (5.23), (5.23a), and (5.24)) where iron oscillates between the ferrous and ferric states, has been challenged.[173]

The classical Fenton or the Fenton-like reaction may be a particularly viable alternative to the treatment of polluted soils and subsurface water reservoirs, where photochemical or photocatalytic remediation approaches (Chapter 6) are precluded and bioremediation may be prohibitively expensive.

5.4.4. *Approaches Relying on pH Manipulation*

In Section 5.3, the pH changes accompanying water electrolysis were shown to be useful for the electroreduction of Cr(VI) and its subsequent immobilization as $Cr(OH)_3$. Further examples include the reduction of water hardness by shifting the equilibrium to the formation of CO_3^{2-} ions to promote the coalescence and precipitation of $CaCO_3$ crystals:

$$HCO_3^- + OH^- \rightarrow CO_3^{2-} + H_2O$$
$$CO_3^{2-} + Ca^{2+} \rightarrow CaCO_3$$

This calcium ion removal can also be enhanced by an applied electrical field, which provokes the coalescence of colloidal particles.[174]

The production of H_3O^+ on an anode during the electrolysis of water can be used to acidify the anolyte, as will be illustrated. This effect has been used, for example, for the neutralization of alkaline water streams[175] and promoting the controlled coagulation of proteins. Strict control of acidity is required to avoid protein denaturation.[176] To reduce the organic load to the furnace in a Kraft pulp and paper process, an electrolytic cell can be used to decrease the pH of the black liquor and precipitate lignin (composed largely of phenolic derivatives), which is then removed before loading the furnace.[177]

5.5. ELECTROFLOTATION, ELECTROCOAGULATION, AND ELECTROFLOCCULATION

Electroflotation involves the electrolytic production of gases (e.g., O_2, H_2) that can be used to attach pollutants (e.g., fats and oils) to the gas bubbles and carry them up to the top of the solution where they can be more easily collected and removed. *Electrocoagulation* refers here to the electrochemical production of destabilization agents that bring about charge neutralization for pollutant removal. *Electroflocculation* is the electrochemical production of agents that promote particle bridging or coalescence.

An increase in bubble size has been observed at cathodes with smaller overvoltages for hydrogen evolution (e.g., Pd, W, Ni) as compared to high overvoltage materials (e.g., Cu, Sn, Pb). Pulsed electrogeneration of bubbles for electroflotation yields optimum-sized bubbles that are independent of solution conditions.[178] Highly durable anodes (estimated lifetime, ~5 years) can be made for O_2 production for the electroflotation of waste water.[179] If hydrogen gas is produced on the cathode and Fe or Al are used as anodes, the Fe^{n+} or Al^{3+} ions resulting from the oxidation of the anode can react with the OH^- ions produced at the cathode and yield insoluble hydroxides that will precipitate, adsorbing pollutants out of the solution (e.g., toxic metal ions like Cr (VI), or Hg) and also will contribute to coagulation–flocculation processes, as discussed later (Figure 5.12).[180] Since Fe(III) hydroxide precipitates more easily than Fe(II)

5.5. Electroflotation, Electroagulation, and Electroflocculation

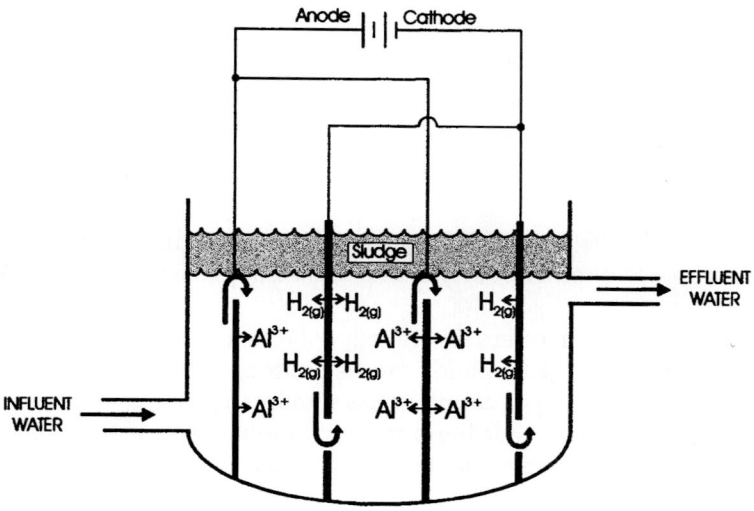

FIGURE 5.12. Schematic diagram of an electrocoagulation process using soluble anodes. (Adapted from Vik et al. [180])

hydroxide, air or oxygen may be injected into the solution to oxidize Fe(II) to Fe(III) and aid in the flotation process.[181] The successful simultaneous removal of oil and heavy metals[182] by a combined chemical precipitation (hydroxides, sulfides)–air flotation process is, in a way, a parallel to its electrochemical counterpart. These electrocoagulation–electroflotation processes can be combined, for example, with microfiltration, to yield irrigation-quality water; in this case, removal of 99% turbidity, 77% COD, and 98% suspended solids can be achieved.[183] Another successful combined process involves the addition of a chemical coagulant, $Fe_2(SO_4)_3$ and application of DC voltage; here, a decrease in the stabilizing zeta potential enables the Van der Waals attraction forces to predominate; oil removal up to 96% has been achieved.[184,184a]

In the case of iron or steel anodes, two mechanisms for the production of the metal hydroxide have been proposed[185–187]: In Mechanism 1,

Anode: $4\ Fe(s) = 4\ Fe^{2+}(aq) + 8\ e^-$
Chemical: $4\ Fe^{2+}(aq) + 10\ H_2O(l) + O_2(g) = 4\ Fe(OH)_3(s) + 8\ H^+(aq)$
Cathode: $8\ H^+(aq) + 8\ e^- = 4\ H_2(g)$
Overall: $4\ Fe(s) + 10\ H_2O(l) + O_2(g) = 4\ Fe(OH)_3(s) + 4\ H_2(g)$

In Mechanism 2,

Anode: $Fe(s) = Fe^{2+}(aq) + 2\,e^-$
Chemical: $Fe^{2+}(aq) + 2\,OH^-(aq) = Fe(OH)_2(s)$
Cathode: $2\,H_2O(l) + 2\,e^- = H_2(g) + 2\,OH^-(aq)$
Overall: $Fe(s) + 2\,H_2O(l) = Fe(OH)_2(s) + H_2(g)$

The insoluble Fe(II) or Fe(III) hydroxides once produced, remove the pollutant via processes rather poorly characterized at present.

Electrocoagulation offers several advantages, including the following:
(1) The smallest charged colloids can be treated, since they can move more easily than their larger counterparts within an electrical field; this facilitates coagulation.[188] Furthermore, such motion avoids the need for mechanical agitation, which (if uncontrolled) may destroy precipitates as soon as they are formed.
(2) The amount of required chemicals is much lower (on the order of 1/10).[180] For example, in conventional lime-neutralization processes, water hardness is increased.
(3) As mentioned previously, a smaller amount of sludge is produced, due to the higher content of dry solids. For example, conventional addition of ferric chloride followed by lime or sodium hydroxide produces up to 30 liters of sludge for every liter of removed oil.[189] In addition, the sludge produced by the electrochemical treatment is more hydrophobic, which leads to more compact residues; this also leads to shorter decantation times.
(4) No mixing of chemicals is required.
(5) The durability of the electrodes translates to low "down times" for maintenance or replacement.
(6) Organic matter removal (including nonbiodegradable organics) is more effective; this facilitates subsequent biological treatment.
(7) Coagulant dosing as well as required overpotentials can be easily calculated and controlled.
(8) These processes are suitable for small water treatment plants.
(9) A pH increase is normally observed, which aids in the removal of heavy metal ions by their precipitation as hydroxides[190] or by adsorption into other flocs or precipitates (e.g., in acid mine waste water).
(10) Often, pH control is not necessary, unless this parameter acquires extreme values; this facilitates process design and operation (e.g., in oil removal, lignin destruction).
(11) High current efficiencies (~90%) can be achieved in well-designed systems.
(12) Short contact times are required. For example, an 8 min treatment suffices for decolorization of typical textile waste water.[185]

5.5. Electroflotation, Electroagulation, and Electroflocculation

(13) Operating costs are much lower when compared with most of the conventional technologies. We shall explore commercial systems in more detail in Chapter 8.

The major challenges for these processes are these:
(1) The production of H_2 at the cathode may prevent precipitated matter from settling properly.
(2) The concentration of aluminum or iron ions in the effluent will most likely be increased (for example, in oil removal, up to 550 mg/liter of dissolved solids can be added by this process; nevertheless, by comparison, chemical coagulation methods generally add 2,000–3,000 mg/liter of dissolved solids).[189] Careful pH control would be needed if Al or Fe contents are outside of the regulatory limits.
(3) The produced insoluble hydroxides may agglomerate between the electrodes, hampering their further production. Alternatives to solve this problem include the use of a moving anode and the promotion of high turbulence by pumping, gas sparging, or mechanical agitation. Fluidized bed electrodes have been used advantageously to clean electrode surfaces by a mild erosive action.[191]
(4) These direct current processes are frequently accompanied by anode passivation and sludge deposition on the electrodes; to prevent this, alternating current with controlled reverse pulses (and with the addition of anode activating ions such as chlorides) has been found to substantially lower the required anodic polarization and facilitate continuous renewal of the electrode surface by redeposition of the metal oxidized in the forward pulse; in addition, hydrogen evolution promotes bond weakening between the sludge and the electrode, enabling its loosening and separation. This procedure has the added benefit of doubling the time required for electrode replacement.
(5) Investment costs are relatively high, although operating costs tend to be smaller than with other techniques.

Other external factors that promote coagulation–sedimentation of the electrogenerated flocs include magnetic fields, that are said to promote sedimentation without interfering in the process, and the addition of small amounts of coagulation–flocculation promoters (e.g., polymer flocculants such as aluminum chloride).[185,186]

<u>Cleanup of Oil-Water Emulsions</u>. Voluntary or involuntary spilling of oil in different kinds of waters is an issue of environmental concern. Oil (as well as hydrocarbons in general) can significantly alter the properties of water and produce optical changes such as in color and opacity (with the concomitant

absorption of the light necessary for photobiological cycles), a negative esthetic impact, foul smell, bad taste, changes in viscosity, conductivity, and the like. Unfortunately, many sources of water-polluting oil, such as mills, refineries, off-shore platforms, cutting machines, oil transportation, distribution, and storage facilities, undergo spills yielding several millions of tons that end up in water reservoirs and the sea every year; about half of this amount contaminates fresh water. Industrial effluents have been found to contain up to 40,000 mg/liter of oil. Furthermore, we use as an average almost 4 liters (about 1 gallon) of hydrocarbons per person each day in the world.[184,192]

Oils can spread and disperse in waste water mainly in the following manners: (a) as a separate phase (supernatant layer), (b) as colloidal droplets (emulsions or coagulates), (c) as dissolved particles, or (d) as adsorbed species on suspended particulate matter. An A/B emulsion consists of one liquid phase (A; for example, oil) dispersed in a continuous phase (B; for example, water) and can be as fluid as water or as viscous as solid fat. It can be produced either in the absence or presence of surface-active substances or finely divided clays; in the former case by the creation of turbulent conditions, and in the latter by the use of asymmetrical molecules (emulsifiers) with hydrophilic and lipophilic segments, which can drastically lower the oil–water interfacial tension and thus facilitate the formation of small droplets (which yield a larger oil–water interfacial area) or else by the addition of finely divided clays (e.g., cocoa, gums).[192]

The main interacting forces between emulsion droplets can be both attractive (Van der Waals) and repulsive (electrostatic) in nature. Oil colloidal particles generally have negative charges on their surface due to frictional charging or ionization of the hydrophilic carboxylic groups. These charges are mainly responsible for the repulsion forces that prevent the particles from colliding and having a chance to flocculate or coalesce. Positively charged cations when added, neutralize the colloids and enable flocculation or coalescence to occur. Organic and inorganic coagulants such as polyamines, H_2SO_4, $AlCl_3$, or $Fe_2(SO_4)_3$ are usually employed. Unfortunately, these chemical addition processes require relatively large amounts of chemicals as well as long periods of time for their handling, produce considerably large volumes of sludge, and are unable to coagulate the smallest colloids.[183] An additional effect produced by the Fe^{3+} ions is the oxidative destruction of the emulsifying agent.[189]

The addition of (polyvalent) metallic ions with high charge densities can also be effected by the controlled electrodissolution of soluble metal anodes like Fe, Al, and steel, as pointed out previously. Here, the potential at the anode must be controlled; if this anode becomes too positive it can become passivated and its dissolution will stop, as can be deduced from the Pourbaix diagram for the M–H_2O system. Careful control of the applied potential and addition of chloride ions can prevent this passivation effect. On the other hand, if the applied potential is not positive enough, the metal would be in the passivation region of its Pourbaix diagram and dissolution will not be feasible.[189]

5.5. Electroflotation, Electroagulation, and Electroflocculation

As mentioned earlier, the corresponding cathodic reaction is normally the reduction of H_3O^+ at an inert cathode. Here, H_2 gas evolves and carries upward flocs or precipitate that provide a large surface area for the adsorption of oil particles and other impurities that may be contained in the aqueous solution. The concomitant pH increase facilitates the flocculation or precipitation of hydroxides of the metallic ions as sludge. The oily layer and the solid sludge produced in this electrocoagulation–electroflotation process can then be mechanically removed. Oil removal efficiencies of up to 96–99% can be achieved.[180-184,189] Dioxygen-based electroflotation of waste water containing dispersed peptides and oils decreases this content by ~99.5%.[182] A simple laboratory demonstration of this technology has been recently described for pedagogical purposes.[193]

Electric fields have also been used for the dewatering of water-in-oil dispersions (*electrical dehydration*). Polarization and alignment of the water droplets result in their coalescence. On the other hand, an electrophoretic effect favors charging the particles, and they can be set in motion by the applied electric field. Reduction of water content by a factor of 200 has been reported.[194]

<u>Dye Removal</u>. Dye fixation effectiveness in the textile industry can be sometimes as low as 60%; this leaves considerable amounts of dyestuffs in waste water. Dyestuffs are also contained in large amounts in printing and dyeing waste water together with significant amounts of suspended solids, dispersing agents, leveling agents, wetting agents, and trace metals. Some of these compounds may be toxic, carcinogenic, mutagenic, or teratogenic in various living organisms. Large amounts of such wastes are produced (e.g., in Canada this number has been estimated to be 8×10^8 liters per day[187]). Usual methods of decolorization include biological, physical, and chemical treatment. Due to the variable contents of the waste water, such methods encounter serious difficulties in meeting environmental discharge requirements. By virtue of their application requirements, dyes are inherently stable to photodegradation as well as microbial attack. Chemical oxidation by chlorine, hypochlorite, and ozone are among the most effective decolorizing methods; however, they are generally rather expensive. Electrochemical approaches such as electrocoagulation and electroflotation appear to be promising.

Aluminum and iron anodes have been used for the electrogeneration of hydroxide flocs that adsorb and remove disperse, metal-complex, and acid dyes and other pollutants (such as metal ions or compounds) from the waste water. The aluminum-based process seems to remove dyes by this adsorption mechanism (e.g., disperse anthraquinone, azo, and quinoline type dyes) without chemically altering their structure, whereas iron anodes were found to additionally yield other degradation products, possibly due to the action of reducing Fe^{2+} ions.[195] Likewise, the iron process seems to be more effective for reactive dyes, whereas the aluminum process is superior for disperse dyes.[196] Addition of oxidizing agents favors the formation of iron (III) hydroxide, which is more effective than iron(II) hydroxide for color removal.[186] Nevertheless, high decolorization is achieved in all these cases as well as a considerable lowering of toxicity, although conditions remain to be found for the simultaneous removal of all the dye additives.

In additon to the pollutant removal mechanisms with electrogenerated metal hydroxides discussed previously, chemical modification may occur here by direct or indirect electron transfer to or from the electrodes. For example, C=C and N=N bonds are susceptible to catalytic hydrogenation or reduction in the presence of H_2 and a catalyst.[186] In fact, a combined coagulation–electroflotation–electrolytic reduction and oxidation method using inert graphite electrodes has been reported to give a COD removal up to ~90%.[197] Advantages include no consumption of electrodes, no addition of chemicals into the solution, smaller doses of coagulating agents needed, and smaller amounts of sludge produced. The nature of the oxidation–reduction products as well as their possible effects on human and animal health remain to be established.

<u>Other Uses.</u> In addition to the uses just discussed, opportunities for the electroflotation–electrocoagulation–electroflocculation processes include:
- Phosphorus removal:[198,199] Eutrophication of rivers and lakes is known to be promoted by high phosphorus content in water. Phosphate can be removed from aqueous media by anodic dissolution of Fe or Al to yield insoluble $AlPO_4$ or $FePO_4$ with simultaneous phosphate adsorption on the corresponding insoluble hydroxides, as discussed already. Phosphorus removal >90% has been obtained. Phosphates and organic nitrogen have been simultaneously removed by this method. Furthermore, this can be combined with chlorine generation in a purification–disinfection system.
- Manufacture of filters by deposition of electrolytically produced metal hydroxides on particulate materials such as sawdust:[191] The gelatinous hydroxides by themselves have the tendency to decrease permeation, percolation, and liquid passage. However, when deposited as earlier, they can be used for a variety of applications, including the removal of phenol, dyes, and pigments, dissolved metals, coffee, ammonia, carbonate sludge, and other colloidal matter. In addition to the Fe and Al hydroxides discussed, hydroxides of Mg, Ba, Sr, Ni, Cu, Cd, and Mn have been used for this application.
- The treatment of waters containing food and protein wastes, synthetic detergents, and fluorides.[196]

5.6. ELECTROCHEMICAL TREATMENT OF GASEOUS POLLUTANTS

As discussed in Chapter 1, the composition of the atmosphere has changed by the incorporation of trace level constituents from industrial operations, automobile exhaust, and other anthropogenic and non-anthropogenic sources (e.g., volcano eruptions). Currently used methods for the treatment of gases include the following:
(1) Physical methods: Gases are adsorbed on solid surfaces (e.g., activated carbon), selectively filtered, electrostatically precipitated, selectively dissolved, and so forth.
(2) Chemical methods: They are incinerated, dissolved in reactive solutions (e.g., oxidation in a ClO^- solution, neutralization of an acid gas stream in a basic solution), adsorbed through chemical reactions (e.g., NH_3 complexation of Cu^{2+} adsorbed on activated carbon), and the like.
(3) Biological methods: Generally speaking, gases are easier to biodegrade than liquids or solids due to their molecular dispersion. Several microbial processes have been utilized in biofiltration for pollutant gas treatment; these include a large amount of rapidly or slowly biodegradable, volatile organic, or inorganic compounds (VIC). For example, for H_2S remediation the following have been utilized: *Thiobacillus thioparus, Beggiatoa, Thiobacillus denitrificans, Chlorobium versutus, Thiobacillus neopolitanus, Thiobacillus thiooxidans,* and others.[200,201]
(4) Electrochemical methods: Polluting gases must generally be transferred by absorption or reaction to the liquid phase (normally aqueous solution) before they can undergo electrochemical oxidation or reduction. This conversion can be effected in two absorption modes: the gas is directly absorbed in an electrochemical cell for treatment (inner-cell process), or the gas is absorbed in a separate reservoir and then transferred to the electrochemical cell (outer-cell process).[202,203] Subsequently, the dissolved pollutant can undergo different reactions (see Figure 5.13),[204] such as homogeneous electron transfer from a dissolved catalyst, $C^+_{(solv)}$, which can then be regenerated either by heterogeneous electron transfer at an anode (path 1) or by homogeneous electron transfer to a dissolved metal ion, $M^{(n+1)+}$ (path 2). Another reaction path (path 3) involves complexation of the dissolved pollutant followed by oxidation of the formed complex, (Red $M)^{n+}$ with $C^+_{(solv)}$ to produce an innocuous (or at least less polluting) substance, O with the concomitant regeneration of M^{n+}. $C^+_{(solv)}$ is then regenerated from $C_{(solv)}$ at the anode of the electrochemical cell. Examples for the choice of M^{n+} include complex-forming transition metal ions (e.g., Cu^{2+}, Pd^{2+}) and for the couple C^+/C, include redox couples with high standard reduction potentials; for example, $Ag^{2+/+}$, $Co^{3+/2+}$, Br_2/Br^-, $Cr_2O_7^{2-}/Cr^{3+}$, VO_2^+/VO^{2+}, or MnO_4^-/MnO_2.

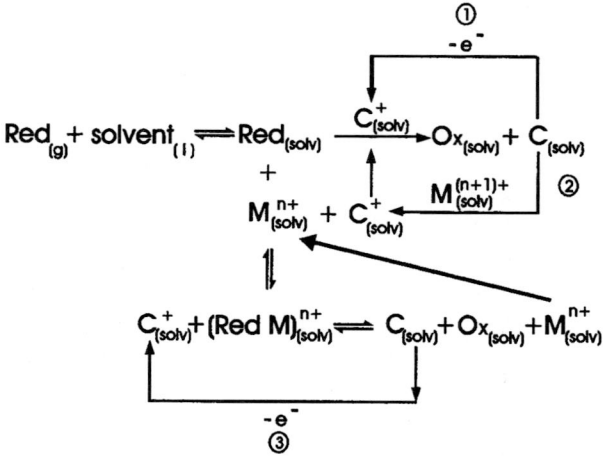

FIGURE 5.13. Pathways for the indirect oxidation of dissolved gaseous pollutants.

The separation-concentration of dissolved gases can also be achieved in some cases by the application of an electrical field across a membrane. Here, the charged species present migrate by virtue of the difference in electrochemical potentials. This produces a separation-concentration effect that is governed by the Nernst equation[205]:

$$E = E° + \frac{RT}{nF} \ln \frac{a_{ia}}{a_{ib}}$$

where a and b refer to the two sides of the separation membrane. Since one is dealing with the same substance on both sides, $E° = 0$ and the ratio of activities (approximated here by partial pressures) is given by the expression

$$\left(\frac{p_{ia}}{p_{ib}}\right) = \exp\left(\frac{nF}{RT}E\right)$$

Hence, for $T = 1{,}000$ K and $n = 1$, a concentration factor of the order of 10^5 can in theory be obtained by the application of a potential difference of just 1 V. In practice, however, charge and mass-transport limitations take up a good portion of this potential. Nonetheless, since the partial pressure ratio is an exponential function of E, even a small potential gradient can yield a large concentration ratio.

5.6. Electrochemical Treatment of Gaseous Pollutants

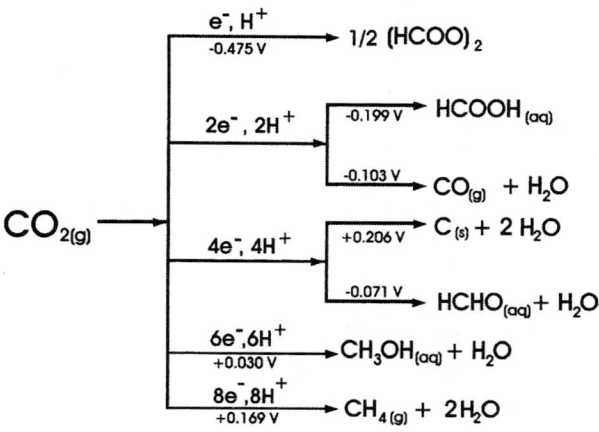

FIGURE 5.14. Pathways for the electroreduction of CO_2. (Adapted from Sullivan et al.[229])

This process has the advantage that a large pressure difference is not required. Reasonably low temperatures and low voltages can also be employed, and high concentration ratios can be obtained. In addition, judicious selection of the applied potential may prevent undesired side reactions.[206] The main disadvantages include the need to first dissolve the gas to be concentrated to produce charged species; each process application is very specific, and favorable economics are achieved only for low pollutant levels.[205] It has been applied under these conditions to SO_2, H_2S, and halogenated acids and the economics have been shown to be favorable versus conventional treatment systems.[207–209]

Electrochemical transformation or removal of carbon, sulfur, and nitrogen oxides and of hydrogen sulfide from gas streams will be discussed in turn.

5.6.1 CO_2 Reduction

Figure 5.14 summarizes the reduction routes discussed earlier in Chapter 3. The mechanism for CO_2 reduction has not been completely understood, but it is clear that it varies with experimental conditions; that is, electrode material, CO_2 partial pressure, solution pH, temperature, applied potential, current density, and nature of the electrolyte. For example, for Pt electrodes with different exposed planes in H_2SO_4, proposed adsorbed intermediates include linear CO, bridged CO, linear COOH and bridged COH,[210] whereas for Cu and Ni electrodes proposed intermediates include CO, -HCO, =CH_2 and H-CO_2 cluster complexes[211–214] (Cu presents high yields for CO_2 reduction to hydrocarbons and alcohols since strong CO adsorption prevents H_2 evolution from occurring[211,212]). The species $CO_2^{\cdot-}{}_{(ads)}$ and $HCO_2^{-}{}_{(ads)}$ seem to be generally accepted as intermediates in the first steps of CO_2 reduction at high hydrogen overpotential cathodes.[213]

The main reduction product in aqueous solutions under normal conditions is usually formic acid or the formate ion, although faradaic efficiencies close to 100% have been reported also for methanol formation. CO_2 reduction to hydrocarbons (e.g., CH_4, C_2H_4, C_2H_6) can be accomplished at many metal electrodes[215,216]; Cu has been shown to be by far the best metal in this regard. The mechanistic and kinetic aspects of the electroreduction of CO_2 in aqueous solutions have been discussed for a wide variety of metals.[217-219] For example, it is known that adsorbed CO is formed during this process on Cu and Ni electrodes.[220] Different factors such as CO_2 partial pressure, solution pH and buffer composition, electrode potential, and solution hydrodynamics play important roles in mechanistic aspects and in product yields. A systematization of the effect of the metallic electrodes upon the nature of the reaction products has led to the proposal of a reaction scheme that considers the differences in the adsorption energies for radical intermediates.[220] Bulk Cu- and Ag-based alloys have been shown to provide electrocatalytic sites to reduce the large overpotentials normally required for CO_2 reduction as well as to facilitate the achievement of high selectivities to form desired reaction products. Alloys tested include Cu–Ni, Cu–Sn, Cu–Pb, Cu–Zn, Cu–Cd, Cu–Ag, Ag–Cd, Ag–Sb. The effect of some of these alloys depends on the crystalline phase used.[220]

Progress in the electroreduction of CO_2 has been reviewed elsewhere.[220-223] Reduction in aqueous media suffers from problems due to the interference from proton reduction and from the low concentrations of CO_2 available in aqueous solution (on the order of 0.033 M under 1 atm at 25°C), which lead to severe mass transfer limitations when high current densities are desired. To overcome the problem of hydrogen evolution interference, materials with high hydrogen overpotential are used (e.g., Pb, Sn, In).[224,225] On the other hand, to overcome the problem of low CO_2 concentration in aqueous solutions, high pressures (up to 60 atm) or low temperatures (ca. 0 °C) or both have been employed[226-228]; substantially higher faradaic efficiencies are obtained due primarily to increased availability of CO_2 on the electrode surface, which can also be enhanced by stirring. For example, at 60 atm of CO_2 its concentration is 1.17 M; that is a 35-fold increase from that at ambient pressure.[228]

This increased CO_2 availability also changes the main reduction products from H_2 to hydrocarbons to CO or HCOOH, depending on current density; that is when the current density is not large enough as to completely reduce all the CO_2 arriving at the electrode, the main products are CO or HCOOH or both, whereas in the opposite case, hydrocarbons (or H_2) are the main products.[227] This is summarized in Figure 5.15. Similarly, alcohol formation is promoted in a supercritical CO_2 electrolyte due to the large concentration of CO_2.[220]

5.6. Electrochemical Treatment of Gaseous Pollutants

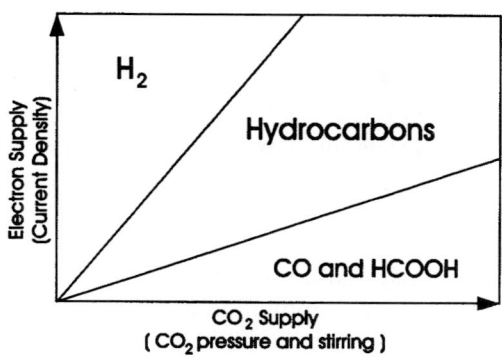

FIGURE 5.15. Product formation as a function of CO_2 supply and electron supply. (Adapted from Hara et al.[227])

Cyclic methods involving selective removal of CO_2 (e.g., molecular sieves, polymer bound amines, dissolution in basic solutions) have been proposed as well as the use of CO_2 carrier molecules, RO (e.g., quinones) that are sequentially reduced and oxidized.[229] This cycle is possible due to the high basicity of RO⁻, which strongly binds CO_2 and removes it from the atmosphere. When $ROCO_2^-$ loses its negative charge on the anode, CO_2 is no longer retained and is desorbed in a high pressure compartment, thereby closing the concentration loop. Further work involves the search for compounds that are not capable of undergoing interfering reactions with ambient dioxygen.[229] A related route involves the permanent electrochemical fixation of CO_2 in organic compounds. For example, the electrolysis of 1, 4-benzoquinone in the presence of CO_2 yields 2, 5-dihydrobenzoic acid.[230,231]

Nonaqueous solvents have also been employed (e.g., acetonitrile, dimethylsulfoxide)[232,233] as well as electrode materials other than metals, including TiO_2 and other oxides.[220,234,235] CuO–ZnO mixed oxide selectively yields ethanol.[220]

In many cases, metal complexes are used as electrocatalysts. Here, two paths have been postulated for the reduction of CO_2 to CO[220,229]: (a) activation of CO_2 by coordination to a metal center and reductive disproportionation into CO and CO_3^-, and (b) insertion of CO_2 into a metal hydride bond and decomposition into CO and a metal hydroxide. Work has to be done to find ways to activate this CO moiety for the further two-electron steps required for the production of useful chemicals such as HCHO, CH_3OH, or CH_4. The picture is much more complicated when attempting to explain the possible reaction paths for the reduction of CO_2 to CH_4; a matrix containing 12 possible intermediates has been proposed.[220,229] Other reduction schemes show routes leading to various gaseous and condensed species.[236–239]

Candidates for promoting CO_2 reduction in aqueous media also include nickel cyclam (cyclam = 1,4,8,11-tetraazacyclotetradecane), pyridinium ions, and metallated porphyrins or phthalocyanines.[220,240–243] Catalysts based on transition metal complexes (e.g., Mo, Co, Fe) have also been immobilized in polymer coatings (e.g. polypyrrole, Nafion, Prussian blue-polyaniline films) on a support electrode (e.g., glassy carbon or Pt) surface and have been shown to yield CO_2 reduction products such as methanol, ethanol, acetone, and lactic acid.[220,243,244] The use of CO_2 reduction in a CO_2–H_2 fuel cell has been proposed and involves a homogeneous catalyst (a Fe or Co complex) and an electron mediator (a Fe complex) for the production of electricity and methanol.[245]

Other promising approaches include the following:
- Microcrystalline metal structures (Cu-Sn and Cu-Zn alloys) have been shown to promote CO_2 reduction to HCOOH and to CO (with faradaic efficiencies of ~80%, depending on alloy composition).[246]
- Metals dispersed in gas diffusion electrodes dramatically improve gas-phase CO_2 reduction as compared to their nondispersed counterparts using electrode fabrication technology developed in the fuel cell area.[220,247] Copper is the only metal studied so far that shows high yields for CO_2 reduction to CO as well as for the subsequent reduction of CO to hydrocarbons. Some metals (e.g., Ag) show good yields for the first reaction but not for the second, whereas others (e.g., Pb) show the opposite behavior. In light of this, bimetallic Cu (e.g., Cu–Ag, Cu–Pb) alloys have been designed and shown to give better results than Cu alone.[248] Copper-containing perovskite-type catalysts yield n-alcohols (C1, C2, and C3) at ambient temperature.[249,250] Hydrogen-storing materials (e.g., Pd) provide hydrogen atoms for the reduction of CO_2 to CO and HCOOH.[220]
- Anion exchange solid polymer electrolytes can be used to produce CO from CO_2 at high current efficiencies (on the order of 85%).[251]
- Recycling of CO_2 using CH_4 and a proton conductor solid electrolyte (e.g., $SrZrO_3$ or $CaZrO_3$) has been proposed in the following modes of an "electrochemical hydrogen pump."[252] For a solid oxide fuel cell (SOFC),

Ni anode:	$CH_4 + CO_2 = 2\,CO + 4\,H^+ + 4\,e^-$
Pt cathode:	$O_2 + 4\,H^+ + 4\,e^- = 2\,H_2O$
Net reaction:	$CH_4 + CO_2 + O_2 = 2\,CO + 2\,H_2O$

5.6. Electrochemical Treatment of Gaseous Pollutants

This cell has been shown to provide stable electric power. For reforming CH_4,

Ni anode:	$CH_4 + CO_2 = 2\,CO + 4\,H^+ + 4\,e^-$
Pt cathode:	$4\,H^+ + 4\,e^- = 2\,H_2$

Net reaction: $CH_4 + CO_2 = 2\,CO + 2\,H_2$

Hydrogen is electrochemically pumped from anode to cathode; the required reaction temperature can be substantially reduced due to the fast kinetics of this reaction. For oxidative coupling of CH_4 with CO_2,

Ag anode:	$2\,CH_4 = C_2H_6 + 2\,H^+ + 2\,e^-$
Pt cathode:	$CO_2 + 2\,H^+ + 2\,e^- = CO + H_2O$

Net reaction: $2\,CH_4 + CO_2 = C_2H_6 + CO + H_2O$

Dimerization of CH_4 is promoted by pumping hydrogen from anode to cathode. Photoelectrochemical methods for the reduction of CO_2 shall be considered in Chapter 6.

5.6.2. H_2S Treatment

This gas is produced in large amounts during the desulfurization of fossil fuels and in coal gasification and liquefaction processes; power plants usually necessitate its removal. In addition, this gas is frequently released to the atmosphere above the waste water in sewers, where it can accumulate in specific spots and then be biologically oxidized to sulfuric acid, producing corrosion in pipes. The gas used in combustion engines must contain only a small amount of H_2S, otherwise corrosion will likely be a problem. Likewise, molten carbonate fuel cells cannot tolerate concentrations of H_2S higher than 1 ppm.[208] H_2S is usually treated by the Claus process[205,253]:

$$H_2S + 3/2\,O_2 \rightarrow SO_2 + H_2O$$
$$H_2S + SO_2 \rightarrow 2\,S + H_2O + 1/2\,O_2$$

However, this process has the following disadvantages[254,255]: (1) hydrogen is essentially wasted, since H_2O is produced from it; (2) the high temperatures and catalysts required do not offer flexible adjustment to varying concentrations of H_2S; (3) pretreatment for the separation of hydrocarbons and H_2 is re-

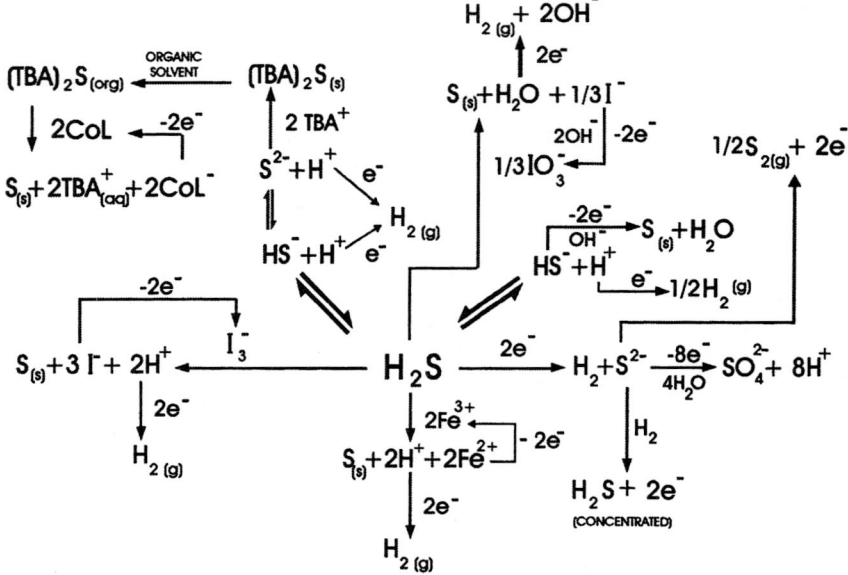

FIGURE 5.16. Electrochemical processes for the treatment of H_2S.

quired; and (4) post-treatment is required since the Claus process converts only 90–98% of the initial H_2S content.

In view of this, electrochemical alternatives have been proposed and tested that eliminate the need for a Claus plant, as can be seen in the summary scheme of Figure 5.16.[208,209,253–261] H_2S removal of ~99% can be achieved in most cases.

All the species in Figure 5.16 are hydrated unless otherwise specified [L = tetraneopentoxyphthalocyanocobalt(III)].[255] In the reaction paths, where redox couple mediators are involved[254–258] (i.e., $Fe^{3+/2+}$, $CoL^{0/+}$, I_3^-/I^-), the cycles are closed by the electrochemical regeneration of the oxidant with the concomitant production of high-purity $H_2(g)$ and of elemental sulfur which can be easily filtered or extracted. In the other reaction paths, sulfide ions can be (a) oxidized to produce gaseous sulfur (using a molten sulfide electrolyte in a porous electrode cell), solid or colloidal sulfur,[207–209,261] or sulfuric acid[258]; (b) reacted with H_2 gas at an anode to produce concentrated H_2S in a true concentration cell[205,262]; or (c) precipitated from the aqueous phase with a large cation (e.g., tetrabutylammonium), forming a compound soluble in the organic phase, where it undergoes oxidation by a Co complex to elemental sulfur.[255] Again, $H_2(g)$ is produced in each case. The HS^- ions in basic media are oxidized to $S(s) + H_2O$;

5.6. Electrochemical Treatment of Gaseous Pollutants 425

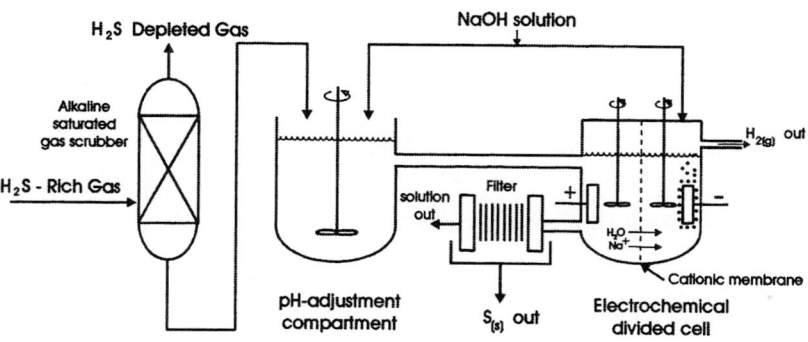

FIGURE 5.17. Schematic flow diagram for the electrochemical production of H_2 and S (outer-cell direct process). (Adapted from Anani et al.[260])

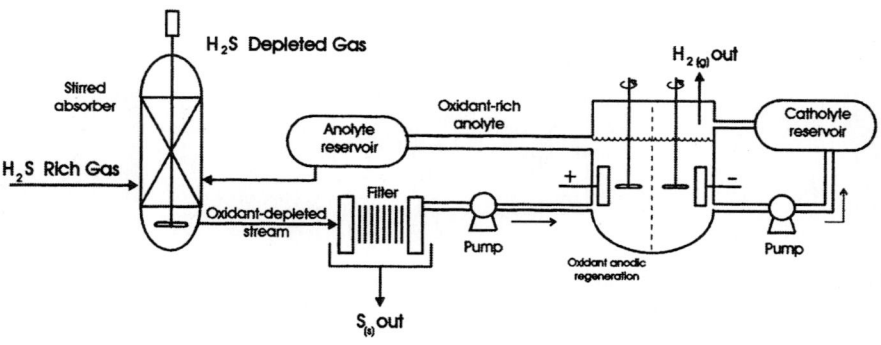

FIGURE 5.18. Same scheme as in Figure 5.17 but for an outer-cell indirect process. (Adapted from Kalina and Maas.[256])

most likely the reaction path involves formation of polysulfide ions (S_n^{2-}) that are further oxidized to $nS^0(s)$.[263] Similarly, another process uses porous, gas diffusion electrodes and aqueous polysulfide electrolyte in a divided cell, where S^{2-} ions are reduced to S_n^{2-}. These ions migrate through the membrane to the anode, where they are oxidized to sulfur.[264] Figures 5.17 and 5.18 show two of these processes in simplified fashion.

An additional electrochemical pathway involves the use of solid oxide fuel cells (yttria or calcium-stabilized zirconia is the solid oxide, W/WS_2 or W/thiospinel is the anode, and $La_{0.89}Sr_{0.11}MnO_3/Pt$ is the cathode), where H_2S is oxidized with air to give sulfur and water at high temperature (ca. 900°C).[253] This solid-oxide fuel-cell (SOFC) approach is schematized in Figure 5.19.

An alternative process is the oxidation of H_2S using vanadium(V) salts and

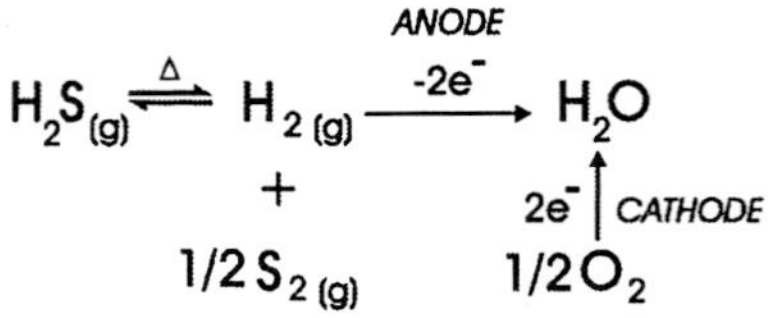

426   Chapter 5

$$H_2S_{(g)} \xrightleftharpoons{\Delta} H_{2(g)} \xrightarrow[\text{ANODE}]{-2e^-} H_2O$$
$$+ \quad\quad 2e^- \uparrow \text{CATHODE}$$
$$1/2\,S_{2(g)} \quad\quad 1/2\,O_2$$

FIGURE 5.19. Electrochemical treatment of H_2S using a solid-oxide fuel cell.

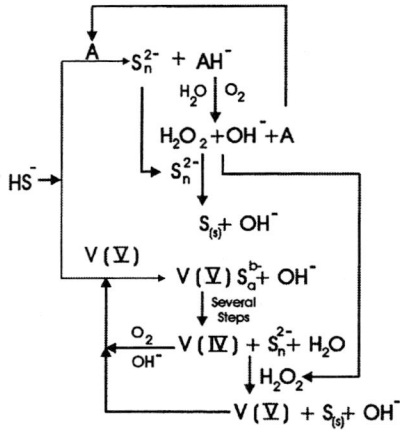

FIGURE 5.20. The British Gas Stretford Process for the treatment of H_2S.

anthraquinone disulphonates (hereby denoted by A), which amounts to the oxidation of HS⁻ by ambient oxygen mediated by redox species of V (V) and A. This is called the *British Gas Stretford Process*.[265–269] Although its chemistry and electrochemistry have been investigated in detail, the reaction mechanism is not well understood. A possible mechanism is shown in Figure 5.20. Due to the complexity of the species involved (e.g., $HV_2O_7^{3-}$, $V_{18}O_{42}^{12-}$, $VO_2S_2^{3-}$, $V_4O_9^{2-}$, $VO_2 \cdot H_2O_2^+$) and the lack of quantitative understanding of the mediating role of these vanadium species, no attempt has been made to write the stoichiometry of each reaction in this simplified scheme.

5.6.3. SO_2 Removal

As mentioned in Chapter 1, gaseous sulfur oxides (SO_2 and SO_3) are implicated in acid rain and photochemical smog. They are produced mainly by combustion of sulfur-containing fuels and coal and as by-products in several industrial processes. Conventional removal processes include the use of basic

5.6. Electrochemical Treatment of Gaseous Pollutants

scrubbing solutions (e.g., lime, limestone, magnesium oxide, amines). Unfortunately, a large amount of sludge is frequently produced and its disposal is expensive in this type of flue gas desulfurization (FGD) processes.

Electrochemical alternatives normally do not produce sludge since the main reagent is the electron. In addition, they usually yield useful by-products and offer competitive economics.[207] Electrochemical sulfur dioxide removal or conversion processes have been classified as follows[270]:

Direct processes:
    Adsorption and adsorbant regeneration by electrochemical oxidation of SO_2.
    Outer-cell or inner-cell absorption and electrochemical conversion.
    Electrochemical reaction at a gas diffusion electrode.

Indirect processes:
    Homogeneous redox mediators in an outer-cell or inner-cell process.
    Heterogeneous redox mediators.
    Catalytic oxidation with oxygen and electrochemical regeneration of the catalyst.

Their main drawbacks are the relatively high energy consumption and the easiness for electrode poisoning by impurities in the flue gas.[270] Some of these processes as well as more recent approaches follow and are depicted in Figure 5.21.

(1) The reduction of a mixture of dissolved ($SO_2 + O_2$) to SO_4^{2-} and further oxidation to $SO_3 + 1/2\ O_2$ at ca. 400°C in a solid membrane (Si_3N_4) electrochemical cell. Here, a molten electrolyte containing a catalyst for the transformation of SO_2 to SO_4^{2-} is retained on porous electrodes and the SO_3 produced is removed as a concentrated stream.[207,220]

(2) The mixed thermal–electrochemical wet scrubbing regenerative FGD process called the *Ispra Mark 13A* that produces H_2SO_4 and H_2 at low temperatures (ca. 60°C) using a dished electrode membrane (DEM) cell in an outer-cell Br_2-based electrochemical cycle.[202,271,272] The undesired reduction of Br_2 at the cathode ("cathodic back reaction") is the culprit for the somewhat less than ideal current efficiencies obtained (ca. 87%). This process is further discussed in Chapter 8.

(3) A process similar to the one just described involves the use of peroxodisulfuric acid (instead of Br_2) as the oxidant for SO_2 and its regeneration in an electrochemical cell.[202]

(4) Fe^{3+} ions are used to oxidize SO_2 to H_2SO_4. The generated Fe^{2+} ions are reoxidized in an electrochemical cell that can be used as an energy generating cell (or electrogenerative cell) since the oxidation of Fe^{2+} at the anode and the reduction of O_2 at the cathode produce a positive thermodynamic cell voltage.[273]

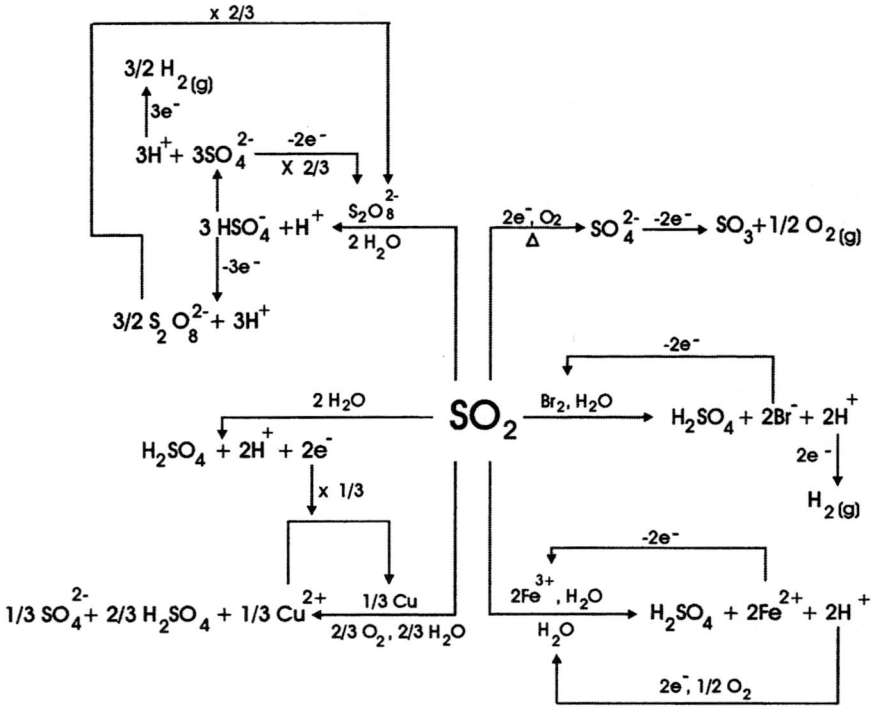

FIGURE 5.21. Electrochemical treatment scheme for SO_2.

(5) Catalytic SO_2 oxidation (by O_2 in aqueous solution containing Cu particles) that produces H_2SO_4 and $CuSO_4$ is combined with an electrochemical reaction. Here, SO_2 is oxidized to H_2SO_4 at the anode whereas the Cu^{2+} ions produced in the catalytic step are reduced to Cu to be recycled into the process.[270] The advantages here are threefold: the cell voltage is reduced (as compared to the process without Cu) by using the reduction of Cu^{2+} as the cathodic reaction instead of H_2 evolution; the current consumption for SO_2 oxidation is decreased by its parallel chemical oxidation; and a bed of Cu particles is used as an absorption column as well as the chemical reaction medium and can be exchanged by the cathode bed after an optimum reaction time. The oxidation of SO_2 has been previously studied for the sulfur cycle hydrogen production process for the electrolytic generation of hydrogen from water.[274]

(6) An absorption column is packed with conductive graphite particles, where SO_2 is both absorbed and electrochemically oxidized to H_2SO_4.[202]

5.6. Electrochemical Treatment of Gaseous Pollutants

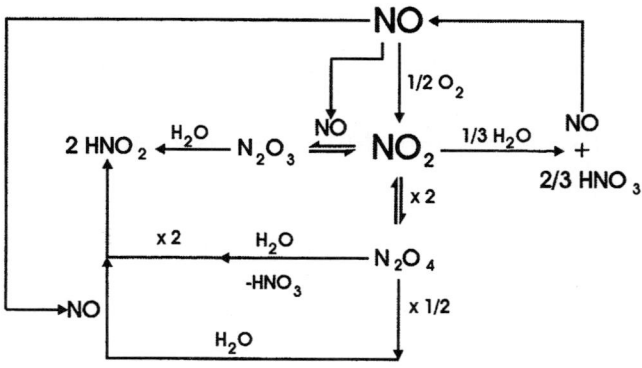

FIGURE 5.22. Various species in the nitrogen oxides system.

5.6.4. NO_x Removal

As mentioned in Chapter 1, nitrogen oxides present in the atmosphere participate in environmentally problematic processes such as the greenhouse effect, acid rain, and photochemical smog. Electrochemical treatment necessitates that these gases be dissolved in an aqueous solution, which may not be a straightforward task as deduced from the following discussion. In addition, complex chemical equilibria exist among the different nitrogen oxides (see Figure 5.22).[275,276]

Simple absorption of NO in alkaline solutions is not effective since it is only physically absorbed.[275,277] Absorption in nitric acid decreases with acid concentration,[276] whereas oxidation with ozone to produce NO_2 facilitates absorption in an alkaline medium.[277] To eliminate the need of an oxidizing agent, absorption of NO with simultaneous oxidation at a gas diffusion electrode has shown that NO is easily absorbed in an alkaline solution due to the following reaction[277]:

$$NO + OH^- = NO_2^- + H^+ + e^-$$

Aqueous absorption of NO is also facilitated by aminopolycarbonato chelates of Fe (II).[202,278] The resulting compounds can then be reduced to yield hydroxylamine, hydrazine, or ammonia, regenerating the chelate. A simplified version of some NO_x reduction processes is given in Figure 5.23. A simultaneous desulfurization–denoxing process (called the Saarberg–Hölter–Lurgi, SHL, process) is based on this principle.[202]

The complex nitrogen oxides N_2O_4 and N_2O_3 undergo chemisorption in aqueous solutions more easily than NO and NO_2.[275]

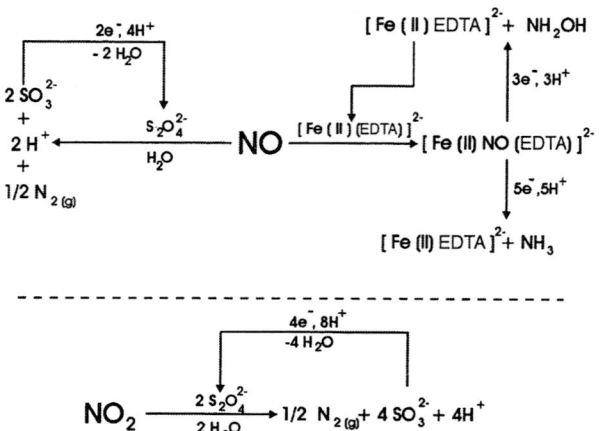

FIGURE 5.23. Catalyzed electroreduction of NO_x.

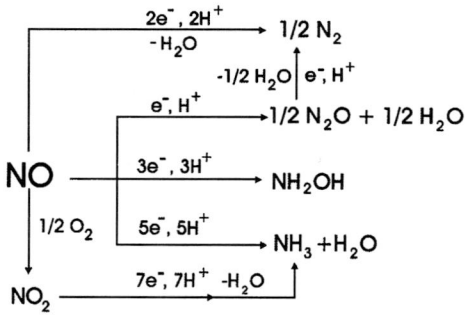

FIGURE 5.24. Electroreduction pathways of NO_x.

The electroreduction of nitrogen oxides in industrial waste gas streams can simultaneously lead to useful or inert compounds, as shown in Table 5.6 as well as in Figures 5.23 and 5.24.[278-285] A major fraction of the current used for reducing NO_x at low potentials results in the formation of ammonia and hydroxylamine in the electrolyte.[281-283]

Ammonia formation is strongly favored in N_2 diluent alone, and a trend toward hydroxylamine formation has been observed in the presence of CO or SO_2.[281-283] The presence of sulfur at porous Pt black gas diffusion cathodes thus favors hydroxylamine production. The effect of sulfur contamination on NO reduction is an important consideration since many flue gases contain significant amounts of sulfurous compounds.

5.6. Electrochemical Treatment of Gaseous Pollutants

Table 5.6. Reduction of NO and SO_2 to Environmentally Less Harmful Products (see Section 3.3.3)

$$2\,NO + 2\,H^+ + 2\,e^- \rightarrow N_2O + H_2O$$

$$2\,NO + 4\,H^+ + 4\,e^- \rightarrow N_2 + 2\,H_2O$$

$$2\,NO + 6\,H^+ + 6\,e^- \rightarrow 2\,NH_2OH$$

$$2\,NO + 10\,H^+ + 10\,e^- \rightarrow 2\,NH_3 + 2\,H_2O$$

$$SO_2 + 4\,H^+ + 4\,e^- \rightarrow S + 2\,H_2O$$

Based upon the thermodynamically favorable reduction of NO to produce ammonia, hydroxylamine and nitrous oxide, an electrogenerative process can be used to react NO in a cell with protons and electrons at one electrode and H_2 at the other to generate current.[280] Note that this configuration resembles that of a fuel cell (although the reactions are not necessarily the same) and the corresponding technology can be borrowed from the fuel cell field.

Electrogenerative processes can be defined as those in which favorable thermodynamic and kinetic factors are utilized for the production of coupled electrode reactions, which take place in separate compartments in a electrochemical cell with simultaneous generation of electricity.[280,284] Removal of NO from flue gases (at the laboratory level) is facilitated by its preferential adsorption on noble metal catalysts. This allows for reduction to occur at high yields (up to ~99.5% conversion) even in dilute gas streams and in the presence of other competitive oxidizers (e.g., O_2, SO_2).[280] Further work is required to inhibit the competitive reaction of O_2 reduction in a broader set of conditions.

N_2O is the most stable nitrogen oxide and its participation is thus inevitable in the greenhouse effect. It can be electroreduced in alkaline and acid media. Under optimum conditions, current efficiencies of 100% [to produce $N_2(g)$] have been reported.[220] Two electroreduction approaches are shown in Figure 5.25, where Pd* represents a Pd catalytic site; Pd-based alloys have been proven useful for this application. An N_2O–H_2 fuel cell was constructed using a Pd–Pt (1:1) catalyst, and an open circuit voltage of up to 1 V was obtained.[220]

5.6.5. Chlorine Removal and Concentration

Chlorine gas present in waste gases can be absorbed in the catholyte compartment of a divided electrochemical cell containing Cu^+ ions; the reduction occurs as follows.

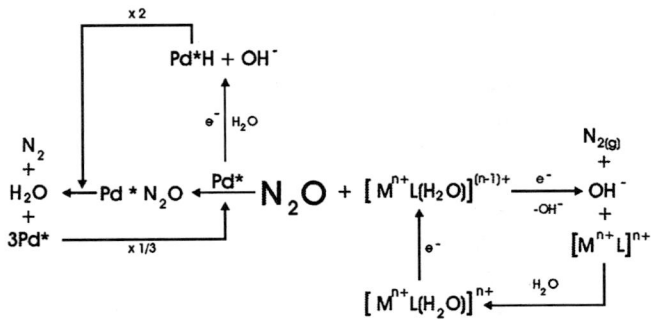

FIGURE 5.25. Catalyzed electroreduction of N_2O.

On the cathode,

$$Cl_2 + 2\,Cu^+ \rightarrow 2\,Cl^- + 2\,Cu^{2+}$$
$$Cl_2 + 2\,e^- \rightarrow 2\,Cl^-$$
$$2\,Cu^{2+} + 2\,e^- \rightarrow 2\,Cu^+$$

On the anode,

$$2\,Cl^- \rightarrow Cl_2 + 2\,e^-$$

This is actually a concentration process whereby undesired Cl_2 is removed from a gas stream by dissolution in the catholyte and the same liquid is passed to the anode, where chloride ions are reconverted to chlorine gas.[203] The process does not require a separation step, since a gas is produced on the anode. A system based on the same principle but without Cu ions has been proposed.[202] Another alternative uses a divided cell and graphite electrodes.[286] The electrolyte consists of $CuCl_2$ + HCl. Chlorine is generated at the anode and the Cu(II) chloro complexes are reduced at the cathode. The resulting catholyte effluent is routed to a packed tower, where the cuprous ions are oxidized by Cl_2 from the waste gas and recycled.

5.7. MEMBRANE-ASSISTED PROCESSES

The reagents, products, and electrode materials involved in the cathodic and anodic electrode reactions of an electrochemical cell can sometimes interact in an undesired manner to foster short-circuits or side reactions that decrease the figures of merit of the system or can even lead to accidents.

5.7. Membrane-Assisted Processes

Electrochemists have devised ways to prevent such interactions by using special separators such as macroseparators, microseparators, and ion selective membranes. Reviews concerning separators, their technology, and applications are available; and this body of literature is listed at the end of this chapter.

The main functions of the macroseparators are to prevent physical contact between electrodes, provide structural support, and promote turbulence to increase the mass-transfer coefficient (examples include polymeric nets and rigid plastic open pore sheets). Microseparators must have passage sites large enough to permit the restricted movement of electrolyte from one side to the other but small enough as to prevent gross exchange of solution due to convective or diffusive mass transport. Separator materials include asbestos diaphragms, porous polymeric materials, porous ceramics, and sintered metals.[287] Unfortunately, an ohmic drop penalty always must be paid for their use, whose magnitude depends primarily on the pore size and tortuosity.

Ion selective membranes or ion exchange membranes (IEMs) are polymeric materials that incorporate within their structure ionic centers capable of binding ions of opposite charge. Such ions can then selectively and directionally migrate under the influence of an electric field and be separated from their counterions.[288] This property is called *permselectivity*. These membrane separators are used in a variety of electrochemical processes and shall be explored in more detail in this section in view of their relevance to the environmental field. The world market for IEM-based technologies is estimated to be on the order of $\$3 \times 10^8$/year, which is an important share of the total membrane technology market estimated to be well over $1 billion.[289,290] Commercial aspects (i.e., membrane producers) are identified in Chapter 8.

5.7.1. Ion Exchange Membranes: General Considerations

Typical cation or anion exchange membranes (CEM, AEM) consist of polymeric chains (e.g., polystyrene, perfluorocarbon) with fixed negatively or positively charged groups covalently bound that act as ion exchange sites (e.g., cation exchange groups, $-SO_3^-$, $-COO^-$, $-PO_3^{2-}$, $-SO_2R$; anion exchange groups, $-NR_3^+$, $-NH_3^+$, $-NH_2R^+$, $-PR_3^+$).[288,291] Cellulose acetate and polyacrylonitrile are also used as membrane materials. In Nafion and other polyelectrolytes based on poly(styrene sulfonate) and poly(vinyl sulfate), the polymer chains have been shown by luminescence spectroscopy to be organized so that the hydrophilic groups can aggregate together, forming hydrated channels that the ions can migrate through.[292-296] An ion–dipole cluster network model has been proposed based on results from X-ray diffraction, neutron scattering, and Mössbauer spectroscopies.[291,297] In this model, fixed ions and their counterions agglomerate together with hydration water forming regions interconnected by a network of microscopic channels. Since the ionized groups render the membranes somewhat soluble, cross-linking groups are frequently inserted to prevent their dissolution (e.g., divinylbenzene).

Membrane hydration is important since too dry a membrane has a very large ohmic resistance, whereas too wet a membrane shows a high degree of swelling and permits the passage of undesired moieties. The number of water molecules absorbed per ionized group decreases exponentially with the group concentration.[291] In addition, the degree of swelling is also affected by the extent of cross-linking. The features sought in an IEM include[288,298]: (a) high selectivity toward the transport of the desired ionic species; (b) fast ionic and negligible electronic transport under the electric field; (c) low ohmic resistance; and (d) high mechanical, chemical, electrochemical, and thermal stability. From the preceding discussion, these properties are seen to be determined by the nature and structure of the cross-linked chain, the nature and concentration of the fixed ionic groups and of the counterions, the degree of hydration and swelling, and so forth.

Composite membranes that combine different characteristics to produce a desired effect have been designed and fabricated. Two important examples are bilayer membranes and bipolar membranes. Bilayer membranes are produced by forming a relatively thin layer of an IEM containing carboxylate groups as the ion exchange centers, on top of an IEM containing sulfonate groups. In this way, the low hydrophilicity and the higher pK_a of the carboxylate membrane are combined with the low electrical resistance and very high chemical stability of the sulfonate membrane.[299] Bipolar membranes are discussed later in this section.

In a cation exchange membrane, mobile cations are attracted to the fixed anionic sites whereas their corresponding counter anions (called *co-ions*) are excluded from it by virtue of electrostatic repulsion. This, called *Donnan exclusion,* creates a concentration gradient between the membrane and the solution. The gradient acts as a driving force for the cations to return to the solution, but due to the electroneutrality required at the membrane, this movement produces a space charge with a concomitant potential gradient called the *Donnan potential*. The potential is of fundamental importance for the distribution of ions at both sides of the interface and therefore for the selectivity of the membrane. A similar rationale can be used for an anion exchange membrane.

As mentioned in Chapter 2, the transport number of an ion is the fraction of current carried by that ion. The closer this number is to unity, the higher the membrane selectivity is to that particular ion. For example, the transport number for Na^+ during sodium chloride electrolysis with a cation exchange membrane is ~0.9, which means that migration of any other ions present (mainly OH^-) is greatly hindered.[288] For cations of different charge, their relative transport numbers are known to be a function of their charge, size, and the ratio of

5.7. Membrane-Assisted Processes

the interaction constants with the polymeric matrix as well as the ratio of the diffusion coefficients in the matrix.[298] Similarly, the selectivity of transport in a mixture of two monovalent anions (ion 1 and ion 2) is known to be proportional to the ratio of ionic mobilities (in the bulk solution). At high current densities, these mobilities approach one another and the membrane becomes less selective. The permselectivity of the membrane for ion 1 relative to ion 2 is given by the expression[300,301]

$$k_{12} = \frac{J_1/J_2}{C_2/C_1} = \frac{\ln\left(\frac{C_{2-0}}{C_{2-t}}\right)}{\ln\left(\frac{C_{1-0}}{C_{1-t}}\right)}$$

where J_1, J_2 are the ionic fluxes and C_{1-0} and C_{1-t} are the ion 1 concentrations at time 0 and time t, respectively (likewise for ion 2). All concentrations pertain to the solution bulk.

The permselectivity of metal ions can vary drastically on complexation. For example, when glycine is added to a solution containing a mixture of Cu^{2+} and Ni^{2+} ions, the selectivity of a CEM to nickel increases several fold. This result can be explained in terms of the much larger complexation constant of glycine with the Cu^{2+} ions that leaves more Ni^{2+} free for cationic transport through the membrane.

Another way of varying the permselectivity consists of membrane surface modification. For example, a thin layer of a cationic polyelectrolyte is spread on the surface of a normal CEM. The higher the positive charge of an approaching cation, the stronger it will experience a repulsion by this polyelectrolyte layer. Thus, monovalent cations will be much less repelled and their transport will be favorable compared to their multivalent counterparts. This principle is used in CEMs (e.g., Neosepta CMS, from Tokuyama Soda Co.) for the recovery of common salt from sea water, which contains substantial amounts of Mg^{2+} and Ca^{2+} ions.[302] Salt of 97% purity can be obtained in this manner and precipitation of Mg and Ca hydroxides or sulfates on the membrane surface is drastically reduced.[303] Similarly, monovalent AEMs with a thin layer of a highly cross-linked matrix are commercially available (e.g., Neosepta ACS, Selemion ASV from Asahi Glass Co.). They have been used, for example, in the separation of monovalent chloride ions from divalent oxalate ions during the electrolytic production of titanium.[303]

The electrical resistance of IEMs is inversely proportional to the number of mobile ions in the membrane, which in turn is a function of the concentration and degree of dissociation of the fixed charges. This is why highly ionizable groups (see the previous examples) are preferred for incorporation into membranes; their concentrations can be determined for example by titration.[298] However, very high concentrations of these groups are undesirable since they lead to excessive swelling with concomitant selectivity reduction and mechanical alteration. Typical concentrations are on the order of 1–2 milliequivalents per gram (of dry matrix); typical measured specific resistances are on the order of 50–100 ohm-cm.[304] The electrical resistance of the IEMs decreases with increasing temperature. For example, the conductivity of a carboxylate cationic exchange membrane layer increases by an order of magnitude with a temperature increase of ~40°C.[305]

The cell resistance can increase sharply if the permeation rate of the selectively transported ions is much greater than their diffusion rate toward the membrane. In this case, the surface concentration of such ions approaches zero, although the reason for the sharp resistance increase seems to be due to a voltage drop instead of an increased membrane resistance due to the surface ion depletion.[306] In spite of these ohmic losses, membrane-based approaches to some processes can be more energy efficient than the conventional ones. For example, during the production of sodium hydroxide and chlorine from brine, the membrane-based process affords an energy reduction up to 37% with respect to the conventional asbestos diaphragm process.[291]

Preparation techniques for ion exchange membranes can be quite varied, depending on the type and the properties of the desired membrane and cannot be comprehensively discussed here. A typical sequence includes the following steps[298]:
- Synthesis of the backbone monomer
- Radical copolymerization of monomer and cross-linker
- Chemical insertion of the ionic groups of interest
- Membrane fabrication—casting, extrusion, lamination (here, a support such as a polymer mesh may be used)
- Surface modification (for example, to increase selectivity, reduce fouling, facilitate gas release, incorporate catalysts)

5.7.2. *Applications of Ion Exchange Membranes: Electrodialysis*

The major electrochemical, membrane-based separation process in commercial use today is electrodialysis. Electrodialysis involves the separation and concentration of electrolytes based on electromigration through IEMs, and was briefly considered in Chapter 1. An array of alternating anion and cation exchange membranes is positioned between two electrodes separated by insu-

5.7. Membrane-Assisted Processes

lating screens. The alternating arrangement of the membranes allows ions to be passed by one membrane and blocked by the next, thus producing in alternate compartments a dilute stream (*diluate*) as well as a concentrated stream (*concentrate* or brine). These solutions are collected and transported out of the membrane stack.

Advantages of using electrodialytic techniques (most of which are also common to other IEM-based separation processes) include the following[290,302,307–316]:

- Essentially no extra reagents are required for the processes.
- Some ionic dissolved substances that cannot be separated by conventional methods can be removed by electrodialysis.
- Electrolyte solutions can be concentrated up to ~20% (or more in some cases).
- Diluates can be obtained with concentrations down to 100 ppm (or less in some cases).
- Concentrate/diluate concentration ratios of ~100 can be obtained.
- Nonelectrolytes can be either separated by exclusion or not affected by the process.
- IEMs are rather durable (typically several years).
- High current efficiencies can be attained (80–90% are common, 100% in optimal cases).
- Unlike ion exchange resins, IEMs do not require a periodic regeneration step.
- Multivalent and complexed ions can be rejected.
- Modular equipment adds versatility to the processes. Up to a hundred or more unit cells can be stacked, and only one anode and one cathode are required.
- Continuous processes with relatively high flow rates can be designed.
- On-site, on-demand production of some substances can be done with attendant reduction in transportation and storage costs.
- No phase changes are required.
- Hybrid processes can be designed to optimize the benefits obtainable from two or more separation processes. For example, a combined electrodialysis–conventional dialysis process has been used for the separation of excess acid from a metal ion solution in an electroplating plant.[313] Ion-exchange materials can be used as packing between membranes to increase both the contact area and the conductivity of the medium. This approach is called *electrodialysis–ion exchange*, EDIX. The electrical resistance in the ion exchange material packed bed can be decreased by applying a suitable binder in an approach called *electrochemical ion exchange* (EIX), see later. Larger decontamination factors can be obtained with EDIX than with EIX, as we shall see later.

Challenges to be faced in these processes include the following[290,307,309,314,317,318]:
- The membrane is fouled. Organic or inorganic insoluble substances may precipitate at the surface of the IEMs thus impeding their proper functioning. Proposed solutions include dissolution by the addition of appropriate chemicals (e.g., acids); turbulence promotion; periodic current reversal; periodic mechanical or chemical cleaning; inclusion of fluid permeable tubules by the membrane surface and continuous fluid pumping as to prevent deposit formation; water pretreatment; and in the case of CEMs, surface modification as described earlier. This last tactic can be used as to prevent polyvalent cations from approaching the membrane surface, where they may encounter precipitating anions (such as OH^- ions).
- Concentration effects. At very high as well as at very low concentrations, these processes become less feasible due to higher energy losses and decreased conductivity, respectively. In addition, concentration gradients work in some cases against the effect of the electrical potential gradient.
- High pH values may adversely affect the membranes.
- Nonionized substances cannot be treated.
- Protons can go through anion exchange membranes from acidic solutions due to their high mobility, small size, and preferential sorption, as well as due to polymeric matrix hydration.
- It is usually necessary to do a multiple pass of the electrolyte to achieve a high removal of its ionic contents. A single pass usually removes less than 50%.

<u>Other Applications of Electrodialysis</u>. Desalination is the most important application of electrodialysis. Here, ionized salts are efficiently removed from solutions by the membrane process described earlier. Since an irreversible energy loss occurs, due to the friction promoted by the passage of ions through the membrane, at high concentrations electrodialysis becomes less profitable and loses ground relative to other techniques.[288,307,319] A rule of thumb is that the lowest cost of desalination of solutions with concentrations smaller than ~0.5 g/liter is attained by conventional ion exchange techniques; for concentrations in the range ~0.5–5 g/liter, by electrodialysis; for ~5–100 g/liter, by reverse osmosis, and for >100 g/liter by distillation. Salt obtained from seawater reaches a production on the order of 1 million tons per year.[303]

The main application is desalination of brackish and seawater for the production of potable water. In addition, this process has been shown to be cheaper in the desalination of circulating water than the supply of a much larger amount of fresh water. The wastewater volume also decreases drastically. Not only from an economical perspective is this process attractive but from an environmental viewpoint as well. Salt removal is also used for facilitating the distillation of organic–salt mixtures.

5.7. Membrane-Assisted Processes

Electrodialysis is used to remove organic matter.[311,320] Strongly basic anion exchange resins are known to be sensitive to organic matter (particularly humic substances), with attendant decreases in their ion exchange capacity, pH, and the active depth as well as increases in the amount of water required for the washing operation and the conductivity of the final solution. Electrodialysis has been successfully used as a pretreatment step that results in a higher degree of removal of these organic substances than with conventional methods of coagulation and lime decarbonization. Another proposed application involving undesirable organic matter consists of coating an ion selective film (Nafion) on a working electrode surface. This prevents it from being poisoned by organic substances during the regeneration of spent oxidants in indirect electrochemical processes (see Section 5.4).

Metal ion removal or recycling by electrodialysis techniques include the following[310,314,321]:
- Rinsing waste waters from electroplating processes can be desalted and reused while the concentrate is recycled into the plating bath. In addition, this electrodialytic process facilitates the regulation of concentration required in the rinse tanks.
- The OH$^-$ ions produced at the cathode of an electrodialytic cell are used for the precipitation and removal of insoluble radioactive metal ion hydroxides.
- Sulfate ions become concentrated during the electroless deposition of copper on printed circuit boards and undergo undesired side reactions. They can be removed by electrodialysis, and sulfuric acid is produced on the anode of the cell, while the OH$^-$ ions produced on the cathode are used to neutralize the plating solution.
- Silver ions are concentrated from spent photographic fixing solutions by electrodialysis and sent to an electrolytic cell for metal recovery.

Anions can be separated from cations.[315] During electroless Ni deposition, hypophosphite ion is used as the reducing agent and is converted to phosphite, which deteriorates the metal coating and lowers the efficiency of the process. An electrodialytic step has been used to separate the nickel ions from the phosphite ions; the phosphite-free metal ion solution is then recycled into the process, minimizing the required solution removal frequency and bettering the quality of the metallic deposit.

Acids can be removed and recovered.[307,310,317] Acidic waste water is generated from a variety of processes, including the production of chlorinated hydrocarbons, regeneration of ion exchange resins, pickling of metals, and other hydrometallurgical processes. Acids that have been recovered by electrodialysis include H_2SO_4, HF, H_2CrO_4, HCl, and H_3PO_4. A problem that frequently has to be addressed is that hydrogen ions are poorly excluded by anion exchange membranes, especially under highly acidic conditions. For this reason, neutralization prior to electrodialysis is recommended.

Bases can be removed and recovered.[322] Waste waters from the textile and molding industries contain considerable amounts of sodium hydroxide. A process has been proposed for its electrodialytic recovery.

5.7.3. Membrane Cell Electrolysis

Electrochemical cells equipped with ion exchange membranes (AEM, CEM, or both) are used for the separation–purification of different substances and process streams. These processes have also been termed *electro-electrodialysis*.[290] Some examples follow:

- Acid recovery: Waste metal-containing acid solutions can be purified and the acid concentrated by simultaneous electrochemical removal of the metal (plating), anion migration and proton production. High recovery (up to 96%) of concentrated waste sulfuric acid with very low cupric ion content (<1 ppm) has been obtained.[290]
- Salt splitting: Waste streams containing dissolved salts are both an economic problem and an environmental problem. On the one hand, a substantial amount of material is wasted and must be replaced by a fresh supply; on the other hand, stricter environmental regulations are placing more pressure on the prevention of materials release to the ecosystem. Recycling or reuse can solve both problems. Here are some examples. [288,290,299,322–325]

Concentrated NaOH, Cl_2 and H_2 are produced from NaCl (Figure 5.26a). Here, the pH of the NaCl solution (pH 2–4) is important, since if it were too low, most of the current would be carried by protons instead of sodium ions, whereas if it were too high, chlorine gas would react with this solution and yield hypochlorite.

H_2SO_4 and NaOH are produced from Na_2SO_4, and $(NH_4)_2SO_4$ and NaOH are produced from Na_2SO_4 and NH_3. Sodium sulfate is a low-value by-product of flue gas desulfurization processes as well as from pulp and paper, plastic recycling, and other chemical industries. Interest in its reuse stems from the possibility of producing its chemical precursors, H_2SO_4 and NaOH. An added benefit is to obtain the caustic while avoiding the attendant coproduction of the environmentally embattled chlorine during the electrolysis of NaCl. This can be achieved by using either a two- or three-compartment cell (Figures 5.26b and 5.26c). In the three-compartment case, hydrogen gas can be fed to a gas diffusion, hydrogen depolarized anode for the production of protons without the need for the high energy consuming O_2 evolution. Undesirable migration of protons across the AEM occurs at low pH values due to their high concentration. A way to prevent

5.7. Membrane-Assisted Processes

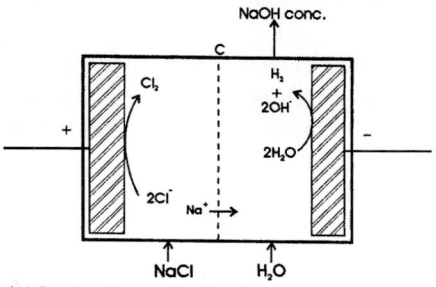

(a) Production of NaOH, Cl_2 and H_2 from NaCl

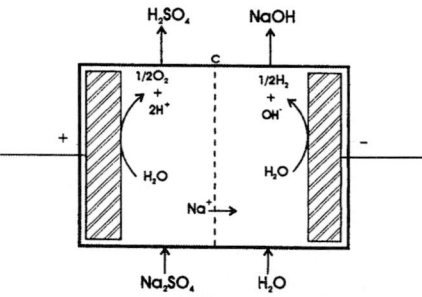

(b) Production of H_2SO_4 and NaOH from Na_2SO_4 (two-compartment cell)

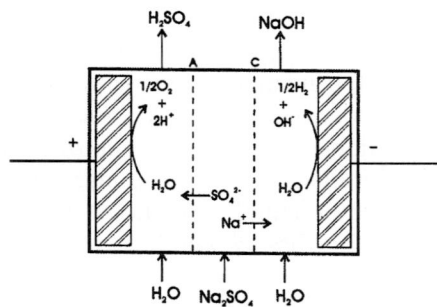

(c) Production of H_2SO_4 and NaOH from Na_2SO_4 (three-compartment cell)

FIGURE 5.26. Schematic diagrams of membrane-based salt-splitting processes.

this is to maintain the pH in the anodic compartment high enough by the addition of ammonia gas. The product is then ammonium sulfate instead of sulfuric acid.

Another alternative that has been shown to be feasible for the production of caustic involves the splitting of Na_2CO_3. Here, a distribution similar to the one shown in Figure 5.26b is used with a gas diffusion anode to

produce protons that react with the carbonate ions and release CO_2; NaOH is produced on the cathodic side.
- Gas removal: As discussed earlier in this chapter (Section 5.6), some polluting gases can be transported across a membrane by virtue of an electrical field to a compartment where they can be either concentrated (e.g., CO_2) or transformed into less harmful substances (e.g., SO_2 into SO_4^{2-}) or even useful products (e.g., H_2S into $S + H_2$). The key factor here is that the pollutant gas must be able to form dissolved ions or undergo electron transfer in solution at lower potentials than the other components of the gas mixture. Compared to pressure-driven gas separation, a potential difference of 60 mV can produce a concentration difference of a bivalent charged species equivalent to that obtained with a pressure difference of 100 atm for a noncharged species.[326]

5.7.4. Bipolar Membrane-Based Processes

A large portion of the potential required in many membrane-based electrochemical cells to bring about a desired process is employed in the water splitting reaction, where hydrogen and oxygen gases are produced at the electrodes with the consequent production of H^+ and OH^- ions. This process requires a (thermodynamic) potential difference of 1.23 V to occur. In practice, however, the required potentials are usually much larger, due to kinetic barriers (overpotentials) (Chapter 2).

This stringent potential requirement (and thus high energy consumption) can be lowered by using composite membranes with an AEM on one side and a CEM on the other. Such a composite, called a *bipolar membrane*, is impermeable to both anions and cations. Its function is to dissociate the water that diffuses to the CEM/AEM interface into its component ions (H^+ and OH^-) with the aid of the external applied potential difference as the driving force. In addition, a catalytic effect for this dissocation due to fixed tertiary ammonium groups in the AEM has been proposed. This has led to a modified design that involves an intermediate layer (between the CEM and AEM layers) incorporating a high concentration of such groups.[304] Another advantage of bipolar membranes is that the electrode reactions are independent of the process. Corrosive acids can then be handled without any harm to the electrodes.[325]

A disadvantage with bipolar membranes is their inability to recover acids (see later) with high concentration of metals that form insoluble hydroxides that foul the membrane. On the other hand, complexing media do not affect their operation.[290] Several applications have followed their invention, some of which are given below.

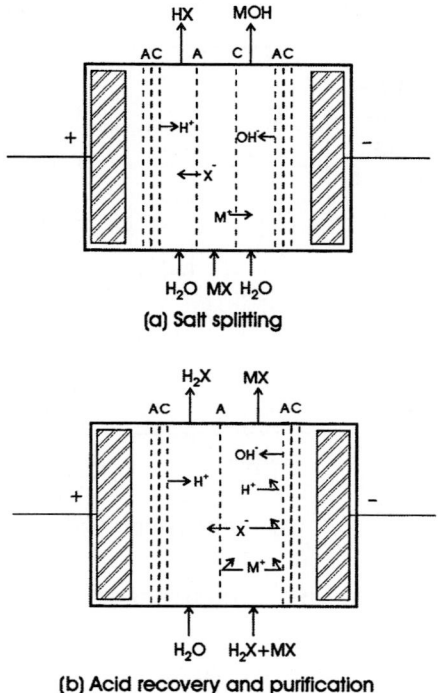

FIGURE 5.27. Salt-splitting and acid recovery with bipolar membranes.

- Salt splitting: This can be achieved using the scheme in Figure 5.27a. Note that the bipolar membranes need to have a specific orientation; in this case, the AEM side must face the anode and the CEM side must face the cathode. The production of the corresponding acid and base has been used for the regeneration of ion exchange resins and the splitting of radioactive waste containing sodium nitrate.[312,316,327]
- Acid recovery: Spent acids from pickling baths contain large amounts of dissolved metal in the form of salts. The acid can be separated from the salts and purified as shown in Figure 5.27b. Recoveries of 80–90% are achievable by this process.[290]
- Organic acid and inorganic base recovery: Sodium hydroxide is used to control the pH in a fermentation process to produce itaconic acid (a common plasticizer). The organic salt thus produced is sent to a bipolar membrane system, where itaconic acid and sodium hydroxide are produced and separated.[304]

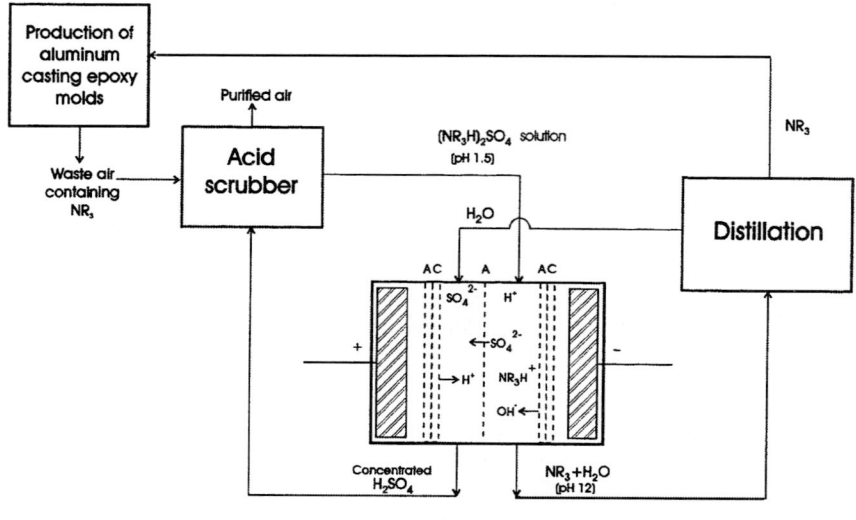

FIGURE 5.28. Regeneration of organic base and inorganic acid with bipolar membranes. (Adapted from Strathmann et al.[304])

- Inorganic acid and organic base recovery: Dimethyl isopropyl amine is used as an epoxy curing agent in the production of metal casting molds. It acts as a catalyst and thus is not consumed during the process, but has to be scrubbed in a sulfuric acid solution from the exit gas stream. A bipolar membrane-based process has been designed and operated for the regeneration of both the tertiary amine (NR_3) and the acid scrubbing solution (Figure 5.28).[304]
- Alcoholate production by alcohol dissociation: Alcohols can be dissociated with the aid of bipolar membranes under the influence of an electric field. For example, sodium methanolate can be produced from methanol (a weak acid and a weak base) by reaction with very strong bases. Milder conditions are required for its production in a bipolar membrane cell, where methanol is ionized in a non-aqueous medium to produce CH_3O^-, which is then reacted with Na^+ fed from another organic salt.[304]

5.7.5. Electrochemical Ion Exchange

The properties of ion exchange materials are combined with electrically driven migration processes to produce *electrochemical ion exchange*, EIX. Here, an ion exchange material is incorporated into the external structure of an electrode by means of a binder.[308,309,328]

5.7. Membrane-Assisted Processes

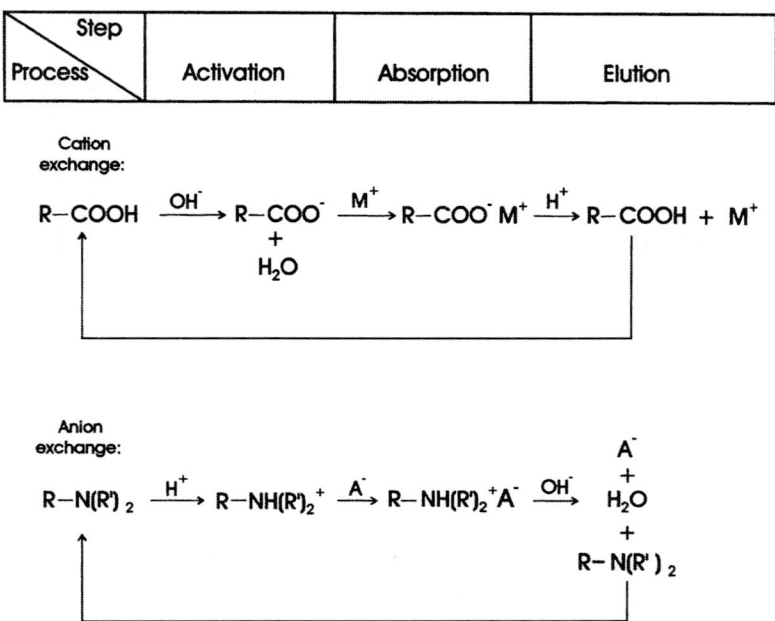

FIGURE 5.29. Basics of an ion exchange process.

In conventional ion exchange, dissolved ions are selectively removed by the ion exchange material according to their cationic or anionic nature (*absorption*). In a later step (*elution*), the ion exchange material is regenerated by releasing the absorbed ions and replacing them with protons (cation exchange) or hydroxyl ions (anion exchange) produced, for example, by the corresponding electrochemical oxidation or reduction of water.[309,328] Finally, the oppositely charged ion from water is partitioned into the ion exchange material to reinitiate the cycle (*activation*). This sequence of steps is illustrated in Figure 5.29.

In EIX, the application of an electric potential promotes ion migration toward the oppositely charged electrode. This migration facilitates or controls both the absorption and elution steps since, by virtue of a deliberate polarity change, the EIX electrode changes its function as a cathode or anode. In other words, a cationic EIX electrode is made the cathode during absorption so that it can attract the positively charged M^+ species, as in Figure 5.29. At the same time, OH^- ions are produced at the EIX electrode due to water reduction and neutralize the H^+ produced at the anode during water oxidation. The polarity is then reversed to facilitate the elution process whereby the positive metal ions are rejected and substituted by the protons produced at the same EIX electrode due to water oxidation. A similar process occurs with an anionic EIX electrode.[309,328]

The efficiency of the EIX cells is proportional to the current and current efficiency and inversely proportional to the electrolyte flow rate and concentration.

Other important advantages of EIX include[308,309,328–330]: (a) the unit size required for comparable performance with an electrodialytic unit is drastically smaller (on the order of 1:5); (b) very low energy is required; (c) there is a high degree of selectivity; (d) high concentrations of eluate can be produced; (e) the kinetics are enhanced; (f) the operating range has a wide pH; (g) small volumes of waste are generated ; and (h) continuous operation is feasible.

Uses of EIX include the following.
- Separation of radionuclides: The concentration of ^{137}Cs in a feed solution has been drastically reduced by a factor of 5,000 using EIX. Also, the radioactivity of a waste water containing a mixture of radionuclides (predominantly ^{58}Co and ^{60}Co) was decreased by a factor of >99%. Likewise, cleanup factors of >11 and 70 were obtained for Pu and Am containing solutions; potential applications in the treatment of Tc and Tu solutions have been demonstrated. When radioactive and metal ion species are removed from a waste stream by a cation exchange resin, they can be migrated toward the cathode, where they can be deposited.[308,309,328–331]
- Water polishing: High-purity water (15 Mohm•cm resistivity) can be obtained by an *electrochemical deionization process* (EDP) called *electrodiaresis polishing* (EDIP) based on the passage of pretreated water (for example, by reverse osmosis) through ion exchange resins in an electrochemical cell with compartments separated by an alternate CEM/AEM arrangement (Figure 5.30). The resin beds increase the electrical conductivity between the electrodes, facilitating treatment of low-conductivity media. In addition, they promote better mass-transfer characteristics and longer residence times. Large flow rates may promote resin crushing as well as too short residence times.[332,333]
- Recycling of chromate ions: The EDIP process just described can also be used for the removal and concentration of anions from an effluent stream. In particular, chromate ions from a simulated plating waste stream can be removed and concentrated up to a factor of ~8 without carrying over other metal residues in the form of cations. Here, the inlet and cathodic compartments are filled with anion exchange resin and the anodic compartment is filled with cation exchange resin. The two membranes are AEMs.[334–336]
- Nitrate removal: Nitrates found in waste waters, in nuclear processing salt cake residues, or in groundwater can be absorbed by an anion exchange resin in a three-compartment electrochemical cell (catholyte, anion exchange

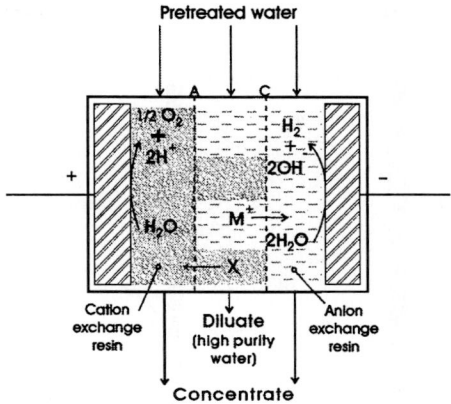

FIGURE 5.30. Schematic diagram of an electrochemical deionization process, electrodiaresis polishing. (Adapted from Dimascio et al.[333])

resin, and anolyte compartments). Under the influence of an electrical field, the nitrate ions migrate toward the anode while OH⁻ ions produced at the cathode (from water reduction) replace them in the resin, regenerating it. The nitrate containing anolyte is then sent to a tortuous path electrochemical cell for its conversion into N_2, O_2, and H_2O. Alternatively, the nitrate ions can be routed to the anolyte compartment to be combined with protons forming nitric acid.[331]

- Other applications: These include the removal of toxic metal ions, the recovery of precious metals, and corrosive anion removal.[309,328]

5.7.6. Other Membranes and Adsorbents

<u>Solid Polymer Electrolytes (SPEs)</u>. An IEM can be "sandwiched" between two porous or wettable electrodes to act as the electrolyte in an electrochemical cell. This approach, developed in the early 1960s in the fuel cell field, greatly reduces the electrolyte resistance and ohmic losses, increases the space–time yields, and facilitates product separation, since there is no need for a supporting electrolyte. Very low-conductivity solutions can thus be employed, which are of special interest in the water treatment field.

<u>Proton Exchange Membranes</u>. These electrolytes constitute a subset of the SPE technology. Highly stable PEM-based cells have been employed in the destruction of organic and bacterial pollutants with the aid of electrocatalysts (e.g., Pt/Ir) integrated into the electrode materials (TOC has been reduced from

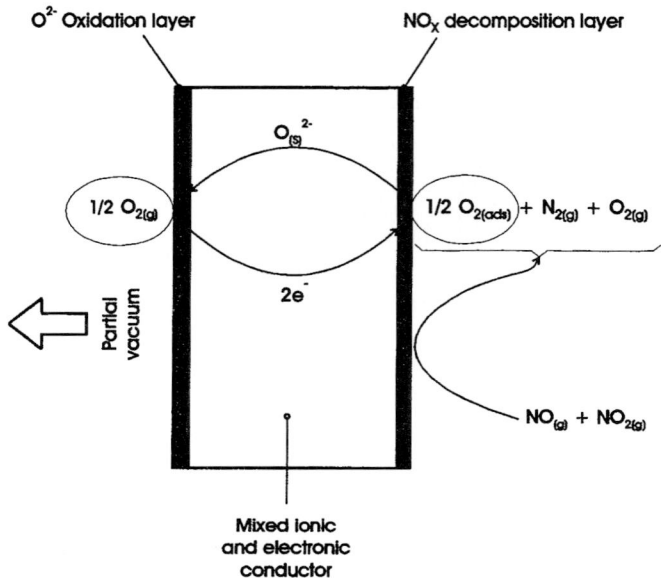

FIGURE 5.31. Spontaneous decomposition of NO_x at a biconductive (mixed ionic and electronic) electrode. (Adapted from Sammells.[340])

100 ppm down to 500 ppb). The rate constant and the activation energy barrier for this process are very similar to those obtained during the photocatalytic destruction of organics on TiO_2 (see Chapter 7).[337,338]

SPEs have also been used for gas separation, concentration, or conversion processes; water and brine electrolysis; fuel cells; and electro-organic synthesis because of the simplicity of a supporting electrolyte-free environment.[291,305,339]

<u>Mixed Ionic and Electronic Conducting Catalytic Membranes</u>. Several spontaneous reactions with synthetic or environmental significance can be carried out at these membranes, which act as short-circuited devices, where ions and electrons are simultaneously transferred from one side to the other. Some perovskite-type mixed ceramic oxides are suitable for this application. On each side of the membrane, specific catalysts may be inserted. A proposed scheme for the spontaneous decomposition of NO_x gaseous pollutants is shown in Figure 5.31. A similar scheme has been proposed for H_2S decomposition into $S + H_2$.[340]

5.8. ELECTROKINETIC PROCESSING OF SOIL

Electric fields as well as electron transfer processes have been used for the treatment of soil and groundwater containing organic or inorganic pollutants. The underlying electrokinetic mechanisms have been discussed in Chapter 2 (Section 2.10) and include electro-osmosis (the movement of a liquid in a pore due to an electric field), electrophoresis (the movement of a charged particle in an electric field), streaming potential (the production of an electric field due to the movement of an electrolyte under a hydraulic potential), and sedimentation potential (the production of an electric field due to the movement of charged particles caused by gravity).[341,342] When suitable anodes and cathodes (e.g., made of graphite or Ebonex) are strategically buried in the ground or placed in contact with a slurry and an electric field from a DC source is applied (typically in the range 40–200 V), one or more of these phenomena occur and are used for the removal of the pollutant(s). The general technique has been called *electroreclamation, electro-osmotic purging, electroremediation, electrorestoration, or electrokinetic processing* in various contexts.

This applied electric field produces inside a charged soil pore filled with a liquid, a drag interaction between the charged outer layer of the liquid, and its bulk. The liquid then moves along the potential gradient to wells or reservoirs, where it can be collected and removed. This phenomenon is called *electro-osmotic transport*. As can be expected, the velocity of the liquid due to this transport is proportional to the applied electric field and to the zeta potential of the pore surface that arises from physical and chemical interactions as well as lattice imperfections; this potential has been found to be negative in wet silts and clays.[341,343]

Simultaneously, water or a purging solution is fed into the soil to aid in the removal of the undesired species by flushing them out by virtue of the electro-osmotic effect. This occurs in soils with low hydraulic conductivities like kaolin, sand–clay mixtures, and the like, and to prevent discontinuities in the soil that would create sites with very high electrical resistivities, thus rendering this technique ineffective. This purging stream can also be used to increase the acidity or basicity of the soil, to increase or decrease the solubility of a given species, to form complexes, and so forth.[344]

In addition, chemical reactions occurring at the electrodes primarily produce $H_2(g)$ and OH^- (aq) at the cathode and $O_2(g)$ and H^+(aq) at the anode as it normally occurs during water electrolysis. These charged species (H^+ and OH^-), along with other ions encountered in the medium, are attracted to the electrodes of opposite polarity and migrate, creating an acid front and a basic front. The movement of these fronts is aided by concentration gradients that promote diffusion and is known to be dominated by the transport of the proton, which neutralizes the base front and impedes its advance toward the anode.[343]

This acid front can be used to inject acidity to soils; solubilize basic metal hydroxides, salts, or adsorbed species; and protonate electron-rich organic functional groups in order to give molecules a more cationic character and promote their migration across the electric field, thus facilitating their removal. Since metals are commonly sorbed on soils by surface complexation or ion exchange, another effect produced by this acid front is the promotion of their desorption by the occupation of the sites available for metal ions,[342,343,345] as expressed by the following reaction:

$$n\text{H}^+ + M^{n+}[\text{soil}]^{n-} \rightarrow \text{H}_n^{n+}[\text{soil}]^{n-} + M^{n+}$$

The movement of the base front often produces the precipitation of cations in the form of hydroxides before they can reach the cathode to be reduced or otherwise removed. This can occur in a very narrow zone of sharp pH change. Although this is not the optimum situation, this "focusing" effect may in some cases facilitate their removal by concentration of the pollutant. In other cases, though, this may clog soil pores.[343] Enhancement techniques have been devised in which a highly soluble and environmentally safe acid (e.g., acetic acid) is injected in the purge solution at the anode to avoid such precipitation. This injection must be carefully controlled, since a high acid concentration will neutralize the negative zeta potential of the pores and significantly slow down the electro-osmotic flow.

Natural buffering produced, for example, by cation exchange capability of the mineral or buffers added in the purge solution tend to balance this acid excess, should it occur. In fact, a linear relationship has been found between the zeta potential and the negative of the logarithm of pH; a low pH also decreases the electro-osmotic conductivity or permeability (see later). Alternatively, when a basic purge solution is used, the flow rate nearly doubles compared to that with the acid purge solution.[343] In the case of mixed pollutants, a stepwise procedure has been envisaged. For example if metal ions and neutral organics are present, the metal ions that neutralize the negative zeta potential of the soil can be removed by electromigration first, and then the organics by electro-osmotic flow.[341] The combined effect due to electric, chemical, and hydraulic potentials results in what is known as *electrokinetic* or *electro-osmotic remediation*, schematized in Figure 5.32.

The main advantages of electrokinetic remediation can be summarized as follows[341,343–347]:
(1) This combined effect is more powerful than hydraulic pressure alone, which is used for *in situ* bioremediation or chemical treatment. These latter treatments require the transport of nutrients or chemicals through the soil, which is particularly difficult to achieve in fine-grained solids with low hydraulic

5.8. Electrokinetic Processing of Soil

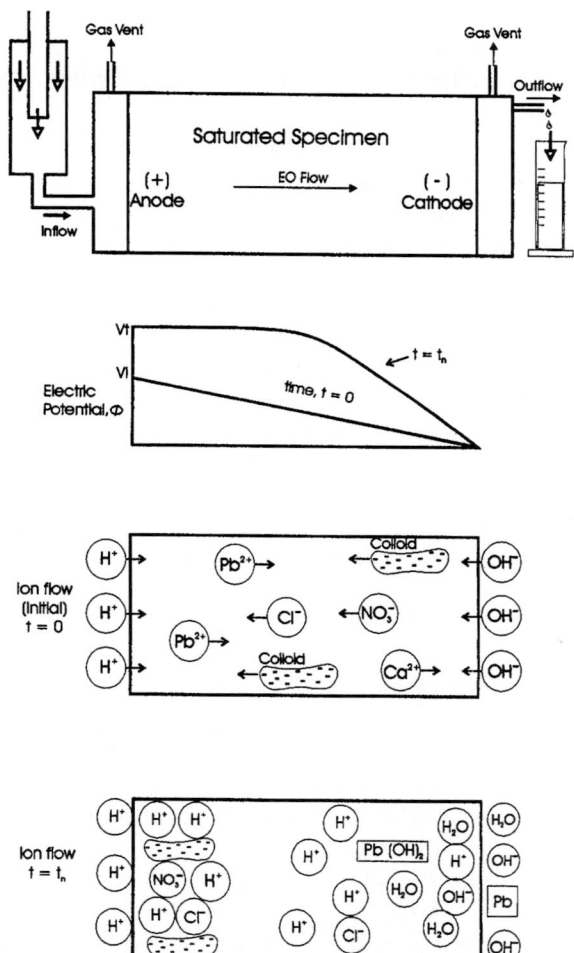

FIGURE 5.32. Electrokinetic soil processing. (Adapted from Acar et al.[349])

conductivities (nitrogen nutrient transport is feasible, but phosphorous transport is not). The electrokinetic approach has no such limitation, since the electro-osmotic flow is normally independent of pore size and grain size.
(2) No channeling is produced, whereas hydraulic pressures often lead to channeling of the fluid.
(3) The *in situ* transport promotes flow direction control.
(4) The approach is effective for low-permeability soils.

(5) Metal ions can normally be recovered in useful form; that is, plated onto the cathode or as concentrated compounds.
(6) Treatment can often be carried out with minimal disruption of normal activities at the site.
(7) Costs are comparable (or in some cases, lower) than conventional treatment techniques.
(8) This method is suitable for combination with other techniques such as pumping, vacuum extraction, or bioremediation.[347] This last combination has been found to be particularly promising and has been termed *electrochemically assisted bioremediation*.[348]

As can be expected, site characterization must be performed before deciding the suitability of this technique in each specific case. Pretreatment is required in some cases. For example, the presence of metallic objects distorts the electrical field path and may consume electricity for their corrosion or even impede the process due to side reactions and product formation.[346,347] Many variables affect the efficiency of this process, including chemical reactions at the electrodes and in the soil, concentration of species, pH (which affects precipitation, dissolution, and the zeta potential of the soil), buffering capacity, hydration of ions and charged particles, saturation, moisture, adsorption capacity, ion exchange properties, mobility of species, permeability, viscosity, dielectric constant, and temperature.[341,349–354]

Clearly, once the polluting species has been removed, the next task is to recover it by treating the resulting concentrated solution (or the plated electrode or the concentrated soil portion) with physical or chemical methods such as chemical or electrochemical ion exchange, adsorption on activated carbon, chemical precipitation, and the like.[341,343,345]

Studies and applications in electrokinetic processing of soils include removal of metal ions (e.g., Cd, Cr, Pb, Hg, Ni, Cu, Zn), anionic pollutants (e.g., Cl^-, NO_3^-, SO_4^{2-}, CN^-), chemical warfare agents, nuclear wastes, coal tar constituents, arsenic, acetate, and organics such as phenol, benzene, toluene, trichloroethylene, or *m*-xylene. Removal of up to 90–99% of some pollutants has been observed under optimum conditions.[342,344,345]

The interested reader is referred to the original literature listed at the end of this chapter for further details on electrokinetic processing of soil and groundwater, particularly on process modeling and simulation. Commercialization efforts are also underway in several countries, and we defer a discussion of these to Chapter 8.

Table 5.7. Commonly Used Industrial Polymers

Monomer	Name	Polymer	Uses	Amt produced in 1975 (millions of kilograms)
$H_2C=CH_2$	Ethylene	Polyethylene	Bags, coatings, toys	3,000
$H_2C=\overset{H}{\underset{CH_3}{C}}$	Propylene	Polypropylene	Beakers, milk cartons	1,000
$H_2C=\overset{H}{\underset{Cl}{C}}$	Vinyl Chloride	Polyvinyl chloride PVC	Floor tile, raincoats, pipe, phonograph records	1,600
$H_2C=\overset{H}{\underset{CN}{C}}$	Acrylonitrile	Polyacrylonitrile PAN	Rugs; Orlon, Acrilan, and Dynel are copolymers with vinyl chloride or other polar monomers	300
$H_2C=\overset{H}{\underset{C_6H_5}{C}}$	Styrene	Polystyrene	Cast articles requiring a transparent plastic	1,200
$H_2C=\overset{CH_3}{\underset{O-C(=O)-CH_3}{C}}$	Methyl methacrylate	Polymethyl methacrylate, Plexiglas	High-quality transparent cast objects	20
$F_2C=CF_2$	Tetrafluoroethylene	Teflon	Gaskets, insulation, bearings, pan coatings	10

Source: Reproduced with permission from W.L. Masterton and E.J. Slowinski, "Chemical Principles." 4th ed., Saunders Co., Philadelphia, 1977.

5.9. EMERGING MATERIALS FOR THE ELECTROCHEMICAL TREATMENT OF POLLUTANTS

The preceding sections in this chapter have highlighted several instances where new electrode materials and membranes have played a key role in the electrochemical treatment of pollutants. In this section, we further elaborate on this aspect and also consider recent developments in new electrolytes (e.g., molten salts, supercritical fluids). Materials science and engineering has much to offer to its electrochemical counterpart in the common goal of pollution abatement. Hence developments in each of these two disciplines will ideally have to mesh with one another, "hand-in-glove", so that new solutions to pollution

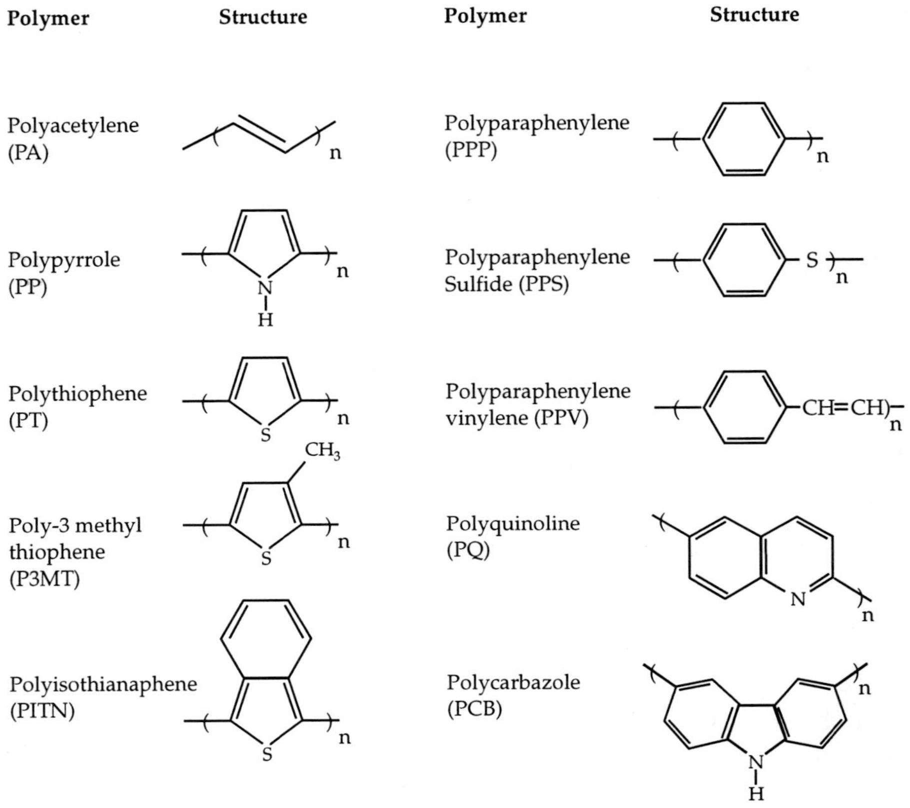

Polymer	Structure	Polymer	Structure
Polyacetylene (PA)		Polyparaphenylene (PPP)	
Polypyrrole (PP)		Polyparaphenylene Sulfide (PPS)	
Polythiophene (PT)		Polyparaphenylene vinylene (PPV)	
Poly-3 methyl thiophene (P3MT)		Polyquinoline (PQ)	
Polyisothianaphene (PITN)		Polycarbazole (PCB)	

FIGURE 5.33. Electronically conducting polymers.

problems can be found. This in turn entails effective communication between the practitioners in each of the three disciplines and efficient cross-fertilization of ideas. In this regard, it is significant that the number of environment-related papers and symposium topics in electrochemistry and materials-related journals and conferences is on the upswing. This trend is likely to continue. Individual classes of materials are considered next.

5.9.1. Electrode Materials

Electronically Conductive Polymers. Polymers, as we know them, are generally electrically insulating. A wide range of polymers (or plastics) have technological importance (Table 5.7). On the other hand, the 1970s and 1980s have heralded a few families of *electroactive* polymers. Electroactive polymers fall

5.9. Emerging Materials for Treatment of Pollutants

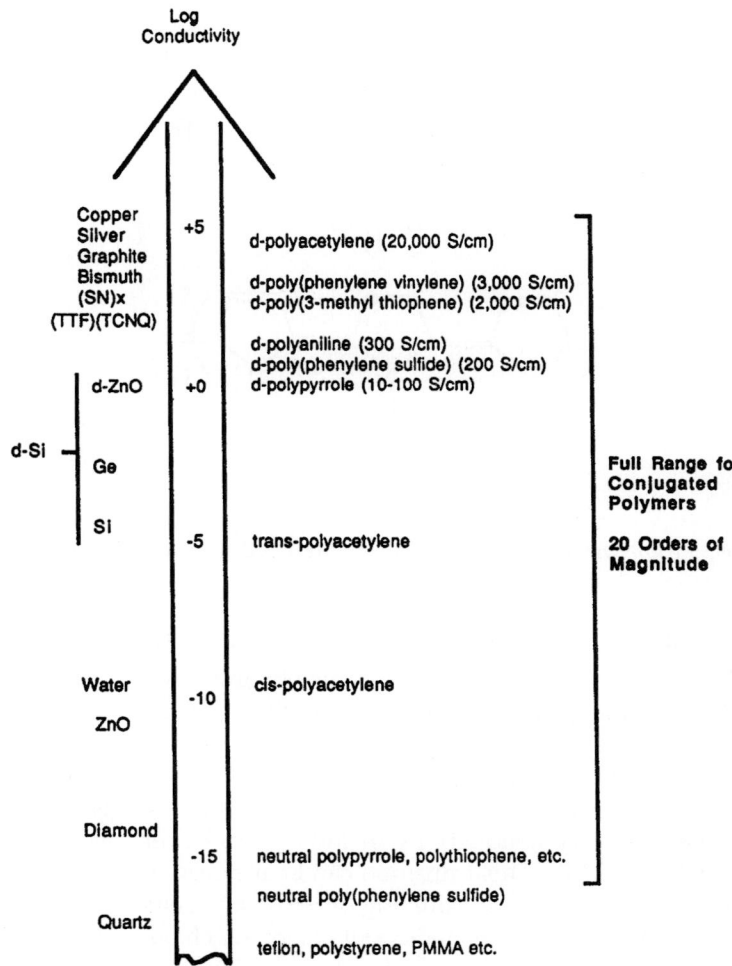

FIGURE 5.34. Conductivity modulation in electronically conductive polymers on doping and comparison with the conductivities of selected materials.

into three broad categories: π-conjugated, electronically conducting polymers; polymers with covalently linked redox groups; and ion-exchange polymers. The polymers considered in Section 5.7 are of the last category. These materials exhibit *ionic* conductivity. On the other hand, the π-conjugated materials are *electronically conducting* because of their delocalized electronic state manifold. The conductive state(s) in these materials are usually induced by oxida-

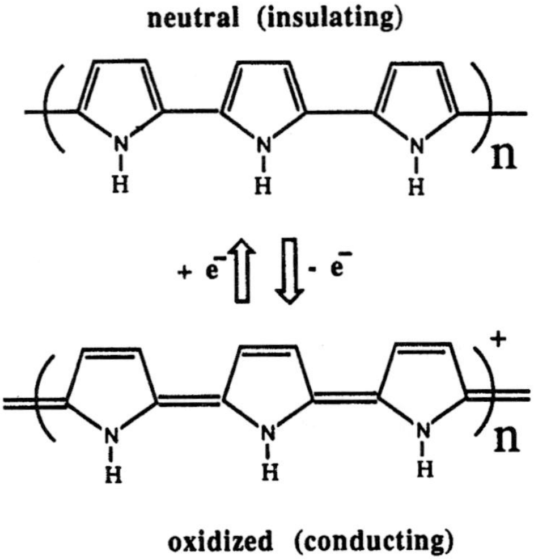

neutral (insulating)

[chemical structure of neutral polypyrrole]

$+e^- \Updownarrow -e^-$

[chemical structure of oxidized polypyrrole]

oxidized (conducting)

FIGURE 5.35. Redox switch in polypyrrole between the doped and neutral states.

tive doping (analogous to a *p*-type semiconductor), although under certain circumstances, *n*-type semiconductivity can also be induced. Figure 5.33 illustrates such *conducting polymers* and representative examples in this category. These materials were briefly considered from a sensor application perspective in Section 4.4.4. Rather remarkable is the property of these polymers to *reversibly* switch from one redox state to the other. Accompanying this transition is a change in the electronic conductivity of the material over several orders of magnitude (Figure 5.34). This transition can be induced either electrochemically or via the use of chemical oxidizing or reducing agents.

Figure 5.35 illustrates the two redox states in one such polymer, polypyrrole, that has been extensively studied in recent years. Unlike the early predecessors in the electronically conducting polymer family such as polyacetylene, polypyrrole has the virtue of excellent stability even in relatively harsh environments as long as the material is not held in the reduced state for an extended period (hr) in contact with O_2. The property of polymers such as polypyrrole to switch from one redox state to another *in a reversible manner* has underpinned a number of technological applications, including secondary rechargeable batteries, fuel cells, chemical sensors, controlled drug delivery, electrochromics, and corrosion protection. They also appear to show promise in environmental pollution control applications.

It has been shown[355] that spontaneous electron transfer occurs from an elec-

5.9. Emerging Materials for Treatment of Pollutants

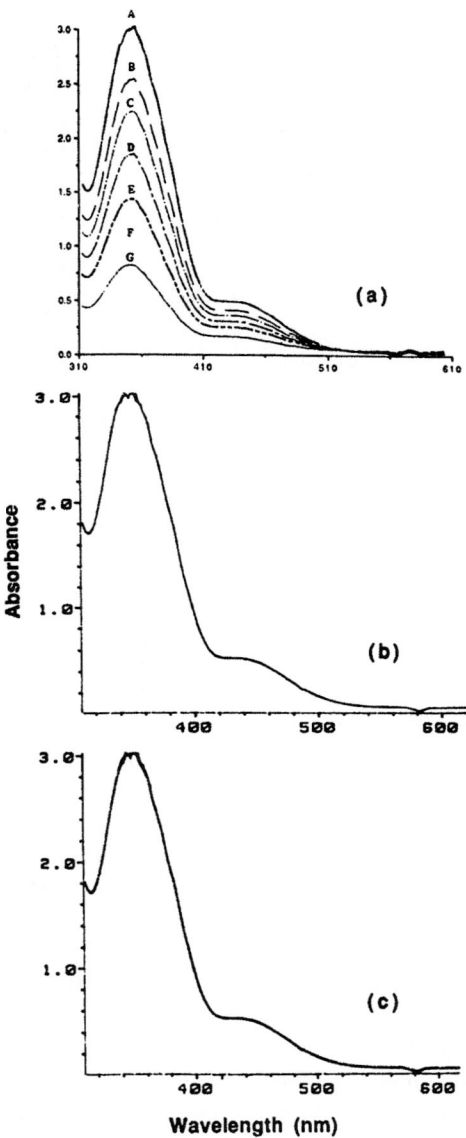

FIGURE 5.36. (a) UV-visible spectra of 1 mM $K_2Cr_2O_7$ in 0.1 M H_2SO_4 solution as a function of time after treatment with Pt-supported polypyrrole at open circuit (~0.40 V): (A) fresh; (B) 2 min; (C) 5 min; (D) 10 min; (E) 20 min; (F) 30 min; and (G) 40 min. (b) Four overlapped spectra of a similar solution treated with a Pt electrode for periods up to 40 min. (c) Five overlapped spectra of a similar solution treated with a polyvinyl chloride pellet for 90 min. The polypyrrole film was grown until a charge of 16 C had accumulated. (Reproduced with permission from Wei et al.[355])

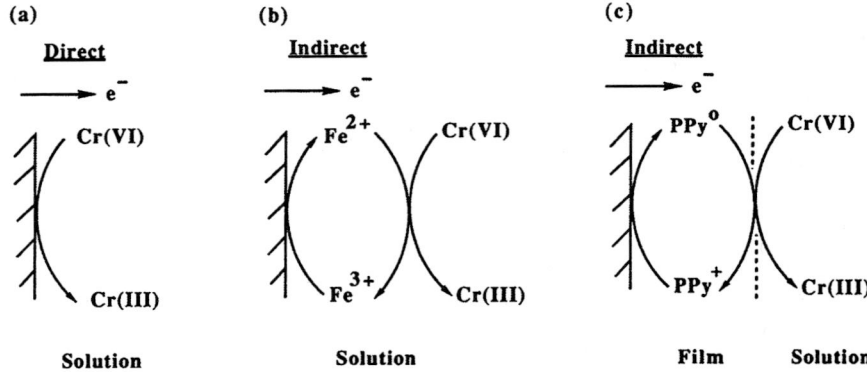

FIGURE 5.37. Schematic model for the electrochemical remediation of Cr(VI): (a) direct electrolysis; (b) indirect electrolysis using $Fe^{2+/3+}$ redox mediator in solution; and (c) modified approach using a polypyrrole (PPy) "catalyst" film that is immobilized at the electrode surface.

trochemically prereduced polypyrrole sample to Cr(VI) consistent with a reaction scheme as follows:

$$Cr_2O_7^{2-} + 6\,PPy^0 + 14\,H^+ \rightarrow 2\,Cr^{3+} + 6\,PPy^+ + 7\,H_2O \tag{5.25}$$

In the above equation, PPy denotes polypyrrole and the superscripts 0 and + correspond to the reduced and oxidized states of the redox polymer, respectively. Taking values for the standard reduction potential of -0.20 V and +1.16 V, respectively, for the two redox couples in this scheme, namely, $PPy^{+/0}$ and $Cr^{6+/3+}$, the process represented by Reaction (5.25) has a standard free energy change (at pH 0) of -376 kJ/mol Cr^{6+}. Thus, electron transfer from PPy to Cr(VI) (reducing the latter to Cr(III)) is spontaneous; that is, thermodynamically "downhill." This is indeed observed experimentally. Contact of a prereduced PPy film with an acidic Cr(VI) solution results in systematic discoloration of the latter (Figure 5.36a). No change in Cr(VI) levels was observed in control experiments where the PPy film was replaced with either a Pt electrode or a polyvinyl chloride specimen (see Figures 5.36b and 5.36c).

Figure 5.37 illustrates how this phenomenon can be used in a practical Cr(VI) treatment application. This illustration also compares the new concept with the *direct* reduction of Cr(VI) (see Section 5.3), which, however, is coulombically inefficient. The new scheme has features in common with the indirect Cr(VI) → Cr(III) reduction approach using a Fe(II/III) mediator redox couple (Section 5.41), except that the PPy "mediator" film now is *immobilized* on a support electrode structure. Hence, the catalyst (such as iron) is not continually consumed in the process.

5.9. Emerging Materials for Treatment of Pollutants

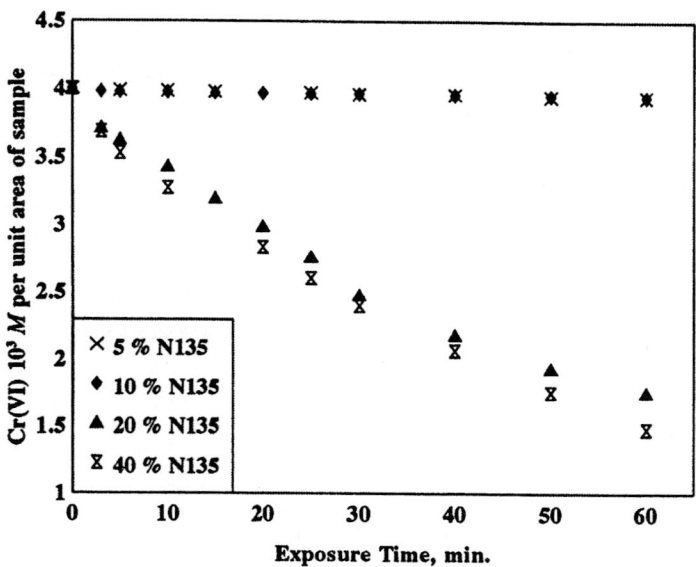

FIGURE 5.38. Influence of carbon black level in the PVC–carbon black composite on the Cr(VI) reduction ability of the composite. The percolation threshold in the composite occurs at ~20% (w/w) of the black. N135 refers to the particular grade of carbon black used. (Reproduced with permission from Wampler et al.[356])

Interestingly, materials such as carbon black also exhibit similar behavior toward Cr(VI); that is, they spontaneously reduce the latter on contact. Figure 5.38 contains representative data for the former material.[356] While carbon black by itself can be compacted into a pellet, it is much too fragile for routine handling and use in this form. Therefore, PVC was chosen as an inert polymeric "binder" for these experiments. Figure 5.39 contains the results from another experiment, where the potential of a carbon black–PVC composite electrode was continuously monitored during contact with Cr(VI). The electrode was first prereduced; on opening the circuit in 0.1 M H_2SO_4, the electrode potential gradually relaxed to the "rest" value. When the solution was subsequently dosed with Cr(VI), the potential underwent a sharp excursion in the positive direction prior to attainment of a plateau. This excursion is symptomatic of the oxidation of the electrode surface by Cr(VI). To examine whether the oxidation resulted in introducing oxygen functionalities to the carbon black surface, X-ray photoelectron spectra of the carbon black samples were compared before and after Cr(VI) treatment. The C 1s signal at 284.6 eV was unaffected by exposure to Cr(VI). Introduction of C-O surface functionalities in the black (as a

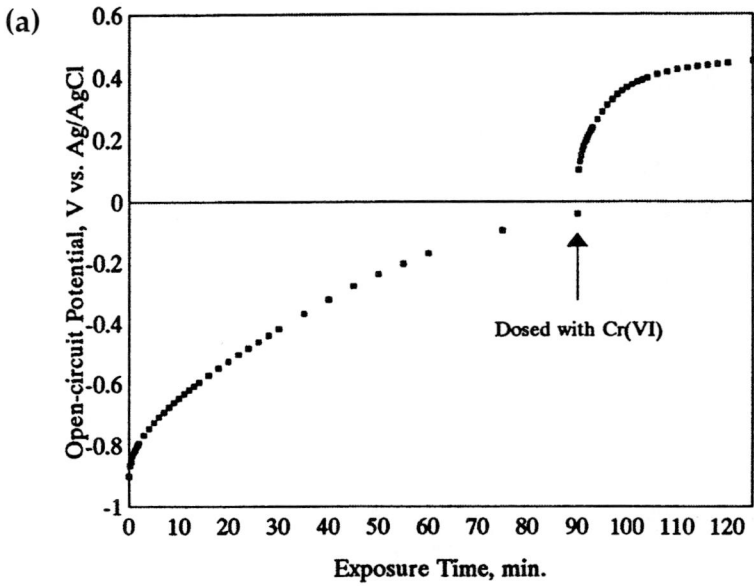

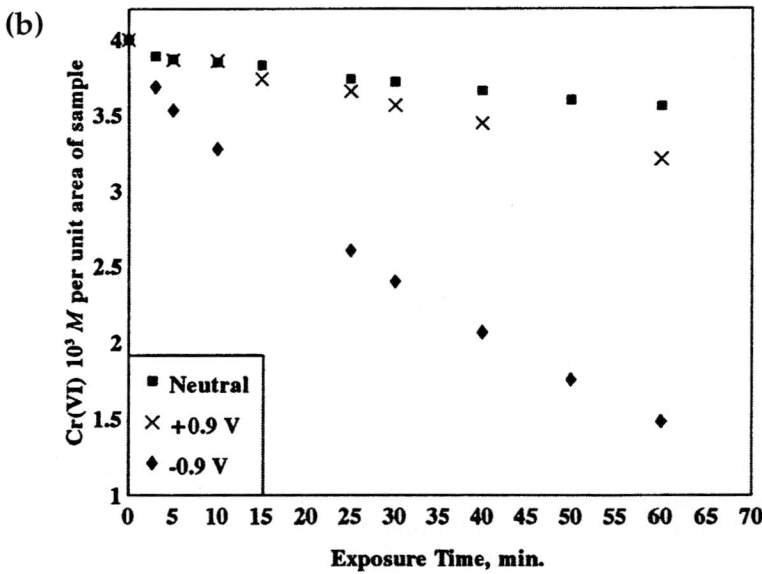

FIGURE 5.39. (a) Variations of the open-circuit potential of a PVC–carbon black composite electrode (30% w/w of carbon black) before and after contact with the Cr(VI) solution. The electrode was prereduced at -0.9 V (vs. Ag/AgCl reference) in 0.1 M H_2SO_4 prior to opening the circuit. (b) Effect of electrochemical pretreatment (in 0.1 M H_2SO_4) on the ability of a PVC–carbon black composite (30% w/w of carbon black) to reduce Cr(VI). (Reproduced with permission from Wampler et al.[356])

5.9. Emerging Materials for Treatment of Pollutants

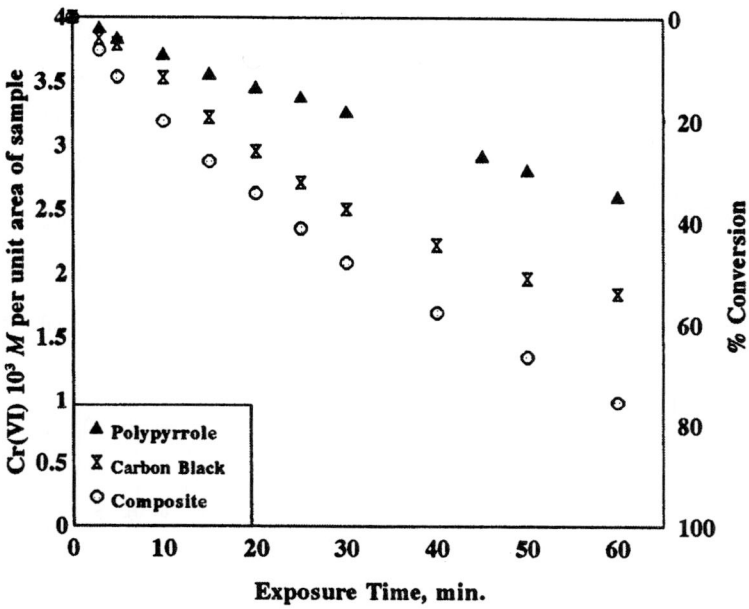

FIGURE 5.40. Comparison of the relative efficacy of polypyrrole, carbon black, and the polypyrrole–carbon black composite for Cr(VI) remediation. The composite contained ~43% (w/w) of the black. In all the three cases, the active material was diluted to 40% by mixing with PVC (refer to text). (Reproduced with permission from Wampler et al.[356])

result of Cr(VI) exposure) should have resulted in a shift in the C 1s peak to higher binding energy. The mechanistic aspects of the Cr(VI) reduction thus are unclear at present for carbon black, particularly at the molecular level.

Composites of PPy and carbon black can be prepared by a variety of methods.[357] Of course, composites of carbon black and a *natural polymer*, namely rubber, have had a long and impressive history in the automobile tire industry. Interestingly, both carbon black and PPy are being considered for very similar technological applications. Figure 5.40 contains a comparison of the relative efficacy of polypyrrole and carbon black for Cr(VI) reduction.[358] These data were generated on three types of samples: a PPy–carbon black composite, a reduced PVC-carbon black composite, and a PPy–PVC composite. The PPy–carbon black composite contained ~43% (w/w) of carbon black. To facilitate the comparison, all these samples were "diluted" to the same extent with the PVC binder to afford 40% (w/w) of the active material. That is, the PPy–carbon black composite in this particular experiment has a composition of 22.8% (w/w) PPy, 17.2% (w/w) carbon black, and 60% (w/w) PVC in the final pelletized sample.

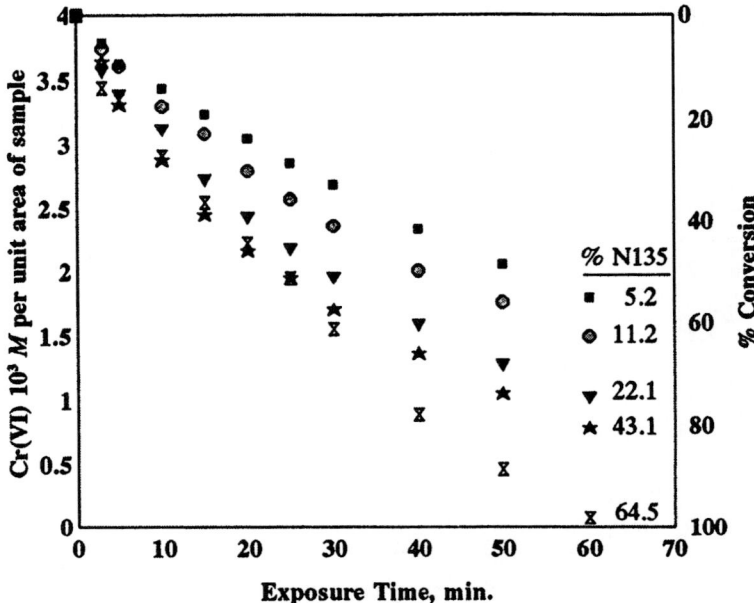

FIGURE 5.41. The influence of carbon black content in a polypyrrole–carbon black composite on its ability to reduce Cr(VI). (Reproduced with permission from Wampler et al.[356])

Interestingly, the composite containing *both* PPy and carbon black as the active material exhibits the highest activity for Cr(VI) reduction. Further, carbon black (at least the grade used in this study) outperforms polypyrrole in its ability to reduce Cr(VI). Figure 5.41 illustrates the influence of carbon black level in the PPy–carbon black composite on the ability of the latter to reduce Cr(VI). Contrary to the experiments considered earlier, the composite samples in this case were prepared *without* the PVC binder. A systematic improvement in the composite performance is noted with an increase in the carbon black content although the rate of improvement tends to saturate at the higher loadings.

While these results are encouraging, it must be noted that the tests to date have been done only at a laboratory scale. Process scale-up as well as important aspects related to the *long-term* stability of PPy on contact with Cr(VI) remain future challenges. Further, experiments are required using reactors that facilitate the treatment of a flow stream containing Cr(VI).

<u>Carbon-Based Electrode Materials</u>. Carbon black was considered in the preceding section as a promising electrode material. Other emerging carbon-derived materials include carbon aerogel composites and carbon foam com-

5.9. Emerging Materials for Treatment of Pollutants

posites. The former are prepared by the sol–gel condensation of resorcinol and formaldehyde in a basic medium followed by drying and pyrolysis.[359] The electrodes possess a high porosity, a surface area between 400 and 1,000 m² g⁻¹, a nominal pore size of less than 50 nm, and a solid matrix composed of interconnected colloid-like particles or fibrils with characteristic diameters of ~10 nm. The porosity and particle size can be varied over a wide range by appropriate tuning of the variables in the fabrication procedure.

Composite metal–carbon electrodes have been recently described.[360,361] Stainless steel fibers (2 μm dia) and carbon fiber bundles (20 μm dia) are combined with cellulose fibers (used as a binder) into a continuous interwoven paper preform. The composite paper preform is then cut into desired dimensions and geometry and sintered to a stainless steel foil substrate. These materials possess specific surface areas (as measured by the BET gas adsorption technique) of ca. 750 m²/g. The high surface area allows easy accessibility to gases and electrolytes while providing adjustable porosities and void volumes. The electrodes can be prepared with other metal fibers including nickel.

Another promising carbonaceous electrode material is diamond. Attractive features of the polycrystalline diamond structure include (a) extreme hardness, (b) corrosion resistance, (c) optical transparency, (d) heat and radiation resistance, and (e) high thermal conductivity. Normally the insulating character of virgin diamond (with resistivity $>10^{12}$ ohm-cm) would preclude the use of this material in electrochemical systems. However, the resistivity of chemical vapor deposited diamond thin films can be made as low as 0.01 ohm-cm by suitable impurity doping. Thus boron-doped diamond thin films have been electrochemically characterized in aqueous media.[362] These materials are chemically inert and microstructurally stable in a variety of acidic and alkaline electrolytes containing fluoride and chloride ions. A recent paper[363] on the electrochemical reduction of nitrate to ammonia using conductive diamond film electrodes exemplifies the promise of this material for environmental applications. However, fundamental studies of the correlation between electrocatalytic behavior and structural–electronic properties are lacking at present for this material. Even for other carbon-derived electrode materials, such information continues to emerge,[364] and this progress is likely to continue in the near future.

<u>Ceramic Electrode Materials</u>. Refractory materials have already revolutionized the microelectronics industry. A new chemically inert ceramic (Ebonex) is finding use in effluent treatment processes.[365,366] The following discussion derives largely from the material contained in Refs. 365 and 366.

The Magneli phase titanium suboxides, of which Ebonex is a member, are homologous compounds of the general formula: Ti_nO_{2n-1}, where n is a number between 4 and 10. Figure 5.42 illustrates a plot of the electronic conductivity versus the compound stoichiometry coefficient, x in the Ti-O binary system.

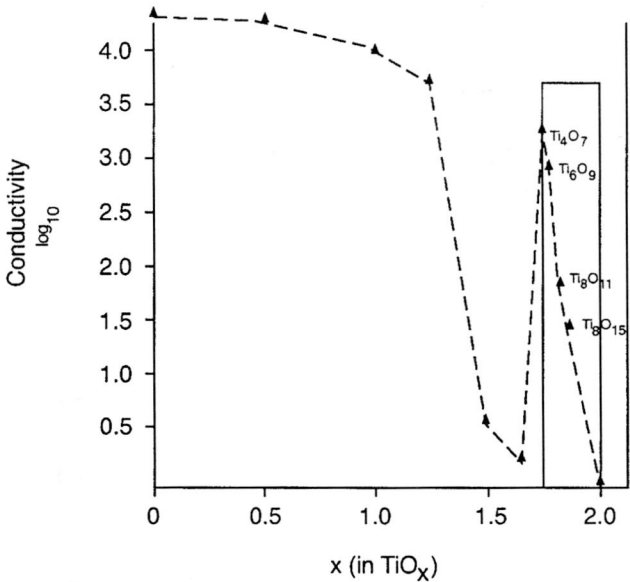

FIGURE 5.42. Electronic conductivity vs. the composition in the titanium–oxygen compound system. The boxed region contains the Magneli (suboxide) phases. (Reproduced with permission from Clarke and Pardoe.[365])

The end members of this system are titanium metal ($x = 0$) and TiO_2 ($x = 2$), which is a semiconductor. We will have much more to say about TiO_2 in the next chapter. The Ebonex ceramic is made from the more conductive members of this series. The compound Ti_4O_7 ($n = 4$, $x = 1.75$) is the most highly conductive phase with a single crystal conductivity of 1,500 mho/cm.[365]

Figure 5.43 contains representative data illustrating the stability of Ebonex electrodes in a rather harsh environment; namely, a H_2SO_4–HF mixture. This remarkable stability is apparently related to the crystal structure of the Magneli phase titanium suboxides. Thus, the oxygen deficient TiO layer (one such layer for every three TiO_2 layers) is protected by a sandwich-type structure made of (chemically inert) TiO_2 layers on either side. In fact, the suboxides appear to be even more chemically inert than titania itself. For example, Ebonex ceramic is resistant to the F⁻ ions that normally dissolve TiO_2 and titanium even in dilute solutions. The reasons for this unusual stability are not clear at present.

Electrochemical studies on Ebonex electrodes have recently appeared.[367] Environmental applications have also been described.[366]

5.9. Emerging Materials for Treatment of Pollutants

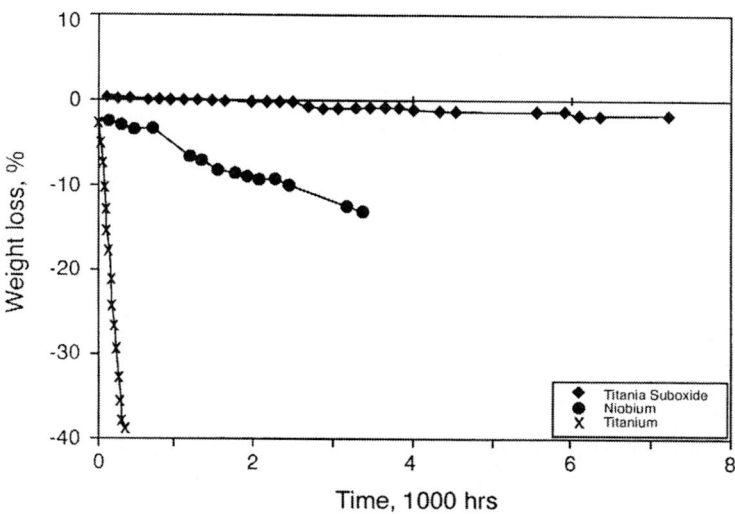

FIGURE 5.43. Weight loss vs. time for three types of electrodes immersed in 15% H_2SO_4 + 0.1% HF at ambient temperature. (Reproduced with permission from Clarke and Pardoe.[365])

5.9.2. Membranes and Electrode Surface Modification Agents

Several examples of this category of materials were already considered from a pollution sensor perspective in Chapter 4. Other examples appeared in Section 5.7 in this chapter.

Inclusion compounds such as cyclodextrins have a wide range of applications as ion exchangers, catalysts or for the microencapsulation of targeted substances. It is in this last context that these compounds may find an application niche in the pollutant treatment field. The most important property of inclusion (or *clathrate*) compounds is that a "host" component can admit "guest" specie(s) into its cavity without any covalent bonds being formed. Other examples in this category include clays, graphite, transition metal chalcogenides and crown-ethers. Cyclodextrins are oligosaccharides constructed from six (α), seven (β), and eight (φ) glucose units. These were first isolated in 1891 although their structural characterization followed much later.[368] Unlike crown ethers, cell toxicity tests have shown that orally administered cyclodextrin is harmless; hence, their widespread use as sweetening agents in the pharmaceutical and food industries.

The molecular shapes of cyclodextrins are truncated cones with internal diameters in the range 0.42–0.88 nm for the α-modification, 0.56–1.08 nm in the β-form, and 0.68–1.20 nm in the φ-form. A recent electrochemical study[369] has

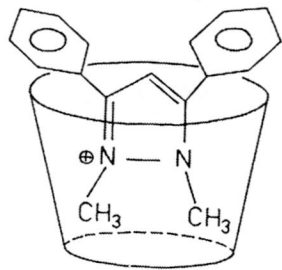

FIGURE 5.44. Structure of the difenzoquat–cyclodextrin complex as deduced from proton NMR. (Reproduced with permission from Pospisil et al.[369])

shown that the herbicide, difenzoquat (1,2-dimethyl-3,5-diphenyl-pyrazolium cation) forms a weak inclusion complex with β-cyclodextrin having the stoichiometry 1:1. On the other hand, the inclusion of this herbicide into α- and φ-cyclodextrins was found to be negligible. Proton NMR data allowed the authors to tentatively assign the inclusion complex structure to the schematic in Figure 5.44.[369] Thus, the NMR data were consistent with a "pyrazole ring in" rather than the "phenyl in" complex structure. Further studies aimed at exploiting the selective encapsulation properties of the cyclodextrin host are needed, particularly with model pollutants as hosts.

5.9.3. Electrolytes

Molten salts, supercritical fluids, and surfactant–oil–water microemulsions constitute three categories of emerging media for carrying out the electrolysis of pollutants. Each of these classes of electrolysis media is discussed next. We already encountered the use of molten salts and emulsions in the treatment of gaseous pollutants (Section 5.6) and oily waste water (Section 5.5).

Molten salt electrochemistry dates back to the time of Sir Humphrey Davy and Michael Faraday. Molten salts are almost always mixtures rather than pure compounds, and therefore an infinite array of solvents is possible. The main advantage of these systems relates to their aprotic nature. This affords the stabilization of unusual oxidation states and reaction intermediates. On the other hand, difficulties associated with their use include the following: (a) they require rather high operating temperatures although ambient-temperature molten salts have recently emerged; (b) they are highly corrosive; and (c) purity presents a problem because the salts are used at a much higher concentration than in the usual case when supporting electrolytes are present only at a level of 1 M or lower (Chapter 2). The molten salt systems include oxides, carbonates, chlorides, nitrates, and fluorides. Mixtures of inorganic and or-

5.9. Emerging Materials for Treatment of Pollutants

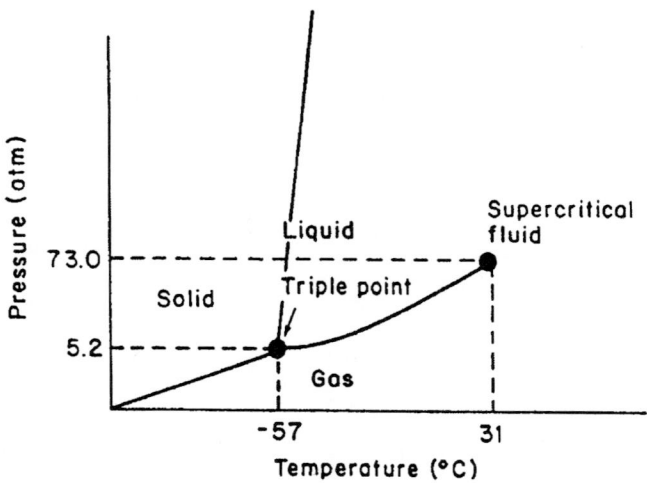

FIGURE 5.45. Phase diagram for carbon dioxide showing supercritical region above critical point at 31°C and 73.0 atm. (Reproduced with permission from Suprex Corp.)

ganic compounds have been used as exemplified by the ambient-temperature molten salt system $AlCl_3$-n-butyl pyridinium chloride.

The treatment of gaseous pollutants is facilitated in molten salts because of their higher solubility in these media and the availability of a wider potential window for electrolysis. The latter advantage is related to the aprotic nature of these solvents. Therefore, the electrochemical reduction of SO_2 has been described in a molten salt electrolyte derived from acetamide.[370] The reduction was shown to proceed via a 1 e⁻ irreversible process to form the radical anion $SO_2^{-\cdot}$. The latter undergoes rapid dimerization to form dithionite, $S_2O_4^{2-}$. It must be noted that this study was *not* performed from an environmental perspective. It remains to be seen whether the positive features, noted here with the use of molten salt electrolytes for the electrolysis of pollutants, outweigh the difficulties associated with their routine use.

A supercritical fluid is produced when a compound (e.g., CO_2) is above its critical pressure and temperature (Figure 5.45). When a gas is raised above its critical temperature and then the pressure is increased, it will not undergo a phase change and condense to form a liquid. Instead, it will become increasingly more dense and, above the critical pressure, will form a single supercritical fluid phase. The properties of a supercritical fluid lie intermediate between those of a liquid and a gas and vary considerably with the applied pressure. Generally, they have viscosities and diffusion coefficients lower than liquids but densities and solubilizing properties similar to liquids. The higher the den-

Table 5.8. Critical Pressures and Temperatures of Selected Compounds

Compound	Critical Temperature (°C)	Critical Pressure (atm)
Xenon	16.5	57.6
Carbon dioxide	31.1	72.9
Nitrous oxide	36.4	71.5
Sulphur hexafluoride	45.5	37.1
Ammonia	132.4	111.3
Pentane	196.5	33.3
Hexane	234.5	29.3

Source: R. M. Smith, "Gas and Liquid Chromatography in Analytical Chemistry." p. 361. Wiley, London, 1988.

sity, the more liquid-like are the properties and vice versa. A limited range of compounds can be easily converted to supercritical fluids at practical temperatures and pressures (Table 5.8). However, many of these compounds are unsuitable for routine use because of toxicity, fire hazard, and corrosion problems. Carbon dioxide has been widely used (particularly in chromatographic separations) because of its low critical values and low toxicity. Supercritical media are intriguing for electrolysis applications, although reports of electrochemistry in such media have only recently begun to appear.[220,371] As with the molten salts discussed previously, the increased solubilization possibility that these media offers is attractive, especially for the efficient treatment of streams containing high levels of organic pollutants.

Polychlorinated biphenyl (PCB) mixtures have been used worldwide as nonflammable oils and lubricants for high-temperature applications (Chapter 1). In spite of the fact that the use of PCBs has ceased because of adverse environmental effects, new methodology for combating PCB pollution continues to be of interest. As discussed in Chapter 1, PCBs are a group of 209 chlorinated aromatic congeners. Chemical and photolytic dechlorination require 2 mol of chemical reductant for each mol of Cl removed.[372] Oxidative procedures have been discussed, although the environmental persistence of PCBs derives from their resistance to oxidative degradation. The more highly chlorinated the congener, the greater is its resistance to oxidation and the longer it will persist in the environment. Further, oxidation can lead to the formation of toxic dioxins. In contrast, reduction of a PCB congener should yield no toxic by-products

5.9. Emerging Materials for Treatment of Pollutants

and, in fact, should be facilitated by high chlorine content. Catalytic reduction using nickel salts and sodium borohydride has been described.[373] However, the removal of the catalyst from treated soils or other solids would be problematic, and many nickel compounds are toxic. A simple titanium catalyst system[374] on the other hand, yields only readily biodegradable organic components and NaCl, TiO_2, and sodium borate as by-products. This latter process thus appears to be more environmentally friendly.

Electrochemical catalytic dehalogenation of PCBs has been described in water-based surfactant media. Micelles of the surfactant cetyltrimethylammonium bromide (CTAB) in water were found to afford efficient electrochemical catalytic dehalogenation.[375,376] Such micellar solutions increase the solubility of the (nonpolar) organohalides over that in water and, in this regard, mimic the function of the molten salt electrolytes discussed earlier. Significant improvements in solubilities and dehalogenation rates of PCBs were achieved by the use of dispersions of the more hydrophobic surfactant didodecyldimethylammonium bromide (DDAB).[377] Dechlorination rates 10-fold larger than those in CTAB micelles were attributed to the improved substrate (organohalide)–DDAB coadsorption on the carbon electrodes employed. A disadvantage, however, is that the pH of the dispersions must be controlled, and added salt is necessary to lower viscosity and increase the conductivity of the medium.

More recently, the same group has explored the use of conductive *bicontinuous microemulsions* as media for catalytic electrochemical dehalogenation.[372,378–380] Microemulsions are clear, thermodynamically stable mixtures of water, oil, and surfactant. Bicontinuous microemulsions are formed of intertwined, dynamic networks of oil and water with the surfactant residing in the interfacial regions.[381,382] Mass transport in these media is much faster than dictated by the *bulk* viscosity of the medium, because molecules and ions can travel along the oil and water conduits in the network. Fast mass transport obviously is an important consideration for electrolysis (Section 2.8).

Catalytic dechlorination of PCBs has been recently reported in such media using constant current at lead cathodes.[372] Zinc phthalocyanine was used as a catalyst. The maximum current efficiency was 20% for 4,4'-dichlorobiphenyl but increased to 42% for the most heavily chlorinated PCB mixture. Nearly total dechlorination of Arochlor 1260 (60% Cl) was reported after 18 hr of electrolysis. Importantly, no salts or buffers were added as in the previous micellar electrolyte cases.[375,383] The particular microemulsion employed was a mixture of DDAB, dodecane, and water. Current problems include side reactions such as water reduction and surfactant decomposition. For the system to be cost effective for large-scale use, the surfactant content of the microemulsion must be decreased severalfold. Nonetheless, these studies do illustrate that the confluence of electrochemistry and colloid science offers powerful solutions to environmental problems.

5.10. SUMMARY

Electrolysis technology for the treatment of a wide variety of environmental pollutants is already at a mature stage of development. In fact, commercialization efforts have been underway during the past few years, and these developments will be reviewed in Chapter 8. Concurrent with this is the continuing discovery of new approaches and new materials for pollution abatement. It will be important and interesting to see how this emerging body of information and technologies will help achieve the realization of even more efficient and stable electrochemical pollutant treatment systems.

REFERENCES

1. D. Golub and Y. Oren, Removal of Chromium from Aqueous Solutions by Treatment with Porous Carbon Electrodes: Electrochemical Principles. *J. Appl. Electrochem.* **19**, 311 (1989).
2. M. Abda, Z. Gavra, and Y. Oren, Removal of Chromium from Aqueous Solutions by Treatment with Fibrous Carbon Electrodes. *J. Appl. Electrochem.* **21**, 734 (1991).
3. C. Comninellis and E. Plattner, Electrochemical Waste Water Treatment. *Chimia* **42**, 250 (1988).
4. C. Comninellis and C. Pulgarin, Anodic Oxidation of Phenol for Waste Water Treatment. *J. Appl. Electrochem.* **21**, 703 (1991).
5. C. Comninellis and C. Pulgarin, Electrochemical Oxidation of Phenol for Wastewater Treatment Using SnO_2 Anodes. *J. Appl. Electrochem.* **23**, 108 (1993).
6. C. Pulgarin, N. Adler, P. Peringer, and C. Comninellis, Electrochemical Detoxification of a 1,4-Benzoquinone Solution in Wastewater Treatment. *Water Res.* **28**, 887 (1994).
7. C. Comninellis, Electrochemical Treatment of Wastewater Containing Organic Pollutants. *In* "Chemical Oxidation: Technologies for the Nineties" (W. W. Eckenfelder, A. R. Bowers and J. A. Roth, eds.), Vol. 3, p. 190. Technomic Publishing, Lancaster and Basel, 1993.
8. V. Smith de Sucre and A. P. Watkinson, Anodic Oxidation of Phenol for Waste Water Treatment. *Can. J. Chem. Eng.* **59**, 52 (1981).
9. C. Seignez, C. Pulgarin, P. Peringer, C. Comninellis, and E. Plattner, Degradation des Pollutants Organiques Industriels. Traitements Electrochimique, Biologique, et leur Couplage. *Swiss Chem.* **14**, 25 (1992).
10. J. R. Akse, J. E. Atwater, L. J. Schussel, J. O. Thompson, and C. E. Verostko, Development and Fabrication of a Breadboard Electrochemical Water Recovery System. 23rd Int. Conf. on Environ. Syst. Colorado Springs, CO, Pap. No. 932032, 1993.

11. M. A. Walsh and R. S. Morris, Electrolytic Reactor and Method for Treating Fluids. U. S. Pat. 4,690,741 (1987).
12. N. Weinberg, The Destruction of Organic Pollutants. *6th Int. Forum Electrolysis Chem. Ind.* The Electrosynthesis Company, Fort Lauderdale, FL, 1992.
13. H. Sharifian and D. W. Kirk, Electrochemical Oxidation of Phenol. *J. Electrochem. Soc.* **133**, 921 (1986).
14. K. F. Kawagoe and D. C. Johnson, Electrocatalysis of Anodic Oxygen-Transfer Reactions. Oxidation of Phenol and Benzene at Bismuth-Doped Lead Dioxide Electrodes in Acidic Solutions. *J. Electrochem. Soc.* **141**, 3404 (1994).
15. M. Gatrell and D. W. Kirk, A Study of the Oxidation of Phenol at Platinum and Preoxidized Platinum Surfaces. *J. Electrochem. Soc.* **140**, 1534 (1993).
16. A. B. Boscoletto, F. Gottardi, L. Milan, P. Pannocchia, V. Tartari, M. Tavan, R. Amadelli, A. de Battisti, A. Barbieri, D. Patracchini, and G. Battaglin, Electrochemical Treatment of Bisphenol-A Containing Wastewaters. *J. Appl. Electrochem.* **24**, 1052 (1994).
17. D. W. Kirk, H. Sharifian, and F. R. Foulkes, Anodic Oxidation of Aniline for Waste Water Treatment. *J. Appl. Electrochem.* **15**, 285 (1985).
18. J. Casado, E. Brillas, and R. M. Bastida, Electrodegradation Kinetics of Aniline and 4-Chloroaniline in Basic Aqueous Media. *In* "Water Purification by Photocatalytic, Photoelectrochemical and Electrochemical Processes" (T. L. Rose, E. Rudd, O. Murphy, and B. E. Conway, eds.) Vol. 94-19, p. 87. The Electrochemical Society, Pennington, NJ, 1994.
19. P. J. van Duin and J. van Erkel, Process for the Detoxification of Chemical Waste Materials. U. S. Pat. 4,443,309 (1984).
20. F. Beck, H. Schulz, and B. Wermeckes, Anodic Dehalogenation of 1,2-Dichloroethane in Aqueous Electrolytes. *Chem. Eng. Technol.* **13**, 371 (1990).
21. L. Kaba, G. D. Hitchens, and J. O'M. Bockris, Electrochemical Incineration of Wastes. *J. Electrochem. Soc.* **137**, 1341 (1990).
22. A. Kowal, J. Kowal, and J. Haber, Electrochemical Oxidation of Formaldehyde, Trioxane and Urotropin for Waste Water Treatment. *Int. Conf. Mod. Electrochem. Ind. Prot. of Environ.* Krakow, Poland, 1993.
22a. W. Meissner and G. Hartel, Degradation of Organics in Waste Water by Electrolysis. 42nd Meet. Int. Soc. Electrochem. Montreaux, Switzerland, 1991.
23. G. S. Zenin, Y. V. Vodolazhkii, L. Y. Markitanova, V. N. Filatova, and E. B. Balakireva, Electrochemical Degradation of Organic Pollutants in Wastewaters with Variable Mineral Content. *J. Appl. Chem. USSR* (Eng. Transl.) **64**, 831 (1991).
24. A. C. Almon and B. R. Buchanan, Electrochemical Oxidation of Organic Waste. *In* "Electrochemistry for a Cleaner Environment" (J. D. Genders

and N. Weinberg, eds.), Chapter 15 The Electrosynthesis Co., East Amherst, NY, 1992.
25. D. T. Hobbs, Electrochemical Treatment of Liquid Nuclear Wastes. *8th Int. Forum Electrolysis Chem. Ind.*, The Electrosynthesis Co., Lake Buena Vista, FL, 1994.
26. F. Hine, M. Yasuda, T. Iida, and Y. Ogata, On the Oxidation of Cyanide Solutions with Lead Dioxide Control Anode. *Electrochim. Acta* **31**, 1389 (1986).
27. U. B. Ogutveren, E. Toru, S. Koparal, and Z. Poyraz, Removal of Cyanide by Anodic Oxidation for Wastewater Treatment. *44th Meet. Int. Soc. Electrochem.* Berlin, 1993.
28. P. Tissot and M. Fragniere, Anodic Oxidation of Cyanide on a Reticulated Three-Dimensional Electrode. *J. Appl. Electrochem.* **24**, 509 (1994).
29. A. Socha and E. Kusmierek, Electrochemical Oxidation of Cyanide Complexes with Copper at Carbon Fibre. Int. Conf. Mod. Electrochem. Ind. Prot. Environ. Krakow, Poland, 1993.
30. K. Scott, Electrochemical Treatment of Aqueous Solutions Containing Organic and Inorganic Effluents. *In* "Water Purification by Photocatalytic, Photoelectrochemical and Electrochemical Processes" (T. L. Rose, E. Rudd, O. Murphy, and B. E. Conway, eds.) Vol. 94-19, The Electrochemical Society, Pennington, NJ, 1994.
31. F. Beck and W. Gabriel, Heterogeneous Redox Catalysis on Ti/TiO$_2$ Cathodes - Reduction of Nitrobenzene. *Angew. Chem. Intl. Ed. Engl.* **24**, 771 (1985).
32. H. J. Byker, Halogenated Aromatic Compound Removal and Destruction Process. U. S. Pat. 4,659,443 (1987).
33. D. J. Mazur and N. L. Weinberg, Methods for Electrochemical Reduction of Halogenated Organic Compounds. U. S. Pat. 4,702,804 (1987).
34. A. Watanabe, H. Takahashi, and Y. Horimoto, The Electrochemical Treatment of Wastewaters Containing o-Chlorobenzoic Acid. Anodic Oxidation After the Cathodic Conversion of o-Chlorobenzoic Acid into a Readily Oxidizable Form. *Denki Kagaku* **53**, 207 (1985).
35. D. Schmal, P. J. van Duin and A. M. C. P. de Jong, Electrochemical Dehalogenation of Organic Compounds in Industrial Waste Water. *Dechema Monogr.* **124**, 241 (1991).
36. D. T. Hobbs, Electrochemical Treatment of Nuclear Waste at the Savannah River Site. *In* "Electrochemistry for a Cleaner Environment" (D. Genders and N. Weinberg, eds.), Chapter 12. The Electrosynthesis Co., East Amherst, NY, 1992.
37. A. B. Mindler and S. B. Tuwiner, Electrolytic Reduction of Nitrate from Solutions of Alkali Metal Hydroxides. U. S. Pat. 3,542,657 (1970).
38. J. J. Kaczur and D. W. Caulfield, Oxyhalide and Oxynitrogen Specie Removal from Aqueous Solutions by Electrochemical Reduction. *8th Int.*

Forum Electrolysis Chem. Ind., The Electrosynthesis Co., Lake Buena Vista, FL, 1994.
39. V. D. Stankovic and A. A. Wragg, A Particulate Vortex Bed Cell for Electrowinning: Operational Modes and Current Efficiency. *J. Appl. Electrochem.* **14**, 615 (1984).
40. R. Bakshi and P. S. Fedkiw, Optimal Time-Varying Potential Control. *J. Appl. Electrochem.* **23**, 715 (1993).
41. R. Bakshi and P.S. Fedkiw, Optimal Time-Varying Cell-Voltage Control of a Parallel-Plate Reactor. *J. Appl. Electrochem.* **24**, 1116 (1994).
42. M. Matlosz and J. Newman, Experimental Investigation of a Porous Carbon Electrode for the Removal of Mercury from Contaminated Brine. *J. Electrochem. Soc.* **133**, 1850 (1986).
43. D. Pletcher, I. Whyte, F. C. Walsh, and J. P. Millington, Reticulated Vitreous Carbon Cathodes for Metal Ion Removal from Process Streams. Part I: Mass Transport Studies. *J. Appl. Electrochem.* **21**, 659 (1991).
44. A. J. Chaudhary and S. M. Grimes, Heavy Metals in the Environment: Part I: Removal of Cobalt from Dilute Effluent Streams by Fluidised Bed Electrolysis. *J. Chem. Tech. Biotechnol.* **56**, 15 (1993).
45. J.-L. Fang, N.-J. Wu, and Z.-W. Wang, Activation Effect of Halides on Chromium Electrodeposition from Chromic Acid Baths. *J. Appl. Electrochem.* **23**, 495 (1993).
46. J. W. Robinson and I. A. L. Rhodes, Development of an Electrochemical Technique for the Removal of Ultratrace Levels of Heavy Metals from Water Using Accelerated Electrodeposition. *Spectrosc. Lett.* **13**, 69 (1980).
47. C. Zur and M. Ariel, The Use of Graphite Cloth Electrodes for the Recovery and Separation of Gold. *J. Appl. Electrochem.* **11**, 639 (1981).
48. D. Simonsson, A Flow-by Packed-bed Electrode for Removal of Metal Ions from Waste Waters. *J. Appl. Electrochem.* **14**, 595 (1984).
49. K. Nishimura, H. Fukushima, T. Akiyama and K. Higashi, Effect of Sulfate and Nickel on Chromium Electrodeposition. *Met. Finish.* **85**, 45 (1987).
50. J. P. Hoare, On the Mechanisms of Chromium Electrodeposition. *J. Electrochem. Soc.* **126**, 190 (1979).
51. F. C. Walsh, The Design and Performance of Electrochemical Reactors for Metal Ion Removal from Dilute Solutions. *8th Int. Forum Electrolysis Chem. Ind.* The Electrosynthesis Co., Orlando, FL, 1994.
52. F. C. Walsh and G. W. Reade, Electrochemical Techniques for the Treatment of Dilute Metal-Ion Solutions. *In* "Environmental Oriented Electrochemistry" (C. A. C. Sequeira, ed.), p. 3. Elsevier, Amsterdam, 1994.
53. A. Yelshin, Dealing with Pollution in Electrochemical Processing and Elsewhere. *Filtr. Sep.* **28**, 14 (1991).
54. F. C. Walsh, The Role of the Rotating Cylinder Electrode Reactor in Metal Ion Removal. *In* "Electrochemistry for a Cleaner Environment" (J. D. Gen-

ders and N. Weinberg, eds.), Chapter 4, p. 101. The Electrosynthesis Co., East Amherst, NY 1992.
55. J. M. Graham, III and N. S. Hammond, Electrolytic Apparatus for Reclaiming Dissolved Metal from Liquid. U. S. Patent 4,269,690 (1981).
56. R. Alkire and A. A. Mirarefi, Current Distribution in a Tubular Electrode Under Laminar Flow: Two Electrode Reactions. *J. Electrochem. Soc.* **124**, 1214 (1977).
57. F. C. Walsh and D. R. Gabe, The Application of Rotating Cylinder Electrode Reactors to Metal Recovery from Industrial Process Liquors. *Trans. Inst. Chem. Eng.* **68**, 107, (1990).
58. F. C. Walsh, The Performance of a 500 Amp Rotating Cylinder Electrode Reactor. Part 3: Methods for the Determination of Mass Transport Data and the Choice of Reactor Model. *Hydrometallurgy* **33**, 367 (1993).
59. D. Robinson and F. C. Walsh, The Performance of a 500 Amp Rotating Cylinder Electrode Reactor. Part 2: Batch Recirculation Studies and Overall Mass Transport. *Hydrometallurgy* **26**, 115 (1991).
60. D. Robinson and F.C Walsh, The Performance of a 500 Amp Rotating Cylinder Electrode Reactor. Part 1: Current-Potential Data and Single Pass Studies. *Hydrometallurgy* **26**, 93 (1991).
61. A. Radwan, A. El-Kiar, H. A. Farag and G. H. Sedahmed, The Role of Mass Transfer in the Electrolytic Reaction of Hexavalent Chromium at Gas Evolving Rotating Cylinder Electrodes. *J. Appl. Electrochem.* **22**, 1161 (1992).
62. C.-D. Zhou and D.-T. Chin, Copper Recovery and Cyanide Destruction with a Plating Barrel Cathode and a Packed-bed Anode. *Plat.Surf. Finish.* **80**, 69 (1993).
63. C.D. Zhou and D.T. Ching, Continuous Electrolytic Treatment of Complex Metal Cyanides with Rotating Barrel Plater as the Cathode and a Packed Bed as the Anode. *Plat. Surf. Finish.* **81**, 70, June (1994).
64. P. M. Maitino and E. J. Maitino, Electrolytic Recovery Unit. U. S. Pat. 5,057,202, (1991).
65. E. Avci, Electrolytic Recovery of Copper from Dilute Solutions Considering Environmental Measures. *J. Appl. Electrochem.* **18**, 288 (1988).
66. D. R. Gabe and F. C. Walsh, The Rotating Cylinder Electrode: A Review of Development. *J. Appl. Electrochem.* **13**, 3 (1983).
67. F. Coeuret, A Flow-through Porous Electrode Patented 100 Years Ago: The Hulin Process. *J. Appl. Electrochem.* **23**, 853 (1993).
68. J. M. Fenton, D. T. Grasso and E. J. Podlaha, A Laboratory Porous Flow-through Electrode Reactor with Interchangeable Electrodes. *J. Electrochem. Soc.* **137**, 2809 (1990).
69. S. Langlois and F. Coeuret, Flow-through and Flow-by Porous Electrodes of Nickel Foam. I. Material Characterization. *J. Appl. Electrochem.* **19**, 43 (1989).

70. S. Langlois and F. Coeuret, Flow-through and Flow-by Porous Electrodes of Nickel Foam. Part III. Theoretical Electrode Potential Distribution in the Flow-by Configuration. *J. Appl. Electrochem.* **20**, 740 (1990).
71. D. Pletcher, I. Whyte, F. C. Walsh, and J. P. Millington, Reticulated Vitreous Carbon Cathodes for Metal Ion Removal from Process Streams. Part II: Removal of Copper(II) from Acid Sulphate Media. *J. Appl. Electrochem.* **21**, 667 (1991).
72. J. V. Zee and J. Newman, Electrochemical Removal of Silver Ions from Photographic Fixing Solutions Using a Porous Flow-through Electrode. *J. Electrochem. Soc.* **124**, 706 (1977).
73. P. S. Fedkiw, Transient Analysis of a Flow-through Porous Electrode at the Limiting Current. *J. Appl. Electrochem.* **11**, 145 (1981).
74. K. Scott, The Role of Diffusion on the Performance of Porous Electrodes. *J. Appl. Electrochem.* **13**, 709 (1983).
75. B. B. Ateya, Effect of Radial Diffusion on the Polarization at Porous Flow-through Electrodes. *J. Appl. Electrochem.* **13**, 417 (1983).
76. J. Jorne and E. Roayaie, Experimental Studies of Flow-through Porous-graphite Chlorine Electrodes. *J. Electrochem. Soc.* **133**, 696 (1986).
77. Y. Sharon and J. R. Goldstein, Mixed Conduction in an Active Porous Electrode. *J. Electrochem. Soc.* **140**, 1655 (1993).
78. J. Qi and R. F. Savinell, Development of the Flow-through Porous Electrode Cell for Bromide Recovery from Brine Solutions. *J. Appl. Electrochem.* **23**, 887 (1993).
79. R. E. Sioda and T. Z. Fahidy, The Performance of Porous and Gauze Electrodes in Electrolysis with Parabolic Velocity Distribution. *J. Appl. Electrochem.* **18**, 853 (1988).
80. J. Salles, J. F. Thovert, and P. M. Adler, Deposition in Porous Media and Clogging. *Chem. Eng. Sci.* **48**, 2839 (1993).
81. B. Fleet, Electrochemical Reactor Systems for Pollution Control and the Removal of Toxic Metals from Industrial Wastewaters. *Collect. Czech. Chem. Commun.* **53**, 1107 (1988).
82. J. Wang, Reticulated Vitreous Carbon – A New Versatile Electrode Material. *Electrochim. Acta* **26**, 1721 (1981).
83. D. Pletcher, I. Whyte, F. C. Walsh, and J. P. Millington, Reticulated Vitreous Carbon Cathodes for Metal Ion Removal from Process Streams. Part III. Studies of a Single Pass Reactor. *Electrochim. Acta.* **23**, 82 (1993).
84. F. C. Walsh, D. Pletcher, I. Whyte, and P. Millington, Electrolytic Removal of Cupric Ions from Dilute Liquors Using Reticulated Vitreous Carbon Cathodes. *J. Chem. Tech. Biotechnol.* **55**, 147 (1992).
85. I. C. Agarwal, A. M. Rochon, H. D. Gesser, and A. B. Sparling, Electrodeposition of Six Heavy Metals on Reticulated Vitreous Carbon Electrodes. *Water Res.* **18**, 227 (1984).

86. T. M. Hung and C. J. Lee, Electrochemical Reduction of Eu(III) Using a Flow-through Porous Graphite Electrode. *J. Appl. Electrochem.* **22**, 865 (1992).
87. V. Tricoli, N. Vatistas and P. F. Marconi, "Removal of Silver Using Graphite-felt Electrodes. *J. Appl. Electrochem.* **23**, 390 (1993).
88. M. Astruc, P.-Y. Guyomar, and C. Lestrade, Modified Carbon or Graphite Fibrous Percolating Porous Electrode, Its Use in Electrochemical Reactions. U. S. Pat. 4,396,474, (1983).
89. B. Fleet and S. D. Gupta, Novel Electrochemical Reactor. *Nature* (London) **263**, 122 (1976).
90. M. Abda and Y. Oren, Removal of Cadmium and Associated Contaminants from Aqueous Wastes by Fibrous Carbon Electrodes. *Water Res.* **27**, 1535 (1993).
91. Y. Oren and A. Soffer, Graphite Felt as an Efficient Porous Electrode for Impurity Removal and Recovery of Metals. *Electrochim. Acta* **28**, 1649 (1983).
92. D. Pletcher and F. C. Walsh, Three Dimensional Electrodes. In "Electrochemistry for a Cleaner Environment" (J. D. Genders and N. L. Weinberg, eds.), Chapter 3, p. 51. The Electrosynthesis Co., East Amherst, NY, 1992.
93. K. Kinoshita and S. C. Leach, Mass-transfer Study of Carbon Felt, Flow-through Electrode. *J. Electrochem. Soc.* **129**, 1993 (1982).
94. J. Qi and R. F. Savinell, Analysis of Flow-through Porous Electrode Cell with Homogeneous Chemical Reactions: Application to Bromide Oxidation in Brine Solutions. *J. Appl. Electrochem.* **23**, 873 (1993).
95. M. Abda, Y. Oren and A. Soffer, The Electrodeposition of Trace Metallic Impurities: Dependence on the Supporting Electrolyte Concentration — A Comparison Between Bipolar and Monopolar Porous Electrodes. *Electrochim. Acta* **32**, 1113 (1987).
96. D. Schmal, J. van Erkel, and P. J. van Duin, Mass Transfer at Carbon Fibre Electrodes. *J. Appl. Electrochem.* **16**, 422 (1986).
97. N. Vatistas, P. F. Marconi and M. Bartolozzi, Mass-transfer Study of the Carbon Felt Electrode. *Electrochim. Acta* **36**, 339 (1991).
98. K. Kusakabe, H. Nishida, S. Morooka, and Y. Kato, Simultaneous Electrochemical Removal of Copper and Chemical Oxygen Demand Using a Packed-bed Electrode Cell. *Electrochim. Acta* **16**, 121 (1986).
99. M. Sankaranarayanan, K. Murugan, C. A. Basha, and R. Vijayavalli, Electrochemical Removal of Nickel from Industrial Effluents. *Bull. Electrochem.* **7**, 75 (1991).
100. G. Kreysa and C. Reynvaan, Optimal Design of Packed Bed Cells for High Conversion. *J. Appl. Electrochem.* **12**, 241 (1982).
101. C.-D. Zhou, E. J. Taylor, R. P. Renz, and E. C. Stortz, Continuous Electrolytic Metal Recovery with a Packed-bed Cathode. AESF/EPA Conf., Orlando, FL, 1995.

102. C.-D. Zhou, E. J. Taylor, R. P. Renz, and M. K. Sunkara, Electrolytic Metal Recovery with a Packed-bed Cathode. AESF SUR/FIN Conf., Indianapolis, IN, 1994.
103. R. H. Davis and H. A. Stone, Flow Through Beds of Porous Particles. *Chem. Eng. Sci.* **48**, 3993 (1993).
104. Y. Bingkun, Improvement of the Performance of Porous Electrodes Using Ionic Conducting Particles: Application to Silver Recovery. *J. Appl. Electrochem.* **20**, 974 (1990).
105. G. Kreysa, K. Jüttner, and J. M. Bisang, Cylindrical Three-dimensional Electrodes Under Limiting Current Conditions. *J. Appl. Electrochem.* **23**, 707 (1993).
106. S. Ehdaie, M. Fleischmann, and R. E. W. Jansson, Application of the Trickle Tower to Problems of Pollution Control. I. The Scavenging of Metal Ions. *J. Appl. Electrochem.* **12**, 59 (1982).
107. E. A. El-Ghaoui and R. E. W. Jansson, Application of the Trickle Tower to Problems of Pollution Control. III. Heavy-Metal Cyanide Solutions. *J. Appl. Electrochem.* **12**, 75 (1982).
108. B. G. Ateya, A. A. Ateya, and M. E. El-Shakre, Applications of Porous Flow Through Electrodes. II. Hydrogen Evolution and Metal Ion Removal at Packed Metal Wool Electrodes. *J. Appl. Electrochem.* **14**, 357 (1984).
109. B. G. Ateya and M. E. El-Shakre, Applications of Porous Flow-through Electrodes. III. Effect of Gas Evolution on the Pore Electrolyte Resistance. *J. Appl. Electrochem.* **14**, 367 (1984).
110. E. C. W. Wijnebelt and L. J. J. Janssen, Reduction of Chromate in Dilute Solution Using Hydrogen in a GBC-Cell. *J. Appl. Electrochem.* **24**, 1028 (1994).
111. G. Lacoste, Influence de la Pulsation Liquide sur le comportment d'un Reacteur Electrochimique Constitue par un Lit de Particules Conductrices. I. Transfert de Matérie. *J. Appl. Electrochem.* **18**, 394 (1988).
112. H. Bergmann and K. Hertwig, Neue Ergenbnisse Der Modellierung Von Reaktoren Mit Partikelelektroden Für Die Elektrochemische Abproduktbehandlung. *Chem. Tech. (Leipzig)* **45**, 77 (1993).
113. R. Alkire and P. K. Ng, Studies on Flow-by Porous Electrodes Having Perpendicular Directions of Current and Electrolyte Flow. *J. Electrochem. Soc.* **124**, 1220 (1977).
114. J. M. Marracino, F. Coeuret, and S. Langlois, A First Investigation of Flow-through Porous Electrodes Made of Metallic Felts or Foams. *Electrochim. Acta* **32**, 1303 (1987).
115. A. Montillet, J. Comiti, and J. Legrand, Application of Metallic Foams in Electrochemical Reactors of Filter-Press Type. Part I: Flow Characterization. *J. Appl. Electrochem.* **23**, 1045 (1993).
116. A. Montillet, J. Comiti, and J. Legrand, Application of Metallic Foams in

Electrochemical Reactors of the Filter-press Type. Part II: Mass Transfer Performance. *J. Appl. Electrochem.* **24**, 384 (1994).

117. L. Chunhua, Fluidized-bed Electrolysis/Ion Exchange Process for the Recovery of Copper from Copper Nitrate Wastewater. *Water Treat.* **6**, 66 (1991).
118. F. Coeuret, The Fluidized Bed Electrode for the Continuous Recovery of Metals. *Water Treat.* **10**, 687 (1980).
119. K. Scott and E. M. Paton, An Analysis of Metal Recovery by Electrodeposition from Mixed Metal Ion Solutions. Part II. Electrodeposition of Cadmium from Process Solutions. *Electrochim. Acta.* **38**, 2191 (1993).
120. F. Goodridge and C. J. Vance, Copper Deposition in a Pilot-plant-scale Fluidized Bed Cell. *Electrochim. Acta.* **24**, 1237 (1979).
121. K. Scott, Metal Recovery Using a Moving-bed Electrode. *J. Appl. Electrochem.* **11**, 339 (1981).
122. K. Scott, A Consideration of Circulating Bed Electrodes for the Recovery of Metal from Dilute Solutions. *J. Appl. Electrochem.* **18**, 504 (1988).
123. R. P. Tison, Copper Recovery Using a Tumbled-bed Electrochemical Reactor. *J. Electrochem. Soc.* **128**, 317 (1981).
124. R. G. Hickman, J. C. Farmer, and Z. Chiba, Mediated Electrochemical Oxidation Techniques as an Alternative to Incineration. *6th Int. Forum Electrolysis Chem. Ind.*, The Electrosynthesis Co., Fort Lauderdale, FL, 1992.
125. T. M. Mill, C. C. D. Yao, S. Smedley, B. Dougherty, and P. Cox, Electrochemical Catalytic Oxidation of Trace Organics in Water for a Space Station Life Support System. *6th Int. Forum Electrolysis Chem. Ind.*, The Electrosynthesis Co., Fort Lauderdale, FL, 1992.
126. D. F. Steele, Electrochemistry and Waste Disposal. *Chem. Br.* **27**, 915 (1991).
127. J. C. Farmer, F. T. Wang, P. R. Lewis, L. J. Summers, and L. Foiles, Electrochemical Treatment of Mixed and Hazardous Wastes: Oxidation of Ethylene Glycol and Benzene by Silver. *J. Electrochem. Soc.* **139**, 659 (1992).
128. C. Zawodzinski, W. Smith, and K. Martinez, Studies of the Electrochemical Generation of Ag(II) and the Catalytic Oxidation of Selected Organics. *6th Int. Forum Electrolysis Chem. Ind.*, The Electrosynthesis Co., Fort Lauderdale, FL, 1992.
129. C. Zawodzinski, W. Smith, and K. Martinez, Kinetic Studies of Electrochemical Generation of Ag(II) Ion and Catalytic Oxidation of Selected Organics. *In* "Environmental Aspects of Electrochemisry and Photoelectrochemistry" (M. Tomkiewicz, H. Yoneyama, R. Haynes and Y. Hori, eds.), Vol. 93-18, p. 170. The Electrochemical Society, Pennington, NJ, 1993.
130. D. E. Kurath, L. A. Bray, and J. L. Ryan, Electrochemical Oxidation of Organic Complexants in High Level Nuclear Wastes. *6th Int. Forum Electrolysis Chem. Ind.*, The Electrosynthesis Co., Fort Lauderdale, FL, 1992.

131. P. M. Dhooge, Catalyzed Mediated Electrochemical Oxidation of Organic Compounds with Iron(III). *185th Meet. Electrochem. Soc.*, San Francisco, 1994. Ext. Abstr. No. 1146.
132. P. M. Dhooge, Catalyzed, Mediated Electrochemical Oxidation of Organic Compounds with Iron(III). *In* "Water Purification by Photocatalytic, Photoelectrochemical and Electrochemical Processes" (T. L. Rose, E. Rudd, O. Murphy and B. E. Conway, eds.), Vol. 94-19, p. 152. The Electrochemical Society, Pennington, NJ, 1994.
133. R. L. Clarke, Iron(III) as a Catalyst for the Anodic Destruction of Carbonaceous Waste. *In* "Electrochemistry for a Cleaner Environment" (J. D. Genders and N. Weinberg, eds.), Chapter 13. The Electrosynthesis Co., East Amherst, NY, 1992.
134. C. A. Koval, S. M. Drew, T. Spontarelli, and R. D. Noble, Concentration and Removal of Nitrogen and Sulfur Containing Compounds from Organic Liquid Phases Using Electrochemically Reversed Chemical Complexation. *Sep. Sci. Technol.* **23**, 1389 (1988).
135. S. B. Gale and P. P. O'Donnell, Method of Treating Contaminant Ions in an Aqueous Medium with Electrolytically Generated Ferrous Ions and Apparatus Therefore. U. S. Pat. 4,693,798, (1987).
136. S. Mizuta, W. Kondo, K. Fujii, H. Iida, S. Isshiki, H. Noguchi, T. Kikuchi, H. Sue, and K. Sakai, Hydrogen Production from Hydrogen Sulfide by the Fe-Cl Hybrid Process. *Ind. Eng. Chem. Res.* **30**, 1601 (1991).
137. J. C. Farmer, F. T. Wang, P. R. Lewis, and L. J. Summers, Destruction of Chlorinated Organics by Co(III)-mediated Electrochemical Oxidation. *J. Electrochem. Soc.* **139**, 3025 (1992).
138. C. Zawodzinski, W. H. Smith, and K. R. Martinez, Cobalt(III) Catalyzed Electrochemical Oxidation of Selected Organics. *8th Int. Forum Electrolysis Chem. Ind.*, The Electrosynthesis Co., Orlando, FL, 1994.
139. J. Farmer and Z. Chiba, Fundamental Studies of the Mediated Electrochemical Oxidation of Wastes. *In* "Water Purification by Photocatalytic, Photoelectrochemical and Electrochemical Processes" (T. L. Rose, E. Rudd, O. Murphy, and B. E. Conway, eds.), Vol. 94-19, p. 144. The Electrochemical Society Proceedings, Pennington, NJ.
140. O. M.-R. Chyan and K. Rajeshwar, Heterojunction Photoelectrodes. II. Electrochemistry at Tin-Doped Indium Oxide/Aqueous Electrolyte Interfaces. *J. Electrochem. Soc.* **132**, 2109 (1985).
141. R. Kötz, S. Stucki and B. Carcer, Electrochemical Waste Water Treatment Using High Overvoltage Anodes. Part I. Physical and Electrochemical Properties of SnO_2 Anodes. *J. Appl. Electrochem.* **21**, 14 (1991).
142. S. Stucki, R. Kötz, B. Carcer, and W. Suter, Electrochemical Waste Water Treatment Using Overvoltage Anodes. Part II. Anode Performance and Applications. *J. Appl. Electrochem.* **21**, 99 (1991).

143. F. Beck and H. Schulz, Heterogeneous Redox Catalysis with Titanium/Chromium(III) Oxide + TiO_2 Composite Anodes. *Electrochim. Acta.* **29**, 1569 (1984).
144. C. Comninellis, Electrocatalysis in the Electrochemical Conversion/Combustion of Organic Pollutants for Wastewater Treatment. *Electrochim. Acta.* **39**, 1857 (1994).
145. F. Hine, M. Yasuda, T. Iida, and Y. Ogata, On the Oxidation of Cyanide Solutions with Lead Dioxide Coated Anode. *Electrochim Acta.* **31**, 1389 (1986).
146. T. Matsue, M. Fujihira, and T. Osa, Oxidation of Alkylbenzenes by Electrogenerated Hydroxyl Radical. *J. Electrochem. Soc.* **128**, 2565 (1981).
147. R. Tomat and A. Rigo, Electrochemical Oxidation of Toluene Promoted by OH Radicals. *J. Appl. Electrochem.* **14**, 1 (1984).
148. H. Chang and D. C. Johnson, Electrocatalysts of Anodic Oxygen-transfer Reactions. Ultrathin Films of Lead Oxide on Solid Electrodes. *J. Electrochem. Soc.* **137**, 3108 (1990).
149. J. Feng and D. C. Johnson, Electrocatalysts of Anodic Oxygen-transfer Reactions. Fe-doped Beta-lead Dioxide Electrodeposited on Noble Metals. *J. Electrochem. Soc.* **137**, 507 (1990).
150. B. Wels and D. C. Johnson, Electrocatalysts of Anodic Oxygen-transfer Reactions. Oxidation of Cyanide at Electrodeposited Copper Oxide Cathodes in Alkaline Media. *J. Electrochem. Soc.* **137**, 2785 (1990).
151. D. C. Johnson, H. Chang, J. Feng, and W. Wang, New Anodes for Electrochemical Incineration: Concepts of Atomic Engineering. *In* "Electrochemistry for a Cleaner Environment" (J. D. Genders and N. Weinberg, eds.), Chapter 17. The Electrosynthesis Co., East Amherst, NY, 1992.
152. R. C. Rhees and B. B. Halker, Process for Producing a Lead Dioxide Coated Anode from a Lead Electrolyte Which Contains Dissolved Bismuth. U. S. Pat. 4,101,390, (1978).
153. T. C. Franklin, W. K. Adeniyi, and R. Nnodimele, The Electrooxidation of Some Insoluble Inorganic Sulfides, Selenides and Tellurides in Cationic Surfactant-aqueous Sodium Hydroxide Systems. *J. Electrochem. Soc.* **137**, 480 (1990).
154. T. C. Franklin, J. Darlington, T. Solouki, and N. Tran, The Use of the Oxidation of Barium Peroxide in Aqueous Surfactant Systems in the Electrolytic Destruction of Organic Compounds. *J. Electrochem. Soc.* **138**, 2285 (1991).
155. T. C. Franklin, G. Oliver, R. Nnodimele and K. Couch, Destruction of Halogenated Hydrocarbons Accompanied by Generation of Electricity. *J. Electrochem. Soc.* **139**, 2192 (1992).
156. T. C. Franklin, R. Nnodimele, and J. Kerimo, Electrochemical Oxidations of Several Organic Compounds in Aqueous Surfactant Suspensions Us-

ing Higher Valence Oxides in Intermediates. *J. Electrochem. Soc.* **140**, 2145 (1993).
157. E. A. El-Ghaoui, R. E. W. Jansson, and C. Moreland, Application of the Trickle Tower to Problems of Pollution Control. II. The Direct and Indirect Oxidation of Cyanide. *J. Appl. Electrochem.* **12**, 69 (1982).
158. J.-S. Do and C.-P. Chen, In Situ Oxidative Degradation of Formaldehyde With Electrogenerated Hydrogen Peroxide. *J. Electrochem. Soc.* **140**, 1632 (1993).
159. J.-S. Do and C.-P Chem, Kinetics of In Situ Degradation of Formaldehyde with Electrogenerated Hydrogen Peroxide. *Ind. Eng. Chem. Res.* **33**, 387 (1994).
160. P. Tatapudi and J. M. Fenton, Synthesis of Hydrogen Peroxide in a Proton Exchange Membrane Electrochemical Reactor. *J. Electrochem. Soc.* **140**, L55 (1993).
161. P. Tatapudi and J.M. Fenton, Simultaneous Synthesis of Ozone and Hydrogen Peroxide in a Proton Exchange Membrane Electrochemical Reactor. *J. Electrochem. Soc.* **141**, 1174 (1994).
162. J. Feng, D. C. Johnson, S. N. Lowery, and J. J. Carey, Electrocatalysis of Anodic Oxygen-transfer Reactions. Evolution of Ozone. *J. Electrochem. Soc.* **141**, 2708 (1994).
163. P. C. Foller and C. W. Tobias, The Anodic Evolution of Ozone. *J. Electrochem. Soc.* **129**, 506 (1982).
164. P. C. Foller and M. L. Goodwin, The Electrochemical Generation of High Concentration Ozone for Small-Scale Applications. *Ozone: Sci. Eng.* **6**, 29 (1984).
165. P. C. Foller and G. H. Kelsall, Ozone Generation via the Electrolysis of Fluoboric Acid Using Glassy Carbon Anodes and Air Depolarized Cathodes. *J. Appl. Electrochem.* **23**, 996 (1993).
166. C. Walling, Fenton's Reagent Revisited. *Acc. Chem. Res.* **8**, 125 (1975).
167. E. Koubek, Photochemically Induced Oxidation of Refractory Organics with Hydrogen Peroxide. *Ind. Eng. Chem., Process. Des. Dev.* **14**, 348 (1975).
168. R. G. Zepp, B. C. Faust, and J. Hoigné, Hydroxyl Radical Formation in Aqueous Reactions (pH 3-8) of Iron(II) with Hydrogen Peroxide. The Photo-Fenton Reaction. *Environ. Sci. Technol.* **26**, 313 (1992).
169. Y.-L. Hsiao and K. Nobe, Hydroxylation of Chlorobenzene and Phenol in a Packed Bed Flow Reactor with Electrogenerated Fenton's Reagent. *J. Appl. Electrochem.* **23**, 943 (1993).
170. A. Safarzedeh-Amiri, Photocatalytic Method for Treatment of Contaminated Water. U. S. Pat. 5,266,214, (1993).
171. D. F. Bishop, G. Stern, M. Fleischmann, and L. S. Marshall, Hydrogen Peroxide Catalytic Oxidation of Refractory Organics in Municipal Waste Waters. *Ind. Eng. Chem., Proc. Des. Dev.* **7**, 110 (1968).

172. A. P. Murphy, W. J. Boegli, M. K. Price, and C. D. Moody, A Fenton-like Reaction to Neutralize Formaldehyde Waste Solutions. *Environ. Sci. Technol.* **23**, 166 (1989).
173. M. L. Kremer, Oxidation Reduction Step in Catalytic Decomposition of Hydrogen Peroxide by Ferric Ions. *Trans. Faraday Soc.* **59**, 2535 (1963).
174. J. Nanzer and F. Coueret, Adoucissement De L'eau A L'aide Du Reacteur Electrochimique A Film Ruisselant. *J. Appl. Electrochem.* **22**, 364 (1992).
175. V. A. Kolesnikov, E. A. Shalyt, and P. K. Aarinola, Complex Technology of Electrochemical Water Treatment with Regeneration of Valuable Components in Electroplating Production. *181st Meet. Electrochem. Soc.*, St. Louis, MO, 1992.
176. R. Labrecque, A. Theoret, F. Lamarche, and M. Boulet, Application of Electrodialysis for the Extraction of Cytoplasmic Leaf Proteins from Alfalfa. *In* "Electrochemistry for a Cleaner Environment" (J. D. Genders and N. Weinberg, eds.), Chapter 9. The Electrosynthesis Co., East Amherst, NY, 1992.
177. J. N. Cloutier, M. K. Azarniouch, and D. Callender, Electrolysis of Weak Black Liquor. Part II. Effect of Process Parameters on the Energy Efficiency of the Electrolytic Cell. *8th Int. Forum Electrolysis Chem. Ind.*, The Electrosynthesis Co., Lake Buena Vista, FL, 1994.
178. N. K. Khosla, S. Venkatachalam, and P. Somasundaran, Pulsed Electrogeneration of Bubbles for Electroflotation. *J. Appl. Electrochem.* **21**, 986 (1991).
179. R. Mraz and J. Krysa, Long Service Life IrO_2/Ta_2O_5 Electrodes for Electroflotation. *J. Appl. Electrochem.* **24**, 1262 (1994).
180. E. A. Vik, D. A. Carlson, A. S. Eikum, and E. T. Gjessing, Electrocoagulation of Potable Water. *Water Res.* **18**, 1355 (1984).
181. M. H. Weintraub, M. A. Dzieciuch, and R. L. Gealer, Method for Breaking an Oil-in-Water Emulsion. U. S. Pat. 4,194,972, (1980).
182. R. W. Peters and G. F. Bennet, The Simultaneous Removal of Oil and Heavy Metals from Industrial Wastewaters Using Hydroxide or Sulfide Precipitation Coupled with Air Flotation. *Hazardous Wastes and Hazardous Materials* **6**, 327 (1989).
183. M.-F. Pouet, F. Persin, and M. Rumeau, Intensive Treatment by Electrocoagulation-flotation-tangential Flow Microfiltration in Areas of High Seasonal Population. *Water Sci. Technol.* **25**, 247 (1992).
184. N. Biswas and G. Lazarescu, Removal of Oil from Emulsions Using Electrocogulation. *Int. J. Environ. Stud.* **38**, 65 (1991).
184a. J.J. McKetta, "Encyclopaedia of Chemical Processing and Design," Vol. 32. Marcel Dekker, NY, 1990.
185. S. H. Lin and C. F. Peng, Treatment of Textile Wastewater by Electrochemical Method. *Water Res.* **28**, 277 (1994).

186. A. Wilcock, M. Brewster, and W. Tincher, Using Electrochemical Technology to Treat Textile Wastewater: Three Case Studies. *Am. Dyestuff. Rep.* **81**, 15 (1992).
187. A. E. Wilcock and S. P. Hay, Recycling of Electrochemically Treated Disperse Dye Effluent. *Can. Text. J.* **108**, 37 (1991).
188. F. Persin and M. Rumeau, Le Traitement Electrochimique des Eaux et des Effluents. *Trib. Eau.* **82**, 45 (1989).
189. M. H. Weintraub, R. L. Gealer, A. Golovoy, M. A. Dzieciuch, and H. Durham, Development of Electrolytic Treatment of Oily Wastewater. *Environ. Progr.* **2**, 32 (1983).
190. D. R. Jenke and F. E. Diebold, Electroprecipitation Treatment of Acid Mine Wastewater. *Water Res.* **18**, 855 (1984).
191. K. Moeglich, Sorbent Particulate Material and Manufacture Thereof. U. S. Pat. 4,048,028, (1977).
192. H. Roques and Y. Aurelle, Oil-water Separations in Oil Recovery and Oily Wastewater Treatment. *In* "New Developments in Industrial Wastewater Treatment" (A. Turkman and O. Uslu, eds.), p. 155. Kluwer Academic Publishers, Dordrecht, The Netherlands, 1991.
193. J. G. Ibanez, M. M. Takimoto, R. C. Vasquez, S. Basak, N. Myung, and K. Rajeshwar, Laboratory Experiments on Electrochemical Remediation of the Environment: Electrocoagulation of Oily Wastewater. *J. Chem. Educ.* **75**, 1050 (1995).
194. G. L. Sjoblom and S. L. Goren, Effect of D. C. Electric Fields on Liquid-liquid Settling. *Ind. Eng. Chem. Fund.* **5**, 520 (1966).
195. A. Wilcock, Spectrophotometric Analysis of Electrochemically Treated, Simulated, Dispersed Dyebath Effluent. *Text. Chem. Color.* **24**, 29 (1992).
196. J.-S. Do and M.-L. Chen, Decolourization of Dye-containing Solutions by Electrocoagulation. *J. Appl. Electrochem.* **24**, 785 (1994).
197. Z. Youchun, L. Dunwen, Z. Yongqi, L. Jianmin, and L. Meiqiang, The Study of the Electrolysis Coagulation Process Using Insoluble Anodes for Treatment of Printing and Dyeing Wastewater. *Water Treat.* **6**, 277 (1991).
198. O. Groterud and L. Smoczynski, Phosphorus Removal from Water by Means of Electrolysis. *Water Res.* **20**, 667 (1986).
199. S. Lech and Z. Beata, Wastewaters Treatment by Direct Electrocoagulation. Int. Conf. Mod. Electrochem. Ind. Prot. Environ., Krakow, Poland, 1993.
200. P. Cadenhead and K. L. Sublette, Oxidation of Hydrogen Sulfide by *Thiobacilli. Biotechnol. Bioeng.* **35**, 1150 (1990).
201. C. Ongcharit, K. L. Sublette, and Y. T. Shah, Oxidation of Hydrogen Sulfide by Flocculated *Thiobacillus Denitrificans* as a Continuous Culture. *Biotechnol. Bioeng.* **37**, 497 (1991).
202. D. van Velzen, Electrochemical Processes in the Protection of the Envi-

ronment. In "Electrochemistry for a Cleaner Environment" (D. Genders and N. Weinberg, eds.), Chapter 21. The Electrosynthesis Co., East Amherst, NY, 1992.
203. G. Kreysa, Elektrochemische Umwelttechnik. *Chem. Ing. Tech.* **62**, 357 (1990).
204. N. L. Weinberg, J. D. Genders and A. O. Minklei, Methods for Purification of Air. U. S. Pat. 5,009,869 (1991).
205. J. Winnick, Electrochemical Waste Treatment: Gas Clean-up. In. "Electrochemical Solid State Science Education at the Graduate and Undergraduate Level" (W. H. Smyrl and F. McLarnon, eds.), p. 303. The Electrochemical Society, Pennington, NJ, 1987.
206. D. Weaver and J. Winnick, Performance of an Electrochemical Membrane H_2S Separator. *J. Electrochem. Soc.* **139**, 492 (1992).
207. J. Winnick, Electrochemical Membrane Gas Separation. *8th Int. Forum Electrolysis Chem. Ind.*, The Electrosynthesis Co., Lake Buena Vista, FL, (1994).
208. S. R. Alexander and J. Winnick, Electrochemical Polishing of Hydrogen Sulfide from Coal Synthesis Gas. *J. Appl. Electrochem.* **24**, 1092 (1994).
209. H. S. Lim and J. Winnick, Electrochemical Removal and Concentration of Hydrogen Sulfide from Coal Gas. *J. Electrochem. Soc.* **131**, 562 (1984).
210. S. Taguchi and A. Aramata, Surface-structure Sensitive Reduced CO_2 Formation on Pt Single Crystal Electrodes in Sulfuric Acid Solution. *Electrochim. Acta* **39**, 2533 (1994).
211. O. Koga and Y. Hori, Reduction of Adsorbed CO on a Ni Electrode in Connection with the Electrochemical Reduction of CO_2. *Electrochim. Acta* **38**, 1391 (1993).
212. Y. Hori, A. Murata, T. Tsukamoto, H. Wakebe, O. Koga and H. Yamazaki, Adsorption of Carbon Monoxide at Copper Electrode Accompanied by Electron Transfer Observed by Voltammetry and IR Spectroscopy. *Electrochim. Acta* **39**, 2495 (1994).
213. G. Z. Kyriacou and A. K. Anagnostopoulos, Influence of CO_2 Partial Pressure and the Supporting Electrolyte Cation on the Product Distribution in CO_2 Electroreduction. *J. Appl. Electrochem.* **23**, 483 (1993).
214. D. W. DeWulf, T. Jin, and A. J. Bard, Electrochemical and Surface Studies of Carbon Dioxide Reduction to Methane and Ethylene at Copper Electrodes in Aqueous Solutions. *J. Electrochem. Soc.* **136**, 1686 (1989).
215. R. L. Cook, R. C. MacDuff, and A. F. Sammells, Ambient Temperature Gas Phase CO_2 Reduction to Hydrocarbons at Solid Polymer Electrolyte Cells. *J. Electrochem. Soc.* **135**, 1470 (1988).
216. D. P. Summers and K. W. Frese, Jr., The Electrochemical Reduction of Aqueous Carbon Monoxide and Methanol to Methane at Ruthenium Electrodes. *J. Electrochem. Soc.* **135**, 264 (1988).

217. J. Ryce, T. N. Anderson, and H. Eyring, The Electrode Reduction Kinetics of Carbon Dioxide in Aqueous Solution. *J. Phys. Chem.* **76**, 3278 (1972).
218. Y. B. Vassilev, V. S. Bagotzky, N. V. Osetrova, O. A. Khazova, and N. A. Mayorova, Electroreduction of Carbon Dioxide. Part I. The Mechanism and Kinetics of Electroreduction of CO_2 in Aqueous Solution on Metals with High and Moderate Hydrogen Overvoltages. *J. Electroanal. Chem. Interfacial Electrochem.* **189**, 271 (1985).
219. Y. B. Vassilev, V. S. Bagotzky, N. V. Osetrova, and A. A. Mikhailova, Electroreduction of Carbon Dioxide. Part III. Adsorption and Reduction of CO_2 on Platinum Metals. *J. Electroanal. Chem. Interfacial Electrochem.* **189**, 311 (1985).
220. M. Tomkiewicz, H. Yoneyama, R. Haynes, and Y. Hori, eds., "Environmental Aspects of Electrochemistry and Photoelectrochemistry." Vol. 93-18. The Electrochemical Society, Pennington, NJ, 1993.
221. Y. Taniguchi, *in* "Modern Aspects of Electrochemistry" (J. O'M. Bockris, R. E. White and B. E. Conway, eds.), Vol. 20, p. 327. Plenum, New York 1989.
222. H. L. Chum and M. M. Baizer, eds., "The Electrochemistry of Biomass and Derived Materials." Chapter 3, p. 141. American Chemical Society, Washington, DC, 1985.
223. M. Halmann, *in* "Energy Resources Through Photoelectrochemistry and Catalysis" (M. Gratzel, ed.), Chapter 15. Academic Press, New York, 1983.
224. P. G. Russell, N. Kovac, S. Srinivasan, and M. Steinberg, The Electrochemical Reduction of Carbon Dioxide, Formic Acid and Formaldehyde. *J. Electrochem. Soc.* **124**, 1329 (1977).
225. S. Kapusta and N. Hackerman, The Electroreduction of Carbon Dioxide and Formic Acid on Tin and Indium Electrodes. *J. Electrochem. Soc.* **130**, 607 (1983).
226. S. Nakagawa, A. Kudo, M. Azuma, and T. Sakata, Effect of Pressure on the Electrochemical Reduction of CO_2 on Group VIII Metal Electrodes. *J. Electroanal. Chem. Interfacial Electrochem.* **308**, 339 (1991).
227. K. Hara, A. Tsuneto, A. Kudo, and T. Sakata, Electrochemical Reduction of CO_2 on a Cu Electrode Under High Pressure: Factors that Determine the Product Selectivity. *J. Electrochem. Soc.* **141**, 2097 (1994).
228. A. Kudo, S. Nakagawa, A. Tsuneto, and T. Sakata, Electrochemical Reduction of High Pressure CO_2 on Nickel Electrodes. *J. Electrochem. Soc.* **140**, 1541 (1993).
229. D. L. Dubois, A. Miedaner, W. Bell. and J. C. Smart, Electrochemical Concentration of Carbon Dioxide. *In* "Electrochemical and Electrocatalytic Reactions of Carbon Dioxide" (B. P. Sullivan, K. Krist, and H. E. Guard, eds.), Chapter 4. Elsevier, Amsterdam 1993.

230. L. O. Bulhoes and A. Zara, The Effect of Carbon Dioxide on the Electroreduction of 1,4-Benzoquinone. *J. Electroanal. Chem.* **248**, 159 (1988).
231. K. Sugimura, S. Kuwabata, and H. Yoneyama, Electrochemical Fixation of Carbon Dioxide in Oxoglutaric Acid Using an Enzyme as an Electrocatalyst. *J. Am. Chem. Soc.* **111**, 2361 (1989).
232. Y. B. Vassilev, V. S. Bagotzky, O. A. Khazova, and N. A. Mayorova, Electroreduction of Carbon Dioxide. Part II. The Mechanism of Reduction in Aprotic Solvents. *J. Electroanal. Chem. Interfacial Electrochem.* **189**, 295 (1985).
233. A. Naitoh, K. Ohta, T. Mizuno, H. Yoshida, M. Sakai and H. Noda, Electrochemical Reduction of Carbon Dioxide in Methanol at Low Temperature. *Electrochim. Acta.* **38**, 2177 (1993).
234. A. H. A. Tinnemans, T. P. M. Koster, D. H. M. W. Thewissen, C. W. de Kreuk, and A. Mackor, On the Electrolytic Reduction of Carbon Dioxide at TiO_2 and Other Titanates. *J. Electroanal. Chem. Interfacial Electrochem.* **145**, 449 (1983).
235. K. Tanaka, K. Miyahara, and I. Toyoshima, Adsorption of CO_2 on TiO_2 and Pt/TiO_2 Studied by X-ray Photoelectron Spectroscopy and Auger Electron Spectroscopy. *J. Phys. Chem.* **88**, 3504 (1984).
236. K. Ogura and M. Takagi, Electrocatalytic Reduction of Carbon Monoxide with a Photocell. *J. Electroanal. Chem. Interfacial Electrochem.* **195**, 357 (1985).
237. G. Silvestri, S. Gambino, G. Filardo, G. Spardo, and L. Palmisano, The Electrochemistry of Carbon Monoxide Reductive Cyclotetramerization to Squarate Anion. *Electrochim. Acta.* **23**, 413 (1978).
238. F. A. Uribe, P. R. Sharp, and A. J. Bard, Electrochemistry in Liquid Ammonia. Part VI. Reduction of Carbon Monoxide. *J. Electroanal. Chem. Interfacial Electrochem.* **152**, 173 (1983).
239. B. Beden, C. Lamy, N. R. de Tacconi, and A. J. Arvia, The Electrooxidation of CO: A Test Reaction in Electrocatalysis. *Electrochim. Acta* **35**, 691 (1990).
240. M. Beley, J. P. Collin, R. Ruppert, and J. P. Sauvage, Electrocatalytic Reduction of CO_2 by Ni Cyclam^{2+} in Water: Study of the Factors Affecting the Efficiency and the Selectivity of the Process. *J. Am. Chem. Soc.* **108**, 7461 (1986).
241. C. M. Lieber and N. S. Lewis, Catalytic Reduction of CO_2 at Carbon Electrodes Modified with Cobalt Phthalocyanine. *J. Am. Chem Soc.* **106**, 5033 (1984).
242. S. Kapusta and N. Hackerman, Carbon Dioxide Reduction at a Metal Phthalocyanine Catalyzed Carbon Electrode. *J. Electrochem. Soc.* **131**, 1511 (1984).
243. A. Y. Breikss and H. D. Abruña, Electrochemical and Mechanistic Studies of [Re(CO)$_3$(dmbpy)Cl] and Their Relation to the Catalytic Reduc-

tion of CO_2. *J. Electroanal. Chem. Interfacial Electrochem.* **201**, 347 (1986).
244. K. Ogura, H. Sugihara, J. Yano, and M. Higasa, Electrochemical Reduction of Carbon Dioxide on Dual-Film Electrodes Modified With and Without Cobalt(II) and Iron(II) Complexes. *J. Electrochem. Soc.* **141**, 419 (1994).
245. K. Ogura, C. T. Migita, and H. Imura, Catalytic Reduction of Carbon Dioxide with a Hydrogen Fuel Cell. *J. Electrochem. Soc.* **137**, 1730 (1990).
246. A. Katoh, H. Uchida, M. Shibata, and M. Watanabe, Design of Electrocatalyst for CO_2 Reduction. V. Effect of the Microcrystalline Structures of Cu-Sn and Cu-Zn Alloys on the Electrocatalysis of CO_2 Reduction. *J. Electrochem. Soc.* **141**, 2054 (1994).
247. R. L. Cook, R. C. MacDuff, and A. F. Sammells, High Rate Gas Phase CO_2 Reduction to Ethylene and Methane Using Gas Diffusion Electrodes. *J. Electrochem. Soc.* **137**, 607 (1990).
248. M. Schwartz, M. E. Vercauteren, and A. F. Sammells, Fischer-Tropsch Electrochemical CO_2 Reduction to Fuels and Chemicals. *J. Electrochem. Soc.* **141**, 3119 (1994).
249. M. Schwartz, R. L. Cook, V. M. Kehoe, R. C. MacDuff, J. Patel, and A. F. Sammells, Carbon Dioxide Reduction to Alcohols Using Perovskite-Type Electrocatalysts. *J. Electrochem. Soc.* **140**, 614 (1993).
250. A. F. Sammells and R. L. Cook, Electrocatalysis and Novel Electrodes for High CO_2 Reduction Under Ambient Conditions. In "Electrochemical and Electrocatalytic Reactions of Carbon Dioxide" (B. P. Sullivan, K. Krist, and H. E. Guard, eds.), Chapter 7. Elsevier, Amsterdam, 1993.
251. Y. Nishimura, M. Mizuhata, K. Asaka, K. Oguro, and H. Takenaka, Solid Polymer Electrolyte CO_2 Reduction. *186th Meet. Electrochem. Soc.*, Miami Beach, FL, 1994.
252. T. Hibino, S. Hamakawa, T. Suzuki, and H. Iwahara, Recycling of Carbon Dioxide Using a Proton Conductor as a Solid Electrolyte. *J. Appl. Electrochem.* **24**, 126 (1994).
253. N. U. Pujare, K. J. Tsai, and A. F. Sammells, An Electrochemical Claus Process for Sulfur Recovery. *J. Electrochem. Soc.* **136**, 3662 (1989).
254. S. Mizuta, W. Kondo, K. Fujii, H. Iida, S. Isshiki, H. Noguchi, T. Kikuchi, H. Sue, and K. Sakai, Hydrogen Production from Hydrogen Sulfide by the Fe-Cl Hybrid Process. *Ind. Eng. Chem. Res.* **30**, 1601 (1991).
255. S. Petrovic, J. C. Donini, J. Szynkarczuk, S. Thind, O. E. Hileman, and A. B. P. Lever, Liquid Phase Hydrogen Sulfide Electrolysis, *6th Intl. Forum Electrolysis Chem. Ind.*, The Electrosynthesis Co., Fort Lauderdale, FL, 1992.
256. D. W. Kalina and E. T. Maas, Jr., Indirect Hydrogen Sulfide Conversion. 1. An Acidic Electrochemical Process. *Int. J. Hydrogen Energy* **10**, 157 (1985).
257. D. W. Kalina and E. T. Mass, Jr., Indirect Hydrogen Sulfide Conversion. 2. A Basic Electrochemical Process. *Int. J. Hydrogen Energy* **10**, 163 (1985).
258. C. Oloman, Electrochemical Synthesis and Separation Technology in the

Pulp and Paper Industry. *6th Int. Forum Electrolysis Chem. Ind.* The Electrosynthesis Co., Fort Lauderdale, FL, 1992.
259. N. L. Weinberg, A New Air Purification Technology: The Electrocinerator System. *In* "Electrochemistry for a Cleaner Environment" (D. Genders and N. Weinberg, eds.), Chapter. 16. The Electrosynthesis Co., East Amherst, NY, 1992.
260. A. A. Anani, Z. Mao, R. E. White, S. Srinivasan, and A. J. Appleby, Electrochemical Production of Hydrogen and Sulfur by Low-Temperature Decomposition of Hydrogen Sulfide in an Aqueous Alkaline Solution. *J. Electrochem. Soc.* **137**, 2703 (1990).
261. D. Weaver and J. Winnick, Electrochemical Removal of H_2S from Hot Gas Streams. *J. Electrochem. Soc.* **134**, 2451 (1987).
262. K. A. White, III and J. Winnick, Electrochemical Removal of H_2S from Hot Coal Gas: Electrode Kinetics. *Electrochim. Acta* **30**, 511 (1985).
263. J. Szynkarczuk, P. G. Komorowski, and J. C. Donini, Redox Reactions of Hydrosulphide Ions on the Platinum Electrode. I. The Presence of Intermediate Polysulfide Ions and Sulfur Layers. *Electrochim. Acta* **39**, 2285 (1994).
264. J. Winnick, Electrochemical Separation and Concentration of Sulfur-Containing Gases from Gas Mixtures. U. S. Pat. 4,772,366, (1988).
265. G. H. Kelsall, Y. Thompson, and P. A. Francis, Redox Chemistry of H_2S Oxidation by the British Gas Stretford Process. Part I: Thermodynamics of Sulphur-Water Systems at 298K. *J. Appl. Electrochem.* **23**, 279 1993.
266. G. H. Kelsall, Y. Thompson, and P.A. Francis, Redox Chemistry of H_2S Oxidation by the British Gas Stretford Process. Part II: Electrochemical Behavior of Aqueous Hydrosulphide (HS^-) Solutions. *J. Appl. Electrochem.* **23**, 287 (1993).
267. G. H. Kelsall, Y. Thompson, and P.A. Francis, Redox Chemistry of H_2S Oxidation by the British Gas Stretford Process. Part III: Electrochemical Behavior of Anthraquinone 2,7-Disulphonate in Alkaline Electrolytes. *J. Appl. Electrochem.* **23**, 296 (1993).
268. G. H. Kelsall, Y. Thompson, and P.A. Francis, Redox Chemistry of H_2S Oxidation by the British Gas Stretford Process. Part IV: V-S-H_2O Thermodynamics and Aqueous Vanadium(V) Reduction in Alkaline Solutions. *J. Appl. Electrochem.* **23**, 417 (1993).
269. G. H. Kelsall, Y. Thompson, and P.A. Francis, Redox Chemistry of H_2S Oxidation by the British Gas Stretford Process. Part V: Aspects of the Process Chemistry. *J. Appl. Electrochem.* **23**, 427 (1993).
270. G. Kreysa, J. M. Bisang, W. Kochanek, and G. Linzbach, Fundamental Studies on a New Concept of Flue Gas Desulphurization. *J. Appl. Electrochem.* **15**, 639 (1985).

271. D. van Velzen, H. Langekamp, and A. Moryoussef, HBr Electrolysis in the Ispra Mark 13A Flue Gas Desulphurization Process: Electrolysis in a DEM Cell. *J. Appl. Electrochem.* **20**, 60 (1990).
272. D. Hughes, Reviving Electrochemistry. *Chem. Eng. J.* September p.17 (1987).
273. J. Lee, H. B. Darus, and S. H. Langer, Electrogenerative Oxidation of Ferrous Ions with Graphite Electrodes. *J. Appl. Electrochem.* **23**, 745 (1993).
274. P. W. T. Lu and R. L. Ammon, An Investigation of Electrode Materials for the Anodic Oxidation of Sulfur Dioxide in Concentrated Sulfuric Acid. *J. Electrochem. Soc.* **127**, 2610 (1980).
275. W. Weisweiler and K.-H. Deib, Influence of Electrolytes on the Absorption of Nitrogen Oxide Components N_2O_4 and N_2O_3 in Aqueous Absorbents. *Chem. Eng. Technol.* **10**, 131 (1987).
276. W. Weisweiler, K. Eidam, M. Thiemann, E. Scheibler, and K. W. Wiegand, Absorption of Nitric Oxide in Dilute Nitric Acid. *Chem. Eng. Technol.* **14**, 270 (1991).
277. N. Furuya and T. Okada, Electrooxidation of NO on Gas-Diffusion Electrode. *179th Meet. Electrochem. Soc.*, Washington, DC, 1991. Ext. Abst. No. 728.
278. S. Uchiyama and G. Muto, Dependence of pH and Chelate Ligand on the Electroreduction Mechanism of a Nitrosyl-Aminopolycarbonato-Ferrous Complex in Weakly Acidic Media. *J. Electroanal. Chem. Interfacial Electrochem.* **127**, 275 (1981).
279. S. Langer and K. T. Pate, Electrogenerative Reduction of Nitric Oxide. *Nature* **284**, 434 (1980).
280. K. Pate and S. Langer, Electrogenerative Reduction of Nitric Oxide for Pollution Abatement. *Environ. Sci. Technol.* **19**, 371 (1985).
281. S. H. Langer, M. J. Foral, J. A. Colucci, and K. T. Pate, Electrogeneration and Related Electrochemical Methods for NO_X and SO_X Control. *Environ. Prog.* **5**, 277 (1986).
282. M. J. Foral and S. H. Langer, The Effect of Preadsorbed Sulfur on Nitric Oxide Reduction at Porous Platinum Black Electrodes. *Electrochim. Acta* **33**, 257 (1988).
283. M. J. Foral and S. H. Langer, Sulfur Coverage Effects on the Reduction of Dilute Nitric Oxide at Platinum Black Gas Diffusion Electrodes, *Electrochim. Acta* **36**, 299 (1991).
284. S. Langer, Electrochemical Processing Without a Power Source: Benefits and Applications. *8th Int. Forum Electrolysis Chem. Ind.*, The Electrosynthesis Co., Lake Buena Vista, FL, 1994.
285. N. Furuya and K. Murase, Electroreduction of Nitric Oxide to Nitrogen Using a Gas Diffusion Electrode Loaded with Noble Metals. *186th Meet. Electrochem. Soc.*, Miami Beach, FL, 1994.

286. F. Hine, M. Nozaki, and Y. Kurata, Bench Scale Experiment of Recovery of Chlorine from Waste Gas. *J. Electrochem. Soc.* **131**, 2834 (1984).
287. F. C. Walsh, "A First Course in Electrochemical Engineering." The Electrochemical Consultancy, Ramsey, England, 1993.
288. H. Strathman, Electrodialytic Membrane Processes and Their Practical Application. *In* "Environmentally Oriented Electrochemistry" (C. A. C. Sequeira, ed.), Elsevier, Amsterdam, 1994.
289 A. S. Michaels, Membranes, Membrane Processes and Their Applications: Needs, Unsolved Problems and Challenges of the 1990's. *Desalination* **77**, 5 (1990).
290. A. P. Velin, Overview of the Application of Dialysis, Electrodialysis and Membrane Electrolysis for the Recovery of Waste Acids. *6th Int. Forum Electrolysis Chem. Ind.*, The Electrosynthesis Co., Fort Lauderdale, FL, 1992.
291. P. Gallone, L. Giuffre, and G. Modica, Developments in Separator Technology for Electrochemical Reactors. *Electrochim. Acta* **28**, 1299 (1983).
292. Y. Kurimura, H. Yokota, K. Shigehara, and E. Tsuchida, The Interaction of the Tris(2,2'-bipyridine) Ruthenium(II) Ion with Poly (*p*-styrenesulfonate). *Bull. Chem. Soc. Jpn.* **55**, 55 (1982).
293. D. Meisel and M. S. Matheson, Quenching and Quenching Reversal of Tris(2,2'-bipyridine) Ruthenium(II) Emission in Polyelectrolyte Solutions. *J. Am. Chem. Soc.* **99**, 6577 (1977).
294. P. C. Lee and D. Meisel, Luminescence Quenching in the Cluster Network of Perfluorosulfonate Membrane. *J. Am. Chem. Soc.* **102**, 5477 (1980).
295. N. Prieto and C. R. Martin, Luminescence Probe Studies of Nafion Polyelectrolytes. *J. Electrochem. Soc.* **131**, 751 (1984).
296. E. Blatt, A. Launikonis, A. W-H. Mau, and W. H. F. Sasse, Luminescence Probe Studies of Pyrene and Two Charged Derivatives in Nafion. *Aust. J. Chem.* **40**, 1 (1987).
297. A. Eisenberg and H. L. Yeager, eds., "Perfluorinated Ionomer Membranes," ACS Symp. Ser. No. 180. American Chemical Society, Washington, DC, 1982. R. S. Yeo and H. L. Yeager, Structural and Transport Properties of Perfluorinated Ion Exchange Membranes. *Mod. Aspects Electrochem.* **16**, 437 (1985).
298. D. Pletcher, Membrane Processes. *In* "Membranes and Their Applications," Short Course, The Electrosynthesis Co., Lake Buena Vista, FL, 1994.
299. J. D. Genders, Electrochemical Membrane Separation for Salt Splitting Processes. In *8th Int. Forum Electrolysis Chem. Ind.*, The Electrosynthesis Co., Lake Buena Vista, FL, 1994.
300. P. R. Shah, A. S. Kovvali, G. Y. Khan, and A. A. Khan, Prediction of Permselectivity of Nitrate and Acetate Ions in the Electrodialysis of Aqueous Solutions. *Chem. Eng. J. (Lausanne)* **51**, 1 (1993).
301. T.-C. Huang and J.-K. Wang, Preferential Transport of Nickel and Cupric

Ions Through Cation Exchange Membrane in Electrodialysis with a Complex Agent. *Desalination* **86**, 257 (1992).
302. T. A. Davis, Electrodialysis and its Applications. *In* "Electrochemistry for a Cleaner Environment" (J. D. Genders and N. Weinberg, eds.), Chapter 6. The Electrosyntheis Co., East Amherst, NY 1992.
303. G. Saracco, M. C. Zanetti and M. Onofrio, Novel Application of Monovalent-ion-permselective Membranes to the Recovery Treatment of an Industrial Wastewater by Electrodialysis. *Ind. Eng. Chem. Res.* **32**, 657 (1993).
304. H. Strathmann, B. Bauer, and H. J. Rapp, Better Bipolar Membranes. *Chemtech*, p. 17, June (1993).
305. J. D. Genders, Application of Membrane Technology to Electrosynthesis. *In* "Membranes and Their Applications," Short Course. The Electrosynthesis Co., Lake Buena Vista, FL, 1994.
306. S. J. Judd, G. S. Solt, and T. Wen, Polarization and Back E.M.F. in Electrodialysis. *J. Appl. Electrochem.* **23**, 1117 (1993).
307. T. A. Davis, Applications of Electrodialysis and Bipolar Membranes. *In* "Membranes and Their Applications," Short Course. The Electrosynthesis Co., Lake Buena Vista, FL, 1994.
308. J. Bontha, Evaluation of Electrochemical Ion Exchange for Cesium Removal from Hanford Tank Waste Supernates. *8th Int. Forum Electrolysis Chem. Ind.*, The Electrosynthesis Co., Lake Buena Vista, FL, 1994.
309. N. J. Bridger, C. P. Jones, and M. D. Neville, Electrochemical Ion Exchange. *J. Chem. Tech. Biotechnol.* **50**, 469 (1991).
310. T. A. Davis, Electrodialysis for Resource Recovery. *6th Int. Forum Electrolysis Chem. Ind.*, The Electrosynthesis Co., Fort Lauderdale, FL, 1992.
311. I. V. Dobrevsky and S. T. Pavlova, Organic Matter Removal from Natural Waters by Electrodialysis Demineralization. *Desalination* **86**, 43 (1992).
312. H. R. Bolton, Use of Bipolar Membranes for Ion Exchange Resin Regenerant Production. *J. Chem. Tech. Biotechnol.* **54**, 341 (1992).
313. J. Haggin, Hybrid Systems Improve Membrane Technology. *Chem. & Eng. News*, February 25, p. 23 (1991).
314. S. Itoi, I. Nakamura, and T. Kawahara, Electrodialytic Recovery Process of Metal Finishing Waste Water. *Desalination* **32**, 383 (1980).
315. K. Okuno and Y. Kuboi, A Process to Reduce Spent Solution of Electroless Nickel by Electrodialysis Method Using Ion Exchange Diaphragms. *Proc. Symp. Environ. Aspects Electrochem. and Photoelectrochemistry*, Vol. 93-18, p. 230. The Electrochemical Society, Pennington, NJ, 1993.
316. J. Farmer, D. Fix, L. Summers, R. Hickman, and M. Adamson, Regeneration of Acids and Bases by Electrodialysis. *In* "Water Purification by Photocatalysis, Photoelectrochemistry and Electrochemistry" (T.L. Rose, E. Rudd, O. Murphy, and B.E. Conway, eds.) Vol. 94-19, p. 184. The Electrochemical Society, Pennington, NJ, 1994.

317. P. M. Shah and J. F. Scamehorn, Use of Electrodialysis to Deionize Acidic Wastewater Streams. *Ind. Eng. Chem. Res.* **26**, 269 (1987).
318. A. L. Goldstein, Electrodialysis Apparatus and Process. U. S. Pat. 4,608,140 (1986).
319. M. Turek, J. Mrowiec, and W. Gnot, Electrodialytic Desalination of Water in a Closed System. *Int. Conf. Mod. Electrochem. Ind. Prot. Environ.*, Krakow, Poland, 1993.
320. J.-H. Ye and P. S. Fedkiw, Electroregeneration of Spent Oxidants Using a Nafion-coated Electrode. *185th Meet., Electrochem. Soc.*, San Francisco, 1994.
321. D. Wedman, Electrodialysis for the Controlled Precipitation/Removal of Neodymium(III) and Thorium(IV) Species from HCl Feeds. *8th Int. Forum Electrolysis Chem. Ind.*, The Electrosynthesis Co., Lake Buena Vista, FL, 1994.
322. J. S. Thompson, Process for Producing Sodium Hydroxide and Ammonium Sulfate from Sodium Sulfate. U. S. Pat. 5,098,532, (1992).
323. C. Kot, Production of Sodium Hydroxide and Sulfuric Acid from Sodium Sulfate Solutions by Electromembrane Techniques. *8th Int. Forum Electrolysis Chem. Ind.*, The Electrosynthesis Co., Lake Buena Vista, FL, (1994).
324. J. D. Genders, D. Hartsough and J. Thompson, Novel Approaches to Salt Splitting. *185th Meet., Electrochem. Soc.*, San Francisco, 1994.
325. C. Kot and M. M. Letord, Production of Sodium Hydroxide and Sulfuric Acid from Sodium Sulfate Solutions: Comparisons Between Membrane Electrolysis and Bipolar Membrane Electrodialysis. *185th Meet. Electrochem. Soc.*, San Francisco, 1994.
326. J. Winnick, Electrochemical Membrane Gas Separation. *Chem. Eng. Prog.* January, **41** (1990).
327. J. Farmer, D. Fix, R. Hickman, V. Oversby, and M. Adamson, Regeneration of Acids and Bases by Electrodialysis. *185th Meet. Electrochem. Soc.*, San Francisco, 1994.
328. C. P. Jones, M. D. Neville, and A. D. Turner, Electrochemical Ion Exchange. In "Electrochemistry for a Cleaner Environment" (J. D. Genders and N. Weinberg, eds.), Chapter 8. The Electrosynthesis Co., East Amherst, NY, 1992.
329. A. D. Turner, W. R. Bowen, N. J. Bridger, and K. T. Harrison, Electrical Processes for the Treatment of Medium Active Liquid Wastes: A Laboratory-scale Evaluation. *Chem. Abstr.* **103**, 28991t (1985).
330. A. D. Turner, W. R. Bowen, N. J. Bridger, and K. T. Harrison, Electrical Processes for the Treatment of Medium Active Liquid Wastes: A Laboratory-scale Evaluation, Final Report, January 1981 - March 1983. *Chem. Abstr.* **102**, 86292f (1985).
331. G. Kalinauskas, Electrochemical Ion Exchange for Removal and Destruc-

tion of Nitrates in Water. *6th Int. Forum Electrolysis Chem. Ind.* The Electrosynthesis Co., Fort Lauderdale, FL, 1992.

332. F. DiMascio, A. Jha, D. Hobro, and J. Fenton, Electrochemical Deionization Process (EDP). *185th Meet., Electrochem. Soc.* San Francisco, 1994.

333. F. DiMascio, A. Jha, D. Hobro, and J. Fenton, Electrodiaresis Polishing (An Electrochemical Deionization Process), *In* "Water Purification by Photocatalysis, Photoelectrochemistry and Electrochemistry" (T.L. Rose, E. Rudd, O. Murphy, and B.E. Conway, eds.), Proc. Vol. 94-19, p. 164. The Electrochemical Society, Pennington, NJ, 1994.

334. D. Hobro, Recycling of Chromium from Metal Finishing Waste Waters Using Electrochemical Ion Exchange (EIX). *8th Int. Forum Electrolysis Chem. Ind.*, The Electrosynthesis Co., Lake Buena Vista, FL, 1994.

335. D. Hobro and J. Fenton, Recycling of Chromium from Metal Finishing Waste Waters Using Electrochemical Ion Exchange (EIX). *185th Meet. Electrochem. Soc.*, San Francisco, 1994.

336. D. Hobro, J. Fenton, F. Dimascio, and A. Jha, Recycling of Chromium from Metal Finishing Waste Waters Using Electrochemical Ion Exchange (EIX). *In* "Water Purification by Photocatalysis, Photoelectrochemistry and Electrochemistry" (T.L. Rose, E. Rudd, O. Murphy, and B.E. Conway, eds.), Proc. Vol. 94-19, p. 173. The Electrochemical Society, Pennington, NJ, 1994.

337. O. J. Murphy, G. D. Hitchens, L. Kaba, and C. E. Verostko, Direct Electrochemical Oxidation of Organics for Wastewater Treatment. *Water Res.* **26**, 443 (1992).

338. L. M. Kaba, S. Srinivasan, and A. J. Appleby, Proton-exchange Membrane Reactor for Removal of Organic and Bacterial Contaminants from Reclaimed Water. *181st Meet. Electrochem. Soc. St.* Louis, MO, 1992.

339. M. Inaba, Z. Ogumi, and Z.-I. Takehara, Application of the Solid Polymer Electrolyte Method to Organic Electrochemistry. *J. Electrochem. Soc.* **141**, 2579 (1994).

340. A. Sammells, Catalytic Membrane Reactors for Chemicals Upgrading and Environmental Control. *8th Int. Forum Electrolysis Chem. Ind.* The Electrosynthesis Co., Lake Buena Vista, FL, 1994.

341. R. F. Probstein and R. E. Hicks, Removal of Contaminants from Soils by Electric Fields. *Science* **260**, 498 (1993).

342. R. J. Gale, H. Li, and Y. B. Acar, Soil Decontamination Using Electrokinetic Processing. *In* "Environmentally Oriented Electrochemistry," (C. A. C. Sequeira, ed.), p. 621. Elsevier, Amsterdam, 1994.

343. Y. B. Acar and A. N. Alshawabkeh, Principles of Electrokinetic Remediation, *Environ. Sci. Technol.* **27**, 2638 (1993).

344. B. A. Segall and C. J. Bruell, Electroosmotic Contaminant – Removal Processes. *J. Environ. Eng.* **118**, 84 (1992).

345. S. Smedly, Electrochemical Treatment of Contaminated Soils. *In 8th Int. Forum Electrolysis Chem. Ind.*, The Electrosynthesis Co., Lake Buena Vista, FL, 1994.
346. J. Trombly, Electrochemical Remediation Takes to the Field. *Environ. Sci. Technol.* **28**, 289A (1994).
347. R. Lageman, Electroreclamation. *Environ. Sci. Technol.* **27**, 2648 (1993).
348. F. G. Hill, ed., "Proceedings of the EPRI Workshop on In Situ Electrochemical Soil and Water Remediation," *Doc. TR-104170*. Electric Power Research Institute, Palo Alto, CA, 1994.
349. Y. B. Acar, H. Li, and R. J. Gale, Phenol Removal from Kaolinite by Electrokinetics. *J. Geotech. Eng.* **118**, 1837 (1992).
350. J. Hamed, Y. B. Acar, and R. J. Gale, Pb(II) Removal from Kaolinite by Electrokinetics. *J. Geotech. Eng* **117**, 241 (1991).
351. D. Cabrera-Guzman, J. T. Swartzbaugh, and A. W. Weisman, The Use of Electrokinetics for Hazardous Waste Site Remediation. *J. Air Waste Management Assoc.* **40**, 1670 (1990).
352. A. P. Shapiro and R. F. Probstein, Removal of Contaminants from Saturated Clay by Electroosmosis. *Environ. Sci. Technol.* **27**, 283 (1993).
353. G. Sposito, Chemical Models of Inorganic Pollutants in Soils. *CRC Criti. Rev. Environ. Control* **15**, 1 (1985).
354. H. E. Allen and P.-H. Chen, Remediation of Metal Contaminated Soil by EDTA Incorporating Electrochemical Recovery of Metal and EDTA. *Environ. Prog.* **12**, 284 (1993).
355. C. Wei, S. German, S. Basak, and K. Rajeshwar, Reduction of Hexavalent Chromium in Aqueous Polypyrrole. *J. Electrochem. Soc.* **140**, L60 (1993).
356. W. A. Wampler, S. Basak, and K. Rajeshwar, Composites of Polypyrrole and Carbon Black. 4. Use in Environmental Pollution Abatement of Hexavalent Chromium. *Carbon* **34**, 747 (1996).
357. W. A. Wampler, Conducting Polymer-Carbon Black Composites. Ph.D. Dissertation, The University of Texas at Arlington (1995).
358. W. A. Wampler, K. Rajeshwar, R. G. Pethe, R. C. Hyer, and S. C. Sharma, Composites of Polypyrrole and Carbon Black. 3. Chemical Synthesis and Characterization. *J. Mater. Res.* **10**, 1811 (1995).
359. J. Wang, L. Angnes, H. Tobias, R. A. Roesner, K. C. Hong, R. S. Glass, F. Kong, and R. W. Pekala, Carbon Aerogel Composite Electrodes. *Anal. Chem.* **65**, 2300 (1993).
360. D. Kohler, J. Zabasajja, A. Krishnagopalan, and B. J. Tatarchuk, Metal-Carbon Composite Materials from Fiber Precursors. I. Preparation of Stainless Steel Carbon Composite Electrodes. *J. Electrochem. Soc.* **137**, 136 (1990).
361. D. Kohler, J. Zabasajja, F. Rose, and B. J. Tatarchuk, Metal-Carbon Composite Materials from Fiber Precursors. II. Electrochemical Characteriza-

tion of Stainless Steel-Carbon Structures. *J. Electrochem. Soc.* **137**, 1750 (1990).
362. G. M. Swain and R. Ramesham, The Electrochemical Activity of Boron-Doped Polycrystalline Diamond Thin Film Electrodes. *Anal. Chem.* **65**, 345 (1993).
363. R. Tenne, K. Palal, K. Hashimoto and A. Fujishima, Efficient Electrochemical Reduction of Nitrate to Ammonia Using Conduction Diamond Film Electrodes. *J. Electroanal. Chem. Interfacial Electrochem.* **347**, 409 (1993).
364. R. L. McCreery, Carbon Electrodes: Structural Effects on Electron Transfer Routes. *Electroanal. Chem.* **17**, p. 221 (1991).
365. R. Clarke and R. Pardoe, Applications of Ebonex^R Conductive Ceramic Electrodes in Effluent Treatment. *In* "Electrochemistry for a Cleaner Environment" (J. D. Genders and N. L. Weinberg, eds.), Chapter. 18, p. 349. The Electrosynthesis Co., East Amherst, NY, 1992.
366. M. Mayr, W. Blatt, B. Busse, and H. Heinke, Treatment of Fluoride Containing Electrolytes by New Electrolytic Systems. *In* "Electrochemistry for a Cleaner Environment" (J. D. Genders and N. L. Weinberg, eds.), Chapter. 19, p. 365. The Electrosynthesis Co., East Amherst, NY, 1992.
367. V. B. Baez, J. E. Graves, and D. Pletcher, The Reduction of Oxygen on Titanium Oxide Electrodes. *J. Electroanal. Chem. Interfacial Electrochem.* **340**, 273 (1992).
368. W. Saenger, Cyclodextrin Inclusion Compounds in Research and Industry. *Angew. Chem. Intl. Ed. Engl.* **19**, 344 (1980).
369. L. Pospisil, J. Hanzlik, R. Fuoco, and M. P. Columbini, Electrochemical and Spectral Evidence of the Inclusion of the Herbicide Difenzoquat by Cyclodextrins in Aqueous Solution. *J. Electroanal. Chem. Interfacial Electrochem.* **368**, 149 (1994).
370. V. Mathew and R. Narayan, Electrochemical Reduction of Sulphur Dioxide from Molten Amides at Glassy Carbon Electrodes. *J. Electroanal. Chem. Interfacial Electrochem.* **364**, 215 (1994).
371. D. W. Tallman and S. A. Olsen, Electrochemistry at Microelectrodes in Supercritical Chlorodifluoromethane. *184th Meet. Electrochem. Soc.*, New Orleans, LA, 1993.
372. S. Zhang and J. F. Rusling, Dechlorination of Polychlorinated Biphenyls by Electrochemical Catalysis in a Bicontinuous Microemulsion. *Environ. Sci. Technol.* **27**, 1375 (1993).
373. J. A. Roth, S. R. Dakoji, R. C. Hughes, and R. E. Carmody, Hydrolysis of Polychlorinated Biphenyls by Sodium Borohydride with Homogeneous and Heterogeneous Nickel Catalysts. *Environ. Sci. Technol.* **28**, 80 (1994).
374. Y. Liu, J. Schwartz and C. L. Carallaro, Catalytic Dechlorination of Polychlorinated Biphenyls. *Environ. Sci. Technol.* **29**, 836 (1995).
375. J. F. Rusling, C.-N. Shi, D. K. Gosser, and S. S. Shukla, Electrocatalytic

Reactions in Organized Assemblies. Part I. Redution of 4-Bromobiphenyl in Cationic and Non-ionic Micelles. *J. Electroanal. Chem. Interfacial Electrochem.* **240**, 201 (1988).
376. J. F. Rusling, Controlling Electrochemical Catalysis with Surfactant Microstructures. *Acc. Chem. Res.* **24**, 75 (1991).
377. M. O. Iwunze and J. F. Rusling, Aqueous Lamellar Surfactant System for Mediated Electrolytic Dechlorination of Polychlorinated Biphenyls. *J. Electroanal. Chem. Interfacial Electrochem.* **266**, 197 (1989).
378. Q. Huang and J. F. Rusling, Formal Reduction Potentials and Redox Chemistry of Polyhalogenated Biphenyls in a Bicontinuous Microemulsion. *Environ. Sci. Technol.* **29**, 98 (1995).
379. M. O. Iwunze, A. S. Sucheta, and J. F. Rusling, Bicontinuous Microemulsions as Media for Electrochemical Studies. *Anal. Chem.* **62**, 644 (1990).
380. G. N. Kamau, N. Hu, and J. P. Rusling, Rate Enhancement and Control in Electrochemical Catalysis Using a Bicontinuous Microemulsion. *Langmuir* **8**, 1042 (1992).
381. S. E. Friberg, Applications of Amphiphilic Association Structures. *Adv. Coll. Interface Sci.* **32**, 167 (1990).
382. F. D. Blum, S. Pickup, B. W. Ninham, S. J. Chen, and D. F. Evans, Structure and Dynamics in Three-component Microemulsions. *J. Phys. Chem.* **89**, 711 (1985).
383. J. F. Rusling, C.-N. Shi, and S. L. Suib, Electrocatalytic Reactions in Organized Assemblies. Part V. Dehalogenation of 4,4'-Dibromobiphenyl in Cationic Micelles at Bare and Clay-modified Carbon Electrodes. *J. Electroanal. Chem. Interfacial Electrochem.* **245**, 331 (1988).

SUPPLEMENTARY READING

Reviews

1. R. W. Peters, Y. Ku and D. Bhattacharya, Evaluation of Recent Treatment Techniques for Removal of Heavy Metals from Industrial Wastewaters. *AIChE Symp. Ser.*, p. 165 (1985).
2. B. Fleet, Electrochemical Reactor Systems for Pollution Control and the Removal of Toxic Metals from Industrial Wastewaters. *Collect. Czech. Chem. Commun.* **53**, 1107 (1988).
3. A. M. Cooper, D. Pletcher, and F. C. Walsh, Electrode Materials for Electrosynthesis. *Chem. Rev.* **90**, 837 (1990).
4. D. F. Steele, Electrochemistry and Waste Disposal. *Chem. Br.* October p. 915 (1991).

5. D. Pletcher and N. L. Weinberg, The Green Potential of Electrochemistry. Part 1: The Fundamentals. *Chem. Eng.*, August, p. 98 (1992).
6. D. Pletcher and N. L. Weinberg, The Green Potential of Electrochemistry. Part 2: The Applications. *Chem. Eng.* November, p. 132 (1992).
7. F. Walsh and G. Mills, Electrochemical Techniques for a Cleaner Environment. *Chem. Ind., (London)*, August, p. 576 (1993).
8. F. Walsh and G. Reade, Design and Performance of Electrochemical Reactors for Efficient Synthesis and Environment Treatment. Part 1. Electrode Geometry and Figures of Merit. *Analyst (London)* **119**, 791 (1994).
9. F. Walsh and G. Reade, Design and Performance of Electrochemical Reactors for Efficient Synthesis and Environmental Treatment. Part 2. Typical Reactors and Their Performance. *Analyst (London)*. **119**, 797 (1994).
10. J. O'M. Bockris, R. C. Bhardwaj and C. L. K. Tennakoon, Electrochemistry of Waste Removal. A Review. *Analyst (London)*. **119**, 781 (1994).
11. K. Rajeshwar, J. G. Ibanez and G. Swain, Electrochemistry and the Environment. *J. Appl. Electrochem.* **24**, 1077 (1994).

Books and monographs

1. B. A. Bolto and L. Pawlowski, "Wastewater Treatment by Ion Exchange." Spon, London and New York, 1987.
2. S. D. Faust and O. M. Aly, "Adsorption Processes for Water Treatment." Butterworth, Boston, 1987.
3. D. Pletcher and F. C. Walsh, "Industrial Electrochemistry." 2nd ed. Chapman & Hall, London, 1990.
4. D. Genders and N. Weinberg, eds., "Electrochemistry for a Cleaner Environment." The Electrosynthesis Co., East Amherst, NY, 1992.
5. C. A. C. Sequeira, ed., "Environmental Oriented Electrochemistry." Elsevier, Amsterdam, 1994.
6. P. M. Bersier, L. Carlsson, and J. Bersier, "Electrochemistry for a Better Environment." *Top. Curr. Chem.*, Vol. 170. Springer-Verlag, Berlin and Heidelberg, 1994.
7. B.P. Sullivan, K. Krist, and H.E. Guard, "Electrochemical and Electrocatalytic Reactions of Carbon Dioxide." Elsevier, Amsterdam, 1993.

CHAPTER SIX

6.1. INTRODUCTION

In this chapter, we shall discuss photoassisted methods for treating polluted air and water. The treatment can be done either via *direct* photoexcitation of the pollutant (for example by irradiating it with UV light) or via *mediated* processes. We saw several examples of the latter in the preceding chapter, where the pollutant decomposition was accomplished by *in situ* electrochemical generation (in the "dark") of agents such as H_2O_2, O_3, and the hydroxyl radical, ˙OH. The dark reactions involving oxidants such as H_2O_2 and O_3 are most often exoergic, that is thermodynamically spontaneous; however, they are kinetically slow. For example, reactions involving *molecular* O_3 are highly selective but have bimolecular rate constants only on the order of 1–10^3 M^{-1} s^{-1}.[1] In the presence of *initiators* such as ˙OH, much faster reaction rates, often approaching the diffusion-controlled limit (ca. 1×10^{10} M^{-1} s^{-1}) can be attained. Dark (electrochemical) methods for generating radical initiators (such as ˙OH) have been discussed in Chapter 5 and include the Fenton reaction. In this chapter, *photoassisted* methods for generating highly reactive radicals will be discussed. There are two broad strategies[2] for photoassisted generation of radicals such as ˙OH:

- Homogeneous photolysis: In this approach, homogeneous solutions containing H_2O_2, O_3, or even a combination of the two are subjected to UV irradiation. Photolysis of the peroxide in these cases generates the highly reactive ·OH.

- Heterogeneous photolysis: Here, colloidal particles of a semiconductor (e.g., TiO_2) absorb the UV light to generate ·OH at the particle/solution interface. This process can be alternatively termed *heterogeneous photocatalysis*, because the semiconductor functions as a *photocatalyst*. Interestingly enough, the photoexcitation is shifted toward *longer* wavelengths (*lower* energy) by the use of the semiconductor as a photoreceptor. For example with TiO_2, light of ~300 nm wavelength can be used instead of the ~250 nm required for sustaining a reasonable photolysis rate of H_2O_2 or O_3 (see later).

Hydroxyl radicals are not the only species of interest in the photoassisted treatment of pollutants. Other species such as e^-_{aq}, the superoxide anion, O_2^-, and the hydroperoxy radical $HO_2·$ can be photogenerated (as will be discussed) and made to react with the pollutant. In the case of UV/TiO_2, in fact, electron–hole pairs are simultaneously generated within the semiconductor on bandgap photoexcitation. A fraction of these migrate to the particle surface where they are available for photo-oxidation and photo-reduction reactions.

Radiolytic methods for generating radicals in water rely on a source of high-energy radiation (e.g., a linear accelerator) and are useful for applications related to drinking water treatment (e.g., disinfection). A brief discussion of these methods is contained in Chapter 7.

6.2. PHOTOLYSIS OF H_2O_2 AND O_3 AND GENERATION OF e^-_{aq}

Figure 6.1 contains absorption spectra of H_2O_2 and O_3^{2-}; the spectral data are shown both for the gas phase and for aqueous solutions of these compounds. The extinction coefficients are generally larger in aqueous solution than in the gas phase. Concomitantly, the quantum yields for photolysis (i.e., the number of molecules photolyzed per photon of light incident in a fixed period of time) are also reduced in the condensed phase because of solvent cage recombination.[2] We shall now consider each compound in turn.

As Figure 6.1a shows, H_2O_2 absorbs fairly weakly in the UV region with increasing absorption at lower wavelengths. For example, at 254 nm (the low-pressure Hg emission wavelength), the molar extinction coefficient (ε in units of M^{-1} cm^{-1}) is 18, whereas at 200 nm, it is 190.[2] The primary photolysis reaction is

$$H_2O_2 \xrightarrow{h\nu} 2·OH \quad (6.1)$$

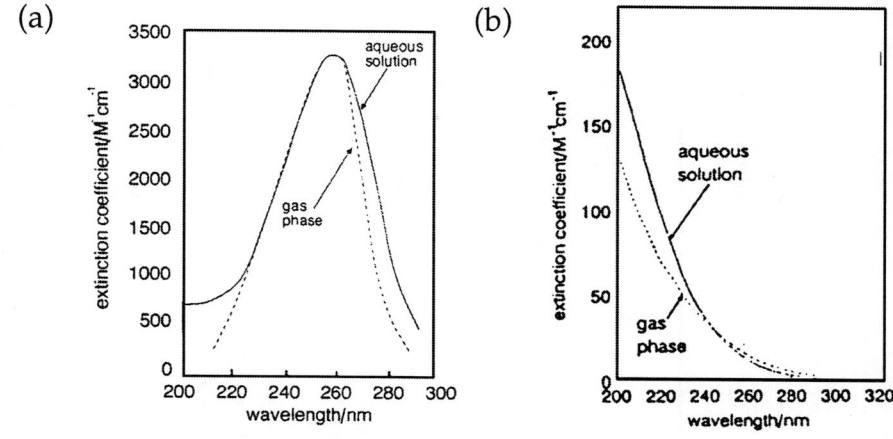

FIGURE 6.1. Absorption spectra of H_2O_2 (a) and O_3 (b) in the gas phase and in aqueous solution. (Reproduced with permission from Bolton and Cater.[2])

The peroxide anion (the conjugate base) is reported to have a higher ε value, accounting for the increase in the photolysis rate with increasing pH.[3] A dismutation reaction (Eq. (6.3))

$$H_2O_2 + H_2O \rightleftharpoons H_3O^+ + HO_2^- \quad (6.2)$$

$$H_2O_2 + HO_2^- \rightarrow H_2O + O_2 + {}^{\bullet}OH \quad (6.3)$$

has been invoked to explain this trend with the maximum rate being located at the pK_a value of the base (11.6).

Much research has focused on the so-called Haber–Weiss mechanism in which the $^{\bullet}$OH radicals initially formed from the photolysis of H_2O_2 (Eq. (6.1)) initiate a radical chain mechanism wherein the propagation cycle is predicted to afford high quantum yields ($\phi \gg 1$). Experimentally, however, the situation is more complex, and high quantum yields are not always obtained. The Haber–Weiss mechanism and other aspects of H_2O_2 photolysis have been reviewed.[4]

At a high (local) concentration of H_2O_2, the back reaction to Eq. 6.1 begins to exert its influence with a consequent degradation of the quantum yield:

$$^{\bullet}OH + {}^{\bullet}OH \rightarrow H_2O_2 \quad (6.4)$$

In the presence of an excess of H_2O_2, hydroxyl radicals are *consumed* by the reaction

$$H_2O_2 + {}^{\bullet}OH \rightarrow H_2O + HO_2^{\bullet} \quad (6.5)$$

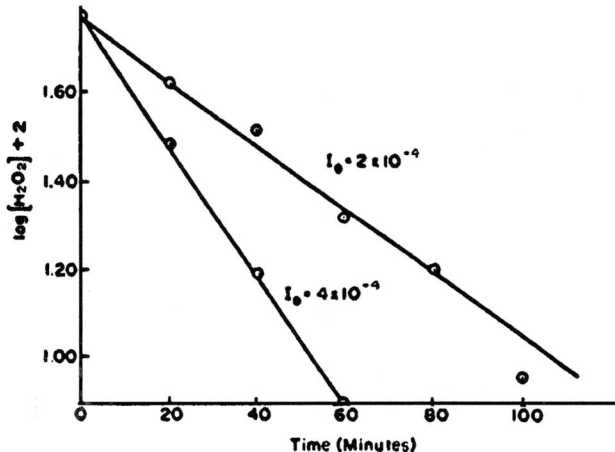

FIGURE 6.2. The photolysis of 0.5 M H_2O_2 in aqueous solution in the presence of sodium acetate. Light intensities (I_o) in units of einsteins/liter·min; temp. = 45°C. (Reproduced with permission from Koubek.[9])

The photolysis of H_2O_2 is first order in peroxide concentration, and the rate is proportional to the square root of the incident light intensity.[4] Figure 6.2 contains data illustrating these points. The major disadvantage with H_2O_2 photolysis as a method of generating ·OH is associated with the low absorptivity of the peroxide. Thus, in waters of high inherent UV absorption, the photolysis efficiency will be low unless prohibitively large peroxide concentrations are used, with attendant problems, as outlined earlier. The transport and storage of H_2O_2 solutions are also not free of practical difficulties. Nonetheless, the UV–H_2O_2 route has the important benefit of good electrical efficiency and immunity from the air emission problems associated with the UV/O_3 approach, to be discussed next.

The use of ozone is established practice in the drinking water treatment industry (Chapter 7). Ozone has a strong absorption band centered at 260 nm with an ε_{max} of ~3,000 M^{-1} cm^{-1} (Figure 6.1b).[2] Photolysis at this wavelength affords quantitative generation of H_2O_2 (Ref. 5):

$$O_3 \xrightarrow{h\upsilon} O(^1D) + O_2 \qquad (6.6)$$
$$O(^1D) + H_2O \rightarrow H_2O_2 \qquad (6.7)$$

6.2. Photolysis of H_2O_2 and O_3 and Generation of e^-_{aq}

Hydroxyl radicals are subsequently formed either via Reaction (6.1) or by the reaction of O_3 with the conjugate base of H_2O_2 (Eq. (6.2))[2]:

$$HO_2^- + O_3 \rightarrow O_3^- + HO_2^\bullet \tag{6.8}$$
$$O_3^- + H_2O \rightarrow HO_3^\bullet + OH^- \tag{6.9}$$
$$HO_3^\bullet \rightarrow {}^\bullet OH + O_2 \tag{6.10}$$

Therefore the net result of UV–O_3 is the conversion of O_3 to H_2O_2. In this respect, this approach appears to be a rather expensive route to H_2O_2 generation! For example, assuming a 30% (electrical) efficiency of generating photons below 300 nm and an ozone generation efficiency of 10 kWh per pound of ozone from air, the energy consumption per mole of $^\bullet OH$ generated has been computed[2] to be 1.05 kWh for UV–O_3 and 0.54 kWh for UV–H_2O_2. This situation can change when the H_2O_2 photolysis becomes inefficient (as, e.g., for opaque water). In this instance, the UV–O_3 approach can become more energy efficient.

The hydrated electron, e^-_{aq} was discussed earlier in Chapter 2. Aside from the metal photoemission route discussed in that chapter for e^-_{aq} generation, certain solutes undergo *photoionization* to produce this specie[2]:

$$Fe(CN)_6^{4-} \xrightarrow{h\upsilon} Fe(CN)_6^{3-} + e_{aq}^-$$

Hydrated electrons react rapidly with dioxygen to produce the superoxide anion $O_2^{\bullet-}$ (in basic media) or the hydroperoxy radical $HO_2^\bullet$. Interestingly enough, as pointed out in Chapter 2, the *reducing* radical e^-_{aq} can be converted to $^\bullet OH$ (a strong oxidant, see later) by the use of N_2O via a reaction that occurs close to the diffusion-controlled limit[2]:

$$e^-_{aq} + H^+ + N_2O \rightarrow N_2 + {}^\bullet OH$$

Table 6.1 lists the redox properties of radicals that are germane to pollutant decomposition scenarios. Specifically, $^\bullet OH$ is seen to be a strong oxidant, second in oxidizing power only to fluorine. It is superior as an oxidant to permangate, Cr(VI), O_3, and H_2O_2—all considered to be strong oxidants in the chemist's arsenal. On the other hand, e^-_{aq} is a powerful reducing agent approaching, in its reducing power, the alkali metals. The hydroperoxy radical and the superoxide anion are rather mild reducing agents.

Table 6.1. Redox Potentials of Some Radical Species and Commonly Used Chemical Oxidants and Reductants

Species	Standard Reduction Potential V vs. SHE
•OH	1.77
$HO_2^\bullet$	-0.30
$O_2^{\bullet -}$	-0.33
e^-_{aq}	-2.90
Oxidants	
F_2	3.03
O_3	2.07
H_2O_2	1.68
Permanganate	1.68
Cl_2	1.36
Cr(VI)	1.33
Reductants	
Lithium	-3.05
Sodium	-2.71
Magnesium	-2.36

6.3. DESTRUCTION OF ORGANICS AS MEDIATED BY •OH AND e^-_{aq}

The •OH radicals attack organics primarily via hydrogen abstraction:

$$RH_2 + {}^\bullet OH \rightarrow RH^\bullet + H_2O$$

Another (less prevalent) route to oxidative decomposition is either electron transfer or electrophilic addition:

$$RX + {}^\bullet OH \rightarrow RX^{\bullet +} + OH^-$$
$$PhX + {}^\bullet OH \rightarrow HOPhX^\bullet$$

6.3. Destruction of Organics as Mediated by •OH and e^-_{aq}

The first type of mechanism is exemplified by methanol.[2] Here, successive hydrogen abstraction steps interspersed with reaction with (molecular) O_2 lead to formaldehyde, formic acid, and ultimately to CO_2:

$$CH_3OH + {}^\bullet OH \longrightarrow {}^\bullet CH_2OH + H_2O$$

$$\downarrow O_2$$

$$HO-CH_2-OH \underset{H_2O}{\rightleftharpoons} H-\overset{O}{\underset{\|}{C}}-H + HO_2^\bullet$$

$$\downarrow {}^\bullet OH$$

$$HO-\overset{\bullet}{C}H-OH + H_2O \xrightarrow{O_2} \longrightarrow H-\overset{O}{\underset{\|}{C}}-OH + HO_2^\bullet$$

$$\downarrow {}^\bullet OH$$

$$CO_2 + H_2O \xleftarrow{O_2} \longleftarrow {}^\bullet \overset{O}{\underset{\|}{C}}-OH + H_2O$$

In the case of aromatic substrates (e.g., benzene), initial •OH attack leads to mono-, then di-, and finally trihydroxy derivatives or quinone intermediates (the electrophilic addition route, see later). Ultimately, the aromatic ring is cleaved to yield first 6-carbon aldehydes and acids followed by fragmentation to smaller derivatives (e.g., glyoxal, oxalic acid, maleic acid). Subsequent oxidation leads to formic acid and CO_2.[2]

Figure 6.3 contains a generic reaction pathway for the UV–H_2O_2 system.[3] The initial hydrogen atom abstraction step and the subsequent reaction with O_2 to yield the organic peroxy radical (see later) can be recognized in both these schemes. These radicals subsequently undergo four possible pathways: (a) heterolysis generating the organic cation and $O_2^{-\bullet}$ (d, Figure 6.3); (b) 1,3-hydrogen shift and homolysis to generate carbonyl derivatives and •OH (e, Figure 6.3); (c) back reaction to RH• and O_2 (f, Figure 6.3); and (d) decomposition of the radical, RH• to polymeric and other products (h, Figure 6.3).

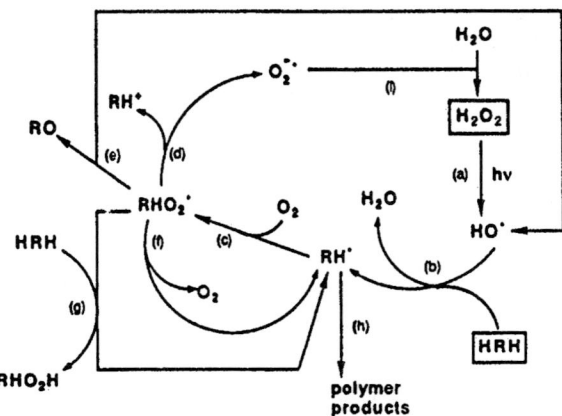

FIGURE 6.3. Reaction pathways in the UV/H_2O_2 process. (Reproduced with permission from Legrini et al.[3])

The last pathway would be particularly important in O_2-starved reactors.

The electron transfer pathway is of particular interest when hydrogen abstraction and electrophilic addition may be inhibited by multiple halogen substitution or steric hindrance. The electrophilic addition mechanism is important in the facile dehalogenation reaction typical, for example, of chlorinated phenols.

Table 6.2 provides a compilation of rate constants for the reaction of both •OH and e^-_{aq} with organic compounds. More extensive compilations are available in the literature.[2,6,7] Of particular interest are the facts that (a) in some cases, the reactions are exceedingly fast, often surpassing the diffusion-controlled rate limit (ca. $1 \times 10^{10}\ M^{-1}\ s^{-1}$); and (b) for highly halogenated derivatives (e.g., chloroform, trichloroethylene), the H-atom abstraction route is inefficient, with consequent deterioration in the overall *oxidative* reaction rate. Note, however, that the corresponding *reductive* reaction rate constants are orders of magnitude higher. This trend has been exploited for the treatment of substrates such as $CHCl_3$ via a UV photo-reduction route.[2]

6.4. DIRECT PHOTODISSOCIATION OF THE POLLUTANT

As indicated in the introductory paragraph, some pollutants can be dissociated by direct UV excitation. For this to happen, the pollutant must have a high absorption cross-section for the excitation light, and also have a decent

6.4. Direct Photodissociation of the Pollutant

Table 6.2. Bimolecular Rate Constants for the Reaction of •OH and e_{aq}^- with Organic Substrates

Substrate	$k/10^9\ M^{-1}\ s^{-1}$ [a]	
	•OH	e_{aq}^-
HCHO	~1 (1)	~1.01
HCOOH	0.13 (1)	0.14 (5)
CH_2Cl_2	0.058 (10)	6.3
$CHCl_3$	~0.005 (5.7)	30
$CHCl=CCl_2$	4.2	19
Tetrachloroethylene	2.3	—
Benzene	7.8	0.009 (10)
Phenol	6.6	0.02
Chlorobenzene	5.5 (9.0)	0.50
Nitrobenzene	3.9	37
Vinyl chloride	7.1	—

[a] Rate constants are for pH = 7 unless otherwise indicated by the number in parentheses.
Source: Bolton and Cater[2] and Ross et al.[6,7]

quantum yield for photodissociation. Unfortunately, most organics absorb strongly only at wavelengths below 250 nm (Table 6.3). The net result of photodissociation is usually oxidation, since initial photoexcitation is almost always followed by reaction with dissolved O_2 in the water:

$$R \xrightarrow{h\nu} R^*$$
$$R^* + O_2 \rightarrow R^{\cdot +} + O_2^{\cdot -}$$
$$R^{\cdot +} \rightarrow products$$

or
$$R-X \xrightarrow{h\nu} R^\bullet + X^\bullet$$
$$R^\bullet + O_2 \rightarrow RO_2$$

Table 6.3. Molar Absorptivities of Selected Organics as a Function of Wavelength

	ε, M^{-1} cm^{-1}		
Compound	254 nm	200 nm	175 nm
Benzene	250	8,500	
Toluene	180	7,000	
Perchloroethylene	44	4,300	2,200
Carbon tetrachloride	0.17	170	2,000
Methylchloroform	<0.1	81	>100
p-Nitrophenol	1,000	9,300	
m-Nitrophenol	3,200	9,000	
o-Nitrophenol	2,000	10,000	
p-Nitroaniline	2,000	10,500	
2,6-Dinitrophenol	9,700	15,000	

Source: Glaze[8]

Homolysis of the C–X bond is an efficient route to decomposition of highly fluorinated or chlorinated saturated aliphatics (e.g., freons), especially since these compounds react only slowly with •OH. Excitation of the C–F bond requires vacuum UV light (<190 nm) but the C–Cl bonds are somewhat less demanding in terms of energy requirements (210–230 nm).

Photolysis of the medium (e.g., O_2 in air or H_2O) can assist in the decomposition of the organics. For example, O_2 absorbs strongly in the vacuum UV region (120–200 nm), forming oxygen atoms and subsequently O_3. Thus, irradiation with a VUV lamp can decompose organics in the gas phase, even though the absorption cross-section of the organics is low. Similarly, the absorption coefficient of H_2O at 185 nm is high and the quantum yield for formation of H and OH is respectable.[8] Thus, polluted moist gas streams can be treated with VUV sources.

In general, however, few applications of direct photolysis have gained widespread acceptance. The optical transmissitivity of most wastewater streams is too low, especially if solids are present. Another practical problem is that associated with lamp technology. High efficiency lamps that provide photons of high energy and at a fast rate are not currently available, especially at low cost.

6.5. Reaction Kinetics and Mechanisms of the UV–H_2O_2 System

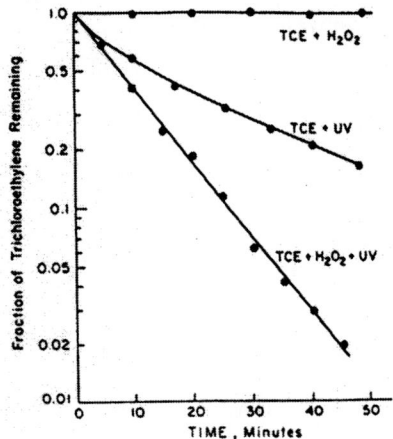

FIGURE 6.4. Effect of H_2O_2 alone, UV alone, and H_2O_2 plus UV on decomposition of trichloroethylene (TCE) at 20°C, pH 6.8. Initial TCE = 58 ppm. Initial H_2O_2/TCE = 4.5 mols/mol. (Reproduced with permission from Sundström et al.[10])

6.5. REACTION KINETICS, MECHANISMS AND EXAMPLES OF APPLICATION OF THE UV–H_2O_2 SYSTEM

An early study[9] showed that the rate of removal of H_2O_2 was four times the rate of organic compound (acetate) decomposition, which suggested the following stoichiometry:

$$4\,H_2O_2 + CH_3CO_2^- \rightarrow 2\,CO_2 + OH^- + 5\,H_2O$$

The reaction was also first order in peroxide (see Figure 6.2) and the rate was *independent* of the initial concentration of acetate. This observation is at variance with a later study[10] on the decomposition of trichloroethylene (TCE), where the initial TCE concentration does indeed affect the reaction rate. Another point of conflict between the two sets of results concerns the effect of temperature: no effect of temperature was noted in the early study[9] while the time needed to reach any specified TCE level was approximately halved for each 10°C increase in the reaction temperature.[10] The initial molar ratio of H_2O_2 to TCE was also found to play a substantial role on the reaction kinetics.[10] As Figure 6.4 indicates, the "dark" reaction of TCE with H_2O_2 is negligible; however, the direct photolysis of TCE plays a significant role. We shall return to this point later in the discussion.

Table 6.4. Rate Constants of •OH in the UV/H_2O_2 System

Species	Reaction	Concentration, μM	Rate Constant k (M^{-1} s^{-1})	10^{-3} k (s^{-1})
H_2O_2	(6.5)	100	2.7×10^7	2.7
HA	(6.11)	25	3.0×10^9	75.0
HCO_3^-	(6.12)	100	8.5×10^5	0.85
		1,000	8.5×10^6	8.5
		15,000	8.5×10^6	127.5

Source: Yao et al.[11]

The organic substrate is not the only entity in the water that can react with the photogenerated •OH. Humic acids (HA) and the bicarbonate–carbonate ion found in most natural waters also consume •OH via reactions such as:

$$HA + {}^{\bullet}OH \rightarrow H_2O + A^{\bullet} \qquad (6.11)$$

$$HCO_3^- + {}^{\bullet}OH \rightarrow H_2O + CO_3^{\bullet -} \qquad (6.12)$$

$$CO_3^{2-} + {}^{\bullet}OH \rightarrow OH^- + CO_3^{\bullet -} \qquad (6.13)$$

For example, it has been estimated,[11] that in the absence of organic substrates, the HCO_3^- ion scavenges 24, 75 and 98% of •OH in 100 μM, 1 mM and 15 mM HCO_3^- solutions, respectively. Similarly, 96% of photogenerated •OH is scavenged by 25 μM HA solution. Table 6.4 contains a listing of the bimolecular rate constants of reactions involving •OH and these radical scavengers. The corresponding *pseudo-first-order* rate constants (in units of s^{-1}) are also contained in this tabulation.

It must be noted that the carbonate radical ion ($CO_3^{\bullet -}$) is rather unreactive toward organics in spite of its oxidizing power. Thus, the reactivity ratio of $CO_3^{\bullet -}/{}^{\bullet}OH$ toward the acetate ion is less than 10^{-5}.[7] A similar trend also exists for secondary radicals such as $HO_2^{\bullet}$, which are generated in the photolysis. In fact, analyses[11] of reactivity ratios for pairs of organic substrates in a variety of matrices point to •OH as the dominant oxidant in these systems. We thus have a steady state kinetics situation in the UV–H_2O_2 system, where the rate of the photogeneration of •OH is balanced by the rate of its consumption via several competitive routes:

6.5. Reaction Kinetics and Mechanisms of the UV–H_2O_2 System

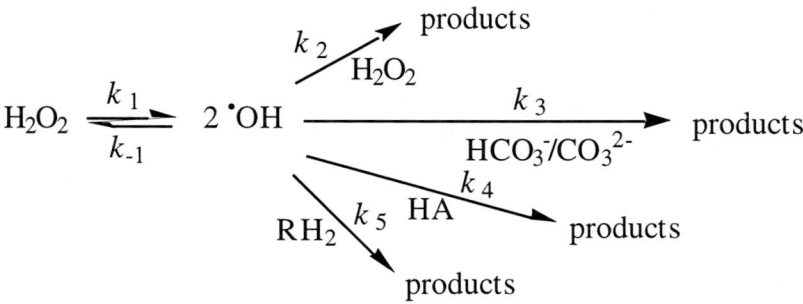

Hence, for optimal photo-decomposition of the organic substrate, k_1 and k_5 must be enhanced, and the magnitudes of k_{-1}, k_2, k_3 and k_4 are to be suppressed. The situation is complicated by the fact that the rate of photolysis of H_2O_2 is first order in H_2O_2. However, too high a concentration of H_2O_2 will cause a deleterious increase in the rate of Reaction (6.5), which *consumes* •OH (i.e., the branch denoted by k_2 in the scheme). Nonetheless, the photo-decomposition of the organic substrate (RH_2) can be sustained at a reasonable rate even with a steady state •OH radical concentration in the range 10^{-10}–10^{-12} M.

For example, consider the prototype substrate in many studies: TCE with an associated rate constant for reaction with •OH of 4.2×10^9 M^{-1} s^{-1} (see Table 6.2). The *pseudo first-order rate constant* is given by

$$k_5' = -d \ln [S]/dt$$
$$= (4.2 \times 10^9) [•OH]_{ss}$$

In this expression, S is the organic substrate and the subscript ss denotes the steady-state condition. With $[•OH]_{ss} = 10^{-11}$ M, k_5' will be 0.042 s^{-1} and the half-life of the substrate will be 16.5 s. The goal of process optimization, therefore, is to maintain a steady-state •OH concentration at least of this magnitude.

With the incident photon flux and the concentrations of all the reactants as known variables, it is possible to model the reaction kinetics in the UV–H_2O_2 system and to verify the efficacy of the model with experimental measurements of the steady-state concentration of •OH and k_1, k_{-1}, k_2, k_3, k_4 and k_5. This has been attempted for a system containing various amounts of HCO_3^- and 25 μM humic acid by modeling and measuring the H_2O_2 loss in the presence and absence of butyrate or propionate ions, respectively.[11]

The rate constants can be computed from the slopes of first-order kinetics plots of ln C_t vs. time (C_t = concentration at any time t) such as those shown in Figures 6.4 and 6.5. Consider the data in Figure 6.5.[11] Two aspects are to be noted: First the presence of HA does not significantly perturb the primary photolysis rate of H_2O_2. This has been interpreted by the authors[11] to mean that

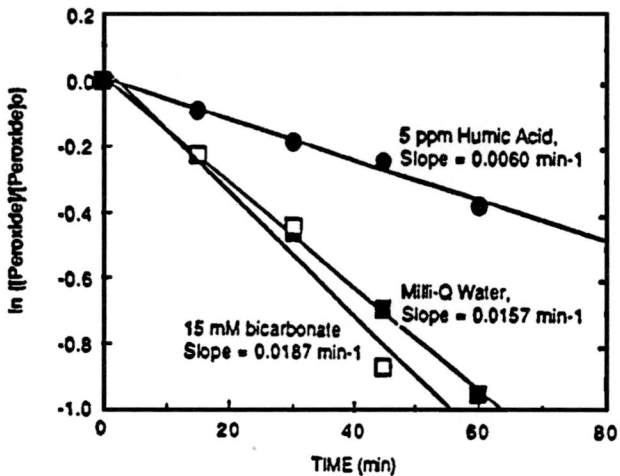

FIGURE 6.5. The measured hydrogen peroxide loss in the absence of butyrate and propionate ions under various conditions. (Reproduced with permission from Yao et al.[11])

Reaction (6.11) does not *regenerate* any H_2O_2. Second, the further addition of HCO_3^- has an insignificant effect over and above the baseline bicarbonate level (100 μM) found in the particular water sample (milli-Q) used in this study. This has been interpreted by the authors[11] to be diagnostic of the efficient scavenging of all the ·OH by HCO_3^- (via Reaction (6.12)) such that Reaction (6.5) is suppressed to a large degree. Further, the $CO_3^{-\cdot}$ generated via Reaction (6.12) can also oxidize H_2O_2,

$$H_2O_2 + CO_3^{-\cdot} \rightarrow HCO_3^- + HO_2^\cdot \tag{6.14}$$

resulting in the same net loss of H_2O_2. Good agreement was observed between the experimentally measured H_2O_2 loss with the model-generated values in the presence of butyrate or propionate ions respectively.[11]

The peroxide photolysis rate was also recently examined[12] under different conditions of solution sparging and in the presence and absence of radical scavengers. Figure 6.6 contains the data from this study.[12] In agreement with the study,[11] the presence of HCO_3^- causes little perturbation (for example, the apparent rate constants are ~8.3×10^{-4} s^{-1} for H_2O_2 only and 9.42×10^{-4} s^{-1} with 0.1 M HCO_3^-) on the H_2O_2 photolysis rate, pointing to the influence of Reaction (6.14). On the other hand, the lower rate constant obtained for acetate with N_2 sparging (5.0×10^{-4} s^{-1}), which is further lowered in the presence of O_2 (Figure 6.6), shows that the presence of O_2 has a significant effect on the extent of peroxide generation during photolysis.[4]

6.5. Reaction Kinetics and Mechanisms of the UV–H_2O_2 System

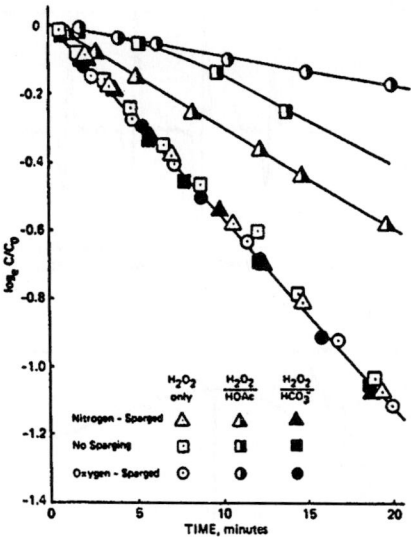

FIGURE 6.6. First-order plot of hydrogen peroxide photolysis data in the presence and absence of radical scavengers. (Reproduced with permission from Payton and Glaze.[12])

In the preceding discussion, and in the kinetics scheme, we have not considered the competition between the direct photolysis and the •OH-mediated organic photo-decomposition routes. Nor have the aforementioned studies specifically examined the temporal profile of the •OH radical concentration in the solution. The primary rate constants for the attack of •OH radicals on pollutant molecules (see Table 6.2) have been experimentally determined[13] using a competition method based on the technique of spin-trapping[14] coupled with detection by electron paramagnetic resonance (EPR) spectroscopy. The spin-trap molecule, 5,5'-dimethylpyroline-N-oxide (DMPO) reacts rapidly with radicals such as •OH to yield relatively long-lived spin adduct nitroxide radicals that can be detected readily by EPR. Figure 6.7 (top) contains a representative EPR spectrum of the DMPO–OH spin adduct recorded during UV irradiation of a 0.20 mM solution of H_2O_2 at pH = 7.0.[13] Other species such as the DMPO—HO_2 adduct would have shown a six-line EPR spectrum.[15] Thus, the HO_2• radical (see Reaction (6.5)) is not a significant intermediate in this system.

In the experiment, the field is set at the maximum of the second peak and the EPR signal amplitude is monitored as a function of time before and during the UV irradiation of the sample (Figure 6.7, bottom). The maximum amplitude of the EPR signal ($A°$) in the control run (i.e., with no pollutant in solution), and the initial slope (Figure 6.7, bottom) (i.e., the initial rate, $R_o°$) are used

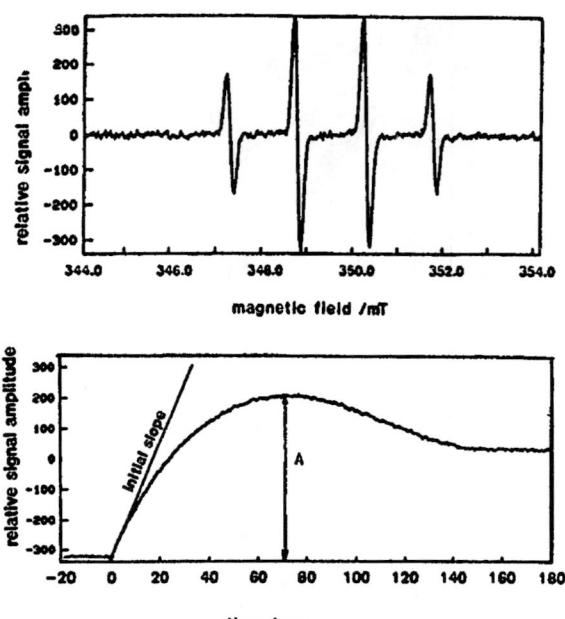

FIGURE 6.7. (top) EPR spectrum of the DMPO–OH spin adduct recorded during UV irradiation of an aqueous solution (pH = 7.0) of H_2O_2 and DMPO (both at a concentration of 0.20 mM); (bottom) Time course of the DMPO–OH signal maximum (low-field central peak) before and during UV irradiation of the same solution as in (top). (Reproduced with permission from Kochany and Bolton.[13])

as reference values. The organic substrate concentration is then varied and the ratios $A°/A$ and $R_o°/R_o$ are monitored to yield sets of data as shown in Figure 6.8 for benzene and three of its derivatives. The results were analyzed within the framework of the following kinetic model[13,16]:

$$H_2O_2 \xrightarrow{\Phi J_{ph}/h\nu} 2\,{}^{\bullet}OH \tag{6.1}$$

$${}^{\bullet}OH + DMPO \xrightarrow{k_6} DMPO-OH \tag{6.15}$$

$${}^{\bullet}OH + \text{pollutant} \xrightarrow{k_5} \text{products} \tag{6.16}$$

$$\text{pollutant} \xrightarrow{\Phi' J_{ph}'/h\nu} \text{products} \tag{6.17}$$

The inclusion of Reaction (6.17) in this scheme takes into account the competition from the direct photolysis route (see Figure 6.4). In the scheme, J_{ph} and J_{ph}' are the rates of photon absorption by H_2O_2 and pollutant, respectively, and Φ

6.5. Reaction Kinetics and Mechanisms of the UV–H_2O_2 System

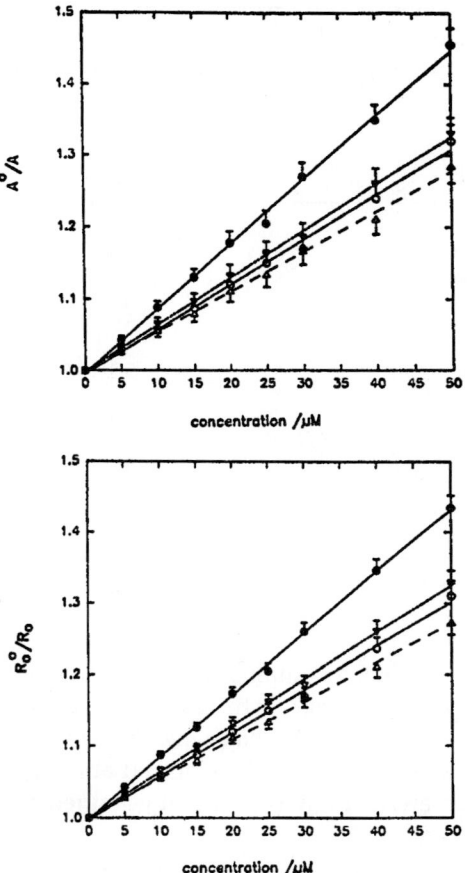

FIGURE 6.8. (top) Inverse amplitude ratio $A°/A$ and (bottom) inverse initial slope ratio $R_o°/R_o$ vs. substrate concentration for the photooxidation of benzene (circles), chlorobenzene (triangles), bromobenzene (circles), and iodobenzene (inverse triangles) with H_2O_2 at pH = 7.0. Error bars are ±1 standard deviation. (Reproduced with permission from Kochany and Bolton.[13])

and Φ' are the corresponding quantum yields. A steady-state analysis[16] yields the following expression:

$$R_o° / R_o = 1 + \frac{k_5 \text{[pollutant]}_o}{k_6 \text{[DMPO]}_o} \quad (6.18)$$

with $R_o° = 2\Phi J_{ph}$. A plot of $R_o°/R_o$ vs. [pollutant]$_o$ (see Figure 6.7a) should be a straight line with intercept 1, from the slope of which k_5 can be computed if k_6 is known. This rate constant is well known[13] and the authors used a value for k_6 of 4.3×10^9 M^{-1} s^{-1} for their computations. A similar analysis can be performed

Table 6.5. Rate Constants (k_5) for the Reaction of •OH Radicals with Benzene and Halobenzenes as Determined by the EPR Spin-Trap Technique

	$k_5/10^{-9}$ M^{-1} s^{-1}	
Compound	$A°/A$ Method	$R_o°/R_o$ Method
Benzene	7.7 ± 0.4	7.5 ± 0.4
Chlorobenzene	4.6 ± 0.3	4.3 ± 0.3
Bromobenzene	5.0 ± 0.3	4.8 ± 0.3
Iodobenzene	5.5 ± 0.3	5.3 ± 0.3
o-Dichlorobenzene	3.9 ± 0.3	3.9 ± 0.3
m-Dichlorobenzene	5.9 ± 0.4	5.4 ± 0.4
p-Dichlorobenzene	5.3 ± 0.5	5.3 ± 0.4

Source: Kochany and Bolton.[13]

for the $A°/A$ data.[13] Table 6.5 contains representative values for k_5 from this study, using both types of analyses. The results for benzene are in good agreement with the value determined by *pulse radiolysis* (see Table 6.2). The corresponding value for chlorobenzene was acquired at a higher solution pH (9.0) (Table 6.2) and is somewhat higher than that obtained from the EPR method (Table 6.5).

This study[13] also permitted an assessment of the importance of the direct photolysis route. This was done using a Pyrex filter, which blocks light with λ < 300 nm. The absorbance of the test compounds is negligible at wavelengths above 300 nm. No difference was seen in the k_5 values (within experimental error) with or without the Pyrex filter. This shows that the rates of reaction of •OH radicals with benzene and its derivatives are much larger than the rates of photolysis of these compounds. In other words, the direct photolysis route is "shunted" or bypassed in the presence of UV–H_2O_2 in the medium.

An important issue in UV–H_2O_2 concerns the degree of mineralization of the organic substrate and the generation of intermediates. An early study on simple anions such as acetate revealed up to 97% COD removal; hence no stable intermediates were detected.[9] Similar studies on dissolved organic impurities (not identified) have revealed reduction of the TOC content of distilled water by about 88% and tap water by 98%.[17] The residual 12% organics in the former case were detected by GC analysis of the hexane extracts after UV–H_2O_2 treatment. Other studies of 2-chlorophenol have reported catechol and

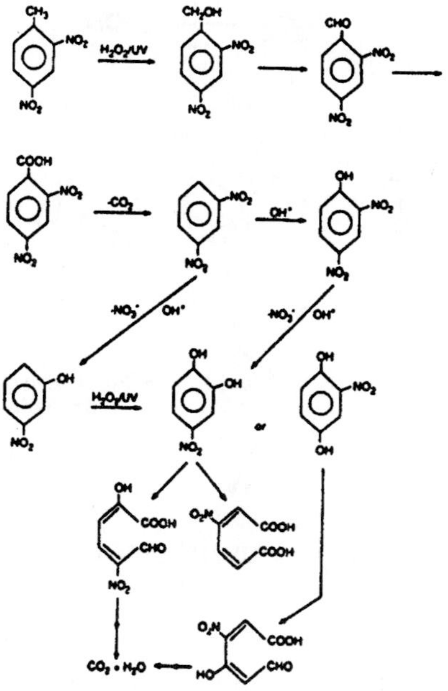

6.5. Reaction Kinetics and Mechanisms of the UV–H_2O_2 System

FIGURE 6.9. Mechanism of UV/H_2O_2 oxidation of 2,4-dinitrotoluene.

2-chlorohydroquinone as final products,[18] and in yet another study of the same compound, more extensive oxidation of catechol to pyrogallol and 1,2,4-trihydroxybenzene was noted.[19] Treatment of benzene yielded phenol, catechol, hydroquinone, and resorcinol as intermediates, all of which decomposed to mineralized products (CO_2 and H_2O) with continued treatment.[20] Compounds similar in structure to HA (humic acids) and resistant to photo-oxidation were reported for the UV-H_2O_2 treatment of 4-bromodiphenyl-ether.[21] Finally, the mechanistic scheme in Figure 6.9 is representative of the sort of intermediates to be expected from the UV–H_2O_2 treatment of 2,4-dinitrotoluene, a waste product from military munitions facilities which is particularly hazardous to biota.[22]

A problem with the use of a steady-state UV light source for identification of reaction intermediates is associated with their photolysis. That is, a suspected "intermediate" may really be the photolytic daughter fragment of a parent (real) intermediate. This problem is particularly severe at the low pollutant levels typical of environmental scenarios. On the other hand, the use of higher substrate concentrations (to achieve significant product yield at low conversion) leads to bimolecular coupling reactions (and even polymerization) that often do not really mimic the environmental condition.

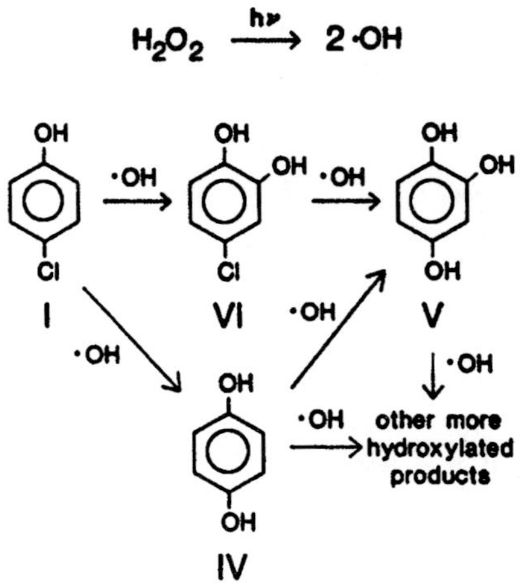

$$H_2O_2 \xrightarrow{h\nu} 2 \cdot OH$$

FIGURE 6.10. Mechanistic scheme for the UV/H_2O_2 photolysis of 4-chlorophenol. (Reproduced with permission from Lipczynska-Kochany and Bolton.[23])

A method based on HPLC detection of flash photolysis products has been recently developed[23] to avoid the possibility of photolysis of intermediates. With this approach, intermediates that are longer-lived than ~15 min. (the time required for a HPLC run) can be monitored. The technique was applied to both the direct photolysis and the •OH-radical mediated photodecomposition of 4-chlorophenol (4-CP).[23-25] Interestingly enough, p-benzoquinone (II) was the only primary photoproduct in the absence of H_2O_2. When the concentration of H_2O_2 was increased to greater than 20 times the amount of 4-CP, the major product was 4-chlorocatechol (VI) with smaller amounts of hydroquinone (IV) and 1,2,4-trihydroxybenzene (V). Figure 6.10 illustrates the reaction scheme.[23] When multiple flashes were applied, IV and V became the dominant products, diagnostic of the further photolysis of VI. However, the overall yield of the products declined suggesting that more massive degradation (and probably mineralization) was occurring. This new technique supplements the conventional configuration, where flash photolysis is coupled with *optical* detection of the intermediates. A major distinction between the two detection approaches relates to the time resolution. That is, shorter-lived intermediates can be accessed with the optical detection approach. Nonetheless, the flash photolysis–HPLC technique ought to be of general utility in further mechanistic studies of UV–H_2O_2 reaction systems.

Table 6.6 provides a compilation of representative examples of the application of UV–H_2O_2 to organic pollutant treatment. The reader is referred to fur-

6.5. Reaction Kinetics and Mechanisms of the UV–H$_2$O$_2$ System

Table 6.6. Listing of Organic Compound Types That Have Been Treated by the UV–H$_2$O$_2$ Approach

Chlorinated Aliphatics
  Dichloroethane
  Trichloroethane
  Tetrachloroethane
  Pentachloroethane
  Carbon tetrachloride
  Chloroform
  Dichloromethane
  1,2-Dibromopropane

Halogenated Ethylenes
  Trichloroethylene
  Dichloroethylene
  Perchloroethylene
  Dibromoethylene

Aromatics
  Benzene
  Toluene
  Ethylbenzene
  Xylenes

Phenols
  Phenol
  2,5-Dimethylphenol
  m-Cresol
  Nitrophenols
  Chlorophenols
  Chloronitrobenzenes
  Trichlorophenol
  Dichlorophenols

Nitroaromatics
  2,4-Dinitrotoluene
  Trinitrotoluene
  1,2-Dimethyl-3-nitrobenzene

Pesticides and Related Compounds
  Chlordane
  Dioxane
  Atrazine
  Malathion
  Carbetamide
  Metoxuron

Cont. Table 6.6.   Organic Compound Types

Alcohols and Ketones
  Methanol
  Butanol
  2-Pentanol
  Methyl isobutyl ketone
  Methyl ethyl ketone

Acids
  Acetic acid
  Butyric acid
  Propionic acid
  Formic acid
  Naphthenic acids
  Fatty acids

Other Miscellaneous Organics
  4-Bromodiphenylether
  Dimethylhydrazine
  Freon-TF
  Dimethyl phthalate
  Diethyl phthalate
  Cumene
  Diethyl malonate

ther extensive reviews of the scope of the UV–H_2O_2 approach.[2,3,26] As seen in this tabulation, the range of organics that can be effectively treated is extensive. A variety of wastewater media have been treated, including boiler feedwater, explosives-containing effluent, tannery wastes and keratin solutions, textile plant waste, paper and pulp bleaching waste water, and cutting oil waste water.[26] In many cases, however, it must be noted that complete mineralization or decolorization is not obtained. Commercialization efforts of this technology are discussed in Chapter 8.

6.6. REACTION KINETICS, MECHANISMS, AND EXAMPLES OF APPLICATION OF THE UV–O_3 SYSTEM

The discussion in Section 6.2 shows that •OH radicals can be generated along two different routes in photolytic ozonation, one involving the direct photolysis of ozone (to •OH) or of the H_2O_2 generated via Reactions (6.6) and

6.6. Reaction Kinetics and Mechanisms of the UV–O$_3$ System

Table 6.7. Important Reactions and Associated Bimolecular Rate Constants in the UV–O$_3$ System (see Figure 6.11)

Reaction	No.	Rate Constant M^{-1} s^{-1} (or s^{-1})	Reference(s)
$HO_2^- + O_3 \rightarrow O_3^- + HO_2^\bullet$	(6.8)	5.5×10^6	27
$O_3^- + H^+ \rightarrow HO_3^\bullet$	(6.9a)	5.2×10^{10}	11,28
$HO_3^\bullet \rightarrow {}^\bullet OH + O_2$	(6.10)	1.4×10^5	11,28
$HO_2^\bullet \rightleftarrows H^+ + O_2^-$	(6.22)	(pK_a = 4.8)	—
$O_3 + O_2^- \rightarrow O_3^- + O_2$	(6.19)	1.5×10^9	29,30
${}^\bullet OH + O_3 \rightarrow O_2 + HO_2^\bullet$	(6.23)	1.1×10^8	31
$O_3 + H_2O \xrightarrow{h\upsilon} O_2 + H_2O_2$	(6.20)	—	—
$H_2O_2 \xrightarrow{h\upsilon} 2\, {}^\bullet OH$	(6.1)	—	—
$H_2O_2 \rightleftarrows HO_2^- + H^+$	(6.2)	(pK_a = 11.6)	—
$H_2O_2 + HO_2^- \rightarrow H_2O + O_2 + {}^\bullet OH$	(6.3)	—	—
$HO_2^\bullet + O_3 \rightarrow {}^\bullet OH + 2\, O_2$	(6.24)	$>10^4$	31
$2\, HO_2^\bullet \rightarrow H_2O_2 + O_2$	(6.25)	8.3×10^5	—
$HO_2^\bullet + O_2^- \rightarrow O_2 + HO_2^-$	(6.26)	9.7×10^7	—
$H_2O_2 + {}^\bullet OH \rightarrow HO_2^\bullet + H_2O$	(6.5)	2.7×10^7	11,32

(6.7), and the other involving ozone and the superoxide radical anion formed by the reaction of organic radicals with O_2 (see Figure 6.3):

$$O_2^- + O_3 \rightarrow O_3^- + O_2 \qquad (6.19)$$

followed by Reactions (6.9) and (6.10). Reaction (6.19) is facile and has an associated rate constant of 1.6×10^9 M^{-1} s^{-1}.[27] Other chain reactions in the UV–O$_3$ system and available rate data are summarized in Table 6.7.[27–32] Indeed, this system is exceedingly complex, and some 84 different reactions have been included in an attempt to kinetically model it.[11] The situation is complicated by

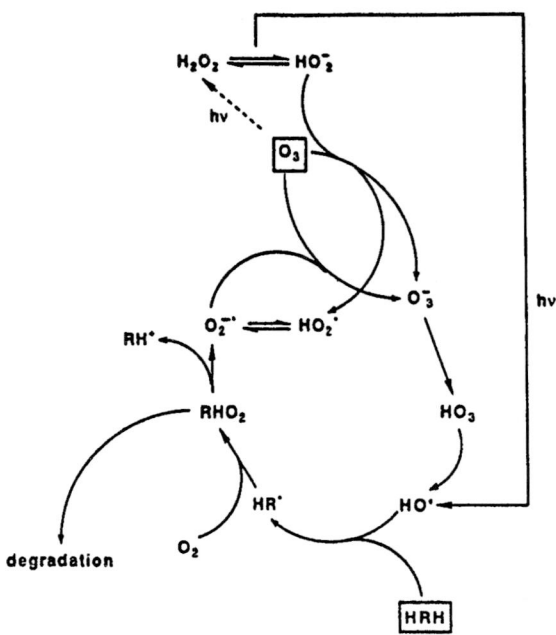

FIGURE 6.11. Reaction pathways in the ozone–UV and ozone–peroxide systems. (Reproduced with permission from Legrini et al.[3])

the fact that ozone decomposes in the dark by a chain reaction that is initiated by OH⁻

$$O_3 + OH^- \rightarrow HO_2^{\bullet} + O_2^-$$

and propagated by O_2^-; the peroxide anion (HO_2^-) continues the chain.

The relative importance of the two ("direct") •OH generation routes,

$$O_3 + H_2O \xrightarrow{h\nu} O_2 + H_2O_2 \quad (6.20)$$
$$O_3 + H_2O \xrightarrow{h\nu} O_2 + 2\,{}^{\bullet}OH \quad (6.21)$$

has also been addressed in a recent study[12] by monitoring the H_2O_2 generated via the photolysis of aqueous ozone. The conclusion of this study was that •OH radical generation is initiated both via Reaction (6.20) coupled with Reaction (6.1) and by Reaction (6.8) coupled with Reactions (6.9) and (6.10). Figure 6.11 summarizes the reaction scheme.[12] Once the cycle is initiated, it continues

6.6. Reaction Kinetics and Mechanisms of the UV–O$_3$ System

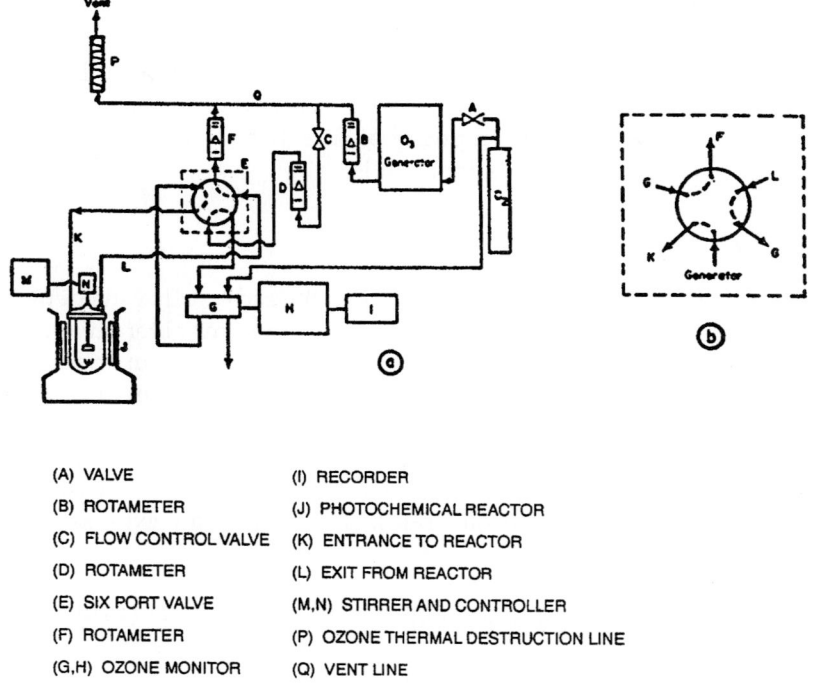

(A) VALVE	(I) RECORDER
(B) ROTAMETER	(J) PHOTOCHEMICAL REACTOR
(C) FLOW CONTROL VALVE	(K) ENTRANCE TO REACTOR
(D) ROTAMETER	(L) EXIT FROM REACTOR
(E) SIX PORT VALVE	(M,N) STIRRER AND CONTROLLER
(F) ROTAMETER	(P) OZONE THERMAL DESTRUCTION LINE
(G,H) OZONE MONITOR	(Q) VENT LINE

FIGURE 6.12. (a) Schematic diagram of experimental system for ozone/UV studies. (b) Alternative flow pattern for ozone off-gas measurement. (Reproduced with permission from Peyton et al.[33])

by the pathway indicated by the lower perimeter of the circle in Figure 6.11, provided the organic radical adduct (RO$_2$H) can facilitate the elimination of O$_2^-$. When this part of the cycle is dominant, the •OH radical yield per O$_3$ molecule may approach unity and ozone photolysis may become relatively unimportant. If, on the other hand, O$_2^-$ elimination is not possible, the organic adduct decays by other (slower) reaction routes. Many of these reactions result ultimately in the regeneration of H$_2$O$_2$. Under these conditions, photolysis of H$_2$O$_2$ may be an important source of •OH throughout the reaction. Ozone or H$_2$O$_2$ will then accumulate until the short-circuiting reactions of O$_3$ and H$_2$O$_2$ with •OH (Reactions (6.23) and (6.5), respectively) become significant.

This last scenario is unfavorable in that the available •OH flux (for reaction with the pollutant) may be drastically diminished. A similar situation may prevail when the pollutant level is very low relative to the ozone input rate. Clearly, the efficiency of photolytic ozonation for destruction of organic substrates can vary widely, depending on the relative ozone input, the UV photon flux, substrate concentration, and medium variables (e.g., carbonate–bicarbonate alkalinity).

The kinetics of the destruction of tetrachloroethylene (PCE) by UV–O_3 has been studied[33] using the experimental setup illustrated in Figure 6.12. The substrate disappearance was monitored by GC with electron capture detection. The rate was modeled as a composite of four terms—purging, photolysis, ozonation, and photolytic oxidation; that is,

$$k_{tot} = k_{purge} + k_p I_o + k_o [O_3]_l + k_1 I_o D \qquad (6.27)$$

where k_{tot} is the composite rate constant, I_o is the radiant flux incident on the reactor (in units of W/liter), D is the ozone dose rate (mg/liter•min) and $[O_3]_l$ is the concentration of ozone in the liquid phase. The first term arises from the volatilization loss from the liquid phase; the second term, from direct photolysis of the substrate; and the third term is the "dark" reaction component between the ozone and the organic substrate. In this particular study, the purge rate was low although, at high linear gas velocities, this term will become important, especially for volatile substrates. The overall rate expression then is given by

$$-d[S]/dt = k_{purge}[S] + k_p I_o^a [S]^b + k_o [O_3]_l^c [S]^d + k_1 I_o^e D^f [S]^g \qquad (6.28)$$

For a first-order process, the exponents b, d, and g are each equal to unity and the rate expression reduces to

$$-d[S]/dt = k_{tot}[S] \qquad (6.29)$$

In Eq. (6.27), it was assumed that the reaction is first order in ozone concentration and the light flux and that $f = 1$ (see later). The component rate constants were individually evaluated in this study[33] from separate experiments involving only purging (no ozone, no UV), purging and ozonation (no UV), and photolysis (no ozone). The corresponding pseudo first-order rate constants were obtained in the usual manner, after correcting the data for the purging loss "background." Additionally, the exponents a, c and f were determined from log–log plots of the rate constant vs. the UV flux, ozone level in the liquid phase, and D, respectively. Two different water samples, purified water and lake water, were used as the matrix for PCE. Interestingly enough, the exponents were all unity for the purified water while $a = 0.67$ and $e = 0.42$ in the lake water. Figure 6.13 contains representative data for the purified water.[33] The solid lines in each case are fits to the data using the preceding kinetics model.

Figure 6.14 contains additional data on chloroform from the same study.[33]

6.6. Reaction Kinetics and Mechanisms of the UV–O$_3$ System

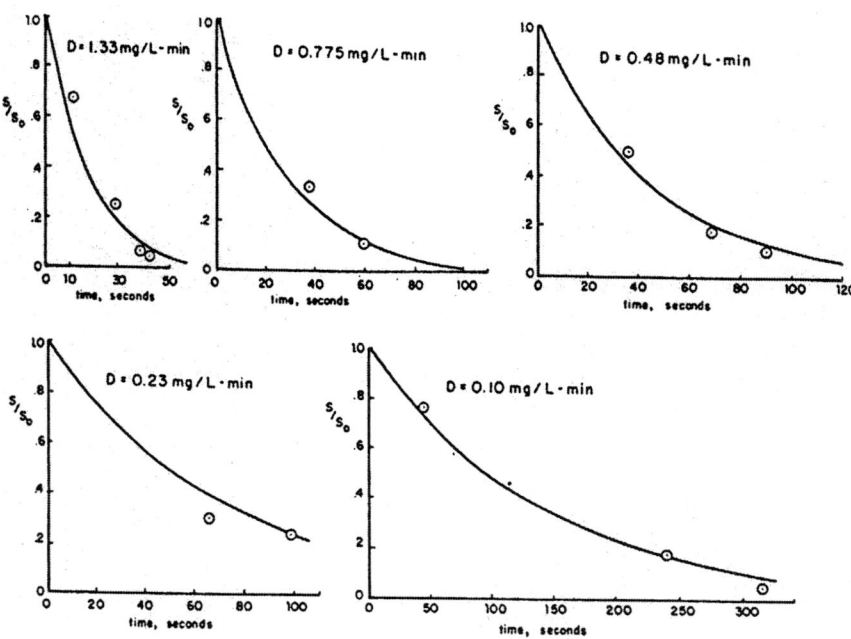

FIGURE 6.13. Normalized kinetics plots for the photolytic ozonation of tetrachloroethylene in purified water: ozone dose rate shown; UV dose, 0.375 W/liter; initial concentration 100 μg/liter. Solid line is predicted by the kinetics model. (Reproduced with permission from Peyton et al.[33])

Figure 6.14a shows the relative importance of the purge and the (dark) ozonation components along with the rather dramatic effect of the UV irradiation on the substrate degradation rate. Of course, the direct photolysis and the UV–O$_3$ photolysis routes are convoluted with one another in the latter case. The accompanying data in Figure 6.14b illustrate the matrix effect. In this particular experiment, a fresh aliquot of chloroform was spiked into the *same* matrix at the end of the first run (A). Note that the pseudo first-order plot is now linear and has a slope near the limiting slope in Run A (dashed line). The authors[33] attribute this to the destruction of interfering substances in the first run to yield the correct matrix-free kinetics in Run B.

The applications of the UV–O$_3$ technology have been as diverse as in the corresponding UV–H$_2$O$_2$ case summarized in Table 6.6. This technology was first used in the 1970's for the treatment of cyanide wastes.[34] Subsequently, it was shown that ozone in combination with UV radiation enhanced the oxidation of chlorinated solvents, pesticides and miscellaneous group parameters such as COD and BOD.[35–37]

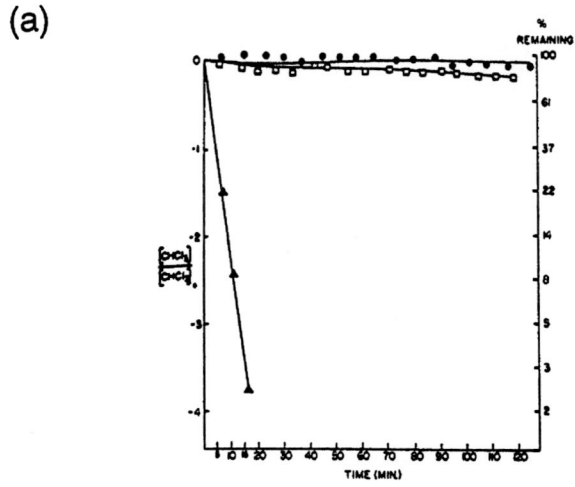

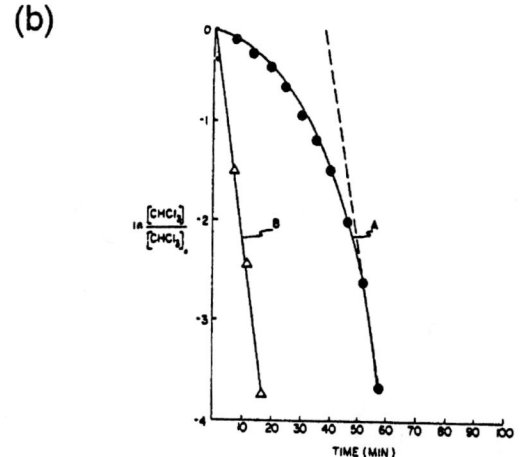

FIGURE 6.14. (a) Example of the effect of UV irradiation on the oxidation of chloroform in purified water matrix: (solid circles) pH 7, N_2 purge at 25 mL/min; (squares) pH 7, O_3 dose rate 1.3 mg/(liter min); (triangles) pH 6.5, O_3 1.4 mg/(liter min). 0.42 W/liter 254-nm UV irradiation (transferred). (b) First-order kinetics plot for the photolytic ozonolysis of chloroform. Initial concentration, 100 µg/liter; ozone dose rate, 3 mg/(liter min); UV dose, 0.4 W/liter; pH 6.5. Run B: Water reused from Run A with second spike of $CHCl_3$. (Reproduced with permission from Peyton et al.[33])

The dark reactions involving O_3 and organic solutes in aqueous media have been analyzed in detail[28-38] because of their relevance to water-treatment processes (Chapter 8) and also because they are important to an understanding of the chemistry of atmospheric water droplets. A reaction model comprising initiation, propagation, and termination steps and expressions for describing the kinetics of the various chain reactions have been presented.[28]

Finally, an example of the use of UV–O_3 is a study on the oxidation of 4-chloro-2-methyl-phenoxyacetic acid.[42] This chemical is widely used as a herbicide, and traces of it have been found in surface water. Two distinct ozonolysis pathways were identified in this study: ring hydroxylation (see Figure 6.9) and cleavage by molecular O_3 in the dark, and side-chain oxidation by ·OH under irradiation.

6.7. THE UV–H_2O_2–O_3 PROCESS

It has been demonstrated[43-46] that the combination of H_2O_2, UV radiation, and ozone increases the fraction of available ozone, and the possibility exists that the rate of ozone mass transfer into the liquid phase is higher than in the treatment process based on ozone alone (see later). This process has been commercialized, and we shall further discuss it in Chapters 7 and 8.

6.8. UV–H_2O_2 AND UV–O_3 SYSTEMS: PRACTICAL CONSIDERATIONS

Further process improvements with both these technologies will hinge on developments in lamp technology. Four types of lamps have significant radiant output from 200 to 300 nm:[2]
- Low-pressure Hg vapor lamps
- Medium-pressure Hg vapor lamps
- Pulsed Xenon flashlamps
- Proprietary designs

Lasers (e.g., KrCl excimer) have been used, although it is doubtful whether they will be practical for large-scale use. Specially tuned Hg arcs and Xe-doped Hg arcs are undergoing development.[2] The UV region has been subdivided into UV-A; 320–380 nm, UV-B; 280–320 nm, and UV-C, 210–280 nm. Mercury has a line at 253.7 nm, and this wavelength has been commonly employed. The spectral output of the low-pressure Hg vapor lamp is mostly at this wavelength with ~1% of the output at 185 nm. Often, however, the 185 nm line is absorbed by the quartz envelope of the lamp. This line can be used if the quartz is replaced with Supracil, which is transparent to 185 nm radiation. The lower

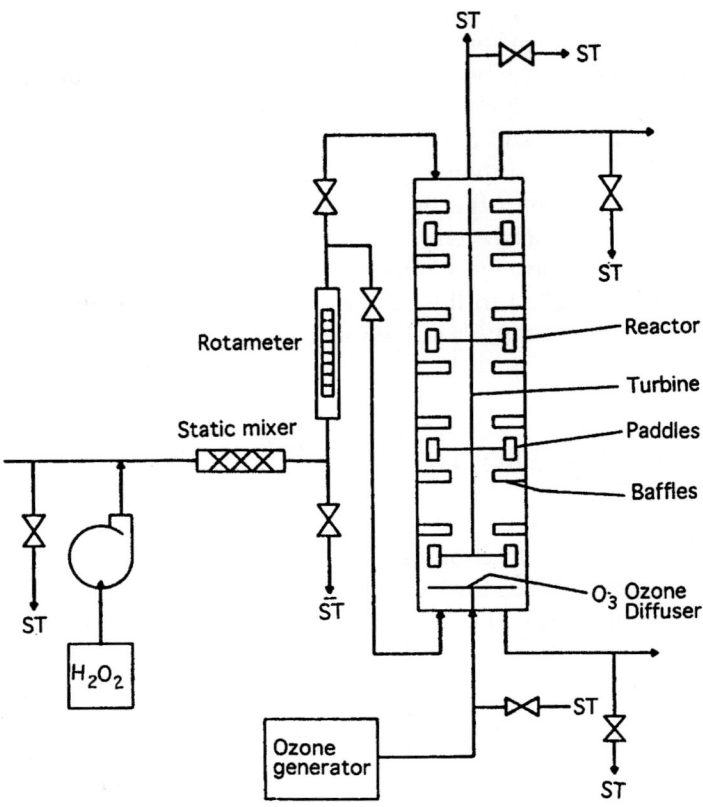

FIGURE 6.15. Pilot-scale reactor system for peroxide–ozone AOP; ST is sample top. (Reproduced with permission from Aieta et al.[47])

wavelength light is more energetic than the higher wavelength light—1 mole of photons (1 einstein) at 254 nm equals 471 kJ/mole, whereas an einstein of 185 nm light equals 647 kJ/mole.[8] This can be especially important for direct photolysis and for H_2O_2 photolysis applications. Until recently, low-pressure UV lamps were available only with relatively low power output. More powerful lamps in the 10–20 kW range are becoming available.[8]

It is worth noting that ozone is most efficiently photolyzed at a wavelength near 254 nm (see Figure 6.1b) with the result that the low-pressure Hg vapor lamp is eminently suited to the UV–O_3 technology. For UV–H_2O_2 (and for direct photolysis), on the other hand, a broad-spectrum UV lamp with significant output at lower wavelengths is to be preferred for reasons noted earlier.

6.8. UV–H_2O_2 and UV–O_3 Systems: Practical Considerations

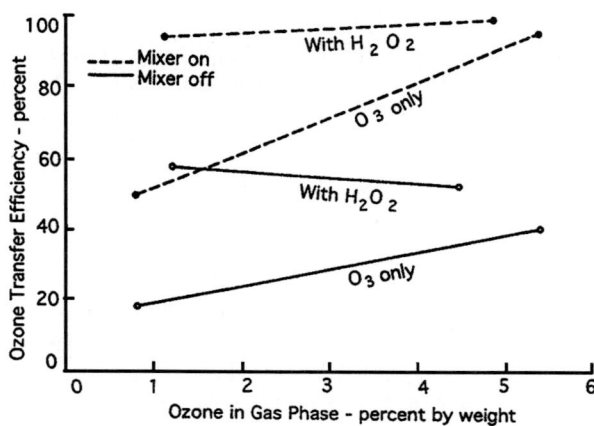

FIGURE 6.16. Ozone transfer efficiency as a function of gas-phase ozone concentration. (Reproduced with permission from Aieta et al.[47])

A serious problem with the UV–O_3 approach is the limited solubility of ozone in water. On the other hand, H_2O_2 is available as an easily handled solution that can be metered into the aqueous waste water to afford a wide range of accurate, reproducible concentrations. The addition of gas-phase oxidants such as O_3 to water containing volatile organics also causes air stripping of the latter, as noted in Section 6.6. The mass-transport limitations imposed by the limited solubility of O_3 have prompted special attention to reactor design in the UV–O_3 technology.

Reactors of the CSTR type (Chapter 2) have proven to be effective in maintaining an adequate O_3 flux in the system. Figure 6.15 contains a schematic of a reactor design utilized in a pilot study,[47] where the liquid flow can be either cocurrent or countercurrent to the gas flow. A four-stage turbine mixer is shown in the column, with each turbine stage buffered by stators attached to the column. The turbine is driven by a variable-speed drive. Ozone enters the reactor from the bottom while the H_2O_2 for the UV–O_3–H_2O_2 process is injected at the top. Sampling parts are located in the piping before and after the reactor. Figure 6.16 contains the results of the O_3 mass-transfer experiments, where the transfer efficiency is expressed in the following manner:

$$\text{Transfer efficiency} = \frac{\text{Feed-gas } O_3 - \text{Off-gas } O_3}{\text{Feed-gas } O_3}$$

The results are shown both with and without the mixer turned on. Clearly, adding H_2O_2 to the feedwater increases the O_3 transfer efficiency significantly, especially at lower gas-phase O_3 levels. The transfer efficiency of the H_2O_2 experiments is nearly independent of the gas-phase O_3 concentration. For O_3 only, the transfer efficiency increases with the latter parameter.

These results have been explained[47] within the framework of the following equation:

$$d[O_3]/dt = k_m A_e ([O_3]_s - [O_3]) - \Sigma k_j [S_j] [O_3] \tag{6.30}$$

where $[O_3]$ is the molar concentration of ozone in the liquid phase, $[O_3]_s$ is the saturation value of ozone in equilibrium with the gas-phase level, $\Sigma[S_j]$ is the sum of all the liquid-phase components that consume ozone (organics, H_2O_2, etc.), and k_j are the corresponding bimolecular rate constants, k_m is the mass transfer coefficient (cm•s^{-1}), A_e is the volumetric surface area (cm^2•cm^{-3}), and $k_m A_e$ is the volumetric mass-transfer coefficient (s^{-1}). There are two limiting scenarios to Eq. (6.30). The first one occurs when k_j is very small or zero. For a fixed contact time, the amount of O_3 transferred into the liquid depends on $k_m A_e$ and $[O_3]_s$. Therefore mechanical mixing, an increased gas flow rate, or an increase in the gas-phase O_3 concentration will all contribute to an increased mass transfer. This is the situation for the "O_3 only" experiments in Figure 6.16.

The second case is when the reactions in the liquid phase are so rapid (high k_j) as to consume all the O_3 transferred from the gas phase. The penetration theory of mass transfer predicts mass-transfer enhancement when the dissolving gas reacts in the liquid phase.[48] In this case, the rate of increase of O_3 in the liquid phase is zero and the bulk phase O_3 concentration, $[O_3]$ is also zero or very small because of the rapid reaction at the interface. The net result is that the transfer efficiency becomes independent of the gas-phase concentration and depends only on the $k_m A_e$ term. This is consistent with the trend in Figure 6.16 in that the transfer efficiency increases significantly with the mixer turned on, although it still maintains a relatively "flat" profile.

Ozone generators are commercially available and have long been used in the drinking water industry (Chapters 7 and 8). In these commercial units, O_3 is usually generated by the silent electric (corona) discharge method. The generator consists of a pair of electrodes separated by a gas-filled space and a layer of dielectric insulation such as glass. Air or pure O_2 is bled into the empty space and high voltage (7,500–20,000 V) AC power (50–2,000 Hz) is applied. Corona discharge occurs in the gas, and a portion of the reactant is converted to O_3. Typical yields vary from 1 to 6%, depending on whether the feed gas is air or O_2. A large fraction of the input energy is dissipated as heat; therefore, efficient cooling (either by air or water) of the generator is critical.

The final aspect to be discussed in this section concerns the analytical chemistry for monitoring the UV–H_2O_2, UV–O_3, and UV–O_3–H_2O_2 processes. Hydrogen peroxide in the liquid phase can be determined either colorimetrically by complexation with Ti(IV)[49] or by using fluorometry.[50] Ozone in the aqueous phase is commonly measured by the indigo method.[51] Other methods are available including iodometry[52] and UV spectrophotometry.[53] Gas-phase ozone levels can be monitored using commercial instruments. The borate-buffered potassium iodide (BKI) test for aqueous ozone,[53] however, is sensitive to oxidants (other than ozone) that have oxidation potentials sufficiently high to oxidize the iodide. Similarly, ozone has been reported[12] to interfere negatively with the titanium method for H_2O_2 assay.[49] It is thus necessary to remove the O_3 from solution (e.g., by sparging) prior to a peroxide determination. The *total* oxidant level can be measured by the iodometric method[52] with a small amount of catalyst added (e.g., ammonium molybdate) to accelerate the reaction with the peroxide.[12]

6.9. HETEROGENEOUS PHOTOCATALYSIS

This method, introduced in Section 1.6.9, uses a semiconductor catalyst, usually in particulate form, to photogenerate e^-–h^+ (electron–hole) pairs at the catalyst/solution interface. As discussed later, these photogenerated carriers can then be used to convert reducible and oxidizable pollutants to less harmful products or to immobilize them on the catalyst surface. A photocatalytic system is a subset of *photoelectrochemical* (PEC) phenomena, where electrochemical processes are induced by irradiation of a semiconductor electrode. The principles underpinning photoelectrochemistry were discussed in Section 2.11.

In a historical context, it is interesting to note that much of the research in the 1970s and the early 1980s was oriented toward the photovoltaic conversion and energy storage possibilities with PEC devices. The late 1980s and plummeting oil prices prompted a reorientation of PEC studies to materials synthesis, processing, and characterization, particularly with the electronics industry as a target consumer. Concomitant with this trend was the realization that PEC methods, and specifically photocatalytic techniques, can play a useful role in hazardous waste treatment. In this respect, the field of photoelectrochemistry will perhaps carve out a niche in the next few years in environment-related technologies. This is already borne out by the fact that TiO_2-based photocatalysis is generating commercial interest, as discussed in more detail later.

From a technological perspective, the photocatalysis methodology will have to be evaluated against air and water remediation methods currently employed: incineration, air stripping, and granular activated carbon adsorption. This will be done in Chapter 8. The photocatalysis method may also be considered under the umbrella of a number of new technologies, collectively known as advanced oxidation processes (AOPs). As discussed earlier, these technologies almost all rely on the generation of very reactive free radicals (e.g., ·OH) that are subsequently used to destroy the organic pollutants or microorganisms.

Table 6.8. Organic Pollutants Successfully Treated by TiO_2-based Photocatalysis

Pollutant category	Reference(s)
BTEX	54,55
Phenols	56–60
TCE and other halogenated hydrocarbons	61–75
DDT, dioxins, other pesticides, and herbicides	76–79
Kraft lignin	80,81
Dye stuffs, organophosphates, and surfactants	82–91
Gasoline	92

Note: Also see Refs. 3, 26, 93–98.

6.9.1. Operating Principle and Scope of the Photocatalysis Method

The absorption of bandgap light by the semiconductor (SC) at the particle/electrolyte interface results in the formation of an electron—hole pair:

$$SC \xrightarrow{h\nu > E_g} e^-_{CB} + h^+_{VB} \qquad (6.31)$$

E_g represents the SC energy bandgap and the subscripts attached to the (photogenerated) electronic carriers specify their location in the conduction band (CB) and valence band (VB) in the semiconductor. A major distinction can be drawn with regard to the interphasial situation at semiconductor slurries–suspensions and their immobilized counterparts. Specifically, when the latter are attached to a conductive support, the possibility exists for imposing a bias potential across the semiconductor/electrolyte interface, thus modifying the electrostatics to bring about better electron–hole separation. In the former case involving semiconductor suspensions, the extent to which the (deleterious) electron–hole pair recombination is suppressed is determined solely by the dynamics of carrier transport across each semiconductor particle (see Section 6.9.5), and by the efficacy with which the carrier-intercepting surface reactions occur.

In terms of applicability to pollutant treatment strategies, the reactions involving photogenerated electrons and holes are best discussed separately. Thus,

6.9. Heterogeneous Photocatalysis

Table 6.9. Inorganic Pollutants Treated by Photocatalytic Techniques

Pollutant(s)	Reference(s)
Cyanide	100–106
Hydrogen sulfide	107–109
Mercury and cadmium	10, 111
Chromium	99, 110, 112–116
Sulfite	101
Manganese	117

the photogenerated holes oxidize water or adsorbed OH^- (on the photocatalyst surface) to hydroxyl radicals:

$$OH^- + h_{VB}^+ \xrightarrow[h\nu]{SC} {}^\bullet OH \qquad (6.32)$$

These highly reactive radicals can then be used to mineralize or at least partially oxidize most organic pollutants. Indeed, the range of compounds that have been successfully treated is impressive, as the compilation in Table 6.8 illustrates. More extensive listings may be found in the literature.[3,26,93–98] The $^\bullet OH$ radicals are also nonselective in their attack of microorganisms and cause biological cell inactivation in many cases. The applicability of the photocatalytic approach for air and water disinfection scenarios will be considered in the next chapter.

Paralleling Reaction (6.32) is the possibility of *direct* attack of an organic substrate by the photogenerated hole. This pathway appears to be particularly important in gas-phase photocatalysis.

Inorganic pollutants may be either reduced to the elemental form and thus removed from the process stream;

$$M_{(aq)}^{n+} + ne_{CB}^- \xrightarrow[h\nu]{SC} M_{(s)} \qquad (6.33)$$

or converted to an environmentally more benign element in a different oxidation state as exemplified by the $Cr(VI) \rightarrow Cr(III)$ system. Strategies incorporating simultaneous pH adjustment also immobilize the pollutant as an (insoluble) oxide onto the photocatalyst, to be subsequently regenerated in a separate step. This approach has been recently demonstrated using the $Cr(VI) \rightarrow Cr(III)$ model

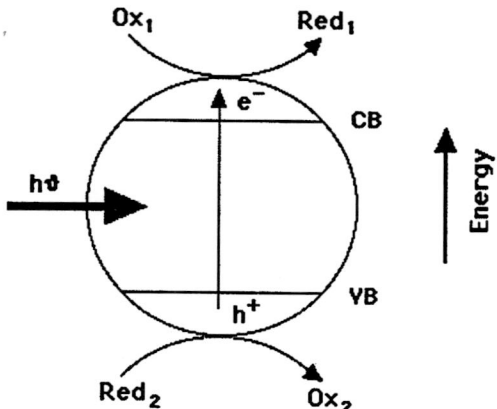

FIGURE 6.17. Schematic diagram of photoresponse of an irradiated semiconductor particle. Spherical particle geometry is assumed.

system.[99] Table 6.9 contains examples of inorganic pollutants that have been treated successfully using the photocatalysis methodology.[100–117]

Indeed, the capability exists with the photocatalysis approach for *simultaneously* treating a water (or air) stream contaminated with all three types of pollutants; namely, organic, biological, and inorganic. This is particularly relevant in the drinking water industry, where the raw water often contains odor- or color-producing organic chemicals, inorganic species such as Pb(II) or As(III), and microorganisms such as bacteria, protozoa, or viruses (see Chapter 7).

Figure 6.17 summarizes the photodriven events at a single semiconductor particle. Within an environmental context, Ox_1 could be a metal ion (e.g., Pb^{2+}) and Red_2 could be either an organic substrate or a microorganism.

The *simple* recombination of the e^-–h^+ pairs often involves a mediating "surface state" (located in the bandgap of the semiconductor) and *results in no net chemical change in the system*. The concept of *conjugate* reactions is useful for analyzing the photocatalysis process. Thus Reaction (6.32) may be alternately expressed as

$$2\,H_2O + 4h^+ \xrightarrow[h\upsilon]{SC} O_2 + 4H^+ \tag{6.34}$$

Now, O_2 functions as an electron acceptor to *chemically reverse* this process:

$$O_2 + 4H^+ + 4e^- \xrightarrow[h\upsilon]{SC} 2\,H_2O \tag{6.35}$$

If the carrier fluxes due to Reactions (6.34) and (6.35) are exactly balanced, the holes and electrons will have been effectively recombined as can be demon-

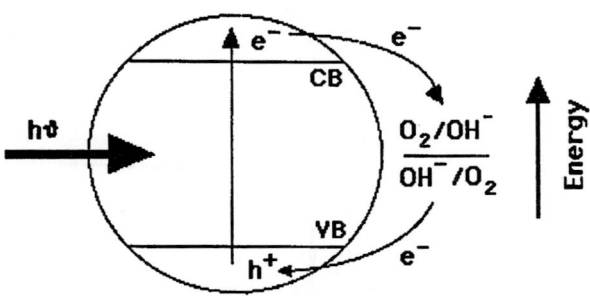

FIGURE 6.18. Short-circuiting of an irradiated semiconductor particle via redox processes mediated by O_2.

strated by simple addition of Reactions (6.34) and (6.35). Again there is no net chemistry and O_2 has functioned in the role of a "chemical surface-state" for mediating the e^-–h^+ recombination process (see Figure 6.18). More complicated chemical scenarios can be envisaged that involve the surface hydroxyl groups at the semiconductor surface as the carrier recombination mediators, but the conclusion remains the same.

The foregoing discussion underlines the fact that net chemistry can occur at the semiconductor surface only if either Reaction (6.34) or (6.35) is intercepted at an intermediate stage. *Alternatively, the e^- and h^+ at the semiconductor surface must react with different redox couples in the contacting medium, as illustrated in Figure 6.17.*

In pollutant decomposition, either Reaction 6.34 or Reaction 6.35 constitutes one-half of the conjugate reaction pair, the other half of the pair or partner is the pollutant molecule, ion or microorganism.

To illustrate, Reactions (6.36) and (6.35) form a conjugate pair in the photocatalytic oxidation of an organic substrate:

$$4\,OH^- + 4\,h^+ \longrightarrow 4\,{}^\bullet OH \quad (6.36a)$$

$$4\,{}^\bullet OH + \text{organic substrate} \longrightarrow \text{products} \quad (6.36b)$$

$$O_2 + 2\,H_2O + 4\,e^- \longrightarrow 4\,OH^- \quad (6.35')$$

Net reaction: organic substrate $+ O_2 + 2\,H_2O \xrightarrow[h\nu]{SC}$ products (6.37)
[(6.35') + (6.36)]

A second example involves the photo-reduction of a toxic metal ion, such as Pb^{2+}, on the semiconductor surface. In this instance, Reactions (6.34) and

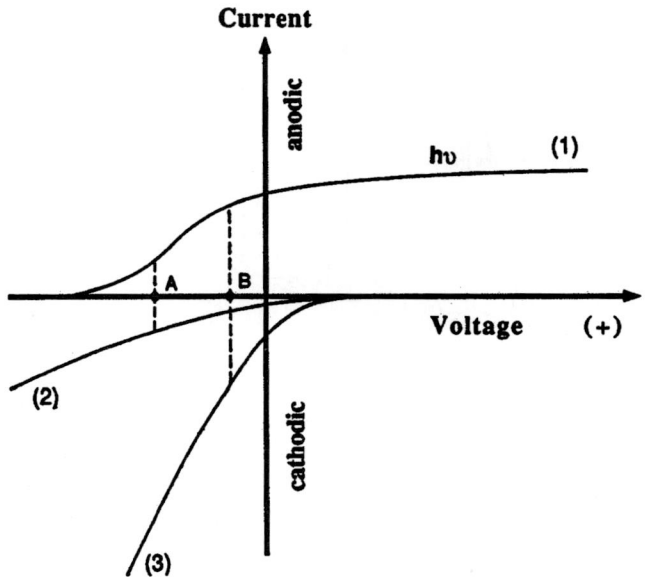

FIGURE 6.19. Component (photo) anodic and cathodic processes at an irradiated n-type semiconductor particle "microelectrode." Refer to the text for notation.

(6.38) form a conjugate pair:

$$2\,H_2O + 4\,h^+ \longrightarrow O_2 + 4\,H^+ \quad (6.34)$$

$$2\,Pb^{2+} + 4\,e^- \longrightarrow 2\,Pb \quad (6.38)$$

Net reaction: $\quad 2\,Pb^{2+} + 2\,H_2O \xrightarrow[h\upsilon]{SC} 2\,Pb + O_2 + 4\,H^+ \quad (6.39)$
[(6.34) + (6.38)]

A useful model framework for discussing conjugate reactions on a photocatalytic particle can be borrowed from the corrosion field.[118,119] The underlying idea is that, on a "microelectrode" particle in a steady state, the net rates of the oxidation and the reduction components must be the same; that is, the anodic and the cathodic current branches must have the same magnitude. This is illustrated in Figures 6.19 and 6.20. For an n-type semiconductor such as TiO_2, the anodic current branch reflects the flux of the minority carriers; that is, the holes. On the other hand, the cathodic component originates in the flow of the majority carriers; that is, the electrons in this case.

6.9. Heterogeneous Photocatalysis 537

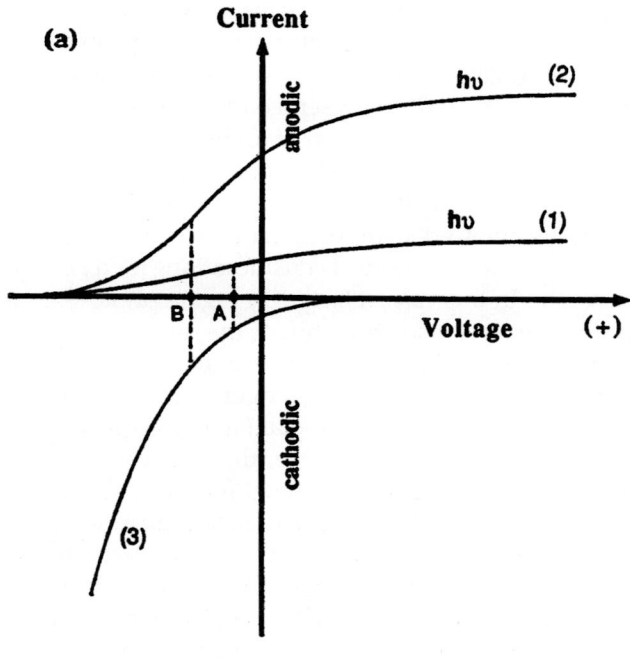

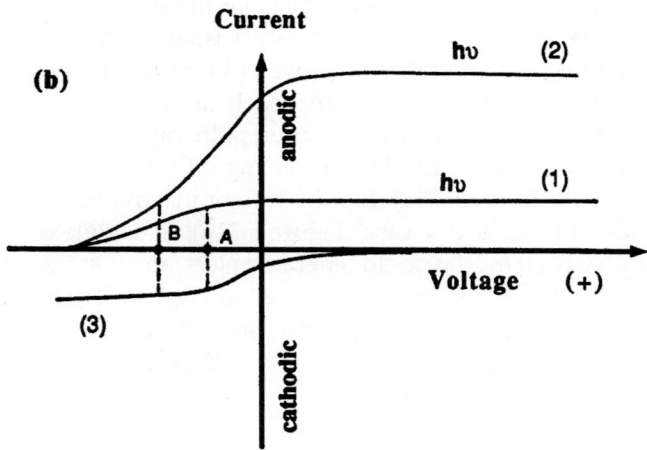

FIGURE 6.20. As in Figure 6.19 but for a photo-reduction target system, such as metal ion reduction at TiO_2. Refer to the text for notation.

Suppose that the desired reaction is the photoanodic decomposition of a pollutant (see Reaction (6.37)). Two scenarios are shown in Figure 6.19 where this oxidation reaction (curve 1) is coupled with a sluggish (curve 2) or facile (curve 3) cathodic conjugate reaction. In the former case, the particle in a steady state attains a "mixed potential," represented by point A. In the latter case, this potential shifts in the desirable direction of higher (i.e, more positive) reaction overpotential (point B). Thus, the key lesson here is that the rate of the conjugate reaction partner (Reaction (6.35)) in the preceding example) must be optimized to afford faster photo-decomposition of the organic pollutant.

Figure 6.20a contains symmetrical considerations for a *photo-reduction* target system (curve 3). In this case, a faster conjugate photo-oxidation reaction (curve 2 instead of curve 1) leads to a shift of the reaction overpotential in the desired negative direction. Thus, if the removal of a toxic metal ion is the goal (see Reaction (6.39)), an efficient hole scavenger can be added to accelerate the cleanup process. However, if the toxic metal ion is present only in low concentration (ppm levels) as is usually the case, then the cathodic branch will be mass-transport limited and exhibit a plateau (curve 3, Figure 6.20b). In this case, the available hole flux may be more than sufficient to balance the maximum cathodic current that can be sustained at the particle surface. Economically speaking, the cases illustrated in Figure 6.20a may thus be more appropriate to a metal recovery or "photoelectrowinning" scheme, where higher levels (millimolar) of metal ions exist in the waste stream. Silver recovery from a photoprocessing operation is such a case.

The modeling strategy of balancing carrier fluxes can be used for explaining the beneficial influence of metal catalyst islands on semiconductor particles,[120,121] and for analyzing the influence of O_2 reduction kinetics on the organic degradation pathway.[122] This approach can also be used for analyzing photon flux effects and competitive reaction pathways.[123] For example, Figure 6.21 contains two examples for the contrasting influence of O_2 on the photoreduction of Cr(VI) (Figure 6.21a) and the rate of inactivation of *E. coli* (Figure 6.21b) by TiO_2. In the former case (Figure 6.21a), O_2 is deleterious because it competes with Cr(VI) for the photogenerated electrons. On the other hand, the photoanodic (hole) reaction rate is enhanced in Figure 6.21b by improving the kinetics of the (cathodic) reaction conjugate, as in the model case in Figure 6.19. This is done by increasing the concentration of the electron acceptor (O_2 in this case) in the electrolyte.[124]

6.9.2. Thermodynamic Aspects

Let us now consider the energetics of Reactions (6.37) and (6.39) as written. Reaction (6.37) is photocatalytic because it involves a negative Gibbs free energy change ($\Delta G° < 0$). Thus the reaction is driven in the spontaneous direction

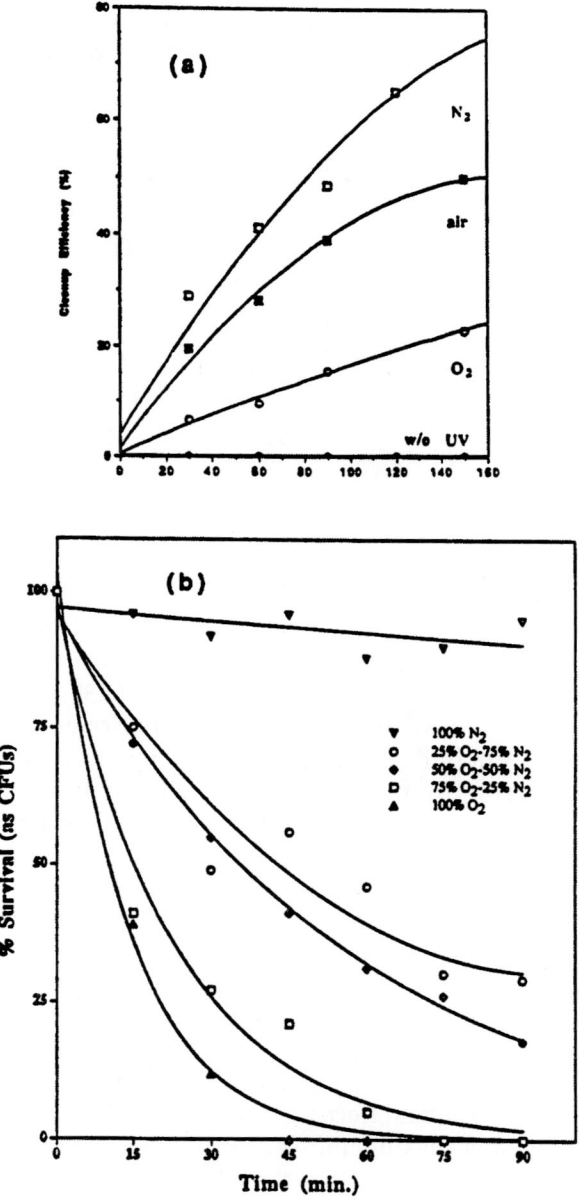

FIGURE 6.21. Effect of ambient gas composition on the photocatalytic rate for Cr(VI) → Cr(III) (a), and for inactivation of *E. coli* (b). Both processes were mediated by irradiated TiO_2 particles; CFUs are colony-forming units. (Reproduced with permission from Lin *et al.*[99] (Figure 6.21a) and Wei *et al.*[124] (Figure 6.21b))

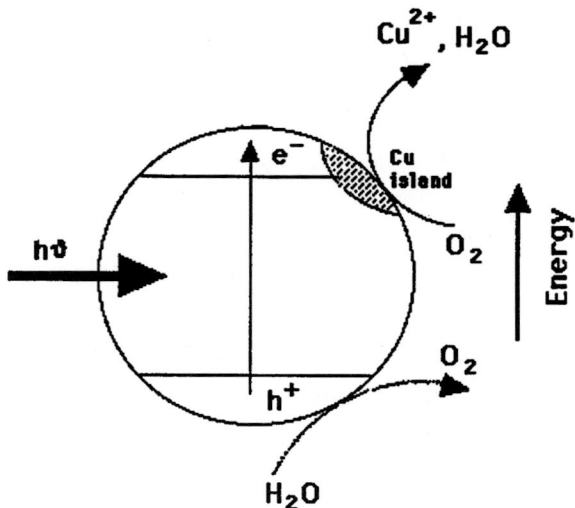

FIGURE 6.22. Photocatalytic deposition and dissolution of copper on TiO_2.

by the light, and the radiant energy simply serves to overcome the activation barrier for the process. On the other hand, Reaction (6.39) involves a positive $\Delta G°$ of +48 kJ and is photosynthetic. Here, the light drives the reaction in the thermodynamically "uphill" direction. Indeed, most reactions of this type are photosynthetic except when the metal ions are reduced at very positive potentials; that is when they have standard reduction potentials that are more positive relative to the O_2–H_2O redox couple.

An important ramification of the preceding discussion is that the *back-reaction* is important only for photosynthetic systems. For example, consider the deposition of copper on TiO_2. The back-reaction between Cu and O_2 (which has a *negative* $\Delta G°$) chemically reverses (the forward) Reaction (6.34). In other words, a situation arises akin to that in Figure 6.18 except that the $Cu^{2+/0}$ redox couple has acted as a *mediator* or *electron relay* (see Figure 6.22). Alternately stated, the Cu^{2+} ions have "short-circuited" the semiconductor microelectrode. Experimental examples of this effect are available.[56,125] In many instances, however, the thermodynamically downhill ($\Delta G° < 0$) back-reaction is *kinetically* slow.

A second important point is that the conjugate reaction partner plays a pivotal role in dictating whether a given reaction is photocatalytic or photosynthetic. For example, the reaction

$$2\,Cu^{2+} + 2\,H_2O \xrightarrow[h\nu]{SC} 2\,Cu + O_2 + 4\,H^+ \tag{6.40}$$

is photosynthetic ($\Delta G° = +164$ kJ). On the other hand, addition of a hole scavenger such as acetic acid to the medium causes the resultant reaction

$$Cu^{2+} + 2\, CH_3COO^- \xrightarrow[h\upsilon]{SC} Cu + C_2H_6 + 2\, CO_2 \qquad (6.41)$$

to be photocatalytic ($\Delta G° = -136$ kJ/mole).[126]

6.9.3. Choice of Semiconductor Photocatalyst, Photocatalyst Configuration and other Materials-Related Aspects

In terms of criteria for choosing an appropriate semiconductor photocatalyst, the most important appears to be factors related to the location of the band edges at the surface. That is, the conduction and valence band edges have to be located at energies (potentials) such that Reactions (6.34) and (6.35) may be photodriven. Hence, the valence band has to be placed at potentials that are at least +2.85 V (with respect to the standard hydrogen electrode). Only then will the photogenerated holes (see Reaction (6.31)) have sufficient energy to oxidize water to •OH. Similarly, if O_2 is used as the electron acceptor (see Reaction (6.35)), the conduction band would have to lie at a value negative of the standard potential for the reduction of O_2. In addition to surface energetics, the stability of the semiconductor particle to photocorrosion is a crucial factor. Finally, in terms of practical considerations, cost is also an issue.

Titanium dioxide has been by far the most popular photocatalyst. Both the anatase and rutile modifications have been used, although commercial samples (e.g., Degussa P-25) often contain a mixture of the two. The energy bandgaps of anatase (3.23 eV, 384 nm) and rutile (3.02 eV, 411 nm) combine with the valence-band positions to create a favorable situation for the photogeneration of highly energetic holes at the interface. However, anatase is superior to rutile for photocatalytic applications. First, the conduction band location for anatase is more favorable for driving conjugate reactions involving electrons. Other variant reasons have been given. For example, the poorer photocatalytic activity of rutile was attributed to its high e^-–h^+ recombination rate and its lower oxygen photoadsorption capacity.[127] A more recent study concludes that very stable surface peroxo groups can be formed at the anatase during photo-oxidation reactions but not at the rutile surface.[128] The implication is that the oxidation of organic compounds such as 4-chlorophenol proceeds through an *indirect* pathway involving these surface species (see later). The decrease noted in the photocatalytic activity when titania (prepared by a precipitation method) is annealed at temperatures higher than 600°C[129] appears to have a similar mechanistic origin in the extent of hydroxylation of the oxide surface. Other semiconductors have also been employed, and a representative listing appears

Table 6.10. Semiconductors That Have Been Employed in Photocatalysis Studies

Material	Bandgap Energy/eV
Si	1.1
TiO_2	3.0 (rutile) 3.2 (anatase)
ZnO	3.2
WO_3	2.7
CdS	2.4
ZnS	3.7
$SrTiO_3$	3.4
SnO_2	3.5
WSe_2	1.2
Fe_2O_3	2.2

in Table 6.10. However, none of these appear to match (at least to date) the attributes of TiO_2.

The bandgaps as listed in Table 6.10 are for the semiconductors in "massive" form or for colloidal particles of several hundred nanometer size. Size quantization, however, can cause sizable shifts in these bandgap energies.[130–132] For example, TiO_2 microcrystallites as small as 2 nm have been prepared in Nafion and clay interlayers with a corresponding bandgap energy as high as 3.95 eV.[133] Similar trends have been observed for Fe_2O_3[134] and ZnS.[135] Very high activities have been observed for Fe_2O_3 nanosized particles (relative to bulk α-Fe_2O_3 powder) for model photoreactions such as the decomposition of saturated carboxylic acids.[136]

The vast majority of the semiconductor photocatalysts listed in Table 6.10 have rather high bandgaps. Large bandgap semiconductors in general tend to be more stable against photocorrosion. Oxidation of many pollutants, especially organic species, requires high potentials, with the result that the valence band location at the semiconductor/electrolyte interface has to be rather positive as exemplified by TiO_2 and CdS. In these instances, the photogenerated holes will have sufficient energy to oxidize the organics either directly or via the generation of hydroxyl radical intermediates (see later). Nonetheless, some of the candidates listed in Table 6.10 do not have long-term stability in aqueous media, notably Si, CdS, and ZnO. The photo-oxidation kinetics are also poor in some instances, as for example, for n–Si.[137] Thus the positive attributes of this material in terms of its good match with the solar spectrum and the advanced technology which exists on it, are offset unfortunately by these other handicaps.

Consider next the physical form of the semiconductor electrode or the photocatalyst. Single crystal materials can safely be discarded as serious candidates for remediation purposes both in terms of economic considerations and, perhaps more important, from a technical perspective, because of finite reaction cross-section with the (pollutant) substrate. Semiconductor thin films are attractive for use in solar (photovoltaic) devices but are inadequate for PEC or photocatalytic waste treatment applications unless special efforts are made to tailor their surface morphology (see later). This then leaves the option of photocatalytic reactors containing semiconductor slurries or suspensions. These do offer the advantage of high surface dispersion, and consequently optimization of the encounter frequency of the active surface with the pollutant substrate. However, photocatalyst recovery after use becomes an issue of major practical concern. Immobilized photocatalysts in the form of highly microporous particles that are attached to a solid support represent an effective compromise incorporating to a degree, the positive features with the use of a semiconductor suspension. Thus TiO_2 has been immobilized on beads,[92,138] hollow tubes,[139–141] Vycor glass,[142,143] woven fabric,[144] silica gel,[58] optical fibers,[145–147] and even sand.[59] Porous thin films of TiO_2 have been immobilized on conductive SnO_2 glass.[148] Such "particulate" films have been synthesized by thermally fusing ZnO or TiO_2 particles onto conducting glass (e.g., indium tin oxide)[149,150] or by dispersing the powder in a polymeric binder.[151] Sol-gel technology can also be used for fabricating porous films onto suitable supports.[148,152]

Other than the ease of handling the photocatalyst after use in the reactor, another crucial advantage with this type of immobilization approach is that a bias potential may be applied to the photocatalyst film to separate the photogenerated carriers, thereby improving the quantum yield.[148–150]

In general, however, immobilized photocatalyst reactor configurations result in lower photocatalytic activity relative to their slurry counterparts (see below). A major factor here relates to substrate mass transport, an aspect discussed in more detail later. Photocatalyst regeneration (e.g., with an acid wash) is perhaps easier with the use of suspensions, and there are indications that the thin-film photocatalyst configuration may be more susceptible to poisoning and permanent deactivation. Incomplete mineralization has also been observed with the use of porous supports such as silica gel.[153] This has been attributed to the competitive adsorption of organics at TiO_2 and the (unreactive) SiO_2 sites. The partially hydroxylated decomposition products apparently escape further encounters with the TiO_2 sites by partitioning at the silica/ solution interface. Interestingly enough, this effect was not observed at *compact* supports such as glass.[153]

For a given semiconductor, factors such as the dopant ion, crystal modification, surface area, and surface morphology can be expected to play an important role in the photocatalytic efficiency, although detailed studies on semiconductors other than TiO_2 appear to be rare. The influence of surface chemistry was already mentioned within the context of the superior performance of anatase relative to rutile.

Using a flame reactor to produce nonporous anatase samples of varying surface area, the degradation rate of 3-chlorophenol has been observed to increase with surface area up to ~150 m^2 g^{-1}.[154] A decrease in the rate was observed for finer particles thereafter, and this trend was attributed to defect-induced e$^-$–h$^+$ recombination. Some dopant ions such as Cr(III) were found to have a deleterious influence on the photocatalytic activity of the parent oxide for similar reasons.[155] The influence of dopant ions on the e$^-$–h$^+$ recombination rate in TiO_2 has been also recognized by other authors.[156]

It was recognized early in the history of photoelectrochemistry that noble metal islands on a photocathode surface accelerated reactions such as the reduction of water. Parallel efforts have gone into modifying semiconductor particles and to show that metal or metal oxide centers on these particles can indeed promote their photocatalytic activity.[120,157-173] Apart from Pt, other metals such as Pd,[157-171] Au,[163] and Ag[167-169] on TiO_2 have been shown to accelerate the photo-oxidation of organic compounds such as 2,2-dichloropropionate,[157] 1,4-dichlorobenzene,[170] 2-propanol,[168] and salicylic acid.[172] These catalyst centers may channel the photogenerated e$^-$–h$^+$ pairs (Eq (6.31)) into physically separated reaction sites. It is possible that oxides such as RuO_2 and NiO serve a similar role. For example, in the case of NiO–$SrTiO_3$, the NiO has been proposed[173] as an e$^-$-scavenging site:

$$2 H^+(ads) + 2 e^- \rightarrow 2 H^\bullet(ads) \rightarrow H_2(ads) \rightarrow H_2(g)$$

Electron transfer kinetics to oxidants such as O_2 are also improved in the presence of the catalyst, and the important role that the kinetics of the conjugate Reaction (6.35) plays in the photocatalytic decomposition of organic compounds has already been discussed. Research also has focused on improving the visible light response of wide bandgap semiconductors such as TiO_2. These studies have been carried out mainly from a photovoltaic application perspective, although there is at least one instance wherein TiO_2 modified with a sensitizer such as zinc tetraphenylporphyrin has been used for the photo-decomposition of PCBs.[72]

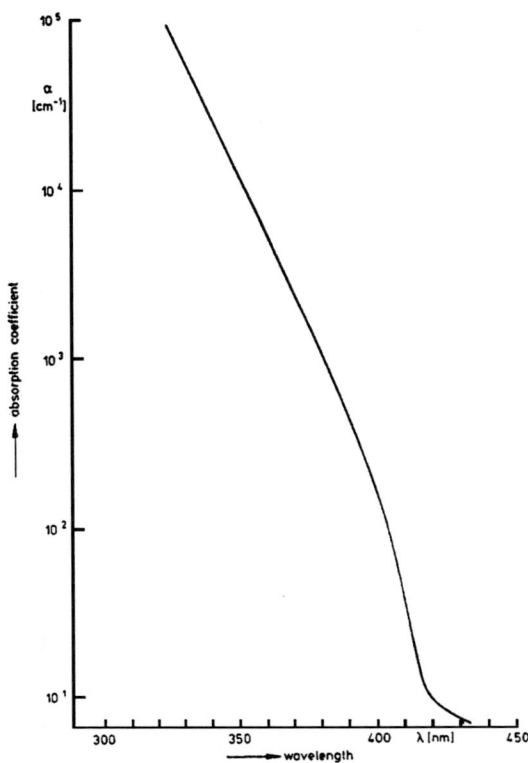

FIGURE 6.23. Absorption coefficient *vs.* wavelength for TiO$_2$ layers. (Reproduced with permission from Möllers *et al.*[175])

6.9.4. *Absorption of Light by the Photocatalyst and Quantum Yield*

Given that the vast majority of studies to date on heterogeneous photocatalysis have used TiO$_2$, we shall focus on this material in the rest of this chapter (unless otherwise indicated). As shown in Table 6.10, the energy bandgap in TiO$_2$ is different in the rutile and anatase modifications. The optical properties of rutile single crystals have been analyzed by a variety of techniques (electrical conductivity, photoconductivity, and optical absorption), and an average value for E_g of 3.05 eV has been quoted.[174] On the other hand, for TiO$_2$ layers prepared by a chemical vapor deposition technique, the absorption sets in at ~420 nm, corresponding to an energy gap of 3.2 eV (Figure 6.23).[175] The near-edge absorption can be analyzed via the expression[176]

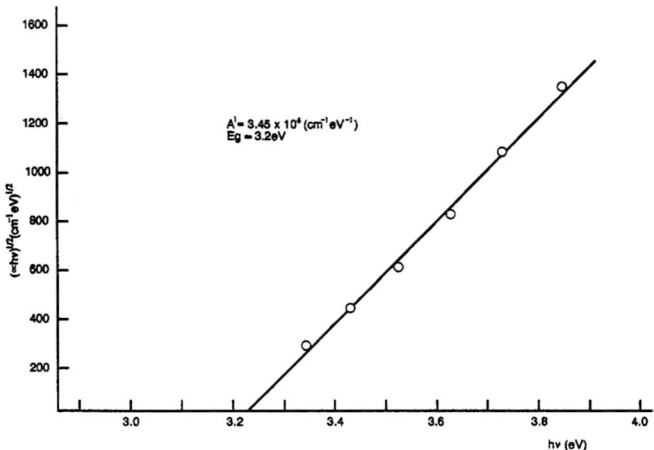

FIGURE 6.24. Calculation of the optical absorption constant A' from the spectral dependence of the TiO$_2$ absorption coefficient using Eq. (6.42). (Reproduced with permission from Salvador et al.[178])

$$\alpha = A'(h\upsilon - E_g)^m / h\upsilon \tag{6.42}$$

where α is the absorption coefficient (cm^{-1}), A' is a constant and the exponent, m is related to the nature of the fundamental optical transition ($m = 1/2$ for direct allowed transitions and $m = 2$ for indirect, i.e. phonon assisted, processes). Figure 6.24 contains an analysis of available optical absorption data[177] on TiO$_2$ according to Eq. (6.42),[178] showing that the transition in TiO$_2$ is indirect. Nonetheless, the α values are high[174] and an average value of ~10^4 cm^{-1} can be taken in the wavelength range from 360 to 400 nm.

The preceding discussion pertains to TiO$_2$ in *massive* (e.g., single crystal or thin film) form. The corresponding optical behavior of TiO$_2$ particles is conveniently considered in two size ranges; namely, nanometer sized particles and larger particles of micrometer (μm) dimensions (typical of "powder" suspensions). In either case, a major distinction must be drawn between the behavior of semiconductor *electrodes* (of the sort considered in Chapter 2) and the corresponding *particles* of the same material.

The absorption length, L_ε is given by

$$L_\varepsilon = 1/\alpha \tag{6.43}$$

Therefore, an α value of 10^4 cm^{-1} corresponds to an absorption length of 1,000 nm, which is smaller than the usual thickness of a TiO$_2$ single crystal. Hence

6.9. Heterogeneous Photocatalysis

Table 6.11. Approximated Light Absorption of TiO$_2$ Particles of Different Size in Photons/Particles with $I_0 = 2 \times 10^{15}$ photons cm^{-2} s^{-1} and $\alpha = 10^4$ cm^{-1}

R/cm	$\pi R^2 I_0$/particles	g/particles
10^{-6}	6.3 × 10^3	4.1 × 10^1
3 × 10^{-6}	5.6 × 10^4	1.5 × 10^3
10^{-5}	6.3 × 10^5	6.0 × 10^4
3 × 10^{-5}	5.6 × 10^6	2.0 × 10^6
10^{-4}	6.3 × 10^7	4.4 × 10^7

Source: Gerischer.[180]

the charge carriers are created mainly near the surface in this latter case. On the other hand, the absorption length is comparable to the size of the TiO$_2$ particles in the case of nanometer size particle suspensions. Therefore *uniform* irradiance (I_0) inside the particle and *uniform* photogeneration of holes and electrons can be assumed.

Two types of theoretical models have been employed for the analysis of the optical behavior of TiO$_2$ particles in the two size ranges: a geometrical optics model for the µm particles and the Mie scattering theory for the nm-sized TiO$_2$ particles.[179,180] An exact calculation was not performed because of the variability of the optical constants of TiO$_2$ in the UV range; instead, a first-order approximation with an averaged α value of 10^4 cm^{-1} (see earlier) yielded the results shown in Table 6.11.[179,180] According to these calculations, the absorption of light with the particle radius, $R < 10^{-5}$ cm is rather weak. The parameter g in Table 6.11 is the (volumetric) carrier photogeneration rate:

$$g = I_0/L_\varepsilon \tag{6.44}$$

The inward photon flux that is *absorbed* by each particle is given by

$$J_{ph} = \frac{1}{3}\frac{I_0 R}{L_\varepsilon} = \frac{1}{3}gR \tag{6.45}$$

The weak absorption of ultrafine (50–150 Å) particles is favorable for surface-limited photoreactions for two reasons. First, the photocurrent density (i_{ph}) decreases *linearly* with R as the ratio of the surface area to the volume increases. Second, the photogeneration rate per unit volume decreases with a decrease in R. Thus, at sufficiently small particle size, and when the photoprocesses are surface limited, recombination losses are *superlinearly* reduced with a decrease in R. We shall return to this point later.

The other major advantage with ultrafine particles is an optical one; namely, that light traverses many particles before becoming completely extinct. In other words, colloidal suspensions of such particles are essentially transparent with negligible scattering cross-sections to the incoming light. This contrasts with the situation discussed earlier (see Table (6.11)) for larger particles (spanning hundreds of nm), where the particle dimensions become comparable to the excitation wavelength.

Two aspects with the optical absorption of ultrafine (quantum-size or Q-size) particles have not been considered thus far. The first concerns *size quantization* effects.[130-132] These can cause sizable shifts in the bandgap energy. For example, as mentioned earlier, TiO_2 microcrystallites as small as 2 nm have been prepared in Nafion and clay interlayers with a corresponding E_g as high as 3.95 eV.[133] For an anisotropic layered crystallite, the bandgap shift, ΔE_g, is given by[181]

$$\Delta E_g = \frac{h^2}{4\, m'_{xy} L^2_{xy}} + \frac{h^2}{8\, m'_z L^2_z} \qquad (6.46)$$

where m'_{xy} and m'_z are the reduced effective masses of electron–hole pairs in the plane of the layer and perpendicular to it, and L_{xy} and L_z are the corresponding dimensions. Quantum-size TiO_2 layers are relatively rare because of the high effective mass of electrons in TiO_2; further, the usual preparative methods yield particles of larger (5–20 nm) dimensions. Size-quantization effects manifest as a blue shift of the absorption edge. In addition to the aforementioned study,[133] there are only a few reports of the preparation of Q-size TiO_2 both in particulate[182,183] and layer[184] form.

A second origin for the blue shift in the absorption edge is the so-called Burstein–Moss effect.[185] This arises from photocharging of the semiconductor particles and the onset of carrier degeneracy. The origin of this effect is schematized in Figure 6.25, and the difference between $E_{g,opt}$ and E_g (called the *Burstein shift*) is given by

$$E_{g,opt} = E_g + [1 + (m_e/m_h)](E_F - E_c - 4\,kT_e) \qquad (6.47)$$

with

$$E_F - E_c = h^2/2\,m_e\,(3n_e/8\pi)^{2/3} \qquad (6.48)$$

$$n_e \gg 1/2\,\pi^2\,(2\,m_e\,kT_e/\hbar^2)^{3/2} \qquad (6.49)$$

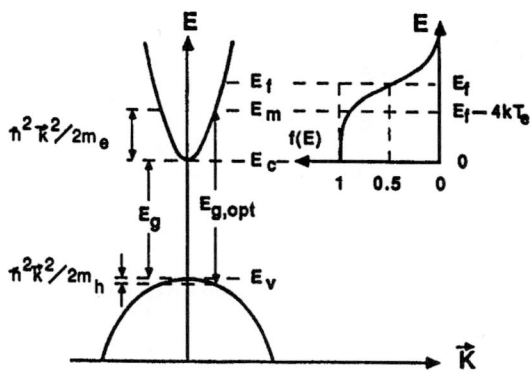

FIGURE 6.25. Energy band model of a degenerate (direct gap) semiconductor and the Fermi–Dirac distribution. (Reproduced with permission from Liu and Bard.[185])

where m_e and m_h are the effective masses of the electron and hole respectively, E_F is the Fermi level (Chapter 2), $(E_F - E_c)$ defines the height of the Fermi level above the bottom of the conduction bond (i.e., the extent of carrier degeneracy), T_e is the electron temperature, n_e is the electron density and $\hbar = h/2\pi$. The Burstein shift will be zero when $(E_F - E_c) = 4\ kT_e$. The corresponding electron density, n_{crit}, can be calculated (from available physical constants) to be 1.5×10^{19} cm^3 for n–CdS.[185] Hence, $E_{g,opt} \simeq E_g$ when $n_e < n_{crit}$ and will be approximately equal to E_g +1.26 $(E_F - E_c - 0.1)$ when $n_e > n_{crit}$ for a value of $T_e = 300$ K. Figure 6.26 contains the results of such calculations[185] on n–CdS for the dependence of ΔE_g on the extra electrons per particle (i.e., above n_{crit}) (Figure 6.26a) and of the parameter n_{crit} on the particle diameter (Figure 6.26b). An extra electron density of 0.1 corresponds to $\sim 7 \times 10^{19}$ cm^{-3} and causes a blue shift of 0.25 eV in E_g. The major point is that in small particles, the electron concentration can be very high even with few excess electrons per particle. Finally, Figure 6.26c shows ΔE_g as a function of the particle diameter for n–CdS. The Burstein shift is sizeable for values of $2R < \sim 30$Å (3 nm).[185]

It must again be emphasized that the Q-size and Burstein–Moss effects generally will not be important for photocatalyst particles synthesized under "nominal" conditions. As mentioned earlier, these particles are in the size range $> \sim 50$ nm. Similarly, commercial TiO_2 samples (e.g., Degussa P-25) have particle sizes typically in the μm dimension range.

The optics of sunlight collection in hollow glass microbead-attached TiO_2 particles floating on oil slicks, have been analyzed.[92] Figure 6.27 contains a schematic diagram of the trajectories of several rays after they are incident at an angle of 45° to the surface. For their calculations the authors assume mono-

550   Chapter 6

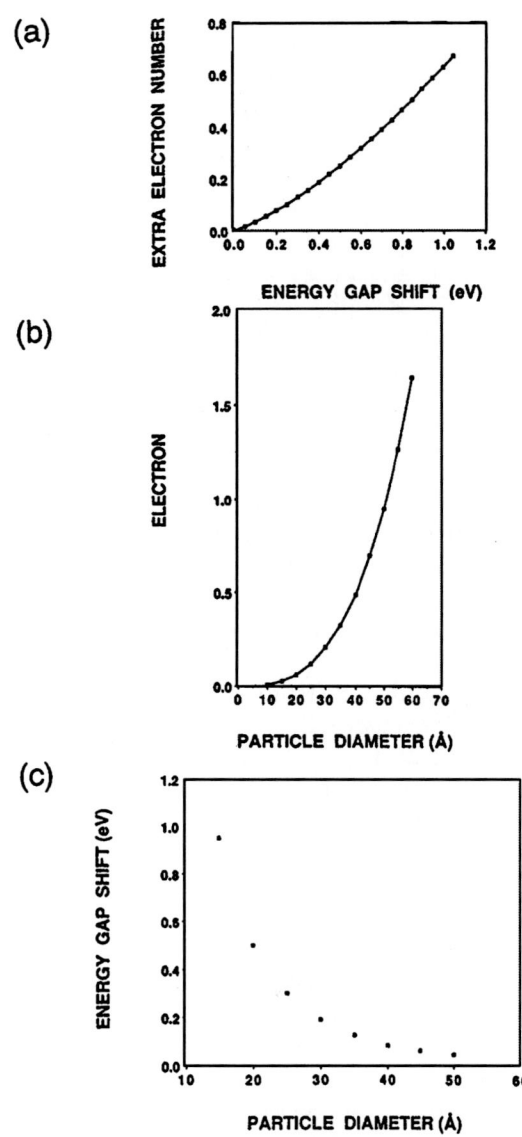

FIGURE 6.26. (a) Dependence of the energy gap shift on number of extra electrons in CdS semiconductor particle (diameter 15 Å). (b) Dependence of n_{crit}, the critical number of electrons per particle for ($|E_F - E_c| = 4kT_e$), on the particle diameter, for CdS, for $T_e = 300$ K, $m_c = 0.21\, m_o$, $m_h = 0.80\, m_o$. (c) Increase in energy gap ($E_{g,opt} - E_g$) as a function of particle diameter for CdS with an extra electron number of 0.58 e$^-$/particle. CdS parameters as in (b). (Reproduced with permission from Liu and Bard.[185])

6.9. Heterogeneous Photocatalysis

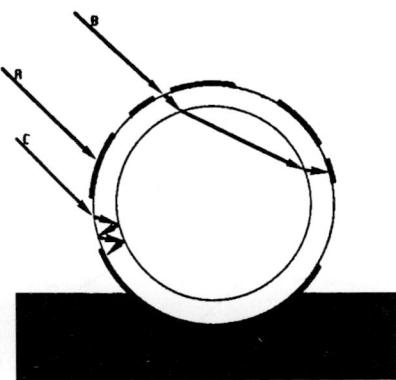

FIGURE 6.27. Schematic diagram of the trajectories of several rays absorbed by bead-attached TiO_2 crystallites. (Reproduced with permission from Rosenberg et al.[92])

chromatic radiation of unit intensity at 360 nm. The light collection is analyzed for beads at varying photocatalyst coverages (the shaded areas in the bead surface in Figure 6.27), and for varying immersion of the bead in the strongly UV-light absorbing crude oil. The outer flux of the incident light involves only the photons that pass through the atmosphere, such as ray A in Figure 6.27. The outer flux is composed of photons refracted into the bead at the air/glass interface and are subsequently absorbed by the attached TiO_2; for example, ray B or ray C in Figure 6.27. The model calculations[92] show that the efficiency of UV light collection by a TiO_2 layer on an aluminosilicate glass microbead peaks near 70% layer coverage, most of this is the inner flux.

In general, very few of the studies on the photocatalytic behavior of TiO_2 suspensions report on the optical details within the reactor. Figure 6.28 contains data from one study[186] where the optical absorption of the TiO_2 suspension was indeed analyzed; the figure shows the absorption spectra of a water blank (A), a 0.05% (w/v) TiO_2 suspension in water (B) and a 0.10% (w/v) sample (C). Two types of excitation lamps were used in this study,[186] a germicidal lamp (emitting monochromatic light) at 254 nm) and a blacklight fluorescent lamp with output from 300 to 425 nm and a peak at 350 nm. There is a discontinuity in the optical spectra at ~325 nm beyond which the solution absorbance steadily decays. Based on the data, it was estimated that any TiO_2 beyond an annular depth of ~2.25 nm within the reactor would be photocatalytically inert because of the negligible light throughput arising from absorption and scattering by the TiO_2 particles within the top layer.

This particular example underlines the need for careful attention to the light intensity profiles *within the reactor*. Another factor that must be consid-

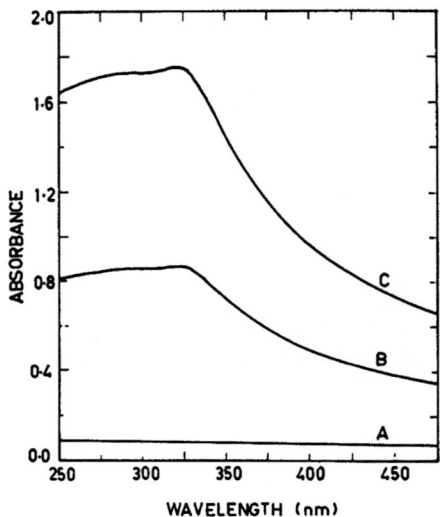

FIGURE 6.28. Optical absorption spectra for TiO_2 suspensions in 1 mm path length cell: A, water; B, 0.025% (w/v) TiO_2; C, 0.05% (w/v) TiO_2. (Reproduced with permission from Matthews and McEvoy.[186])

ered is the progressive coating of the inner lamp walls by fine TiO_2 particles and the consequent attenuation of the light emanating from the lamp. Similar considerations apply to *solar* reactors—in this case, the light trajectory obviously is in the *opposite direction* from the UV case; that is, from the outside to the interior of the reactor.

Quantum Yield. The concept of quantum yield is of great value in photochemistry. The quantum yield is generally defined as the number of events occurring per photon absorbed. The difficulty with defining such a parameter for heterogeneous systems is related to uncertainties arising from light scattering by the photocatalyst particles. As seen earlier, the catalyst particle size is comparable in many cases to the incident photon wavelength. Accordingly, an *apparent* quantum yield can be quoted wherein the inherent assumption is that all the incident light is absorbed by the semiconductor particle. The quantum yield (Φ) can be defined in terms of the ratio of the rate of product generation (v_p) and the rate of carrier generation.

$$\Phi = \frac{v_p}{g} \tag{6.50}$$

Alternatively, a charge "collection" parameter equivalent may be defined in terms of the ratio of photogenerated electrons (or holes) leaving the particle compared to the number of photons absorbed by the particle:[15]

$$N' = \frac{4\pi R^2 i_{ph}}{(4/3)\pi R^3 g} = \frac{i_{ph}}{1/3\, R\, g}$$
$$= \frac{i_{ph}}{J_{ph}} \qquad (6.51)$$

where the last equality uses Eq. (6.45.) In Eq. (6.51), i_{ph} is the photocurrent density. As mentioned earlier, the collection efficiency is seen to scale inversely with the particle radius, R; that is, the smaller the particle, the more efficient is the charge collection. Parameters such as N' are accessible from slurry photoelectrochemical cell measurements, as will be seen shortly. The quantum yield defined in terms of *product* collection can be readily measured within the constraints associated with the uncertainty in the knowledge of g arising from light reflection and scattering losses and the like.

Before discussing the magnitude of the quantum yields that have been measured for TiO_2 suspensions, it is necessary to first consider the carrier dynamics and the electron transfer rates at the TiO_2/solution interface. This is done next.

6.9.5. Carrier Dynamics in Irradiated TiO_2 Particles

<u>Potential Distribution in the Dark.</u> In the preceding section, the assumptions of uniform irradiance and a uniform electron–hole pair photogeneration rate were predicated on the comparable dimensions of the TiO_2 particle and the absorption length (see Eq. (6.43)). It is instructive to first probe the potential distribution within a TiO_2 particle and the corresponding *macroelectrode* of the same material *in the dark*. Spherical geometry can be assumed as an approximation in the former case, and the potential distribution has been mapped[187] by solution of the Poisson–Boltzmann equation for a sphere. The key result is that for a larger particle such that the particle dimension is much larger than the depletion layer width, that is, for a thin depletion layer close to the surface of a sphere, the spherical geometry is unimportant. The depletion layer width, is given by

$$W = \left[\frac{2\varepsilon\varepsilon_0 \Delta\phi}{e_0 N}\right]^{1/2} \qquad (6.52)$$

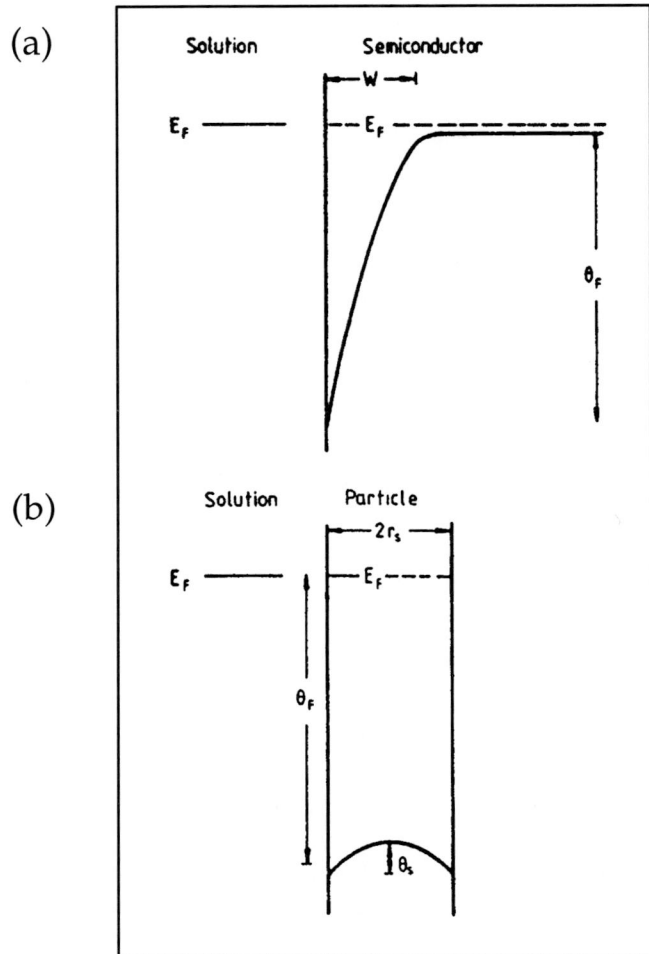

FIGURE 6.29. Comparison of the potential distribution and Fermi levels for a macroscopic planar electrode (a) and for a spherical particle (b). See text for definition of symbols. (Reproduced with permission from Albery and Bartlett.[187])

where ε is the static dielectric constant (173 for TiO_2, see Ref. 188), ε_o is the permittivity of free space (8.85×10^{-12} F/m), $\Delta\phi$ is the potential drop across the depletion layer, and N is the charge carrier density. For particles of dimension such that $R \sim 10^{-6}$ cm, the particle radius may become *smaller* than the depletion layer width. In this scenario, the particle essentially has no *field-free* region

6.9. Heterogeneous Photocatalysis

unlike the situation at a macrosized planar electrode (Figure 6.29).[187] That is, the entire particle is depleted of majority carriers. Figure 6.29b also shows however, that the particle dimension may be too small for the entire potential drop (at an equivalent planar electrode, Figure 6.29a) to be developed across the particle. Therefore, θ_S in the particle case may be much smaller than θ_F, the shift in the Fermi level from its position at flat band; that is, the extent of "band bending" at a planar electrode (see Figure 6.29). The maximum value of θ is

$$\theta_S = \frac{1}{6}\left(\frac{R}{L_D}\right)^2 \tag{6.53}$$

The diffusion length, L_D, depends on a variety of parameters such as carrier density and material quality, but it varies between 1 μm and 1 nm for TiO_2.[189] Equation (6.53) thus predicts that $\theta_S < 1$! Under these conditions, the bulk particle may be treated as field free.[187] *This is the situation in the dark.* Carrier accumulation at the *surface* will give rise to a surface charge component and a consequent potential drop across the Helmholtz layer, as we shall see later.

Oxide semiconductor particles such as TiO_2 have a *surface charge* associated with amphoteric surface hydroxyl groups[190]:

$$Ti-OH + H_2O \rightleftharpoons Ti-OH_2^{\delta+} + OH^- \tag{6.54}$$

$$Ti-OH + OH^- \rightleftharpoons Ti-O^{\delta-} + H_2O \tag{6.55}$$

Thus the surface charge varies with the solution pH, becoming more negative at the rate of −59 mV/pH unit at 25°C. This surface charge change manifests in the electrophoretic mobility of the particle (see Section 2.10) as a function of solution pH. Figure 6.30 contains representative electrokinetic data for ~ 50 Å TiO_2 particles.[191] The intersection of the curve with the abscissa corresponds to PZC or PZZP, the point of zero charge or zero zeta potential, a parameter of central importance in colloid chemistry. For this particular sample and solution conditions, the PZZP is 4.7. This is the particular pH at which the positively charged and negatively charged surface groups counterbalance each other, and the net surface charge is zero. At a pH > PZZP, the *net* surface charge is negative and vice versa. We shall see later that variation of the solution pH affords a powerful route to tuning of the interfacial charge–transfer reaction kinetics.

<u>Carrier Transport.</u> Again assuming spherical geometry for the semiconductor (TiO_2) particles, the master differential equation describing carrier generation, recombination, and bulk may be written in radial coordinates as

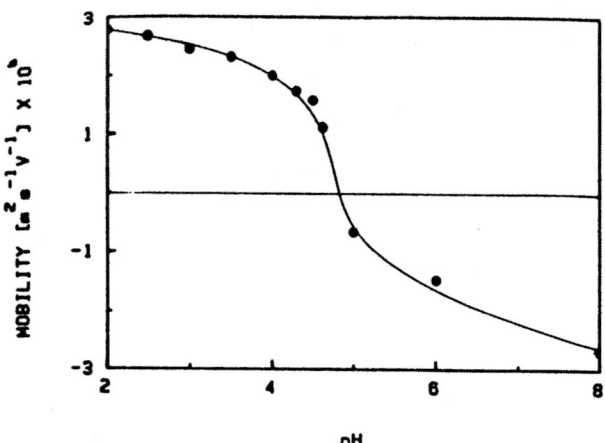

FIGURE 6.30. Electrophoretic measurements of the mobility of colloidal TiO_2 particles as a function of pH ($[TiO_2]$ = 20 mg/liter, ionic strength = 2×10^{-1} M). (Reproduced with permission from Moser and Grätzel.[191])

$$\frac{\partial C}{\partial t} = D\left(\frac{\partial^2 C}{\partial r^2} + \frac{2}{r}\frac{\partial C}{\partial r}\right) + v \tag{6.56}$$

Note (via the last equality) that this is for the *steady-state condition*; that is, the photogeneration term is balanced by the loss of carriers either via recombination or surface reactions. Equation (6.56) has been written for the simplified case, where (field-assisted) migration of the carriers can be neglected. More complete expressions incorporating carrier transport by both migration and diffusion have been developed.[187] However, the overall conclusions to be drawn in the following are not essentially altered by this simplification. In Eq. (6.56), v is the carrier recombination term. Equation (6.56) can be solved with the boundary conditions

$$r = 0; \quad \frac{\partial C}{\partial r} = 0 \tag{6.57}$$

and $\quad r = R; \quad D\dfrac{\partial C}{\partial r} = k\, C(R) \tag{6.58}$

where k is the rate constant for carrier consumption and $C(R)$ is the surface carrier concentration. It must be noted that k is a heterogeneous rate constant (units of cm s^{-1}), and the carrier concentration (density) is expressed in volume basis (cm^{-3}). Thus, the expression in Eq. (6.58) describes the carrier flux (di-

6.9. Heterogeneous Photocatalysis

mension of number per unit area per unit time). Before considering the solution of Eq. (6.56), it is instructive to examine the competition between *bulk* carrier recombination and escape (transit) to the surface. To a first approximation, the carrier transit time, is given by[192]

$$\tau = \mu_e^2 / \pi^2 D \tag{6.59}$$

The electron mobility (μ_e) in TiO$_2$ has been determined to be 0.5 cm^2/Vs at room temperature.[188] We then have the Stokes–Einstein equation (see Eq. (2.8)):

$$D = \mu_e kT / e_o \tag{6.60}$$

which leads to a diffusion coefficient of 1.2×10^{-2} cm^2/s, a relatively small value because of the heavy effective mass ($m_{eff} = 30 \times m_e$) for electrons in the conduction band. Nonetheless, the transit time for a range of particle dimensions in the Å–nm range is only in the picosecond range. Bulk recombination in semiconductors is a function of the majority carrier density, N. However, a value of 100 ps has been estimated [193] for $N = 2 \times 10^{19}$ cm^{-3}. Clearly, ultrafine particles afford efficient collection of the photogenerated carriers as was qualitatively seen earlier. Note that τ scales with the square of the particle radius (Eq. (6.59)). Thus τ increases quickly to ~100 ns for $R = $ ~1.1 µm, a situation typical of TiO$_2$ powder suspensions. In these instances, the quantum yield will be correspondingly small, and this prediction is borne out by experimental data from many laboratories.

It must be noted that if an electric field is present within the semiconductor particle, the transit time will be reduced because of the migration component. Under these conditions, it may be more appropriate to use L_D^2 / D for the transit time (instead of Eq. (6.59)). For example, for a value of $L_D = 1$ nm, a τ value of ~ 1 ps is predicted.

Returning to Eq. (6.56), the differential equation has been solved[180] for the case where the bulk recombination is neglected (i.e., $v = 0$). As seen earlier, this is not strictly valid for the larger particles; nonetheless, the simple calculations permit some useful conclusions to be drawn. The result is a parabolic concentration profile within the particle[180]

$$C(r) = \frac{g}{3k}R + \frac{g}{6D}R^2\left[1-\left(\frac{r}{R}\right)^2\right] \tag{6.61}$$

This expression is valid for both types of particles, electrons and holes. The carrier distribution is shown in Figure 6.31 in dimensionless form for the distance from the center. An average concentration $\overline{C}$ can be defined as

$$\overline{C} = \frac{g}{3k}R + \frac{g}{15D}R^2 \tag{6.62}$$

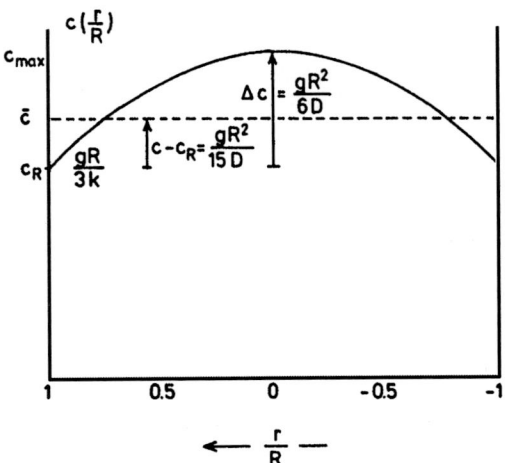

FIGURE 6.31. Concentration profile of charge carriers in an irradiated semiconductor particle in the steady state. (Reproduced with permission from Gerischer.[180])

As mentioned earlier, the surface carrier concentrations are controlled both by the interfacial reaction kinetics and by *surface* recombination losses. The latter can be modeled by the Hall–Shockley–Read mechanism. Suppose that $k_p \gg k_n$ (we shall later show experimental evidence that this is indeed the case), where k_p and k_n are the heterogeneous rate constants for the hole and electron reactions at the interface respectively (see Figure 6.17). All the recombination traps will be occupied by electrons, and the rate of surface recombination, v_{rs} will be first order with respect to holes and zero order with respect to electrons, that is,

$$v_{rs} = k_{rs} p_s \tag{6.63}$$

where k_{rs} is the surface recombination rate constant. The boundary conditions in Eq. (6.58) must now be rewritten as

$$D_n \frac{dn}{dr}\bigg|_{r=R} = k_n n_s + k_{sr} p_s$$

for electrons and

$$D_p \frac{dn}{dr}\bigg|_{r=R} = \left(k_p + k_{sr}\right) p_s$$

for holes.

The surface concentrations, n_s and p_s of electrons and holes are now given by

$$n_s = \frac{gR}{3\,k_n\left(1+\dfrac{k_{sr}}{k_p}\right)} \tag{6.64a}$$

$$p_s = \frac{gR}{3\left(k_p + k_{sr}\right)} \tag{6.64b}$$

Note that when $k_{sr} \ll k_p$, these expressions collapse to the first term in Eq. (6.62).

More complicated models have been considered,[187] where the equation for the Hall–Shockley–Read model is expressed in kinetic terms as

$$\frac{1}{v_{rs}} = \frac{1}{k_I n} + \frac{1}{k_{II} np} + \frac{1}{k_2 p} \tag{6.65}$$

where n and p are given by the Boltzmann distributions

$$n = n_s \exp(\theta - \theta_s) \tag{6.66a}$$

$$p = p_s \exp(\theta - \theta_s) \tag{6.66b}$$

In Eqs. 6.66, θ is a dimensionless potential and θ_s is defined in Figure 6.29b. These models lead to a number of different cases wherein kinetic control is exerted by one or more of the terms in Eq. (6.65). These cases have been analyzed via kinetic case diagrams.[187]

Both these models[180,187] lead to the same general conclusion; namely, that the carrier concentration and their fate inside the particle are critically dependent on the interfacial rate constants k_p and k_n and on the particle radius (dimension).

6.9.6. Photo-charging of the TiO_2 Particles

If the rate constants (k_n and k_p) for the interfacial electron–transfer reactions involving electrons and holes are not the same, charges of one polarity will accumulate within the particle, thereby causing charge buildup. That this is indeed the case for UV-irradiated TiO_2 is shown by the potential–time transient in Figure 6.32.[194] These particular experiments were done for platinized TiO_2 but a similar trend appears to hold for the unmodified TiO_2 as well; namely, that the particles acquire a net negative surface charge. This is a direct indica-

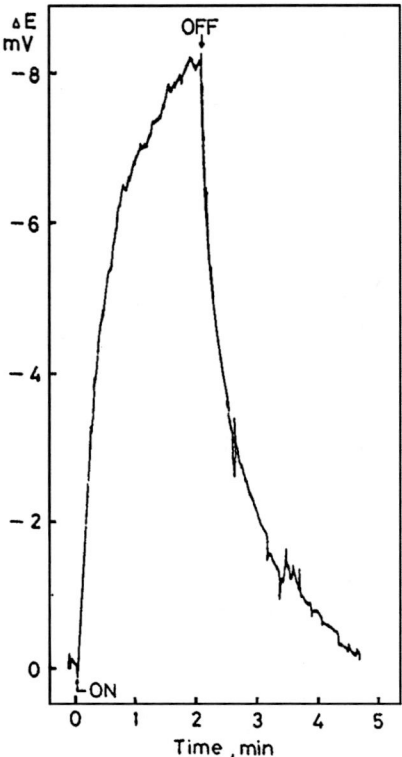

FIGURE 6.32. The potential–time relation of Pt/TiO$_2$ catalyst in 1 N NaOH under pulse illumination with 360 nm monochromatic light. (Reproduced with permission from Uosaki and Kita.[194])

tion that $k_p \gg k_{n'}$ namely, that electrons accumulate within the particle. The photo-charging situation is schematized in Figure 6.33.[194] The excess negative charge on the particle surface is balanced by the counterions in the Helmholtz double layer at the particle/solution interface (provided that the ionic strength is large enough). Calculations of the corresponding potential drop[180] appear to be in reasonable agreement with the experimental data in Figure 6.32. Note that this photopotential develops *in addition* to the interfacial potential drop caused by ionization of surface groups at TiO$_2$ (see Eqs. (6.55) and (6.56)), as discussed earlier.

In chemical terms, the photo-charging phenomenon may be represented as

$$(TiO_2)OH^- \xrightarrow{h\upsilon} TiO_2^- + {}^{\bullet}OH \qquad (6.67)$$

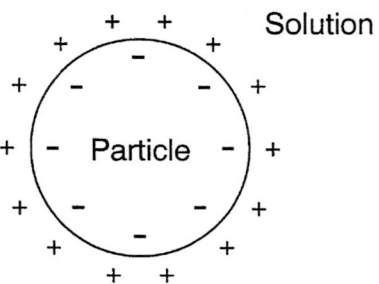

FIGURE 6.33. Photo-charging of an irradiated TiO_2 particle. Spherical particle geometry is assumed.

There is evidence from EPR[195] and optical spectroscopy[196,197] that the electrons accumulated within the irradiated particles (in the presence of hole scavengers) are not free electrons in the conduction band. Rather they are trapped at surface sites, identified as Ti^{3+} (small polaron) species. Photoacoustic spectroscopy on TiO_2 powders in a pump–probe arrangement[198] confirms the formation of these species (estimated to be generated to an extent of ~10^{20} cm^{-3}) and are thought to be stabilized by local anion distortion. Interestingly enough, two factors appear to be needed for observation of this photochromic effect: moisture and an O_2-free atmosphere. The presence of physisorbed or weakly H-bonded H_2O molecules at the TiO_2 surface would reduce the e^-–h^+ recombination rate by rapidly removing the photogenerated holes. Similarly, O_2 is expected to capture the photogenerated electrons (thus precluding their accumulation at surface traps) and assist in surface recombination via the mechanism illustrated in Figure 6.18. Trapped holes have also been identified in TiO_2 colloids by the EPR technique.[199]

When the colloidal TiO_2 particles are supported on a conductive electrode (e.g., Sn-doped In_2O_3 or ITO glass), potentiostatic control of the Fermi level within the particle is possible. Concomitant monitoring of the near-IR absorbance (either transiently by stepping the potential or at a steady state) affords a probe of both the (free) conduction band electrons and electrons localized as small polarons.[200]

An effective strategy for studying the charge–separation process in particulate *films* is to compare the photoaction spectra obtained from front- *vs.* back-surface illumination. The Gärtner-Butler model[201] can be used to analyze the photoaction spectra in the two cases. For front-side illumination, where the electron–hole pairs are excited within the depletion layer, the action spectrum would be expected to match the shape of the absorption spectrum. On the other hand, with rear-side illumination, a decrease in photoresponse (quantum yield) and the appearance of a maximum in the photoaction spectra are expected. This can be explained as follows.

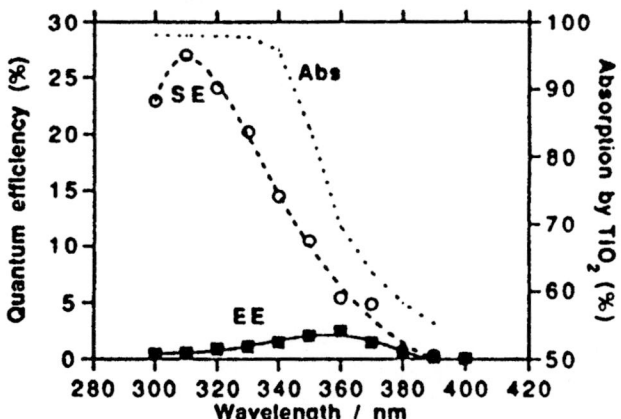

FIGURE 6.34. The action spectra for back-side (SE) and front-side (EE) irradiation of a TiO_2 electrode at -0.3 V (vs. SCE reference) in 0. 1 M KSCN/water, pH = 5. The absorption spectrum (Abs) of the TiO_2 electrode is also shown. (Reproduced with permission from Hagfeldt et al.[202])

Short-wavelength light (with high α) is absorbed in the field-free region, where no charge separation occurs. As the optical mean free path increases with longer wavelengths, the light can penetrate deeper into the TiO_2 layer, and the quantum yield begins to increase because of the built-in field. At very long wavelengths (photon energies close to the bandgap), the α begins to decrease and the quantum yield decreases again. Interestingly enough, this expected trend was *not* observed experimentally on TiO_2 particulate films.[202] For a thick (7 µm) film, the trend was just the opposite, as shown in Figure 6.34. These results indicate that the most efficient charge collection takes place close to the back contact. The lower-quantum yield for front-side illumination is explained by recombination losses during the transport of electrons through the colloidal film to the back contact. These observations reaffirm the conclusions arrived at earlier, namely that depletion layers are relatively unimportant in describing the carrier dynamics in colloidal particles (Figure 6.29).

A variety of electrical techniques exist for the study of charge separation and accumulation within irradiated semiconductor particles. Not all of these have attained a high state of refinement, at least at the time of this writing. Two such promising candidates include time-domain reflectometry (TDR),[203] and AC conductometry.[204] In the former, a voltage pulse travels from a generator to the sample cell that is equipped with a special type of electrode to sense the changes in the electrical properties of the TiO_2 suspension. The voltage pulse is reflected at this electrode by impedance mismatch, resulting in an echo signal that is propagated back toward the generator. Both TDR and conductometry are in their infancy in terms of their application to the study of photocatalytic systems.

Electrophoretic measurements under illumination ("photoelectrophoretic" experiments[205,206]) of the TiO_2 particles show a shift consistent with the buildup of negative charge on the particles.[207] Interestingly enough, this charge appears to persist for a time duration up to several minutes after irradiation is terminated. This charge can be collected at an inert electrode as an anodic photocurrent.[207,208] We shall return to a more detailed discussion of these slurry PEC cells later in this chapter.

The dynamics of charge carrier trapping and recombination can also be studied by picosecond or nanosecond laser flash photolysis.[209] The absorption spectra of the trapped electron appears within the leading edge of the 30 ps laser pulse, while hole trapping is much slower, requiring ~250 ns. A rate constant for electron–hole pair recombination of $\sim 3.2 \times 10^{-11}$ cm^3 s^{-1} has been derived in this study. The kinetics switches from second order to first order at very low charge carrier occupancy (low irradiance), the lifetime of a single electron–hole pair (exciton) in a 120 Å sized TiO_2 particle being ~30 ns.

Luminescence decay measurements offer yet another route to the study of electron–hole pair recombination.[210,211] In general, both the absorption and luminescence decay profiles are multiexponential and require distributed (stochastic) kinetics formulations (e.g., Williams–Watt model). Non-monoexponential kinetics behavior has also been correlated with the distribution of particle sizes in TiO_2 colloids.[212]

We shall next consider the dynamics of charge transfer when electron and hole scavengers are present at the TiO_2/solution interface.

6.9.7. Dynamics of Interfacial Electron and Hole Transfer at Irradiated TiO_2 (and Other Semiconductor) Suspensions

Viologen derivatives are effective probes of electron transfer dynamics at the TiO_2/solution interface because of the strong optical absorption of the cation radicals that are thus generated. For example, the methyl viologen dication, MV^{2+};

$$CH_3-\overset{+}{N}\underset{}{\bigcirc}-\underset{}{\bigcirc}\overset{+}{N}-CH_3$$

is an efficient scavenger of electrons with an $E°$ value of -0.44 V, and a λ_{max} for the cation radical, $MV^{+\bullet}$ of ~600 nm (Figure 6.35).[213] The molar absorptivity at this wavelength is in the range 11,000–12,000 M^{-1} cm^{-1}.[214–216] Quantum yields close to unity have been measured for the reduction of MV^{2+} at TiO_2 using a thin-layer spectroelectrochemical technique[213] and laser photolysis.[214]

Laser flash photolysis has been effective for unraveling the intricacies of electron transfer at TiO_2/solution interfaces containing MV^{2+} as the electron

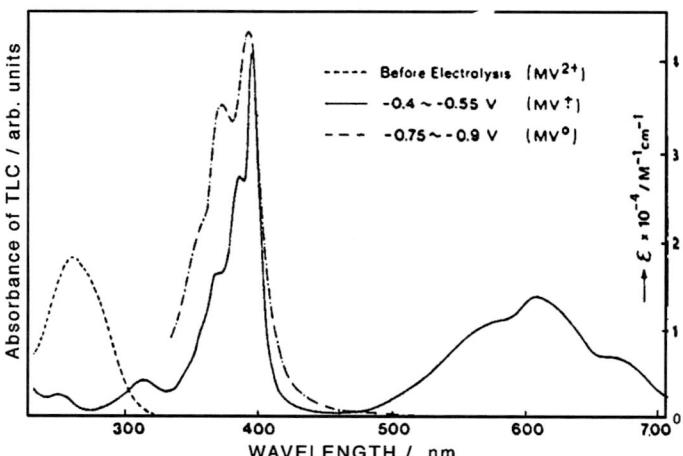

FIGURE 6.35. Spectra of the electrolyte solution in the thin-layer cell (TLC) at different electrode potentials $[MV^{2+}]_0 = 2$ mM in C_2H_5OH; [LiCl] = 0.1 M. (Reproduced with permission from Watanabe and Honda.[213])

acceptor.[192,214] Similar studies have been performed on CdS suspensions.[210,214,217,218] Figure 6.36 illustrates representative data for the former case.[219] The growth of the 602 nm absorption is concomitant with the appearance of blue coloration in the solution because of the MV$^{+\bullet}$ species. Kinetic analysis of these data shows pseudo-first-order behavior; the plot of k vs. MV^{2+} concentration (for [MV^{2+}] ≥ 2.5 × 10^{-5} M) is a straight line, from the slope of which a second-order rate constant of 1.2 × 10^7 M^{-1} s^{-1} can be estimated for the reduction of MV^{2+} by excited TiO$_2$ particles; that is,

$$TiO_2 \xrightarrow{h\nu > E_g} e^- + h^+$$

$$MV^{2+} + e^- \rightarrow MV^{+\bullet}$$

$$R + h^+ \rightarrow \text{products}$$

where, R denotes a reducing agent. Tafel analysis (Section 2.5.4) of the flash photolysis data[192] yield a heterogeneous electron-transfer rate constant (k_n) of 4 × 10^{-3} cm/s. The driving force (and hence the dynamics) of electron transfer can be tuned in this system by varying the solution pH since the flat-band potential (or equivalently, the conduction band edge) of TiO$_2$ shifts negatively

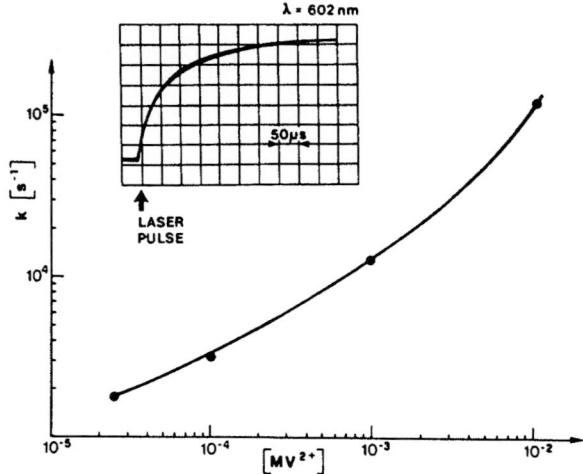

FIGURE 6.36. The 347.1 nm laser photolysis of colloidal TiO_2 (500 mg/liter, pH 5). The observed rate constant for MV^+ formation is plotted against MV^{2+} concentration. Insert: oscilloscope trace illustrating time course of 602 absorbance for $[MV^{2+}] = 10^{-3}$ M. (Reproduced with permission from Duonghong et al.[219])

at the rate of 59 mV/pH, whereas the $E°$ value of the $MV^{2+/+•}$ redox couple is pH invariant.

Evidence has been presented[191,220] for multiple electron transfer after band-gap excitation of TiO_2. An amphiphilic viologen derivative, $C_{14}MV^+$;

$$CH_3-(CH_2)_{13}-\overset{+}{N}\bigcirc\!\!\!-\!\!\!\bigcirc\overset{+}{N}-CH_3$$

and a cofacial dimeric viologen, DV^{4+};

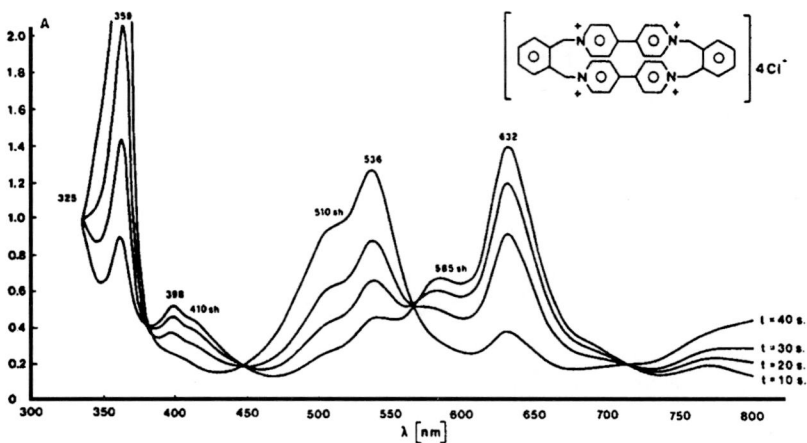

FIGURE 6.37. Spectral changes observed under irradiation of deaerated TiO_2 solutions (500 mg/liter) by $\lambda > 330$ nm light in the presence of 2×10^{-4} M DV^{4+}. Spectra of the products were recorded after 10, 20, 30, and 40 s of irradiation time. The spectra were measured against colloidal TiO_2 (500 mg/liter) as a reference solution; optical pathlength, 1 cm. Note the clean isosbestic points at 714, 517, 447, 380, and 325 nm. (Reproduced with permission from Moser and Grätzel.[191])

were used in these studies. Consecutive reduction of $C_{14}MV^{2+}$ to the radical cation $C_{14}MV^{+\bullet}$ and neutral $C_{14}MV^{\circ}$ was facilitated by surface adsorption of the compound and completed within 100 μs.[220] Similarly, the 2 e⁻ reduction of DV^{4+} to DV^{3+} and to DV^{2+} was probed by the optical absorption of the products at 632 and 536 nm, respectively (Figure 6.37).[191] Cyclic voltammetry yields the $E°$ values for these electron transfer steps as $E_1°$ ($DV^{4+/3+}$) = 0 V and $E_2°$ ($DV^{3+/2+}$) = -0.07 V. As with the MV^{2+} reduction discussed earlier, the corresponding second-order rate constants are in the ~10^5 M^{-1} s^{-1} range, values significantly lower than the diffusion limit.

The dynamics of the corresponding valence-band processes (hole transfer) can be similarly studied by using optical probes such as methyl orange[83] or thiocyanate.[219,221,222] Alternatively, halide species such as iodide can be used.[223] A value for k_p of 15 cm s⁻¹ has been estimated in the latter case.[223] This is significantly larger than the corresponding rate constants for *electron* transfer, as noted earlier, and supports the notion that, in general, hole transfer is faster at irradiated TiO_2 than the corresponding electron-transfer pathway.

Among other studies related to interfacial electron transfer at colloidal semiconductor particles, mention may be made of a picosecond resonance Raman scattering study[224] and picosecond laser flash photolysis studies[217,218] all on CdS and employing methyl viologen as the electron transfer probe. In the former case, the rise time of the $MV^{+\bullet}$ radical was found to be between 5 ns and 20 ps,[224] whereas the transient absorption experiments indicate the reduction of MV^{2+} to occur within 18–100 ps after the excitation.[217,218] Interestingly enough, the oxidation of diethyldithiocarbamic acid on CdS was found to be *slower* and occurs within 250 ps with a quantum yield around $\Phi = 0.2$.[218] A disturbing aspect is that the electron transfer rates reported in various studies vary widely. For example, time-resolved absorbance spectra on CdS show formation of $MV^{+\bullet}$ (and its dimer at $\lambda_{max} = 530$ nm) in ~1 ns, obviously a rise time orders of magnitude variant from previous estimates. These discrepancies probably reflect the critical influence of colloid preparation and surface variables on the measured electron transfer rates.

Slurry PEC cells offer another route to the study of interfacial electron transfer on TiO_2 (and other semiconductor) suspensions. Both intentionally added electron acceptors (e.g., Cu^{2+}, methyl viologen) and electron donors (acetate) or a combination of the two can be utilized for this purpose. Figures 6.38a and 6.38b contain the results from a kinetic model including seven distinct steps[225]:

(1) $S \xrightarrow{h\upsilon} S^*$ (e⁻–h⁺ pair generation)

(2) $S^* \xrightarrow{k_1} S$ (recombination)

(3) $S^* + A \xrightarrow{k_2} S^+ + A^-$ (electron transfer to acceptor, A)

(4) $A^- + S^+ \xrightarrow{k_3} A + S$ (back reaction)

(5) $S^+ + D \xrightarrow{k_4} S + D^+$ (hole transfer to donor, D)

(6) $S^+ + SC \xrightarrow{k_5}$ products   (semiconductor photo-decomposition)

(7) $A^- \xrightarrow{p} A + e^-$ (current collection)

Step 6 can be ignored for a stable semiconductor such as TiO_2 but must be considered for a photocatalyst material such as CdS (see later). In this scheme, p is the coulometry cell constant, in s⁻¹ [$p = k_m$ (electrode area)$/V_s$, where k_m is the mass-transfer coefficient]. The steady-state current I_{ss} can be calculated with the equation

$$I_{ss}/nFV_s = p[A^-] \tag{6.68}$$

Figures 6.38a and 6.38b show the computed variation of I_{ss} vs. C^* [C^* is the total (analytical) concentration of A].[225] The rate parameters in the two cases are $B = k_3/(k_4D + k_5)$ and $R = k_1/k_2$, respectively, and describe the influence of the back reaction and recombination. To generate these model curves, the steady-state concentration of A^- (e.g., $MV^{+\bullet}$) was calculated as a function of the experimental parameters I_o/pV and C^*, where I_o is proportional to the irradiation flux.

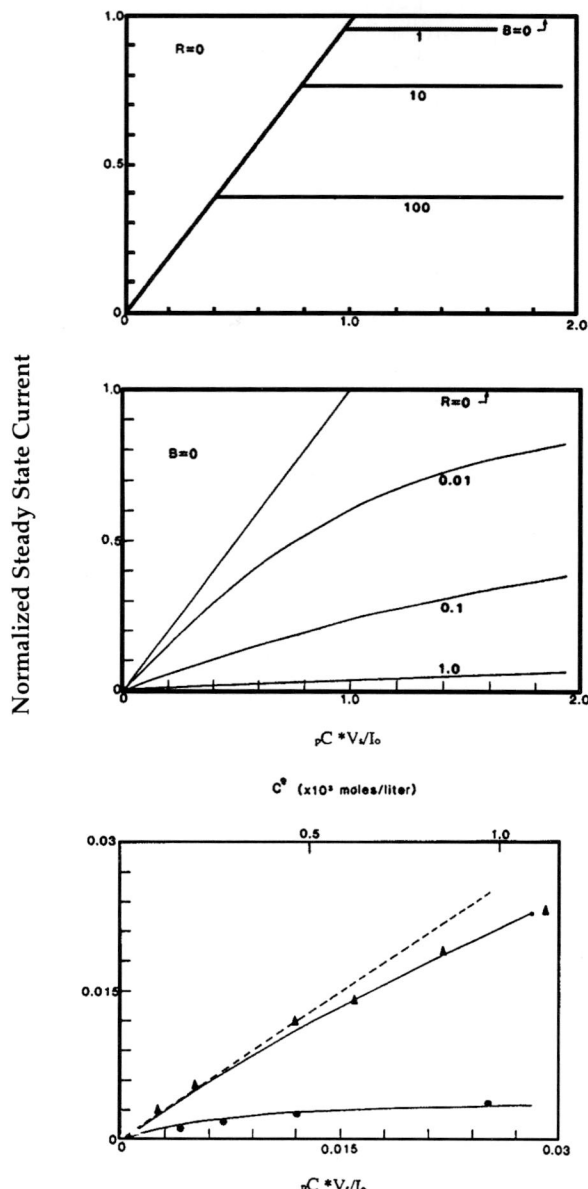

FIGURE 6.38. Dependence of the normalized steady-state current (see Eq. (6.68)) on C^*. Refer to the text for details. The experimental data in (c) were generated under the following conditions: 1.0 M NaOH, 1.0 mM MV^{2+}, 100 mg CdS; triangles, with 0. 1 M tartrate; dots, no tartrate, solid lines in all cases are model predictions. (Reproduced with permission from White and Bard.[225])

6.9. Heterogeneous Photocatalysis

In the absence of the back reaction ($B = 0$) and recombination ($R = 0$), the normalized current, I_{ss}/nFI_o increases linearly (with a slope of unity) with the normalized initial concentration, pC^*V_s/I_o for values less than 1.0. Under these conditions, the current is mass-transfer limited. When $pC^*V_s/I_o > 1$, there is a sufficient excess of oxidized mediator that I_{ss} is limited by the light flux. Under these conditions, I_{ss}/nFI_o reaches a constant, light-limited value. An increase in the parameter B, has little effect at low values of pC^*V_s/I_o. The slope of the curve in this region is similar to that in the absence of kinetic limitations. On the other hand, the recombination parameter affects the initial slope as well as the limiting current (Figure 6.38b).

In the absence of a donor, best fits to the experimental data[225] yield values of $R = 10^{-3}$ and $B = 10^6$. Hence $k_5 = 10^{-6} k_3$. In the presence of tartrate as a donor, $B = 7.5 \times 10^3$ and $R = 10^{-3}$. Note that the same value of R is obtained in both cases. This interesting result shows that the *initial* hole capture occurs at a rate fast enough (even in the absence of a donor) such that the recombination kinetic limitations are not altered. Now, $k_3 = 7.5 \times 10^3 k_4 D$ so that $k_4 D \gg k_5$. Thus, the presence of the donor effectively suppresses the photocorrosion reaction (CdS was employed in this particular study[225]). Figure 6.38c shows a comparison of experimental data and model predictions for the CdS/MV^{2+}/tartrate system.[225] Under the conditions corresponding to the first three data points in the upper curve, the maximum collection efficiency is obtained, and the losses are caused by the mass-transfer limitations of the cell.

Similar experiments have been performed with methyl viologen/acetate[226] and on platinized TiO_2 particles.[227] Slurry cell experiments have also been reported for other semiconductors (iron oxide polymorphs,[228] WO_3,[227]), and the reader is referred to the original literature on these. Figure 6.39 contains more recent slurry PEC cell data on TiO_2 suspensions.[229] The photocurrent is initially cathodic and rapidly (within 25 ms) becomes anodic. Once a steady state is reached, the *net* current is anodic. However, this is still a composite of anodic and cathodic current components, as shown by the anodic spike that occurs when the light is shut off. Reactions (6.69) and (6.70) generate the anodic and cathodic currents, respectively, in the slurry PEC cell:

$$TiO_2^- \rightarrow TiO_2 + e^- \quad (6.69)$$

$$\cdot OH \rightarrow OH^- + h^+ \quad (6.70)$$

Photoluminescence (PL) is a useful probe of interfacial electron-transfer processes. Such studies have been performed on ZnO,[230] CdS,[231,232] and ZnS.[233] For example, changes in the PL quantum yields have been correlated to the effect of surface-adsorbed species[232] and added electron acceptors such as methyl viologen.[231] The PL spectrum of ZnO powder exhibits a sharp band (~380 nm) and a broad (~500 nm) band, both emissions being quenched by added

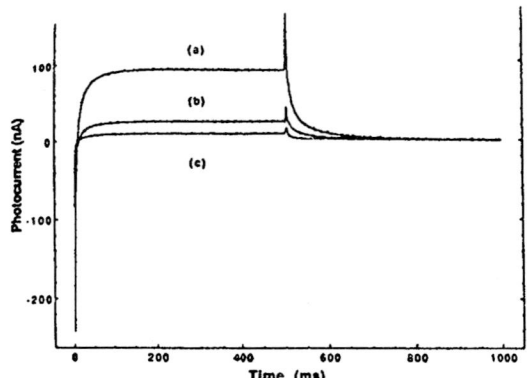

FIGURE 6.39. Slurry PEC cell transient response for equal concentrations of (a) P-25, (b) Fisher, and (c) Aldrich TiO_2 in pH 1 aqueous solution. (Reproduced with permission from Peterson et al.[229])

O_2.[230] The former has been assigned to the annihilation of bound excitons (electron–hole pairs) formally represented as $[Zn^+–O^-]_s^*$ species. The emission around 500 nm appears to be associated with the presence of oxygen anion vacancies near the surface.[230]

Similar PL studies have been performed on TiO_2, both on films anchored onto porous Vycor glass[142] and on "standard" catalyst powder.[234] Quenching of PL was observed with O_2 or N_2O with a concomitant blue shift in the absorption spectrum.[142] This was attributed to electron transfer from the excited catalyst to either O_2 or N_2O. On the other hand, the addition of unsaturated hydrocarbons (e.g., $C_2H_5C\equiv CH$) caused an increase in the PL intensity.[234] The extent of the PL enhancement appears to correlate with the ionization potential of the organic compound; that is, the lower the ionization potential, the higher is the PL intensity. These data suggest that the formation of negatively charged adducts (via electron transfer) on the TiO_2 causes PL quenching while the formation of positive adducts (via hole trapping) enhances the PL process. Additional mechanistic research is required on this topic.

Changes in the microwave conduction characteristics have been used as a probe of electronic processes in suspensions of TiO_2, CdS, and Se in a nonpolar medium (e.g., *trans*-decalin).[235] There are two problems associated with electrical measurements on micrometer sized particles. We have seen earlier that the transit time for electronic carriers within an excited particle is much shorter than a microsecond. Thus very fast perturbation and detection techniques are needed. A second problem, also due to the small dimensions involved, is that the probing electric field, E, must be applied in any given direction only for a time, Δt_E, much shorter than the time required for the charge carriers to un-

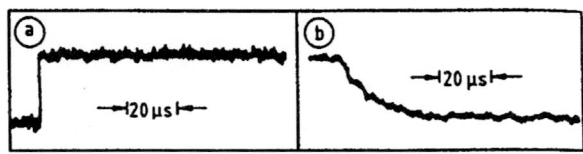

FIGURE 6.40. Time profile of the 392 nm absorption (a) and the conductivity (b) of a TiO_2 solution containing 5×10^{-5} M methyl viologen, pH 10. (Reproduced with permission from Bahnemann et al.[221])

dergo drift in the field direction. This distance roughly translates to the particle dimension. This produces the condition $\Delta t_E \propto d \times 10^{-6}/\mu E$ so that for a field strength of ~10^4 V m^{-1} and a mobility of 0.1 m^2 V^{-1} s^{-1}, $\Delta t_E \ll 1$ ns.[235] Thus, the field modulation has to be in the microwave frequency (GHz) range or higher. Such measurements have been made on TiO_2, albeit in low-loss suspension media such as hydrocarbons. Transient measurements indicate a fast rise in the microwave absorption followed by an initially fast decay and a much longer-lived conducting state.[235] The short lifetime has been ascribed to a combination of bulk electron–hole recombination and rapid diffusion (migration) to the surface. The longer-lived component is the more interesting one from the perspective of surface (interfacial electron transfer) processes. More studies in this area would undoubtedly contribute to a better understanding of the molecular details of events subsequent to the initial e$^-$–h$^+$ pair generation within the semiconductor particle.

Laser flash photolysis can be usefully combined with conductivity measurements. For example, Figure 6.40 contains the results from simultaneous laser flash photolysis and conductivity measurements on a TiO_2 solution at pH 10.[221] Note that the colloidal particles are negatively charged at this pH (see Eqs. (6.55) and (6.56)). In the presence of MV^{2+} as an electron scavenger, the radical cation (MV$^{+\bullet}$) absorption is present immediately after the laser flash (Figure 6.40a). The conductivity of the solution decreases with a half-life of 8.5 μs (Figure 6.40b). With increasing OH$^-$ concentration, this time becomes shorter. The decrease in conductivity is attributed to the oxidation of OH$^-$ by the photogenerated holes in TiO_2. That this reaction takes place *after* the laser flash is taken as an indication for the existence of a long-lived oxidizing intermediate (possibly adsorbed •OH). At low laser dose, the quantum yield for OH$^-$ consumption was ~35%.[221]

We turn next to a discussion of the dynamics of electron transfer from colloidal TiO_2 to O_2, a reaction of central importance in environmental photocatalysis (see Figure 6.21). The situation is schematized in Figure 6.41a. In this model and as borne out by the experimental data discussed earlier, the hole transfer reaction(s) are assumed to be very fast. Electron transfer from the

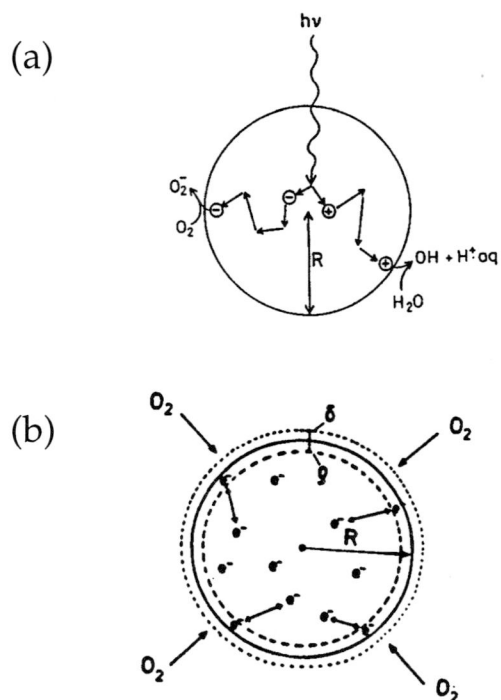

FIGURE 6.41. Two different models for the dynamics of electron transfer from irradiated TiO_2 to O_2. Refer to the text for details. (Reproduced with permission from Gerischer and Heller.[179])

photocharged TiO_2 particle to the O_2 acceptor is asumed to be the rate-determining step. Figure 6.42 illustrates slurry PEC cell data showing the depolarization of the TiO_2 particles in the presence of O_2 in the solution.[157] In the N_2-saturated solution, the charge accumulated on the TiO_2 particles is collected at the collector electrode as an anodic photocurrent. The rate of particle depolarization was shown to be accelerated by the catalytic modification of TiO_2 with Pd islands.[157] The small reverse current in Figure 6.42 is assigned to the back reaction associated with electron transfer *to* the TiO_2 particle.

Different models have been considered for electron transfer from irradiated TiO_2 particles to O_2.[179,180,236] The model contained in Figure 6.41b perhaps is the most realistic.[180] Here only electrons in a hemisphere of radius ρ from the particle surface, are supposed to interact strongly with an O_2 surface for the electron transfer step. The other critical distance is δ outward from the TiO_2 particle surface, which defines the locus of the O_2 molecules that are positioned close enough for electron transfer to occur. This leads to an expression for the

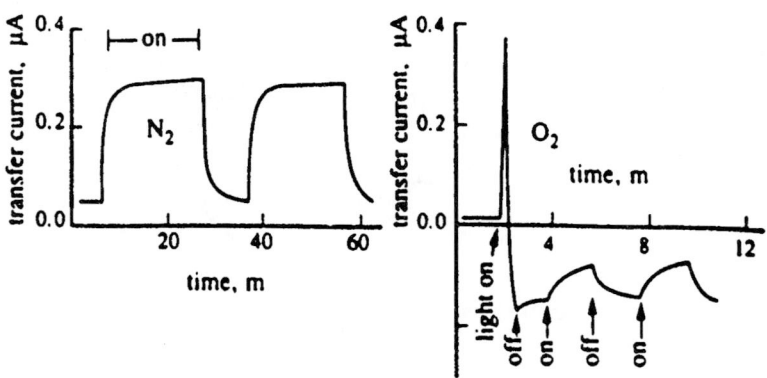

FIGURE 6.42. Depolarization of TiO_2 particles by dissolved O_2. Charge collection following particle irradiation is shown for N_2-saturated 1.6 M methanol-0.3 M NaCl (left) and the same O_2-saturated solution (right). The particles were modified with Pd catalyst (3 wt %); the charge collection potential was 0.35 V (vs. Ag/AgCl reference). (Reproduced with permission from Wang et al.[157])

pseudo-first-order rate constant k_3 (i.e., $v_3 = k_3 n_s$, where k_3 incorporates the O_2 concentration, which is assumed to be constant):

$$k_3 = \frac{2\pi}{3} \rho^3 \delta N_{O2} k_{et} \qquad (6.71)$$

where N_{O2} is the concentration of O_2 in water ($\sim 10^{17}$ cm^{-3}). The rate constant k_{et} (s^{-1}) has been modeled[180] using the Marcus theory and has been estimated to be in the range: 3×10^9–3×10^8 s^{-1}. With values of $\rho = 10^{-7}$ cm, $\delta = 3 \times 10^{-8}$ cm, and $N_{O2} = 1.7 \times 10^{17}$ cm^{-3}, Eq. (6.71) yields the following approximation[180]:

$$k_3 \sim 10^{-12} k_{et}$$

Unfortunately, experimental data on k_3 are scarce. One report quotes a value of 10^{-7} cm s^{-1} [236] which translates to a corresponding k_{et} of $\sim 10^5$ s^{-1}. This lower value (than that predicted by Marcus theory) indicates that the electron transfer may be non-adiabatic. However, it is worth noting that the use of Marcus's theory for modeling the heterogeneous electron transfer reaction to O_2 may not be justified in the absence of more insights into the molecular details of this process.

Upper bounds for k_3 have been computed[180] for three different particle dimensions as follows: $R = 10^{-6}$ cm, $k_3^{min} = 4 \times 10^{-4}$ cm s^{-1}, $R = 10^{-5}$ cm; $k_3^{min} = 4 \times 10^{-1}$ cm s^{-1}, and $R = 10^{-4}$ cm; and $k_3^{min} = 4 \times 10^2$ cm s^{-1}. Therefore, the larger the TiO_2 particles, the faster the interfacial electron transfer has to be so that the quantum yields are not deleteriously affected by electron–hole recombination. With

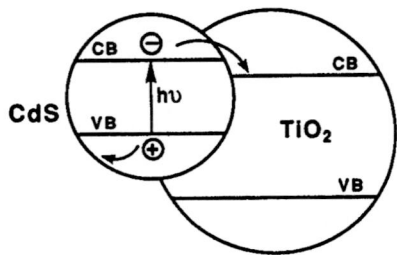

FIGURE 6.43. Coupled ("sandwich") semiconductor particles. (Reproduced with permission from Gopidas et al.[239])

these optimized conditions, electron–hole pair generation will assume the role of determining the overall rate of the process. It is interesting to note that the electron transfer rate–constant experimentally determined from laser-flash photolysis (4×10^{-3} cm s^{-1}, see above) for *one-electron* acceptors such as MV^{2+} may be fast enough to compete with the e$^-$–h$^+$ generation rate. However, it is unlikely that fast enough reactions can be sustained at larger (μm-sized) TiO$_2$ particles for *multielectron* processes (e.g., O$_2$ reduction) without catalytic modification. Even under these conditions, the available data indicate[157] that electron accumulation at the particle surface is not completely suppressed (i.e., the particle is not depolarized fast enough).

The preceding model has been criticized on the grounds[122] that the hole-transfer kinetics is not explicitly taken into account. This criticism was based partly on the observations at TiO$_2$ single-crystal electrodes of rather slow hole-transfer kinetics for CHCl$_3$ oxidation.[122] However, it is not clear to what extent this trend can be extrapolated to the behavior of TiO$_2$ *colloids*. As indicated earlier, much of the available data are consistent with facile hole-transfer kinetics at the TiO$_2$ particle/water interface.

Finally, we close the discussion in this section with a brief examination of mixed or "sandwich" colloidal suspensions, containing two different semiconductors (e.g., CdS and TiO$_2$). These sandwich structures can be used for efficient charge separation, as very schematically illustrated in Figure 6.43. Bandgap excitation of CdS is accompanied by electron transfer to TiO$_2$. However, the holes are "trapped" in CdS because of the location of the valence band in TiO$_2$ relative to CdS. In other words, we have created a favorable situation for charge separation. It has been shown[237] that the quantum yield for charge injection from CdS to TiO$_2$ can be close to unity, provided a large number of TiO$_2$ particles are present per CdS particle. Fluorescence decay of CdS was used as a probe in this particular study. Other combinations have been studied including CdS–ZnO,[237,238] CdS–AgI,[239] Cd$_3$P$_2$–TiO$_2$,[240] AgI–Ag$_2$S,[241] and CdSe–TiO$_2$.[242] One practical problem associated with the use of these mixed colloids is the susceptibility of one (or both) components to photocorrosion. Thus, the rate of photoanodic corrosion of CdS in aerated solution is drastically increased in the presence of TiO_2. The discovery of a mixed colloid system in which *both* the semiconductor components are stable against photocorrosion would open up a promising avenue in terms of its use in environmental remediation applications.

6.9.8 Kinetics and Mechanistic Aspects of Photocatalytic Reactions at TiO_2

In this section, we consider specifically the kinetics of product evolution in aqueous suspensions containing UV-irradiated TiO_2. We also consider the issue of reaction intermediates and finally our present state of knowledge as regards the mechanism of heterogeneous photocatalysis.

The general expression describing the rate of decomposition for a number of organic substrates on irradiated TiO_2 is of the Langmuir–Hinshelwood (L-H) kinetics form:

$$-\frac{dC}{dt} = k_1 \cdot \frac{KC}{1+KC} \tag{6.72a}$$

where k_1 is a reactivity constant, and K is an adsorption equilibrium (binding) constant. When the product $KC \ll 1$, Eq. (6.72a) collapses to an expression similar to that applicable to first-order kinetics:

$$-\frac{dC}{dt} = k_1 \cdot KC = k'C \tag{6.72b}$$

Indeed, such "first-order" reaction kinetics have been repeatedly observed for both inorganic and organic solutes. Figure 6.44 contains three representative plots for the photocatalytic reaction of H_2PtCl_6,[243] CN^-,[105] and salicyclic acid[244] at irradiated TiO_2 in aqueous media. In the first case, the photocatalytic reaction involves the *reduction* of Pt(IV) by the photogenerated electrons, and in the latter (more common) instances, photogenerated holes in TiO_2 oxidize the inorganic (CN^-), and organic (salicyclic acid) substrates, respectively. These observations are not without conflict, however. For example, a recent study[155] on the photooxidation of 4-chlorophenol at TiO_2 reports that plots of log C_t/C_0 time for initial concentrations ranging from 5.0 to 120.0 ppm are *not* straight lines.

Obviously, sweeping generalizations are hazardous, and the reaction kinetics often are crucially dependent on the substrate chemistry and the molecular details of the processes at the TiO_2 particle/solution microinterface(s).

Returning to the data in Figure 6.44c, the slopes of the log C_t plots *vs.* time are seen to decrease with increasing C_0. This is typical of L-H kinetics behavior, that is, at low substrate concentrations, roughly first-order behavior is observed.

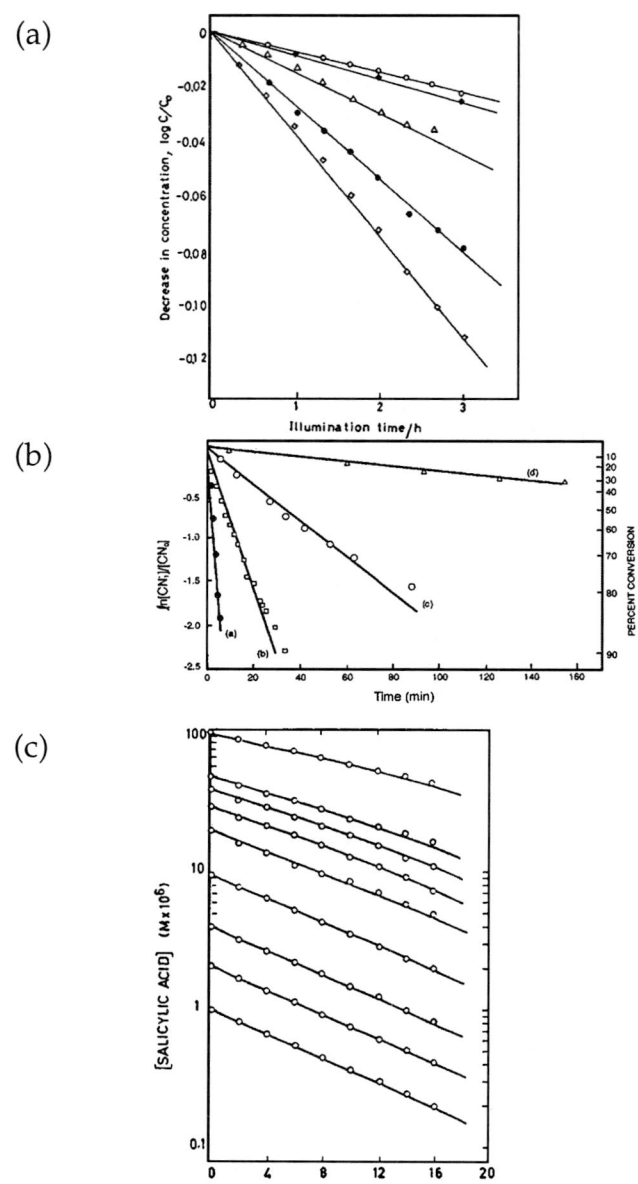

FIGURE 6.44. First-order kinetics plots for the photo-reduction of H_2PtCl_6 (a), photo-oxidation of CN^- (b) and salicyclic acid (c) in all cases mediated by illuminated TiO_2 suspensions. (Reproduced with permission from Doménech and Andres[111] (Figure 6.44a), Rose and Nanjundiah[105] (Figure 6.44b) and Matthews[244] (Figure 6.44c).)

6.9. Heterogeneous Photocatalysis

The reaction trends to zero order at high substrate concentrations, and the rate constant k_1 decreases with increasing concentration.[129]

The rate parameter k_1 in the L-H expression (Eq. (6.72a)) is usually interpreted as reflecting the limiting rate at high solute concentrations, that is, when $KC \gg 1$:

$$-\frac{dC}{dt} = k_1 \qquad (6.72c)$$

In this situation, essentially all the surface sites available for adsorbing the substrate are occupied. At the other limit, the parameter K reflects the ability of the substrate to be "bound" or adsorbed at the TiO_2 surface. Thus, a substrate often difficult to oxidize at high concentrations may be decomposed relatively easily at low levels, where it appears to be strongly adsorbed at the TiO_2 surface. Unfortunately, the *molecular* significance of the parameter K is not yet clear. For example, serious discrepancies have been noted[245,246] when parameters from *dark* adsorption isotherms are introduced into the L-H expression and the computed rates are compared with the measured photodecomposition rates. It is possible that such discrepancies can be resolved by invoking photoadsorption or other processes in the TiO_2 particle/solution double layer (termed micro-interfaces, see Ref. 246). Another complication arises from the possibly different reaction cross-sections of mono- vs. multilayer adsorbates with the photogenerated ·OH at these interfaces.[246] It is clear that further research is needed on these issues and much can be learned from the catalysis community. What can be concluded at this juncture is that the extracted kinetics parameters have a *phenomenological* rather than *fundamental* significance.

An alternative format for convenient analysis of kinetics data within the L-H model framework uses the reciprocal expression:

$$1/\text{rate} = \frac{1}{k_1} + \frac{1}{k_1 K} \cdot \frac{1}{C} \qquad (6.72d)$$

Thus, k_1 and K can be computed from the intercept and slope respectively of a plot of (rate)$^{-1}$ vs. C^{-1}. Representative plots of this type for benzene and perchloroethylene (PCE) are contained in Figure 6.45.[247] Interestingly, the intercepts for two variant substrates appear to converge to a single point; we shall return to this point later in this discussion. However, it is worth noting that this method of analysis is contingent on obtaining reliable values for the *initial rate*. Initial rates are often notoriously prone to scatter and uncertainty. Consequently, a nonlinear least-squares fit of Eq. (6.72a) to the *entire* rate vs. concentration curve may be a preferred route to reliable computation of k_1 and K.

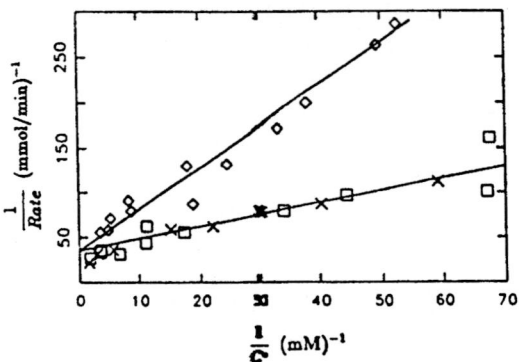

FIGURE 6.45. Reciprocal initial rate vs. reciprocal initial concentration for (squares, x) benzene and (diamonds) perchloroethylene (PCE). Catalyst 1.0 g/L Degussa P25 TiO_2. (Reproduced with permission from Turchi and Ollis.[247])

Analysis of reaction half-lives, $t_{1/2}$ (or other characteristic times) provides another route to kinetics analyses of photocatalytic reactions. Thus, the L-H expression (Eq. (6.72a)) can be integrated to yield:[129,248]

$$t_{1/2} = \frac{0.693}{k_1 K} + \frac{0.5 \, C_0}{k_1} \tag{6.72e}$$

A plot of $t_{1/2}$ vs. C_o should yield a straight line whose slope is $0.5/k_1$ and whose intercept is $0.693/k_1 K$. Hence, the solid lines in Figure 6.44c were calculated by back-inserting the values of k_1 and K into the integrated L-H expression.[129] Good agreement with the experimental data (circles) is seen in Figure 6.44c.[129]

We can also start with the master rate expression

$$-\frac{dC}{dt} = kC^n$$

and obtain an expression for $t_{1/2}$ as

$$t_{1/2} = \frac{(2^{n-1} - 1) \, C^{-n+1}}{k(n-1)}$$

or
$$\log_{10} t_{1/2} = \log_{10}\left[\frac{(2^{n-1} - 1)}{k(n-1)}\right] + (1-n)\log_{10} C \tag{6.73}$$

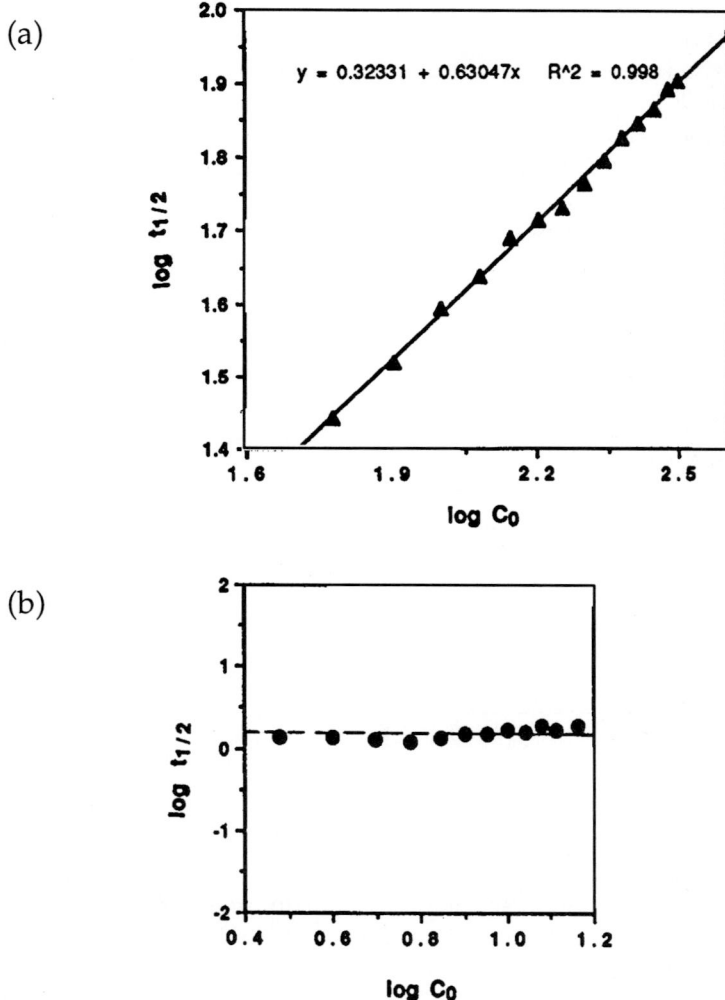

FIGURE 6.46. Analysis of photocatalysis data according to Eq. (6.73) for formaldehyde (a) and TCE (b) as the substrates.

Thus a log–log plot of $t_{1/2}$ vs C yields values for the reaction order, n, and the rate constant from the slope and intercept respectively. Such an analysis is contained in Figure 6.46a for formaldehyde as the substrate.[249] This analytical method obviously is not applicable to a first-order reaction for which $t_{1/2}$ is constant (Figure 6.46b). For this latter case, k can be measured by the usual method of plotting ln C vs. t (Figure 6.44).

The attractive feature of this approach based on the use of $t_{1/2}$ is that only a *single* experiment is needed to quickly establish values for k and n. In the usual method, the initial substrate concentration is varied from run to run and the conversion is mapped as a function of time in each case. This is obviously a tedious exercise. On the other hand, it must be noted that analyses based on Eqs. (6.72e) and (6.73) generally will not be valid in the presence of reaction intermediates that may alter the kinetics during the course of the reaction.

Product inhibition has been noted, as for example, via the addition of HCl to a solution containing TCE and UV-irradiated TiO_2.[66] This has been modeled within the L-H framework by competitive adsorption of the substrate and the product; that is, Eq. (6.72a) now becomes

$$-\frac{dC}{dt} = k_1 \sigma_s \qquad (6.74)$$

and

$$\sigma_s = \frac{K_s [C]}{1 + K_s C + K_p [P]} \qquad (6.74a)$$

More generally, reaction intermediates can also be included:

$$\sigma_s = \frac{K_s C}{1 + K_s C + K_I [I] + K_p [P]} \qquad (6.74b)$$

In Eqs. (6.74a) and (6.74b), σ_s is a TiO_2 surface coverage parameter (i.e., fraction of surface sites occupied by S) and the letters I and P stand for reaction intermediate(s) and product(s), respectively. Representative inhibition plots are contained in Figure 6.47 for PCE and dichloroacetic acid, respectively.[250] Plots such as those in Figure 6.47 afford a route to the determination of K_p.

A more complicated two-site Langmuirian model has been considered[73] to account for observations on chloroform at high substrate levels (up to 63 mM). Now, σ_s takes the form

$$\sigma_{CHCl_3} = \frac{x_1 K_1 [CHCl_3]}{1 + K_1 [CHCl_3]} + \frac{x_2 K_2 [CHCl_3]}{1 + K_2 [CHCl_3]} \qquad (6.74c)$$

where x_1 and x_2 are the fractions of surface sites 1 and 2, respectively; and K_1 and K_2 are the corresponding binding constants for the adsorption of $CHCl_3$ at these sites. According to these authors,[73] at low-substrate levels, adsorption is dominated by the tight binding site (x_1), while at higher concentrations, ~75%

6.9. Heterogeneous Photocatalysis

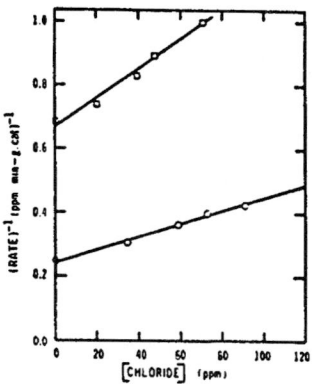

FIGURE 6.47. Inhibition by HCl. Inverse initial rate $vs.$ initial chloride (as HCl) concentration: (circles) perchloroethylene [$(Cl_2C=CCl_2)_o$ = 69.6 ppm]; (squares) dichloroacetic acid [$(Cl_2HCCOOH)_o$ = 67 ppm]. (Reproduced with permission from Ollis et al.[250])

of the observed rate is accommodated by the loosely binding site (x_2). The *molecular* nature of these two types of surface sites remains yet to be established.

The reaction rate dependence on O_2 has also been modeled[56,73,251,252] by the L-H expression as follows:

$$\text{rate} = \frac{K_{O_2}[O_2]}{1 + K_{O_2}[O_2]}$$

The value quoted for K_{O_2} (13 ± 7 × 10⁴ M^{-1}, Ref. 73) appears to support the notion that O_2 binds rather strongly to the TiO_2 surface sites. For comparison, the binding constants of organic substrates are in the ~10^3 M^{-1} range.[129]

The reaction between the organic substrate and the photogenerated •OH has four possible routes: (a) the reaction occurs between the two species in the adsorbed state; (b) a nonbound •OH radical reacts with an adsorbed organic molecule; (c) an adsorbed •OH radical reacts with a freely mobile organic molecule arriving at the catalyst surface; or (d) reaction occurs between the two species in the fluid phase. Unfortunately, all four mechanisms yield rate expressions that are indistinguishable within the L-H model framework.[247] Back-of-the-envelope calculations[247] indicate that •OH does not diffuse very far into solution even for very low concentrations of the substrate. On the other hand, this expectation appears to be in conflict with the analysis of slurry PEC cell data on TiO_2.[229]

The *direct* attack of the organic substrate by the photogenerated holes appears to be an unlikely scenario even though this process is often thermodynamically possible. This is borne out by the lack of photocatalytic reactivity observed in water-free organic solvents. As indicated earlier, the present state of knowledge as regards molecular details of the adsorption of organics at the TiO_2 surface is fragmentary. To add to the difficulty, the measurement at a single-crystal TiO_2 surface in an ultrahigh-vacuum environment often cannot be extrapolated to the colloidal suspension case. Nonetheless, new information is gradually emerging on this important aspect. For example, some aromatics appear to adsorb to surface hydroxyls rather than directly to the TiO_2 surface.[253] This would tend to favor the reaction route (c).

Attempts have been made[73,98,254] to correlate the reactivity trends for organic substrates at TiO_2 with the corresponding *homogeneous* solution reaction:

$$·OH + S \rightarrow products$$

As we have seen earlier in this chapter, bimolecular reaction rate constants are available for ·OH radical attack on a variety of organics. For example, a rate constant of $7 \times 10^3 \ M^{-1} \ s^{-1}$ has been estimated[73,98] for the photocatalytic decomposition of $CHCl_3$ at TiO_2. This is considerably lower than the value ($\sim 10^7 \ M^{-1} \ s^{-1}$)[6] measured for the corresponding homogeneous reaction. On the other hand, good agreement appears to exist between the two cases as exemplified by the comparison in Table 6.12.[254] This comparison was facilitated[254] by setting up a competition for the photogenerated ·OH radical as follows:

$$C_6H_5COOH + ·OH + O_2 \xrightarrow{k_1} C_6H_4OHCOOH + HO_2^·$$

$$RH + ·OH \xrightarrow{k_2} products$$

Suspensions of TiO_2 in water containing sodium benzoate and benzoic acid were dosed with known ·OH radical scavengers (RH) as shown in Table 6.12.

The form of the dependence of the photocatalytic reaction rate (or the quantum yield) on the incident photon flux offers mechanistic insights. A linear dependence has been observed especially at low photon fluxes.[73] This gives way to a square-root dependence at higher light intensities, and even ultimately to a photon flux-independent regime at very high rates of photon incidence. This has ramifications in the use of solar concentrator technology for photocatalytic reactor designs, as discussed later. The quantum yield is constant in the linear regime, varies as $I_o^{-0.5}$ (I_o is the photon flux) in the intermediate regime and degrades to I_o^{-1} in the plateau region. The transition between these boundaries depends on the catalyst material, although it appears that the

6.9. Heterogeneous Photocatalysis

Table 6.12. Comparison Between Rate Constants Determined with Illuminated Aqueous Slurries of TiO_2 (k_1) and Bimolecular OH Radical Reaction Rate Constants (k_2)

Solute	pH	$k_1/M^{-1}\,s^{-1}$	$k_2/M^{-1}\,s^{-1}$
Benzoic acid	3	3.4×10^9 [a]	$(3.4 \pm 1.9) \times 10^9$
Ethanol	3	$(1.0 \pm 0.1) \times 10^9$	$(1.6 \pm 0.4) \times 10^9$
Propan-2-ol	3	$(1.0 \pm 0.1) \times 10^9$	$(2.2 \pm 1.2) \times 10^9$
Methanol	3	$(4.1 \pm 0.7) \times 10^8$	$(9.2 \pm 4.0) \times 10^8$
Benzoate	7	4.8×10^9	$(4.8 \pm 1.3) \times 10^9$
Iodide	7	$(1.0 \pm 0.1) \times 10^{10}$	$(1.6 \pm 0.9) \times 10^{10}$
Formate	7	$(2.4 \pm 0.3) \times 10^9$	$(3.0 \pm 0.8) \times 10^9$
Bromide	7	$(8.6 \pm 1.9) \times 10^8$	$(1.3 \pm 0.8) \times 10^9$
Hydrogen carbonate	7	$(3.7 \pm 0.7) \times 10^7$	$(4.0 \pm 2.5) \times 10^7$
Chloride	7	$<2.4 \times 10^7$	10^6

[a] Assumed value from which other rates were calculated.
Source: Matthews.[254]

linear regime may be extended to much higher I_o by adding better electron acceptors (than O_2).[94] Two explanations have been put forward for the square-root dependence. In the first, carrier recombination (i.e., $e^- + h^+ \rightarrow$ heat) is thought to dominate at the higher light intensities.[225] On the other hand, a kinetics model[73] shows that the bimolecular recombination of $^{\bullet}OH$;

$$2\,^{\bullet}OH \xrightarrow{k_r} H_2O_2 \qquad (6.1a)$$

can also account for the square-root dependence. Slurry PEC cell data also indicate that the back reaction of $^{\bullet}OH$ with the semiconductor surface; that is,

$$(TiO_2)OH^- \xrightarrow{h\nu > E_g} TiO_2^- + \,^{\bullet}OH \qquad (6.67)$$

$$TiO_2^- + \,^{\bullet}OH \rightarrow TiO_2 + OH^- \rightarrow (TiO_2)OH^- \qquad (6.67a)$$

is significant.[229] A half-life of 1.6 ms has been estimated for the oxidizing intermediate [(TiO$_2^-$)·OH] in Reaction (6.67).[98]

Other studies[252,254,256] have addressed the issue of mixed reactants and kinetic effects of the position of chlorine atoms in the photodecomposition of di- and trichlorophenols.[257] The nonlinear nature of semilogarithmic plots of conversion vs. time [i.e., ln (C/C_o) vs. t] has been modeled in terms of *dispersed kinetics*.[258] That is, a reaction mechanism involving two or more consequent steps in heterogeneous media warrants a Gaussian distribution of rate constants. The reader is referred to the original literature for further information on these specialized topics.

Finally, intense illumination with solar radiation (even without concentration) causes a heating of the reactant solution unless infrared filters are used. Therefore, the temperature dependence of the photocatalytic degradation rate is of both fundamental and practical significance. However, the trends reported to date have been contradictory. An increase in the photocatalytic rate with increasing temperature was reported for salicyclic acid and phenol as the model substrates.[57,140] These data obeyed the Arrhenius law. On the other hand, a decrease in the photocatalytic reaction rate was noted for chloroform with increasing temperature.[98] A recent study on 4-chlorophenol also reveals adherence to Arrhenius behavior with an activation energy (E_a) of ca. 5.5 kJ mol^{-1}.[259] The low value of E_a indicates that thermally activated steps are negligible; that is, the adsorption–desorption processes are almost temperature independent. A similarly low value for E_a has been quoted earlier for phenol as the substrate.[57]

6.9.9 Reaction Intermediates

Two mechanistic aspects associated with photocatalysis at TiO$_2$ have not been thoroughly explored as yet in this chapter: The first concerns the molecular nature of the oxidizing intermediate at the TiO$_2$/solution interface. Second, in many cases, the extent of mineralization of the organic substrate is not complete, and intermediates—often more toxic than the reactant—are formed. We shall examine each of these issues in turn.

The importance of surface hydroxylation in photocatalytic activity was mentioned earlier (Section 6.9.3) within the context of the variant performance of TiO$_2$ in anatase vs. rutile form. The recognition of the role of chemisorbed water on TiO$_2$ (hydroxyl groups and undissociated molecular water) has instigated a number of studies based on the use of specific chemical probes,[260] calorimetry,[261,262] thermal desorption measurements,[263,264] NMR spectroscopy,[265] and IR absorption spectroscopy.[266–272] When H$_2$O undergoes dissociative adsorption at a TiO$_2$ surface, two distinctive hydroxyl groups are formed (Figure 6.48).[247] The theoretical maximum surface coverage has been estimated to be 5–15 OH$^-$/

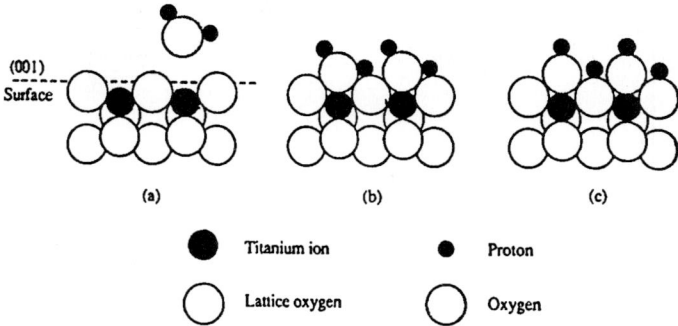

FIGURE 6.48. Surface hydroxyl groups on TiO_2: (a) hydroxyl-free surface; (b) physical adsorption of water; (c) dissociation of water, giving rise to two distinct OH⁻ groups. (Reproduced with permission from Turchi and Ollis.[247])

nm², depending on the specific crystallographic plane. Infrared evidence along with models[266-273] indicate that certain crystal faces [(100), (101)] bond only molecular H_2O, while others [(110), (001)] produce OH⁻ group pairs as in Figure 6.49. Anatase particles obviously contain a mixture of these planes. Thus a surface coverage of 4–10 OH⁻/nm² appears to be a reasonable estimate, both based on models and experiment.

There appears to be little doubt at this point that surface hydroxylation is a key factor in the photocatalytic activity of oxide semiconductors such as TiO_2. For example, a gradual decrease in the photocatalytic activity with time has been noted for gas-phase reactions at TiO_2.[61,274] However, the activity is restored when humid air is flowed over the photocatalyst. Second, anatase exhibits Brönsted acidity towards NH_3 molecule but rutile does not.[272] Third, stable peroxo species are observed at the anatase modification of TiO_2 but not at the rutile surface:[128,275]

$$\begin{array}{c} \text{OH} \quad \text{OH} \\ | \quad\quad | \\ Ti_s\text{—}O^-\text{—}Ti_s + h^+ \end{array} \longrightarrow \begin{array}{c} \text{OH} \quad \text{OH} \\ | \quad\quad | \\ Ti_s\text{—}O^\bullet\text{—}Ti_s \end{array}$$

$$\begin{array}{c} \text{OH} \quad \text{OH} \\ | \quad\quad | \\ Ti_s\text{—}O^\bullet\text{—}Ti_s + h^+ \end{array} \longrightarrow \begin{array}{c} \quad O \quad\quad \text{OH} \\ /\;\backslash \quad | \\ Ti_s\text{—}O\text{—}Ti_s + H^+_{aq} \end{array}$$

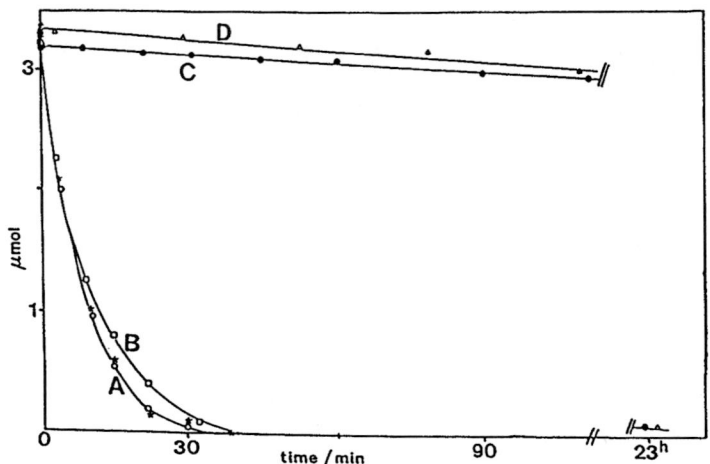

FIGURE 6.49. Kinetics of disappearance of benzamide under the following conditions: (A) H_2O_2 ($\lambda \geq 220$ nm) (open circles) or TiO_2 ($\lambda \geq 220$ nm) (stars), pH = 6.5; (B) H_2O_2 - TiO_2 ($\lambda \geq 220$ nm), pH = 6.5; (C) direct photolysis ($\lambda \geq 220$ nm); (D) Fenton reagent (dark), pH = 3.4. (Reproduced with permission from Maillard et al.[299])

Thus, oxidation is initiated at an acidic oxygen site (see later) on the surface. In a subsequent step, H_2O_2 is formed on the TiO_2 surface (but see later), or alternatively, the (chemisorbed) H_2O_2 species are further oxidized to O_2. The fact that ·OH accumulates at the surface indicates that these subsequent steps must be kinetically slow.[276]

In localized terms, the initial photoinduced charge transfer step can be written as

$$[M^{n+}-O^{2-}] \xrightleftharpoons{h\nu} [M^{(n-1)+}-O^-]^*$$

The O⁻ hole centers react with O_2 to form (unstable) O_3^- species, which are implicated in the photo-oxidation of alkenes and in the photoinduced $^{16}O_2-^{18}O_2$ isotope exchange process.[277,278] The O_3^- anion radical has been detected via its EPR signal.[278] Interestingly, addition of CO leads to the disappearance of the EPR signal and the formation of CO_2^-. From these observations, it was concluded[278] that O_2 in the gas phase (and not lattice oxygen) is involved in the formation of O_3^- and in the photocatalytic oxidation of alkenes. That is, O_2 undergoes photoactivation (weakening of the O-O bond) by interaction with the O⁻ hole center. This defect has also been implicated in the photoassisted production of $^{13}CO_2$ and H_2 at TiO_2.[279]

The radical species, ·OH, $O_2^{-·}$, and $HO_2^·$ have been detected by EPR on TiO_2,[280] ZnO,[281] and CdS.[282] The methyl viologen cation radical, $MV^{+·}$ (formed as a result of electron transfer from the semiconductor to MV^{2+}), has also been detected by this technique.[282,283] The formation of hydrocarbon free radicals as a result of the photo-Kolbe decarboxylation reaction:

$$RCO_2^- \xrightarrow[-e^-]{h\nu} [RCO_2^\bullet] \rightarrow R^\bullet + CO_2$$

has been confirmed by EPR spectroscopy.[284] The technique of spin trapping (see Section 6.5) has also enabled[281] the detection of intermediates such as $^\bullet CO_2^-$ which are formed at ZnO via the reaction of $^\bullet OH$ with the formate anion:

$$HCOO^- + {}^\bullet OH \rightarrow H_2O + {}^\bullet CO_2^-$$

The *direct* oxidation of the latter by the photogenerated hole

$$HCOO^- + h^+ \rightarrow {}^\bullet CO_2^- + H^+$$

also cannot be ruled out.

The *net* formation of H_2O_2 via the hole route (see above) does not appear to be an important mechanism. For example, H_2O_2 formation was not observed[285] when O_2 was precluded from the reaction medium. This indicates that the superoxide radical anion pathway must be the dominant route to H_2O_2 generation[286]:

$$O_2 + e^- \rightarrow O_2^{-\bullet}$$
$$O_2^{-\bullet} + H^+ \rightarrow HO_2^\bullet$$
$$HO_2^\bullet + H^+ + e^- \rightarrow H_2O_2$$

Other reports exist on photoadsorption–desorption equilibria involving O_2 and H_2O_2 at TiO_2,[287–289] and the reader is referred to this original literature for further details. The photocatalytic activity of TiO_2 for the reduction of silver ions, the oxidation of isopropanol, and O_2 evolution has been correlated with the density of surface hydroxyl groups.[290–292] These hydroxyl groups can be silylated as follows:

M-OH + X_3SiR → M-O-SiX$_2$R + HX
(hydrophilic)   (hydrophobic)

In this scheme, X and R are halide and alkyl groups respectively. When the silane-bonded TiO_2 particles in suspension are illuminated with UV light, the silane groups are decomposed, causing the particles to sink to the bottom. This "photo-sinking" phenomenon has been correlated with the degree of photocatalytic activity of various types of TiO_2 anatase preparations, a rutile powder type, and oxides such as SiO_2 and Al_2O_3.[293]

Turning next to reaction intermediates that are generated during the photocatalytic decomposition of organic substrates, the parameter k_1, K (or k') in the L-H expression (Eq. (6.72)) can be used as a measure of the reactivity. That is, the higher this value, the more easily mineralized the compound will be. The halogenated hydrocarbons, aromatics, and chlorophenols appear to have been the most thoroughly investigated in this regard. A general conclusion that can be drawn is that halogenated alkenes (e.g., TCE and PCE) are relatively easily mineralized. On the other hand, species such as CCl_4 and chloromethanes (e.g., $CHCl_3$) and chlorinated benzenes will be difficult candidates for photocatalytic treatment. It must be noted that this trend is quite similar to that noted earlier for the treatment of organics via the $UV–H_2O_2$ process (Section 6.5), a conclusion that must be hardly surprising given the fact that the •OH radicals are the causative agents in both the cases.

Even for relatively "simple" substrates such as acetate, it was noted fairly early in the evolution of the heterogeneous photocatalysis field that complete mineralization was not obtained. For example, methane, ethane, and CO_2 were all observed during the photocatalytic decomposition of acetate, and the yield of CO_2 exceeded that expected from the sum of methane and ethane produced.[294] This was rationalized on the basis of the generation of ethanol and acetaldehyde as reaction intermediates.[294] The formation of methane and ethane can be explained by the photo-Kolbe decarboxylation mechanism[120]:

$$CH_3COO^- + h^+ \rightarrow {}^\bullet CH_3 + CO_2$$
$$ {}^\bullet CH_3 + {}^\bullet CH_3 \rightarrow C_2H_6$$
$$ {}^\bullet CH_3 + {}^\bullet H \rightarrow CH_4$$

The hydrogen atom is formed via the reduction of H^+ by the photogenerated electrons at the TiO_2 surface sites. (This is favored on anatase but not on rutile particles.) Alternatively, H-atom abstraction is possible from the substrate leading to another route for methane production:[294]

$$ {}^\bullet CH_3 + CH_3COOH \rightarrow CH_4 + {}^\bullet CH_2COOH$$

Hole reactions are not the sole causative agents for organic decomposition. For example, both one-electron and two-electron *reductions* of halothane (2-bromo-2-chloro-1,1,1-trifluoroethane) have been noted in aqueous colloidal suspensions of platinized TiO_2.[76] These were detected by the products, namely, Br^- and F^-, generated:

6.9. Heterogeneous Photocatalysis

$$F-\underset{\underset{F}{|}}{\overset{\overset{F}{|}}{C}}-\underset{\underset{Cl}{|}}{\overset{\overset{H}{|}}{C}}-Br + e^- \longrightarrow F-\underset{\underset{F}{|}}{\overset{\overset{F}{|}}{C}}-\underset{\underset{Cl}{|}}{\overset{\overset{H}{|}}{C}}^{\bullet} + Br^-$$

$$F-\underset{\underset{F}{|}}{\overset{\overset{F}{|}}{C}}-\underset{\underset{Cl}{|}}{\overset{\overset{H}{|}}{C}}^{\bullet} + e^- \longrightarrow \underset{F}{\overset{F}{\diagdown}}C=C\underset{Cl}{\overset{H}{\diagup}} + F^-$$

CDFE

The other reaction product, 2-chloro-1,1-difluoroethylene (CDFE) was also detected and identified by GC-MS.[295] The reaction of the carbon-centered radical, $CF_3CHCl^{\bullet}$ with O_2 leads to trifluoroacetate and other products in a complex series of reactions.

In another study,[296] hole attack of TCE and PCE has been shown to yield trichlorinated acetic acid products, whereas the parallel reaction initiated by the photogenerated electrons (e_{CB}^-) appears to yield dichlorinated by-products. The reduction reaction is postulated to occur on the TiO_2 surface (i.e., in the adsorbed state). Interestingly, it has been noted[296] that the reaction of TCE and PCE with the hydrated electron ($k_{TCE} = 1.9 \times 10^{10}$ M^{-1} s^{-1}; $k_{PCE} = 1.3 \times 10^{10}$ M^{-1} s^{-1}, Ref. 7) proceeds at a comparable rate to that of the reduction of O_2 ($k_{O_2} = 1.9 \times 10^{10}$ M^{-1}, Ref. 7), with which the organic substrates must compete. Indeed, the yield of the dichlorinated byproducts was found to increase in O_2-deficient systems, consistent with a dual e^-–h^+ decomposition scheme.

Table 6.13 contains a summary of the reaction intermediates that have been observed during the photocatalytic oxidation of organic substrates.[297-299] The mechanistic pathways leading to the by-products listed in this table are discussed in the original literature cited there.

In closing this section, it is once again emphasized that, while complete mineralization has been observed in many instances for the photocatalytic oxidation of organics, highly noxious by-products are formed in some cases (as exemplified by the examples in Table 6.13). In situations where these reaction intermediates are strongly bound to the TiO_2 surface or can be suitably trapped at the TiO_2/solution interface (e.g., within the microporous framework of an immobilized particulate film photocatalyst), further conversion and ultimately mineralization would be facilitated. However, it does appear that the

Table 6.13. Reaction Intermediates in the TiO_2-Mediated Photocatalytic Oxidation of Organic Substrates

Substrate(s)	Intermediate(s)	Comments	Reference(s)
Chlorobenzenes	Chlorophenols; benzoquinones; and hydroquinones	Ortho and para substituted by-products are formed because of the propensity of the chloro-substituent as ortho–para director	250
1,2-Dibromoethane	Vinyl bromide	The 1,1-dibromo isomer, by contrast, is completely mineralized to HBr and CO_2	259
Benzene	Phenol; 1,4-benzoquinone	Traces of other species (e.g., 2-hydroxy-1,4-benzoquinone) are also generated	57, 252
2-Chlorophenol	Chlorohydroquinone (CHQ); hydroxyhydroquinone (HHQ); and 1,2-dihydroxybenzene	A total of eight intermediates identified by HPLC	297
3-Chlorophenol	CHQ; HHQ	The sequence 3 CP $\rightarrow$ CHQ $\rightarrow$ HHQ is proposed implying that C-Cl bond scission occurs in the second step	297
4-Chlorophenol	Hydroquinone (HQ)	1,4-benzoquinone (formed from HQ) and 4-chloro-1,2-dihydroxybenzene also detected	155
Di- and tri-chlorophenols	Para benzoquinones with 0-2 Cl atoms and 0 or 1 OH group	Intermediates with two aromatic or quinoid rings also detected	257
Pentachlorophenol	Chloranil; tetrachlorohydroquinone; H_2O_2	Hydroxyl radical attachment to the para position of the aromatic ring yields the semi-quinone radical, which then disproportionates; high-intensity radiation yields HCO_2^-, $CH_3CO_2^-$, CO_2, H^+, and Cl^- (i.e., more complete mineralization)	74

Cont. Table 6.13. Reaction Intermediates

Substrate(s)	Intermediate(s)	Comments	Reference(s)
TCE	Dichloroacetaldehyde (DCAAD); dichloroacetic acid (DCAA); dichloroacetyl chloride; and phosgene	Traces of trichlorinated by-products (TCAAD and TCAA) also formed	62,296
PCE	DCAA; TCAA	—	296
Pyridine	Acetate; formate; seven aliphatic intermediates with 1-5 C atoms; 2-hydroxy-pyridine is a major intermediate	Dipyridyl and carbamoyl pyridine isomers detected at high initial concentrations of substrate	258
Benzamide	4-hydroxybenzamide	—	298
Chloromethane	CO; CH_2Cl_2	On a single-crystal (rutile) surface, formaldehyde also is detected	299

Note: These studies were conducted both in the gas phase and in solution; no distinction is made here.

photocatalytic decomposition products often are less toxic than those generated via direct photolysis of the organic substrate. These comparative aspects form the topic of the next section.

6.9.10. Comparison of the Direct Photolysis, UV–H_2O_2, Fenton Reaction, and Heterogeneous Photocatalysis Approaches for the Destruction of Organic Pollutants

This comparison is not straightforward because the efficiency of each method depends, in a complex manner, on a number of process variables. Nonetheless, data are available under comparable conditions to enable some general conclusions to be made on the relative efficacy of these AOPs.

It appears that, in some cases, the reaction intermediates differ in the case of direct photolysis and in the TiO_2-mediated pathway. For example, direct photolysis of 3-chlorophenol (at $\lambda > 290$ nm) yields 1,3-dihydroxybenzene (resorcinol).[297] This has been postulated to occur via polarization of the C-Cl bond in the excited state leading to its cleavage and subsequent hydroxylation of the *meta*-ring position.[297] We have seen (Table 6.13) that the photocatalytic route leads to *para*-hydroxy derivatives. Direct photolysis of 4-chlorophenol yields hydroxylated biphenyl derivatives that are exceedingly toxic.[155] On the other hand, HQ is favored as a major reaction intermediate in the photocatalytic pathway (Table 6.13), leading eventually to complete mineralization.

It has been observed[249] that the direct photolysis pathway has an activation threshold in the case of formaldehyde. However, when TiO_2 is also present in the reaction medium, the photocatalytic reaction is initiated at the very outset, and therefore this route bypasses the direct photolysis pathway.

Figure 6.49 contains a comparison of the UV–H_2O_2 (A), UV–TiO_2 (A), UV–H_2O_2–TiO_2 (B), direct photolysis (C), and Fenton reagent (D), data for the rate of disappearance of benzamide.[299] The last two approaches clearly are ineffective for the treatment of this organic pollutant. Hydroxybenzamides are the major intermediates in both the UV–H_2O_2 and UV–TiO_2 routes; however, two more (unidentified) intermediates were detected by HPLC in the former case.[299] Interestingly, the formation of nitrate (favored as a final product of benzamide in the photocatalytic treatment) was not observed after UV–H_2O_2 treatment.[299]

The Fenton reagent approach is especially effective in the presence of electron-donating substituents in the substrate and with hydroxylated compounds (e.g., phenols). A recent comparison[300] of this approach with the direct photolysis and the UV–H_2O_2 alternatives for six test substrates (Table 6.14) is consistent with this notion. As in the case of benzamide, direct photolysis is an inefficient route for all these substrates. The decomposition of nitrobenzene and *N,N*-dimethylaniline via UV/H_2O_2 is slower than the Fenton reaction route.[300] This has been attributed to protonation of these compounds in the singlet excited state leading to radically altered reactivity, relative to the dark.

Finally, it must be noted that the reactivity of pollutants such as halogenated alkanes toward •OH is not especially good. In these instances the dark electrochemical reductive dehalogenation route (Chapter 5) may indeed prove to be the preferred remediation route. Alternatively, such pollutants can be attacked using either solvated electrons (e_{aq}) or reactions mediated by conduction band electrons in UV-irradiated TiO_2. The latter possibility was already discussed in the preceding section.

6.9.11. Photocatalytic Oxidation of Organic Pollutants: Miscellaneous Aspects

Attempts have been made to systematically study the influence of anions such as Cl^-, SO_4^{2-}, NO_3^-, ClO_4^-, HCO_3^-, and PO_4^{3-} on the photocatalytic activity of TiO_2 toward selected organic substrates.[73,154,301,302] These studies are in agreement that anions such as NO_3^- and ClO_4^- have no effect. On the other hand, the inhibitory effect of the other ionic species was attributed to either specific adsorption[154] or an alternative mechanism[301,302] involving competitive adsorption on the TiO_2 surface of the anions followed by radical anion generation by reaction with the photogenerated holes. We have already seen the ad-

Table 6.14. Rate Constants for the Destruction of Six Test Organic Substrates by Various AOPs

Substrate[a] A_{254}, A_{300}, Φ	Rate constant, k^b		
	Fenton reaction	Direct photolysis	UV/H_2O_2
Nitrobenzene $A_{254} = 0.42$ $A_{300} = \sim 0.15$ $\Phi = \sim 10^{-5}$	0.035	Not observed	0.017
Chlorobenzene $A_{254} = 0.01$ $A_{300} = <0.001$ $\Phi = 0.1$	0.077	0.0029	0.099
Toluene $A_{254} = 0.01$ $A_{300} = <0.01$	0.116	0.0058	0.154
Phenol $A_{254} = 0.125$ $A_{300} = <0.001$ $\Phi = 0.12$	0.247	0.0019	0.199
N,N-dimethylaniline $A_{254} = \sim 0.6$ $A_{300} = \sim 0.22$	0.105	Not observed	0.057

[a] A_{254}, A_{300} refer to the optical absorption of each compound at 254 nm, and at 300 nm for 1 cm path length; Φ is the quantum yield.
[b] Pseudo-first-order rate constant in min^{-1}.

Source: Lipczynska-Kochany.[300] The reaction medium in all the cases was water.

verse effects of species such as carbonate–bicarbonate and fulvic acids in terms of consuming •OH (Section 6.5).

Electrolyte pH shifts the standard potentials of Reactions (6.34) and (6.35) at the rate of -59 mV/pH at 25°C. The surface energetics at the TiO_2/solution (and indeed at other oxide semiconductor/solution) interfaces are also affected in a similar manner. As is well known in colloid science, this is a consequence of the varying surface charge at the semiconductor surface as a function of electrolyte pH. This in turn could have an impact on the adsorption of the organic substrate at the semiconductor (e.g., TiO_2) surface and consequently its reactivity. Rather variant trends have been observed for the influence of electrolyte pH on the photocatalytic reaction rate and yield. For example, the pH reaction dependence of the photocatalytic rate was found to be weak in the range from ~1.3 to ~6.3 for nitrobenzene, benzamide, 3-chlorophenol, and 1,2-dimethoxybenzene.[154] A similar trend was observed in a field test for BTEX.[55] On the other hand, the photon flux dependence of the photocatalytic reaction rate (see later) was observed to switch from linear at pH 2.6, 7, and 11 to a power (1/2) law at pH 5.[73,98] This was attributed to a "zero proton condition" on the TiO_2 surface at the last pH, which was estimated by the authors to correspond to the point of zero charge of TiO_2. A strong effect of pH was also noted by the same group in the photocatalytic treatment of biologically pretreated landfill effluent.[303]

These contrasting results are perhaps not entirely surprising, given that *heterogeneous* processes such as specific adsorption can be expected to be crucially dependent on the surface characteristics of the photocatalyst used and the chemical structure of the adsorbing substrate molecule (or ion).

Cations such as Cu(II), Fe(III), and Ag(I) have been shown to enhance the oxidation rates of organic substrates.[56,110,125,304-308] A likely mechanism is the reduction in the extent of $e^-–h^+$ recombination via the trapping of the photogenerated species by these electron acceptors. The beneficial influence of enhancing the rate of the conjugate process was discussed earlier.

Finally, the influence of additives such as H_2O_2 on the photocatalytic reaction rates at TiO_2 deserves mention. The photocatalytic oxidation rates were enhanced severalfold for substrates such as chloroethylenes, phenol, and chloral hydrate.[77,129,308] The H_2O_2 presumably increases the oxidation rate in these cases via its role as an electron scavenger. Comparisons between TiO_2–UV and H_2O_2–UV processes have also been made.[154] The synergistic effect observed in the TiO_2–H_2O_2–UV case shows that the photoaction spectrum of H_2O_2 has deleterious consequences, as noted earlier in this chapter.

Photochemical reactors employing conventional chemical oxidants (e.g., H_2O_2, ozone) employ short-wavelength UV-B or UV-C light. On the other hand, photocatalysis experiments with TiO_2 are usually conducted with UV-A (i.e., long-wavelength) light. A variety of illumination conditions have been employed in these studies ranging from high to medium-high pressure Hg arc lamps (major emission at ~365 nm), broadband solar simulators, "black light" fluorescent tubes, germicidal lamps, and outdoor sunlight. In attempting a systematic examination of the influence of irradiation wavelength, a comparison of 254 and 350 nm excitation of TiO_2 using phenol and salicylic acid as model water contaminants[309] has revealed the shorter wavelength to be considerably more effective. As the authors comment, however, it is debatable whether the rate enhancement is sufficient to offset the negative aspects associated with the practical use of short-wavelength UV light.

6.9.12. Photocatalytic Treatment of Inorganic Pollutants

A comparison of Tables 6.9 and 6.10 reveals that, in contrast to their organic counterparts, documented instances for the photocatalytic treatment of inorganic pollutants appear to be fewer in variety. In the case of the treatment of metal ions, it is also noteworthy that the conceptual basis is the same as in metal recovery or catalytic modification of the semiconductor, two processes with very different objectives. In some instances, however, there *is* an environmental relevance in metal recovery applications as in the treatment of metal plating bath formulations, many of which also contain pollutants such as cyanide.[100] Organic pollutants were also present in addition to the inorganic species[110,310]. As pointed out earlier, photocatalysis offers an elegant approach to the waste treatment in these cases by using both the conjugate electron- and hole-driven photoprocesses at the semiconductor particle–electrolyte interphase.

As in the organic systems discussed earlier, substrate (photo) adsorption on the semiconductor surface appears to be a fairly common trend, as exemplified by chromium,[99] manganese,[117] and silver.[311] The kinetics of the disappearance of these species have also been analyzed by the Langmuir–Hinshelwood model.[99,117,311] An important point to note with the inorganic pollutants is that their toxicity may be highly dependent on the oxidation state, such as in chromium and arsenic. Thus reduction of the metal ion to the elemental state is often not required, and solubility differences can be exploited to "immobilize" the pollutant. This is exemplified by the chromium system, where remediation is accomplished by Cr(VI) → Cr(III) reduction at TiO_2 followed by precipitation of Cr(III) as the hydroxide.[99]

Again, as with the organic counterparts, the vast majority of the species contained in Table 6.10 have been treated with TiO_2 photocatalyst, although, in a few instances, the other candidates listed in Table 6.11 have also been utilized.

6.9.13. Gas-Phase Photocatalysis and the Treatment of Air-Borne Pollutants

Air-borne pollutants (e.g., VOCs or volatile organic compounds) can be treated in a photocatalytic reactor either directly in the gas phase or after partitioning them in a fluid phase. The results to date[61,62,312] indicate that the photocatalytic reaction rates for some compounds (e.g., TCE) are orders of magnitude faster in the gas phase than in aqueous media. In general, substrate mass transport will be much more efficient in the gas phase than in condensed media. Second, oxidant starvation (see later) may be less of a problem with gas phase photocatalytic reactors. On the other hand, the available data for TCE indicate the lack of complete mineralization and the formation of by-products such as phosgene and dichloroacetyl chloride.[62] However, as the authors themselves point out, this in itself should not be a limitation since a caustic scrubber can be incorporated at the "end of the pipe." Thus, a two-step treatment scheme has been considered,[312] where the (contaminated) groundwater is first air stripped in a packed tower, then the contaminant-laden air from the stripper is routed to the photocatalytic reactor where the pollutant subsequently is degraded.

Another difficulty with gas-phase photocatalysis is the slow degradation in the activity of the photocatalyst.[61,274] This appears to be associated with the depletion of hydroxyl groups or chemisorbed water at the semiconductor (e.g., TiO_2)/gas interface. Support for this notion accrues from the restoration of photocatalytic activity when humid air is flowed over the photocatalyst surface after a "run."

Air-borne pollutants such as ozone and CO_2 have been treated after partitioning into the liquid phase. The conversion and fixation of CO_2 has both environmental and energy implications. The global accumulation of CO_2 has been associated with the "greenhouse" effect (Chapters 1 and 5). On the other hand, the reduction of CO_2 yields energy-rich fuels. Recent advances in this area have been reviewed.[313–316]

There are two basic approaches for converting CO_2 to useful chemicals: direct reduction of CO_2 to CO, formic acid, formaldehyde, methanol, methane, and the like and the fixation of CO_2 in organic and biological molecules. It was first demonstrated that CO_2 could be photoreduced at p–GaP photocathodes.[317] Subsequent work from the same group as well as others has shown that other materials such as $SrTiO_3$ single crystals[318] as well as semiconductor powders including $SrTiO_3$, WO_3, and TiO_2[319,320] can be used. (It must be noted that the photoreduction of CO_2, like the splitting of water, is an energetically "uphill" process, i.e., positive ΔG, contrasting with many of the organic photo-oxidation reactions discussed in a preceding section, which are *photocatalytic*.) Size-quantized semiconductor particles (e.g., ZnS) have also been used, and the quantum efficiencies for the formation of formic acid (a reduction product) have been found to increase with particle dimension in the range from 3.4 to 5.3 nm.[321] Catalysts such as crown ethers have been found to assist in the photoelectrochemical reduction of CO_2 on p–GaP in lithium carbonate electrolytes.[322]

A main difficulty in achieving high coulombic efficiency in this process as well as the reduction of CO[327] is the prevalence of competing side reactions such as hydrogen evolution. Another problem is the low solubility of substrates such as CO in aqueous media. Macrocycles such as metallated tetraphenylporphyrins have been effectively used to counter this by reversibly binding CO, thus enhancing the effective surface concentration of the substrate.[323]

A second approach has been to use organic acids such as pyruvic acid to fix CO_2 in a scheme involving an electron relay such as methyl viologen and malic enzyme (ME) along with ferredoxin–NADP⁺–reductase, FNR (NADP =

nicotinamide adenine dinucleotide phosphate).[324] The net reaction has an associated Gibbs free-energy charge of +57.7 kJ mol^{-1}. This approach appears to be promising in the search for the conversion of CO_2 under mild conditions. However, some problems remain to be solved in terms of the unfavorable effect of substrates such as lactic acid on the activities of ME and FNR.[324]

Although molecules such as N_2 have been fixed via PEC means,[325] we are not aware of documented attempts to photo-reduce other air-borne pollutant species such as NO_x and SO_x. However, an early study describes the fixation of NO on an illuminated TiO_2 catalyst in $HClO_4$,[326] the main product was found to be N_2.

6.9.14. Photocatalytic Reactor Design Considerations

The major issues are whether to use (a) a suspended or supported photocatalyst, (b) solar or UV light, and (c) if solar energy is used, whether to use concentrated or nonconcentrated sunlight. We shall address each of these in turn.

The relative merits of slurry versus immobilized photocatalytic reactor configurations were discussed earlier in this chapter. From a reactor design standpoint, the low pressure drop attendant with the use of slurry reactors is an added advantage. Strategies to remove the catalyst particles after use can be devised using commercially available membrane filtration units. These units can be operated either in direct-flow (also called *dead-end*) or cross-flow modes. A difficulty with the cross-flow mode is that a large volume of treated water is constantly recycled with the catalyst particles, thereby reducing the effective capacity of the reactor by an equivalent amount. Slurry and supported TiO_2 photocatalysts have been compared,[144] using TCE as a model pollutant; the slurry system was found to outperform the suspended photocatalyst almost by a factor of 2. This has been the trend without exception in the tests conducted by a number of laboratories to date, namely, the slurry systems are more efficient than their immobilized photocatalyst reactor counterparts.

The obvious trade-off in solar *vs.* UV photocatalysis technology is treatment cost. Unless the locale is favorable in terms of electricity cost (e.g., proximity to a hydroelectric generation site), the high cost of electrically generated UV light will always be a handicap relative to its solar counterpart. On the other hand, the direct UV (especially in conjunction with oxidants) technology is relatively well advanced (see earlier), and many elements of it could be borrowed for use with the photocatalysis approach. This may indeed prove to be the preferred route in certain specialized (especially indoor) applications. Annular and spiral reactors are generally used for photocatalysis with UV radiation.

Solar photoreactors for water detoxification were originally configured with parabolic trough concentrators.[144] This is due partly to the extant technology base for solar thermal conversion. The concentrator approach has the advantage of compactness. On the other hand, nonconcentrating or "one-sun" photoreactors can capture diffuse sunlight as well as the direct radiation. The diffuse component can amount to ~50% of the total available UV light even on a clear day.[144] Under cloudy conditions only diffuse light is available, and a nonconcentrator design could be operated even under these conditions (albeit at low conversion rates) while the trough unit would have to be shut down. Another major disadvantage with concentrator units is the loss in quantum efficiency at high photon fluxes (see previously). Other hybrid types of solar collectors (e.g., the compound parabolic concentrator), which are a cross between the trough concentrators and the one-sun units, have been considered for photocatalytic treatment applications.[327]

The system design for a solar photocatalytic reactor has to incorporate provisions for intermittent operation. In particular, the system design is simplified (and attendant costs lower) if only daylight operation is needed. Otherwise, the solar unit will have to be coupled with a nonsolar backup such as UV lamps or carbon adsorption. Other factors to be taken into account in the solar system design include those related to the diurnal and seasonal variations in the UV flux to the surface of the earth. The effect of these on the decomposition rate of phenol has been considered, using a borosilicate glass dish reactor.[59]

Using the slurry approach, a variety of photocatalytic reactor designs have been modeled,[328] and falling film or slit fluid flow geometries have been shown to yield superior performance. Two types of reactors utilizing thin liquid films can be envisioned: spinning disk or falling film reactors. The former afford high liquid throughput, and ozonolysis reactors of this type have demonstrated good mass transfer characteristics. The characteristics of a falling film reactor have been established using a 2.5 kW medium pressure UV lamp and several model pollutants.[329] It must be noted that this type of design can accommodate both slurries and immobilized photocatalyst reactor configurations. It is also compatible with both solar and the electric (UV lamp) approaches.

Another reactor design consideration relates to continuous flow *vs.* batch operation. In the latter, shallow panels or tanks hold the water for several days to allow for complete destruction of the pollutant (or microorganisms). Geometries simulating shallow lagoons have been tested.[59] Indeed operating concepts incorporating both of these designs have been envisioned. For example, in the Lazy River concept,[327] the fluid continuously enters and exits a long serpentine channel. This channel, which serves the dual functions of photoreactor and holding tank, is dimensioned such that the fluid residence time is adequate for complete detoxification.

Progress is being made in modeling the fluid hydrodynamics and flow inside photocatalysis reactors.[312] In most cases, reactant mass transport from

6.9. Heterogeneous Photocatalysis

the bulk fluid phase to the "wall" must be considered, since the photoreaction mainly occurs at the latter. The tube geometry is obviously important here, and results have been presented[330] for mass transfer to a coiled tube geometry of the type used in conjunction with UV lamps.[331] Other aspects to be yet considered in the development of a process model include oxidant (e.g., O_2) mass transport, homogeneous reaction chemistry (i.e., reaction of ˙OH radicals with the substrate in the bulk fluid phase away from the wall), and the effect of interparticle distance on the convective–diffusive substrate mass transport in reactors using immobilized photocatalyst. For further discussion on the reactor design issues, the reader is referred to the original literature cited above and in Ollis and Turchi[328] and Wyness et al.[332]

6.9.15. The Photocatalytic Pollutant Treatment Approach: Prospects and Problems

While notable advances have been made in this field since the early 1980s, many challenges remain. First, the quantum yields have remained rather low for TiO_2-based photocatalysis reactors. On the other hand, gas-phase photocatalysis systems have shown very high quantum yields (close to 1). Improvements in the conjugate reaction kinetics or oxidant supply and strategies for suppression of carrier recombination in the photocatalyst remain challenges. Second, the solar photocatalytic approach suffers from the poor overlap of the absorption profile of TiO_2 with the solar spectrum. (Only ca. 5% of the available energy is utilized.) Attempts to extend the photoresponse of TiO_2 have yielded rather disappointing results. This is an area that warrants further careful research. Third, the poor performance of immobilized TiO_2 vis-à-vis its slurry reactor counterpart remains a vexing problem. There are also indications that photocatalyst film-based reactors may be more prone to poisoning and inactivation. The surface chemistry in these deleterious situations must be properly understood so that steps may be taken to avoid catalyst fouling. We can learn much from the heterogeneous catalysis community in this area. Finally, ingenuity is required to devise reactor designs for overcoming the mass transfer limitations with the immobilized photocatalyst. Approaches incorporating turbulent fluid flow remain to be tested.

Notwithstanding these shortcomings, the photocatalytic treatment approach has the potential to fare well against competing technologies in the next few years and into the next millenium. In particular, this approach is compatible with many of the water treatment technologies currently used.

The commercial prospects of this technology will be discussed in Chapter 8.

6.10. SUMMARY

Various photochemical methods for pollution abatement have been reviewed in this chapter. Notwithstanding the intangible factors that underlie the practical and widespread adoption of any new technology, it is likely that photochemical technologies will soon pose stiff competition to established methods of combating environmental pollution. Importantly, these technologies have many similarities with Nature's own self-repair mechanisms. They can also be adapted for use in a wide range of areas in developed and developing countries alike, as we shall see in Chapter 8.

REFERENCES

1. J. Hoigné and H. Bader, Rate Constants of Reactions of Ozone with Organic and Inorganic Compounds in Water. I. Non-Dissociating Organic Compounds. *Water Res.* **17**, 173 (1983). Rate Constants of Reactions of Ozone with Organic and Inorganic Compounds in Water. II – Dissociating Organic Compounds. *ibid.* p. 85.
2. J. R. Bolton and S. R. Cater, in "Aquatic and Surface Photochemistry" (G. R. Helz, R. G. Zepp and D. G. Crosby, eds.), Chapter. 33, pp. 467–490. Lewis Publishers, Boca Raton, FL, 1994.
3. O. Legrini, E. Oliveros, and A. M. Braun, Photochemical Processes for Water Treatment. *Chem. Rev.* **93**, 671 (1993).
4. S. Lunak and P. Sedlak, Photoinitiated Reactions of Hydrogen Peroxide in the Liquid Phase. *J. Photochem. Photobiol. A: Chem.* **68**, 1 (1992).
5. H. Taube, Photochemical Reactions of Ozone in Solution. *Trans. Faraday Soc.* **53**, 656 (1957).
6. P. C. Farhataziz and A. B. Ross, Selected Specific Rates of Reactions of Transients from Water in Aqueous Solution. III. Hydroxyl Radical and Perhydroxyl Radical and Their Radical Ions. Rep. NSRDS-NBS 59. National Bureau of Standards, Washington, DC, 1977.
7. G. V. Buxton, C. L. Greenstock, W. P. Helman, and A. B. Ross, Critical Review of Rate Constants for Reactions of Hydrated Electrons, Hydrogen Atoms and Hydroxyl Radicals in Aqueous Solution. *J. Phys. Chem. Ref. Data* **17**, 513 (1988).
8. W. H. Glaze, in "Chemical Oxidation: Technologies for the Nineties" (W. W. Eckenfelder, A. R. Bowers and J. A. Roth, eds.), Vol. 3, pp. 1-11. Technomic Publishing, Lancaster and Basel, 1993.
9. E. Koubek, Photochemically Induced Oxidation of Refractory Organics with Hydrogen Peroxide. *Ind. Eng. Chem. Process Des. Dev.* **14**, 348 (1975).

10. D. W. Sundstrom, H. E. Klei, T. A. Nalette, D. J. Reidy, and B. A. Weir, Destruction of Halogenated Aliphatics by Ultraviolet Catalyzed Oxidation with Hydrogen Peroxide. *Hazard. Waste Hazard. Mater.* **3**, 101 (1986).
11. C. C. D. Yao, W. R. Hag, and T. Mill, *in* "Chemical Oxidation: Technologies for the Nineties," W. W. Eckenfelder, A. R. Bowers, and J. A. Roth, eds., Vol. 2, pp. 112-139. Technomic Publishing, Lancaster and Basel, 1992.
12. G. R. Payton and W. H. Glaze, Destruction of Pollutants in Water with Ozone in Combination with Ultraviolet Radiation. 3. Photolysis of Aqueous Ozone. *Environ. Sci. Technol* **22**, 761 (1988).
13. J. Kochany and J. R. Bolton, Mechanism of Photodegradation of Aqueous Organic Pollutants. 2. Measurement of the Primary Rate Constants for Reaction of •OH Radicals with Benzene and Some Halobenzenes Using an EPR Spin-Trapping Method Following the Photolysis of H_2O_2. *Environ. Sci. Technol.* **26**, 262 (1992).
14. R. P. Mason and K. M. Morehouse, *in* "Free Radicals – Methodology and Concepts" (C. Rice-Evans and B. Halliwell, eds.), pp. 157-167. Richelieu Press, London, 1988.
15. J. R. Harbour, V. Chan, and J. R. Bolton, An Electron Spin Resonance Study of the Spin Adducts of OH and HO_2 Radicals with Nitrones in the Ultraviolet Photolysis of Hydrogen Peroxide Solutions. *Can. J. Chem.* **52**, 3549 (1974).
16. J. Kochany and J. R. Bolton, Mechanism of Photodegradation of Aqueous Organic Pollutants. 1. EPR Spin-Trapping Technique for the Determination of •OH Radical Rate Constants in the Photooxidation of Cholorophenols Following the Photolysis of H_2O_2. *J. Phys. Chem.* **95**, 5116 (1991).
17. M. Malaiyandi, M. H. Sadar, P. Lu, and R. O'Grady, Removal of Organics in Water Using Hydrogen Peroxide in Presence of Ultraviolet Light. *Water Res.* **14**, 1131 (1980).
18. P. N. Moza, K. Fytsanos, V. Samanidou, and F. Korte, Photodecomposition of Chlorophenols in Aqueous Medium in Presence of Hydrogen Peroxide. *Bull. Environ. Contam. Toxicol.* **41**, 678 (1988).
19. K. Omura and T. Matsumura, Photo-induced Reactions – IX. The Hydroxylation of Phenols by the Photodecomposition of Hydrogen Peroxide in Aqueous Media. *Tetrahedron* **24**, 3475 (1968).
20. B. A. Weir, D. W. Sundstrom, and H. E. Klei, Destruction of Benzene by Ultraviolet Light-Catalyzed Oxidation with Hydrogen Peroxide. *Hazard. Waste Hazard. Mater.* **4**, 165 (1989).
21. J. C. Milano, S. Yassin-Hussan, and J. L. Vernet, Photochemical Degradation of 4-Bromodiphenylether: Influence of Hydrogen Peroxide. *Chemosphere* **25**, 353 (1992).

22. P. C. Ho, Photooxidation of 2,4-Dinitrotoluene in Aqueous Solution in the Presence of Hydrogen Peroxide. *Environ. Sci. Technol.* **20**, 260 (1986).
23. E. Lipczynska-Kochany and J. R. Bolton, Flash Photolysis/HPLC Applications. 2. Direct Photolysis vs. Hydrogen Peroxide Mediated Photodegradation of 4-Chlorophenol as Studied by a Flash Photolysis/HPLC Technique. *Environ. Sci. Technol.* **26**, 259 (1992).
24. E. Lipczynska-Kochany and J. R. Bolton, Flash Photolysis/HPLC Method for Studying the Sequence of Photochemical Reactions: Applications to 4-Chlorophenol in Aerated Aqueous Solution. *J. Photochem. Photobiol.* **58**, 315 (1991).
25. E. Lipczynska-Kochany and J. R. Bolton, Flash-Photolysis-HPLC Method Applied to the Study of Photodegradation Reactions. *J. Chem. Soc., Chem. Commun.* p. 1596 (1990).
26. R. Venkatadri and R. W. Peters, Chemical Oxidation Technologies: Ultraviolet Light/Hydrogen Peroxide, Fenton's Reagent and Titanium Dioxide-Assisted Photocatalysis. *Hazard. Waste Hazard. Mater.* **10**, 107 (1993).
27. P. Neta, R. E. Huie, and A. B. Ross, Rate Constants for Reactions of Inorganic Radicals in Aqueous Solution. *J. Phys. Chem. Ref. Data* **17**, 1027 (1988).
28. J. Staehelin and J. Hoigné, Decomposition of Ozone in Water in the Presence of Organic Solutes Acting as Promoters and Inhibitors of Radical Chain Reactions. *Environ. Sci. Technol.* **19**, 1206 (1985).
29. K. Sehested, J. Holeman, E. Bjergbakke, and E. J. Hart, Formation of Ozone in the Reaction of OH with O_3^-, and the Decay of the Ozonide Ion Radical. *J. Phys. Chem.* **88**, 269 (1984).
30. K. Sehested, J. Holeman, and E. J. Hart, Rate Constants and Products of the Reaction of e_{aq}^-, O_2^- and H with Ozone in Aqueous Solutions. *J. Phys. Chem.* **87**, 1951 (1983).
31. K. Sehested, J. Holeman, E. Bjergbakke, and E. J. Hart, A Pulse Radiolytic Study of the Reaction: $OH + O_3$ in Aqueous Medium. *J. Phys. Chem.* **88**, 4144 (1984).
32. H. Christensen, K. Sehested, and H. Corfitsen, Reactions of Hydroxyl Radicals with Hydrogen Peroxide at Ambient and Elevated Temperatures. *J. Phys. Chem.* **86**, 1588 (1982).
33. G. R. Peyton, F. Y. Huang, J. L. Burleson, and W. H. Glaze, Destruction of Pollutants in Water with Ozone in Combination with Ultraviolet Radiation. 1. General Principles and Oxidation of Tetrachloroethylene. *Environ. Sci. Technol.* **16**, 448 (1982).
34. R. L. Garrison, C. E. Mauk, and H. W. Prengle, Jr., in "First International Symposium on Ozone for Water and Wastewater Treatment" (R. G. Rice and M. E. Browning, eds.), p. 551. International Ozone Institute, Syracuse, NY 1975.

35. H. W. Prengle, Jr., C. E. Mauk, R. W. Legan, and C. G. Hewes III, Ozone/UV Process Effective Wastewater Treatment. *Hydrocarbon Process.* **54**, 82 1975.
36. H. W. Prengle, Jr., C. G. Hewes, III, and C. E. Mauk, Oxidation of Refractory Materials by Ozone with Ultraviolet Radiation. *In* "Second International Symposium on Ozone Technology" (R. G. Rice, P. Pichet, and M.-A. Vincent, eds.), p. 224. International Ozone Institute, Syracuse, NY 1976.
37. H. W. Prengle, Jr., and C. E. Mauk, Ozone/UV Oxidation of Pesticides in Aqueous Solution. *In* Ozone/Chlorine Dioxide Products of Organic Materials, (R. G. Rice and J. A. Cotruro, eds.), p. 302. Ozone Press International, Cleveland, OH 1978.
38. J. Hoigné and H. Bader, The Role of Hydroxyl Radical Reactions in Ozonation Processes in Aqueous Solutions. *Water Res.* **10**, 377 (1976).
39. J. Staehelin and J. Hoigné, Decomposition of Ozone in Water: Rate of Initiation of Hydroxide Ions and Hydrogen Peroxide. *Environ. Sci. Technol.* **16**, 676 (1982).
40. M. Pelig, The Chemistry of Ozone in the Treatment of Water. *Water Res.* **10**, 361 (1976).
41. J. Hoigné and H. Bader, Ozonation of Water: Selectivity and Rate of Oxidation of Solutes. *Ozone: Sci. Eng.* **1**, 73 (1979).
42. J. L. Benoit-Guyod, D. G. Crosby, and J. B. Bowers, Degradation of MCPA by Ozone and Light. *Water Res.* **20**, 67 (1986).
43. S. Nakayama, K. Esaki, K. Namba, Y. Taniguchi, and N. Tabata, Improved Ozonation in Aqueous Systems. *Ozone: Sci. Eng.* **1**, 119 (1979).
44. R. Brunet, M. M. Bourbigot, and M. Dore, Oxidation of Organic Compounds Through the Combination Ozone-Hydrogen Peroxide. *Ozone: Sci. Eng.* **6**, 163 (1984).
45. J. P. Duguet, E. Brodard, B. Dussert, and J. Mallevialle, Improvement in the Effectiveness of Ozonation of Drinking Water Through the Use of Hydrogen Peroxide. *Ozone: Sci. Eng.* **7**, 241 (1985).
46. W. H. Glaze, J. W. King, and D. H. Chapin, The Chemisry of Water Treatment Processes Involving Ozone, Hydrogen Peroxide and Ultraviolet Radiation. *Ozone: Sci. Eng.* **9**, 335 (1987).
47. E. M. Aieta, K. M. Reagan, J. S. Lang, L. McReynolds, J.-W. Kang, and W. H. Glaze, Advanced Oxidation Processes for Treating Groundwater Contaminated with TCE and PCE: Pilot-Scale Evaluations. *J. Am. Water Works Assoc.* **80**, 64 (1988).
48. P. V. Danckwerts, "Gas-Liquid Reactions," McGraw-Hill, San Francisco, 1970.
49. C. N. Satterfield and A. H. Bonnell, Interferences in the Titanium Sulfate Method for Hydrogen Peroxide. *Anal. Chem.* **27**, 1174 (1955).

50. A. L. Lazrus, G. L. Kok, S. N. Gitlin, J. A. Lind, and S. E. McLaren, Automated Fluorometric Method for Hydrogen Peroxide in Atmospheric Precipitation. *Anal. Chem.* **57**, 917 (1985).
51. H. Bader and J. Hoigné, Determination of Ozone in Water by the Indigo Method. *Water Res.* **15**, 449 (1981).
52. D. L. Flamon, Analysis of Ozone at Low Concentrations with Boric Acid Buffered KI. *Environ. Sci. Technol.* **11**, 978 (1977).
53. E. J. Hart, K. Sehested, and J. Holeman, Molar Absorptivities of Ultraviolet and Visible Absorption of Ozone in Aqueous Solutions. *Anal. Chem.* **55**, 46 (1983).
54. R. Beyerle-Pfnur, K. Hustert, P. N. Moza, and M. Voget, The Role of Oxygen Species in the Degradation of Selected Environmental Chemicals. *Toxicol. Environ. Chem.* **20-21**, 129 (1989).
55. C. S. Turchi, J. F. Klausner, D. Y. Goswami, and E. Marchand, in "Chemical Oxidation: Technologies for the Nineties" (W.W. Eckenfelder, A.R. Bowers, and J.A. Roth, eds.), Vol. 3, NREL/RP-471-5345, Technomic Publishing, Lancaster and Basel, 1993.
56. K. Okamoto, Y. Yamamoto, H. Tanaka, M. Tanaka, and A. Itaya, Heterogeneous Photocatalytic Decomposition of Phenol over Anatase Titanium Dioxide Powder. *Bull. Chem. Soc. Jpn.* **58**, 2015 (1985).
57. K. Okamoto, Y. Yamamoto, H. Tanaka, and A. Itaya, Kinetics of Heterogeneous Photocatalytic Decomposition of Phenol over Anatase Titanium Dioxide Powder. *Bull. Chem. Soc. Jpn.* **58**, 2023 (1985).
58. R. W. Mathews, An Adsorption Water Purifier with In Situ Photocatalytic Regeneration. *J. Catal.* **113**, 549 (1988).
59. R. W. Mathews and S. R. McEvoy, Destruction of Phenol in Water with Sun, Sand, and Photocatalysis. *Solar Energy* **49**, 507 (1992).
60. R. Terzian, N. Serpone, C. Minero, and E. Pelizzetti, Photocatalyzed Mineralization of Cresols in Aqueous Media with Irradiated Titania. *J. Catal.* **128**, 352 (1991).
61. L. A. Dibble and G. B. Raupp, Fluidized Bed Photocatalytic Oxidation of Trichloroethylene in Contaminated Air Streams. *Environ. Sci. Technol.* **26**, 492 (1992).
62. M. R. Nimlos, W. A. Jacoby, D. M. Blake, and T. A. Milne, Direct Mass Spectrometric Studies of the Destruction of Hazardous Wastes. 2. Gas Phase Photocatalytic Oxidation of Trichloroethylene over TiO_2: Products and Mechanisms. *Environ. Sci. Technol.* **27**, 732 (1993).
63. J. H. Carey, J. Lawrence, and H. M. Tosine, Photodechlorination of PCB's in the Presence of Titanium Oxide in Aqueous Suspensions. *Bull. Environ. Contam. Toxicol.* **16**, 697 (1976).
64. B. G. Oliver, E. G. Cosgrove, and J. H. Carey, Effect of Suspended Sediments on the Photolysis of Organics in Water. *Environ. Sci. Technol.* **13**, 1075 (1979).

65. J. H. Carey and G. B. Oliver, The Photochemical Treatment of Wastewater by Ultraviolet Irradiation of Semiconductors, *Water Pollut. Res. J. Can.* **15**, 157 (1980).
66. A. L. Pruden and D. F. Ollis, Photoassisted Heterogeneous Catalysis: The Degradation of Trichloroethylene in Water. *J. Catal.* **82**, 404 (1983).
67. C. Hsiao, C. Lee, and D. F. Ollis, Heterogeneous Photocatalysis: Degradation of Dilute Solutions of Dichloromethane, Chloroform, and Carbon Tetrachloride with Illuminated Titania Photocatalyst. *J. Catal.* **82**, 418 (1983).
68. R. W. Matthews, Photooxidation of Organic Material in Aqueous Suspensions of Titanium Dioxide. *Water Res.* **20**, 569 (1986).
69. R. W. Matthews and D. F. Ollis, Photocatalytic Oxidation of Chlorobenzene in Aqueous Suspensions of Titanium Dioxide. *J. Catal.* **97**, 565 (1986).
70. M. B. Borup and E. J. Middlebrooks, Photocatalyzed Oxidation of Toxic Organics. *Water Sci. Technol.* **19**, 381 (1987).
71. A. E. Hussain and N. Serpone, Kinetic Studies in Heterogeneous Photocatalysis. I. Photocatalytic Degradation of Chlorinated Phenols in Aerated Aqueous Solutions over TiO_2 Supported on a Glass Matrix. *J. Phys. Chem.* **92**, 5726 (1988).
72. P. E. Menassa, M. K. S. Mak, and C. H. Langford, A Study of the Photodecomposition of Different Polychlorinated Biphenyls by Surface Modified Titanium(IV) Oxide Particles. *Environ. Technol. Lett.* **9**, 825 (1988).
73. C. Kormann, D. W. Bahnenann, and M. R. Hoffmann, Photolysis of Chloroform and Other Organic Molecules in Aqueous TiO_2 Suspensions. *Environ. Sci. Technol.* **25**, 494 (1991).
74. G. Mills and M. R. Hoffmann, Photocatalytic Degradation of Pentachlorophenol on TiO_2 Particles: Identification of Intermediates and Mechanism of Reaction. *Environ. Sci. Technol*. **27**, 1681 (1993).
75. R. W. Matthews, Purification of Water with Near UV-illuminated Suspensions of Titanium Dioxide. *Water Res.* **24**, 653 (1990).
76. D. W. Bahnemann, J. Mönig, and R. Chapman, Efficient Photocatalysis of the Irreversible One-electron and Two-electron Reduction of Halothane on Platinized Colloidal Titanium Dioxide in Aqueous Suspension. *J. Phys. Chem.* **91**, 3782 (1987).
77. K. Tanaka, T. Hisanaga, and K. Harada, Efficient Photocatalytic Degradation of Chloral Hydrate in Aqueous Semiconductor Suspensions. *J. Photochem. Photobiol. A: Chem.* **48**, 155 (1989).
78. R. Borello, C. Minero, E. Pramauro, E. Pelizzetti, N. Serpone, and H. Hidaka, Photocatalytic Degradation of DDT Mediated in Aqueous Semiconductor Slurries by Simulated Sunlight. *Environ. Toxicol. Chem.* **8**, 997 (1989).
79. E. Pelizzetti, M. Borgarello, C. Minero, E. Pramauro, E. Borgarello, and N. Serpone, Photocatalytic Degradation of Polychlorinated Dioxins and

Polychlorinated Biphenyls in Aqueous Suspensions of Semiconductors Irradiated with Simulated Solar Light. *Chemosphere* **17**, 499 (1988).
80. K. Kobayakawa, Y. Sato, S. Nakamura, and A. Fujishima, Photodecomposition of Kraft Lignin Catalyzed by Titanium Dioxide. *Bull. Chem. Soc. Jpn.* **62**, 3433 (1989).
81. R. A. Sierka and C. W. Bryant, *in* "Proceedings of the First International Conference on TiO_2 Photocatalytic Purification and Treatment of Water and Air. (D. F. Ollis and H. Al-Ekabi, eds.), Elsevier, Amsterdam, 1993.
82. G. K. Grätzel, M. Jirousek, and M. Grätzel, Decomposition of Organophosphorous Compounds on Photoactivated TiO_2 Surfaces. *J. Mol. Catal.* **60**, 375 (1990).
83. G. T. Brown and J. R. Darwent, Methyl Orange as a Probe for Photooxidation Reactions of Colloidal TiO_2. *J. Phys. Chem.* **88**, 4955 (1984).
84. J. R. Darwent and A. Lepre, Photooxidation of Methyl Viologen Sensitized by Zinc Oxide. Part I. Mechanism. *J. Chem. Soc., Faraday Trans. 2* **82**, 1457 (1986).
85. R. W. Matthews, Photocatalytic Oxidation and Adsorption of Methylene Blue on Thin Films of Near-Ultraviolet Illuminavd TiO_2. *J. Chem. Soc., Faraday Trans. I* **85**, 1291 (1989).
86. R. W. Matthews, A Comparison Between UV-Illuminated TiO_2 and Co(60) Gamma Rays for the Destruction of Organic Impurities in Waste. *Appl. Radiat. Isot.* **37**, 1247 (1986).
87. C. K. Grätzel, M. Girousek, and M. Grätzel, Accelerated Decomposition of Active Phosphates on TiO_2 Surfaces. *J. Mol. Catal.* **39**, 347 (1987).
88. K. Harada, T. Hisanaga, and K. Tanaka, Photocatalytic Degradation of Organophosphorous Compounds in Semiconductor Suspension. *New J. Chem.* **11**, 597 (1987).
89. H. Hidaka, K. Ihara, Y. Fujita, S. Yamada, E. Pelizzetti, and N. Serpone, Photodegradation of Surfactants. IV: Photodegradation of Non-Ionic Surfactants in Aqueous Titanium Dioxide Suspensions. *J. Photochem. Photobiol. A: Chem.* **42**, 375 (1988).
90. E. Pelizzetti, C. Minero, V. Maurino, A. Sclafani, H. Hidaka, and N. Serpone, Photocatalytic Degradation of Nonylphenol Ethoxylated Surfactants. *Environ. Sci. Technol.* **203**, 1380 (1989).
91. H. Hidaka, J. Zhao, Y. Satoh, K. Nohara, E. Pelizzetti, and N. Serpone, Photodegradation of Surfactants. Part XII: Photocatalyzed Mineralization of Phosphorus-Containing Surfactants at TiO_2/H_2O Interfaces. *J. Mol. Catal.* **88**, 239 (1994).
92. I. Rosenberg, J. R. Brock, and A. Heller, Collection Optics of TiO_2 Photocatalyst on Hollow Glass Microbeads Floating on Oil Slicks. *J. Mol. Catal.* **96**, 3423 (1992).

93. D. F. Ollis, Contaminant Degradation in Water. *Environ. Sci. Technol.* **19**, 480 (1985).
94. D. F. Ollis, E. Pelizzetti, and N. Serpone, Photocatalyzed Destruction of Water Contaminants. *Environ. Sci. Technol.* **25**, 1523 (1991).
95. M. A. Fox, Photocatalysis: Decontamination with Sunlight. *Chemtech* p. 680, December (1992).
96. Y. Zhang, J. C. Crittenden, and D. W. Hand, The Solar Photocatalytic Decontamination of Water. *Chem. Ind.* **18**, 714, (1994).
97. K. Rajeshwar, Photoelectrochemistry and the Environment. *J. Appl. Electrochem.* **25**, 1067 (1995).
98. D. Bahnemann, D. Bockelmann, and R. Goslich, Mechanistic Studies of Water Detoxification in Illuminated TiO_2 Suspensions. *Sol. Energy Mater.* **24**, 564 (1991).
99. W.-Y. Lin, C. Wei, and K. Rajeshwar, Photocatalytic Reduction and Immobilization of Hexavalent Chromium at Titanium Dioxide in Aqueous Basic Media. *J. Electrochem. Soc.* **140**, 2477 (1993).
100. N. Serpone, E. Borgarello, M. Barbeni, E. Pelizzetti, P. Pichat, J.-M. Herrmann, and M. A. Fox, Photochemical Reduction of Gold(III) on Semiconductor Dispersions of Titanium Dioxide in the Presence of Cyanide Ions: Disposal of Cyanide by Treatment with Hydrogen Peroxide. *J. Photochem.* **36**, 373 (1987).
101. S. N. Frank and A. J. Bard, Heterogeneous Photocatalytic Oxidation of Cyanide and Sulfite in Aqueous Solutions at Semiconductor Powders. *J. Phys. Chem.* **81**, 1481 (1977); Heterogeneous Photocatalytic Oxidation of Cyanide Ion in Aqueous Solutions at Titanium Dioxide Powder. *J. Am. Chem. Soc.* **99**, 303 (1977).
102. K. Kogo, H. Yoneyama, and H. Tamura, Photocatalytic Oxidation of Cyanide on Platinized TiO_2. *J. Phys. Chem.* **84**, 1705 (1980).
103. A. J. Bard, Semiconductor Particles and Arrays for Photoelectrochemical Utilization of Solar Energy. *Ber. Bunsenges. Phys. Chem.* **92**, 1187 (1988).
104. J. Peral and J. Doménech, Photocatalytic Cyanide Oxidation from Aqueous Copper Cyanide Solutions over Titania and Zinc Oxide. *J. Chem. Tech. Biotechnol.* **53**, 93 (1992).
105. T. L. Rose and C. Nanjundiah, Rate Enhancement of Photooxidation of CN^- with TiO_2 Particles. *J. Phys. Chem.* **89**, 3766 (1985).
106. D. Bhakta, S. S. Shukla, M. S. Chandrasekhariah and J. L. Margrave, A Novel Photocatalytic Method for Detoxification of Cyanide Wastes. *Environ. Sci. Technol.* **26**, 265 (1992).
107. N. Serpone, E. Borgarello, and M. Grätzel, Visible Light Induced Generation of Hydrogen from Hydrogen Sulfide in Mixed Semiconductor Dispersions: Improved Efficiency through Inter-particle Electron Transfer. *J. Chem. Soc., Chem. Commun.* 342 (1984).

108. E. Borgarello, K. Kalyanasundaran, M. Grätzel, and E. Pelizzetti, Visible-Light Induced Generation of Hydrogen from H_2S in CdS Dispersions, Hole Transfer Catalysis by RuO_2. *Helv. Chim. Acta* **65**, 243 (1982).
109. M. Matsumura, Y. Sato, and H. Tsubomura, Photocatalytic Hydrogen Production from Solutions of Sulfite Using Platinized Cadmium Sulfide Powder. *J. Phys. Chem.* **87**, 3807 (1983).
110. M. R. Prairie, L. R. Evans, B. M. Stange, and S. L. Martinez, An Investigation of TiO_2 Photocatalysis for the Treatment of Water Contaminated with Metals and Organic Chemicals. *Environ. Sci. Technol.* **27**, 1776 (1993).
111. J. Doménech and M. Andres, Elimination of Hg(II) Ions from Aqueous Solutions by Photocatalytic Reduction over ZnO Powder. *New J. Chem.* **11**, 443 (1987).
112. J. Muñoz and J. Doménech, TiO_2 Catalyzed Reduction of Cr(VI) in Aqueous Solutions Under Ultraviolet Illumination. *J. Appl. Electrochem.* **20**, 518 (1990).
113. J. Doménech and J. Muñoz, Photochemical Elimination of Cr(VI) from Neutral-Alkaline Solutions. *J. Chem. Technol. Biotechnol.* **47**, 101 (1990).
114. H. Yoneyama, Y. Yamashita, and H. Tamura, Heterogeneous Photocatalytic Reduction of Dichromate on n-Type Semiconductor Catalysts. *Nature* (London) **282**, 817 (1979); M. Miyake, H. Yoneyama, and H. Tamura, The Correlation Between Photoelectrochemical Cell Reactions and Photocatalytic Reactions on Illuminated Rutile. *Bull. Chem. Soc. Jpn.* **50**, 1492 (1977).
115. J. Doménech and J. Muñoz, Photocatalytical Reduction of Cr(VI) over ZnO Powder. *Electrochim. Acta* **32**, 1383 (1987).
116. Y. Xu and X. Chen, Photocatalytic Reduction of Dichromate over Semiconductor Catalysts. *Chem. Ind.* (London) p. 492 (1990).
117. A. Lozano, J. Garcia, J. Doménech, and J. Casado, Heterogeneous Photocatalytic Oxidation of Manganese(II) over Titania. *J. Photochem. Photobiol. A: Chem.* **69**, 237 (1992).
118. C. Wagner and W. Traud, Über Die Deutung Von Korrosionsvorgängen Durch Überlagerung Von Elektrochemischen Teilvorgängen Und Über Die Potentialbildung An Mischelektroden. *Z. Elektrochem.* **44**, 391 (1938).
119. R. F. Steigerwald, Electrochemistry of Corrosion. *Corrosion* (Houston) **24**, 1 (1968).
120. B. Kraeutler and A. J. Bard, Heterogeneous Photocatalytic Decomposition of Saturated Carboxylic Acids on Titanium Dioxide Powder. Decarboxylative Route to Alkanes. *J. Am. Chem. Soc.* **100**, 5985 (1978).
121. A. J. Bard, Photoelectrochemistry. *Science* **207**, 139 (1980).
122. J. M. Kesselman, A. Kumar, and N. S. Lewis, in "Proceedings of the First International Conference on TiO_2 Photocatalytic Purification and Treatment of Water and Air" (D. F. Ollis and H. Al-Ekabi, eds.), Elsevier, Amsterdam, 1993.

123. K. Rajeshwar and J. G. Ibanez, Electrochemical Aspects of Photocatalysis: Application to Detoxification and Disinfection Scenarios. *J. Chem. Educ.* **72**, 1044 (1995).
124. C. Wei, Z. Zainal, W.-Y. Lin, N. Williams, R. L. Smith, K. Rajeshwar, and A. Kruzic, Bactericidal Activity of TiO_2 Photocatalyst in Aqueous Media: Towards a Solar Assisted Water Disinfection System. *Environ. Sci. Technol.* **28**, 934 (1994).
125. M. Fujihira, Y. Satoh, and T. Osa, Heterogeneous Photocatalytic Reactions on Semiconductor Materials. III. Effect of pH and Copper(2+) Ions on the Photo-Fenton Type Reaction. *Bull. Chem. Soc. Jpn.* **55**, 666 (1982).
126. H. Reiche, W. W. Dunn and A. J. Bard, Heterogeneous Photocatalytic and Photosynthetic Deposition of Copper on TiO_2 and WO_3 Powders. *J. Phys. Chem.* **83**, 2248 (1979).
127. A. Sclafani, L. Palmisano, and M. Schiavello, Influence of the Preparation Methods of TiO_2 on the Photocatalytic Degradation of Phenol in Aqueous Solution. *J. Phys. Chem.* **94**, 829 (1990).
128. J. Augustynski, The Role of the Suface Intermediates in the Photoelectrochemical Behavior of Anatase and Rutile TiO_2. *Electrochim. Acta* **38**, 43 (1993).
129. R. W. Matthews, *in* "Photochemical Conversion and Storage of Solar Energy" (E. Pellizzetti and M. Schiavello, eds.), Kluwer Academic Publishers, Dordrecht, The Netherlands, 1991.
130. Y. Wang and N. Herron, Nanometer-Sized Semiconductor Clusters: Materials Synthesis, Quantum Size Effects, and Photophysical Properties. *J. Phys. Chem.* **95**, 525 (1991).
131. L. E. Brus, Nanometer-Sized Semiconductor Clusters: Materials Synthesis, Quantum Size Effects, and Photophysical Properties. *Annu. Rev. Mater. Sci.* **19**, 471 (1989).
132. A. Henglein, Small-Particle Research: Physicochemical Properties of Extremely Small Colloidal Metal and Semiconductor Particles. *Chem. Rev.* **89**, 1861 (1989).
133. H. Miyoshi, S. Nippa, H. Uchida, H. Mori, and H. Yoneyama, Photochemical Properties of TiO_2 Microcrystallites Prepared in Nafion. *Bull. Chem. Soc. Jpn.* **63**, 3380 (1990).
134. H. Yoneyama, Electrochemical Aspects of Light Induced Heterogeneous Reactions on Semiconductors. *Res. Chem. Intermed.* **15**, 101 (1991).
135. H. Inoue, T. Torimoto, T. Sakata, H. Mori, and H. Yoneyama, Effects of Size Quantization of Zinc Sulfide Microcrystallites on Photocatalytic Reduction of Carbon Dioxide. *Chem. Lett.* p. 1483 (1990).
136. H. Miyoshi, H. Mori and H. Yoneyama, Light-Induced Decomposition of Saturated Carboxylic Acids in Iron Oxide Incorporated Clay Suspended in Aqueous Solutions. *Langmuir* **7**, 503 (1991).

137. H. Yoneyama, N. Matsumoto, and H. Tamura, Photocatalytic Decomposition of Formic Acid on Platinized n-Type Silicon Powder in Aqueous Solution. *Bull. Chem. Soc. Jpn.* **59**, 3302 (1986).
138. N. Serpone, E. Borgarello, R. Harris, P. Cahill, M. Borgarello, and E. Pellizzetti, Photocatalysis over Titanium Dioxide Supported on Glass Substrate. *Sol. Energy Mater.* **14**, 121 (1986).
139. R. W. Matthews, Solar-Electric Water Purification Using Photocatalytic Oxidation with TiO_2 as a Stationary Phase. *Sol. Energy* **38**, 405 (1987).
140. R. W. Matthews, Photooxidation of Organic Impurities in Water Using Thin Films of Titanium Dioxide. *J. Phys. Chem.* **91**, 3328 (1987).
141. R. W. Matthews, M. Abdullah, and G. K.-C. Low, Photocatalytic Oxidation for Total Organic Carbon Analysis. *Anal. Chim. Acta* **233**, 171 (1990).
142. M. Anpo, N. Aikawa, Y. Kubokawa, M. Che, C. Louis, and E. Giamello, Photoluminescence and Photocatalytic Activity of Highly Dispersed Titanium Oxide Anchored onto Porous Vycor Glass. *J. Phys. Chem.* **89**, 5017 (1985).
143. M. Anpo, M. Sunamoto and M. Che, Preparation of Highly Dispersed Anchored Vanadium Oxides by Photochemical Vapor Deposition Method and Their Photocatalytic Activity for Isomerization of *Trans*-2-Butene. *J. Phys. Chem.* **93**, 1187 (1989).
144. C. S. Turchi and M. S. Mehos, in "Chemical Oxidation: Technology for the Nineties" (W.W. Eckenfelder, A.R. Bowers, and J.A. Roth, eds.), Vol. 2. Technomic Publishing, Lancaster and Basel, 1992.
145. R. E. Marinangeli and D. F. Ollis, Photoassisted Heterogeneous Catalysis with Optical Fibers. I. Isolated Single Fiber. *AICh.E. J.* **23**, 415 (1977).
146. R. E. Marinangeli and D. F. Ollis, Photoassisted Heterogeneous Catalysis with Optical Films. II. Nonisothermal Single Fiber and Fiber Bundle. *AICh.E. J.* **26**, 1000 (1980).
147. R. E. Marinangeli and D. F. Ollis, Photoassisted Heterogeneous Catalysis with Optical Fibers. III. Photoelectrodes. *AICh.E. J.* **28**, 945 (1982).
148. D. H. Kim and M. A. Anderson, Photoelectrocatalytic Degradation of Formic Acid Using a Porous TiO_2 Thin-Film Electrode. *Environ. Sci. Technol.* **28**, 479 (1994).
149. S. Hotchandani and P. V. Kamat, Photoelectrochemistry of Semiconductor ZnO Particulate Films. *J. Electrochem. Soc.* **139**, 1630 (1992).
150. K. Vinodgopal, S. Hotchandani, and P. V. Kamat, Electrochemically Assisted Photocatalysis. TiO_2 Particulate Film Electrodes for Photocatalytic Degradation for 4-Chlorophenol. *J. Phys. Chem.* **97**, 9040 (1993).
151. R. E. Hetrick, Powder Layer Photoelectrochemical Structure. *J. Appl. Phys.* **58**, 1397 (1985).
152. M. A. Anderson, M. J. Gieselman, and Q. Xu, Titania and Alumina Ceramic Membranes. *J. Membrane Sci.* **39**, 243 (1988).

153. R. W. Matthews, in "Proceedings of the First International Conference on TiO$_2$ Photocatalytic Purification and Treatment of Water and Air" (D. F. Ollis and H. Al-Ekabi, eds.), p. 121. Elsevier, Amsterdam, 1993.
154. P. Pichat, C. Guillard, C. Maillard, L. Amalric, and J.-C. D'Oliveira, in "Proceedings of the First International Conference on TiO$_2$ Photocatalytic Purification and Treatment of Water and Air" (D. F. Ollis and H. Al-Ekabi, eds.), Elsevier, Amsterdam, 1993.
155. G. Al-Sayyed, J.-C. D'Oliveira, and P. Pichat, Semiconductor-Sensitized Photodegradation of 4-Chlorophenol in Water. *J. Photochem. Photobiol. A: Chem.* **58**, 99 (1991).
156. M. Grätzel and R. F. Howe, Electron Paramagnetic Resonance Studies of Doped TiO$_2$ Colloids. *J. Phys. Chem.* **94**, 2566 (1990).
157. C.-M. Wang, A. Heller, and H. Gerischer, Palladium Catalysis of O$_2$ Reduction by Electrons Accumulated on TiO$_2$ Particles During Photoassisted Oxidation of Organic Compounds. *J. Am. Chem. Soc.* **114**, 5230 (1992).
158. N. Jaffrezic-Renault, P. Pichat, A. Foissy, and R. Mercier, Effect of Deposited Pt Particles on the Surface Charge of TiO$_2$ Aqueous Suspensions by Potentiometry, Electrophoresis and Labeled Ion Adsorption. *J. Am. Chem. Soc.* **90**, 2733 (1986).
159. P. Keller and A. Moradpour, Is There a Particle-Size Dependence for the Mediation by Colloidal Redox Catalysts of the Light-Induced Hydrogen Evolution from Water? *J. Am. Chem. Soc.* **102**, 7193 (1980).
160. J.-M. Lehn, J.-P. Sauvage, and R. Ziessel, Photochemical Hydrogen Production: Development of Efficient Heterogeneous Redox Catalysts. *Nouv. J. Chim.* **5**, 291 (1981).
161. H. Yoneyama, N. Nishimura, and H. Tamura, Photodeposition of Palladium and Platinum onto Titanium Dioxide Single Crystals. *J. Phys. Chem.* **85**, 268 (1981).
162. P.-A. Brugger, P. Cuendet, and M. Grätzel, Ultrafine and Specific Catalysts Affording Efficient Hydrogen Evolution from Water Under Visible Light Illumination. *J. Am. Chem. Soc.* **103**, 2923 (1981).
163. M. Koudelka, J. Sanchez, and J. Augustynski, Electrochemical and Surface Characteristics of the Photocatalytic Platinum Deposits on TiO$_2$. *J. Am. Chem. Soc.* **86**, 4277 (1982).
164. J. Disdier, J.-M. Herrmann, and P. Pichat, Platinum/Titanium Dioxide Catalysts. A Photoconductivity Study of Electron Transfer from the Ultraviolet-Illuminated Support to the Metal and of the Influence of Hydrogen. *J. Chem. Soc., Faraday Trans. I* **79**, 651 (1983).
165. J.-M. Herrmann, J. Disdier, and P. Pichat, Photoassisted Platinum Deposition on TiO$_2$ Powder Using Various Platinum Complexes. *J. Phys. Chem.* **90**, 6028 (1986).

166. D. N. Furlong, D. Wells, and W. H. F. Sasse, Colloidal Semiconductors in Systems for the Sacrificial Photolysis of Water. 1. Preparation of a Pt/ TiO_2 Catalyst by Heterocoagulation and its Physical Characterization. *J. Phys. Chem.* **89**, 626 (1985).
167. M. M. Kondo and W. F. Jardin, Photodegradation of Chloroform and Urea Using Ag-Loaded Titanium Dioxide as Catalyst. *Water Res.* **25**, 823 (1991).
168. A. Sclafani, M.-N. Mozzanega, and P. Pichat, Effect of Silver Deposits on the Photocatalytic Activity of Titanium Dioxide Samples for the Dehydrogenation or Oxidation of 2-Propanol. *J. Photochem. Photobiol. A: Chem.* **59**, 181 (1991).
169. S.-I. Nishimoto, B. Ohtani, H. Kajiwara, and T. Kagiya, Photoinduced Oxygen Formation and Silver Metal Deposition in Aqueous Solutions of Various Silver Salts by Suspended Titanium Dioxide Powder. *J. Chem. Soc., Faraday Trans. I* **79**, 2685 (1983).
170. J. Papp, H.-S. Shen, R. Kershaw, K. Dwight, and A. Wold, Titanium(IV) Oxide Photocatalysts with Palladium. *Chem. Mater.* **5**, 284 (1993).
171. N. Serpone, E. Borgarello, M. Barberi, and E. Pelizzetti, Effect of CdS Preparation on the Photo-Catalyzed Decomposition of Hydrogen Sulfide in Alkaline Aqueous Media. *Inorg. Chim. Acta* **90**, 191 (1984).
172. A. Wold, Photocatalytic Properties of TiO_2. *Chem. Mater.* **5**, 280 (1993).
173. K. Domen, S. Naito, T. Onishi, K. Tamura, and M. Soma, Study of the Photocatalytic Decomposition of Water Vapor Over a $NiO-SrTiO_3$ Catalyst. *J. Phys. Chem.* **86**, 3657 (1982).
174. D. C. Cronemeyer, Electrical and Optical Properties of Rutile Single Crystals. *Phys. Rev.* **87**, 876 (1952).
175. F. Möllers, H. J. Tolle, and R. Memming, On the Origin of the Photocatalytic Deposition of Noble Metals on TiO_2. *J. Electrochem. Soc.* **121**, 1160 (1974).
176. E. J. Johnson, *in* "Semiconductors and Semimetals" (R. K. Willardson and A. C. Bear, eds.), Vol. 3, Chapter. 6. Academic Press, NY, 1967.
177. D. M. Eagles, Jr., Polar Modes of Lattice Vibration and Polaron Coupling Constants in Rutile TiO_2. *J. Phys. Chem. Solids* **25**, 1243 (1964).
178. P. Salvador, Analysis of the Physical Properties of TiO_2-Be Electrodes in the Photoassisted Oxidation of Water. *Sol. Energy Mater.* **6**, 241 (1982).
179. H. Gerischer and A. Heller, Photocatalytic Oxidation of Organic Molecules at TiO_2 Particles by Sunlight in Aerated Water. *J. Electrochem. Soc.* **139**, 113 (1992).
180. H. Gerischer, Photoelectrochemical Catalysis of the Oxidation of Organic Molecules by Oxygen on Small Semiconductor Particles with TiO_2 as an Example. *Electrochim. Acta* **38**, 3 (1993).
181. C. Sandorf, S. Kelly, and D. M. Hwang, Clusters in Solution: Growth and Optical Properties of Layered Semiconductors with Hexagonal and Honeycombed Structures. *Chem. Phys.* **85**, 5337 (1986).

182. C. Kormann, D. W. Bahnemann, and M. R. Hoffmann, Preparation and Characterization of Quantum-Size Titanium Dioxide. *J. Phys. Chem.* **92**, 5196 (1988).
183. M. Anpo, T. Shima, S. Kodama, and Y. Kubokawa, Photocatalytic Hydrogenation of CH_3CCH with H_2O on Small-Particle TiO_2: Size Quantization Effects and Reaction Intermediates. *J. Phys. Chem.* **91**, 4305 (1987).
184. L. Kavan, T. Stoto, M. Grätzel, D. Fitzmaurice, and V. Shklover, Quantum Size Effects in Nanocrystalline Semiconducting TiO_2 Layers Prepared by Anodic Oxidative Hydrolysis of $TiCl_3$. *J. Phys. Chem.* **97**, 9493 (1993).
185. C.-Y. Liu and A. J. Bard, Effect of Excess Charge on Band Energetics (Optical Absorption Edge and Carrier Redox Potentials) in Small Semiconductor Particles. *J. Phys. Chem.* **93**, 3232 (1989), and references therein.
186. R. W. Matthews and S. R. McEvoy, A Comparison of 254 nm and 350 nm Excitation of TiO_2 in Simple Photocatalytic Reactors. *J. Photochem. Photobiol. A: Chem.* **66**, 355 (1992).
187. W. J. Albery and P. N. Bartlett, The Transport and Kinetics of Photogenerated Carriers in Colloidal Semiconductor Electrode Particles. *J. Electrochem. Soc.* **131**, 315 (1986).
188. R. G. Breckenridge and W. R. Hosler, Electrical Properties of Titanium Dioxide Semiconductors. *Phys. Rev.* **91**, 793 (1953). Y. Yahia, Dependence of the Electrical Conductivity and Thermoelectric Power of Pure and Aluminum-Doped Rutile on Equilibrium Oxygen Pressure and Temperature. *ibid.* **130**, 1711 (1963).
189. P. Salvador, Hole Diffusion Length in n-TiO_2 Single Crystals and Sintered Electrodes: Photoelectrochemical Determination and Comparative Analysis. *J. Appl. Phys.* **55**, 2977 (1984).
190. H. Gerischer, Neglected Problems in the pH Dependence of the Flatband Potential of Semiconducting Oxides and Semiconductors Covered with Oxide Layers. *Electrochim. Acta* **34**, 1005 (1989).
191. J. Moser and M. Grätzel, Light-Induced Electron Transfer in Colloidal Semiconductor Dispersions: Single *vs.* Dielectronic Reduction of Acceptors by Conduction-Band Electrons. *J. Am. Chem. Soc.* **105**, 6547 (1983).
192. M. Grätzel and A. J. Frank, Interfacial Electron-Transfer Reactions in Colloidal Semiconductor Dispersions. Kinetic Analysis. *J. Phys. Chem.* **86**, 2964 (1982).
193. F. Williams and A. J. Nozik, Irreversibilities in the Mechanism of Photoelectrolysis. *Nature (London)* **271**, 137 (1979).
194. K. Uosaki and H. Kita, Photopotential Behavior of Platinized TiO_2 Particles. *J. Electrochem. Soc.* **129**, 1752 (1982).
195. R. F. Howe and M. Grätzel, EPR Observation of Trapped Electrons in Colloidal TiO_2. *J. Phys. Chem.* **89**, 4495 (1985).

196. B. O'Regan, M. Grätzel, and D. Fitzmaurice, Optical Electrochemistry I: Steady-State Spectroscopy of Conduction-Band Electrons in a Metal Oxide Semiconductor Electrode. *Chem. Phys. Lett.* **183**, 89 (1991).
197. U. Köhle, J. Moser, and M. Grätzel, Dynamics of Interfacial Charge-Transfer Reactions in Semiconductor Dispersions. Reduction of Cobalticeniumdicarboxylate in Colloidal TiO_2. *Inorg. Chem.* **24**, 2253 (1985).
198. J. G. Highfield and M. Grätzel, Discovery of Reversible Photochromism in Titanium Dioxide Using Photoacoustic Spectroscopy. Implications for the Investigation of Light-Induced Charge-Separation and Surface Redox Processes in Titanium Dioxide. *J. Phys. Chem.* **92**, 464 (1988).
199. O. I. Micic, Y. Zhang, K. R. Cromack, A. D. Trifunac, and M. C. Thurnauer, Trapped Holes on TiO_2 Colloids Studied by Electron Paramagnetic Resonance. *J. Phys. Chem.* **97**, 7277 (1993).
200. B. O'Regan, M. Grätzel, and D. Fitzmaurice, Optical Electrochemistry. 2. Real-Time Spectroscopy of Conduction Band Electrons in a Metal Oxide Semiconductor Electrode. *J. Phys. Chem.* **95**, 10525 (1991).
201. W. W. Gärtner, Depletion-Layer Photoeffects in Semiconductors. *Phys. Rev.* **116**, 84 (1959). M. A. Butler, Photoelectrolysis and Physical Properties of the Semiconducting Electrode WO_3. *J. Appl. Phys.* **48**, 1914 (1977).
202. A. Hagfeldt, U. Bjorkstin, and S. E. Lindquist, Photoelectrochemical Studies of Colloidal TiO_2 Films. The Charge Separation Process Studied by Means of Action Spectra in the UV Region. *Sol. Energy Mater. Sol. Cells* **27**, 293 (1992).
203. S. Nakakayashi, A. Fujishima, and K. Honda, Single Charge Accumulation Dynamics on Photocatalytic TiO_2 Particles in Ethanol Slurries by Time-Domain Reflectometry. *J. Am. Chem. Soc.* **107**, 250 (1985).
204. A. Henglein and J. Lilie, Storage of Electrons in Aqueous Solution: The Rates of Chemical Charging and Discharging the Colloidal Silver Microelectrode. *J. Am. Chem. Soc.* **103**, 1059 (1981).
205. A. Takahashi, Y. Aikawa, Y. Toyoshima, and M. Sukigara, Photoelectrophoresis and Photoelectrohydrodynamic Instability of TiO_2 Particle Suspension Systems. *J. Phys. Chem.* **83**, 2854 (1979).
206. C. Boxall, The Electrophoresis of Semiconductor Particles. *Chem. Soc. Rev.* 137 (1994).
207. W. W. Dunn, Y. Aikawa, and A. J. Bard, Characterization of Particulate Titanium Dioxide Photocatalysts by Photoelectrophoretic and Electrochemical Measurements. *J. Am. Chem. Soc.* **103**, 3456 (1981).
208. W. W. Dunn, Y. Aikawa, and A. J. Bard, Semiconductor Electrodes. XXXV. Slurry Electrodes Based on Semiconductor Powder Suspensions. *J. Electrochem. Soc.* **128**, 222 (1981).
209. G. Rothenberger, J. Moser, M. Grätzel, N. Serpone, and D. K. Sharma, Charge Carrier Trapping and Recombination Dynamics in Small Semi-

conductor Particles. *J. Am. Chem. Soc.* **107**, 8054 (1985).
210. N. Serpone, D. K. Sharma, M. A. Jamieson, M. Grätzel, and J. J. Ramsden, Photophysical and Photochemical Primary Events in Semiconductor Particulate Systems. Colloidal CdS in Methylviologen. *Chem. Phys. Lett.* **115**, 473 (1985).
211. M. O'Neill, J. Marohn, and G. McLendon, Dynamics of Electron-Hole Pair Recombination in Semiconductor Clusters. *J. Phys. Chem.* **94**, 4356 (1990).
212. G. T. Brown, J. R. Darwent, and P. D. I. Fletcher, Interfacial Electron Transfer in TiO_2 Colloids. *J. Am. Chem. Soc.* **107**, 6446 (1985).
213. T. Watanabe and K. Honda, Measurement of the Extinction Coefficient of the Methyl Viologen Cation Radical and the Efficiency of Its Formation by Semiconductor Photocatalysis. *J. Am. Chem. Soc.* **86**, 2617 (1982).
214. J. R. Darwent, Photoreduction of Methyl Viologen in Micellar Solutions Sensitized by Zinc Phthalocyanine. *J. Chem. Soc., Chem. Commun.*, p. 805 (1980).
215. R. J. Crutchley and A. B. P. Lever, Ruthenium(II) Tri(bipyrazyl) Dication - A New Photocatalyst. *J. Am. Chem. Soc.* **102**, 7129 (1980).
216. D. Meisel, W. A. Mulac, and M. S. Matheson, Catalysis of Methyl Viologen Radical Reactions by Polymer-Stabilized Gold Sols. *J. Phys. Chem.* **85**, 179 (1981).
217. Y. Nosaka, H. Miyama, M. Terauchi, and T. Kobayashi, Photoinduced Electron Transfer from Colloidal Cadmium Sulfide to Methylviologen: A Picosecond Transient Absorption Study. *J. Phys. Chem.* **92**, 255 (1988).
218. P. V. Kamat, T. W. Ebbesen, N. M. Dimitrijevic, and A. J. Nozik, Primary Photochemical Events in CdS Semiconductor Colloids as Probed by Picosecond Laser Flash Photolysis. *Chem. Phys. Lett.* **157**, 384 (1989).
219. D. Duonghong, J. Ramsden, and M. Grätzel, Dynamics of Interfacial Electron-Transfer Processes in Colloidal Semiconductor Systems. *J. Am. Chem. Soc.* **104**, 2977 (1982).
220. M. Grätzel and J. Moser, Multielectron Storage and Hydrogen Generation with Colloidal Semiconductors. *Proc. Natl. Acad. Sci. U.S.A.* **80**, 3129 (1983).
221. D. Bahnemann, A. Henglein, J. Lilie, and L. Spanhel, Flash Photolysis Observation of the Absorption Spectra of Trapped Positive Holes and Electrons in Colloidal TiO_2. *J. Phys. Chem.* **88**, 709 (1984).
222. R. B. Draper and M. A. Fox, Titanium Dioxide Photooxidation of Thiocyanate $(SCN)_2^{\bullet-}$ Studied by Diffuse Reflectance Flash Photolysis. *J. Phys. Chem.* **94**, 4628 (1990).
223. D. J. Fitzmaurice, M. Eschle, H. Frei, and J. Moser, Time-Resolved Rise of I_2^- Upon Oxidation of Iodide at Aqueous TiO_2 Colloid. *J. Phys. Chem.* **97**, 3806 (1993).

224. R. Rossetti and L. E. Brus, Picosecond Resonance Raman Scattering Study of Methylviologen Reduction on the Surface of Photoexcited Colloidal CdS Crystallites. *J. Phys. Chem.* **90**, 558 (1986).
225. J. R. White and A. J. Bard, Electrochemical Investigation of Photocatalysis at CdS Suspensions in the Presence of Methylviologen. *J. Phys. Chem.* **89**, 1947 (1985).
226. M. D. Ward, J. R. White, and A. J. Bard, Electrochemical Investigation of the Energetics of Particulate Titanium Dioxide Photocatalysts. The Methyl Viologen-Acetate System. *J. Am. Chem. Soc.* **105**, 27 (1983).
227. M. D. Ward and A. J. Bard, Photocurrent Enhancement via Trapping of Photogenerated Electrons of TiO_2 Particles. *J. Phys. Chem.* **86**, 3599 (1982).
228. J. K. Leland and A. J. Bard, Photochemistry of Colloidal Semiconducting Iron Oxide Polymorphs. *J. Phys. Chem.* **91**, 5076 (1987). J. K. Leland and A. J. Bard, Electrochemical Investigation of the Electron Transfer Kinetics and Energetics of Illuminated Tungsten Oxide Colloids. *ibid.* **91**, 5083 (1987).
229. M. W. Peterson, J. A. Turner, and A. J. Nozik, Mechanistic Studies of the Photocatalytic Behavior of TiO_2 Particles in a Photoelectrochemical Slurry Cell and the Relevance to Photodetoxification Reactions. *J. Phys. Chem.* **95**, 221 (1991).
230. M. Anpo and Y. Kubokawa, Photoluminescence of Zinc Oxide Powder as a Probe of Electron-Hole Surface Processes. *J. Phys. Chem.* **88**, 5556 (1984).
231. A. Henglein, Photochemistry of Colloidal Cadmium Sulfide. 2. Effects of Adsorbed Methyl Viologen and of Colloidal Platinum. *J. Phys. Chem.* **86**, 2291 (1982).
232. R. Rossetti and L. Brus, Electron-Hole Recombination Emission as a Probe of Surface Chemistry in Aqueous CdS Colloids. *J. Phys. Chem.* **86**, 4470 (1982).
233. W. G. Becker and A. J. Bard, Photoluminescence and Photoinduced Oxygen Adsorption of Colloidal Zinc Sulfide Dispersions. *J. Phys. Chem.* **87**, 4888 (1983).
234. M. Anpo, M. Tomonari, and M. A. Fox, In Situ Photoluminescence of TiO_2 as a Probe of Photocatalytic Reactions. *J. Phys. Chem.* **93**, 7300 (1989).
235. J. M. Warman, M. P. de Haas, M. Grätzel, and P. P. Infetta, Microwave Probing of Electronic Processes in Small Particle Suspensions. *Nature (London)* **310**, 306 (1984).
236. H. Gerischer and A. Heller, The Role of Oxygen in Photooxidation of Organic Molecules on Semiconductor Particles. *J. Phys. Chem.* **95**, 5261 (1991).
237. L. Spanhel, H. Weller, and A. Henglein, Photochemistry of Semiconductor Colloids. 22. Electron Injection from Illuminated CdS into Attached TiO_2 and ZnO Particles. *J. Am. Chem. Soc.* **102**, 6632 (1987).

238. S. Hotchandani and P. V. Kamat, Charge-Transfer Processes in Coupled Semiconductor Systems. Photochemistry and Photoelectrochemistry of the Colloidal CdS-ZnO System. *J. Phys. Chem.* **96**, 6834 (1992).
239. K. R. Gopidas, M. Bohorquez, and P. V. Kamat, Photophysical and Photochemical Aspects of Coupled Semiconductors. Charge-Transfer Processes in Colloidal CdS-TiO$_2$ and CdS-AgI Systems. *J. Phys. Chem.* **94**, 6435 (1990).
240. L. Spanhel, A. Henglein, and H. Weller, Photochemistry of Colloidal Semiconductors. 24. Interparticle Electron Transfer in Cd$_3$P$_2$-TiO$_2$ and Cd$_3$P$_2$-ZnO Sandwich Structures. *Ber. Bunsenges. Phys. Chem.* **91**, 1359 (1987).
241. A. Henglein, M. Gutiérrez, H. Weller, A. Fojtik, and J. Jirkovsky, Photochemistry of Colloidal Semiconductors. 30. Reactions and Fluorescence of AgI and AgI-Ag$_2$S Colloids. *Ber. Bunsenges. Phys. Chem.* **93**, 593 (1989).
242. D. Liu and P. V. Kamat, Electrochemical Rectification in CdSe + TiO$_2$ Coupled Semiconductor Films. *J. Electroanal. Chem. Interfacial Electrochem.* **347**, 451 (1993).
243. H. Yoneyama, N. Nishimura, and H. Tamura, Photodeposition of Palladium and Platinum onto Titanium Dioxide Single Crystals. *J. Phys. Chem.* **85**, 268 (1981).
244. R. W. Matthews, Photooxidation of Organic Impurities in Water Using Thin Films of Titanium Oxide. *J. Phys. Chem.* **91**, 3328 (1987).
245. J. Cunningham and G. Al-Sayyed, Factors Influencing Efficiencies of TiO$_2$-Sensitized Photodegradations. Part I – Substituted Benzoic Acids: Discrepancies with Dark-Absorption Parameters. *J. Chem. Soc., Faraday Trans.* **86**, 3935 (1990).
246. J. Cunningham and P. Sedlak in "Proceedings of the First International Conference on TiO$_2$ Photocatalytic Purification and Treatment of Water and Air" (D. F. Ollis and H. Al-Ekabi, eds.), p. 67. Elsevier, Amsterdam, 1993.
247. C. S. Turchi and D. F. Ollis, Photocatalytic Degradation of Organic Water Contaminants: Mechanisms Involving Hydroxyl Radical Attack. *J. Catal.* **122**, 178 (1990).
248. R.W. Matthews, Kinetics of Photocatalytic Oxidation of Organic Solutes Over Titanium Dioxide. *J. Catal.* **111**, 264 (1988).
249. E. M. Shin, R. Senthurchelvan, J. Muñoz, S. Basak, K. Rajeshwar, G. Beneglas-Smith, and B. C. Howell, III, Heterogeneous Photocatalytic Degradation of Formaldehyde in UV-Irradiated TiO$_2$ Suspension in Water: Degradation Kinetics, Direct *vs.* TiO$_2$-Mediated Photolysis and Long-Term Photocatalyst Stability. *J. Electrochem. Soc.* **143**, 1562 (1996).
250. D. F. Ollis, C-Y. Hsiao, L. Budiman, and C.-L. Lee, Heterogeneous Photoassisted Catalysis: Conversions of Perchloroethylene, Dichloroethane, Chloroacetic Acids and Chlorobenzene. *J. Catal.* **88**, 89 (1984).

251. V. Augugliara, L. Palmisano, A. Sclafani, C. Minero, and E. Pelizzetti, Photocatalytic Degradation of Phenol in Aqueous Titanium Dioxide Dispersions. *Toxicol. Environ. Chem.* **16**, 89 (1988).
252. C. S. Turchi and D. F. Ollis, Mixed Reactant Photocatalysis: Intermediates and Mutual Rate Inhibition. *J. Catal.* **119**, 483 (1989).
253. M. Nagamo and Y. Suda, Molecularly Adsorbed H_2O on the Bare Surface of TiO_2 (Rutile). *Langmuir* **3**, 786 (1987).
254. R. W. Matthews, Hydroxylation Reactions Induced by Near-Ultraviolet Photolysis of Aqueous Titanium Dioxide Suspensions. *J. Chem. Soc., Faraday Trans. I* **80**, 457 (1984).
255. T. A. Egerton and C. J. King, The Influence of Light Intensity on Photoactivity in TiO_2 Pigmented Systems. *J. Oil Colour Chem. Assoc.* **62**, 386 (1979).
256. H. Al-Ekabi, N. Serpone, E. Pelizzetti, C. Minero, M. A. Fox, and R. B. Draper, Kinetic Studies in Heterogeneous Photocatalysis. 2. TiO_2-Mediated Degradation of 4-Chlorophenol Alone and in a Three-Component Mixture of 4-Chlorophenol, 2,4-Dichlorophenol, and 2,4,5-Trichlorophenol in Air-Equilibrated Aqueous Media. *Langmuir* **5**, 250 (1989).
257. J.-C. D'Olivera, C. Minero, E. Pelizzetti, and P. Pichat, Photodegradation of Dichlorophenols and Trichlorophenols in TiO_2 Aqueous Suspensions: Kinetic Effects of the Positions of the Cl Atoms and Identification of the Intermediates. *J. Photochem. Photobiol. A: Chem.* **72**, 261 (1993).
258. C. Maillard-Dupuy, C. Guillard, H. Courbon, and P. Pichat, Kinetics and Products of the TiO_2 Photocatalytic Degradation of Pyridine in Water. *Environ. Sci. Technol.* **28**, 2176 (1994).
259. T. Nguyen and D. F. Ollis, Complete Heterogeneously Photocatalyzed Transformation of 1,1- and 1,2-Dibromoethane to CO_2 and HBr. *J. Phys. Chem.* **88**, 3386 (1984).
260. H. P. Boehm, Acid and Basic Properties of Hydroxylated Metal Oxide Surfaces. *Disc. Faraday Soc.* **52**, 264 (1971).
261. T. Morimoto, M. Nagao, and T. Omari, Heat of Immersion of Titanium Dioxide in Water. I. The Effect of the Hydration Treatment of Titanium Dioxide. *Bull. Chem. Soc. Jpn.* **42**, 943 (1969).
262. T. Omari, J. Imai, M. Nagao, and T. Morimoto, Heat of Immersion of Titanium Dioxide in Water. II. The Effect of the Crystallinity of Anatase. *Bull. Chem. Soc. Jpn.* **42**, 2198 (1969).
263. G. Munuera and F. S. Stone, Adsorption of Water and Organic Vapours on Hydroxylated Rutile. *Discuss. Faraday Soc.* **52**, 205 (1971).
264. M. Egashira, S. Kawasumi, S. Kagawa, and T. Seiyama, Temperature-Programmed Desorption Study of Water Adsorbed on Metal Oxides. I. Anatase and Rutile. *Bull. Chem. Soc. Jpn.* **51**, 3144 (1978).

265. C. Daremieux-Morin, M.-A. Enriquez, J. Sanz, and J. Fraissard, Rigid Lattice Proton NMR Study of the Constitutive Water of Titanium Oxides (Rutile, Anatase, Amorphous Oxides). *J. Colloid Interface Sci.* **95**, 502 (1983).
266. D. J. C. Yates, Infrared Studies of the Surface Hydroxyl Groups in Titanium Dioxide, and of the Chemisorption of Carbon Monoxide and Carbon Dioxide. *J. Phys. Chem.* **65**, 746 (1961).
267. M. Primet, P. Pichat, and M-V. Mathieu, Infrared Study of the Surface of Titanium Dioxides. I. Hydroxyl Groups. *J. Phys. Chem.* **75**, 1216 (1971).
268. P. Jackson and G. D. Parfitt, Infra-red Study of the Surface Properties of Rutile. Water and Surface Hydroxyl Species. *Trans. Faraday Soc.* **67**, 2469 (1971).
269. P. Jones and J. A. Hockey, Infra-red Studies of Rutile Surfaces. Part 3. Adsorption of Water and Dehydroxylation of Rutile. *J. Chem. Soc., Faraday Trans. I* **68**, 907 (1972).
270. G. Munuera, V. Rives-Arnau, and A. Saucedo, Photo-adsorption and Photo-desorption of Oxygen on Highly Hydroxylated TiO_2 Surfaces. Part 1 - Role of Hydroxyl Groups in Photo-adsorption. *J. Chem. Soc., Faraday Trans. I* **75**, 736 (1979).
271. K. Tanaka and J. M. White, Characterization of Species Adsorbed on Oxidized and Reduced Anatase. *J. Phys. Chem.* **86**, 4708 (1982).
272. K. Morishige, F. Kanno, S. Ogawara, and S. Sasaka, Hydrated Surfaces of Particulate Titanium Dioxide Prepared by Pyrolysis of Alkoxide. *J. Phys. Chem.* **89**, 4404 (1985).
273. T. Bredow and K. Jug, SINDO1 Study of Photocatalytic Formation and Reaction of OH Radicals at Anatase Particles. *J. Phys. Chem.* **99**, 285 (1995).
274. J. Peral and D. F. Ollis, Heterogeneous Photocatalytic Oxidation of Gas-Phase Organics for Air Purification: Acetone, 1-Butanol, Butyraldehyde, Formaldehyde, and m-Xylene Oxidation. *J. Catal.* **136**, 554 (1992).
275. E. Pelizzetti and C. Minero, Mechanism of the Photo-oxidative Degradation of Organic Pollutants over TiO_2 Particles. *Electrochim. Acta* **38**, 47 (1993).
276. M. L. Garcia Gonzalez and P. Salvador, The Influence of Oxygen Vacancies on the Kinetics of Water Photoelectrolysis at (001) n-TiO_2 Rutile. *J. Electroanal. Chem. Interfacial Electrochem.* **325**, 369 (1992).
277. H. Courbon, M. Formenti, and P. Pichat, Study of Oxygen Isotopic Exchange over Ultraviolet Irradiated Anatase Samples and Comparison with the Photooxidation of Isobutane into Acetone. *J. Phys. Chem.* **81**, 550 (1977).
278. M. Anpo, Y. Kubokawa, T. Fujii, and S. Suzuki, Quantum Chemical and $^{18}O_2$ Tracer Studies of the Activation of Oxygen in Photocatalytic Oxidation Reactions. *J. Phys. Chem.* **88**, 2572 (1984).
279. J. Cunningham, J. P. J. Tobin, and P. Meriaudeau, Oxygen Migration and Isotope Exchange During Photoassisted Splitting of Water Vapour on TiO_2-Carbon Mixtures. *Surf. Sci.* **108**, L465 (1981).

280. C. D. Jaeger and A. J. Bard, Spin Trapping and Electron Spin Resonance Detection of Radical Intermediates in the Photodecomposition of Water at TiO_2 Particulate Systems. *J. Phys. Chem.* **83**, 3146 (1979).
281. J. R. Harbour and M. L. Hair, Radical Intermediates in the Photosynthetic Generation of H_2O_2 with Aqueous ZnO Dispersions. *J. Phys. Chem.* **83**, 652 (1979).
282. J. R. Harbour and M. L. Hair, Superoxide Generation in the Photolysis of Aqueous Cadmium Sulfide Dispersions. Detection by Spin Trapping. *J. Phys. Chem.* **81**, 1791 (1977).
283. J. R. Harbour, R. Wolkow, and M. L. Hair, Effect of Platinization on the Photoproperties of CdS Pigments in Dispersion. Determination of H_2 Evolution, O_2 Uptake, and Electron Spin Resonance Spectroscopy. *J. Phys. Chem.* **85**, 4026 (1981).
284. B. Kraeutler, C. D. Jaeger, and A. J. Bard, Direct Observation of Radical Intermediates in the Photo-Kolbe Reaction—Heterogeneous Photocatalytic Radical Formation by Electron Spin Resonance. *J. Am. Chem. Soc.* **100**, 4903 (1978).
285. R. Cai, K. Hashimoto, A. Fujishima, and Y. Kubota, Conversion of Photogenerated Superoxide Anion into Hydrogen Peroxide in TiO_2 Suspension System. *J. Electroanal. Chem. Interfacial Electrochem.* **326**, 345 (1992).
286. V. Rao, K. Rajeshwar, V. Pai Verneker, and J. DuBow, Photosynthetic Production of H_2 and H_2O_2 on Semiconducting Oxide Grains in Aqueous Solutions. *J. Phys. Chem.* **84**, 1987 (1980).
287. A. H. Boonstra and C. A. H. A. Mutsaers, Relation Between the Photoadsorption of Oxygen and the Number of Hydroxyl Groups on a Titanium Dioxide Surface. *J. Phys. Chem.* **79**, 1694 (1975).
288. A. H. Boonstra and C. A. H. A. Mutsaers, Adsorption of Hydrogen Peroxide on the Surface of Titanium Dioxide. *J. Phys. Chem.* **79**, 1940 (1975).
289. G. Munuera, A. R. Gonzalez-Felipe, J. Soria, and J. Sanz, Photo-adsorption and Photo-desorption of Oxygen on Highly Hydroxylated TiO_2 Surfaces. 3. Role of H_2O_2 in Photo-desorption of O_2. *J. Chem. Soc., Faraday Trans. I* **76**, 1535 (1980).
290. K. Kobayakawa, Y. Nakazawa, M. Ikeda, Y. Sato, and A. Fujishima, Influence of the Density of Surface Hydroxyl Groups on TiO_2 Photocatalytic Activities. *Ber. Bunsenges. Phys. Chem.* **94**, 1439 (1990).
291. Y. Oosawa and M. Grätzel, Enhancement of Photocatalytic Oxygen Evolution in Aqueous TiO_2 Suspensions by Removal of Surface-OH Groups. *J. Chem. Soc., Chem. Commun.* p. 1629 (1984).
292. Y. Oosawa and M. Grätzel, Effect of Surface Hydroxyl Density on Photocatalytic Oxygen Generation in Aqueous TiO_2 Suspensions. *J. Chem. Soc. Faraday Trans. I* **84**, 197 (1988).

293. K. Patel, S. Yamagata, A. Fujishima, B. H. Loo, and T. Kato, Photo-sinking Phenomenon: Photodecomposition Rate of Silane Bonded on TiO_2 Powders. *Ber. Bunsenges. Phys. Chem.* **95**, 176 (1991).
294. H. Yoneyama, Y. Takao, H. Tamura, and A. J. Bard, Factors Influencing Product Distribution in Photocatalytic Decomposition of Aqueous Acetic Acid on Platinized TiO_2. *J. Phys. Chem.* **87**, 1417 (1983).
295. K. D. Asmus, D. Bahnemann, K. Krischer, M. Lal, and J. Mönig, One-Electron Induced Degradation of Halogenated Methanes and Ethanes in Oxygenated and Anoxic Aqueous Solutions. *Life Chem. Rep.* **3** 1 (1985).
296. W. H. Glaze, J. F. Kenneke, and J. L. Ferry, Chlorinated Byproducts from the TiO_2-Mediated Photodegradation of Trichloroethylene and Tetrachloroethylene in Water. *Environ. Sci. Technol.* **27**, 177 (1993).
297. J.-C. D'Olivera, G. Al-Sayyed and P. Pichat, Photodegradation of 2- and 3-Chlorophenol in TiO_2 Aqueous Suspensions. *Environ. Sci. Technol.* **24**, 990 (1990).
298. C. Maillard, C. Guillard, P. Pichat, and M. A. Fox, Photodegradation of Benzamide in Titanium Dioxide Aqueous Suspensions. *New J. Chem.* **16**, 821 (1992).
299. C. Maillard, C. Guillard, and P. Pichat, Comparative Effects of the TiO_2-UV, H_2O_2-UV, H_2O_2-Fe^{2+} Systems on the Disappearance Rate of Benzamide and 4-Hydroxybenzamide in Water. *Chemosphere* **24**, 1085 (1992).
300. E. Lipczynska-Kochany, in "Chemical Oxidation. Technologies for the Nineties" (W. W. Eckenfelder, A. R. Bowers and J. A. Roth, eds.), Vol. 3, pp. 12-27. Technomic Publishing., Lancaster and Basel, 1993.
301. M. Abdullah, G. K.-C. Low, and R. W. Matthews, Effects of Common Inorganic Anions on Rates of Photocatalytic Oxidation of Organic Carbon over Illuminated Titanium Dioxide. *J. Phys. Chem.* **94**, 6820 (1990).
302. G. K.-C. Low, S. R. McEvoy, and R. W. Matthews, Formation of Nitrate and Ammonium Ions in Titanium Dioxide Mediated Photocatalytic Degradation of Organic Compounds Containing Nitrogen Atoms. *Environ. Sci. Technol.* **25**, 460 (1991).
303. D. W. Bahnemann, D. Bockelmann, R. Goslich, M. Hilgendorff, and D. Weichgrebe, in "Proceedings of the First International Conference on TiO_2 Photocatalytic Purification and Treatment of Water and Air" (D. F. Ollis and H. Al-Ekabi, eds.), Elsevier, Amsterdam, 1993.
304. M. Fujihira, Y. Satoh, and T. Osa, Heterogeneous Photocatalytic Reactions on Semiconductor Materials. Part 1. Heterogeneous Photocatalytic Oxidation of Aromatic Compounds on Semiconductor Materials: The Photo-Fenton Type Reaction. *Chem. Lett.* p. 1053 (1981).
305. S-I. Nichimoto, B. Ohtani, H. Kajiwara, and T. Kagiya, Photoinduced Oxygen Formation and Silver-metal Deposition in Aqueous Solutions of Vari-

ous Silver Salts by Suspended Titanium Dioxide Powder. *J. Chem. Soc., Faraday Trans. I* **79**, 2685 (1983).
306. K. Hashimoto, T. Kawai, and T. Sakata, Photocatalytic Reaction of Hydrocarbons and Fossil Fuels with Water. Hydrogen Production and Oxidation. *J. Phys. Chem.* **88**, 4083 (1984).
307. M. Fujihira, Y. Satoh, and T. Osa, Heterogeneous Photocatalytic Oxidation of Aromatic Compounds on Titanium Dioxide. *Nature (London)* **293**, 206 (1981).
308. K. Tanaka, T. Hisanaga, and K. Harada, Photocatalytic Degradation of Organohalide Compounds in Semiconductor Suspension with Added Hydrogen Peroxide. *New J. Chem.* **13**, 5 (1989).
309. R. W. Matthews and S. R. McEvoy, A Comparison of 254 nm and 350 nm Excitation of Titania in Simple Photocatalytic Reactors. *J. Photochem. Photobiol. A: Chem.* **66**, 355 (1992).
310. J. Peral, J. Casado, and J. Doménech, Competitive Processes in Photocatalysis. Phenol-Sulphide and Phenol-Cyanide Competitive Photooxidation over ZnO. *Electrochim. Acta* **34**, 1335 (1989).
311. J-M. Herrmann, J. Disdier, and P. Pichat, Photocatalytic Deposition of Silver on Powder Titania: Consequences for the Recovery of Silver. *J. Catal.* **113**, 72 (1988).
312. C. S. Turchi, E. J. Wolfrum, and M. Nimlos, 5th Annu. Symp. Emerging Technol. Hazard. Waste Manage. American Chemical Society, Atlanta, GA (1993).
313. M. Halmann, *in* "Energy Resources Through Photoelectrochemistry and Catalysis" (M. Gratzel, ed.), Chapter. 15. Academic Press, NY, 1983.
314. I. Taniguchi, *in* "Mod. Aspects Electrochem," **20**, 327 (1989).
315. N. S. Lewis and G. A. Shreve, *in* "Electrochemical and Electrocatalytic Reactions of Carbon Dioxide" (B. P. Sullivan, K. Krist and H. E. Guard, eds.), Chapter. 8, p. 263. Elsevier, Amsterdam, 1993.
316. M. Tomkiewicz, H. Yoneyama, R. Haynes, and Y. Hori, eds., "Environmental Aspects of Electrochemistry and Photoelectrochemistry," Vol. 93-18. The Electrochemical Society, Pennington, NJ, 1993.
317. M. Halmann, Photoelectrochemical Reduction of Aqueous Carbon Dioxide on p-Type Gallium Phosphide in Liquid Junction Solar Cells. *Nature (London)* **275**, 115 (1978).
318. J. C. Hemminger, R. Carr, and G. A. Somorjai, The Photoassisted Reaction of Gaseous Water and Carbon Dioxide Adsorbed on the $SrTiO_3$(III) Crystal Face to Form Methane. *Chem. Phys. Lett.* **57**, 100 (1978).
319. T. Inoue, A. Fujishima, S. Konishi, and K. Honda, Photoelectrocatalytic Reduction of Carbon Dioxide in Aqueous Suspensions of Semiconductor Powders. *Nature (London)* **277**, 637 (1979).

320. R. Aurian-Blajeni, M. Halmann, and J. Manassen, Photoreduction of Carbon Dioxide and Water into Formaldehyde and Methanol on Semiconductor Materials. *Sol. Energy* **25**, 165 (1980).
321. H. Inoue, T. Torimoto, T. Sakata, H. Mori, and H. Yoneyama, Effects of Size Quantization of Zinc Sulfide Microcrystallites on Photocatalytic Reduction of Carbon Dioxide. *Chem. Lett.* p. 1483 (1990).
322. Y. Taniguchi, H. Yoneyama, and H. Tamura, Photoelectrochemical Reduction of Carbon Dioxide at p-Type Gallium Phosphide Electrodes in the Presence of Crown Ethers. *Bull. Chem. Soc. Jpn.* **55**, 2034 (1982).
323. H. Yoneyama, K. Wakamoto, N. Hatanaka, and H. Tamura, Photoelectrochemical Reduction of Carbon Monoxide on Iron(II) Tetraphenylporphyrins Coated p-Type GaP Electrodes. *Chem. Lett.* p. 539 (1985).
324. H. Inoue, M. Yamashita, and H. Yoneyama, Photocatalytic Conversion of Lactic Acid to Malic Acid Through Pyruvic Acid in the Presence of Malic Enzyme and Semiconductor Photocatalysts. *J. Chem. Soc., Faraday Trans. I* **88**, 2215 (1992).
325. G. N. Schrauzer and T. D. Guth, Photolysis of Water and Photoreduction of Nitrogen on Titanium Dioxide. *J. Am. Chem. Soc.* **99**, 7189 (1977).
326. H. Yoneyama, H. Shiota, and H. Tamura, Heterogeneous Reactions of Nitrogen Monoxide on Titanium Dioxide Photocatalysts in Solutions. *Bull. Chem. Soc. Jpn.* **54**, 1308 (1981).
327. C. Turchi, M. Mehos, and J. Pacheco *in* "Proceedings of the First International Conference on TiO_2 Photocatalytic Purification and Treatment of Water and Air" (D. F. Ollis and H. Al-Ekabi, eds.), Elsevier, Amsterdam, 1993.
328. D. F. Ollis and C. Turchi, Heterogeneous Photocatalysts for Water Purification: Contaminant Mineralization Kinetics and Elementary Reactor Analysis. *Environ. Progr.* **9**, 229 (1990).
329. H. C. Yatmaz, C. R. Howarth, and C. Wallis, *in* "Proceedings of the First International Conference on TiO_2 Photocatalytic Purification and Treatment of Water and Air" (D. F. Ollis and H. Al-Ekabi, eds.), Elsevier, Amsterdam, 1993.
330. C. Turchi and D. F. Ollis, Photocatalytic Reactor Design: An Example of Mass-Transfer Limitations with an Immobilized Catalyst. *J. Phys. Chem.* **92**, 6852 (1988).
331. R. W. Matthews, Response to Comment in Turchi and Ollis. *J. Phys. Chem.* **92**, 6853 (1988).
332. P. Wyness, J. F. Klausner, D. Y. Goswami, and K. S. Schanze, Performance of Nonconcentrating Solar Photocatalytic Oxidation Reactors; Part I: Flat-Plate Configuration. *J. Sol. Energy Eng.* **116**, 2 (1994).

CHAPTER SEVEN

7.1. INTRODUCTION

Disinfection of water can be traced back to ca. 2000 BC to ancient Sanskrit writings that prescribed that water should be exposed to sunlight and filtered through charcoal and that "foul water" be treated by boiling and "by dipping seven times into it a piece of hot copper and then filtering it."[1] Other very early references to boiling water and storage in silver flagons and other containers (Figure 7.1)[2] exist as are ancient efforts of water disinfection based on the use of copper, silver and electrolysis. The first U.S. patent on chlorination of water dates back to May 22, 1898, and was awarded to Albert R. Lieds. The low cost and high potency of chlorine as a water disinfectant promoted its usage since the mid-18th century. However, the practice of continuous addition was not initiated until the early 1900s and it is still the main water disinfectant used throughout the world.

A distinction must be made here between disinfection and sterilization procedures. *Disinfection* relates to the killing of disease-related organisms, whereas *sterilization* involves the killing of all organisms present.[3,4]

The primary purpose of water disinfection, of course, is the prevention of waterborne disease. Table 7.1 contains a summary of disease-causing microorganisms in water; a more extensive compilation is available.[1] Of particular concern is that in a sizeable fraction of the disease outbreaks recorded

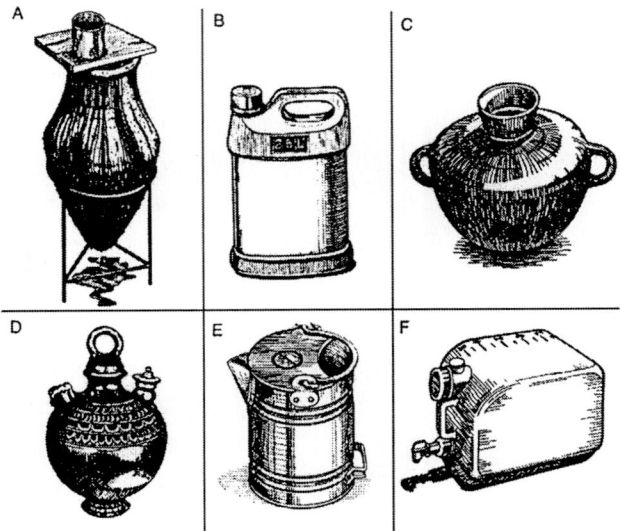

FIGURE 7.1. Traditional water storage vessels and water storage vessels that have been modified to reduce contamination during storage. Storage vessel A; traditional Egyptian zir; B; plastic container used to sell vegetable oil in Zambia; C; traditional cantero from El Salvador; D; sorai used in an intervention trial in India; E; tin bucket used in an intervention trial in Malawi; F; plastic container meeting the Centers for Disease Control and Prevention/Pan American Health Organizafion design criteria and used in an intervention trial in Bolivia. (Reproduced with permission from Mintz et al.[2])

in the United States and elsewhere, the etiologic agent could not be identified. However, in many of these cases, enteric viruses are believed to be the causative agents.

The promulgation of a number of new regulations for the control of microbiological and chemical pollutants in drinking water has prompted the search for suitable, cost-effective alternative methods for primary disinfection. Of particular concern are the disinfection by-products (DBPs) of chlorination and that groundwaters high in natural organic matter (NOM) may be incompatible with the more traditional chemical disinfectants. Even the alternatives currently considered such as chloramines may be inappropriate because they are weak virucides and would be unlikely to meet primary disinfection requirements.

We shall discuss alternative technologies for water disinfection in this chapter including those based on electrolysis, γ- and UV irradiation, O_3–UV, O_3–H_2O_2–UV, and photoelectrochemical (PEC) methods.

Table 7.1. Waterborne Diseases Caused by Bacteria, Viruses, and Other Microorganisms

Agent	Disease/Symptoms
Bacteria	
Salmonella typhosa	Typhoid fever/headache, nausea, diarrhea, increasing fever
S. ponatyphii	Paratyphoid fever/symptoms same as above
S. enteritis	Gastroenteritis/diarrhea, nausea, and dehydration
Shigella flexneri	Bacillary dysentry/diarrhea, fever, tenesmus, and stool with mucus and blood
Sh. dysenteriae	Same as above
Sh. sonuei	Same as above
Enterovirus	
Poliovirus (three subtypes)	Muscular paralysis, aseptic meningitis, fever
Coxsackie virus (>24 subtypes)	Respiratory disease, muscular paralysis, hepatitis, infantile diarrhea, and the like
Reovirus	
Six subtypes	Symptoms not well known
Protozoan cysts	
Giardia muris	Giardiasis/diarrhea, vomiting, and the like
Giardia lamblia	

7.2. WATER DISINFECTION: BACKGROUND AND PRINCIPLES

7.2.1. General Considerations

The methods for water disinfection may be classified as follows.[3]

- Chemical action: A variety of chemical agents can be used to inactivate microorganisms. These include halogens and derivatives (Cl_2, Br_2, I_2, $HOCl$, OCl^-, ClO_2, $HOBr$, HOI, polyiodide anion exchange resins, etc.), oxygenated and highly oxidizing compounds (ozone, hydrogen peroxide, phenols, alcohols, persulfate and percarbonate, peracetic acid, potassium perman-

ganate, etc.), metal ions (Ag^+, Cu^{2+}, etc.), dyes, quaternary ammonium compounds, strong acids and bases, and enzymes.
- Physical action: Electromagnetic radiation (ultrasonic waves, heat, visible light, UV light, gamma radiation, X-rays), particle radiation (electron beam), and electrical current.

The mechanisms for microbial inactivation include the following:
- Laceration of the cell wall
- Modification of cell permeability
- Modification of the nature of the protoplasm
- Alteration of nucleic acids
- Disruption of protein synthesis
- Induction of abnormal redox processes
- Inhibition of enzyme activity

A variety of factors influence the disinfection efficiency, including the contact time, chemical nature, and concentration of the disinfecting agent as well as the initial mixing mode and point of injection, nature and intensity of the physical agents, temperature, type, concentration and age of the microorganisms, and the nature of the liquid carrier. We shall explore a few of these in the sections that follow.

7.2.2. Chemical Disinfection

Common attributes of the important chemical disinfectants (Cl_2, OCl^-, ClO_2, and O_3) are the following:[3]
- Highly potent microorganism inactivation and relatively high toxicity to humans and animals.
- Active interaction (normally oxidation or addition) with organic matter and with inorganic reducing agents.
- Sufficient solubility in aqueous media (except the dihalogens due to their nonpolar nature).
- Penetration capability through surfaces and cell membranes.
- Moderate to good deodorizing ability.

The relative stabilities of these chemicals follow the order $Cl_2 > OCl^- > ClO_2$, O_3, and their relative costs are $O_3 > ClO_2$, $OCl^- > Cl_2$. Undesirable characteristics include DBP production, corrosivity to metallic materials, membrane attack, and discoloration of dyes and tints.[5] In fact, it has been stated that it would be ideal to *separate* the oxidation and disinfection functions in the water treatment system.[6] In this regard, it is worth noting that another major realm of use of these chemicals is in the pulp and paper industry.

7.2. Water Disinfection: Background and Principles

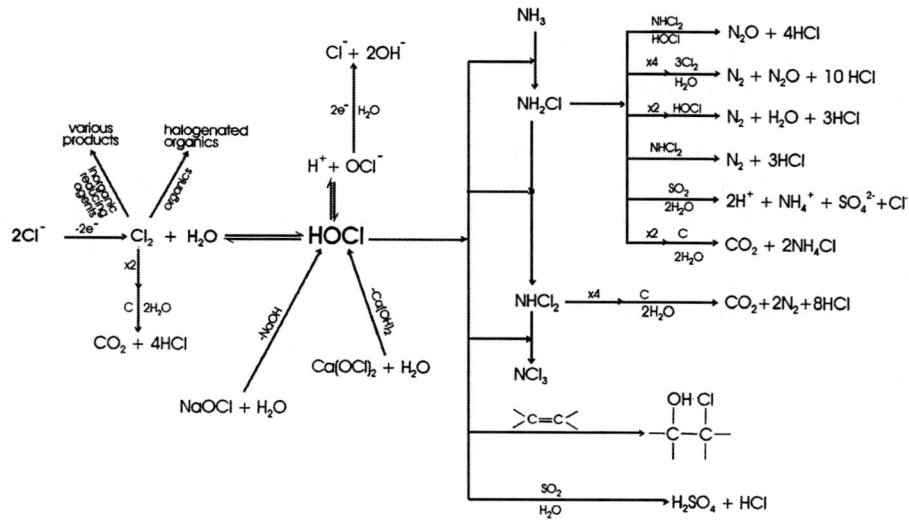

FIGURE 7.2. Production and reactions of Cl_2-based water disinfectants other than ClO_2.

We next review individual classes of chemicals for water disinfection.

<u>Chlorine, Hypochlorous Acid, and Hypochlorite.</u> As mentioned earlier, most water treatment plants in the world use some form of chlorine as a disinfectant. For example, ~60% of the treatment facilities in the Atlantic region in Canada use Cl_2 gas, $Ca(OCl)_2$, or $NaOCl$.[7] By the same token, in the United States, more than 98% of the plants in a major survey reported usage of chlorine-based disinfectants.[8] Other halogen-based disinfectants are also used but in much smaller quantities.

Disinfection by chlorination is simple, inexpensive, and effective. In addition, chlorine-based disinfectants can be solid, liquid, or gas. Their applications include the treatment of potable, cooling, bathing, brackish, and waste water. Biofouling in marine environments (e.g., algae growth and barnacle formation) is avoided by continuously treating the problem surface with chlorine or hypochlorite generated *in situ*.[9]

The main drawbacks of chlorine are the production of DBPs (see later), relatively high electric power requirement, its short-lasting residual action and its ability to react with a myriad of organic and inorganic pollutants, which reduces its available concentration for disinfection.

Dissolved chlorine is in equilibrium with hypochlorous acid ($K^{25°C} = 4 \times 10^{-4}$), another powerful disinfectant (Figure 7.2). Likewise, HOCl is in equilibrium with hypochlorite ions ($K^{25°C} = 2.9 \times 10^{-8}$).[3,4] These equilibrium constants mean that, at pH ≥ 4, chlorine is present mainly in the form of HOCl and, at pH ≥ 6, it is converted to the OCl⁻ ion. Aqueous hypochlorite solutions are much safer to handle than chlorine gas.[10] Hypochloric acid and hypochlorite ions are also produced by the dissolution of sodium or calcium hypochlorite. According to the equilibrium constants just given and the equations in Figure 7.2, the dissolution of Cl_2 in water lowers the pH, whereas dissolution of sodium or calcium hypochlorite increases its value.[4] HOCl is a much better disinfectant than OCl⁻ ions or monochloramine. For example, for 99% *E. coli* inactivation, HOCl is better than OCl⁻ by a factor of ~70 in terms of the required time and almost by a factor of 300 better than monochloramine.[4]

Ammonia when present or added (a widespread practice in water treatment facilities) reacts with hypochlorous acid to form the chloramines, NH_2Cl, $NHCl_2$, or NCl_3. The first two are known to have a disinfecting action that lasts longer than that of chlorine (and thus are called *combined chlorine residuals*),[6] which explains the ammonia addition practice.[3,4,8,11] In addition, this can reduce the formation of trihalomethanes (THMs).

At the same time, long chlorine residence times are to be avoided since Cl_2 has more opportunity to form DBPs. The formation of DBPs may be prevented by adding SO_2 (which reacts with residual HOCl and chloramines) or activated carbon (that reacts with residual Cl_2 as well as with chloramines and also removes DBP precursors)[3,4] as shown in Figure 7.2. Residual chlorine can also be removed with the aid of a bipolar membrane.[12]

It is worthy of note that, introduction of desalinated water into a distribution network in coastal areas, necessitates some form of mineralization or disinfection. The first is achieved by the addition of CaO and CO_2 (or $CaCO_3$ and CO_2); the second objective can be fulfilled by the introduction of electrolytically generated Cl_2 (see later).

<u>Chlorine Dioxide.</u> Chlorine dioxide is a strongly oxidizing gas. It is unstable in the presence of light and explosive in concentrations higher than 10% v/v under certain circumstances.[13] For this reason, it must be prepared on site before use. Unlike chlorine, it does not undergo hydrolysis in water.[10] In fact, it is a better virus inactivator than Cl_2.[14] It is also used for taste and odor reduction, to control algae growth and to remove iron and manganese.[6]

The main reaction for its production as a disinfectant is the chlorination of a chlorite solution. Chlorine dioxide may react violently with organic matter or reducing agents. It does not react with ammonia or produce halogenated organics, which is a major advantage with the use of ClO_2 as compared to chlorine.[4] At high pH it hydrolyzes to chlorite and chlorate, its main DBPs.[6] Chlorite ions may further react under acidic conditions to yield chloride ions. Dissolved SO_2 reacts with residual ClO_2 and removes it from solution as HCl.[3]

7.2. Water Disinfection: Background and Principles

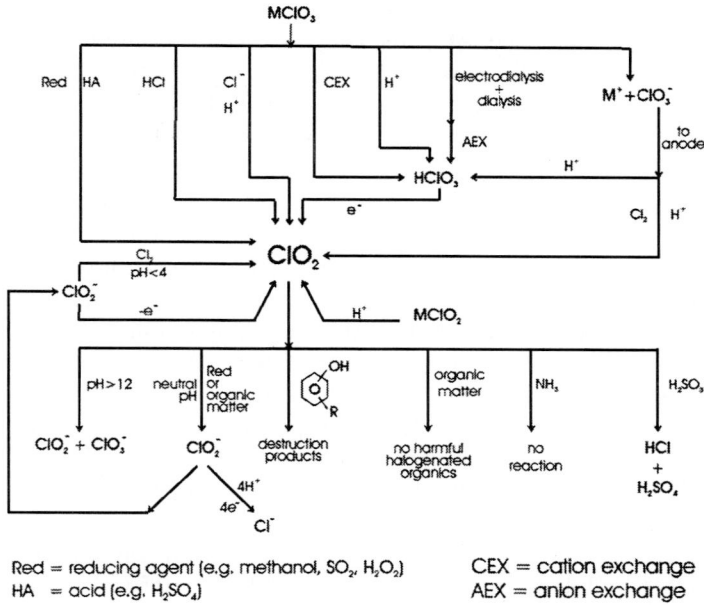

FIGURE 7.3. Production and reactions of ClO_2. The equations are representative and are not balanced.

These reactions are summarized in Figure 7.3. An alternative method for the production of ClO_2 is via the oxidation of sodium chlorite.[10]

A major drawback is that chlorine dioxide cannot achieve the 4-log (99.99%) reduction of *Giardia* required under the U.S. Surface Water Treatment Rule,[8] and therefore the addition of chlorine is required. In fact, the combination of ClO_2 and Cl_2 has been shown to improve the inactivation of pathogenic microorganisms including total coliforms, fecal coliforms, fecal streptococci, and *E*-coliphages. In addition, this combination produces a relatively stable high residual concentration of both disinfectants.[15] This practice is gaining acceptance in the United States.[8]

<u>Bromine-Based Disinfectants.</u> Bromine chloride is a promising alternative to chlorine-based disinfection. Its disinfection-related chemistry is very similar to that of Cl_2. It adsorbs on the microorganism and affects normal enzyme activity, hydrolyzes to produce HOBr and HCl, and reacts with ammonia to produce the three bromamines.[3]

Advantages with respect to chlorine disinfection include the following[3]: smaller contact times are required for similar disinfection results, bromamines are more effective than chloramines, brominated organic DBPs are more susceptible to hydrolysis and photolysis, and in general, the environmental effects are less dangerous than those related to chlorine disinfection. The main drawback seems to be the lack of an extensive research and database to support more widespread use.

<u>Iodine-Based Disinfectants</u>. Aqueous iodine undergoes hydrolysis to hypoiodous acid, HOI, which is unstable at pH $\geq$ 8 and decomposes into iodate, IO_3^- and iodide, I^-. Iodide can form the triiodide complex, I_3^- with I_2. The first two species (I_2 and HOI) are good disinfectants, but the last three species (IO_3^-, I^- and I_3^-) do not show significant germicidal action.[16] The main differences between iodine and chlorine as disinfectants include the lower pH sensitivity of I_2 as well as slower reaction rates with organic matter. In addition, I_2 does not react with ammonia. These properties entail that iodine does not considerably lose its disinfecting power due to side reactions.

It has been suggested that, other factors being equal, the standard reduction potential can be used as an explanation for the relative disinfection effectiveness between two agents in the presence of oxidizable material. For example, HOI ($E° = 1.482$ V) is less efficient than I_2 ($E° = 0.535$ V). This is due to the enhanced thermodynamic driving force for reacting with any oxidizable material present, which leaves a smaller amount of disinfectant for microorganism inactivation.[16]

Anion exchange resins based on quaternary ammonium strong bases in the polyiodide form (triiodide, pentaiodide) afford agents for broad-spectrum microorganism inactivation on demand. The mechanisms proposed include iodine release and iodine hydrolysis to hypoiodous acid.[17,18]

<u>Ozone</u>. The use of ozone for drinking water treatment is not a new concept; for example, the city of Nice has used ozone since 1907.[19] Ozone is a popular disinfectant in the water treatment industry in Europe.

Ozone is used for microorganism inactivation both in water sterilization and disinfection procedures.[20] In adition to its extremely high oxidizing power ($E°$ for $O_3–H_2O = 2.07$ V) (Section 3.3.3, Chapter 3), its self-decomposition in aqueous solutions involves highly oxidizing radical intermediates that can interact with microorganisms in a lethal way. UV light as well as hydrogen peroxide greatly enhance this decomposition as we have already seen in Chapter 6.

Compared to chlorine, ozone leaves fewer residues and highly reduced levels of toxicity. For this reason, it is preferred by swimmers as a disinfectant.[21] The introduction of ozone into a water treatment plant decreased the average THM levels by a factor of 3, and the monthly volume of water needed to be flushed from hydrants to avoid retention time was reduced by a factor of 10.[22] Ozone can also be used for taste and odor removal.

Ozone is safer to store and use since only 1–2 atm are required, compared with >30 atm required for chlorine. Accidental exposure to ozone is rarely of serious consequence. The action of ozone is less pH-sensitive than that of chlorine. It shows a longer half-life at lower pH. It is also capable of precipitating undesirable metal ions (e.g., Fe^{2+}, Mn^{x+}) by oxidizing them to a higher oxidation state in which they form the corresponding insoluble oxides or hydroxides.[3,23]

Interestingly, algae develop immunity to chlorine with time, but ozone controls algae growth by an indirect mechanism (it oxidizes the organic matter that serves as their nutrient).[24] Microbial inactivation by ozone is known to be much faster than that by chlorine,[25] and biological regrowth potential is minimized.[22] Chlorine has a longer decomposition time than ozone, which allows for better residual disinfecting action together (unfortunately) with a higher probability of DBP formation. Chloramines are frequently added to ozonation processes to compensate for this lack of residual disinfecting action.[25] For ultrapure water sterilization applications (for example, in the semiconductor industry), ozone has been shown to be quite effective and to decrease the frequency required for regular sterilization.[20] Residual ozone destruction or removal is achieved catalytically (MnO_2, $Pt-Al_2O_3$, Pt-activated C, CoO_x-activated carbon),[22,24,26] thermally (300°C),[24] or physically (activated carbon, bipolar membranes).[12,24]

A major drawback of ozone is its low solubility in water. However, several processes have been developed to circumvent this problem. These are based on dispersors, injectors, mixers, distributors, sprayers, packed towers, bubble columns, and so forth.[21,24,27,28] Another drawback is that its high oxidizing power promotes strong interaction (usually leading to degradation) with surrounding materials. Stainless steel does not undergo appreciable degradation.[20]

Hydrogen Peroxide. The properties of this important disinfection chemical have already been summarized in Chapter 6.

Other Miscellaneous Chemical Disinfectants. A "cocktail" of disinfection chemicals is often used in the water treatment field. The combination of ClO_2 + Cl_2 was discussed earlier. Other combinations that are in use in the United States include Cl_2 + NH_3, Cl_2+OCl^-, ClO_2 + Cl_2 + NH_3, Cl_2 + OCl^- + NH_3, and ClO_2 + Cl_2 + OCl^-.[8]

Potassium permanganate is an effective oxidant for the removal of undesired compounds such as THMs and other taste- and odor-causing substances. However, its disinfecting action is limited due to its lack of residual action.[8]

As mentioned earlier, the use of metal ions as disinfectants has a long history. Silver ions are poisonous to most microorganisms in trace amounts. Further, Ag^+ and Cu^{2+} show a synergistic disinfection effect.[29-31] A combination of

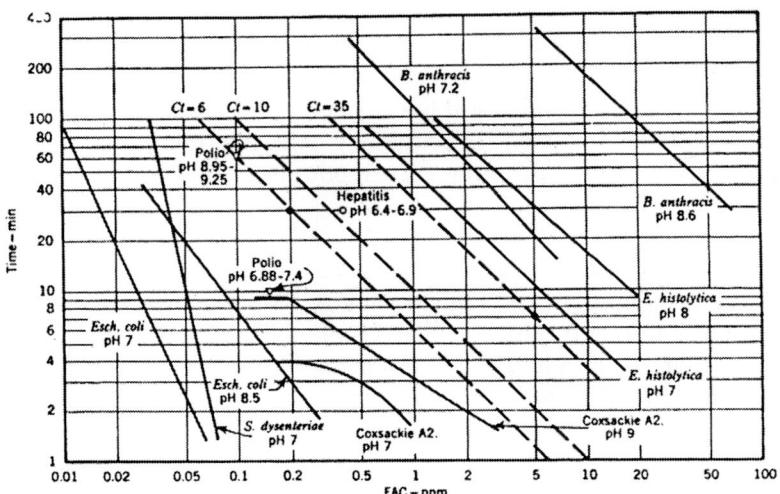

FIGURE 7.4. Disinfection vs. free available chlorine (FAC) residuals. Time scale is for 99.6–100% kill. Temperature was in the range 20–29°C, with pH as indicated. [Reproduced with permission from E. R. Baumann and D. D. Ludwig, *J. Am. Water Works Assoc.* **54**, 1379 (1962).]

Cu^{2+}, Ag^+, and Cl_2 attains 4-log reduction of poliovirus in well and tap water, although the infectivity of adenovirus, rotavirus, and hepatitis A virus is not completely removed.[32]

Chemical Dose. A major factor in disinfection studies and practices is the relationship between the disinfectant dosage and contact time to effect a specified level of "kill." An empirical equation has been developed[33]:

$$t = aC^b \qquad (7.1)$$

where t is the contact time(s), C is the disinfectant dosage (mg/liter), and a and b are constants. Representative plots of Eq. (7.1) for several organisms are shown in Figure 7.4.[34] These plots generally have negative slope equal to -1. Hence, it follows that:

$$a = Ct \qquad (7.1a)$$

The constant a therefore can be used as a figure of merit to compare disinfection efficiencies and organismal response as long as $b = 1$ and the survival is less than ~log 3. The slope parameter b has been called the *coefficient of dilution*. When $b > 1$, the effectiveness of the disinfectant decreases rapidly with dilution. On the other hand, when $b < 1$, time of contact is more important than dosage. Finally, when $b = 1$, C and t are weighted equally.[1]

   Mechanisms of Microorganism Inactivation. As mentioned at the outset of this chapter, the disinfection mechanism(s) have been the subject of much research. The mechanisms of Cl_2 and O_3 disinfection in particular have been extensively studied.[34a-d] Chlorine can alter cell permeability resulting in leakage of the cytoplasm. It can also diffuse into the cell to oxidize nucleic acids and proteins.[34a] Chlorine inactivates viruses by oxidation of the capsid or nucleic acids. Ozone disrupts the function of the bacterial cell membrane.[34b,c] It also oxidizes the viral capsid, which subsequently interferes with its ability to invade host cells. Aside from these routes, *radical-induced* oxidation and inactivation of cellular components are also important.[35]

7.2.3. Disinfection By-Products

   After the initial reports on the formation of trihalomethanes in chlorinated natural surface water,[36-38] much research has gone into factors underlying the formation and control of disinfection by-products in water. The "maximum contaminant level" (MCL) as set by the U.S. EPA for THMs is 100 µg/liter, and there are strong indications that the standard will be significantly lowered to 50 µg/liter or even 20 µg/liter. This has spurred the search for alternative disinfectants to chlorine and for advanced methods of DBP removal in treated water.[39-41] Another viable THM control strategy is predicated on precursor removal prior to conventional disinfection practice. This latter strategy has the obvious advantage of not necessitating radical changes in the disinfection equipment, personnel training, and the like from that practiced currently by the water treatment industry.

   Two questions arose after early work on the chemical characterization of the THMs: What is the precursor? What is the source of the brominated derivatives that were observed[38,42] such as $CHCl_2Br$, $CHClBr_2$, $CHBr_3$, and CH_2Br_2? Both questions have led to topics of active research and reviews are available.[1,6] The possibility of chlorine gas contamination with elemental bromine and formation of Br_2 from naturally occurring Br^- in the surface water during chlorination have been raised as explanations for the second question. The second possibility appears to be more tenable at present. Thus, chlorine in the form of $HOCl-OCl^-$ can oxidize Br^- to $HOBr-OBr^-$, which in turn can lead to the formation of the four THM species ranging from chloroform to bromoform ($CHBr_3$).

   It must be pointed out that disinfectants other than Cl_2 also lead to the production of DBPs as long as the water influent contains natural organic matter. In fact, the measurement of total organic carbon was previously considered as a surrogate for THMs, but now many studies are underway to characterize the NOM in water.[6] As will be discussed in further detail, many organic compounds commonly found in water such as lignin, tannin, humic substances, fulvic acids, and various other organics strongly absorb UV radiation. Water utilities across the United States are gathering data for correlating UV absorption with organic carbon content, color, THM formation potential, and other DBPs.[6,42]

It has been shown that O_3 in the presence of Br^- alone can lead to the formation of $CHBr_3$.[44,45] Ozone oxidizes Br^- to $HOBr$ under water treatment conditions. The OBr^- derived from the ionization of the weak acid $HOBr$ further reacts to form BrO_3^-.[46] Eventually, all of the original bromide may be quantitatively converted to BrO_3^-.[39] Animal experiments indicate the carcinogenic nature of BrO_3^-. Other research has also shown the formation of BrO_3^- from Br^- on UV irradiation.[6]

Processes for DBP control and THM precursor removal currently use granular activated carbon (GAC) or powdered activated carbon (PAC), adsorption ultrafiltration (UF), and ozonation.[39-41] However, the efficacy of these methods can be highly variable. For example, UF when used in isolation is relatively ineffective for DBP removal because of its high molecular weight (ca. 100,000 daltons) cutoff.[41] Removal efficiency of TOC using PAC adsorption has been shown to vary from 29 to 85% depending on the type of PAC used, the PAC dosage, and the quality of the source water. Combination of UF with PAC has been shown to result in enhanced removal of THM and trihaloacetic acid precursors.[41] Finally, neither this method nor activated carbon adsorption applied alone appears to result in bromide removal. The net result often is an increase in the ratio of Br^-/DOC (DOC is dissolved organic carbon) and an increase in the relative abundance of brominated THMs after treatment.[39] Clearly, further research is warranted on the treatment of source waters containing the bromide ion.

Advanced methods for DBP and THM precursor removal based on UV–O_3 and photoelectrochemistry will be discussed in later sections in this chapter.

7.2.4. Taste and Odor Removal

The potability of drinking water is judged by its *organoleptic* and esthetic qualities; that is, taste and odor. Off-tasting or malodorous water is thought to be unsafe for consumption, although this association can often be misleading. That is, taste and odor are not always indicative of the bacteriological and pathogenic quality of the particular water.

Sources of tastes and odors are many and varied. They range from biological sources such as algae and actinomycetes to organic compounds such as chlorophenols. Several attempts have been made to chemically characterize the taste and odor agents from biological sources. Thus compounds that have

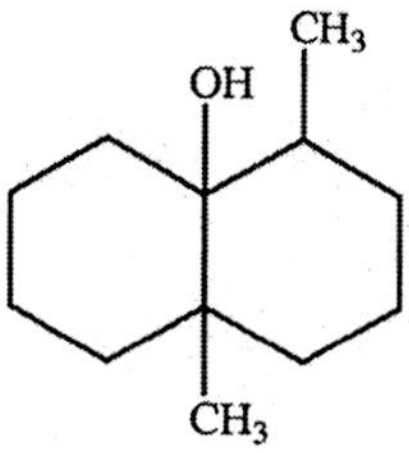

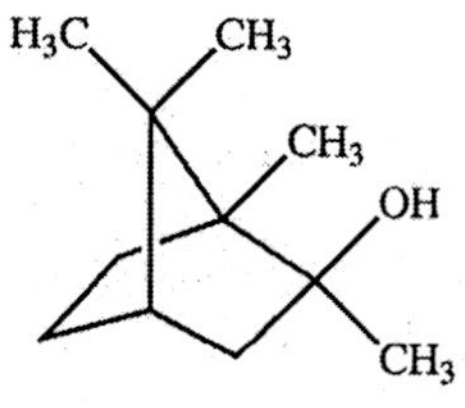

7.2. Water Disinfection: Background and Principles

FIGURE 7.5. Taste and odor causing agents in water.

been isolated and identified include mucidone (derived from the Latin *mucid* meaning musty and *one* denoting a ketone), geosmin (from the Greek *ge* meaning earth and *osme* meaning odor), and methylisoborneol (MIB) (Figure 7.5).

Drinking water frequently may have an "iodine" or a "medicinal" taste. This is attributed to the result of chlorination of the phenolic compounds that may be present in the water. Dissolved inorganic substances also impart an undesirable taste to drinking water. Examples include the "mineral" taste that often correlates with the total dissolved solids (TDS) of the particular water and the "bitter" or "metallic" taste from metals such as Cu, Fe, Mn, and Zn. The latter are believed to originate from the corrosion of pipes through which the water is distributed.

Methods are available for quantifying taste and odor.[1] For example the odor threshold concentration (OTC) is frequently used as an indicator of the manifestation of a certain agent or group of compounds. In fact, the American Society for Testing and Materials (ASTM) prescribes a method for classifying odors by chemical types.[47] Flavor profile analyses[48] can be used for detecting, controlling, and understanding off-flavors in drinking water.

Advanced methods for the removal of taste and odor compounds from water will be reviewed later in this chapter.

7.2.5. Indicator Organisms

The most desirable index of pathogenic microorganism pollution is its own presence or absence. However, practical constraints (e.g., time, cost, laboratory capability) have prompted the use of surrogates or indicator organisms for assessing the water quality and treatment efficacy.

Current microbial indicators are based on total coliforms and fecal coliforms (see Section 1.5.2). However, several studies have shown[49-51] that coliforms are inadequate for indicating the presence of pathogens, especially viruses and parasites. Current research suggests that *Clostridium perfringens* and somatic coliphages are better suited as indicators of viruses and protozoan cysts including *Giardia lamblia* cysts and *Cryptosporidium* oocysts.[52]

7.3. ELECTROCHEMICAL DISINFECTION OF WATER

7.3.1. Introductory Remarks

As with the organic pollutants discussed earlier in Chapter 5, microorganisms can be electrochemically inactivated either directly or via the generation of "killer" agents such as •OH. A third route involves the electrosorption of bacteria and the like on the electrode surface and their subsequent inactivation. Again, as with their organic counterparts, the direct and indirect routes are not always distinguishable; and it is possible that, in many of the studies done to date, both processes play a significant role.

Electrode materials vary widely, depending on the disinfecting agent desired. Cathode materials include stainless steel, copper, graphite, carbon cloth, and reticulated vitreous carbon (RVC). Anode materials include platinized titanium or niobium, tantalum, graphite, carbon, metal oxides, silver, copper, nickel, monel, dimensionally stable anodes, and combinations thereof. Electrocatalytic materials can be incorporated into electrodes, for example, in the form of coatings, or incorporated into cell separators.[53] Three-dimensional electrodes have also been successfully used.[54]

Narrow gap cell technology involving the use of a solid polymer electrolyte (SPE) (see Chapter 5) has been applied to electrochemical water disinfection to decrease cell resistance and avoid the need for adding supporting electrolytes. Examples of the present application include chlorine and hypochlorite production (see later).[53]

Direct as well as low and high frequency alternating current (AC) have been used for disinfection purposes. Deposit formation may occur on the cathodes, particularly in the case of hard water. This problem has been prevented by periodic current reversal and producing oscillations of the electrode.[55–57] Treatment with DC or AC has been shown to inactivate a large variety of microorganisms including viruses, bacteria, algae, coliforms, fecal streptococci, and relatively large species such as *Euglena*.[56]

7.3.2. Electrosorption of Microorganisms and Direct Electron Transfer

Bacteria show a tendency to adsorb onto surfaces such as activated carbon, fibrous carbon, or ion exchange resins.[58,59] This tendency is driven mainly by electrostatic forces between charged groups on the cell wall (e.g., amino and carboxylic groups) and on the adsorbant. For example, Gram-negative bacilli

7.3. Electrochemical Disinfection of Water 639

concentration was reduced by some five orders of magnitude upon adsorption on activated charcoal.[59]

The potential-induced adsorption of solutes onto the surface of an electrode is called *electrosorption*, and its effectiveness depends on the potential of zero charge of the adsorbate. (This is also the principle used in the removal of suspended solids by electrofiltration.) For example, the application of an external potential (positive with respect to the potential of zero charge of the adsorbate) to a carbon felt electrode promoted a reduction in the concentration of *Escherichia coli* in the suspension passing through the electrode by three or four orders of magnitude.[58] A major reduction in *S. typhimurium* concentration was also observed.[58,59] Interestingly, this method is not effective for microorganism *inactivation* since part of the bacteria can be released back to the suspension.

Even though this phenomenon does not strictly fall within the framework of the definition of *disinfection* given at the beginning of this chapter, it does remove *disease-related* microorganisms and thus can be an effective method for the prevention of infection-related problems. Additional advantages are that chemicals need not be added to the suspension for treatment, the adsorbent can be at least partially regenerated, dead microorganisms do not remain in the treated water, and inexpensive adsorbates can be used. An interesting alternative consists of the adsorption of bactericides onto the electrode.[60] The applicability of the DC electrosorption approach to nonbacterial pathogens (e.g., viruses, protozoan cysts), however, remains to be established.

An interesting electrochemical disinfection process has been described[56,57] based on AC perturbation of the electrode/electrolyte interface. In the anodic part of the cycle, the pathogen adsorbed on the electrode is oxidized. In the cathodic portion, the oxidation products are reductively removed from the electrode surface and a clean electrode surface is regenerated for subsequent disinfection cycles. This system is reported to work not only on bacteria but on larger organisms such as protozoa.

Possible electrocution mechanisms involve the induction of abnormal redox processes or even forced (unnatural) electro-osmotic flow[61] at the cellular level because of the current flow. Direct electrochemical oxidation of intercellular coenzyme A has been claimed for the inactivation of bacteria at electrode surfaces.[62–65] This method is claimed to also reduce microbial fouling of the electrode and other surfaces. Carbon anodes were used in early work.[62–64] This approach has since been extended to the use of large surface area graphite–silicone and carbon–chloroprene electrode surfaces.[65,66]

An important application of the "direct" oxidation approach is prevention of marine biofouling of structures such as water-cooling pipes and ship hulls. The accumulation of biomass on these surfaces causes increased fluid frictional resistance in the case of ship hulls and decreased heat-transfer efficiency for cooling pipes.[67] The use of toxic chemical agents such as copper and organotin is not environmentally safe because of leaching of these species from the surfaces to be protected.

7.3.3. In situ Electrogeneration of Disinfection Agents

<u>Chlorine, Hypochlorous Acid, and Hypochlorite</u>. The disinfection properties of these chemicals were reviewed earlier in Section 7.2.2. In this section, we review the possibilities for electrochemically generating these species for their subsequent use in a water treatment system.

Chlorine production necessitates the use of high O_2 overpotential and low chlorine overpotential anodes, since the standard reduction potential of chlorine ($E° = 1.358$ V) is more positive than that of oxygen ($E° = 1.229$ V), and therefore dioxygen production is thermodynamically favorable (see Chapter 2). Mixtures of Cl_2 and ClO_2 are normally produced.[68] Simultaneous cathodic processes can beneficially occur in certain cases (e.g., metal ion removal, nitrate ion reduction to N_2).[11,69] Chlorine yields are normally a function of chloride ion concentration.[70] Since chlorine produces chloride ions on disinfection or oxidation, a recycling concept can be used for reusing the Cl^- in the spent solution to produce chlorine again.[71] Automated control of Cl_2 production can be based on Cl^- concentration, on Cl_2 concentration or on the amount of charge passed.[72–75]

As shown in Figure 7.2, chlorine can produce hypochlorous acid and hypochlorite. For example, OCl^- has been produced by the application of a DC voltage in a cell containing a chloride salt solution (normally alkali or ammonium chlorides)[69,76–78] or HCl.[79,80] The metal chloride can also be added in the form of tablets, pills, or granules.[81] Seawater is a common raw material for this process.[82] However, inorganic compounds (e.g., $Mg(OH)_2$) may produce scale deposits that lower the efficiency of chlorine production from seawater.[10,55] This problem has been solved by periodically performing an acid cleaning of the anodes.[83] Production of scale can be made beneficial if used for flocculation purposes.[84]

The alkaline solution produced at the cathode in an electrochemical cell during water reduction can be allowed to mix with the Cl_2 solution produced at the anode. The resulting products are OCl^-, H_2O and Cl^-. Hypochlorite has to be used immediately after production since it undergoes photolytic decomposition.[70] As mentioned previously, HOCl is a more powerful disinfectant than OCl^-. For this reason bipolar membranes, direct HCl addition, or proton ion exchange are used to acidify hypochlorite solutions.[85–89]

Sodium or calcium hypochlorite can advantageously replace Cl_2 in many small- to medium-scale applications (e.g., swimming pools). Automated dosing of hypochlorite can be based on pH and redox potential monitoring.[90]

7.3. Electrochemical Disinfection of Water

A variety of anodes are used for chlorine or hypochlorite production for disinfection or bleaching purposes including[53,70,91]

- A mixture of ruthenium, palladium, and titanium oxides: Such a mixture can be produced by the thermal decomposition of a paint solution with chlorides of the three metals.
- Ti, Nb, Ta, and some of their alloys, coated with Pt or PtO_x.
- Graphite, PbO_2, and dimensionally stable anodes (TiO_2–RuO_2).

<u>Chlorine dioxide</u>. An aqueous solution of an alkali metal chlorate, $MClO_3$ is passed through a cation exchange resin (CEX) and electrolyzed to produce a mixture of ClO_2 and $HClO_3$. ClO_2 is then stripped off. Alternatively, the chlorate salt is reacted with an acid stream externally added or else produced at the anode of an electrochemical cell to yield ClO_2. A mixture of the chlorate salt and chloric acid is produced by a similar principle and is suitable for the production of ClO_2 by reduction. The chlorate ions can also be obtained from electrodialysis–dialysis processing of an alkali metal chlorate.[92,93]

Chloric acid for the production of ClO_2 can also be obtained in an electrochemical cell from an alkali metal chlorate by selective migration of the chlorate ions through a diaphragm toward the anode, where they encounter H^+ ions and produce the acid. A simpler chemical route consists of the addition of H_2SO_4 to $NaClO_3$; however, the resulting product contains salt impurities that must be removed.[94]

A cell with electrodes separated by a cation exchange membrane is used for the production of ClO_2 for industrial wastewater and sewage treatment.[95] ClO_2 is also produced by the electrolysis of a saturated NaCl solution to form concentrated Cl_2 which is then reacted with a hypochlorite solution to generate chlorine dioxide.[96] The likely reaction is

$$3/2\, Cl_2 + NaOCl + H_2O = ClO_2 + NaCl + 2\, HCl$$

<u>Ozone and Hydrogen Peroxide</u>. Methods for the electrogeneration of these species were reviewed earlier in this book (Chapter 5). The inactivation of marine bacteria such as *Vibrio anguillarum* and other microorganisms using *in situ* electrogeneration of H_2O_2 has been described.[97–99]

<u>Other Miscellaneous Agents</u>. Potassium permanganate can be used as a bactericidal agent. Its production is based on the chemical oxidation (by O_2) of MnO_2 to MnO_4^{2-} followed by electrochemical oxidation at a Ni or monel anode to MnO_4^-.[100,101] It can also be produced in an electrochemical reactor loaded with manganese metal particles (or one of its alloys) that are made bipolar in an alkaline electrolyte.[102] Its combined action with Ag^+ and Cu^{2+} ions inactivates bacteria, virus, fungi, and parasites.[103]

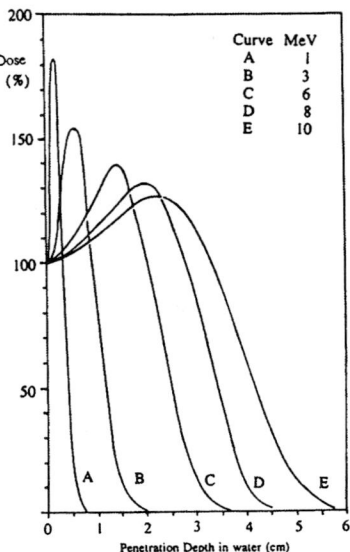

FIGURE 7.6. Depth–dose distribution in water at different electron energies. (Reproduced with permission from Getoff.[110])

Metal ions are particularly amenable to electrogeneration. Precisely metered quantities of these ions can be electrochemically dosed into the water stream on demand. For example, water containing 10^7 microorganisms/mL (mainly *Escherichia coli*) was given a contact time of 4 s in the anodic compartment of a cell containing Ag^+ ions; 99.43% of the microorganisms were inactivated.[104] Since only minute amounts of Ag^+ are required, a recovery unit for the excess concentration can be economically and environmentally attractive.[104] Alkaline solutions of Ag^+ yield AgOH, another powerful disinfectant.[105,106] In a similar fashion, Cu^{2+} ions from consumable Cu anodes are used to disinfect swimming pools and water for drinking purposes.[106] They are also used to prevent marine biofouling.[65,66] Inhibition of normal enzymatic activity by these ions is the most likely inactivation path. Since both Ag and Cu anodes develop an undesirable film on their surfaces after some time, it is advisable to add inert particles to the electrochemical system to slightly erode the surfaces, thus preventing film formation.[107] Other transition metal compounds such as $CoCl_2$, $NiCl_2$, or $(NH_4)_2IrCl_6$ have also been shown to inactivate *E. coli*.[108]

The electrolysis of water can generate nascent oxygen that has been implicated in bactericidal action.[109] These species are produced as precursors to *bulk* O_2 generation at the anode surface:

Table 7.2. Products of the Radiolysis of Water

Primary reactions

$$H_2O \rightsquigarrow \begin{cases} H_2O^* \rightarrow H^\bullet + {}^\bullet OH \\ H_2O^+ + e^- \end{cases}$$

$$H_2O \rightsquigarrow \underset{(2.7)}{e^-_{aq}}, \underset{(0.6)}{H^\bullet}, \underset{(2.8)}{{}^\bullet OH}, \underset{(0.45)}{H_2}, \underset{(0.7)}{H_2O_2}, \underset{(3.2)}{H^+}, \underset{(0.5)^{a)}}{OH^-}$$

Secondary reactions	k (M^{-1} s^{-1})
$H^\bullet + H^\bullet \rightarrow H_2$	1×10^{10}
$H^\bullet + {}^\bullet OH \rightarrow H_2O$	2.5×10^{10}
${}^\bullet OH + e^-_{aq} \rightarrow OH^-$	2.5×10^{10}
${}^\bullet OH + {}^\bullet OH \rightarrow H_2O_2$	6×10^9
$H^+ + e^-_{aq} \rightarrow H^\bullet$	2.3×10^{10}
$O_2 + e^-_{aq} \rightarrow O_2^{-\bullet}$	2×10^{10}
$O_2 + H^\bullet \rightarrow HO_2^\bullet$	1.9×10^{10}

[a] The numbers in parentheses are the G values at pH 7.
Source: From Getoff.[110]

$$2\ OH^- \rightarrow H_2O + [O] + 2\ e^-$$
$$2\ [O] \rightarrow O_2$$

7.4. DISINFECTION BY HIGH-ENERGY RADIATION

High-energy electrons and γ-rays can be used for the production of reactive transients in water. These intermediates in turn can initiate the destruction of organic pollutants (Chapters 5 and 6) as well as the disinfection of water.[110] High-energy electrons are preferred for several reasons. Manipulation with radioactive isotopes (e.g., ^{60}Co) and their disposal are not required. The equipment is easier to operate and maintain in terms of output power regulation, "down time," and so forth. Modern electron accelerators provide electrons with variable energy (e.g., from 0.5 to 2.8 MeV) and rather high output power (80–100 kW) with efficiencies (electron power/incident power) of ~80%. The electron beam (E-beam) penetration depth in the water column depends on the electron energy content. Figure 7.6 shows the dose–depth distribution in water with electron energy as the parameter.[110] The absorbed dose units are commonly quoted in terms of "rad" (1 rad = 100 erg (6.24 × 10^{13} eV)/g H$_2$O).

The interaction between the ionizing radiation and water yields several transients and molecular products, as listed in Table 7.2.[110] The "G-value" is quoted for *radiolytic yield* as the number of converted molecules per 100 eV (1.60 × 10^{-17} J) of absorbed energy. The corresponding conversion factor to SI units is 0.10364 to obtain the yield in terms of µmol•J^{-1}.

As pointed out earlier in Chapter 1, scavenger species such as O$_2$ and N$_2$O are often important in radiochemistry. Via such species, e_{aq}^- and H atoms are converted to •OH and peroxy radicals that can then initiate the disinfection reaction.

While γ-ray induced water disinfection has a long history, its counterpart based on the use of the E-beam is at an experimental stage.[111] Field demonstration tests have been completed using a 1-kW linear accelerator.[111] The system effectiveness for disinfection was found to be very sensitive to flow control configuration and E-beam shape. Further demonstration of this technology with higher-power E-beam generators is being planned at the time of this writing.

7.5. UV DISINFECTION OF WATER

7.5.1. General Considerations

Ultraviolet (UV) disinfection has rapidly become an alternative to chlorination of wastewater effluents. UV radiation has been used commercially for many years in the pharmaceutical, cosmetic, beverage, and electronic industries. As we have seen earlier in Chapter 6, UV photolysis of pollutants is an important member of the emerging family of AOP-based technologies. UV was first used on drinking water in the early 1900s but was abandoned shortly thereafter for a variety of reasons, including high operating costs, poor equipment reliability and maintenance problems. However, a major factor in its decline was the advent of chlorination, which was found to be more efficient and reliable.

From a historical perspective, it is interesting that we are witnessing a cyclical change and a resurgence in the popularity of UV-based *wastewater* disinfection technology, thanks largely due to concern over meeting standards for discharge of chlorinated organics (i.e., DBPs). Further, chlorination produces

7.5. UV Disinfection of Water

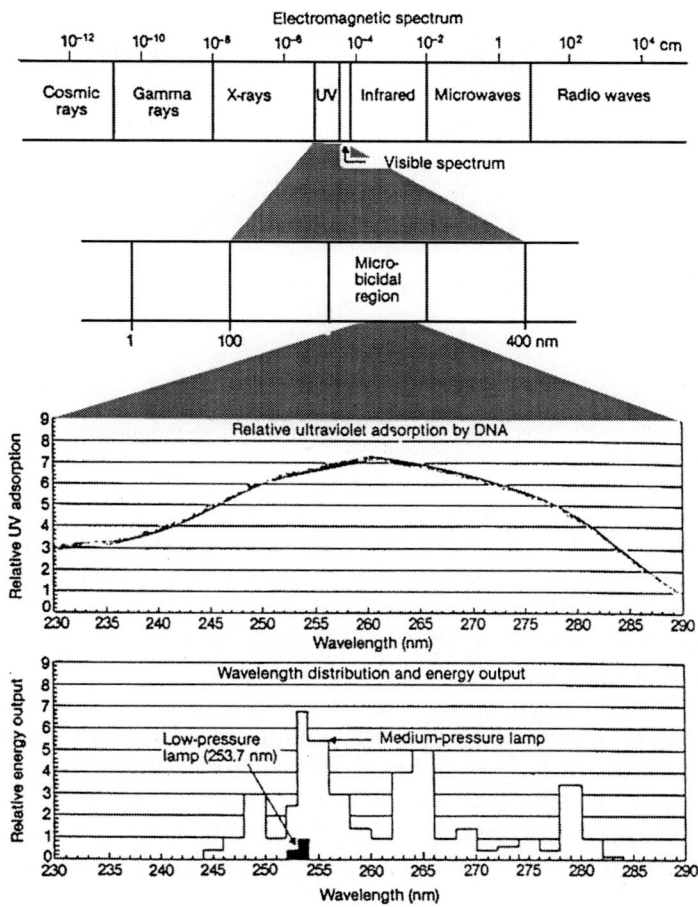

FIGURE 7.7. Microbicidal wavelengths of UV and distribution of energy output for low- and medium-pressure arc lamps. (Reproduced with permission from Wolfe.[49])

effluents that can be toxic to receiving water biota as encountered in the treatment of combined sewer overflows into oceans, rivers, or lakes. Current regulations limit total residual chlorine to 19 ppb for effluents discharged to fresh water and 13 ppb for those discharged to saltwater.[111] Compared with chlorination, UV disinfects waste water without the need for storing or handling dangerous chemicals. The short contact times needed with UV also make it practical to reduce the size of treatment tanks, and the absence of moving parts considerably simplifies plant operation. Finally, equipment reliability has considerably improved during the last decade.

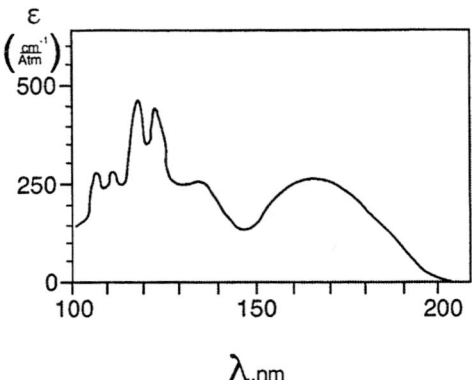

FIGURE 7.8. Absorption spectrum of water. [Reproduced with permission from K. Watanabe and M. Zelikoff, *J. Opt. Soc. Am.* **43**, 753 (1953).]

A number of full-scale UV wastewater disinfection systems have been built or planned, after several pilot tests demonstrated its effectiveness.[112–119] Importantly, full-scale UV disinfection of wastewater has been found to be economically competitive with chlorination.[112,113] Some 2,000 water treatment plants in Europe currently use UV.[49]

As in the case of organics decomposition (Chapter 6), two mechanisms can be envisioned for the inactivation of microorganisms using UV radiation. Electromagnetic radiation in the range from 240 to 280 nm kills microorganisms by causing irreparable damage to their nucleic acid. The most potent wavelength for DNA damage is ~260 nm (Figure 7.7).[49] Currently two types of commercial UV systems are in use: those that use low-pressure Hg vapor lamps and those that use medium-pressure Hg vapor lamps. The spectral output in these two cases is also shown in Figure 7.7.[49] The low-pressure lamp output peaks near the optimal wavelength for DNA damage. Hence, these lamps have been commonly used as a source of *germicidal* UV light. Medium-pressure lamps provide a broader band of UV light (Figure 7.7), and their overall energy output is greater than that of low-pressure lamps. Although both types can be used for water treatment, the medium-pressure systems have a much greater (~25 times) treatment *capacity* because of their higher intensity.

An "indirect" mode of microorganism inactivation can also be envisioned based on the generation of potent radical intermediates by the UV irradiation of water. Figure 7.8 illustrates the absorption spectrum of water.[110] A series of absorption maxima span the vacuum UV (VUV) region ($\lambda < 200$ nm). Low-pressure Hg lamps have emission lines at 184.9 nm while electrodeless discharge lamps emit light at 123.6 and 147 nm. Light absorption at these wavelengths triggers the chain of reactions (see also Table 7.2)[110]

7.5. UV Disinfection of Water

$$H_2O \xrightarrow{h\nu} H_2O^* \rightarrow H^\bullet + {}^\bullet OH$$

$$H^\bullet + O_2 \rightarrow HO_2^\bullet$$

$$2\,H_2O^* \rightarrow H_2O + H_2O^{**} \text{ (super-excited molecules)}$$

$$H_2O^{**} \rightarrow H_2O^+ + e_{aq}^-$$

$$H_2O^+ + H_2O \rightarrow 2\,{}^\bullet OH + 2\,H^+$$

$$e_{aq}^- + O_2 \rightarrow O_2^-$$

$$O_2^-, HO_2^\bullet \rightarrow \rightarrow \rightarrow H_2O_2$$

The quantum yields for generation of $H^\bullet$ and ${}^\bullet OH$ range from 0.33 to 1.0.[110] These free radicals can then initiate the inactivation of microorganisms. Unfortunately, however, practical difficulties are associated with the use and operation of VUV lamps, the strong absorption of this radiation by atmospheric O_2 being one of them. Thus, direct attack on the microorganism DNA by the UV radiation appears to be a more viable route under "normal" irradiation scenarios.

7.5.2. UV Dose and Disinfection Kinetics

Dose is defined as

$$\text{Dose (mW} \cdot \text{s/cm}^2) = \text{intensity (mW/cm}^2) \times \text{contact time (s)} \quad (7.2)$$

Less frequently, the dose is specified in terms of energy units (e.g., joules) per unit area of surface. (Recall that power × time = energy.) The estimation of UV dose is fraught with several problems[120]: (a) radiometer detectors measure intensity on a planar surface and hence are not compatible for use with long tubular UV lamps to measure the *three-dimensional intensity* to which microbial cells may be exposed; (b) a detector placed in the reactor wall cannot be used to estimate the average intensity in an absorbing medium within the reactor; and (c) particles in wastewater scatter UV light so that spectrophotometers tend to overestimate the UV absorbance. The last parameter is needed to take into account (via Beer's law) the light attenuation factor within the absorbing medium. These problems are compounded for flow-through disinfection reactors, as we shall see later.

A bioassay method has been developed[120] to address these problems and indirectly measure the average intensity within the UV disinfection reactor. In this method, the survival of spores of *Bacillus subtilis* is determined as a function of contact time to a beam of UV light that is collimated by a black tube (Figure 7.9). The suspensions containing the microorganism are kept in a stirred

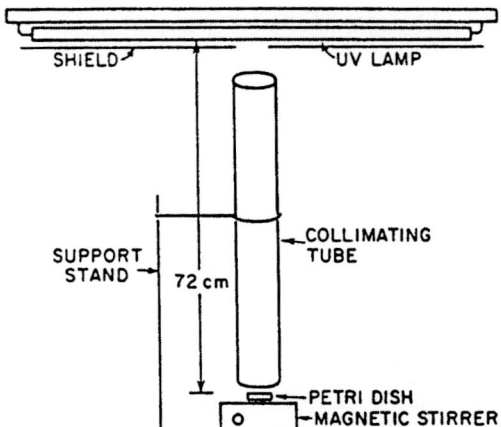

FIGURE 7.9. Collimated beam apparatus for UV disinfection of water. (Reproduced with permission from Qualls and Johnson.[120])

petri dish. The intensity at the surface of the suspension is first measured with a radiometer. A correction for reflection of the light at the water surface can be incorporated (nominally 4%).[121] Since fluid depth and absorbance are minimal, the dose can be calculated based on the measured intensity and the known exposure time. In cases where the absorption is significant, the average intensity is calculated by an integration of Beer's law over the fluid depth.[12] Standard curves of log survival *vs.* dose have been presented by the authors (e.g., Figure 7.10).[120]

The intensity can be determined from such standard curves by determining the survival ratio, reading the dose corresponding to the ratio from the calibration plot, and using the known contact time to calculate the average intensity:

$$\text{intensity} = \frac{\text{dose}}{\text{contact time}} \qquad (7.2a)$$

The authors have validated the bioassay method[120] by modeling the intensity profile around a tubular lamp using the point source summation (PSS) method.[122] The PSS method assumes that the lamp is a line segment source and can be treated as the sum of a number of point sources. This method has also been used by other authors to compute average UV intensities within a pilot plant reactor.[118]

A vexing problem with precise determination of the contact time is associated with the UV lamp warm-up. The lamp has to be equilibrated when expo-

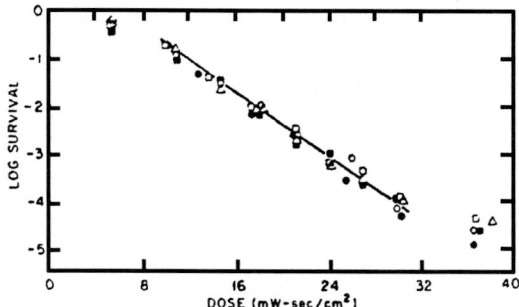

FIGURE 7.10. Log survival of B. subtilis spores vs. UV dose in a collimated beam of known intensity. Different symbols represent five different runs. Data from doses of 10–30.5 mW/cm² appeared to be linear and fit the regression line Y = 0.167 X + 1.01 (r = 0.98). (Reproduced with permission from Qualls and Johnson.[120])

sure begins so that the output does not fluctuate. The authors of the aforementioned study[120] have used an ingenious method based on the use of a slidable (black) paper sleeve that is positioned between the lamp and the quartz envelope. This sleeve can be withdrawn after the lamp is properly "conditioned."

The survival ratio (N_s/N_o) of microorganisms is generally a function of the UV dose:

$$N_s/N_o = f(\text{dose}) \tag{7.3}$$

where N_o and N_s are the density of organisms before and after UV irradiation, respectively.

The disinfection kinetics are usually described by Chick's law,[123] which states that the number of organisms destroyed per unit time (i.e., the rate of kill) is proportional to the organisms remaining, at time t. This law is analogous to first-order reaction kinetics and can be written in the form

$$\frac{N_s}{N_o} = e^{-kt} \tag{7.4}$$

where k is the rate (or decay) constant for disinfection. Combining of Eqs. (7.2)–(7.4) leads to the expectation that a plot of the log survival versus dose will be linear. This is generally borne out for pure cultures of organisms such as *E. coli* (see Figure 7.11).[124] However, departures from Chick's law are frequent and common in both laboratory experiments and treatment plant practice. The observed nonidealities include the following:

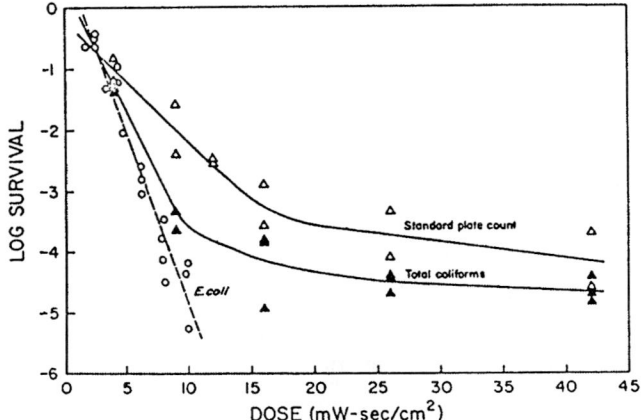

FIGURE 7.11. Survival vs. UV dose for cultured *E. coli*, filtered (10 μm pores) total coliforms, and microorganisms growing on standard plate count agar from a secondary effluent. Each data point represents one irradiation and three replicate platings. (Reproduced with permission from Chang et al.[124])

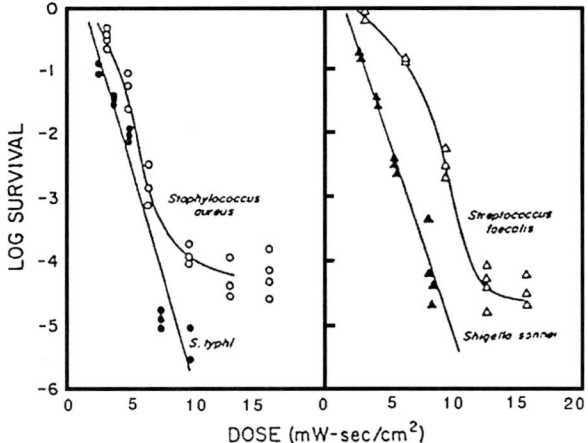

FIGURE 7.12. Survival vs. UV dose for *S. sonnei*, *S. typhi*, *S. faecalis*, and *S. aureus*. Each data point represents one irradiation and three replicate platings. (Reproduced with permission from Chang et al.[124])

(1) Initially the slope or "shoulder" lags. This is exemplified by the data for *S. aureus* and *S. faccalis* in Figure 7.12.[124] Such behavior has been attributed to "multiple-hit" kinetics or related phenomena.[121,125] The multiple-hit theory can apply both to single cells and cell aggregates.

7.5. UV Disinfection of Water

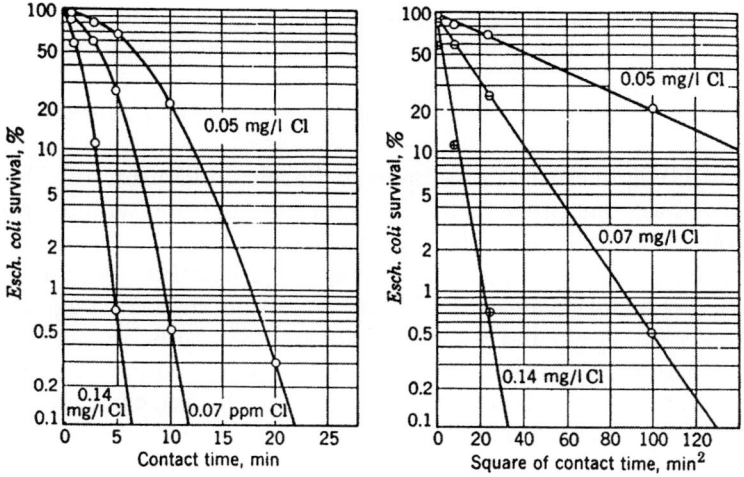

FIGURE 7.13. Length of survival of *E. coli* in pure water at pH 8.5 and 2–5°C. (Reproduced with permission from G. M. Fair et al., "Water Supply and Wastewater Disposal," Wiley, New York, 1954.)

(2) Close examination of the data in Figures 7.11 and 7.12[124] as well as others reported in the literature for enteric viruses[126] shows that the devitalization plots do not pass through the origin (i.e., with zero intercept). Classical *single-event* disinfection phenomena (as embodied in Eq. (7.3)) would yield plots with zero intercept.

Linearization of plots such as those in Figure 7.12 can be achieved[127] by adding empirical coefficients to the t term in the following:

$$\frac{N_s}{N_o} = e^{-kt^m} \tag{7.5}$$

An example is shown in Figure 7.13, where linearization is achieved with $m = 2$.[1] When $m > 1$, the rate of kill increases with time; when $m < 1$, the rate decreases with time. Of course, the classical case corresponds to a constant rate of kill with time. A $t^{1/3}$ inactivation model was reported[112] for a pooled set of data on the disinfection of fecal coliforms in three municipal wastewater effluents.

An intrinsic assumption in the preceding discussion is that inactivation is a function of only the UV dose (see Eq. (7.3)) and is *unaffected by the light intensity*. That is, a short contact time at a high intensity will achieve the same degree of inactivation as a long exposure at low intensity. This is the Bunsen–Roscoe law[121] that has been experimentally verified by other authors.[128]

7.5.3. The Role of Suspended Particles and Chemicals in UV Disinfection

Because UV light must be absorbed into the microorganisms to achieve inactivation, anything that prevents the UV light from reacting with the micro-organisms obviously will impair the degree of disinfection. Chemical substances that interfere with UV transmission in the region of 254 nm have been reviewed.[129] These include phenolic compounds, humic acids from decaying vegetation, lignin sulfonates from pulp and paper mill effluents, dyestuffs, and ferric ions.[129]

Suspended matter in water can affect UV disinfection in two ways. First, colloidal matter can absorb or scatter the UV light. In fact, computation of the UV dose via Eq. (7.2) requires estimation of the UV intensity. This latter parameter in turn is computed from the measured UV *absorbance*, which must be separated from the scattering effect (see later).

The second effect arises from the encapsulation of microbes by the solid matter. In fact, this effect is clearly seen in the survival–dose plots for the standard plate count and the total coliforms in Figure 7.11.[124] The tendency for the survival curves to level out at low levels of survival is caused by a small fraction of cells that are protected by their association with particulate matter. Filtration can remove some, but not all, of the aggregates and particles capable of harboring protected coliform cells. Early work has shown[128,130] that a sample dispersed by ultrasonication is made more sensitive to UV disinfection. Unfortunately, while the particles harboring microbes are only a small percentage of the initial coliform population, they become a limiting factor as disinfection efficiency approaches the levels necessary to meet disinfection standards.[131]

It must be noted that the role of particles in protecting embedded microbes is not a disadvantage unique to UV disinfection. Particles play a similar role in chlorine, chlorine dioxide, and ozone disinfection.[132] Penetration of chemical disinfectants into the particles may be an additional limiting factor in these latter cases.

The encapsulation effect can be clearly discerned by systematic filtering of an initially unfiltered effluent sample.[131] Coliforms are approximately 1–2 µm in size, so that an 8 µm filter would allow only single cells or very small aggregates to pass. The unfiltered effluent is compared with that filtered through 8 and 70 µm filters in terms of the log survival–dose plot in Figure 7.14.[131] The 8 µm filtered sample shows disinfection beyond -4.5 log survival units, where survivors are undetectable otherwise. On the other hand, the unfiltered and 70 µm filtered samples level off after -2 or -3 log survival units. These results are generally borne out by other studies of particle size effects. For example, germicidal efficiencies have been found to differ between irradiated raw waste

7.5. UV Disinfection of Water

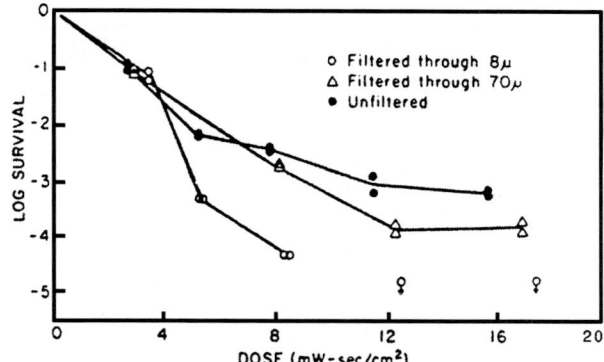

FIGURE 7.14. Effect of filtration on survival of total coliforms in irradiated Sandy Creek effluent with arrows indicating limit of detectability for exposure in which no survivors were found. (Reproduced with permission from Qualls et al.[131])

water, on the one hand, and secondary effluent[128] or sand-filtered activated sludge effluent, on the other.[112] Even in the secondary effluent case, disinfection was significantly better after sand–anthracite filtering than in the unfiltered water.[131] Figure 7.15a shows the results from a pilot plant study[118] on the dependence of effluent fecal coliform counts on the total suspended solids (TSS).

Figure 7.15b contains data from another study[133] that attempts to correlate particle size analyses of the water samples with the coliform count. The particle size distribution was measured with a microscope. The correlation in Figure 7.15b[134] while weak ($r^2 = 0.55$, $p < 0.0001$) does indicate that the number of survivors increases with the number of suspended particles larger than 40 μm diameter. That this is because of the protection effect (rather than simply caused by a greater *initial* number (N_o) of coliforms in samples with a higher number of particles) is indicated by other data trends in this study. Attempts were also made[134] by the authors of this study to correlate the survival count to the nephelometric turbidity unit (NTU) with the UV absorbance at 254 nm, and in both cases the correlations were statistically significant.

The true UV absorbance of the turbid effluent samples can be estimated[131] from the *measured* absorbance by using the bioassay method described earlier[120] and by systematically mixing portions of the filtered and unfiltered samples of the turbid effluent. In this manner, the particulate content of the solution can be varied while the soluble component is maintained constant. Recall that the measured "absorbance" consists of the absorbance due to soluble substances, particle absorbance, and scattering from the particles. The scattered light adds to the transmitted light component and is *incorrectly* read by the spectrophotometer as an absorbance.

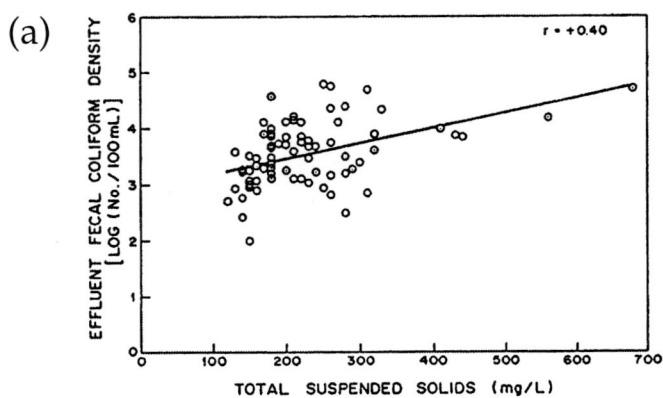

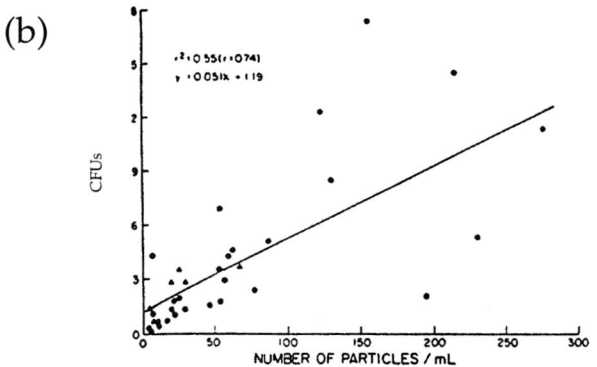

FIGURE 7.15. (a) Effect of TSS on fecal coliform density in UV reactor effluent. (b) Coliform CFUs/mL surviving a dose of 26 mW·cm² as a function of the number of particles (mL⁻¹) greater than 40 μm diameter. Triangles represent sand-filtered effluent samples. [Reproduced with permission from Zukovs et al. [118] (Figure 7.15a) and Qualls et al.[133] (Figure 7.15b).]

Figure 7.16 contains the correlation between the bioassayed absorbance and the spectrophotometric absorbance.[131] The difference between the two quantities is the scattered light component that amounts to a maximum of 12% in this particular case. In the unfiltered wastewater, the soluble absorbance and the particulate absorbance amount to 47 and 41%, respectively. Of the spectrophotometric "absorbance" caused by the particles, about 75% was the true absorbance and ~25% was actually scattering. This fraction of course varied with the NTU of the water sample, which in this particular case was 14 units.[131]

7.5. UV Disinfection of Water

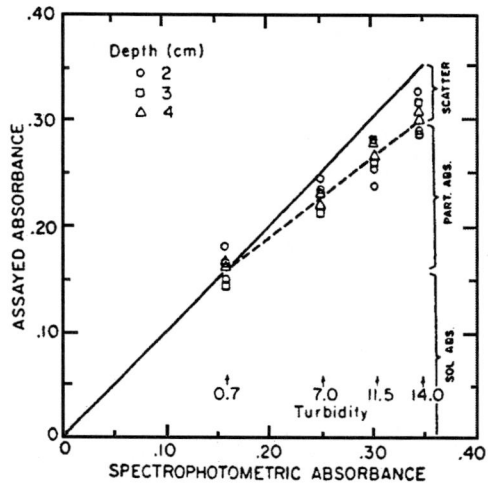

FIGURE 7.16. Spectrophotometric absorbance *vs.* absorbance measured by the bioassay method for a Chapel Hill secondary effluent sample. The soluble UV absorbance was kept constant and the particulate concentrations varied by diluting the unfiltered(14 NTU) sample with filtered (0.07 NTU) sample. The solid line represents an exact correspondence between the two methods. The dashed line is a regression through the data points. The soluble and particulate absorbance and scatter components of the spectrophotometric absorbance of the unfiltered sample are indicated. (Reproduced with permission from Qualls *et al.*[131])

A standard water quality parameter that predicts the UV absorbance (e.g., at 254 nm) is useful for the site-specific engineering of the reactor and other hardware. Presently, UV absorbance is not a standard parameter, although this situation is likely to change, thanks to concern over DBPs.[43] Fecal and total coliform density was found to correlate fairly well with UV absorbance, and this correlation improved with the addition of TSS and COD to the multiple regression.[131] It must be noted that this correlation does not indicate per se, absorption by the bacterial cells themselves. Filtration analyses[131] reveal that most of the UV absorbance is caused by the dissolved component. It is quite likely that these absorbing substances (e.g., humic acids) are closely associated with regions of high bacterial density.

Efforts have been made[134] to correlate UV absorbance with color and turbidity agents that simulate natural UV absorbers in water. These agents were of two general groups. Those in the first category included visibly colored or transparent solutions that absorbed strongly in the UV region, such as lignin sulfonates, methylene blue, ferric chloride, instant tea, and tannic acid. The second group consisted of materials that form particulate suspensions in water such as kaolin, bentonite, and ferric sulfate. Figure 7.17a contains the absorp-

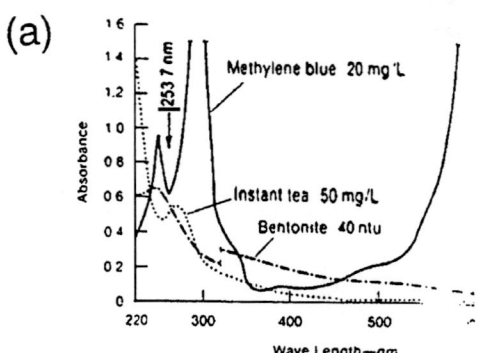

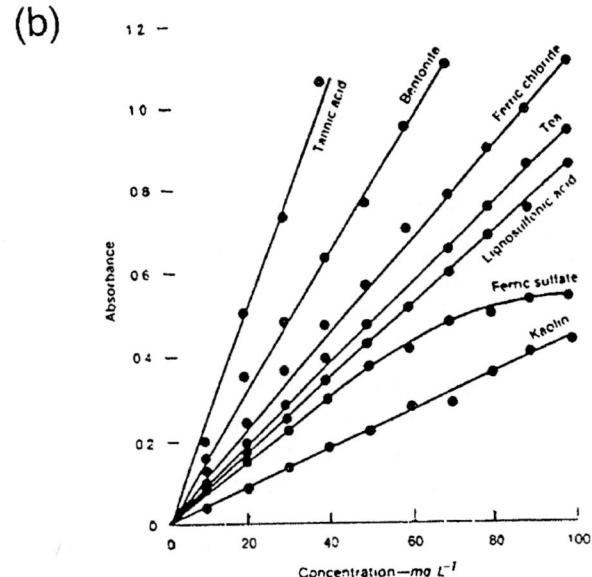

FIGURE 7.17. (a) Absorption spectra of color and turbidity agents. (b) Relationship between ultraviolet light absorbance and concentration of all color and turbidity agents tested. (Reproduced with permission from Tobin et al.[134])

tion spectra of a few of these agents, and Figure 7.17b shows the correlation between the UV absorbance and concentration.[134] All the Beer's law plots in Figure 7.17b are linear except for ferric sulfate at high concentrations (> ~60 ppm). The impetus for this study was to locate a fail-safe UV light monitor for use with disinfection devices. That is, the unit shuts down when the sensor detects a UV dose below the "safe limit." The sensor was set at 16 mW•s/cm² as per a 1966 policy document from the U.S. Department of Health, Education and

Table 7.3. Approximate Dosages for 90% Inactivation of Selected Microorganisms by UV

Microorganism	Dose ($\mu W \cdot s$)/cm^2
Bacteria	
Escherichia coli	3,000
Salmonella typhi	2,500
Pseudomonas aeruginosa	5,500
Salmonella enteritis	4,000
Shigella dysenteriae	2,200
Shigella paradysenteriae	1,700
Shigella flexneri	1,700
Shigella sonnei	3,000
Staphylococcus aureus	4,500
Legionella pneumophila	380
Vibrio cholerae	3,400
Viruses	
Poliovirus 1	5,000
Coliphage	3,600
Hepatitis A virus	3,700
Rotavirus SA 11	8,000
Protozoan cysts	
Giardia muris	82,000
Acanthamoeba castellanii	35,000

Source: Wolfe.[49]

Welfare,[49] although this limit is now known to be inadequate for waters bearing protozoan cysts (see latter).

From a sensor perspective, it is preferable that the compound have a maximum absorbance near the germicidal wavelength, and negligible absorption in the remainder of the UV spectrum. Both instant tea (color agent) and bentonite (turbidity agent) satisfy this criterion (Figure 7.17a). Of course, it is essential that the UV absorbing agent (at the levels required in the test) be not lethal to or inhibit subsequent growth of the microbial population.

7.5.4. UV Dose Sensitivity of Various Microorganisms and Comparison with Chlorine and Other Disinfection Methods

Most bacteria and viruses require relatively low UV dosages for inactivation. Table 7.3 lists the dosage for 90% kill (1 in log survival units) for several types of bacteria, viruses, and protozoan cysts.[49] Protozoan cysts appear to be considerably more resistant to UV inactivation than other types of microorganisms. Figure 7.18 contains UV inactivation data from another study[124] for 99.9% kill (3 log survival units) expressed relative to *E. coli*. The viruses, bacterial spores, and amoebic cysts require about 3–4 times, 9 times, and 15 times, respectively, the dose required for *E. coli* in broad agreement with the trends in Table 7.3. Significantly, it appears that the range of UV dose necessary to disinfect pathogens is narrower than it is for chlorine disinfection.

Available data for the UV inactivation of *Giardia lamblia* cysts are not encouraging.[127] In recent years, giardiasis has been a frequently occurring waterborne disease in the United States, and *G. lamblia* is the etiological agent.[135-137] This is an intestinal parasite transmitted via the water route in the cyst stage. Even at a UV dose of 63 mW•s/cm^2, less than 1-log reduction in cyst survival was noted.[138] This is a significant result, in that the maximum designed dose of many commercial UV units is only 25–35 mW•s/cm^2. Further, the limit of 16 mW•s/cm^2 for the acceptability of UV disinfection units is clearly inadequate when cysts are present in the water stream. Other reports[139,140,140a] indicate that viable *G. lamblia* and *G. muris* cysts can be destroyed by chlorine, although the latter appear to be more resistant. Further, no differences in resistance to chlorination were noted[140] for *G. lamblia* cysts from symptomatic and asymptomatic carriers.

Although we are not aware of UV inactivation data on *Cryptosporidium*, it is likely that the required dosages would be even higher than for *Giardia* given the extreme resistance of *Cryptosporidium* to chlorine.

A comparison study on well-head disinfection of virus-contaminated groundwater by free chlorine and UV shows that UV is a more potent virucide than free chlorine, even when the chorine residual was increased to 1.25 mg/liter at a constant time of 18 min. Other studies on enteric viruses[18,141,142] generally indicate that effective virological disinfection is provided by UV both at the laboratory scale[141] and under field conditions.[142]

A study[119] of the effect of UV light disinfection on antibiotic-resistant coliforms in municipal wastewater effluents explored whether the ratio of antibiotic-resistant bacteria to antibiotic-sensitive strains could be reduced. Total coliforms and total coliforms resistant to streptomycin, tetracycline, or

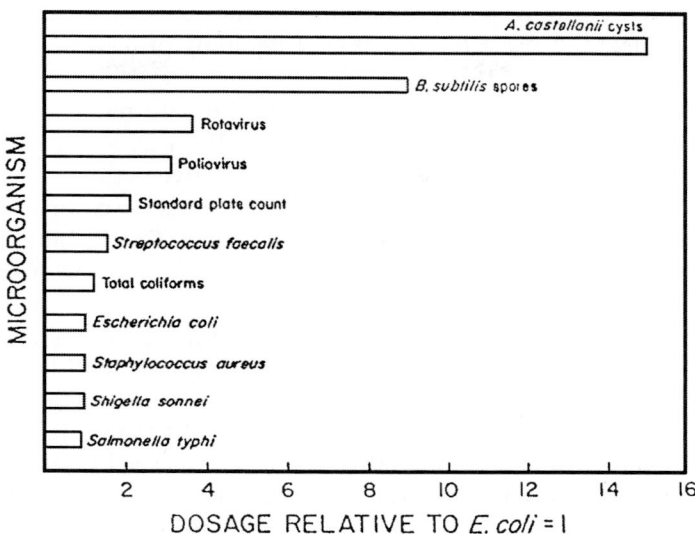

FIGURE 7.18. Relative UV doses required for 99.9% inactivation of various microorganisms compared to that for *E. coli*. (Reproduced with permission from Chang et al.[124])

chloramphenicol were isolated from filtered activated sludge effluents before and after UV light disinfection.[143] Although UV irradiation effectively disinfected the wastewater effluent, the percentage of the total surviving coliform population resistant to tetracycline or chloramphenicol had significantly increased on UV irradiation. This finding was attributed to the mechanism of R-factor mediated resistance to these antibiotics. An R-factor mediating UV resistance had been previously characterized in *E. coli*, K-12,[119] and *S. typhimurium*.[144] As the authors of this study note,[143] there is little information on the effect of other disinfectants (e.g., chlorine, ozone) on antibiotic-resistant organisms.

In concluding this section, it must be noted that UV treatment has no residual disinfection capacity (unlike chlorine, for example). To maintain microbiological integrity in the water distribution system, a post-disinfectant will have to be added, especially for surface waters. The need for a post-disinfectant in groundwater will depend on the potential for bacterial regrowth and cross-contamination.[49]

7.5.5. Photoreactivation and Sublethal UV Damage Repair

The sensitivity of bacteria to a wide variety of inactivating and mutagenic agents is dependent on the activity of various repair enzymes.[145,146] The enzymes for the excision of pyrimidine dimers from UV-damaged deoxyribonucleic acid (DNA),[147] and those capable of rejoining broken DNA strands[148] are examples of such repair enzymes.

Repair of cells that are subjected to sublethal damage manifests in terms of differences in the results of enumeration methods for assaying the survival count. For example, differences have been noted between the most-probable-number (MPN) and membrane filtration (MF) methods for enumerating total and fecal coliforms that survive chlorine disinfection. On the other hand, no significant differences have been noted in the results for the two enumeration methods when ozonation or UV disinfection[149] is used.

Repair of cell damage by UV light can be accelerated by exposure to visible light, a phenomenon known as *photoreactivation*. Under favorable conditions, photoreactivation of inactivated coliforms has been reported to result in an increase in survival of 1 log survival unit, and up to 1.8 log survival units under optimal conditions.[133] Although this may be of concern in wastewater and surface water where UV-treated water in basins is exposed to sunlight, it is likely to be less of a problem in groundwater unless open reservoirs are present. In fact, a careful assessment of the impact of coliform photoreactivation in systems with open reservoirs, has been listed as a high-priority research area in a recent review.[49]

Another manifestation of cell repair is the varying degree of UV sensitivity during the growth and harvest of microbial cells. This can be probed by measuring the sensitivity of indicator organisms harvested at various times during growth in batch cultures. Hence a study of two *E. coli* strains,[150] a resistant variety that repairs UV damage (*E. coli* B/r) and another sensitive *E. coli* B_{2-1} mutant that lacks the ability for repair, shows interesting differences. The cultures entered a period of increased UV sensitivity in the late log phase (just before they entered the stationary phase) in the former case. This increased sensitivity was associated with a decreased shoulder in the log survival–dose plots (see Section 7.2.2). The trend was not found in the radiation-sensitive, repair-deficient mutant.[150] A tentative explanation based on the partial depletion of the medium nutrients during the late log phase was advanced by the authors to explain the increased sensitivity in the former case.[150]

7.5.6. Flow Systems for UV Disinfection of Water

Practical UV reactors are necessarily flow-through systems rather than batch type. Fluid flow may be either parallel or perpendicular to the lamp axis. Alternatively, the fluid flows through tubes interspersed among the UV lamps.[151]

Various models of UV disinfection in a flow-through system have been presented.[152] The applicability of these models (which include CSTR, plug-

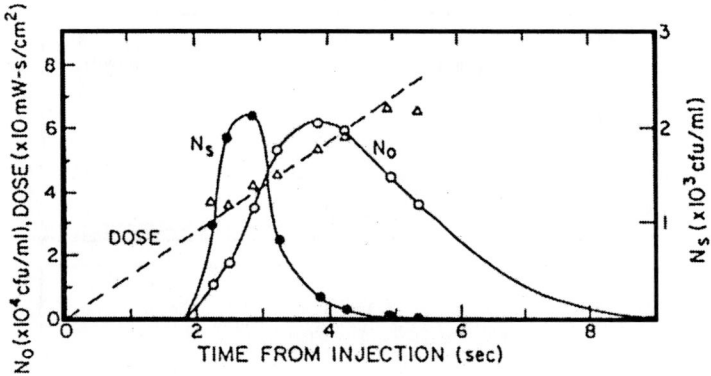

FIGURE 7.19. Assay of dose as a function of time from injection in a flow reactor. (Reproduced with permission from Qualls et al.[154])

flow, and series CSTR; see Section 2.9) has been critically reviewed.[152] An important parameter in these studies is the residence time distribution (RTD). The flow characteristics of a given unit may affect both the exposure time and the intensity to which various flow fractions are subjected within the distribution envelope. To complicate matters further, intensity gradients will exist in any chamber, and ideally, the microorganisms must be well mixed across these gradients. Attempts have been made[153] to improve mixing by adding baffles. Baffles can also aid in improving plug flow.

Another parameter in reactor design is lamp spacing. It has been suggested that closely spaced lamps are needed for the UV disinfection of waste water because of the high UV absorbance of wastewater solutions.[152] The product of average intensity times the volume of the chamber is a factor that is directly proportional to the *capacity* of the unit under *ideal flow conditions*. Thus the volume would serve as a weighting factor to account for the larger capacity of units with larger volume. This approach isolates the trade-off between intensity and volume from the effects of nonideal flow or how well the volume is used.[152] The *efficiency* of a unit expressed in terms of the ratio (intensity × time)/input lamp wattage is another useful figure of merit for reactor engineering purposes.[152]

How are the hydrodynamic characteristics (i.e., RTD) of a given UV flow disinfection reactor established? The very short exposure times typical of UV disinfection (a few seconds) present special problems in measuring the RTD by tracers. The bioassay method described earlier is useful in this application.[120] Thus, spores may be injected into the entrance of the reactor. Samples of the outflow may be collected at precisely timed intervals after injection. (This is best

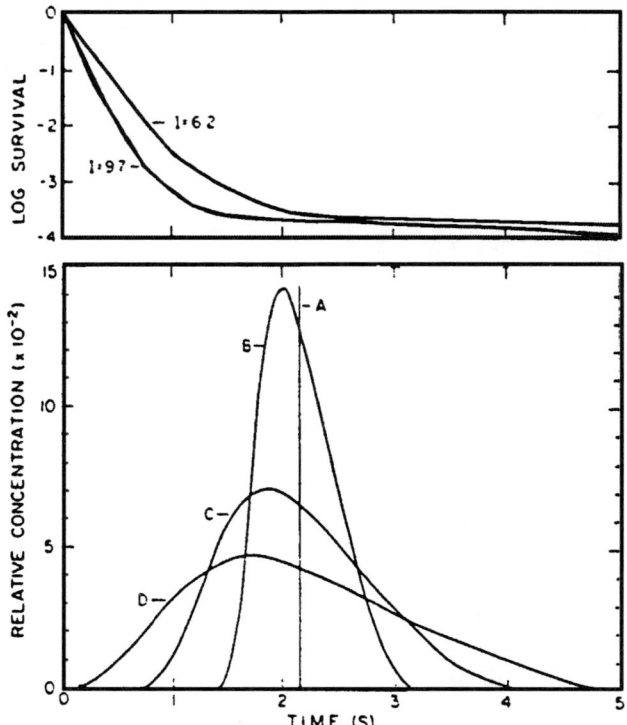

Figure 7.20. Effect of varying amounts of dispersion on average survival in residence time distributions with the same average residence time. Curve C was that measured for the Aquafine system. The other curves (A, B, and D) were generated by transforming the time axis symmetrically about the average residence time. Above the RTDs, the survival curve is shown as a function of residence time for the indicated intensities. Note what portions of the RTDs correspond to the "steep" and "flat" portions of the survival curves. (Reproduced with permission from Qualls and Johnson.[152])

done via an "auto-sampler" unit.) A number of such samples would provide data over the entire RTD curve. Representative data are contained in Figure 7.19.[154]

The effect of flow dispersion on disinfection efficiency is significant when a significant portion of the flow distribution is in a sensitive part of the survival–dose curve. In fact, the spores traversing the unit in the first half of the RTD dominate the average survival because of the logarithmic relationship of survival to dose (Eqs. (7.4) and (7.5)). This can be illustrated by superimposing the survival–dose and dispersion profiles (Figure 7.20).[152] In the flat region of the survival–dose curve, the flow dispersion obviously would exert little effect on the average survival. In this region, increasing dose by increasing either time or intensity would have little effect on survival.

7.5. UV Disinfection of Water

These considerations demonstrate the need for exercising caution in the interpretation of curves of average log survival *vs.* the residence time for practical flow-through systems. This is particularly the case when the residence time is computed from the volume/flow rate ratio. The first fractions of the RTD become disproportionately important in flow systems with large dispersion. Only in relatively few cases are true plug flow conditions achieved.

7.5.7. Photolysis of Aqueous Chlorine

Waters containing chlorine are exposed to sunlight not only in swimming pools but also in natural water receiving chlorinated effluents. Outdoor holding ponds for drinking water production and cooling tower reservoirs also are subjected to UV irradiation. Chlorine photolysis can also be considered to occur in atmospheric water droplets, where chlorine is produced from chloride by slow ozonation. Finally, chlorination and UV photolysis may be combined in a water treatment strategy for specialized applications.

Prompted by these possibilities, the photolysis of aqueous chlorine has been studied at solar and UV wavelengths.[155,156] This process can be approximated by a kinetic law that is first-order in chlorine residual. The pseudo-first order rate constants have been measured to be 2×10^{-4} s^{-1} for HOCl and 1.2×10^{-3} s^{-1} for OCl$^-$. The half-life increases with decreasing pH (from pH = 8) due to the decreasing ratio of OCl$^-$ and HOCl. The most effective photolysis wavelength is ca. 330 nm.

The number of reactive free radicals (·OH and Cl·) produced per molecule of photolytically decomposed aqueous Cl$_2$ can be determined by following the decomposition rate of added organic probes.[156] These probes are chosen such that their reaction rate constants with the generated radicals are previously well established. Only a small fraction of the ·OH radicals (formed from the photolyzed chlorine) will react with the probe, P, when this is present at low concentrations relative to scavengers ($\sum_i S_i$) such as DOM and bicarbonate–carbonate that are present in natural waters (Chapter 6).

This competition can be schematized as follows[156]:

$$Cl^{\cdot} \xrightarrow{h\upsilon} {}^{\cdot}OH \longrightarrow \begin{array}{c} \xrightarrow{P, \, k_p} P_{oxid} \\ \xrightarrow{\sum S_i, \, k_i} S_{i, \, oxid} \end{array}$$

Kinetics analysis leads to the prediction[156] that the logarithm of the relative residual probe concentration declines linearly with the dose of •OH released and subsequently consumed. A constraint for the efficacy of this prediction is that the scavenging efficiency does not vary widely over the course of the reaction. This kinetic model has been successfully tested with the photolysis of H_2O_2,[157] ozone decomposition,[158] photolysis of nitrate[159] and Fenton's reagent.[160] In all these cases, the active species are •OH radicals (Chapter 6). For this competition approach to be successful, several criteria will have to be satisfied: (a) the probes must be resistant both to direct photolysis and to chlorine in the dark; and (b) their reaction rates with radicals *other than* •OH (such as Cl•) must be small.

The yields of •OH thus assayed varied strongly with pH and with the irradiation wavelength.[156] These data indicate that aqueous Cl_2 is a rather less effective agent for •OH formation than aqueous ozone. For example, the •OH yield factor with Cl_2 is ~0.1 for pH > 8 at solar wavelengths, whereas it is ~0.6 for ozone *in the dark*.[156] It must be noted that the yield factor is not significantly changed for UV–O_3 or UV–H_2O_2–O_3.[156]

The preceding study also indicates that chloroform formation in filtered lake water samples is decreased during sunlight exposure because of the consumption of the chlorine residual.[156] On the other hand, an earlier study[161] had shown that UV exposure *increased* the incorporation of chlorine into the DOM. Clearly, further work is warranted to resolve these inconsistencies.

7.5.8. UV–O_3 and Other Advanced Oxidation Processes for DBP Control and Water Treatment

The feasibility of UV–O_3 for removing THM precursors has been evaluated by batch-scale kinetics studies on lake water samples high in THM precursors.[162] The idea here is that ozonation would be capable of oxidizing molecular sites that would otherwise produce THMs on chlorination. The complication is that oxidation of the carbonaceous matrix of natural waters would have the unfortunate consequence of also *producing* THM precursors. Water samples are routinely analyzed in the drinking water industry for their THM *formation potential* (THMFP). Thus, small increases in THMFP have been noted in some instances upon direct oxidation. These trends have been accommodated[162] within a kinetics scheme as follows:

$$P_1 + O_3 \xrightarrow{k_1} X \qquad (7.6)$$

$$C + O_3 \xrightarrow{k_2} P_2 \qquad (7.7)$$

7.5. UV Disinfection of Water

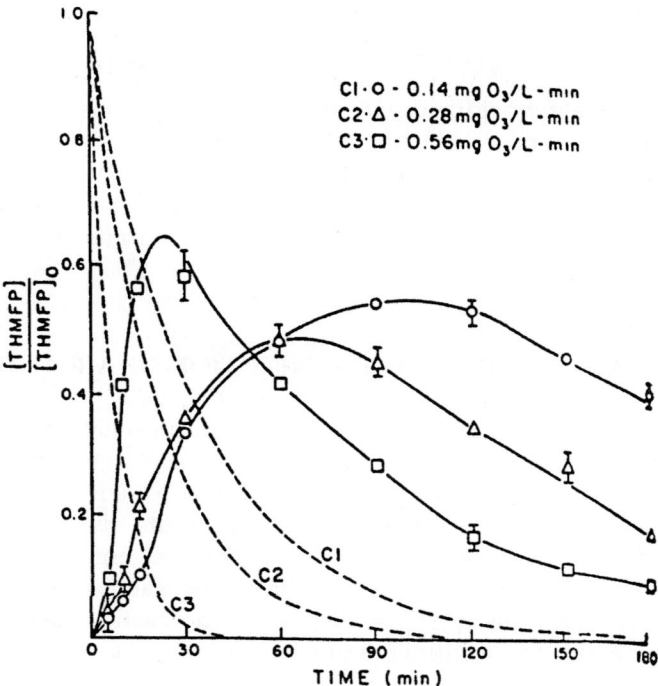

FIGURE 7.21. Destruction of primary precursor P_1 and formation–destruction of secondary precursor P_2 by ozonation of Caddo Lake water. (dashed line) decay of P_1; (solid line) formation and decay of P_2 (error bars shown in the latter). (Reproduced with permission from Glaze et al.[162])

$$P_2 + O_3 \xrightarrow{k_3} X \qquad (7.8)$$

$$C + O_3 \xrightarrow{k_4} X \qquad (7.9)$$

Here, P_1 represents a precursor initially present in the water, P_2 is precursor produced by ozonation of the carbon matrix C, and X is nonprecursor carbon. Figure 7.21 contains reconstructed experimental data for this study,[162] which show that P_2 is significantly less reactive than P_1. This is indicative that the two types of THM precursors are of different chemical nature. Overall, the conclusion was that UV–O_3 was more effective than ozone alone for the destruction of the THM precursors. In particular, in the "ozone-only" systems, the secondary precursors P_2 persisted for long periods; with simultaneous UV, they were destroyed more effectively.

Another study[163] compares peroxone ($H_2O_2 + O_3$) and O_3 for the control of DBPs and for the removal of taste and odor compounds. This pilot test shows that the peroxone process requires a significantly lower applied ozone dosage (relative to O_3 alone) for oxidizing MIB and geosmin. Preoxidation with O_3 or peroxone followed by post-disinfection with chloramines is being considered as a viable alternative by this particular water utility for maintaining a low level of THMs/DBPs and microorganisms in the water. On the other hand, post-disinfection with Cl_2 resulted in >50 µg THMs/liter after one day in this particular series of tests.

7.6. PHOTOELECTROCHEMICAL DISINFECTION OF AIR AND WATER

We have already seen in Chaper 6 that bandgap excitation of TiO_2 produces the highly potent •OH radicals at either the TiO_2/air or TiO_2/water interfaces. These radicals in turn can be used to inactivate microorganisms in the contacting medium. Aside from this "mediated" route, direct attack of the microorganisms by the photogenerated holes in TiO_2 also cannot be discounted. An intriguing practical application of gas-phase photocatalysis is the possibility of coating interior walls and floor tiles with a photoactive material such as TiO_2 and using either natural sunlight or room lighting (e.g., fluorescence) for disinfection and deodorizing purposes.[164] This application is particularly attractive in a hospital setting for example.

The bactericidal properties of irradiated TiO_2 suspensions appear to have been first established in ca. 1985.[165] Since then, a number of microorganisms have been treated by the heterogeneous photocatalysis approach (Table 7.4).[165-174] There are no reports as yet of the successful treatment of protozoan cysts (e.g., *Giardia*, *Cryptosporidium*) although a variety of bacteria have been successfully treated, and there is at least one report[166] on the photocatalytic inactivation of a virus.

Innovations in photocatalytic disinfection include the development of a continuous (flow)-sterilization system using acetyl cellulose membrane-immobilized TiO_2,[167] enhancement of inactivation by added Fe(II),[174] the combined use of UV-TiO_2 and H_2O_2 for microorganism inactivation,[168] the use of "diffuse-light emitting" optical fibers for enhancing the light throughput,[169] the use of metal island-modified TiO_2,[165,170] and the demonstration of a sunlight-driven outdoor water disinfection system.[171]

These advances must be balanced against several mitigating factors with this approach. A study[168] of pond water containing significant amounts of algae and TOC (~20 mg/liter) showed little inactivation (ca. 1 log survival unit as measured by the heterotrophic plate count) although the *total coliform count* reduction was significant. Clearly, as with the organic pollutant cases consid-

7.6. Photoelectrochemical Disinfection of Air and Water

Table 7.4. Microorganisms Inactivated by TiO_2-Based Heterogeneous Photocatalysis

Microorganism(s)	Reference(s)
Escherichia coli	165–173
Lactobacillus acidophilus	165
Saccharomyces cerevisiae	165
Phage MS2	174
Poliovirus 1	166
Fecal coliforms	173

ered earlier in Chapter 6, •OH radical scavengers can have an adverse impact on the efficacy of the process.

Second, the contact times needed for disinfection to the levels needed to meet effluent discharge requirements (i.e., 4 log units) appear to be prohibitively long relative to conventional disinfectants such as Cl_2. For example, 2 log inactivation of coliform bacteria and poliovirus 1 required ca. 150 and 30 min; respectively.[166] This can be compared with 5.5 log reduction of total coliforms with 30 min contact time with 8 mg/liter residual chlorine,[175] or 4 log reduction of poliovirus 1 with 16.2 min contact time with a free chlorine residual even lower at 0.5 mg/liter.[166]

In spite of these deficiencies, the heterogeneous photocatalytic disinfection approach does have several positive attributes. It is particularly amenable to rural and remote (e.g., camping, military) applications for the treatment of potable water. Unlike the ozonation and the direct UV-radiation approaches, this approach does not require high-voltage supplies and can use natural sunlight along with an abundantly available photocatalyst material. Generating disinfection agents such as chlorine, ozone, or H_2O_2 *on site* poses special problems. On the other hand, transporting these chemicals over long distances is neither cost effective nor safe. The UV–TiO_2 approach can also be used for the removal of taste- and odor-causing organics and for lowering the total organic content of the water prior to a chlorination step.

Finally, the UV–TiO_2 disinfection approach shares with the UV photolysis counterpart the handicap that no residual disinfection capacity is inherently built into this method. An assessment of the current state of this technology leads to the conclusion that further process improvements are needed before heterogeneous photocatalysis can be seriously considered for routine adop-

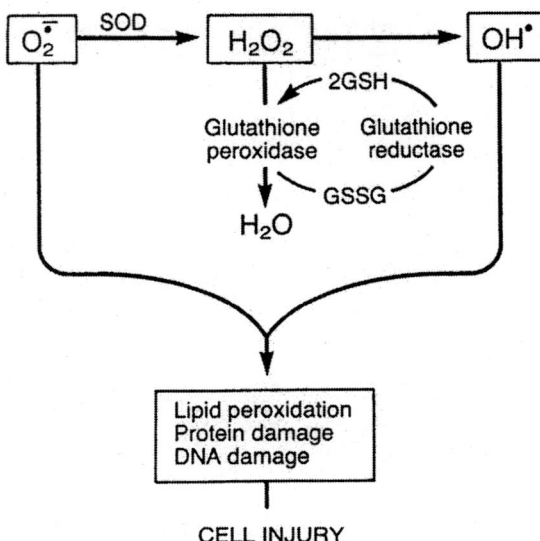

FIGURE 7.22. Schematic diagram of routes involving ·OH-induced inactivation of microbial cells. (Adapted from Cotran et al.[35])

tion. Especially important would be the demonstration that it can be used for the treatment of "difficult" pathogens. Further understanding at a fundamental mechanistic level is also needed.

While ·OH radicals are well-known in their ability to inactivate microorganisms,[35] the importance of other (competitive) routes is not clear. These include the implication of intracellular coenzyme A,[165,167] oxidation of the cell wall or cell membrane, and the importance of diffusion of the disinfecting species. Figure 7.22 contains a schematic of routes involving ·OH-induced inactivation of microbial cells.[35]

7.7. ELECTROCHEMICAL DETECTION AND ENUMERATION OF MICROORGANISMS

The standard methods for coliform assay are the most probable number and membrane filtration tests.[176] Both tests are laborious and time consuming (requiring several hours). Electrochemical methods can play a key role in the development of simple and rapid methods for coliform assay.

The methods developed thus far are based on the potentiometric measurement of a *lag time*.[177-184] This lag time can be due to either gas (H_2) production

7.7. Electrochemical Detection and Enumeration of Micro-Organisms

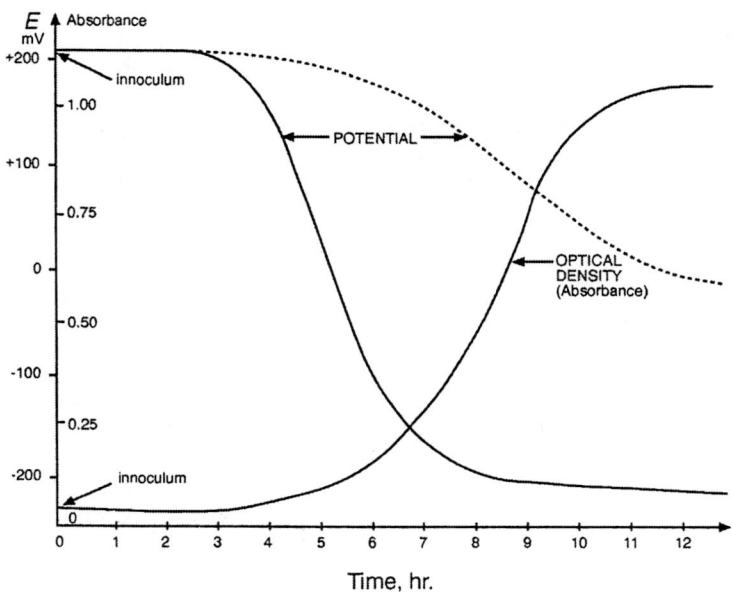

FIGURE 7.23. Evolution with time of the electric potential and of the optical density (absorbance) in the minimal glucose growth medium: (solid line) in presence of lipoic acid (2 mg per liter or 10^{-5} M); (dashed line) without lipoic acid. (Reproduced with permission from Junter et al.[181])

by the microorganism[177] or dissolved O_2 uptake by it, the potential–time profile showing an abrupt discontinuity in either case.

Subsequent work[181-183] has shown that the reduction of lipoic (thioctic) acid by bacteria coupled with O_2 consumption during glucose absorption can be followed potentiometrically with a gold indicator electrode–reference electrode combination in a minimal culture medium containing only salts and glucose. The simplified composition of the culture broth is claimed to limit the number of possible metabolic pathways during bacterial growth and thus enhances the reproducibility of the redox potentials measured *in situ*. Figure 7.23 contains a representative time profile of the potential and the absorbance of a minimal growth medium containing *E. coli*.[181] The influence of adding lipoic acid is also illustrated in this figure. Figure 7.24a contains the time evolution of potential for various cases tested in this work.[181]

Theoretical expressions for the lag time (t_1) have been presented under various assumptions[181]:

$$t_1 = \frac{p}{\ln 2} \ln \left[1 + \frac{[O_2]_o}{Q_m' N_o} \right] \tag{7.10}$$

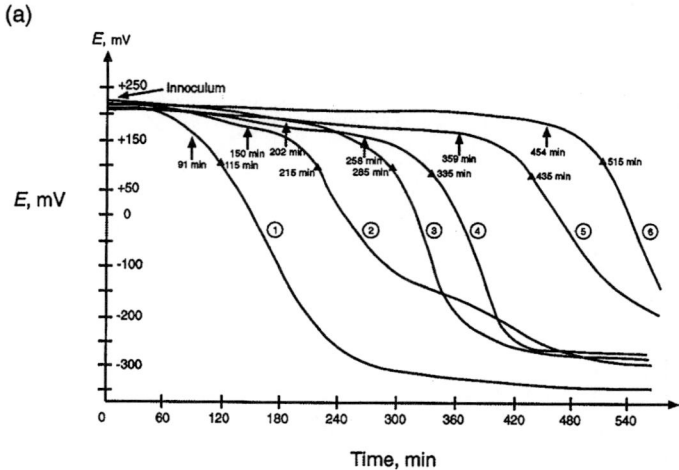

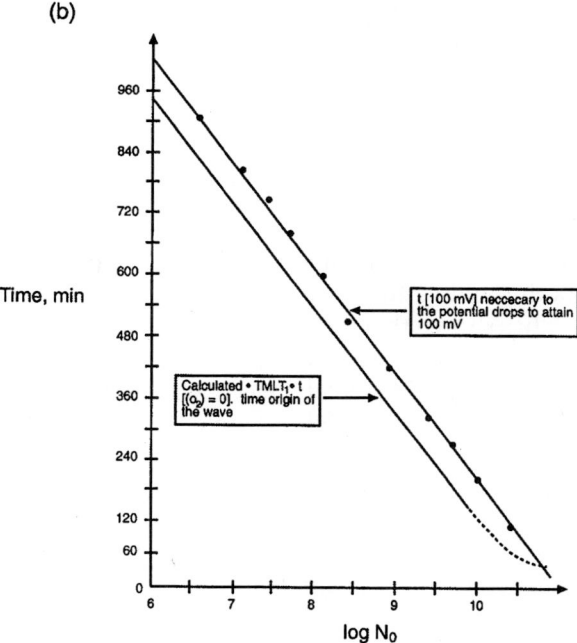

FIGURE 7.24. (a) Time evolution of potential for various initial cell concentrations (x_0) in the minimal medium. Symbols: arrows, calculated TMLT, t_1; triangles, experimental t (100 mV); N_o values (in cells per liter) were as follows: Curve 1, 2.6×10^{10}; curve 2, 1.04×10^{10}; curve 3, 5.2×10^9; curve 4, 2.6×10^9; curve 5, 7.8×10^8; curve 6, 2.6×10^8. (b) Calculated and experimental plots of t_1 vs. the logarithm of the initial cell concentration, N_o in the minimal medium. (Reproduced with permission from Junter et al.[181])

7.7. Electrochemical Detection and Enumeration of Micro-Organisms

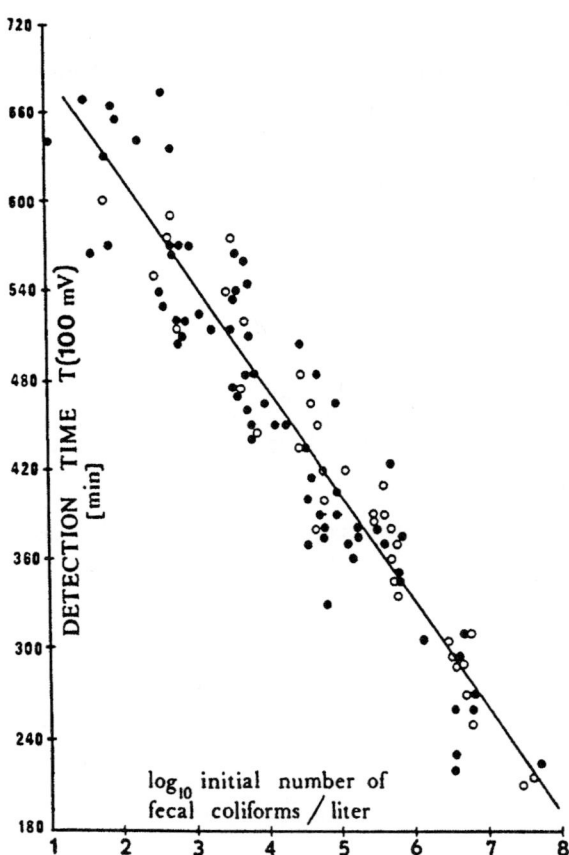

FIGURE 7.25. Potentiometric detection times [t (100 mV)] as a function of the initial concentration of fecal coliform organisms in flasks inoculated with wastewater samples. Symbols: ●, experimental data attributable to *E. coli* alone; ○, experimental data attributable to *E. coli* accompanied by *K. pneumoniae*; solid line, correlation line obtained by least-squares analysis of experimental data attributed to *E. coli* alone or accompanied by *K. pneumoniae*. (Reproduced with permission from Jouenne et al.[183])

where $[O_2]_o$ is the initial concentration of dissolved O_2 in the medium, Q'_m is a corrected specific O_2 uptake rate, N_o is the initial cell concentration in the liquid, and p is a slope factor. The Q'_m value can be determined with an O_2 (Clark) amperometric electrode (see Section 4.6) and has been found to be ~4.5×10^{-15} mol/cell for the particular strain and the minimal growth medium considered in the above work.[181] With a nominal value for $[O_2]_o$ of 2.15×10^{-4} M, this expression becomes

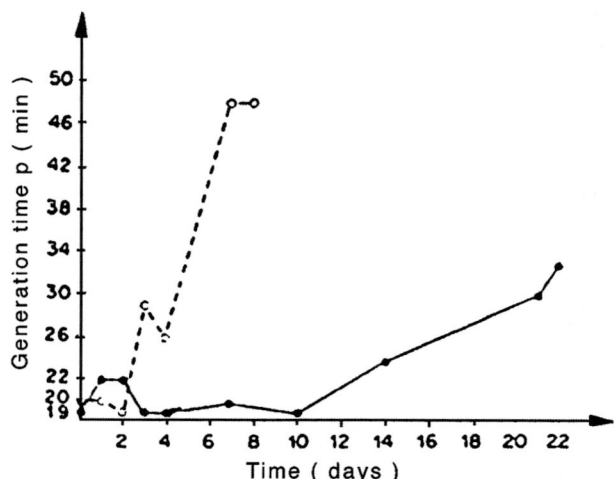

FIGURE 7.26. Changes in potentiometric generation parameter p of strains A (●) and B (○) during the *in situ* experiments. (Reproduced with permission from Jouenne et al.[187])

$$t_1 = \frac{p}{\ln 2} \ln\left[1 + \frac{4.8 \times 10^{10}}{N_o}\right]$$

$$\simeq \frac{p}{\ln 2} \ln\left[\frac{4.8 \times 10^{10}}{N_o}\right] \tag{7.10a}$$

Thus a plot of t_1 vs. log N_o is predicted to be linear; this is borne out by the analysis in Figure 7.24b.

The lag time, t_1, is measured for an arbitrarily chosen 100 mV potential shift from the initial value.[181–183] Multiplying the slope of the t_1 vs. log N_o plot by log 2 yields the "generation time."

This potentiometric method has been extended to various laboratory strains of organisms belonging to aquatic flora[183] and fecal coliforms other than *E. coli* (*Enterobacter cloacae, Klebsiella pneumoniae, Acinetobacter calcoaceticus*, and *Citrobacter*).[182,183] Figure 7.25 contains a t_1 (100 mV) vs. log N_o plot for fecal coliform analysis.[183] This method has been patented both in the United States and in Europe by the authors, and assembly and demonstration of units amenable to automated, on-site analysis of water samples are being considered. A review is also available.[184]

Electrochemical detection methods are also useful for probing the survival patterns and the influence of sublethal stress on microorganisms (Section 7.5.5). An early work demonstrated the influence of salinity on sublethal stress in *E. coli* as measured by electrochemical lag times.[185] The *stress* was defined as the difference between a predicted lag time for unstarved cells (as prepared from a standard curve) and the observed lag time for cells starved in seawater. The higher the salinity, the greater was the stress for all the test media examined.[185] A subsequent study probed the survival of *E. coli* and *Salmonella* spp. in estuarine environments.[186] Finally, the survival patterns of two wild strains of *E. coli* (clinical isolates from human feces) were compared during *in situ* exposure of the organisms to environmental fresh water in dialysis testing.[187] These strains exhibited significant differences as expressed by the plots of the parameter, p (Eqn. 7.10) *vs.* time (Figure 7.26).[187] The lower survivability of strain B in the aquatic environment was confirmed by standard enumeration techniques.

7.8. SUMMARY

In this chapter, electrochemical, photochemical, and photoelectrochemical methods for inactivating and enumerating microorganisms in water have been reviewed. These methods have been discussed against the backdrop of current practice in the water treatment industry. In the next chapter, we will examine to what extent the approaches discussed in Chapters 4 through 7 have attained their commercial potential.

REFERENCES

1. S. D. Faust and O. M. Aly, "Chemistry of Water Treatment," Chapter 10, p. 597. Butterworth, Boston, 1983.
2. E. D. Mintz, F. M. Reift, and R. V. Tauxe, Safe Water Treatment and Storage in the Home. A Practical New Strategy to Prevent Waterborne Disease. *J.A.M.A.* **273**, 948 (1995).
3. Metcalf and Eddy, Inc., "Wastewater Engineering Treatment, Disposal and Re-Use," Chapter 7. McGraw-Hill, New York, 1991.
4. C. N. Sawyer, P. L. McCarty, and G. F. Parkin, "Chemistry for Environmental Engineering," Chapter 19. McGraw-Hill, New York, 1994.
5. R. P. Allison, Electrodialysis Membrane Performance Characteristics. *Proc. Membr. Technol. Conf.*, Am. Water Works Assoc., Denver, CO (1993); *Chem. Abstr.* **121**, 116850 (1993).
6. S. Miller, Disinfection Products in Water Treatment. *Environ. Sci. Technol.* **27**, 2292 (1993).
7. C. A. Garron, Disinfectant Chemical Use in Domestic Water and Wastewater Treatment in the Atlantic Region in 1990. *Surveillance Rep. EPS—*

Can. *(Environ. Prot. Serv.)* **EPS-5-AR-92-2**; *Chem. Abstr.* **118**, 11438 (1990).
8. Water Quality Division Disinfection Committee, Survey of Water Utility Disinfection Practices. *J. Am. Water Works Assoc.* **84**, 121 (1992).
9. J. E. Bennett and J. E. Elliot, Anodically Polarized Surface for Biofouling and Scale Control. U. S. Pat. 4,345,981 (1982).
10. P. M. Bersier, L. Carlsson, and J. Bersier, Electrochemistry for a Better Environment. *Top. Curr. Chem.*, **170**, 144 (1994).
11. L. G. Bompastor, Process for Electrolytic Treatment of Supply Water and Wastewater. *Braz. Pedido PI* **BR 9200133 A** (1993); *Chem. Abstr.* **120**, 143251 (1993).
12. Y. Nakagawa, Bipolar Membrane Used in Tap Water Disinfection. *Jpn. Kokai Tokkyo Koho* **JP 06165989 A2** (1994); *Chem. Abstr.* **121**, 186714 (1994).
13. "The Merck Index," 11th ed. p. 324. Merck, Rahway, NJ, 1989.
14. R. E. Stetler, S. C. Waltrip, and C. J. Hurst, Virus Removal and Recovery in the Drinking Water Treatment Train. *Water Res.* **26**, 727 (1992).
15. A. Katz, N. Narkis, F. Orshansky, E. Friedland, and Y. Kott, Disinfection of Effluent by Combinations of Equal Doses of Chlorine Dioxide and Chlorine Added Simultaneously over Varying Contact Times. *Water Res.* **28**, 2133 (1994).
16. K. V. Ellis, A. P. Cotton, and M. A. Khowaja, Iodine Disinfection of Poor Quality Waters. *Water Res.* **27**, 369 (1993).
17. G. L. Hatch, J. L. Lambert, and L. R. Fina, Some Properties of the Quatenary Ammonium Anion-Exchange Resin-Triiodide Disinfectant for Water. *Ind. Eng. Chem. Prod. Res. Dev.* **19**, 259 (1980).
18. J. L. Lambert, G. T. Fina, and L. R. Fina, Preparation and Properties of Triiodide-, Pentaiodide- and Heptaiodide-Quatenary Ammonium Strong Base Anion-Exchange Resin Disinfectant. *Ind. Eng. Chem. Prod. Res. Dev.* **19**, 256 (1980).
19. M.-M. Bourbigot, Ozone Disinfection in Drinking Water. *J. New Engl. Water Works Assoc.* **103**, 1 (1989).
20. T. Ohmi, T. Isagawa, T. Imaoka, and I. Sugiyama, Ozone Decomposition in Ultrapure Water and Continuous Ozone Sterilization for a Semiconductor Ultrapure Water System. *J. Electrochem. Soc.* **139**, 3336 (1992).
21. "Electrolytic Ozonizer and its Application," Purezone Brochure. Permelec Electrode Ltd., Kanagawa, Japan.
22. W. Dunkieberger and B. A. Beaudet, Ozonation Enhances Water Quality at Florida Utility. *Water/Eng. Manage.* May, Vol. **32** (1994).
23. R. Song, A. Eyring, M. Cowell, R. Minear, P. Westerhoff, and G. Amy, Ozone-Bromide-NOM Interactions in Water Treatment. 3. Formation of Brominated Organic DBP's During Water Ozonation. *208th Am. Chem. Soc. Meet.*, Washington, DC (1994).
24. S. H. Lin and K. L. Yeh, Looking to Treat Wastewater? Try Ozone. *Chem.*

Eng. **100**, 112 (1993).
25. M. S. Stein, W. J. O'Brien, and J. K. Murphy, Strict Water Regulations Call for New Processes. *Water/Eng. Manage.* May, Vol. 20 (1992).
26. N. Seki and K. Ogawa, Apparatus for Decomposition of Dissolved Ozone in Water. *Jpn. Kokai Tokyo Koho* **JP 05050074 A2** (1993); Heisei, *Chem. Abstr.* **119**, 55641 (1993).
27. M. R. Matsumoto, J. N. Jensen, P. McGinley, and B. E. Reed, Physicochemical Processes. *Water Environ. Res.* **66**, 316 (1994).
28. T. Otani, H. Takahashi, and T. Matsunaga, Apparatus for Production of Ozonized Water. *Jpn. Kokai Tokyo Koho* **JP 03254890 A2** (1991); Heisei, *Chem. Abstr.* **116**, 135961 (1991).
29. J. Hayes, Improvements to Ionic Disinfection Systems. *Chem. Abstr.* **121**, 286304 (1994).
30. H. Nishio, M. Ikoma, S. Katsumata, and S. Takagi, Method and Apparatus for Disinfection of Cooling Water in Cooling Towers. *Jpn. Kokai Tokyo Koho* **JP 05245478 A2** (1993); Heisei, *Chem. Abstr.* **120**, 86034 (1993).
31. M. Kimura, Apparatus for Water Purification with Metal Ions. *Jpn. Kokai Tokyo Koho* **JP 05261376 A2** (1993); Heisei, *Chem. Abstr.* **120**, 86039 (1993).
32. A. Bosch, J. M. Diez, and F. X. Abad, Disinfection of Human Enteric Viruses in Water by Copper: Silver and Reduced Levels of Chlorine. *Water Sci. Technol.* **27**, 351 (1993).
33. G. C. White,"Handbook of Chlorination." Van Nostrand-Reinhold, New York, 1972.
34. E. R. Baumann and D. D. Ludwig, Free Available Chlorine Residuals for Small Nonpublic Water Supplies. *J. Am. Water Works Assoc.* **54**, 1379 (1962).
34a. C. N. Hass and R. S. Engelbrecht, Physiological Alterations of Vegetative Microorganisms Resulting from Aqueous Chlorination. *J. Water Pollut. Control Fed.* **52**, 1976 (1980).
34b. E. T. Bryant, G. P. Fulton, and G. C. Budd, Disinfection *In* "Disinfection Alternatives for Safe Drinking Water," pp. 15-42. Van Nostrand-Reinhold, New York, 1992.
34c. V. W. Riesser, J. R. Perrich, B. B. Silver, and J. R. McCammon, Possible Mechanisms of Poliovirus Inactivation by Ozone. *In* "Forum on Ozone Disinfection" (E. G. Fochtman, R. G. Rice, and M. E. Browning, eds.), pp. 186-192. International Ozone Institute, New York, 1977.
34d. D. Ray, P. K. Y. Wong, R. S. Engelbrecht, and E. S. K. Chian, Mechanism of Enterioviral Inactivation by Ozone. *Appl. Environ. Microbiol.* **41**, 718 (1981).
35. R. S. Cotran, V. Kumar, and S. L. Robbins, "Pathologic Basis of Disease," pp. 8-12. Saunders, Philadelphia, 1989.
36. J. J. Rook, Haloforms in Drinking Water. *J. Am.Water Works Assoc.* **68**, 168 (1976).
37. J. J. Rook, Chlorination Reactions of Fulvic Acids in Natural Waters.

Environ. Sci. Technol. **11**, 478 (1977).
38. G. M. Thompson and J. M. Hayes, Trichlorofluoromethane in Groundwater — A Possible Tracer and Indicator of Groundwater Age. *Water Resour. Res.* **15**, 546 (1979).
39. G. L. Amy, L. Tan, and M. K. Davis, The Effects of Ozonation and Activated Carbon Adsorption on Trihalomethane Speciation. *Water Res.* **25**, 191 (1991).
40. A. G. Myers, Evaluating Alternative Disinfectants for THM Control in Small Systems. *J. Am. Water Works Assoc.* **82**, 77, June (1990).
41. J. G. Jacangelo, J.-M. Laine, E. W. Cummings, and S. S. Adham, UF with Pretreatment for Removing DBP Precursors. *J. Am. Water Works Assoc.* p. 100, March (1995).
42. M. Rebhun, J. Manka, and A. Zilberman, Trihalomethane Formation in High-Bromide Lake Galilee Water. *J. Am. Wat. Works Assoc.* **80**, 84 (1988).
43. A. Eaton, Measuring UV-Absorbing Organics: A Standard Method. *J. Am. Water Works Assoc.*, p. 86, February (1995).
44. W. J. Cooper, R. G. Zika, and M. S. Steinhauer, Bromide-Oxidant Interactions and THM Formation: A Literature Review. *J. Am. Water Works Assoc.* **77**, 116 (1985).
45. W. J. Cooper, G. L. Amy, C. A. Moore, and R. G. Zika, Bromate Formation in Ozonated Groundwater Containing Bromide and Humic Substances. *Ozone: Sci. Eng.* **8**, 63 (1986).
46. W. R. Haag and J. Hoigne, Ozonation of Bromide-Containing Waters: Kinetics of Formation of Hypobromous Acid and Bromate. *Environ. Sci. Technol.* **17**, 261 (1983).
47. "Annual Book of ASTM Standards," Part 31. American Society for Testing and Materials, Philadelphia, 1975.
48. S. W. Krasner, M. J. McGuire, and V. B. Ferguson, Tastes and Odors: The Flavor Profile Method. *J. Am. Water Works Assoc.* p. 34, March (1985).
49. R. L. Wolfe, Ultraviolet Disinfection of Potable Water. *Environ. Sci. Technol.* **24**, 768 (1990).
50. S. M.. Goyal, Indicators of Viruses. *In* "Viral Pollution of the Environment" (G. Berg, ed.), pp. 211-230. CRC Press, Boca Raton, FL, 1983.
51. P. Payment and R. Armon, Virus Removal by Drinking Water Treatment Processes. *CRC Crit. Rev. Environ. Control* 31 (1989).
52. P. Payment and E. Franco, *Clostridium perfringens* and Somatic Coliphages as Indicators of the Efficiency of Drinking Water Treatment for Viruses and Protozoan Cysts. *Appl. Environ. Microbiol.* **59**, 2418 (1993).
53. J. M. Hinden, L. M. Ernes, and P. E. Visel, Electrocatalytic Electrode. U. S. Pat. 4,517,068 (1985).
54. N. Goshima, H. Hashimoto, and T. Takahashi, Electrolytic Treatment of

Waters. *Jpn. Kokai Tokyo Koho* **JP 06086981 A2** (1994); Heisei, *Chem. Abstr.* **121**, 42325 (1994).

55. F. Stummer and J. Miller, Electrolytic Cell for Treatment of Water Solutions. U. S. Pat. 4,169,035 (1979).
56. G. E. Stoner, G. L. Cahen, M. Sachyani, and E. Gileadi, The Mechanism of Low Frequency AC Electrochemical Disinfection. *Bioelectrochem. Bioenerg.* **9**, 229 (1982).
57. G. E. Stoner, Electrochemical Inactivation of Pathogens. U. S. Pat. 3,725,226 (1973).
58. Y. Oren, H. Tobias, and A. Soffer, Removal of Bacteria from Water by Electroadsorption on Porous Carbon Electrodes. *Bioelectrochem. Bioenerg.* **11**, 347 (1983).
59. D. Golub, E. Ben-Hur, Y. Oren, and A. Soffer, Electroadsorption of Bacteria on Porous Carbon and Graphite Electrodes. *Bioelectrochem. Bioenerg.* **17**, 175 (1987).
60. Y. Karizume, Y. Saihara, T. Omochi, and A. Kishimoto, Water-Purification Apparatus with Means for Regeneration. *Jpn. Kokai Tokyo Koho* **JP 05228468 A2** (1993); Heisei, *Chem. Abstr.* **119**, 256260 (1993).
61. J. O'M. Bockris, R. C. Bhardwaj, and C. L. K. Tennekoon, Electrochemistry of Waste Removal. *Analyst (London)* **119**, 781 (1994).
62. Y. Kitajima, A. Shigomatsu, N. Nakamura, and T. Matsunaga, Electrochemical Sterilization Using Carbon Fiber Microelectrode. *Denki Kagaku* **52**, 1082 (1988).
63. T. Matsunaga, Y. Namba, and T. Nakajima, Electrochemical Sterilization of Microbial Cells. *Bioelectrochem. Bioenerg.* **13**, 393 (1985).
64. T. Matsunaga, S. Nakasono, and S. Masuda, Electrochemical Sterilization of Bacteria Adsorbed on Granular Activated Carbon. *FEMS Microbiol. Lett.* **93**, 255 (1992).
65. S. Nakasono, N. Nakamura, K. Sode, and T. Matsunaga, Electrochemical Disinfection of Marine Bacteria Attached on a Plastic Electrode. *Bioelectrochem. Bioenerg.* **27**, 191 (1992).
66. S. Nakasono, J. G. Burgess, K. Takahashi, M. Koike, C. Murayama, S. Nakamura, and T. Matsunaga, Electrochemical Prevention of Marine Biofouling with a Carbon-Chloroprene Sheet. *Appl. Environ. Microbiol.* **59**, 3757 (1993).
67. K. C. Marshall, Biofilms. An Overview of Bacterial Adhesion, Activity, and Control at Surfaces. *ASM News* **58**, 202 (1992).
68. C. T. Sweeney, Generation of Chlorine-Chlorine Dioxide Mixtures. U. S. Pat. 4,308,117 (1981).
69. F. Nakamura, Apparatus and Method for Decomposition and Reclamation of Salts and Inorganic Nitrogen Compounds from Wastewater Treatment. *Jpn. Kokai Tokyo Koho* **JP 06182344 A2** (1994); Heisei, *Chem. Abstr.*

121, 2859848 (1994).
70. P. Fabian, M. Gündling, and P. Rössler, Electrolytic Cell. U. S. Pat. 4,422,919 (1983).
71. T. Okazaki, Washing and Disinfection in Apparatus for Manufacture of Ionized Waters by Electrolysis. *Jpn.Kokai Tokyo Koho* **JP 06047381 A2** (1994); Heisei, *Chem. Abstr.* **120**, 330730 (1994).
72. T. Shinohara, K. Tsuruta, J. Nakakubo, K. Nagata, and K. Kanzaki, Apparatus for Disinfection of Drinking Water in Vending Machine. *Jpn. Kokai Tokyo Koho* **JP 03293093 A2** (1991); Heisei, *Chem. Abstr.* **116**, 200826 (1991).
73. K. Kanzaki, Apparatus for Control of Chlorine Generator for Drinking Water. *Jpn. Kokai Tokyo Koho* **JP 04066186 A2** (1992); Heisei, *Chem. Abstr.* **117**, 55626 (1992).
74. F. Nakagawa, Apparatus for Chlorine Generation in Stored Tap Water. *Jpn. Kokai Tokyo Koho* **JP 05269469 A2** (1993); Heisei, *Chem. Abstr.* **120**, 116431 (1993).
75. H. Kayano, Y. Nakakuki, T. Onomi, and H. Watanabe, Apparatus Containing Electrolysis and Bubbling Means for Purifying Seawater. *Jpn.Kokai Tokyo Koho* **JP 06178982 A2** (1994); Heisei, *Chem. Abstr.* **121**, 212582 (1994).
76. R. Fate de Campos and P. E. Cohn, Electrochemical Apparatus for Water Sterilization. *Braz. Pedido PI* **PI BR 9202824 A** (1994); *Chem. Abstr.* **120**, 307009 (1994).
77. H. Iwata, M. Shinagawa, Y. Nishikawa, N. Kawamura, and A. Kishimoto, Apparatus for Water Purification. *Jpn. Kokai Tokyo Koho* **JP 04265192 A2** (1992); Heisei, *Chem. Abstr.* **118**, 27229 (1992).
78. S. Varennes, Electrochemical Bleaching of Kraft and Mechanical Pulp. *6th Int. Forum Electrolysis Chem. Ind.*, The Electrosynthesis Co., Fort Lauderdale, FL (1992).
79. A. Zurbruegg, Water Disinfection with Electrolytically Manufactured Chlorine at the Blaesi Public Baths in Zurich (Switzerland). *Umwelttechnik* **26**, 5 (1992); *Chem. Abstr.* **116**, 262204 (1992).
80. P. Schaetzle, Disinfection of Swimming Pool Water. Gaseous Chlorine and the Legislation on Accidents. *Umwelttechnik* **26**, 10 (1992); *Chem. Abstr.* **118**, 131634 (1992).
81. F. Heurtebise and P. Teissedre, Disinfection of Water by Electrochloration Using a Combination of Alkali Metal Chlorides and a Stabilizer Such as Isocyanuric Acid. *Fr. Demande* **FR 2670198 A1** (1992); *Chem. Abstr.* **118**, 27221 (1992).
82. P. Leroy, Treatment of Desalinated Water. Recommendations for Corrosion Control. *Tech., Sci., Methodes: Genie Urbain-Genie Rural* **1**, 13 (1993); *Chem. Abstr.* **119**, 188185 (1993).
83. R. Bolton, C. Marson, J. W. Bess, Jr., C. D. Ellingson, and D. W. Hill, Main-

taining Efficiency of an Electrochlorination Disinfection System Operating in Adverse Seawater Conditions. *Off. Proc. - 52nd Int. Water Conf.*, p. 344 (1991); *Chem Abstr.* **117**, 76152 (1991).
84. J. O'M. Bockris, "Electrochemistry of Cleaner Environments," Chapter 4. Plenum, New York, 1972.
85. T. Okazaki, Y. Sasaki, H. Kitamur, and K. Oshima, Apparatus for Disinfection of Tap Water in Electrolysis Tank Using Hypochlorites. *Jpn. Kokai Tokyo Koho* **JP 04094787 A2** (1992); Heisei, *Chem. Abstr.* **117**, 137356 (1992).
86. Liquid Disinfectant for Water Purification by Electrolysis. *Jpn. Kokai Tokyo Koho* **JP 04131184 A2** (1992); Heisei, *Chem. Abstr.* **117**, 198160 (1992).
87. T. Okazaki, Disinfection of Water by Electrolysis with Addition of Chlorides. *Jpn. Kokai Tokyo Koho* **JP 06106169 A2** (1994); Heisei, *Chem Abstr.* **121**, 91231 (1994).
88. T. Okazaki, Y. Sasaki, H. Kitamura, and K. Ooshima, Manufacture of Sterilized Water. *Jpn. Kokai Tokyo Koho* **JP 05237478 A2** (1993); Heisei, *Chem. Abstr.* **119**, 256262 (1993).
89. Apparatus for Disinfection of Tap Water Using Hypochlorites in Electrolysis Tank. *Jpn. Kokai Tokyo Koho* **JP 04094788 A2** (1992); Heisei, *Chem. Abstr.* **117**, 137355 (1992).
90. E. Mourgues, Means for Dosing Water Disinfection Chemicals into a Water Basin Such as a Swimming Pool. *Fr. Demande* **FR 2676218 A1** (1992); *Chem. Abstr.* **118**, 219402 (1992).
91. S. Trasatti and G. Lodi, Properties of Conductive Transition Metal Oxides with Rutile-Type Structure. *In* "Electrodes of Conductive Metallic Oxides" (S. Trasatti, ed.), Part A, p. 301. Elsevier, Amsterdam, 1980.
92. J. J. Kaczur, D. W. Cawlfield, K. E. Woodard, Jr., and B. L. Duncan, Chloric Acid-Alkali Metal Chlorate Mixtures for Chlorine Dioxide Generation. U. S. Pat. 5,242,553 (1993).
93. J. J. Kaczur, D. W. Cawlfield, J. F. Watson, C. J. Rolison, S. K. Mendiratta, and R. T. Brooker, Electrolytic Production of Chloric Acid and Sodium Chlorate Mixtures for the Generation of Chlorine Dioxide. U. S. Pat. 5,242,554 (1993).
94. R. J. Coin, J. E. Elliott, E. J. Rudd, and A. R. Sacco, System for Electrolytically Generating Strong Solutions by Halogen Oxyacids. U. S. Pat. 5,242,552 (1993).
95. B. Wang, Apparatus and Process for Disinfecting Wastewater. Can. Pat. Appl. CA 2089961 AA (1993); *Chem. Abstr.* **120**, 85797 (1993).
96. K. Taku, Disinfection of Waters Using Chlorine Dioxide Generated from Chlorine Gas. *Jpn. Kokai Tokyo Koho* **JP 05161890 A2** (1993); Heisei, *Chem. Abstr.* **119**, 256246 (1993).
97. H. P. Dhar, J. O'M. Bockris, and D. H. Lewis, Electrochemical Inactiva-

tion of Marine Bacteria. *J. Electrochem. Soc.* **128**, 229 (1981).
98. K. Shimada and K. Shimahara, Responsibility of Hydrogen Peroxide for the Lethality of Resting *Escherichia coli* B Cells Exposed to Alternating Current in Phosphate Buffer Solution. *Agric. Biol. Chem.* **46**, 1329 (1982).
99. A. Ponta and A. Kulhanek, Process for the Electrochemical Decontamination of Water Polluted by Pathogenic Germs with Peroxide Formed *In Situ*. U. S. Pat. 4,619,745 (1986).
100. P. Tatapudi and J. M. Fenton, Electrochemical Oxidant Generation for Wastewater Treatment. *In* "Environmental Oriented Electrochemistry" (C. A. C. Sequeira, ed.), p. 103. Elsevier, Amsterdam, 1994.
101. D. Pletcher and F. C. Walsh, "Industrial Electrochemistry," Chapter 5. Chapman & Hall, London, 1990.
102. R. I. Agladze, E. A. Manukov, and G. R. Agladze, Electrolyzer for Conducting Electrolysis Therein. U. S. Pat. 4,269,689 (1981).
103. M. T. Yahya and C. P. Gerba, Water Disinfection System and Method. *Chem. Abstr.* **118**, 87342 (1992).
104. V. Eibl, Electrolytic Purification of Aqueous Liquids in the Presence of Silver Ions. U. S. Pat. 4,048,032 (1977).
105. T. Okazaki, Apparatus Having Diaphragm for Formation of Alkaline Electrolyzed Water Containing Silver Hydroxide. *Jpn. Kokai Tokyo Koho* **JP 05115881 A2** (1993); Heisei, *Chem. Abstr.* **119**, 188214 (1993).
106. Apparatus for Tap Water Treatment. *Jpn. Kokai Tokyo Koho* **JP 05317861 A2** (1993); Heisei, *Chem. Abstr.* **120**, 279786 (1993).
107. J. Miller, Electrolytic Cell for Treatment of Water. U. S. Pat. 4,048,030 (1977).
108. B. Rosenberg, L. Van Camp, and T. Krigas, Inhibition of Cell Division in *Escherichia coli* by Electrolysis Products from a Platinum Electrode. *Nature (London)* **205**, 698 (1965).
109. G. Patermarakis and E. Fountoukidis, Disinfection of Water by Electrochemical Treatment. *Water Res.* **24**, 1491 (1990).
110. N. Getoff, Purification of Drinking Water by Irradiation. A Review. *Proc. Indian Acad. Sci., Sect. A* **105 A. (Pt. 1)**, 373 (1993).
111. J. Douglas, Electrotechnologies for Water Treatment. *EPRI J.* **18**, 4 (1993).
112. B. F. Severin, Disinfection of Municipal Wastewater Effluents with Ultraviolet Light. *J. Water Pollut. Control Fed.* **52**, 2007 (1980).
113. O. K. Scheible and C. D. Bassel, "Ultraviolet Disinfection of a Secondary Wastewater Treatment Plant Effluent." EPA 600/S2-81-52. U. S. Environ. Prot. Agency, Washington, DC, 1981.
114. G. E. Whitby, G. Palmatier, W. G. Cook, J. Maarschalkerweird, D. Huber, and K. Flood, Ultraviolet Disinfection of Secondary Effluent. *J. Water Pollut. Control Fed.* **56**, 844 (1984).

115. O. K. Scheible, Development of a Rationally Based Design Protocol for the Ultraviolet Light Disinfection Process. *J. Water Pollut. Control Fed.* **59**, 25 (1987).
116. R. J. Fahey, The UV Effect on Wastewater. *Water/Eng. Manage.* p. 15, December (1990).
117. H. Dizer, W. Burtocha, H. Bartel, K. Seidel, M. J. Lopez-Pila, and A. Grohmann, Use of Ultraviolet Radiation for Inactivation of Bacteria and Coliphages in Pretreated Wastewater. *Water Res.* **27**, 397 (1993).
118. G. Zukovs, J. Kollar, H. D. Monteith, K. W. A. Ho, and S. A. Ross, Disinfection of Low Quality Wastewaters by Ultraviolet Light Irradiation. *J. Water Pollut. Control Fed.* **58**, 199 (1986).
119. E. B. Marsh, Jr. and D. H. Smith, R-factors Improving Survival of *Escherichia coli* K-12 After Ultraviolet Irradiation. *J. Bacteriol.* **100**, 128 (1969).
120. R. G. Qualls and J. D. Johnson, Bioassay and Dose Measurement in UV Disinfection. *Appl. Environ. Microbiol.* **45**, 872 (1983).
121. J. H. Jagger, "Introduction to Research in UV Photobiology." Prentice-Hall, Englewood Cliffs, NJ, 1967.
122. S. M. Jacob and J. S. Dranoff, Light Intensity Probes in a Perfectly Mixed Photoreactor. *AIChE J.* **16**, 359 (1970).
123. H. E. Chick, An Investigation of the Laws of Disinfection. *J. Hyg.* **8**, 92 (1908)
124. J. C. H. Chang, S. Ossoff, D. C. Lobe, M. H. Dorfman, C. M. Dumais, R. G. Qualls, and J. D. Johnson, UV Inactivation of Pathogenic and Indicator Microorganisms. *Appl. Environ. Microbiol.* **49**, 1361 (1985).
125. J. H. Wei and S. L. Chang, *in* "Disinfection: Water and Wastewater" (J. D. Johnson, ed.). Ann Arbor Sci. Publ., Ann Arbor, MI, 1975.
126. W. F. Hill, Jr., F. E. Hamblet, W. H. Benton, and E. W. Akin, Ultraviolet Devitalization of Eight Selected Enteric Viruses in Estuarine Water. *Appl. Microbiol.* **19**, 805 (1970).
127. C. N. Haas, Disinfection. *In* "Water Quality and Treatment" (F. Pontiers, ed.), pp. 877-932. American Water Works Association, Denver, CO.
128. B. G. Oliver and E. G. Cosgrove, The Disinfection of Sewage Treatment Plant Effluents Using Ultraviolet Light. *Can. J. Chem. Eng.* **53**, 170 (1970).
129. R. W. Yip and D. E. Konasewich, Ultraviolet Sterilization of Water — Its Potential and Limitations. *Water Pollut. Control Can.* **110**, 6 (1972).
130. B. G. Oliver and J. H. Carey, Ultraviolet Disinfection: An Alternative to Chlorination. *J. Water Pollut. Control Fed.* **48**, 2619 (1976).
131. R. G. Qualls, M. P. Flynn, and J. D. Johnson, The Role of Suspended Particles in Ultraviolet Disinfection. *J. Water Polut. Control Fed.* **55**, 1280 (1983).

132. O. J. Sproul, "Effect of Particulates on Ozone Disinfection of Bacteria and Viruses in Water." EPA-600/2-79-089. U. S. Environ. Prot. Agency, Cincinnati, OH, 1979.
133. R. G. Qualls, S. F. Ossoff, J. C. H. Chang, M. H. Dorfman, C. M. Dumais, D. C. Lobe, and J. D. Johnson, Factors Controlling Sensitivity in Ultraviolet Disinfection of Secondary Effluents. *J. Water Pollut. Control Fed.* **57**, 1006 (1985).
134. R. S. Tobin, D. K. Smith, A. Horton, and V. C. Armstrong, Methods for Testing the Efficacy of Ultraviolet Light Disinfection Devices for Drinking Water. *J. Water Pollut. Control Fed.* p. 481. September (1983).
135. E. A. Meyer and E. L. Jarroll, Giardiasis. *Am. J. Epidemiol.* **3**, 1 (1980).
136. G. F. Craun, Waterborne Giardiasis in the United States: A Review. *Am. J. Public Health* **69**, 817 (1979).
137. M. S. Wolfe, Giardiasis. *Pediatr. Clin. North Am.* **26**, 295 (1979).
138. E. W. Rice and J. C. Hoff, Inactivation of *Giardia lamblia* Cysts by Ultraviolet Irradiation. *Appl. Environ. Microbiol.* **42**, 546 (1981).
139. E. L. Jarroll, A. K. Bingham, and E. A. Meyer, Effect of Chlorine on *Giardia lamblia* Cyst Viability. *Appl. Environ. Microbiol.* **41**, 483 (1981).
140. E. W. Rice, J. C. Hoff, and F. W. Schaefer, III, Inactivation of *Giardia* Cysts by Chlorine. *Appl. Environ. Microbiol.* **43**, 250 (1982).
140a. J. S. Slade, N. R. Harris, and R. G. Chisholm, Disinfection of Chlorine Resistant Enteroviruses in Ground Water by Ultraviolet Irradiation. *Water Sci. Technol.* **18**, 115 (1986).
141. W. F. Hill, Jr., F. E. Hamblet, W. H. Benton, and E. W. Akin, Ultraviolet Devitalization of Eight Selected Enteric Viruses in Estuarine Water. *Appl. Microbiol.* **19**, 805 (1970).
142. W. F. Hill, Jr., F. E. Hamblet, and W. H. Benton, Inactivation of Poliovirus Type I by the Kelly-Purdy Ultraviolet Seawater Treatment Unit. *Appl. Microbiol.* **17**, 1 (1969).
143. M. C. Meckes, Effect of UV Light Disinfection on Antibiotic-Resistant Coliforms in Wastewater Effluents. *Appl. Environ. Microbiol.* **43**, 371 (1982).
144. W. T. Drabble and B. A. D. Stocker, R. (Transmissable Drug Resistance) Factors in *Salmonella typhimurium*: Pattern of Transduction by Phage P22 and Ultraviolet Protection Effect. *J. Gen. Microbiol.* **53**, 109 (1968).
145. R. H. Haynes, Role of DNA Repair Mechanism in Microbial Inactivation and Recovery Phenomena. *Photochem. Photobiol.* **3**, 429 (1964).
146. R. H. Haynes, R. M. Baker, and G. E. Jones, Genetic Implications of DNA Repair. *In* "Energetics and Mechanisms in Radiation Biology" (G. Phillips, ed.), pp. 425-465. Academic Press, London, 1968.
147. W. L. Carrier and R. B. Setlow, Excision of Pyrimidine Dimers from Irradiated Deoxyribonucleic Acid *In Vitro*. *Biochim. Biophys. Acta* **129**, 318 (1966).

148. M. L. Gefter, A. Becker, and J. Hurwitz, The Enzymatic Repair of DNA. I. Formation of Circular DNA. *Proc. Natl. Acad. Sci. USA* **58**, 240 (1967).
149. R. G. Qualls, J. C. H. Chang, S. F. Ossoff, and J. D. Johnson, Comparison of Methods of Enumerating Coliforms After UV Disinfection. *Appl. Environ. Microbiol.* **48**, 699 (1984).
150. R. A. Morton and R. H. Haynes, Changes in the Ultraviolet Sensitivity of *Escherichia coli* During Growth in Batch Cultures. *J. Bacteriol.* **97**, 1379 (1969).
151. S. C. White, E. B. Jernigan, and A. D. Venose, A Study of Operational Ultraviolet Disinfection Equipment at Secondary Treatment Plants. *J. Water Pollut. Control Fed.* **58**, 181 (1986).
152. R. G. Qualls and J. D. Johnson, Modeling and Efficiency of Ultraviolet Disinfection Systems. *Water Res.* **19**, 1039 (1985).
153. J. R. Cortelyou, M. A. McWhinnie, M. S. Riddiford, and J. E. Semrad, Effects of Ultraviolet Irradiation on Large Populations of Certain Water-Borne Bacteria in Motion. I. The Development of Adequate Agitation to Provide an Effective Exposure Period. II. Some Physical Factors Affecting the Effectiveness of Germicidal Ultraviolet Radiation. *Appl. Microbiol.* **2**, 262 (1954).
154. R. G. Qualls, M. H. Dorfman, and J. D. Johnson, Evaluation of the Efficiency of Ultraviolet Disinfection Systems. *Water Res.* **23**, 317 (1989).
155. L. H. Nowell and J. Hoigné, Photolysis of Aqueous Chlorine at Sunlight and Ultraviolet Wavelengths. I. Degradation Rates. *Water Res.* **26**, 593 (1992).
156. L.H. Nowell, and J. Hoigné, Photolysis of Aqueous Chlorine at Sunlight and Ultraviolet Wavelengths. II. Hydroxyl Radical Production. *Water Res.* **26**, 599 (1992).
157. W. R. Haag and J. Hoigné, Photo-sensitized Oxidation in Natural Water via OH Radicals. *Chemosphere* **14**, 1659 (1985).
158. J. Staehelin and J. Hoigné, Decomposition of Ozone in Water in the Presence of Organic Solutes Acting as Promoters and Inhibitors of Radical Chain Reactions. *Environ. Sci. Technol.* **19**, 1206 (1985).
159. R. G. Zepp, J. Hoigné, and H. Bader, Nitrate-Induced Photooxidation of Trace Organic Chemicals in Water. *Environ. Sci. Technol.* **21**, 443 (1987).
160. R. G. Zepp, B. C. Faust, and J. Hoigné, Hydroxyl Radical Formation in Aqueous Reactions (pH 3-8) of Iron(II) with Hydrogen Peroxide: The Photo-Fenton Reaction. *Environ. Sci. Technol.* **26**, 313 (1992).
161. B. G. Oliver and J. H. Carey, Photochemical Production of Chlorinated Organics in Aqueous Solutions Containing Chlorine. *Environ. Sci. Technol.* **11**, 983 (1977).
162. W. H. Glaze, G. R. Peyton, S. Lin, R. Y. Huang, and J. L. Burleson, Destruction of Pollutants in Water with Ozone in Combination with Ultra-

violet Radiation. 2. Natural Trihalomethane Precursors. *Environ. Sci. Technol.* **16**, 454 (1982).
163. D. W. Ferguson, M. J. McGuire, B. Koch, R. L. Wolfe, and E. M. Aieta, Comparing PEROXONE and Ozone for Controlling Taste and Odor Compounds, Disinfection By-Products, and Microorganisms. *J. Am. Water Works Assoc.* **82**, 161 (1990).
164. A. Fujishima, L. A. Nagahara, H. Yoshiki, K. Ajito, and K. Hashimoto, Thin Semiconductor Films: Photoeffects and New Applications. *Electrochim. Acta* **39**, 1229 (1994).
165. T. Matsunaga, R. Tomoda, T. Nakajima, and H. Wake, Photoelectrochemical Sterilization of Microbial Cells by Semiconductor Powders. *FEMS Microbiol. Lett.* **29**, 211 (1985).
166. R. J. Watts, S. Kong, M. P. Orer, G. C. Miller, and B. E. Henry, Photocatalytic Inactivation of Coliform Bacteria and Viruses in Secondary Wastewater Effluent. *Water Res.* **29**, 95 (1995).
167. T. Matsunaga, R. Tomoda, T. Nakajima, N. Nakamura, and T. Komine, Continuous-Sterilization System that Uses Photosemiconductor Powders. *Appl. Environ. Microbiol.* **54**, 1330 (1988).
168. J. C. Ireland, P. Klostermann, E. W. Rice, and R. M. Clark, Inactivation of *Escherichia coli* by Titanium Dioxide Photocatalytic Oxidation. *Appl. Environ. Microbiol.* **59**, 1668 (1993).
169. T. Matsunaga and M. Okochi, TiO_2-Mediated Photochemical Disinfection of *Escherichia coli* Using Optical Fibers. *Environ. Sci. Technol.* **29**, 501 (1995).
170. R. L. Smith, A. Walker, W.-Y. Lin, and K. Rajeshwar, unpublished work (1996).
171. C. Wei, W.-Y. Lin, Z. Zainal, N. E. Williams, K. Zhu, A. Kruzic, R. L. Smith, and K. Rajeshwar, Bactericidal Activity of TiO_2 Photocatalyst in Aqueous Media: Toward a Solar-Assisted Water Disinfection System. *Environ. Sci. Technol.* **28**, 934 (1994).
172. P. Zhang, R. J. Scrudato, and G. Germano, Solar Catalytic Inactivation of *Escherichia coli* in Aqueous Solutions Using TiO_2 as Catalyst. *Chemosphere* **28**, 607 (1994).
173. R. W. Matthews, Photocatalysis in Water Purification: Possibilities, Problems and Prospects. *In* "Photocatalytic Purification and Treatment of Water and Air" (D. F. Ollis and H. Al-Ekabi, eds.), pp.121-138. Elsevier, Amsterdam, 1993.
174. J. C. Sjögren and R. A. Sierka, Inactivation of Phage MS2 by Iron-Aided Titanium Dioxide Photocatalysis. *Appl. Environ. Microbiol.* **60**, 344 (1994).
175. E. M. Aieta, J. D. Bug, P. V. Roberts, and R. C. Cooper, Comparison of Chlorine and Chlorine Dioxide in Wastewater Disinfection. *J. Wat. Pollut. Control Fed.* **52**, 810 (1980).

176. American Public Health Association, American Water Works Association, Water Pollution Control Federation, "Standard Methods for the Examination of Water and Wastewater."American Public Health Association, Washington, DC, 1992.
177. J. R. Wilkins, G. E. Stoner, and E. H. Boykin, Microbial Detection Method Based on Sensing Molecular Hydrogen. *Appl. Microbiol.* **27**, 949 (1974).
178. J. R. Wilkins and E. H. Boykin, Electrochemical Method for Early Detection and Monitoring of Coliforms. *J. Am. Water Works Assoc.* **68**, 257 (1976).
179. J. R. Wilkins, Use of Platinum Electrodes for the Electrochemical Detection of Bacteria. *Appl. Environ. Microbiol.* **36**, 683 (1978).
180. T. Matsunaga, I. Karube, and S. Suzuki, Electrode Systems for the Determination of Microbial Populations. *Appl. Environ. Microbiol.* **37**, 117 (1979).
181. G.-A. Junter, J.-F. Lemeland, and E. Selegny, Electrochemical Detection and Counting of *Escherichia coli* in the Presence of a Reducible Coenzyme, Lipoic Acid. *Appl. Environ. Microbiol.* **39**, 307 (1980).
182. G. Charriere, T. Jouenne, J.-F. Lemeland, E. Selegny, and G.-A. Junter, Bacteriological Analysis of Water by Potentiometric Measurement of Lipoic Acid Reduction: Preliminary Assays for Selective Detection of Indicator Organisms. *Appl. Environ. Microbiol.* **47**, 160 (1984).
183. T. Jouenne, G.-A. Junter, and G. Charriere, Selective Detection and Enumeration of Fecal Coliforms in Water by Potentiometric Measurement of Lipoic Acid Reduction. *Appl. Environ. Microbiol.* **50**, 1208 (1985).
184. G.-A. Junter, Potentiometric Detection and Study of Bacterial Activity. *Trends Anal. Chem.* **3**, 253 (1984).
185. I. C. Anderson, M. W. Rhodes, and H. Kator, Sublethal Stress in *Escherichia coli*: A Function of Salinity. *Appl. Environ. Microbiol.* **38**, 1147 (1979).
186. M. W. Rhodes and H. Kator, Survival of *Escherichia coli* and *Salmonella* Spp. in Estuarine Environments. *Appl. Environ. Microbiol.* **54**, 2902 (1988).
187. T. Jouenne, L. Bertins, G. Charriere, and G. A. Junter, Electrochemical Assessment of *Escherichia coli* Survival in Natural Water. *Water Res.* **25**, 829 (1991).

CHAPTER EIGHT

8.1. INTRODUCTION

The preceding chapters, we hope, established that electrochemical and photoelectrochemical approaches to the sensing and treatment of environmental pollutants have a sound scientific basis. In this chapter, the commercial applications of many of these approaches will be explored. In the vast majority of cases to be discussed here, the commercial potential has been realized after years of in-house developmental efforts on a laboratory and pilot scale. In particular, many of the technological innovations have been borne out of a need, within the organization, to tackle an environmental or waste disposal problem at hand. Clearly, in these cases, the electrochemical approach proved to be the most viable alternative. A case in point is the "silver bullet" process developed by AEA Technology (Dounreay, Scotland). This process was originally developed for the treatment of non-biodegradable organic solvent waste generated within the nuclear industry.

The discussion here will conform to the order of presentation of the topics in this book. Therefore, after a general review of vendors of electrochemical equipment and accessories, we shall examine, in turn, the commercial status of electrochemical sensors, reactors for the electrochemical remediation of pollutants, systems for performing advanced oxidation processes, and equipment for the *in situ* electrogeneration of disinfectant chemicals. Pilot

Table 8.1. Suppliers of Electrochemical Equipment and Accessories

Company	Country	Cells	Electrodes	Power Sources or galvanostats	Potentiostats
Astris	USA	x	x		
Bioanalytical Systems	USA	x	x		x
Cypress Systems	USA	x	x	x	x
Electron Transfer Technologies	USA			x	
Electrosynthesis	USA	x	x	x	x
Princeton Applied Research (EG&G)	USA	x	x		x
Solartron Instruments	USA				x
Advanced Electronics	Denmark			x	
Tacussel/Radiometer Analytical	France	x	x	x	x
Fairtec	France	x	x		
SRTI	France	x	x		
Sodilec	France			x	
Wenking	Germany				x
Zentro-Elektronik	Germany			x	
Heinzinger Regal	Germany			x	
Sigri Elektrographit	Germany		x		
Ringsdorff-Werke	Germany		x		
Degussa	Germany		x		
Heraeus Elektroden	Germany		x		
Amel	Italy			x	
Goss	UK			x	
Nichia-Keiki-S.	Japan			x	x
Hokuto	Japan			x	x
Metronics	Japan			x	
Ashai Glass	Japan	x			
Tokai Carbon	Japan		x		
Nihon Carbon	Japan		x		

Note: See also Refs. 1–3.

and field test results will also be presented in a few instances as are cost–performance comparisons with alternative technologies.

Most of the commercial activity is in a rapid state of flux, and we realize that the information may be already obsolete by the time this book appears in print. Similarly, the listing of commercial vendors and products in what follows is meant to be only representative rather than all inclusive. In this regard,

8.1. Introduction

Table 8.2. Manufacturers of Membranes

Company	Country	CEM	AEM	BPM	Porous Ceramics	Porous Polymers
Electrosynthesis	USA	x	x	x		x
Ionsep	USA	x	x	x		
DuPont	USA	x				
Sybron	USA	x	x			
Norton	USA					x
Eltron	USA					x
Aqualytics	USA			x		
Rhone Poulenc	France	x	x			
Serva Feinbiochemika	Germany	x	x			
Schumacher'sche	Germany					x
Staatliche-porzellan	Germany				x	
ICI	UK	x	x			
Electro-Cell AB	Sweden	x	x	x		
Ashai Glass	Japan	x	x	x		
Ashai Kasei	Japan	x	x	x		
Tokuyama Soda	Japan	x	x	x		
Mitsui DuPont	Japan	x	x	x		
Nihon-Kagaku-T.	Japan				x	

CEM = cation exchange membrane; AEM = anion exchange membrane; BPM = bipolar membrane.

Table 8.3.A Suppliers of Electrode Materials

Company	Country	RVC	Clothes, Felts	Metallic Foams	Mesh
ERG	USA	x		x	
Electrosynthesis	USA	x	x	x	
Olin	USA		x	x	
E-Tek	USA		x		
Delker	USA				x
Astro Met	USA			x	
Sorapec	France	x	x	x	
EPCI	Sweden			x	

the product and process information is presented in good faith. Finally, it is stressed that the inclusion of a particular product or process for discussion does not at all imply endorsement of the authors or the publisher.

Table 8.3.B Suppliers of Specialty Electrodes[a]

Company	Country	A	B	C	D	E	F
Electrosynthesis	USA		x	x	x	x	x
Electrode Products	USA		x	x	x	x	x
Eltech	USA		x		x		
E-Tek	USA					x	
ABB	Switzerland	x					
Atraverda	UK	x					
Permelec	Japan	x	x	x			

[a] These include Ebonex, DSA, and metal oxide-coated Ti. The major uses of such electrodes are labeled as follows:
  A, oxidation of organics
  B, regeneration/reuse of spent chemicals
  C, metal ion recovery
  D, production of water disinfectants
  E, applications involving gaseous reactants
  F, electrodialysis.

8.2. COMMERCIAL VENDORS OF ELECTROCHEMICAL EQUIPMENT AND ACCESSORIES

Tables 8.1–8.3 list suppliers of electrochemical equipment and electrode materials. It must be noted that companies for general laboratory supplies (e.g., Fisher, Cole-Parmer, Alfa-Ventron) also provide cells, power supplies, and electrodes or electrode materials for laboratory-scale tests.

A directory of commercial vendors appears in alphabetical order in Appendix D.

8.3. ELECTROCHEMICAL SENSORS OF ENVIRONMENTAL POLLUTANTS: COMMERCIALIZATION EFFORTS

Table 8.4 contains a compilation of manufacturers of electrochemical sensors for environmental pollution. The products marketed range from simple potentiometric sensors to complete systems for automated stripping analyses and LC-EC measurements.

It must be noted that many of the companies listed in Table 8.1 also manufacture instrumentation systems that can be configured for environmental electroanalyses.[1-3]

8.4. Electrochemical Reactors for Pollutant Treatment

Table 8.4. Companies Specializing in Electrochemical Sensors of Environmental Pollutants

Company[a]	Product Details
ATI Orion	Potentiomeric sensors, ISEs.
Bioanalytical Systems, Inc.	LC-EC systems, stripping analyses equipment, speciality electrodes, etc.
Radiometer America	Computerized voltammetry and stripping potentiometry instrumentation (e.g., VoltLab™ and TraceLab™)
Universal Sensors, Inc.	Flow cells, gas sensing electrodes, H_2O_2 sensors, etc.

[a] Refer to the company directory in Appendix D for further details.

8.4. ELECTROCHEMICAL REACTORS FOR POLLUTANT TREATMENT AND CHEMICAL RECYCLING

8.4.1. Water treatment systems

Table 8.5 contains a listing of several types of electrochemical cell designs that have seen commercial fruition. These encompass several of the design variations considered in Chapter 2 (Section 2.9) and Chapter 5 (Section 5.3). A brief description of some of the popular cell designs follows. More extensive discussions, including theoretical treatment, are contained in Refs. 4–6.

<u>Chemelec Cell</u>. Figure 8.1 contains a schematic of this device, which features a series of closely spaced gauze or expanded metal electrodes to afford an alternating anode–cathode arrangement. The anodes and cathodes are separated by nonconducting spherical glass particles. This partially fluidized bed of particles results in efficient mass transport within the cell. The effluent solution enters the cell via a flow distributor (not shown in Figure 8.1). The cathodes are designed to have a relatively large surface area, and they can be readily dismantled to recover the deposited metal.

The Chemelec cell may be regarded as a compromise design to afford easy maintenance and operation without realizing the full advantages of a cell with three-dimensional electrodes.

<u>ElectroSyn Cell</u>. This cell design is marketed by ElectroCell AB (Taby, Sweden). With the advent of three-dimensional electrode materials, it has become

Table 8.5. Commercial Cell Designs for Environmental Applications

Company (Country)	Name	M/B[a]	Cell Arrangement	Area Range (m²)	Sample Uses
Bewt (UK)	Chemelec Cell	M/B	Six cathodes/ seven anodes	~3.3	Metal ion removal
Electrocatalytic (USA)	Dished Electrode Membrane Cell	M/B	Dished electrodes, flow-by mode	0.01–1.00	Cathodic–anodic treatment of organic pollutants
Electrosynthesis (USA)	ABC Cell, Aquanautics	B	Flow-by and flow-through	0.0025->1	Various processes
	Gas Diffusion Cell	M	Single cell	0.0025	Processes involving gases
	QC200	M	Single cell	0.0005	Various processes
Electro Cell (Sweden)	MicroFlow	M/B	Filter press	0.001	Electrolysis–electrodialysis, gases
	MP-Cell	M/B	Same as above	0.01–0.20	Same as above
	Syn Cell	M/B	Same as above	0.04–1.04	Same as above
	Prod Cell	M/B	Same as above	0.04–16.0	Same as above
Eltech (USA)	Retec Cell	M/B	G-50 cathodes from foam materials interspersed with DSAs (usually oxide-coated Ti mesh)	Variable	Various pollutant treatment or metal electrowinning processes
EnViro Cell (Germany)	EP, ES	M/B	Packed bed	Variable	Heavy metals removal–recovery, recycling of chemicals, direct oxidation, indirect oxidation, flue gas desulfurization and denitrification
ICI (UK)	FM type and FB type	M/B	Filter press	0.0064–0.20	Pollutant destruction, reagent regeneration, salt splitting
Steetley (UK)	EcoCell	M	Rotating cylinder cascade with six chambers	Variable	Metal ion removal

[a] Monopolar or bipolar connections.

8.4. Electrochemical Reactors for Pollutant Treatment

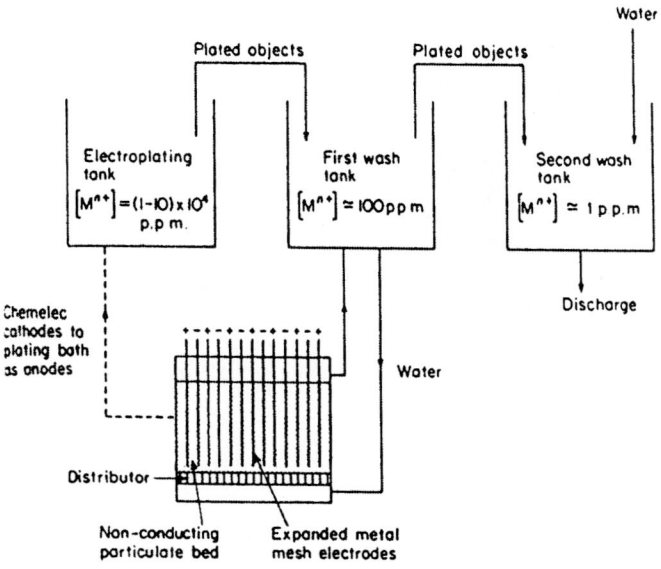

FIGURE 8.1. The Chemelec cell as operated for the treatment of water in a static electroplating rinse tank. (Reproduced with permission from Genders and Weinberg.[4])

commonplace for suppliers of classical plate-and-frame cells to offer modifications of existing designs to allow their use with high surface-area electrodes. One such design utilizes a packed-bed of carbon particles in a stationary, flow-by arrangement, where the original flat plate electrode is used as a current collector.

Thorough characterization of these cells at the laboratory level has been done.[4] Such modifications to existing cell designs have also included the use of three-dimensional nickel electrodes with the EM01-LC electrolyzer marketed by ICI Chemicals and Polymers Ltd. in the United Kingdom.[5]

Retec Cell. This cell design is marketed by Eltech Systems Corp. (Chardon, Ohio) and is illustrated in Figure 8.2. It uses a three-dimensional electrode and contains 6–50 cathodes interspersed with dimensionally stable anodes (usually oxide-coated Ti mesh). The effluent solution flows perpendicularly through the electrodes (i.e., in a flow-through arrangement). This design creates a large cathode volume and a moderate mass-transfer coefficient with air sparging of the cell.

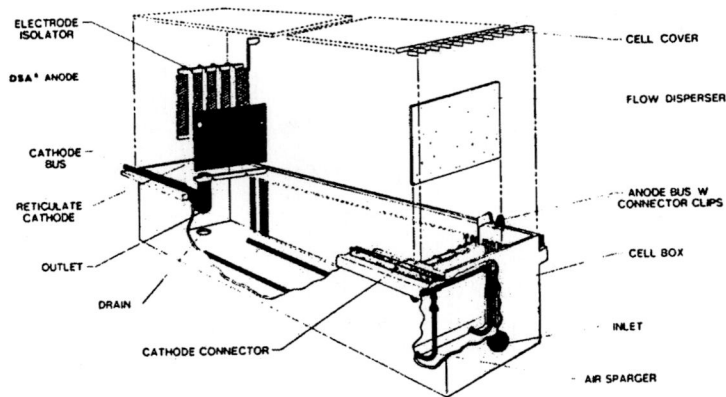

FIGURE 8.2. Schematic drawing of the ReTec cell technology. (Reproduced with permission from Genders and Weinberg.[4])

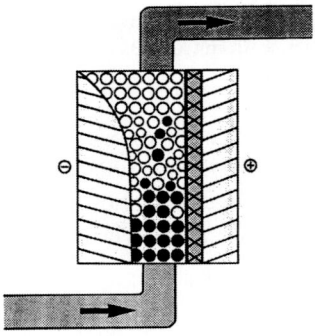

FIGURE 8.3. Schematic drawing of the EnViro cell. (Diagram courtesy of EnViro-cell Umwelttechnik GmbH, Oberursel, Germany.)

EnViro Cell. This device is offered by EnViro-cell Umwelttechnik Gmbh (Oberursel, Germany) and is based on a packed bed of carbon granules in a tapered compartment designed to increase the cathode area toward the cell outlet as illustrated in Figure 8.3. This is done to ensure a constant current distribution along the flow direction.

Unlike the Chemelec and Retec cells, this cell design is compatible with relatively low metal-ion concentrations (10–50 ppm). Reduction to less than 1 ppm can be achieved with a single pass of the effluent through the cell. To

8.4. Electrochemical Reactors for Pollutant Treatment

avoid periodic removal of the cell packing, acid washing or current reversal is recommended on a regular basis.

Eco cell. The rotating cylinder electrode design was considered earlier in this book (Chapters 2 and 5). The Eco cell was one of the earliest designs based on this concept, where the rotating drum cathode was scraped by a wiper to dislodge deposited metal which is then passed through a hydrocyclone to produce a metal powder (see Figure 5.8). A cascade version of this cell, with six chambers separated by baffles, has been demonstrated to reduce a 50 ppm copper inlet stream to ca. 1 ppm at the outlet.[6] Commercial development of this design is now being pursued by Steetley Engineering in the United Kingdom.

Miscellaneous Cell Designs. Two well-established technologies for water treatment are hypochlorite generation and electrodialysis. A number of companies now supply electrolyzers for hypochlorite generation and other applications. Table 8.6 contains a representative listing, and Figures 8.4 and 8.5 contain schematic diagrams of two commercial systems designed to handle waste streams containing sodium sulfate. Such streams are generated by viscose rayon, pulp and paper, and other sectors in the chemical industry. The useful products of the electrolysis are NaOH, H_2SO_4, and H_2. The Retec cell design (Figure 8.2) can also be used for the treatment of Na_2SO_4, for *in situ* acid–base generation from salts (e.g., sodium carbonate, sodium borate, sodium chlorate) and for the oxidation of organics.

Another important category of cells not specifically considered in Table 8.5 includes fluidized-bed designs. The Chemelec cell does use an *inert (insulating)* fluidized bed but simply to enhance mass transport. In the more common designs, the electrode particles themselves constitute the fluidized bed (Chapters 2 and 5). These *conductive* particles are contacted by a porous feeder electrode. The main limitation of this design is that fluidization often causes loss of electrical contact between the particles and attendant *iR* drop problems. After the original work by the University of Newcastle group,[7] much of the commercial development is now developed jointly by Akzo Zout Chimie and the Billiton group of Shell Research in the Netherlands.

8.4.2. Commercial Processes for Water Treatment and Chemical Recycling

In Situ Oxidant Generation. The concept of generating potent oxidants for destroying organics was discussed in Chapter 5.

A process developed by AEA at Dounreay uses a divided, parallel plate reactor to generate Ag(II) species anodically via the oxidation of Ag(I) in a nitric acid solution.[8] This technology was originally developed to solubilize PuO_2 during the treatment of nuclear wastes. It is now being evaluated for the treat-

Table 8.6. Companies (Other Than Those Listed in Table 8.5) Marketing Environmental Electrochemistry Systems

Company/Country	System	Application(s)
AEA/UK	Silver(II) process	Destruction of toxic organics
Andco/USA	Based on soluble iron anodes	Fluoride, arsenic, heavy metal removal, oxidation of organics
Brinecell/USA	Various systems for electrolytic oxidant generation	Diverse
Chemetics/Canada	Electrolyzers for ClO_2 and hypochlorite generation	Diverse
Delphi/USA	Fe(III) mediated oxidation	Indirect oxidation of organics
Dow Chemical Co./USA	H-D Tech Systems for H_2O_2 generation	Diverse
Faratech/USA	Plating barrel cathode and packed bed anode	Copper recovery and cyanide destruction
FSL/UK	Regen	Cupric chloride etchant regeneration and metal recovery
Liquipure/USA	EDP, EIX	Water deionization, recycling of chemicals
Ozonia/Switzerland	Membrel for O_3 production	Diverse
Titalyse/Switzerland	Electrodialysis, Electrowin, Retec, Ceracell and Electrox cells	Regeneration–recycling of etchants, pickling and process liquors, dissolved metal recovery, cyanide oxidation
Trionetics/USA	Rotacat	Dissolved metal recovery
Umpqua/USA	EWRS	Breadboard electrochemical water recovery system

8.4. Electrochemical Reactors for Pollutant Treatment

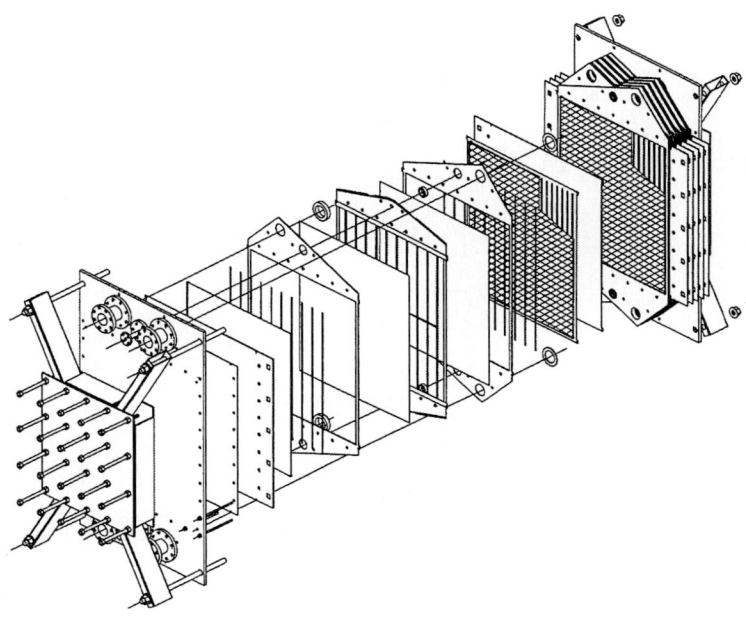

FIGURE 8.4. A three-component electrolyzer marketed by Electrode Corporation. (Diagram courtesy of Electrode Corporation, Chardon, Ohio.)

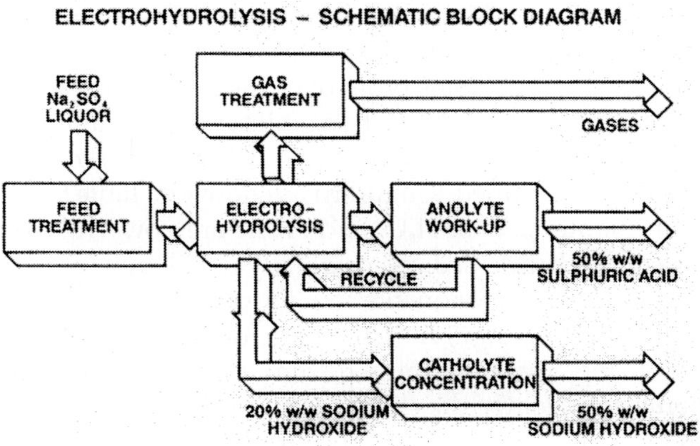

FIGURE 8.5. The electrohydrolysis process for the treatment of Na_2SO_4 liquor. (Diagram courtesy of ICI Electrochemical Technology, UK.)

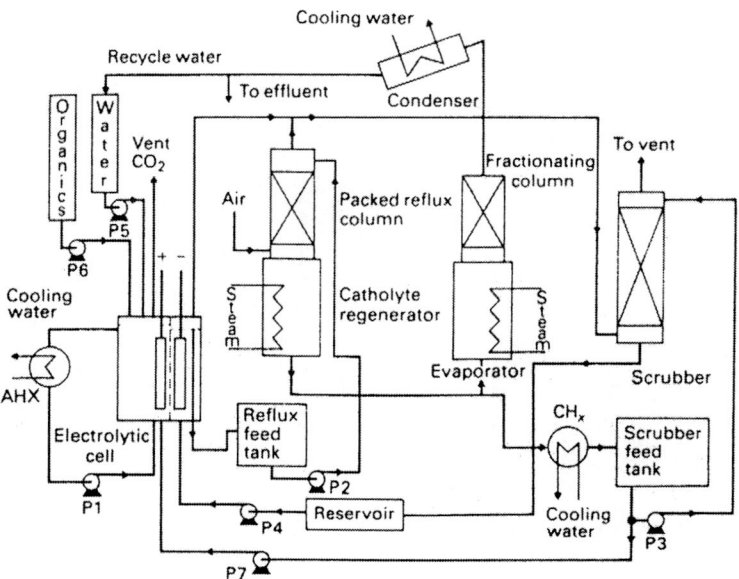

FIGURE 8.6. Schematic diagram of the electrochemical oxidation process developed by AEA, Dounreay. (Reproduced with permission from Steele.[8])

ment of a growing list of industrial organics. A schematic of the oxidation process is contained in Figure 8.6.[8] The low temperature–low pressure operating regime and the ability of the process to accept a wide range of waste and compositions are claimed to make the process an attractive alternative to incineration for toxic and troublesome types of industrial organic wastes (e.g., kerosene, rubber, plastics, polyurethane, hydraulic and lubricating oils).

The Ag(II)-mediated oxidation technology is also marketed by other companies (e.g., FSL in the United Kingdom) as shown in Table 8.6.

As seen in Chapter 2, both H_2O_2 and ozone are powerful oxidants. A diaphragm flow control trickle bed cell originally developed by H-D Tech is now being commercially offered for on-site alkaline peroxide production by the Dow Chemical Co.[9] The cell design, shown schematically in Figure 8.7,[9] is based on the concept of using a cathode formed from a self-draining bed of graphite–Teflon–carbon black "composite chips." The cell operates at essentially ambient pressure with no liquid recycle. Sodium hydroxide is fed into the anode compartment where O_2 evolution occurs. This O_2 can be either vented or recycled to the cathode depending on the site characteristics.

Other commercial systems for *in situ* generation of ozone are listed in Table 8.6.

8.4. Electrochemical Reactors for Pollutant Treatment 699

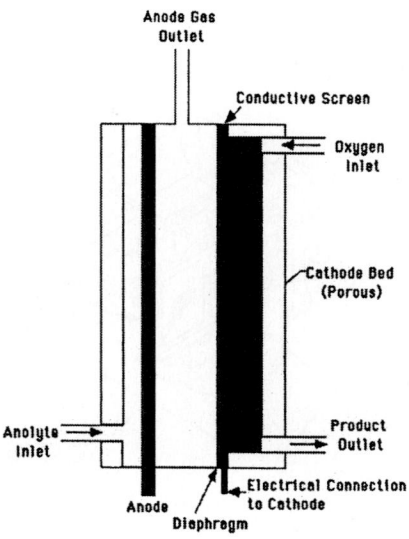

FIGURE 8.7. Diaphragm flow control trickle bed cell for the electrogeneration of H_2O_2. (Reproduced with permission from McIntyre.[9])

Spent Etchant Regeneration and Effluent Treatment. Chemical etchants such as Cr(VI) are used in a variety of industries (e.g., aerospace, microelectronics). Disposal of the spent etchant has increasingly become an expensive proposition for these companies, resulting in a search for in-house technology to regenerate the original chemical. A *zero-emission* process is thus facilitated.

Anodic oxidation of Cr(III) in the spent etchant can be carried out at Sb-doped PbO_2 anodes in a process originally developed by the U.S. Bureau of Mines.[10] Other cations that may be present in the etchant stream (such as Cu^{2+}, Al^{3+}) are sequestered into the catholyte compartment via a cation-selective membrane (Nafion). These can be deposited on the cathode and recovered as value-added chemicals when needed. The cell design can accommodate a variable number of anode–cathode pairs, depending on the plant needs. One version of this system is illustrated in Figure 8.8. This process is also compatible with the regeneration of other etchants such as the Fe(III)-based Aldox formulation. Figure 8.9 contains data from a bench-top reactor illustrating the regeneration of the iron etchant (Aldox V) and the accumulation of Al^{3+} in the catholyte compartment.[11]

Every year, large quantities of sodium chlorate are produced electrolytically for the pulp and paper industry. In the process, Cr(VI) is added to the brine electrolyte to improve the current efficiency. A process developed by

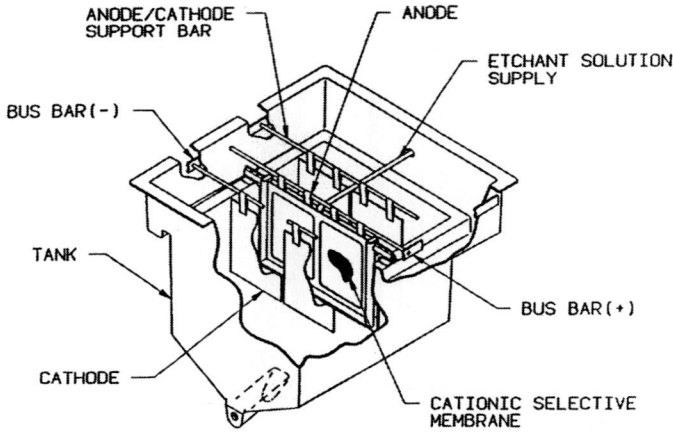

FIGURE 8.8. Schematic drawing of the US Bureau of Mines cell for the electrochemical regeneration of Cr(VI).

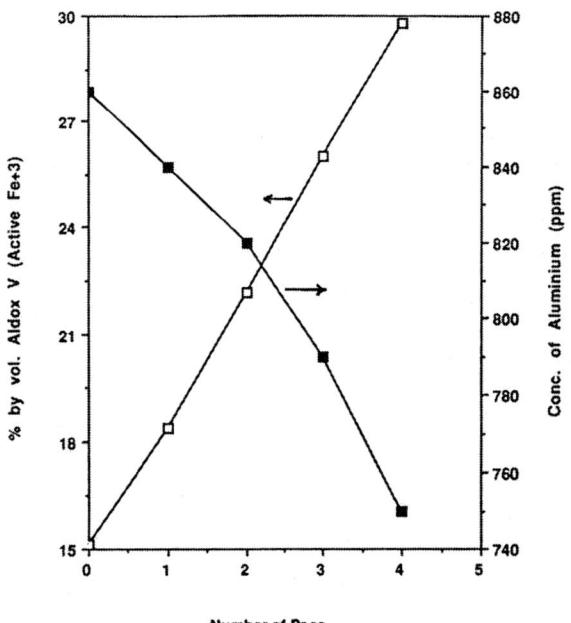

FIGURE 8.9. Change of ferric and aluminum ion concentrations with number of passes, at a fixed flow rate, in a bench-top electrochemical reactor.

8.4. Electrochemical Reactors for Pollutant Treatment

Table 8.7. Comparison of Cr(VI) Treatment Costs Using Various Methods

Method No.	Treatment Method	Cost[a] Capital	Annual Operating
"Standard"			
1	Sodium metabisulfite	1,045,000	122,000
2	Ferrous sulfate	1,060,000	115,000
"Nonstandard"			
3	Electrochemical	490,000	60,000
4	Regenerative carbon adsorption	970,000	74,000
5	Single-usage carbon adsorption		
	Constructed	360,000	45,000
	Leased	200,000	82,000

[a] In 1984 U.S. Dollars.
Source: From Praino and O'Gorman.[12]

Albright and Wilson Americas (Islington, Ontario) uses a carbon-bed cathode cell for the removal of Cr(VI) from sodium chlorate liquors. This process, which is the *opposite* of that just described, reduces the Cr(VI) to Cr(III) that precipitates as a hydroxide within the cathode bed. Periodically, the $Cr(OH)_3$ layer is treated with sodium hypochlorite to regenerate Cr(VI) in a form suitable for recycling back to the sodium-chlorate electrolysis units.

The photography industry generates large quantities of Cr(VI)-containing waste. Polaroid Corp. has conducted a comparative study of various technologies under the umbrella of "non-standard" and "standard" treatment methods (Table 8.7).[12] The standard methods (Methods 1 and 2 in Table 8.7) involve the use of treatment chemicals such as sodium metabisulfite and $FeSO_4$ to reduce Cr(VI) to Cr(III). Both these methods have high capital and annual operating costs. On the other hand, the "nonstandard" treatments have significantly lower capital and operating costs. The higher costs with regenerative carbon are associated with the regenerative chemical and sludge handling steps. However, both types of carbon-based approaches ultimately suffer from problems associated with the disposal of the spent carbon. The electrochemical approach is advantageous in this regard.

702   Chapter 8

Groundwater and Soil Remediation. The treatment of pollutants in soil and groundwater in leaking underground storage tank (LUST) sites presents complex technical challenges. Both *in situ* and *ex situ* treatment technologies were discussed in Section 1.6.10. Most of these technologies are based on the separation of the pollutant from the soil or groundwater. The separated pollutant is then adsorbed on carbon, condensed to a liquid or separated from the extracting solvents by distillation. The disposal or destruction of these waste mixtures is a tedious and expensive proposition. Further, many of these technologies are applicable only to volatile or semivolatile (organic) pollutants.

The efficacy of advanced oxidation processes (AOPs) in the treatment of LUST sites, as borne out by pilot and field experiments, has been reviewed.[13] In this section, we shall consider the applicability of electrochemical and electrokinetic remediation technologies. The use of AOPs will be described in Section 8.5.

The Andco Process. The process was piloted at a Superfund Site in U.S. EPA Region 4. This site was operated as a wood preserving facility from 1963 to 1965. From 1963 until 1980, fluoride–chromate–arsenate–phenol and acid–copper wood preserving processes were used. From 1980 to ca. 1985, a chromate–copper–arsenic wood treatment method was used. After adding this site to the national high-priority list of EPA Superfund Sites (see Chapter 1), a Record of Decision was made to use a "pump and treat" method for toxic waste cleanup. A flow diagram of the proposed electrochemical treatment setup is contained in Figure 8.10. After the electrochemical treatment, a polymer is added to aid in flocculation and settling. The effluent analyses after treatment are reported to be within the site specifications, and processed stream can be discharged to surface water. Details of the Andco process are available.[14]

The Andco process features a "dissolvable iron anode" in an automated system. Thus, Fe(II) species are released on demand from the anode and mediate the reduction of Cr(VI):

$$3\,Fe^{2+} + CrO_4^{2-} + 4\,H_2O \rightarrow 3\,Fe^{3+} + Cr^{3+} + 8\,OH^-$$
$$Cr^{3+} + 3\,OH^- \rightarrow Cr(OH)_3$$

Simultaneously, OH^- ions are released at the cathode from water electrolysis and H_2 dischage; these aid in the precipitation of Cr(III) as the highly insoluble hydroxide. Economic estimates of the process are contained in Table 8.8.[14] The operating costs of the process are primarily the electrical costs and the cost for replacement of electrode materials. The treatment costs for the other methods listed in Table 8.8 have been culled from U.S. EPA documentation.[14]

Other field tests and treatability studies have been conducted for the Andco process, including sites in California[14] and Texas.[15] The Odessa Chromium I

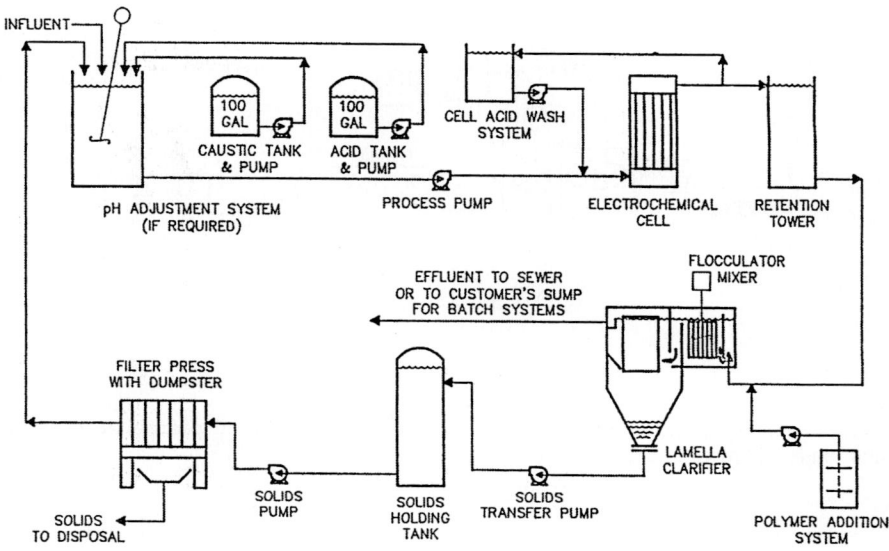

FIGURE 8.10. Flow diagram for the Andco process for heavy metal removal. (Reproduced with permission from Andco Environmental Processes, Inc., Buffalo, New York.)

Table 8.8. Comparative Treatment Costs for Hexavalent Chrome Removal

Method	Relative Cost/1,000 gallons
Ferrous sulfate	$1.50
Ion exchange	0.80
Reverse osmosis	3.00
Electrochemical	0.60

Source: From Reich.[14]

and II Sites consist of a series of chromium-contaminated wells that lie within reach of residential, commercial, and industrial facilities. Site I is an area where several chrome plating businesses operated between 1972 and 1977 and also has at least one currently operating plating facility. The Chromium II Site (a LUST site) originates from a plant manufacturing cooling water additives that contained Cr(VI).

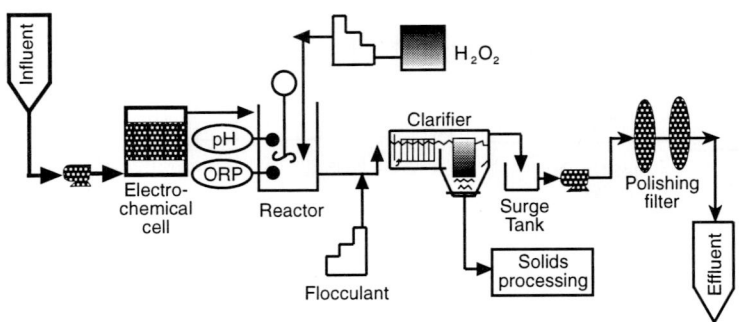

FIGURE 8.11. Electrochemical arsenic removal process. (Reproduced with permission from Andco Environmental Processes, Inc., Buffalo, New York.)

Andco Environmental Processes, Inc., is also marketing (as a licensee) the Boeing technology for the treatment of organics at these sites. Among the chemicals that are used in lumber-treatment operations is pentachlorophenol. The Boeing technology is based on the use of the Fenton reagent, $Fe^{2+} + H_2O_2$ for generating $^\bullet OH$ (see Chapter 5). The Fe^{2+} ions needed for this reaction can also be supplied via the iron anode on demand.

The Andco process has also been developed for the treatment of arsenic-contaminated groundwater (Figure 8.11). The process in this particular application uses iron and H_2O_2 to remove the arsenic as easily precipitated ferric arsenate and an arsenic–hydrous iron oxide complex. The process is claimed to not increase the total dissolved solids (TDS) of the treated stream, unlike conventional chemical treatment (precipitation) schemes that add sulfate or chloride in the form of iron or aluminum salts. This is because the electrochemical process generates only ferrous and OH^- ions, and both are precipitated from the solution phase.

As in the Cr(VI) treatment process, the OH^- ions released at the cathode result in the formation of $Fe(OH)_2$. This in turn is oxidized to $Fe(OH)_3$ by H_2O_2. The latter also oxidizes any arsenite (H_3AsO_3) to arsenate (H_3AsO_4). The pH of the medium is maintained at ~6.5 for optimum ferric arsenate precipitation and adsorption. A small amount of polymer flocculant may be added to the reactor tank to improve the settling characteristics of the solids. The settled solids are routed to a plate and frame filter press for dewatering.

Electrokinetic soil remediation has been reasonably well developed to the point of commercialization. Table 8.9 contains a listing of companies that market this category of technologies. Further details of pilot and field tests and commercialization efforts may be found elsewhere.[16]

8.4. Electrochemical Reactors for Pollutant Treatment

Table 8.9. Companies Marketing Electrokinetic Soil Remediation Technologies

Company	Country	Sample Applications
General Electric	USA	Chromate remediation
Isotron	USA	Heavy metals and anions
Lockheed[a]	USA	Diverse
SRI[a]	USA	Groundwater remediation, *in situ* soil remediation, bioelectroremediation, etc.
Westinghouse (The Savannah River Laboratory)	USA	Diverse
Geokinetics[a]	The Netherlands	Removal of heavy metals, toxic organics, and inorganics from soil and sludge

[a] Members of the Electrochemical Remediation Group.

Table 8.10. Systems/Processes for Gas Treatment and Odor Control

Company/Country	Product/Process
Aquatech Systems/USA	SO_2 treatment
Electrocinerator Technologies, Inc.,/USA	The Electrocinerator
Pepcon/USA	The Odor Master
Joint Research Center of the European Community/Italy	Ispra Mark 13A/flue gas desulfurization

8.4.3. Gas and Air Treatment Systems and Processes

Many of the gaseous pollutants are electroactive (Chapters 3 and 5), so that electrolytic approaches to improving air quality are promising. Table 8.10 contains a listing of air and gas electrochemical treatment systems that have attained commercial maturity. In most of the cases, however, the air to be treated is sorbed into an aqueous solution of a reactive intermediate, which is electrogenerated so that the gases themselves are not electrolyed directly. Included in this category are four products or processes which are briefly described in turn.

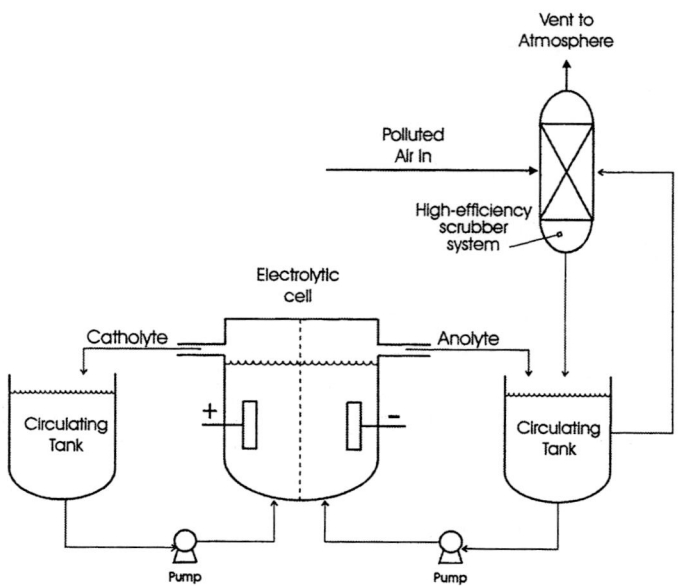

FIGURE 8.12. Flow diagram of the Electrocinerator system. (Adapted from Genders and Weinberg.[4])

<u>The Odor Master</u>. This is a product from Pepcon Systems Inc. (Henderson, Nevada), for the destruction of foul-smelling vapors (e.g., H_2S, amines, thiols) at sewage-treatment facilities. The device is based on an electrolytic cell that generates hypochlorite solution. The solution is then passed into a packed-bed scrubber, where it is mixed with the contaminated air.

<u>The Electrocinerator System</u>. This system, marketed by Electrocinerator Technologies, Inc. (Lancaster, New York), employs powerful redox reagents such as Ag(II), Co(III), Ce(IV), and Cr(VI) see (see Chapter 5). Figure 8.12 contains a schematic of this system, whose economics have been claimed[4] to be very favorable compared to conventional air-treatment technologies, such as carbon adsorption and incineration.

<u>Ispra Mark 13A Process</u>. This process for treating SO_2 in flue-stack emissions, was invented in 1979 at the Joint Research Centre of the European Community in Ispra, Italy. A pilot plant is operating at the Saras Refining Sarroch in Sardinia. Another pilot version uses a stack of dished-electrode membrane (DEM) (see Figure 8.13) cells, marketed by Electrocatalytic, Inc. (Union, New Jersey), to generate Br_2 by the electrolysis of HBr:

8.4. Electrochemical Reactors for Pollutant Treatment 707

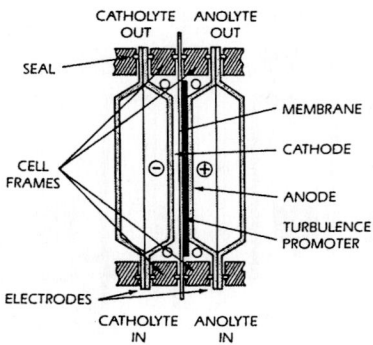

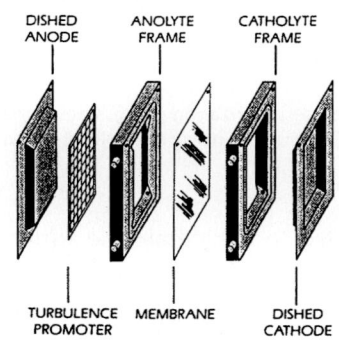

FIGURE 8.13. The dished-electrode membrane (DEM) cell marketed by Electrocatalytic, Inc. (Diagrams courtesy of Electrocatalytic Inc., Union, New Jersey.)

$$2\,HBr \rightarrow H_2 + Br_2$$

This solution is used to scrub the SO_2:

$$SO_2 + 2\,H_2O + Br_2 \rightarrow H_2SO_4 + 2\,HBr$$

The two useful products from the process are sulfuric acid and H_2 gas. The Br_2 intermediate is completely recycled. A process schematic is contained in Figure 8.14.

The Aquatech Process. This process, developed by Aquatech Systems Division of Allied-Signal Inc. (Warren, New Jersey), is also aimed at treating SO_2 emissions. The acid gas is first absorbed into an aqueous base, then a bipolar

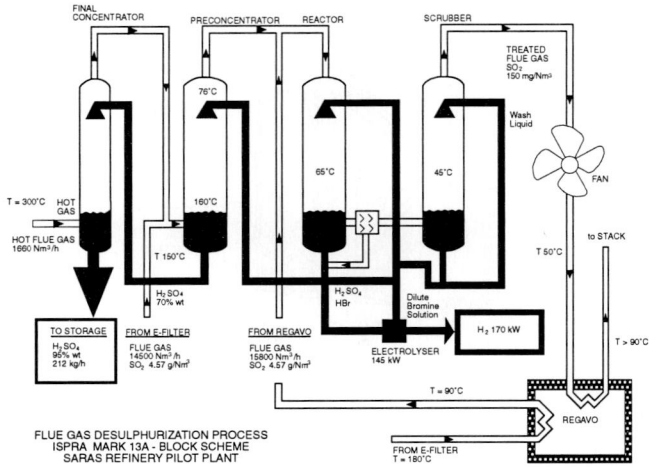

FIGURE 8.14. ISPRA flue gas desulfurization process developed at the Joint Research Centre of the European Community. (Diagram courtesy of the Joint Research Centre, Ispra, Italy.)

membrane system is used to regenerate the aqueous base. The process produces a concentrated aqueous SO_2 solution for conversion to sulfur or sulfuric acid, depending on site specifications.

8.5. ADVANCED OXIDATION PROCESSES: PILOT AND FIELD TESTS AND COMMERCIALIZATION EFFORTS

The $UV-H_2O_2$, $UV-O_3$ and $UV-TiO_2$ technologies for pollutant remediation were discussed in Chapter 6. In this section, their status will be examined from a commercialization perspective. We begin with a listing of vendors of ozonators, UV lamps, and the like, and then proceed to discuss specific systems and processes for implementing $UV-H_2O_2$ and $UV-TiO_2$ technologies. Pilot and demonstration test results on these technologies are then summarized. Finally, the economic aspects of these technologies are briefly discussed. The interested reader is encouraged to consult the original literature, listed at the end of this chapter, for further details of these aspects.

8.5.1. Suppliers of Ozonators, UV Lamps, and Catalysts for $UV-H_2O_2$, $UV-O_3$ and $UV-TiO_2$ Technologies

Table 8.11 contains a listing of vendors of the title items that exist at the time of this writing.

Table 8.11. Vendors of Ozonators, UV Lamps, Hydrogen Peroxide and Catalysts in the United States

Ozone generators
  Capital Controls Co. Inc.
  Griffin Technics Inc.
  Ozone Technology Inc.
  O_3 Associates
  Applied Science and Technology
  Lynntech Inc.

Manufacturers of H_2O_2[a]
  FMC
  Solvay Interox
  Degussa

UV Lamps
  Hanovia
  Emerson Electric Co.
  Voltarc Tubes, Inc.
  Light Sources, Inc.
  General Electric
  Magnum Technology
  Ultrox
  Peroxidation Systems, Inc.
  Solarchem Environmental Systems

Titanium Dioxide[a]
  Degussa
  DuPont
  Tioxide
  Kronos

[a] Most chemical vendors also carry these products.

The ozonators are rated in terms of their capacity in units of kg (or g)/h. The units are either water or air cooled. The feed gas is either air or O_2.

Solar concentrator panels can be used either for solar–TiO_2 AOP or for the direct photolysis of an organic pollutant (see Chapter 6). Several companies in the United States. (e.g., Solar Kinetics, Inc., Dallas, Texas) manufacture such panels and tracking units. The requirements of UV reflector materials for the direct photolysis of hazardous wastes, using solar concentration, have been reviewed.[17] An industry survey is also contained in this review; this survey was conducted to acertain the capabilities and limitations of industry as well as to obtain recommendations regarding designs, materials, and processes relevant to the production of UV reflectors for solar applications.[17]

Table 8.12. Companies Marketing Advanced Oxidation Processes

Company	Process	UV/H_2O_2	UV/O_3	UV/O_3/H_2O_2	UV/TiO_2	Commercialization Status
Energy and Environmental Engineering	LIPOD	x				Laboratory-scale testing completed; not fully commercialized yet
Excalibur Enterprises	—		x (with ultrasound)			Accepted in the SITE program in 1989; is not developed to full commercialization yet
Magnum Technology	CAV-OX	x (with cavitation)				Fully commercialized
Matrix Photo-Catalytic	—				x	Being developed to full commercialization
Peroxidation	Perox-Pure	x	x			Fully commercialized
Solarchem Environmental	Ray-Ox UV/O_3	x	x			Fully commercialized
Solar Kinetics Inc.	Solox	x			x	Being developed to full commercialization
Sun River	—		x			Being developed to full commercialization
Ultrox Intl.	Ultrox	x	x	x		Field-scale demonstration completed in 1989; fully commercialized
VM Technology	—		x			Being developed to full commercialization

8.5.2. Companies Marketing AOP Technologies

Table 8.12 contains a listing of the companies in the United States and Canada. A brief description of each company in this tabulation and its products or services follows next.

8.5. Advanced Oxidation Processes

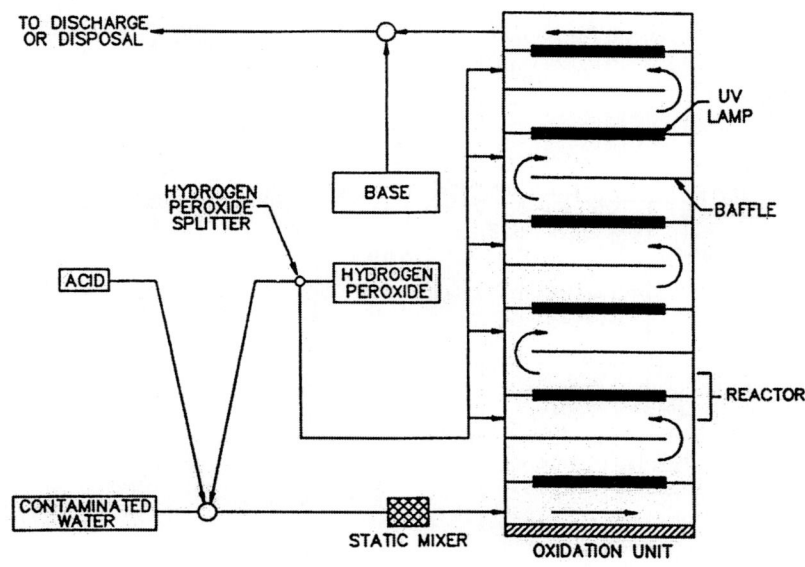

FIGURE 8.15. Schematic diagram of the Perox-Pure chemical oxidation system. (Diagram courtesy of Peroxidation Systems, Inc.)

<u>Peroxidation Systems, Inc.</u> This company developed a patented UV/H_2O_2 system called Perox-Pure in the late 1970s for the destruction of organic pollutants in water. The process utilizes in addition catalyst additives (e.g., iron) if required. Figure 8.15 contains a flow diagram of the Perox-Pure system. The system uses medium-pressure Hg-vapor UV lamps housed in quartz sleeves. The lamps are equipped with a patented tube cleaner (U.S. Patent No. 5,227,140) to clean the surfaces of the quartz tubes. This device maximizes utilization of the radiant output of the UV lamps.

<u>Solarchem Environmental Systems</u>. This company has developed several UV–H_2O_2 processes in the Rayox product line covered by U.S. Patents 5,266,214 (Rayox-A), 5,258,124 (Rayox-R), 5,324,438 (Rayox-F), and also a UV–O_3 process (U.S. Patent 5,043,079). As with the preceding entry, all these processes are directed at the treatment of water-borne pollutants. Each Rayox process differs in the proprietary catalyst formulation used along with H_2O_2. A schematic of the process is contained in Figure 8.16.

Solarchem also holds U.S. Patents 5,133,945 and 5,266,280 for steel-brush quartz tube cleaning assemblies that are pneumatically driven. These "wipers" are claimed to last up to three years of regular service before replacement is required.

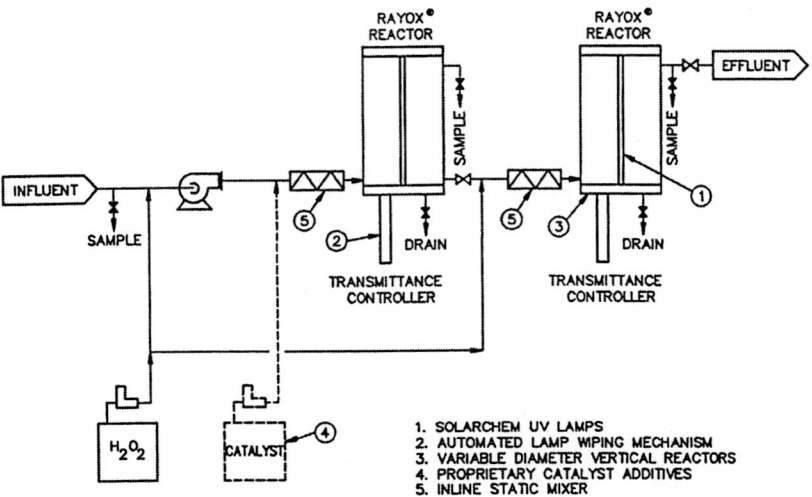

FIGURE 8.16. Schematic diagram of the Rayox Solarchem environmental process. (Reproduced with permission from Solarchem Environmental Systems.)

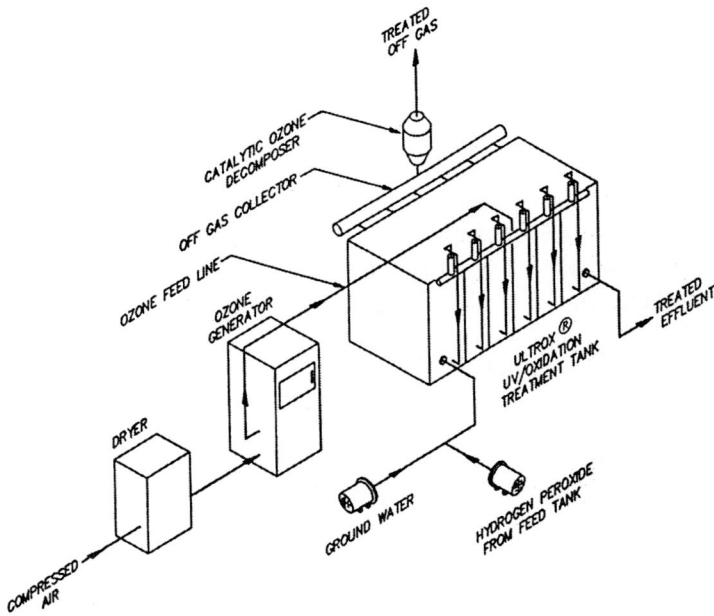

FIGURE 8.17. The Ultrox process. (Diagram courtesy of Ultrox International.)

8.5. Advanced Oxidation Processes

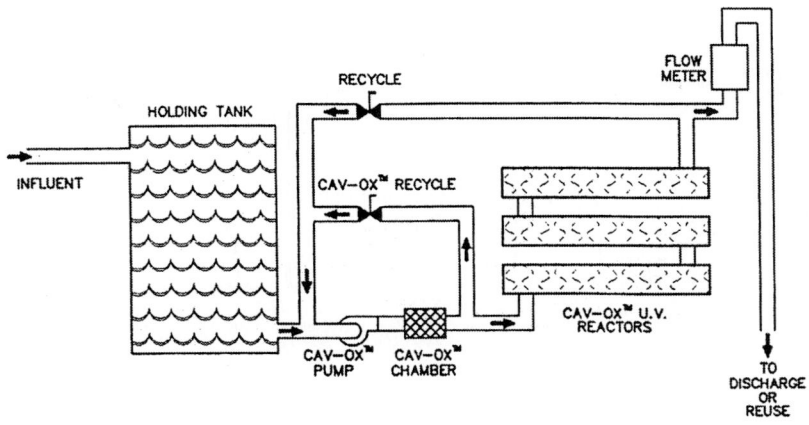

FIGURE 8.18. The Cav-Ox process (Diagram courtesy of Magnum Water Technology.)

Ultrox International. This company was formed in 1984 to develop and commercialize UV–H_2O_2 and UV–O_3 technologies for both waste- and drinking-water treatment. Four U.S. Patents (4,941,957, 4,849,114, 4,792,407, and 4,780,287) cover processes combining UV light, H_2O_2, and O_3. The basic Ultrox process is schematized in Figure 8.17. The UV lamps in the Ultrox systems are of low-pressure 70W Hg/vapor design. They are mounted vertically in the reactor with a gas bubbler. The combination of low heat generation (from the low intensity) and scrubbing action of the bubbler minimizes lamp fouling. The process has also been modified by General Electric (Fairfield, Connecticut) and Nuclear Energy Division (San Jose, California) for the removal of organics from radioactive wastes.[18]

Magnum Water Technology. The Cav-Ox process developed by this company also uses •OH radicals for oxidatively degrading organic pollutants in water. In addition, hydrodynamic cavitation is produced in a specially designed chamber (Figure 8.18). Local temperatures up to 5,000K and pressures of several hundred atmospheres are generated by the cavitation effect, where microdroplets develop, grow, and suddenly collapse because of the pressure transient. These high temperatures and pressures cause *local* dissociation of many liquid molecules generating radical species. According to Magnum Technology, the cavitation feature reduces the H_2O_2 demand for treating the polluted stream.

Energy and Environmental Engineering, Inc. The laser-induced photochemical oxidative destruction (LIPOD) process developed by this company is illus-

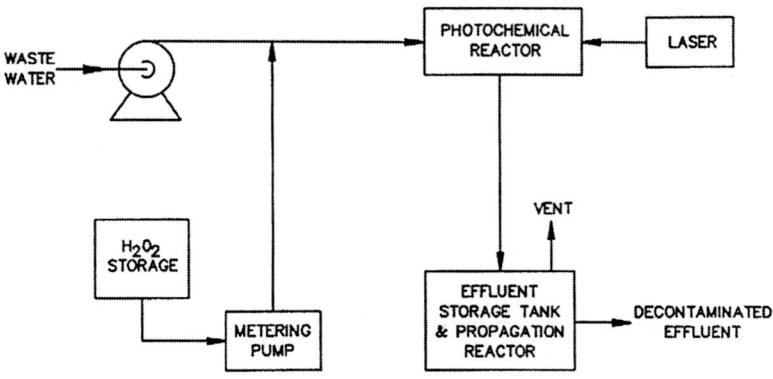

FIGURE 8.19. The laser-induced photochemical oxidative destruction (LIPOD) process. (Diagram courtesy of Energy and Environmental Engineering Inc.)

trated in Figure 8.19. The UV light source is provided by an excimer laser, and the laser light is coabsorbed by the chemical oxidant (H_2O_2) and the organic pollutant. This process is envisioned as a final "polishing" step to reduce organic levels in groundwater and industrial waste water to acceptable levels. The use of the laser does cut down the flow rate of the process stream to ca. 1 gpm.

Excalibur Enterprises, Inc. This company was first formed in 1973 to focus on water purification and later expanded in 1983 to include its service to hazardous waste treatment. The featured process is a UV–O_3–ultrasound system (U.S. Patent 4,548,716) designed for the treatment of both organic and inorganic pollutants in soil. This two-stage process involves a first soil extraction stage and a subsequent oxidation step involving the pollutant-laden extract. The ultrasound agitation is designed to keep the UV lamps clean and improve the mass transport of O_3 in the liquid.

Sun River Innovations, Ltd. This company has developed a process that combines air stripping, UV excitation, and O_3 in one compact unit, called SR2000, for the degradation of organic compounds in water. A four-step process is envisioned, where the water to be purified first enters the unit, cascades through sripper media, and mixes with O_3 gas prior to entry into the UV reaction chamber. An air stream cools the UV reaction chamber and simultaneously passes through the cascading water, stripping the latter of its volatile pollutant(s). The VOC-laden air stream re-enters the UV chamber for direct gas-phase photolysis of the gaseous pollutant. The UV light is then refocused onto the transparent column of moving water to induce further UV-O_3-assisted oxidation of the

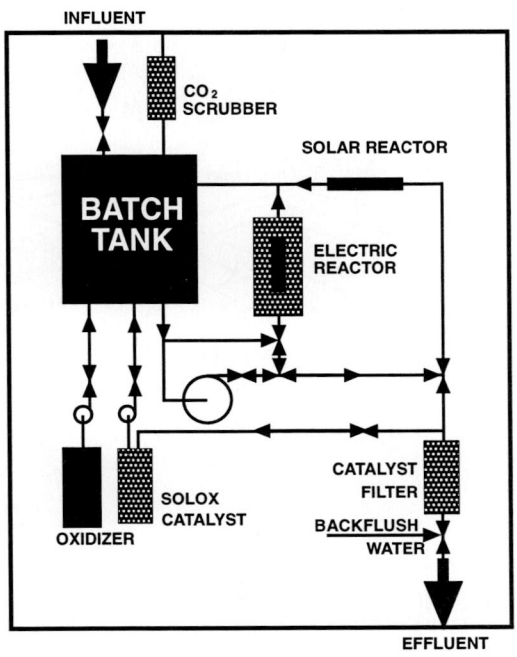

FIGURE 8.20. A hybrid solar–electric wastewater purification system. (Reproduced with permission from Solar Kinetics, Dallas, Texas.)

polluted water. The treated water either exits from the system or is recirculated for a further pass.

Solar Kinetics, Inc. Solox Wastewater Purification Systems of Solar Kinetics, Inc. (Dallas, Texas), offers the Solox-Solar (ST), the Solox-Electric (SE), and a hybrid system for the treatment of water-borne organic pollutants. The hybrid system (Figure 8.20) is designed for the heterogeneous photo-catalytic oxidation process. The catalyst filter removes the TiO_2 particles from the treated water stream. The solar system design is based on a parabolic trough with a line focus. A glass reactor tube is located at the focus and the process stream flows through the reactor and down the trough (Figure 8.21). The electric version is based on the use of UV lamps. A chemical oxidant (e.g., H_2O_2) is also added to the UV/TiO_2 system if needed.

VM Technology. This company offers a variety of environmental pollution abatement equipment and technologies including those based in UV light, ozone, and activated carbon. Two separate systems are offered: one for treating polluted waters and another for air-pollution control. In the UVOX system

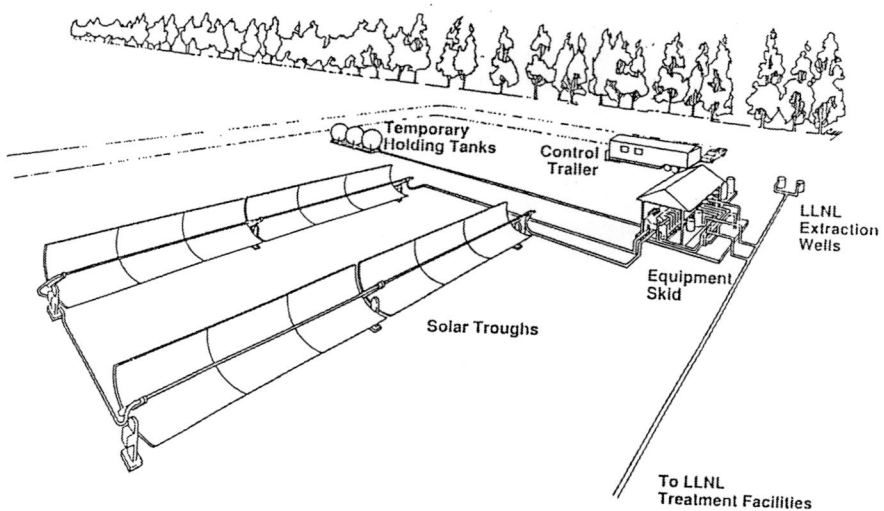

FIGURE 8.21. Schematic diagram of the solar water detoxification field experiment at the Lawrence Livermore National Laboratory site. (Reproduced with permission from National Renewable Energy Laboratory, Golden, Colorado.)

for groundwater treatment, ozone is generated *photolytically* by exposing compressed air to UV light. The photogenerated O_3 is then mixed with the polluted water stream. Unreacted O_3 and other gases exit from the top of the reactor and pass through a coalescer that separates water from air. The air stream is then routed to activated carbon beds. The system uses two alternating carbon beds: one is for adsorbing the VOCs and the other is regenerated with O_3. Recycled O_3 is fed back into the reactor, and the treated air is vented to the atmosphere. Since the carbon is regenerated within the unit, the vendor claims that the carbon life is affected only by its own durability, typically 5–10 years.

8.5.3. *Companies Marketing Direct Photolysis Technology*

At least one commercial enterprise (Purus, Inc.) markets direct photolysis units for the UV-assisted breakdown of volatile organic pollutants. The Purus system is schematized in Figure 8.22. The system uses a pulsed Xe lamp (flashlamp) that emits short-wavelength (<250 nm) light at very high intensities. A feature of this system is that the photolysis parameter can be optimized by shifting the spectral output via control of the peak pulse power. The process

8.5. Advanced Oxidation Processes

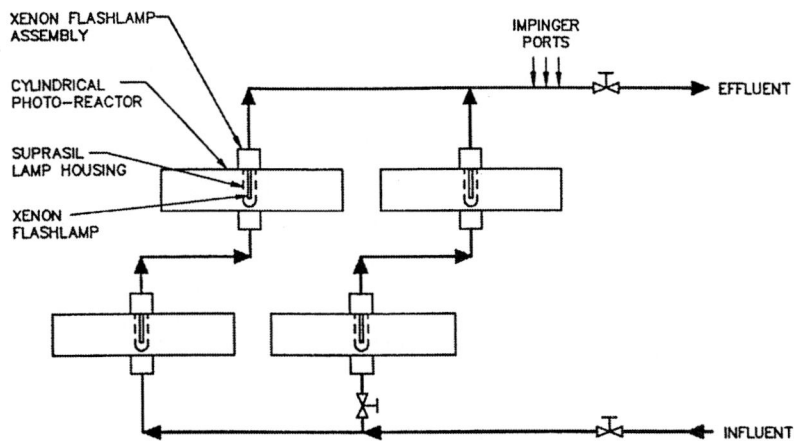

FIGURE 8.22. The Purus direct photolysis system. (Diagram courtesy of Purus, Inc.)

uses vacuum extraction or air stripping to first remove the VOCs from soil or groundwater. The VOCs are then routed into the photolysis reactor.

8.5.4. Pilot, Field, and Demonstration Tests of the AOPs

Generally, the evaluation and selection of a particular AOP from among the various commercial systems that are available, are based on two criteria. The first is *treatability*, that is the effectiveness of a given process to treat the waste stream; the second is *cost*. Treatability studies have been carried out in the United States mainly under the EPA's SITE (Superfund Innovative Technology Evaluation) program. The SITE program is designed to evaluate new treatment methods through technology demonstrations. The latter in turn generate engineering and cost data for further evaluations that can be used as a basis for scale-up and commercial adoption. The demonstrations also facilitate the evaluation of long-term risks.

UV–H_2O_2 and UV–O_3. Tables 8.13[1] and 8.14 contain a compilation of pilot- or bench-scale performance data on UV–H_2O_2 and UV–O_3 systems, respectively. Table 8.15 contains similar data for the UV–H_2O_2–O_3 process. While many factors control the pollutant decomposition kinetics (Chapter 6), it is clear that all three technologies have demonstrated effectiveness in the treatment of waterborne pollutants. However, one problem with comparative assessment of the available information is that, in many cases, the process variables (e.g., light

Table 8.13. Results of Pilot, Bench, and Field Tests on UV–H_2O_2 Systems[a]

Pollutant(s)/matrix	Other Process Details	Treatability Data	Reference
1,4-Dioxane/groundwater	UV dose: 12 kWh/1,000 gallons	Level reduced from 1,800 ppb to 20 ppb in 5.5 min	19
BTEX/waste water	Preoxidation/filtration system added to reduce iron and calcium hardness; this prevents quartz tube fouling	Influent level ranged from 3.3 ppm to 4.6 ppm; corresponding effluent levels were 0.26 ppm and 1.9 ppm respectively; with filtration the effluent levels were below detection limits	20
Methylene chloride, TCE, PCE, toluene, 1,1-DCA, 1,2-DCA/waste water	H_2O_2 dose: 40 ppm	Level reduced from 1–272 ppb to below detection limit in 2 min	18
TCE, 1,2-DCE, chloroform/groundwater	—	Level reduced from 2.1–66.3 ppb range to below detection limits	18
PCE/waste water	H_2O_2 dose: ~20 ppm	Greater than 99.9% removal in less than 1 min	21
Acetone, phenol, aniline, hydrazine, bis-2-ethyl hexyl phthalate/groundwater	—	Influent concentration ranged from 14 ppb to 180 ppm; effluent reduced to below detection limit; the TOC of the groundwater lowered from 31 ppm to 2 ppm; a retention time of 18 min used	22
Trichlorophenol, 2,4-dichloro-4-nitroaniline, 3,3-dichlorobenzidine/waste water	H_2O_2 dose: ~340 ppm and UV dose: 2.31 kWh/gallon	Effluent level reduced to 10 ppb in a 30 min retention time	23
TCE, 1,2-DCE/waste water	Temperature effects also probed	Level reduced from 58 ppm to 1 ppm for TCE in 50 min	24
TCE, 1,2-DCE, vinyl chloride, PAHs/groundwater	—	Greater than 99% destruction achieved	25

Cont. Table 8.13.

Pollutant(s)/matrix	Other Process Details	Treatability Data	Reference
Benzene/waste water	Molar ratio of H_2O_2: benzene in the 6.6–10.6 range	Approximately 98% destruction attained in 40–80 min	18
Acetone/waste water	An iron catalyst used at pH 2	Approximately 85% destruction achieved in 60 min	26
Atrazine/groundwater	Negative effect of bicarbonate and humic substances noted	Influent level ranged from 3.8×10^{-5} to 9.6×10^{-5} M; greater than 99% removal reported	27

BTEX = benzene, toluene, ethylbenzene, xylene; TCE = trichloroethylene; PCE = perchloroethylene; DCA = dichloroethane; DCE = dichloroethylene; PAHs = polycyclic aromatic hydrocarbons.
[a] See also Ref. 18.
[b] The treatment time refers to retention time or residence time in the reactor.

Table 8.14. Results of Pilot, Bench, and Field Tests on UV–O_3 Systems[a]

Pollutant(s)/matrix	Other Process Details	Treatability Data	Reference
Nine herbicides, two insecticides in artificially spiked water	—	Three influent levels of 10, 100, and 1,000 ppm used; the time for 90% destruction increased with influent concentration and were 22 min and 61 min for the lower levels	28
TCE, PCE/groundwater	—	TCE reduced from 330 ppb to 3.2 ppb and PCE reduced from 160 ppb to 5.5 ppb	18
TCE, 1,1,1-TCA, 1,1-DCA/groundwater	—	Greater than 90% removal efficiency on the range of operating conditions	29

Chapter 8

Cont. Table 8.14.

Pollutant(s)/matrix	Other Process Details	Treatability Data	Reference
Free and complex cyanides/waste water	Temperature effects probed	Free cyanide destroyed by O_3 alone without UV light; complex cyanides readily degraded with UV/O_3	30
Hydrazine, alkyl hydrazine, dimethyl nitrosamine/waste water	pH variation probed	Oxidation more rapid in base than in acid; presence of Cu and Fe(III) accelerates the degradation	31
Pink water (explosives-containing water)	Ozone/TOC ratio: 18	TOC level reduced from 15 ppm to 1 ppm	18
Acetic acid, ethanol, glycerol, glycine, palmitic acid/waste water	Temperature effect noted	Compared to ozone alone, the oxidation rate increased 100- to 10,000-fold when UV/O_3 was used	32

Note: Refer to Table 8.14 for compound abbreviations. TCA = trichloroethane.
[a] Also see Ref. 18.

flux) are not completely specified in the original reports and documents on these tests. Nor are they expressed on the same basis. The H_2O_2 and O_3 dose specifications in Table 8.13–8.15 ought to provide cases in point.

Case Studies. The Los Angeles Department of Water and Power undertook a pilot-scale evaluation of the H_2O_2–O_3 AOP, using water from a polluted well in North Hollywood.[34] The process was run on a continuous basis. Treatment efficiency was evaluated as a function of H_2O_2 to O_3 dosage ratio, O_3 dosage and contact time. Laboratory analyses of the water indicated that the optimum H_2O_2:O_3 dosage ratio was 0.4–0.5 by weight. For a constant H_2O_2:O_3 ratio and constant water quality, the rate of oxidation of TCE and PCE depended only on the rate of mass transfer of O_3 from the gas phase to the liquid phase. Thus, the reaction of H_2O_2 with O_3 and the subsequent oxidation of TCE and PCE by •OH are virtually instantaneous.

A cost evaluation[34] of the H_2O_2–O_3 AOP indicates that the process is competitive with more conventional technologies. We shall explore this aspect in a subsequent section (Section 8.5.5).

8.5. Advanced Oxidation Processes

Table 8.15. Results of Pilot, Bench, and Field Tests on UV–H_2O_2–O_3 Systems

Pollutant(s)/matrix	Other Process Details	Treatability Data	Reference
TCE/groundwater	Peroxide and O_3 dose 13 ppm and 110 ppm respectively; influent pH: 7.2	TCE removal efficiency greater than 99% for influent levels ranging from 48 to 85 ppm	18
Methanol/distilled water	H_2O_2 and O_3 dose: 31 millimoles and 39 millimoles	TOC decreased from 75 ppm to ~1 ppm after 30 min; the initial concentration of methanol was 200 ppm	33
Methylene chloride-spiked water	H_2O_2 and O_3 dose: 6.6 and 13.4 millimoles	Level reduced from 100 ppm to 7.6 ppm in 25 min	33
1,4-Dioxane, ethylene-glycol, acetaldehyde/waste water	Ozone dose: 205 mg/min; 35 mL of 30% H_2O_2 added during the first 90 min	Influent concentration range: 700–5,000 ppm; dioxane reduced to 50 ppm in 150 min	33
Phenol, penta-chlorophenol/waste water	System with 72 lamps of 65 W each used	Influent level ranged from 100–500 ppm for phenol to 5–10 ppm for pentachlorophenol; the targeted effluent limits of 40 ppm and 0.1 ppm for these two pollutants were achieved in 60 min	18

The Ultrox technology was demonstrated in 1989 under the auspices of the U.S. EPA SITE program at the Lorentz Barrel and Drum Site in San Jose, California.[35] This site was used primarily for drum recycling operations from ca. 1947 to 1987. The drums contained residual aqueous wastes, organic solvents, acids, metal oxides, and oils. The preliminary site assessment report indicated that both soil and groundwater were polluted with organics and metals. The upper aquifer at the site was selected as the waste stream for evaluation of the Ultrox UV–H_2O_2–O_3 system. Laboratory analyses of the water samples showed high levels of TCE (280–920 ppb), vinyl chloride (51–146 ppb), and 1,2-DCE (42–68 ppb). The pH and alkalinity of the groundwater was 7.2 and 600 mg/liter (as $CaCO_3$), respectively. A total of 13 test runs were performed with five process parameters being varied: (a) influent pH; (b) hydraulic retention time; (c) O_3 dose; (d) H_2O_2 dose; and (e) UV radiant flux. The best operating conditions were determined to be pH 7.2; retention time, 40 min; O_3 dose, 110 ppm; H_2O_2 dose, 13 ppm; and all 24 65W UV lamps operating.

The Peroxidation Systems technology was applied at the Old O-Field site located at the Aberdeen Proving Ground in Maryland.[35] This site was used for the disposal of chemical warfare agents, munitions, and various other hazardous materials during the 1940s and early 1950s. Groundwater samples collected from the site showed the presence of VOCs, organosulfur compounds, and explosives at levels in the 10–500 ppb range. Trace levels of arsenic were also observed in some locations.

Treatability studies were performed in 1991, where 37,000 gallons of groundwater from three wells were used as the influent. After arsenic and other metals (mainly iron and manganese) were removed by a precipitation step, the water was treated with two parallel systems: (a) an air stripping–GAC system developed by Carbonair and (b) the Peroxidation Systems technology. The treated effluent meets the federal MCLs for all the compounds. Details of the demonstration test results are available.[35]

The Peroxidation Systems, Inc. Perox-Pure technology was also demonstrated under the SITE program at the Lawrence Livermore National Laboratory Site 300 in Tracy, California.[36] Over a three-week period in 1992, ca. 40,000 gallons of groundwater polluted with TCE, PCE, and other VOCs were treated in the Perox-Pure system. The system effluent met California drinking water action levels and federal drinking water MCLs for all the pollutants at the 95% confidence level.

<u>UV–TiO_2 and Solar–TiO_2 Case Studies.</u> An early demonstration of the TiO_2-based heterogeneous photocatalysis technology was that conducted jointly by Sandia National Laboratories (Albuquerque, New Mexico) and the National Renewable Energy Laboratory (NREL) (Golden, Colorado) in the U.S.[37-39] A bench-scale test using a UV lamp source was used to guide subsequent engineering-scale experiments using concentrated solar energy and TiO_2. Two different solar systems were used: a line-focusing parabolic trough design and a heliostat–falling-film reactor design.

The model pollutant in these studies was salicyclic acid that was spiked into the water stream to be treated. Parameters such as H_2O_2 concentration, temperature, and catalyst dose were used as variables. The catalyst, TiO_2 (Degussa P-25 grade) was suspended in mixing–holding tanks and then recirculated through the reactors. At the particular salicyclic acid influent level used (30 ppm), the pollutant was reduced to below detection limits in less than 30 s. An important aspect of the systems using solar concentrator technology is that the pollutant mineralization appears to occur before boiling the water stream. This is an important consideration in terms of the possibility of thermal stripping of the pollutant from the aqueous phase.

8.5. Advanced Oxidation Processes

Another field experiment was jointly conducted by NREL and Sandia researchers at the Lawrence Livermore Laboratory Superfund Site in Livermore, California.[40] The system for this field test (Figure 8.21) again used a commercially available line-focusing parabolic solar trough design (see Solar Kinetics Inc., Section 8.5.2). This solar detoxification field experiment was designed to draw water from an existing well pipeline, treat it, and then return it to the pipeline so that it could be processed by the treatment facility for discharge into the aquifer. Four wells connected to the pipeline could be operated individually or in parallel. The TCE level of the influent was in the 80–500 ppb range. Other VOCs were present at <10 ppb level and the bicarbonate alkalinity was 500 ppm with the water pH being in the 6.5–8.0 range. The bicarbonate interference (see Chapter 6) was minimized by acidifying the water stream to ~5.6 by HCl addition. In a single pass experiment, the inlet TCE level of 106 ppb was reduced to <0.5 ppb.

Other field experiments of the solar–TiO_2 technology in the United States have included two separate efforts at the Tyndall Air Force Base in Panama City, Florida.[41,42] The pollutants of concern at this site are the BTEX group of compounds. In one such test, the total concentration of 2 ppm of these mixed aromatics was lowered to less than 0.001 ppm under 6.5 min of contact time, even during a rainy day with a UV irradiance a tenth of the flux at high noon in winter.[42]

Interestingly enough, the cost estimates projected[41] for the treatment of water at this site (U.S. $70–100 per 1,000 gallons) do not appear to be favorable relative to the other estimates currently available for UV–TiO_2 or solar–TiO_2 technology (see later). This high cost has been attributed to the pronounced interference from *nontarget* contaminants in the water stream.

Joint tests between the Plataforma Solar de Almeria in Spain and the Institut für Solarenergieforscheng in Hannover, Germany have used both a parabolic trough reactor and a falling-film design incorporating a fixed bed of immobilized TiO_2 particles.[43,44] Water spiked with dichloroacetic acid and waste water from a phenolic resin factory were studied as the treatment streams. To facilitate a comparison of the efficacy of the two types of reactor designs (which differed in size and hence the residence time), a dimensionless "photon efficiency" parameter $\varepsilon^{h\upsilon}$, was used[43]:

$$\varepsilon^{h\upsilon} = \frac{C_o \dot{V}}{I_o}$$

<u>Gas-Phase Heterogeneous Photocatalysis.</u> Experiments in the laboratory or bench scale (see Section 6.9.13) levels have shown that reaction rates are orders of magnitude faster in the gas phase than in aqueous solution, for the

Table 8.16. Engineering Decisions in a Detoxification Process Design Based on Heterogeneous Photocatalysis[a]

Issue	Pros	Cons
Operation Mode		
Batch recirculation	Unit is compact	Suitable for treating only small quantities of effluent stream
Continuous flow	Can treat large quantities of effluent stream	—
Nature of Process Stream		
Gas	High mass transfer rate translates to good process efficiency	Suited for only volatile pollutants; requires periodic reactivation of photocatalyst surface
Liquid	Process compatible with a wide range of pollutant streams	Process efficiencies generally are poor because of mass-transfer limitations
Photocatalyst Configuration		
Slurry	High dispersion facilitates large contact cross-section with pollutant molecules or ions	Catalyst recovery after use complicates system design
Thin film (or immobilized)	Catalyst recovery after use not an issue	Surface area will not be as high as in the slurry case; photocatalyst may be more prone to fouling
Irradiation Source		
UV (electric)	Higher UV content relative to the solar spectrum translates to better match with photocatalyst (e.g., TiO_2) absorption profile	Electricity requirement may restrict use for remote-site applications
Solar (one sun)	No reflective surface; collects both diffuse and direct sunlight; no tracking needed; low flux results in higher quantum yield	Weather resistant, chemically inert, UV light transmitting, glazing needed

Cont. Table 8.16.

Issue	Pros	Cons
Solar (compound parabolic trough concentrator)	Collects both diffuse and direct sunlight; no tracking needed; low flux results in higher quantum yield	
Solar (concentrating trough with heliostat)	Small receiver–reactor allows more expensive components to be used without adversely affecting overall system cost; established design; easy fluid distribution and containment	High flux results in poor quantum yield; cannot capture diffuse sunlight

[a] Also consult Turchi et al.[39,40]

photoassisted degradation of some (especially volatile) pollutants. Hence, there is much commercial interest in first stripping these VOCs from water or soil and then degrading them via gas-phase photocatalysis. At the time of this writing, NREL and IT Corporation (Knoxville, Tennessee) had entered into a cooperative agreement to spur the commercialization efforts in this area.

Design Issues for Heterogeneous Photocatalysis Systems Based on TiO_2. Table 8.16 contains a list of engineering decisions to be made in the choice of a particular photocatalysis technology based on TiO_2. These have been thoroughly discussed in the literature.[39,40] Some of these aspects were also discussed earlier in this book in Section 6.9.14. Needless to say, the choice of a particular process is largely dictated by the site characteristics. For example, an application involving a remote site and rather small (several liters) quantities of water to be treated would call for a process based on solar–TiO_2 rather than UV–TiO_2. On the other hand, we can look beyond heterogeneous photocatalysis and consider the other AOPs discussed earlier in this chapter. Even combinations of these technologies with more conventional ones, for example GAC adsorption–gas-phase photocatalysis, could be considered. Such "hybrid" designs are not considered per se in Table 8.16. Coupling of a solar system with UV backup can also be envisioned for 24-hr use.

This brings us to the issue of relative treatment costs, which is the topic of the next section.

Table 8.17. Treatment Costs for the Various AOPs

Entry No.	Process	Treatment Cost U.S. $/1,000 gal (3,785 liters)	Reference
1	UV–H_2O_2	4.40	45
2	UV–H_2O_2 (Perox-Pure)	11.00	36
3	UV–H_2O_2–O_3	0.094	34
4	UV–TiO_2	5.22	42
5	Solar–TiO_2 (trough)	13.00	46
6	Solar–TiO_2	6.00	45
7	GAC	6.20	45
8	GAC (liquid phase)	0.397	34
9	GAC (gas phase with air stripping)	0.277	34
10	Air stripping	0.075	34

8.5.5. Economic Considerations

Table 8.17 contains the (annualized) treatment costs for the various AOPs expressed in terms of U.S.$/1,000 gallons. The projects cost estimates are seen to vary widely even for a given process, undoubtedly because of the assumptions involved in the particular costing procedure. However, barring Entries 3, 8, 9, and 10 in Table 8.17, the costs for the other cases fall in the range from ca. $5 to $13 for every 1,000 gallons (3,785 liters) treated. The solar–TiO_2 approach employing the solar concentrator technology appears to be more expensive than the nonconcentrating ("one-sun") approach. Air stripping alone is the lower-cost alternative (Entry 10 in Table 8.17). However, in many areas of the United States, air stripping *alone* would not be an acceptable option to the public or to regulatory agencies.

It must also be borne in mind in considering the data in Table 8.17, that the costs are often very sensitive to the nature of the pollutant and its concentra-

8.5. Advanced Oxidation Processes

Table 8.18. Comparison of Operating Costs Between UV–H_2O_2 and GAC Adsorption Technologies

Pollutant Concentration (ppm)	UV/H_2O_2	Operating Cost ($/1,000 gal)		
		"Poor" Carbon Adsorbers	"Average" Carbon Adsorbers	"Good" Carbon Adsorbers
0.1	1–2.5	>3	0.5–3	<0.5
1	1.5–4	>10	0.7–10	<0.7
10	2–5	>50	1–50	<1

Source: Adapted from Notarfonzo and McPhee.[47]

tion in the effluent. This is illustrated in part in Table 8.18.[47] Note that in the case of GAC adsorption, the cost is inversely proportional to the adsorbability of the pollutant on the carbon phase.

A costing figure of merit for the UV–H_2O_2 process has been presented[47] in terms of a parameter called the *electrical energy per order* (EE/O). It is defined[47] as "the kilowatt hours of electricity required to reduce the concentration of a compound in 1,000 gallons by 1 order of magnitude (or 90%); the unit for EE/O is kWh/1,000 gallons/order." For example, if 10 kWh of electricity is required to reduce the level of a target pollutant from 10 ppm to 1 ppm in 1,000 gallons of groundwater, then the EE/O is 10 kWh/1,000 gal/order for this compound. Table 8.19 shows the EE/O values for selected pollutants.[47]

The costing components for the UV/H_2O_2 system are[47]
(1) UV dose (kWh/1,000 gallons)
(2) Electrical cost ($/1,000 gallons)
(3) Operating cost ($/1,000 gallons) (electrical + peroxide)
(4) Capital cost

Capital cost is computed as a function of system size, which in turn scales with the UV power (kW) required to destroy the targeted pollutant(s). Correlations of capital cost with UV power have been presented.[47] Lamp replacement costs typically range between 40–50% of the electrical cost.[47] Table 8.20 contains a representative costing protocol using the EE/O parameter for TCE as a pollutant.

Table 8.19. Electrical Energy/Per Order (EE/O) for Selected Pollutants

Pollutant	EE/O kWh/1000 gal/order
Atrazine	30
Benzene	5
Chloroform	15[a]
Chlorobenzene	5
1,4-Dioxane	6
DCA	15[a]
DCE	5
Freon	10[a]
Phenol	5
PCE	5
PCP	10
TCE	4
Toluene	5
TNT	12
Vinyl chloride	3
Xylene	5

[a] With reduction catalyst.

Source: From Notarfonzo and McPhee.[47]

Treatment costs for the other AOPs can be similarly computed, once the operating variables (e.g., O_3 dose, TiO_2 dose) and the pollutant profile are established from treatability tests. The UV-based systems tend to be less expensive than the solar alternative for the heterogeneous photocatalysis process. This largely accrues from the much higher process efficiency realized in the former case. An exception to this trend would be when the site locale is such that the electricity costs and other system variables (e.g., UV lamp replacement) are prohibitively high. This could well prove to be the case in Third World countries. Several factors can lead to a drop in the solar unit cost. These include the solar collector component and water storage tanks. For example, it has been estimated[46] that the treatment cost drops by ~30% if 24 hr/day pumping is not required. This scenario can occur where large well pumps can be installed.

Figure 8.23 shows a comparison of the UV–TiO_2 and solar–TiO_2 treatment costs for the water- and air-based systems.[46] The air-stripper–gas-phase treatment concept is significantly less expensive than the aqueous treatment options. The reasons for this include (a) higher reaction rates in the gas-phase; (b) less pretreatment of the stream; and (c) lower catalyst replacement costs. Importantly, inhibiting and catalyst-fouling species such as bicarbonate or other

Table 8.20. Costing Protocol Using the EE/O Parameter for a UV–H_2O_2 Treatment System

Assumptions

Pollutant:	TCE
Flow rate:	125 gpm
Influent:	10 ppm
Effluent:	0.01 ppm
Power cost:	$0.06/kWh
Peroxide demand:	Twice the pollutant level (2 × 10 ppm = 20 ppm in this example)
Peroxide cost:	$0.005/ppm/1,000 gal
Lamp replacement cost:	45% of electrical cost

Electrical Cost

(With reference to Table 8.19, EE/O for TCE is 4 kWh/1000 gal/order)

Electrical cost: $4 \times \log(10/0.01) \times 0.06 = \$0.72/1{,}000$ gal

Operating Cost

Total operating cost: $1.45 \times$ electrical cost + peroxide cost $= 1.45 \times 0.72 + (20 \times 0.005) = \$1.14/1{,}000$ gal

UV Power (kW)

$$\text{Basis: } \frac{\text{UV dose} \times 60 \times \text{flow (gpm)}}{1000}$$

$$\text{or } \frac{\text{EE/O} \times 60 \times \text{flow (gpm)} \times \log(\text{initial/final})}{1000}$$

$$\frac{4 \times 60 \times 125 \times \log(10/0.01)}{1000} = \underline{90 \text{ kW}}$$

Capital Cost

Related to UV wattage required. Thus, 90 kW corresponds to ca. $150,000.

---

[a] Also see Notarfonzo and McPhee.[47] All costs in U.S. dollars.

ions and nonvolatile organics remain in the water phase. The figure also contains the range of cost for the GAC or UV–H_2O_2-based treatment options.

In summation of the discussion contained in this section, it can be stated that the various AOPs, at the current level of refinement, are at least competi-

730   Chapter 8

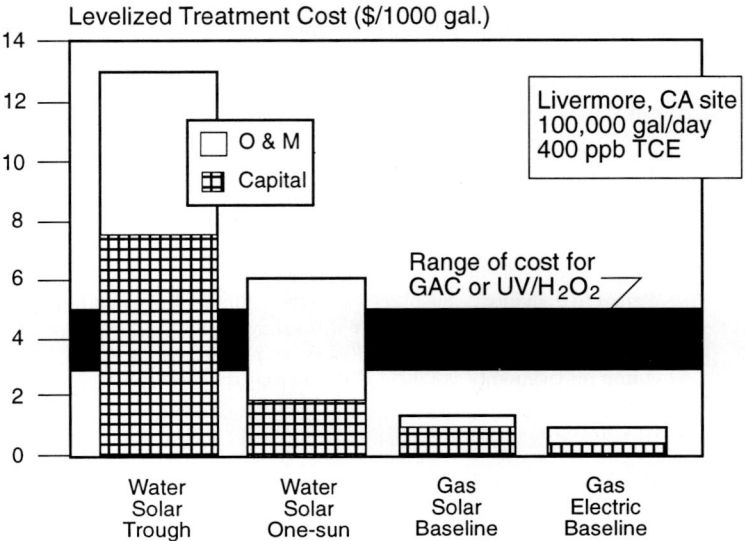

FIGURE 8.23. Comparison of estimated treatment cost (U.S. $/1000 gal) for the Livermore, California site. (Reproduced with permission from the National Renewable Energy Laboratory, Golden, Colorado.)

tive in terms of treatment costs with existing technologies. They may become more attractive as these technologies develop even further, and when some of the "hidden" handicaps of extant treatment methods (e.g., spent GAC disposal) are taken into account. Finally, the AOPs appear to be the technologies of choice and the most "appropriate" in terms of integration into the infrastructure in Third World countries and other developing nations.

8.6. WATER DISINFECTION TECHNOLOGY

Of the various technologies discussed in Chapter 7 for the disinfection of water, only the H_2O_2–O_3 AOP and UV disinfection appear to have attained maturity. We shall review case studies and the economic aspects for each of these technologies in turn.

8.6.1. Case Studies

Table 8.21 contains a representative listing of the pilot and full-scale studies conducted between ca. 1980 and 1993 on UV disinfection of water.

One of the early studies[48] utilized a commercially available system (Ellner Model EP-50 UV water purifier) for the disinfection of fecal coliforms in mu-

Table 8.21. Pilot and Full-Scale Studies on UV Disinfection

Site/System	System Details	Reference
Ellner Model EP-50 UV water purifier	10 UV lamps with a lamp density of 0.17 m of lamp per liter of water; UV dose: 35,00 µW·s/cm^2	48
Tillsonburg, Ontario	Two units, A and B: Unit A; 80 lamps in one channel with 76.2 cm arc length/lamp; Unit B; 120 lamps; 13.8 W of 253.7 nm light; Average water flow rate; 0.03 m^3/s	49
Northwest Bergen County Utilities Authority Plant Waldwich, New Jersey	Open channel system with four banks of lamps, two in each channel; the channels operate in parallel; the lamps can be oriented either parallel or perpendicular to the direction of water flow	50
PEP-Tegel, Berlin	Open-basin of 0.75 × 0.51 × 2.25 m with six frames mounted longitudinally in the basin; each frame carries four UV lamps one above the other; at 60% water transmission, the radiant flux is computed to be 13.3 mW/cm^2; average contact time is 3.54 s with the average dose being 47 mW·s/cm^2	51

nicipal wastewater effluents. Fifteen experiments were conducted using primary clarifier effluent, settled activated sludge effluent, activated sludge effluent with waste-activated solids added, tertiary sand filter effluent, mixed-media filtration effluent, and trickling (roughing) filter effluent. Cost estimates for UV disinfection were also made in this pilot study.[48]

Another study was conducted in Tillsonburg, Ontario, Canada.[49] This project site was a conventional secondary plant that produces an above-average effluent from primarily domestic waste water. Two separate molecular UV systems were tested and compared with the chlorine disinfection procedure using split clarifiers, channels, and contact chambers.

The Northwest Bergen County Utilities Authority wastewater treatment plant, located in Waldwick, New Jersey, converted from chlorination to a UV disinfection system in 1989.[50] This was done by retrofitting existing Cl$_2$ contact tanks with UV lamps. The conversion was necessitated by the imposition of a "zero residual" in the plant's revised permit concurrent with the passage of the state's Toxic Catastrophe Prevention Act.

Full-scale pilot plants have been tested in other countries also, as exemplified by tests done to the outlet of a phosphate elimination plant in Berlin-Tegel.[51] The surface water of a channel in Berlin that is up to 70% polluted by secondary effluents is treated at PEP-Tegel by flocculation and filtration to reduce eutrophication in the Lake Tegel downstream.

In all these pilot and full-scale studies, the UV process was demonstrated to consistently achieve the objective of 200 fecal coliforms per 100 mL of effluent. Importantly, UV-irradiated effluent exhibited no toxicity to aquatic life in the receiving waters. Suspended solids is a major issue both from the perspective of UV transmittivity of the turbid water and because micro-organisms are "encapsulated" by the suspended particles. Full-scale tests, however, have shown[52] that a flocculation and filtration step improves the process efficiency in these cases.

The H_2O_2–O_3 AOP technology was tested by the Metropolitan Water District (MWD) of Southern California as described earlier in this book (Chapter 7).[53] MWD treats two source waters in its system: state project water from northern California and Colorado River water. MWD considered GAC for controlling THMs and taste-and-odor compounds in 1988. These tests showed that GAC is an expensive technology and did not appear to achieve THM levels less than 5 ppb even with extended empty-bed contact times.[53] Subsequent studies with nine different oxidants[53] showed that H_2O_2–O_3 (Peroxone) was the most effective in the removal of MIB and geosmin, and the removal rates of these three compounds were better with Peroxone than with ozone alone. The microbiological results also indicate[53] that for MWD's water sources, Peroxone and ozone are comparable disinfectants at H_2O_2:O_3 ratios of 0.3 or less and 12 min contact time. Demonstration studies (5.5 mgd or 0.24 m^3/s) have been scheduled to evaluate both the ozone and Peroxone processes on a larger scale.

Small-scale hybrid ion exchange–UV disinfection units are currently in use in households in Third World countries such as India.

8.6.2. Economic Aspects

Table 8.22 contains a summation of various cost estimates for AOP-based disinfection technologies and comparisons with existing approaches, such as chlorination. In using these data especially in a comparative sense (i.e., study to study), the initial asumptions as to the electricity cost and the effect of inflation must be carefully taken into account. The figures in Table 8.22 (and indeed the cost estimates presented earlier in Table 8.17), therefore, are presented only as a very rough guide. In this vein, it is perhaps safe to state that the UV-based disinfection approach is a cost-effective alternative to chlorination. The situation with the ozonation (or the H_2O_2–O_3 approach) is murkier, and the available estimates show conflicting trends. (For example, compare Entries 3 and 4, Tables 8.17 and 8.22, respectively.)

Table 8.22. Cost Estimates for AOP-Based Water Disinfection Technologies and Comparison with Chlorination

Entry No.	Cost (per 1,000 gal or 3,785 L)			Reference
	UV	Chlorine	Ozone	
1	0.021–0.043	0.043–0.07	0.053–0.066 (from O_2)	48
2	0.042	0.035	0.071 (from O_2)	48
3	0.0072 (good-quality effluent)[b] 0.0079 (secondary effluent)[b] 0.015 (sand-filled effluent)[b]	0.005–0.013		48
4	1 (normalized)	2.77	8.67 (from air)	49
5	THM Removal 0.55 (60% removal) 0.70 (70% removal) 0.91 (80% removal)			54 54 54

[a] Based on 1 mgd plant capacity.
[b] Cost of lamp cleaning and overdesign to account for lamp deterioration and fouling not included.

8.7. SUMMARY

We hope that the preceding discussion illustrates that electrochemical methods for sensing and control of environmental pollutants have progressed well beyond the laboratory into the commercial sector. To what extent these "new" technologies will supplant (or complement) existing ones on a routine basis is anybody's guess. Nonetheless, their feasibility on both technical and economic grounds bode well for the future.

REFERENCES

1. T. Shono, "Electroorganic Synthesis." Academic Press, London, 1991.
2. "Analytical Chemistry," 1995 Lab Guide Edition. American Chemical Society, Washington, DC, 1994.
3. J. Wang, "Analytical Electrochemistry," Table 3.1, p.73. VCH, New York, 1994.
4. J. D. Genders and N. L. Weinberg, eds., "Electrochemistry for a Cleaner Environment." The Electrosynthesis Co., East Amherst, NY, 1992.
5. C. J. Brown, D. Pletcher, F. C. Walsh, J. K. Hammond, and D. Robinson, Studies of Three-dimensional Electrodes in the FM01-LC Laboratory Electrolyzer. *J. Appl. Electrochem.* **24**, 95 (1994).
6. F. C. Walsh N. A. Gardner, and D. R. Gabe, Development of the Eco-Cascade Cell Reactor. *J. Appl. Electrochem.* **12**, 229 (1982).
7. J. R. Backhurst, J. M. Coulson, F. Goodridge, R. E. Plimley, and M. Fleischmann, A Preliminary Investigation of Fluidized Bed Electrodes. *J. Electrochem. Soc.* **116**, 1600 (1969).
8. D. F. Steele, Electrochemistry and Waste Disposal. *Chem. Br.* October, p. 915 (1991).
9. J. A. McIntyre, An Old Solution Finally Finds an Application. *Interface (Electrochem. Soc.)* Spring, Vol. 29 (1995).
10. D. M. Soboroff, J. D. Troyer, and A. A. Cochran, "Regeneration and Recycling of Waste Chromic Acid-Sulfuric Acid Etchants," Rep. Investi. 8377. U. S. Department of the Interior, Washington, DC, 1979.
11. S. Basak and K. Rajeshwar, unpublished data (1993).
12. R. F. Praino, Jr. and R. O'Gorman, Technology Evaluation, Installation and Performance of a Chromium Removal System for Aqueous Discharges. *Hazard. Waste* **1**, 469 (1984).
13. C. T. Chen, Assessment of the Applicability of Chemical Oxidation Technologies for the Treatment of Contaminants at Leaking Underground Storage Tank (LUST) Sites. *In* "Chemical Oxidation: Technologies for the Nineties" (W. W. Eckenfelder, A. R. Bowers, and J. A. Roth, eds.), Vol. 3, pp. 225-243. Technomic Publishing, Lancaster and Basel, 1993.
14. J. I. Reich, "Groundwater Treatment Technologies for Heavy Metal and Organics Removal." Andco Environmental Processes, Amherst, NY.
15. B. Heineman, U. S. EPA, Region VI, Dallas, TX, private communication (1994).
16. R. Lageman, W. Pool, and G. Sebsinge, Electro-reclamation; Theory and Practice. *Chem. Ind.* **18**, 585 (1989).
17. G. Jorgensen and R. Govindarajan, "Ultraviolet Reflector Materials for Solar Detoxification of Hazardous Waste," Rep. SERI/TP-257-4418. Solar Energy Research Institute, Golden, CO (1991).

18. "Design Considerations for Ultraviolet/Chemical Oxidation Engineering Technical Manual," Doc. No. 9000-020.028. U. S. Army Corps of Engineers, Omaha, NE, 1994.
19. D. Edwards and W. McPhee, "The Design of UV Oxidation Systems for Groundwater Remediation." SolarChem Environmental Systems, Markham, Ontario, Canada.
20. E. M. Froelich, Advanced Chemical Oxidation of Contaminated Water Using the Peroxpure™ Oxidation System. *In* "Chemical Oxidation: Technology for the Nineties" (W.W. Eckenfelder, A.R. Bowers, and J.A. Roth, eds.), Vol. 2. Technomic Publishing, Lancaster and Basel, 1992.
21. R. Sirabian, T. Sanford, and R. Barbour, "UV Peroxidation with Air Stripping for Optimized Removal of VOC's from Groundwater." ERM-Northeast Inc. (as cited in Ref. 18).
22. F. E. Bernadin, Jr., UV/Peroxidation Destroys Organics in Groundwater. *83rd Annu. Meet. Air Waste Manag. Assoc.*, Pittsburgh, PA, 1990.
23. K. W. Yost, "Ultraviolet Peroxidation: An Alternative Treatment Method for Organic Contamination Destruction in Aqueous Waste Streams." *43rd Purdue Ind. Waste Conf.* Lewis Publishers, Boca Raton, FL, 1989.
24. D. W. Sundstrom, B. A. Weir, and H. E. Klei, Destruction of Aromatic Pollutants by UV Light Catalyzed Oxidation with Hydrogen Peroxide. *Environ. Prog.* February, p. 8 (1989).
25. D. M. Thompson and M. A. Wowland, Groundwater Cleanup with UV Light and Hydrogen Peroxide. *Hazmat World*, June (1990).
26. M. E. Zappi, B. C. Fleming, and M. J. Cullinane, "Treatment of Contaminated Groundwater Using Chemical Oxidation." USCOE Waterways Experimental Station, Vicksburg, MI.
27. F. J. Beltran, G. Ovejero, and B. Acedo, Oxidation of Atrazine in Water by Ultraviolet Radiation Combined with Hydrogen Peroxide. *Water Res.* **27**, 1013 (1993).
28. P. C. Kearney, M. T. Muldoon, and C. J. Somich, UV-Ozonation of Eleven Major Pesticides as a Waste Disposal Pretreatment. *Chemosphere* **16**, 2321 (1987).
29. K. V. Topudurti, N. Lewis, and S. R. Hirsh, The Applicability of UV/Oxidation Technologies to Treat Contaminated Groundwater. *Environ. Prog.* **12**, 54 (1993).
30. H. W. Prengle, Jr., New Technology: Evolution of the Ozone/UV Process for Wastewater Treatment. *I01/EPA Colloqu. Wastewater Treat. Disinfect. Ozone*, Cincinnati, OH, 1977.
31. B. J. Jody, M. J. Klein, and H. Judekis, Catalytic O_3/UV Treatment of Wastewater Containing Mixtures of Organic Pollutants. *Proc. 9th. World Ozone Cong.*, New York, 1989.
32. H. W. Prengle, Jr., C. G. Hewes, III, and C. E. Mauk, Oxidation of Refrac-

tory Materials by Ozone with Ultraviolet Radiation. *Proc. 2nd Inter. Symp. Ozone Technol.*, Montreal, Canada, 1975.
33. J. D. Zeff and E. Leitis, Oxidation of Toxic Compounds in Water. U. S. Pat. 4,849,114 (1989).
34. E. M. Aieta, K. M. Reagan, J. S. Lang, L. McReynolds, J.-W. Kang, and W. H. Glaze, Advanced Oxidation Processes for Treating Groundwater Contaminated with TCE and PCE: Pilot-Scale Evaluations. *J. Am. Water Works Assoc.* **80**, 64 (1988).
35. N. M. Lewis and K. Topudunti, Advanced Oxidation Technologies for the Treatment of Contaminated Groundwater. *In* "Chemical Oxidation: Technologies for the Nineties" (W. W. Eckenfelder, A. R. Bowers, and J. A. Roth, eds.), Vol. 2, pp. 406-427. Technomic Publishing, Lancaster and Basel, 1992.
36. "Perox-Pure™ Chemical Oxidation Technology," Peroxidation Systems, Inc., Applications Analysis Report, EPA/540/AR-93/501. U. S. Environ. Prot. Agency, Cincinnati, OH, 1993.
37. J. E. Pacheco and J. T. Holmes, Comparison of Falling-Film and Glass-Tube Solar Photocatalytic Reactors for Destroying Toxic Organic Chemicals in Water. *ACS Conf. Emerging Technol. Hazard. Waste Treat.*, Atlanta, GA (Sandia Rep. SAND 88-2726) (1989).
38. J. V. Anderson, H. Link, M. Dohn, and B. Gupta, Development of Solar Detoxification Technology in the USA — An Introduction. *Sol. Energy Mater.* **24**, 538 (1991).
39. C. Turchi, M. Mehos, and J. Pacheco, Design Issues for Solar-Driven Photocatalytic Systems. *In* "Photocatalytic Purification and Treatment of Water and Air" (D. F. Ollis and H. Al-Ekabi, eds.), p. 789. Elsevier, New York and Amsterdam, 1993.
40. C. S. Turchi and M. S. Mehos, Solar Photocatalytic Detoxification of Groundwater: Developments in Reactor Design. *In* "Chemical Oxidation: Technologies for the Nineties" (W. W. Eckenfelder, A. R. Bowers, and J. A. Roth, eds.), Vol. 2. Technomic Publishing, Lancaster and Basel, 1992.
41. C. S. Turchi, J. F. Klausner, D. Y. Goswami, and E. Marchand, Field Test Results for the Solar Photocatalytic Detoxification of Fuel-Contaminated Groundwater. *In* "Chemical Oxidation: Technologies for the Nineties" (W. W. Eckenfelder, A. R. Bowers, and J. A. Roth, eds.), Vol. 3, NREL/TP 471-5345. Technomic Publishing, Lancaster and Basel, 1993.
42. Y. Zhang, J. C. Crittenden, and D. W. Hand, The Solar Photocatalytic Decontamination of Water. *Chem. Ind.* **18**, 714 (1994).
43. D. Bockelmann, R. Goslich, D. Weichgrebe, and D. Bahnemann, Solar Detoxification of Polluted Water: Comparing the Efficiencies of a Parabolic Trough Reactor and a Novel Thin-Film Fixed-Bed Reactor. *In* "Photocatalytic Purification and Treatment of Water and Air" (D. F. Ollis

and H. Al-Ekabi, eds.), p. 771. Elsevier, New York and Amsterdam, 1993.
44. D. W. Bahnemann, D. Bockelmann, R. Goslich, M. Hilgendorff, and D. Weichgrebe, Photocatalytic Detoxification: Novel Catalysts, Mechanisms and Solar Applications. *In* "Photocatalytic Purification and Treatment of Water and Air" (D. F. Ollis and H. Al-Ekabi, eds.), p.301. Elsevier, New York and Amsterdam, 1993.
45. These cost figures (as cited in Ref. 42) are from the National Renewable Energy Laboratory.
46. C. S. Turchi, E. J. Wolfrum, and M. Nimlos, Cost Estimation for Treating VOCs in an Air Stripper Offgas with Gas-Phase Photocatalysis. *5th. Annu. ACS Symp. Emerging Technol. Hazard. Waste Manag.*, Atlanta, GA, 1993.
47. R. Notarfonzo and W. McPhee, How to Evaluate and Cost UV/Oxidation Systems. Solarchem Environmental Systems, 1993.
48. B. F. Severin, Disinfection of Municipal Wastewater Effluents with Ultraviolet Light. *J. Water. Pollu. Control Fed.* **52**, 2007 (1980).
49. G. E. Whitby, G. Palmateer, W. G. Cook, J. Maarschalkerweerd, D. Huber, and K. Flood, Ultraviolet Disinfection of Secondary Effluent. *J. Water Pollut. Control Fed.* **56**, 844 (1984).
50. R. J. Fahey, The UV Effect on Wastewater. *Water Eng. Manag.*, December p. 15 (1990).
51. H. Dizer, W. Bartocha, H. Bartel, K. Seidel, J. M. Lopez-Pila, and A. Grohmann, Use of Ultraviolet Radiation for Inactivation of Bacteria and Coliphages in Pretreated Wastewater. *Water Res.* 27, 397 (1993).
52. O. K. Scheible and C. D. Bassell, "Ultraviolet Disinfection of a Secondary Wastewater Treatment Plant Effluent," EPA-600/2-81-152. U. S. Environ. Prot. Agency, Cincinnati, OH, 1981.
53. D. W. Ferguson, M. J. McGuire, B. Koch, R. L. Wolfe, and E. M. Aieta, Comparing PEROXONE and Ozone for Controlling Taste and Odor Compounds, Disinfection By-products and Microorganisms. *J. Am. Water. Works Assoc.* **82**, 181 (1990).
54. W. H. Glaze, G. R. Peyton, B. Sohm, and D. A. Meldrum, "Pilot-Scale Evaluation of Photolytic Ozonation for Trihalomethane Precursor Removal," EPA-600/S2-84-136. U. S. Environ. Prot. Agency, Cincinnati, OH, 1984.

APPENDIX A

Abbreviations, Acronyms, Symbols, and Notation

AA	Atomic absorption
AAS	Atomic absorption spectroscopy
ABS	Alkyl–aryl branched sulfonate
AC	Alternating current
ACSV	Adsorption cathodic stripping voltammetry
ADP	Adenosine nicotinamide diphosphate
AEM	Anion exchange membrane
AES	Atomic emission spectroscopy
AOP	Advanced oxidation process
APAD	Activated pulse amperometric detection
AsBet	Arsenobetaine
AsChol	Arsenocholine
ASV	Anodic stripping voltammetry
ATP	Adenosine nicotinamide triphosphate
BET	Brunnauer–Emmett–Teller
BOD	Biological oxygen demand
BPM	Bipolar membrane
BTEX	Benzene, toluene, ethylbenzene, xylene
CA	Cellulose acetate
CB	Conduction band
CE	Capillary electrophoresis (or counterelectrode)
CEM	Cation exchange membrane
CFC	Chlorofluorohydrocarbon
CFU	Colony-forming unit
CHEMFET	Chemical field-effect transistor
CME	Chemically modified electrode
COD	Chemical oxygen demand
4-CP	4-Chlorophenol
CSTR	Continuous stirred tank reactor
CSV	Cathodic stripping voltammetry
CTAB	Cetyl trimethylammonium bromide
CV	Cyclic voltammetry
CZE	Capillary zone electrophoresis

DBCP	Dibromochloropropane
DBP	Disinfection byproduct
DC	Direct current
DCA	Dichloroethane
DCE	Dichloroethylene
DDAB	Dodecyldimethylammonium bromide
DDT	2,2-bis(p-chlorophenyl)-1,1,1-trichloroethane
DEM	Dished electrode membrane
DMA	Dimethyl arsine (or dimethyl arsinic acid; see Figure 1.10)
DME	Dropping mercury electrode
DMF	Dimethyl formamide
DMPO	5,5'-Dimethylpyroline-N-oxide
DNA	Deoxyribonucleic acid
DP	Differential pulse
DSA	Dimensionally stable anode
EA	Electron affinity
EC	Electrochemical
ECAD	Electrochemical array detector
ECD	Electron capture detector
EDIP	Electrodiaresis polishing
EDP	Electrochemical deionization process
EDTA	Ethylenediamine tetraacetate
EE/O	Electrical eenrgy per order
EIX	Electrochemical ion exchange
EMC	Ethyl mercury chloride
EOD	Electrochemical oxygen demand
EOI	Electrochemical oxidability index
EPR	Electron paramagnetic resonance
EQCM	Electrochemical quartz crystal microgravimetry
FAAS	Flame atomic absorption spectroscopy
FAC	Free available chlorine
FGD	Flue gas desulfurization
FIA	Flow injection analysis
FPT	Freeze–pump–thaw
FT-IR	Fourier transform infrared
GAC	Granular activated carbon
GC	Gas chromatography

HA	Humic acid
HER	Hydrogen evolution reaction
HG	Hydride generation
HMDE	Hanging mercury drop electrode
HMO	Hückel molecular orbital
HOMO	Highest-occupied molecular orbital
HPLC	High-performance (pressure) liquid chromatography
IC	Ion chromatography
ICE	Instantaneous current efficiency
ICP	Inductively coupled plasma
IEM	Ion exchange membrane
ISE	Ion-selective electrode
ISFET	Ion-selective field-effect transistor
LC	Liquid chromatography
L-H	Langmuir–Hinshelwood
LIA	Lock-in amplifier
LIPOD	Laser-induced photochemical oxidative destruction
LSV	Linear sweep voltammetry
LUMO	Lowest-unoccupied molecular orbital
LUST	Leaking underground storage tank
MBE	Moving bed electrode
MCL	Maximum contaminant level
MF	Membrane filtration
MIB	Methyl isoborneol
ML	Metal ligand
MMA	Monomethyl arsenate
MO	Molecular orbital
MPN	Maximum probable number
MS	Mass spectrometry
MTFE	Mercury thin-film electrode
MWD	Metropolitan water district
NASICON	Sodium superionic conductor
NMR	Nuclear magnetic resonance
NOM	Natural organic matter
NTU	Nephelometric turbidity unit

ODC	Odor threshold concentration
PAC	Powdered activated carbon
PAD	Pulse amperometric detection
PAH	Polycyclic aromatic hydrocarbon
PCB	Polychlorinated biphenyl
PCDD	Polychlorinated dibenzo-p-dioxin
PCDF	Polychlorinated dibenzofuran
PCE	Perchloroethylene
PCP	Pentachlorophenol
PEC	Photoelectrochemical
PED	Photoelectrochemical detector
PEM	Proton exchange membrane
PFR	Plug flow reactor
PMMA	Polymethyl methacrylate
PP (or PPy)	Polypyrrole
PSA	Potentiomeric stripping analysis
PSS	Point-source summation
PTFE	Polytetrafluoroethylene
PVC	Polyvinylchloride
PZC	Point of zero charge
PZZP	Point of zero zeta potential
QPVP	Quaternized polyvinylpyridine
RCE	Rotating cylinder electrode
RDE	Rotating disk electrode
RE	Reference electrode
RRDE	Rotating ring disk electrode
RTD	Residence time distribution
RVC	Reticulated vitreous carbon
SAM	Self-assembled monolayer
SBE	Spouted bed electrode
SBR	Stirred batch reactor
SC	Semiconductor
SCE	Saturated calomel electrode
SECM	Scanning electrochemical microscope
SERS	Surface-enhanced Raman scattering
SFC	Super-critical fluid chromatography
SHE	Standard hydrogen electrode
SITE	Superfund innovative technology evaluation
SMDE	Static mercury drop electrode

S/N	Signal-to-noise ratio
SOFC	Solid oxide fuel cell
SPE	Solid polymer electrolyte (or screen-printed electrode)
SW	Square wave
TBAP	Tetrabutylammonium perchlorate
TCA	Trichloroethane
TCDD	Tetrachlorodibenzo-p-dioxin
TCE	Trichloroetylene
TDR	Time domain reflectometry
TDS	Total dissolved solids
TEAP	Tetraethylammonium perchlorate
THC	Total hydrocarbon
THM	Trihalomethane
THMFP	Trihalomethane formation potential
TMA	Trimethylarsine
TMAO	Trimethylarsine oxide
TNP	Tris (p-nitrophenyl)phosphate
TNT	Trinitrotoluene
TOC	Total organic carbon
TPP	Triphenyl phosphine
TPPO	Triphenyl phosphine oxide
TSS	Total suspended solids
TU	Thiourea
UF	Ultrafiltration
UME	Ultramicroelectrode
UPD	Underpotential deposition
UV	Ultraviolet
UV-VIS	Ultraviolet visible
VB	Valence band
VBE	Vortex bed electrode
VOC	Volatile organic hydrocarbon
VUV	Vacuum ultraviolet
WE	Working electrode
WJ	Wall jet

Appendix

Environmental regulations in the United States

CAA	Clean Air Act
CERCLA	Comprehensive Environmental Response, Compensation and Liability Act
FEPCA	Federal Environmental Pollution Control Act
FIFRA	Federal Insecticide, Fungicide and Rodenticide Act
FHSA	Federal Health and Safety Act
FWPCA	Federal Water Pollution Control Act
HMTA	Hazardous Materials Transportation Act
OSHA	Occupational Safety and Health Act
RCRA	Resource Conservation and Recovery Act
SARA	Superfund Amendments and Reauthorization Act
SDWA	Safe Drinking Water Act
TSCA	Toxic Substances Control Act

Organizations in the United States

ACS	American Chemical Society
AWWA	American Water Works Association
ASTM	American Society for Testing and Materials
EPA	Environmental Protection Agency
EPRI	Electric Power Research Institute
LLNL	Lawrence Livermore National Laboratory
NREL	National Renewable Energy Laboratory
OSHA	Occupational Safety and Health Administration

Chapter 2

Italic Letters (boldface in the text denotes vector quantities)

E	electrode potential
$E°$	standard reduction potential
$E°'$	formal (or conditional) potential
F	Faraday constant
h	Planck's constant

R	gas constant
T	absolute temperature
a	activity
A	electrode area
A_a	cross-sectional area of a molecule
A_e	specific surface area (i.e., area/volume)
A_R	total area occupied by R
C	concentration
C^*	bulk concentration
C^s	surface concentration
C^o	concentration when oxidized and reduced forms of redox couple are in equal amounts
C_{dl}	double-layer capacitance
D	diffusion coefficient
e_o	elementary charge
E	energy
E_F	Fermi level
E_g	energy bandgap
E_s	energy consumption for electrolysis
g	gravitational acceleration
i	current density
$i^{o\prime}$	(conditional) exchange current density
i_L (or I_L)	limiting current density (or current)
I	current
I_d	diffusion current
I_p	peak current
J	flux
$k^{o\prime}$	conditional rate constant
k_{ij}	potentiometric selectivity coefficient
k_m	mass-transfer coefficient
K	conductivity (or potentiometric constant; see Section 2.6.1)
L (or l)	length (or height)
m	mass flow rate (of Hg)
n	electron stoichiometry
n	electron density in a semiconductor (see Section 2.11)
n_R	number of molecules of R
N	number of moles electrolyzed
N_A	Avogadro's number

$N_A(N_D)$	acceptor (donor) density in a semiconductor
O	oxidized species
p	hole density in a semiconductor
Q	charge
Q_R	reaction quotient
R	reduced species (or resistance)
R_{ct}	charge-transfer resistance
R_s	series resistance
R_u	(uncompensated) series resistance
t	time
u	solution velocity
v	potential sweep rate
V	volume
$\dot{V}$	volume flow rate
V_e	electrode volume
V_R	reactor volume
V_s	solution volume
W_{cell}	electrolytic power
W_R	mass of reactant
x	fraction converted (or Cartesian coordinate)
z	ionic charge

Greek Letters

α	symmetry factor or transfer coefficient
γ	activity coefficient
δ	layer thickness
ε	permittivity
ξ	electrokinetic potential
η	overvoltage (or solution viscosity)
κ	electrical conductivity
ρ_{ST}	space–time yield
μ	mobility
ν	kinematic viscosity of water
υ	frequency
τ	time constant
ϕ	electrical potential

Chapter 3

Italic Letters

$E°$	standard reduction potential
$E°'$	formal (conditional) potential
E_{ox} (E_{red})	potential for oxidation or reduction of a compound
$E_{1/2}$	half-wave potential
E_{ref}	correction for reference electrode
K	equilibrium constant
K_{sp}	solubility product
C	constant
k	(heterogeneous) rate constant for electron transfer
n	electron stoichiometry
O	oxidized species
R	reduced species

Greek Letters

η	overpotential

Chapter 4

Italic Letters

E	electrode potential
E_{act}	activation potential
E_d	deposition potential
E_{det}	detection potential
E_{ox} (E_{red})	potential for oxidative (or reductive) cleaning of the electrode
$E_{1/2}$	half-wave potential
F	Faraday constant
a	tube radius (or inlet diameter)
A	electrode area
b	channel height
B	constant (see Eq. (4.6))
C^*	bulk concentration
C^{max}	maximum concentration (at a FIA peak)

C^o	concentration (without dispersion)
d	MTFE thickness
D	diffusion coefficient
D'	dispersion coefficient
f	correction factor (see Eq. (4.3))
I_d	diffusion current
I_k	kinetic current
I_L	limiting current
k_d, k_f	rate constant for dissociation and formation of ML complex
L	length
r	disk radius (see Eq. (4.8))
R	resistance
t_d	deposition time
t_e	effective measurement time
u	stirring or flow rate of solution
u	solution velocity

Greek Letters

α	exponent (see Eq. (4.6))
δ	layer thickness
ν	kinematic viscosity of water
ω	electrode rotation rate (rad/s)

Chapter 5

Italic Letters

E	electrode potential
E°	standard reduction potential
F	Faraday constant
R	gas constant
T	absolute temperature
a	activity
A	electrode area
A_e	specific surface area
C	concentration
C^*	bulk concentration
d	diameter of cylinder
D	diffusion coefficient

I	current
I_L	limiting current
J	flux
k_m	mass-transfer coefficient
k_{12}	ion selectivity coefficient
K	constant (see Eq. (5.18))
l	height of cylinder
n	electron stoichiometry
p	partial pressure
t	time
u	solution velocity
V	peripheral solution velocity
$\dot{V}$	volume flow rate
V_s	volume of solution
x	exponent (see Eq. (5.18))

Greek Letters

ρ	electrodic parameter
σ	Hammett parameter
τ	time to ICE = 0 (see Figure 5.2)
ν	kinematic viscosity of water

Chapter 6

Italic Letters

$E°$	standard reduction potential
$E°{'}$	formal (conditional) potential
h	Planck's constant
k	Boltzmann constant
T	absolute temperature
T_e	electron temperature
A'	constant (see Eq. (6.42))
C	concentration
$\overline{C}$	average concentration
C_O	initial concentration
d	drift distance
D	diffusion coefficient (or ozone dose rate)
$D_n(D_p)$	diffusion coefficient of electrons (holes)

Appendix

e_o	elementary charge
E	field strength
E_c	conduction band edge
E_F	Fermi energy
E_g	energy bandgap
g	carrier photogeneration rate
$\hbar$	$h/2\pi$
i_{ph}	photocurrent density
I_o	light intensity
I_{ss}	steady-state current
J_{ph}	photon flux
k	rate constant
k_m	mass-transfer coefficient
$k_p(k_n)$	rate constant for hole (electron) transfer
K	equilibrium binding constant (see Eq. (6.72))
L	layer dimension
L_D	Debye length
L_E	absorption length
m	exponent (see Eq. (6.42))
$m_e(m_h)$	effective mass of electron (hole)
m'	reduced effective mass of electron–hole pair
n	reaction order
n_e	electron density
$n_s(p_s)$	surface electron (hole) concentration
N	carrier density
N'	charge collection efficiency (charge out/photons in)
N_{O_2}	number density of O_2 molecules
p	coulometry cell constant
r	radial distance (variable)
r'	normalized radial distance, that is, $r' = r/L_D$ (variable)
R	particle radius
t	time
$t_{1/2}$	half-life of reaction
W	depletion layer width

Greek Letters

α	absorption coefficient
δ	critical distance for electron transfer from semiconductor particle to O_2
ε	permittivity

ε_o	free-space permittivity
θ	dimensionless potential (see Eq. (6.50))
Φ	quantum yield
ϕ	electrical potential
μ_e	electron mobility
υ	frequency
ρ	hemisphere radius (see Figure 6.41b)

Chapter 7

Italic Letters

K	equilibrium constant
p	slope factor (see Eq. (7.10))
a	constant (see Eq. (7.1))
b	constant (see Eq. (7.1))
C	dosage of disinfectant
k	rate constant
m	exponent (see Eq. (7.5))
N_o	initial cell concentration
N_s	number density of surviving organisms
t	contact time
t_1	potentiometric lag time (see Eq. (7.10))

Chapter 8

Italic Letters

C_0	initial concentration
I_0	light intensity
$\dot{V}$	volume flow rate

Greek Letters

$\varepsilon^{h\upsilon}$	photon efficiency

APPENDIX B

Physical Constants, Units and Unit Conversions, and Reference Electrode Potentials

Constants (in SI units)

Elementary charge, e_o	1.60277×10^{19} C
Avogadro's number, N_A	6.02214×10^{23} mol^{-1}
Boltzmann constant, k	1.38066×10^{-23} J K^{-1}
Faraday constant, F	96485.3 C mol^{-1}
Gas constant, R	8.31451 J K^{-1} mol^{-1}
Free-space permittivity, ε_o	8.85419×10^{-12} F m^{-1}
Planck's constant, h	6.62618×10^{-34} J•s
Speed of light in vacuo, c	2.99792×10^8 m/s

Units and Unit Conversions

Use of the SI system of units is by no means universal, and in the environmental field, nonmetric (i.e., English) units are still being used. Some metric units and their English counterparts follow:

Length
1 m = 39.37 in
1 in = 2.54 cm

Mass
1 kg = 2.205 lb
1 lb = 453.6 g
1 metric ton (tonne) = 1.103 tons

Volume
1 L = 1,000 cm^3
1 L ≡ 1 dm^3
1 L = 1.057 quarts
1 U. S. gal = 3.785 liters
1 imperial gal = 5 liters

Other unit conversions

Temperature (The SI system uses the Kelvin scale.)

$$\text{Absolute temperature, K} = °C + 298.15$$
(Kelvin scale) (Celsius scale)

$$°F = 32 + 9/5 °C$$
(Fahrenheit scale)

Pressure (The SI system uses Pa or pascal.)

1 atm = 760 torr (760 mm Hg)

1 atm = 1.01325 bar

1 bar = 10^5 Pa

Energy (The SI system uses J or joule.)

1 cal = 4.184 J

1 erg = 10^{-7} J

1 eV = 1.6021×10^{-19} J

kT (for 25°C) = 4.12×10^{-21} J or 25.7 meV

RT = 2.48 kJ/mol = 592 cal/mol

Reference Electrode Potentials

All potentials in the text are quoted *vs.* the standard hydrogen electrode (SHE) reference, unless otherwise specified. Other reference electrodes used in aqueous media include the saturated calomel electrode (SCE) and Ag–AgCl–(saturated) KCl. Sodium-saturated calomel electrode (SSCE) is also used in selected instances. In nonaqueous media (e.g., acetonitrile, DMF), these reference electrodes generally cannot be used. A Ag–Ag$^+$ ion reference (composed of a silver wire in contact with 0.1 M AgNO$_3$ dissolved in the nonaqueous solvent) can be used in such cases.

The potentials of these ("secondary") reference electrodes on the SHE scale are shown in the following diagram. Further, Figure A1 illustrates the relationship between the SHE and SCE scales for two selected redox couples.

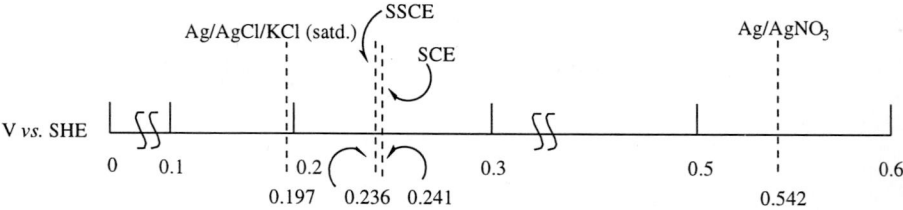

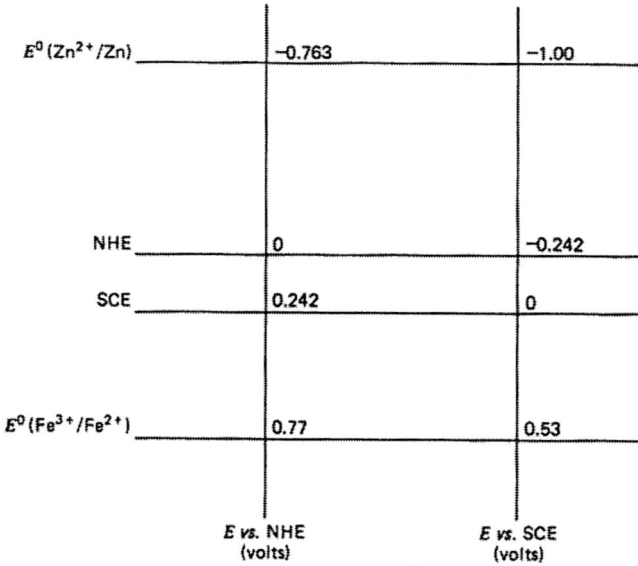

Figure A1. Relationship between potential scales based on the SHE and SCE. (Reproduced with permission from A. J. Bard and L. R. Faulkner, "Electrochemical Methods." Wiley, New York, 1980.)

APPENDIX C

Selected Standard Reduction Potentials in Aqueous Media at 25°C[a]

Reaction	Potential, V
$Ag^+ + e^- = Ag$	0.7996
$Al^{3+} + 3\,e^- = Al$ (0.1 M NaOH)	-1.706
$H_3AsO_4 + 2\,H^+ + 2\,e^- = H_3AsO_3 + H_2O$	0.559
$Br_2(aq) + 2\,e^- = 2\,Br^-$	1.087
$Cd^{2+} + 2\,e^- = Cd$	-0.4026
$Cd^{2+} + 2\,e^- = Cd(Hg)$	-0.3521
$Cl_2(g) + 2\,e^- = 2\,Cl^-$	1.3583
$HClO + H^+ + e^- = \frac{1}{2}Cl_2 + H_2O$	1.63
$2\,CO_2(g) + 2\,H^+ + 2\,e^- = H_2C_2O_4$	-0.49
$Cr^{3+} + 3\,e^- = Cr$	-0.744
$Cr_2O_7^{2-} + 14\,H^+ + 6\,e^- = 2\,Cr^{3+} + 7\,H_2O$	1.33
$Cu^{2+} + e^- = Cu^+$	0.158
$Cu^{2+} + 2\,e^- = Cu$	0.3402
$Cu^{2+} + 2\,e^- = Cu(Hg)$	0.345
$Fe^{3+} + e^- = Fe^{2+}$ (1 M HCl)	0.77
$2\,H^+ + 2\,e^-$	0.0000
$H_2O_2 + 2\,H^+ + 2\,e^- = 2\,H_2O$	1.776
$2\,Hg^{2+} + 2\,e^- = Hg_2^{2+}$	0.905
$Hg_2^{2+} + 2\,e^- = 2\,Hg$	0.7961
$IO_3^- + 6\,H^+ + 5\,e^- = \frac{1}{2}I_2(aq) + 3\,H_2O$	1.178

Appendix

$MnO_2(s) + 4\,H^+ + 2\,e^- = Mn^{2+} + 2\,H_2O$	1.23
$Ni^{2+} + 2\,e^- = Ni$	-0.25
$O_2(g) + 4\,H^+ + 4\,e^- = 2\,H_2O$	1.229
$O_3(g) + 2\,H^+ + 2\,e^- = O_2(g) + H_2O$	2.07
$Pb^{2+} + 2\,e^- = Pb$	-0.126
$S(s) + 2\,H^+ + 2\,e^- = H_2S(g)$	0.141
$SO_4^{2-} + 4\,H^+ + 2\,e^- = H_2SO_3 + H_2O$	0.172
$H_2SeO_3 + 4\,H^+ + 4\,e^- = Se(s) + 3\,H_2O$	0.740
$SeO_4^{2-} + 4\,H^+ + 2\,e^- = H_2SeO_3 + H_2O$	1.15
$Tl^+ + e^- = Tl(s)$	-0.336
$Zn^{2+} + 2\,e^- = Zn$	-0.7628

[a]See also the references cited in Chapter 3 for more extensive compilations.

APPENDIX D

Companies Marketing Electrochemical Technologies and Accessories for Pollution Sensors and Pollution Abatement

Advanced Electronics A/S, Mariendalsvej 55, DK-2000, Frederiksberg, Denmark.
AEA Technology, Dounreay, Thurso, Caithness KW14 7TZ, United Kingdom. Ph. (0847) 802 121, Fax (0847) 802 850.
Amel Instruments, Via Bolzano 30, 20127, Milano, Italy. Ph. (02) 289-3617, Fax (02) 282-8551.
Andco Environmental Processes, Inc., 595 Commerce Drive, Buffalo, New York 14228-2380 USA. Ph. (716) 691-2100, Fax (716) 691-2880.
Aqualytics Systems, Inc., 7 Powder Horn Drive, P. O. Box 4904, Warren, New Jersey 07059 USA. Ph. (908) 563-2800, Fax (908) 563-2816.
Asea Brown Boveri, Corporate Research CRB, CH-5405 Baden, Switzerland.
Ashai Chemical Ind. Co. Ltd., 1-3-2 Yako Kawasaki-ku, Kawasaki-shi 210, Japan. Ph. or Fax (44) 271-2320.
Ashai Glass Co. Ltd., 2-1-2 Marunouchi, Chiyoda-ku, Tokyo, 100, Japan. Fax (045) 334-6187.
Astris, Inc., 318 Pinehurst Drive, Oakville, Ontario L6J 4X5, Canada. Ph. (416) 844-4522.
Astro Met, Inc., 9974 Springfield Pike, Cincinnati, Ohio 45215 USA. Ph. (513) 772-1242, Fax (513) 772-9080.
Atraverda Ltd., Darenth House, Rotherham Road, Eckington, Sheffield S31 9FH, United Kingdom. Ph. (44246) 430 177, Fax (44246) 430 188.
Benham Electrosynthesis, 9400 North Broadway, P. O. Box 20400, Oklahoma City, Oklahoma 73156 USA. Ph. (405) 478-5353, Fax (405) 478-1238.
BEWT (Water Engineers) Ltd., Tything Road, Arden Forest Industrial Estate, Alcester, Warwickshire, United Kingdom. Ph. (0789) 763 669, Fax (0789) 400 274.
Bioanalytical Systems, Inc., 2701 Kent Avenue, West Lafayette, Indiana 47906 USA. Ph. (317) 463-4527, Fax (317) 497-1102.
Brinecell, Inc., 2109 West 2300 South, Salt Lake City, Utah 84119 USA. Ph. (801) 973-6400, FAX (801) 973-6463.
Chemetics International Co. Ltd., 1818 Cornwall Avenue, Vancouver, British Columbia, Canada V6J 1C7. Ph. (604) 737-4517, Fax (604) 734-0340.
Cypress Systems, Inc., 2500 West 31st Street, Suite D, Lawrence, Kansas 66047 USA. Ph. (800) 235-2436, Fax (913) 832-0406.
Danica Supply A/S, Smedevaenget 1-7, DK-5560 Arup, Denmark.
Degussa (see Fa. Degussa AG).

Delker Corporation, Inc., 16 Commercial Street, P. O. Box 427, Branford, Connecticut 06405 USA. Ph. (203) 481-4277, Fax (203) 488-6902.

Delphi Research Inc., 701 Haines Avenue N. W., Albuquerque, New Mexico 87102 USA. Ph. (505) 243-3111, Fax (505) 243-3188.

DuPont Company, Polymer Products Department, Wilmington, Delaware 19898 USA.

EA Technology Ltd., Capenhurst Chester CH1 6ES, United Kingdom. Ph. (051) 339-4181, Fax (051) 357-1581.

E. G. & G., Princeton Applied Research, P. O. Box 2565, Princeton, New Jersey 08543-2565 USA. Ph. (609) 530-1000, Fax (609) 883-7259.

E-Tek, Inc., 6 Mercer Road, Natick Industrial Park, Natick, Massachusetts 01760 USA. Ph. (508) 653-9331, Fax (508) 653-4225.

Ebara Co., 1-6-27 Kohnan, Minato-ku, Tokyo 108, Japan.

Electrocatalytic, Inc., 2 Milltown Court, Union, New Jersey 07083 USA. Ph. (908) 851-2277, Fax (908) 851-6906.

ElectroCell AB, Box 7007, S-183 07 Taby, Sweden. Ph. (468) 732-8965, Fax (468) 732-7159.

Electrochemical Design Associates (EDA), Inc., 74 Muth Drive, Orinda, California 94563 USA. Ph. (510) 254-2335, Fax (510) 254-8451.

Electrode Corporation, 100 Seventh Avenue, Suite 300, Chardon, Ohio 44024-1095 USA. Ph. (216) 285-0339; Fax (216) 285-0302 or 0386.

Electrode Products, Inc., 163 Washington Valley Road, Unit 101, Warren, New Jersey 07059 USA. Ph. (908) 302-1686 or (800) 553-5228, Fax (908) 627-9496.

Electron Transfer Technologies, P. O. Box 160, Princeton, New Jersey 08542 USA. Ph. (609) 921-0070, Fax (609) 683-0079.

Electrosynthesis (see The Electrosynthesis Co.).

Eltech Systems Company, Inc., 6100 Glades Road, Suite 305, Boca Raton, Florida 33434 USA. Ph. (407) 487-3600.

Energy Research and Generation, Inc., 900 Stanford Avenue, Oakland, California 94608 USA. Ph. (510) 658-9785, Fax (510) 658-7428.

Enviro-cell Umwelttechnik GmbH, Gattenhöfer Weg 29, D-6370 Oberursel, Germany. Ph. (06171) 55 096, Fax (06171) 55 095.

EPCI, 14 Avenue des Sports, Ch-1400 Yverdon, Sweden. Ph. (024) 215 079, Fax (024) 210 732.

European Commission, Joint Research Centre, 21010 Ispra (VA) Italy. Ph. (39332) 789 487, Fax (39332) 789 124.

Fa. Degussa Ag., Leipziger Strasse 10, Postfach 1351, D-6450 Hanau 1, Germany.

Fa. Serva Feinbiochemika GmbH & Co., P. O. Box 105260, D-6900 Heidelberg 1, Germany.

Fairtec, 50 rue Carnot, Bp 148, 92154 Suresnes Cedex, France.

Faraday Technology, Inc., 3155 Research Boulevard, Suite 105, Dayton, Ohio 45420-4011 USA. Ph. (513) 252-2113, Fax (513) 252-2131.
Feinbiochemika (see Fa. Serva Feinbiochemika GmbH & Co.).
Finishing Services Limited (FSL), Woburn Rod, Industrial Estate Postley Road, Kempston, Bedfordshire MK42 7BU, United Kingdom. Ph. (0234) 857 004, Fax (0234) 855 712.
G. Bank (Wenking), Werner v. Siemens-str. 3, D-3400 Göttingen, Germany.
General Electric Corporate Research & Development, Environmental Laboratory, P. O. Box 8, K-1, 4B30, Schenectady, New York 12301-0008 USA. Ph. (518) 387-6035, Fax (518) 387-7611.
Geokinetics, Dannenberg 16, 7461 TK Rijssen, The Netherlands. Ph. (315480) 40584, Fax (315480) 40697.
Goss Scientific Instruments, Ltd., Marks Close, Ingatestone, Essex CM4 9AR, United Kingdom.
H-D Tech, Inc., 25 Don Street, Kingston, Ontario, Canada K7L 4V4.
Heinzinger Regel- und Messtechnik, Happinger Strasse 71, D-8200 Rosenheim, Germany.
Heraeus Elektrochemie GmbH, Industriestrasse 17, D-63517, Rodenbach, Germany. Ph. (06184) 598-0, Fax (06184) 598/181-184.
Heraeus Elektroden GmbH, Industriestrasse 9, D-6463, Freigericht 2, Germany.
Hokuto-Denko Co. Ltd., 4-22 Hibundani, Meguroku, Tokyo, 152, Japan.
ICI, Electrochemical Technology, P. O. Box 13, The Heath, Runcorn, Cheshire WA7 4QG, United Kingdom. Ph. (0928) 517 467, 835/6/7, Fax (0928) 569 487.
Isotron Corp., Inc., 13152 Chef Menteur Highway, New Orleans, Louisiana 70129 USA. Ph. (504) 254-4624, Fax (504) 254-5172.
Ionsep Corporation, Inc., P. O. Box 258, Rockland, Delaware 19732 USA. Ph. (302) 798-7431, Fax (302) 798-7402.
Liquipure Technologies, Inc., 8 West Street, Plantsville, Connecticut 06479 USA.
Lockheed Missiles & Space Co., Research and Development Division, 3251 Hanover Street, Org. 93-50, Building 204, Palo Alto, California. Ph. (415) 424-3176, Fax (415) 354-5795.
Lynntech, Inc., 7610 Eastmark Drive, Suite 105, College Station, Texas 77840 USA. Ph. (409) 693-0017, Fax (409) 764-7479.
Metronics Co. Ltd., 1-14-3 Chidori, Ohta-ku, Tokyo, 146, Japan.
Mitsui DuPont Fluoro-Chemicals Co. Ltd., 1-2-3 Ootemachi, Chiyoda-ku, Tokyo, 100, Japan.
Nichia-Keiki-Seisakusho Co. Ltd., 2-4-0 Minamikaneda, Suita, Osaka, 564, Japan.
Nihon Carbon Co. Ltd., 2-6-1 Hacchoubori, Chuo-ku, Tokyo, 104, Japan.
Nihon-Kagaku-Tohgyo Co. Ltd., 2-24 3-Chou, Ono-Chou, Tohsato, Sakai, Osaka, 590, Japan.

Nihon Wacon Co., Ltd., 1-44 Katabira-Cho, Hodogaya-Ku, Yokohama, 240, Japan.
Norton Performance Plastics, 150 Dey Road, Wayne, New Jersey 07470-4699 USA. Ph. (201) 696-4700.
Olin Corporation, Charleston Technology Center, P. O. Box 248, Charleston, Tennessee 37310 USA. Ph. (615) 336-4076, Fax (615) 336-4554.
Oxytech, Inc., 141B Albright Way, Los Gatos, California 95030 USA.
Ozone Research and Equipment Corporation, 4953 West Missouri Avenue, Phoenix, Arizona 85301 USA. Ph. (602) 936-7332, Fax (602) 931-7727.
Ozonia Ltd., Stettbachstrasse 1, CH-8600, Deubendorf, Switzerland. Ph. (411) 801-8650, Fax (411) 801-8501.
Pepcon Systems, Inc., P. O. Box 797, Henderson, Nevada 89015 USA.
Permelec Electrode Ltd., No. 1159, Ishikawa, Fujisawa-shi, Kanagawa Pref., 252, Japan. Ph. (0466) 878 801/841, Fax (0466) 878 850/852.
Peroxidation Systems (PA Vulcan), Stonewood Commons I, 101 Bradford Road, Suite 200, Wexford, Pennsylvania 15090 USA. Ph. (412) 934-2240.
Prosep Technologies, Inc., 817 Brock Road South, Unit 7, Pickering, Ontario, Canada L1W 2L9.
Radiometer America Inc., 810 Sharon Drive, Westlake, Ohio 44145 USA. Ph. (216) 871-8900, Fax (216) 899-1139.
Rhone Poulenc Recherches Activite Membrane, 25, Quai Paul Doumer, 92400 Courbevoie, France.
Ringsdorff-Werke GmbH, Bad Godesberg, D-5300 Bonn 2, Germany.
Schumacher'sche Fabrik, D-7120 Bietigheim, Württ, Germany.
Sigri Elektrographit GmbH, Postfach 1160, D-8901 Meitingen, Germany.
Sodilec, 7 Avenue Louise, 93360 Neuilly-Plaisance, France.
Solar Kinetics Inc. (Solox Division), 10635 King William Drive, Dallas, Texas 75220 USA. Ph. (214) 556-2376, Fax (214) 569-4158.
Solarchem Environmental Systems, 130 Royal Crest Court, Markham, Ontario L3R OA1, Canada. Ph. (905) 477-9242, Fax (905) 477-4511.
Sorapec, 192 Avenue Carnot, 94124 Fontenay-Sous-Bois Cedex, France. Ph. (33148) 774 959, Fax (33148) 770 270.
SRI International, 333 Ravenswood Avenue, Menlo Park, California 94025 USA. Ph. (415) 859-6173, Fax (415) 859-3678.
SRTI, Bd Poincare, Bp 18, 26701 Pierelatte Cedex, France.
Staatliche Porzellan-Manufaktur Berlin, Wegelystr. 1, D-1000 Berlin 12, Germany.
Sun River Innovations Ltd., 2255 SW 28th Street, Coconut Grove, Florida 33133 USA. Ph. (606) 268-4200, Fax (606) 268-4089.
Tacussel (Solea-Tacussel), 72 rue d'Alsace, Lyon F69627 Villeurbanne, France. Ph. (337) 868-0122, Fax (337) 868-8812.

The Electrosynthesis Co., Inc., 72 Ward Road, Lancaster, New York 14086-9779 USA. Ph. (716) 684-0513, Fax (716) 684-0511.
Titalyse SA, 29 Rue Lect-CH 1217 Meyrin, Geneve, Switzerland. Ph. (022) 785-5777, Fax (022) 785-5779.
Tokai Carbon Co. Ltd., 1-2-3 Kita-Aoyama, Minato-ku, Tokyo, 105 Japan.
Tokuyama Soda Co. Ltd., 4-5, 1-chome Nishi-Shimbashi, Minato-ku, Tokyo 105, Japan. Ph. (03) 597-5120, Fax (03) 597-5126.
Trionetics, Inc., 2021 Midway Drive, Twinsburg, Ohio 44087 USA. Ph. (216) 425-2846, Fax (216) 425-2908.
Ultrox (Zimpro), 2435 South Anne Street, Santa Anna, California 92704-5308 USA. Ph. (714) 545-5557.
Umpqua Research Co., P. O. Box 791, 125 Volunteer Way, Myrtle Creek, Oregon 97457 USA. Ph. (503) 863-7770, Fax (503) 863-7775.
Universal Sensors, Inc., 5258 Veterans Boulevard, Suite D, Metairie, Louisiana 70006 USA. Ph./Fax (504) 885-8443.
Westinghouse Savannah River Co., SRTC 773-A, B-132, Aiken, South Carolina 29808 USA. Ph. (803) 725-5276, Fax (803) 725-4704.
Zentro-Elektronik GmbH & Co. KG, Sandweg 20, Postfach 2070, D-7530 Pforzheim, Germany.

Index

A

Absorption cross-section 506, 508
Absorption length 546, 548, 555
AC electrical impedance 76
Acetarsone 182
Acetone 150, 422
Acid front 449, 450
Acid rain 6, 8, 10, 426, 429
Acid recovery 440, 443
Advanced oxidation processes (AOPs) 46, 531, 591, 702, 709, 710, 717, 720, 725, 726, 728-730, 732
Air stripping 22, 36, 38, 529, 531, 714, 717, 722, 726, 735
Alcohol dissociation 444
Alcoholate production 444
Aldehydes 137, 148-151, 153, 505
Aldox formulation 699
Algae 13, 39, 46, 244, 629, 630, 633, 636, 638, 666
Alkyl hydrazine destruction 720
Amperometry 131, 220, 221, 231, 242, 247, 249, 250, 253-256, 265, 266, 270, 272, 273, 275, 276, 282, 295, 304, 313, 318
Anatase 541, 544, 545, 584, 585, 588, 604
Andco process 702, 704
Aniline 145, 147, 314, 372, 373
Anion exchange 433, 434, 438, 440
  See also Ion Exchange.
Anodic dehalogenation 374
Anodic oxidation 149, 165, 185, 190, 191, 194, 197, 232, 363, 365, 368, 369, 371, 372, 374, 375, 378, 391, 399, 470-472, 489, 699
Anolyte 98, 104, 373, 402, 420, 465, 472
Antibiotic-resistant coliforms 658
Aquatech process 707
Aromatic amines 253, 265, 266, 372
Array detector 226, 229, 230
Arsenic 27-29, 31, 32, 38, 42, 46, 47, 147, 167, 169, 170, 181, 184, 298, 452, 595, 704, 722
Arsenic removal 704
Atomic absorption spectroscopy 31
Average current efficiency 364

B

Bacillus subtilis 647
Band structure 117
Bandgap shift 548
Barium peroxide 404
Basic front 449
Batteries 24, 49, 57, 65, 379, 391, 393, 461
Bed voidage 395
Benzene 8, 19, 20, 22, 133, 139, 141, 265, 268, 365, 374, 505, 514, 516, 517, 577, 588
Bicontinuous microemulsions 469
Bimolecular rate constants 403, 499, 507, 510, 521, 530
Biodegradation 11, 14, 22, 40, 41, 370
Biological oxygen demand 13
Biphenyls 133, 141, 142, 154, 253

Bipolar membranes 434, 442-444, 633, 640
Bipolar trickle tower 376, 391, 394
Bisulfite 175, 176
Boeing technology 704
Boltzmann distribution 559
British Gas Stretford Process 426, 488, 491
Bromine-based disinfectants 631

C

Cadmium 24, 38, 49, 96, 157-160, 248, 249, 285, 286, 292, 294, 299, 379, 388, 396
Capacitance 92, 118, 123, 281, 310, 311
Capillary electrophoresis 32, 112, 201, 250, 318
Carbon adsorption 22, 370, 531, 598, 636, 706
Carbon black 240, 310, 459-462, 698
Carbon cloth 638
Carbon dioxide 178, 468
Carbon felt 275, 370, 391, 392, 639
Carbon fiber 230, 233, 235, 264, 374, 376, 378, 388, 392, 463
Carbon monoxide 10, 178
Carbon paste electrode 89, 171, 231, 234, 235, 304
Carbonaceous materials 44, 232, 234, 370, 401
Carbonate radical ion 510
Carbonyls 149-151
Carboxylic acid anions 375
Carboxylic acids 148, 151, 542
Carrier degeneracy 548, 549
Carrier dynamics 553, 562
Carrier transport 532, 555, 556
Cathode materials 376, 378, 379, 638
Catholyte 98, 104, 373, 402, 431, 432, 446, 699

Cation exchange membranes 436
Cav-Ox process 713
Cellulosic materials 401
Ceramic electrode materials 463
Chattering control of potential 382
Chemelec cell 691-693, 695
Chemical disinfection 628
Chemical dose 634
Chemical oxygen demand 13, 364
Chemical stripping analysis 343
Chemically modified electrode 133, 189, 210, 224, 236
Chick's law 649
Chloramines 626, 630, 632, 633, 666
Chlorinated organics 47, 377, 644
Chlorination 29, 36, 253, 625, 626, 629, 630, 635, 637, 644-646, 658, 663, 664, 667, 731-733
Chlorine 9, 30, 42, 49, 105, 625, 629, 630, 631-635, 638, 640, 641, 645, 652, 658-660, 663, 664, 667, 731, 733
Chlorine dioxide 270, 630, 631, 641, 652
Chlorine photolysis 663
Chlorine removal 431
Chlorobiphenyls 143
Chlorofluorocarbon 7, 8, 46
Chlorophenols 137, 253, 317, 378, 519, 584, 588, 590, 636
Chromate 24, 192, 218, 384, 394, 395, 446, 477, 702, 705
Chromatography 31, 32, 52, 201, 203, 216, 220, 253, 266, 270, 276, 309, 314, 315, 318
Chromium 7, 25, 27, 38, 46, 160-163, 279, 298, 299, 384, 389, 402, 470, 533, 595, 703
Circulating bed electrodes 396, 397
Clark electrode 271, 272
Claus process 423, 424

CO$_2$ reduction 419, 436-439
Coagulation 21, 28, 36, 37, 41, 42, 410, 412, 413, 416, 439
Cobalt 243, 317, 382, 383, 396, 403
Coliform 35, 36, 652-655, 659, 660, 666-668, 672
Combined chlorine residuals 630
Complexation 99, 157, 160, 170, 181, 188, 189, 243, 376, 384, 388, 417, 435, 450, 530
Complexing capacity 294-296
Conductor 58, 64, 422
Conjugate reactions 534, 536, 541
Continuous stirred tank reactor 105
Convective diffusion 100, 111, 221, 222
Copper 23, 24, 27, 44, 65, 66, 81, 82, 83, 155-157, 188, 212, 218, 243, 279, 281, 286, 290, 292, 294-296, 298-300, 302, 317, 376, 379, 387-389, 392, 394, 396, 404, 422, 439, 472, 540, 625, 638, 640, 695, 696, 702
Coupled chemical reactions 88, 131, 171, 383
Crown ethers 243, 465, 596
Cryptosporidium 29, 658, 666
Current efficiency 113, 160, 163, 373, 378, 382, 384, 388, 407, 408, 446, 469, 699
Cyanate 176, 376
Cyanide 7, 27, 47, 159, 175, 176, 217-219, 374, 376, 381, 387-389, 394, 404, 525, 533, 595, 696, 720
Cyclic voltammetry 88, 89, 95, 182, 185, 188, 304, 307, 566. *See also* Voltammetry.
Cyclohexane 139, 374, 375
Cylindrical packed bed 394
Cysteine 187-189, 301
Cystine 187-189, 301

D

Daniell cell 64-66, 70
DBP control 636, 664
DDT 16, 19, 22, 41, 532
Dechlorination 42, 378, 468, 469
Dendrites 382
Depletion layer 553, 554, 561, 562
Desalination 438
Desulfurization 423, 427, 429, 440, 692, 705, 708
Detection limit 30, 32, 33, 95, 202, 203, 218, 219, 225, 226, 231, 243, 244, 253, 255, 256, 264-270, 274, 306, 314, 718, 722
Differential pulse voltammetry 93, 95, 305, 306
Dimensionless numbers 102
Dimethyl arsinic acid 181
Direct electrolysis 361, 363, 367, 369, 371, 373, 375, 458
Disinfection 30, 50, 270, 362, 405, 416, 500, 533, 625-635, 637-641, 643, 644, 646, 647, 649, 651, 652, 656, 658, 659, 660, 662, 666, 667, 730-733
Disinfection by-products 15, 29, 271, 379, 626, 635
Dispersed kinetics 584
Dispersion 204-207, 235, 240, 417, 543, 662, 663, 724
Dissociative adsorption 151, 584
Distillation 2, 36, 45, 253, 438, 702
Disulfide 184, 187, 188, 401
DNA 646, 660
Donnan potential 86, 434
Doping 116, 117, 236, 310, 311, 404, 455, 456, 463
Double-layer 92, 163, 286, 295, 306
Dye removal 415

E

E. coli 29, 538, 539, 630, 642, 649-651, 658-660, 669, 671-673. *See also* Coliform.
Ebonex 370, 449, 463, 464, 690
Eco cell 695
Electrical conductance 62
Electrical conductivity 62, 109, 446, 545
Electrical dehydration 415
Electrical energy per order 727
Electro-osmosis 449
Electro-osmotic flow 112, 114, 450, 451
Electro-osmotic purging 449
Electroactive polymers 454. *See also* Polymers.
Electrochemical disinfection 361, 638, 639, 641
Electrochemical equipment 36, 687, 688, 690
Electrochemical hydride generation 309
Electrochemical ion exchange 380, 437, 444, 452
Electrochemical lag times 673. *See also* Lag time.
Electrochemical oxidability index 364
Electrochemical oxygen demand 365, 367
Electrochemical pH control 393
Electrochemical reactor 105-107, 109-113, 363, 381, 385, 386, 390, 407, 641, 691, 693, 695, 697, 699-701, 705, 707
Electrochemical reactor design 104, 105, 107-109, 111
Electrochemical reactor engineering 57, 112
Electrocinerator 705, 706
Electrocoagulation 361, 363, 410-412, 415, 416
Electrocution 639
Electrode configurations 104, 111, 222, 223, 249, 271, 282
Electrode materials 105, 112, 153, 171, 220, 229, 231, 232, 236, 240, 242, 256, 274, 287, 315, 362, 369, 370, 376, 390, 391, 407, 421, 432, 447, 453, 454, 462, 463, 638, 690, 691, 702
Electrode materials, Suppliers 689
Electrodeposition 97, 383, 384
Electrodialysis 44, 45, 436-440, 641, 690, 692, 695, 696
Electroflocculation 361, 410, 411, 413, 415, 416
Electroflotation 361, 363, 410, 411, 415, 416
Electrogenerated chemiluminescence 313, 316
Electrogenerative processes 431
Electrohydrolysis process 697
Electrokinetic phenomena 112, 113, 115, 216
Electrokinetic processing of soils 452
Electrokinetic soil remediation technologies 705
Electrolysis 49, 62, 66, 68, 95, 97, 98, 113, 120, 131, 147, 181, 228, 247, 276, 284, 290, 307-309, 362, 364, 367, 374
Electrolytic cell 111, 410, 439, 706
Electron accelerators 644
Electron beam 644
Electron beam penetration depth 644
Electron density 116, 139, 364, 549
Electron mediator 363, 422
Electron mobility 557
Electron relay 540, 596
Electron transfer kinetics 69, 544

Electron-donating group 364
Electron-hole pairs 312
Electron-withdrawing group 151, 364
Electronic current 58
Electronically conductive polymers 211, 239, 454, 455
Electrooxidation 135, 136, 145, 147, 148, 155, 163, 175, 190, 480. *See also* Anodic oxidation.
Electrophoresis 112, 216, 225, 318, 321, 324, 449
Electrophoretic mobility 555
Electroreclamation 449
Electroreduction 132, 133, 138, 140-142, 146, 151, 152, 178, 179, 195, 409, 419, 420, 430, 431
Electroremediation 449, 705
Electrorestoration 449
ElectroSyn cell 691
Electrothermal atomization 32
Energy consumption 113, 376, 378, 388, 393, 427, 503
Energy efficiency 362
Enterovirus 627
EnViro cell 692, 694
Environmental compatibility 362, 369
Escherichia coli 639, 642, 657, 667. *See also* E. coli and Coliform.
Ethylene glycol 42, 149, 151, 310, 401, 403
Euglena 638
Excitons 570
Extrinsic semiconductor 116

F

Falling-film reactor 722
Fats 21, 401, 410
Fecal coliform counts 653

Fecal coliforms 36, 631, 637, 651, 660, 667, 672, 730, 732. *See also* Coliform.
Fenitrothion 263, 264
Fenton reaction 408, 409, 499, 591, 592
Fermi level 117, 118, 549, 555, 561
Ferredoxin–$NADP^+$–reductase 596
Field-effect transistor 214
Field-free region 554, 562
Figures of merit 111-113, 363, 384, 432
Fin-type current feeders 390
Flat band 555
Flow cell 49, 68, 207, 220, 228, 249, 379
Flow-by mode 390, 393, 692
Flow-injection analysis 201
Flow-through mode 393
Fluid flow 58-60, 224, 231, 598, 599, 660
Fluidized bed electrodes 396-398
Fluorides 416, 466
Formal potential 164, 384
Formaldehyde 7, 129, 137, 149, 151, 235, 403, 407, 409, 463, 505, 579, 591, 592, 596

G

G-value 644
G. muris 658
Galvanic cell 66
Gamma rays 606
Gas-phase heterogeneous photocatalysis 533, 595, 596, 666, 723, 735
Geosmin 637, 666, 732
Giardia lamblia 627, 638, 658

Glassy carbon 96, 158, 171, 176, 210, 225, 231-236, 243-245, 249, 255, 256, 265, 280, 284, 289, 304, 306, 391, 422
Glutathione 187-189, 301
Graphite cloth 392, 473
Graphite felt 378, 392, 393, 476
Greenhouse effect 6, 10, 429, 431
Groundwater transport 11

H

H_2S treatment 423
Halogenated derivatives 506
Hanging mercury drop electrode 95, 202, 234, 304, 342, 347-350
Herbicide 13, 19, 147, 148, 190, 466, 495, 527
Heteroatom-containing pollutants 402
Hexavalent chrome removal 703
Hg vapor lamps 527, 646
Hg(II) 284, 379, 608
Hole scavenger 538, 541
Hollow perforated cylinders 388
Homogeneous photolysis 46, 499
Homolysis 505, 508
Humic acids 44, 510, 517, 652, 655
Hydrated electrons 46, 503, 600
Hydrazine 177, 243, 429, 718, 720
Hydrodynamic cavitation 713
Hydrodynamic layer 100, 101
Hydroperoxy radical 500, 503
Hydroxyl radical 9, 46, 405, 406, 480, 481, 499, 542, 590, 600, 603, 617, 683
Hypochlorite 374, 388, 405, 415, 440, 629, 630, 638, 640, 641, 695, 696, 701, 706
Hypochlorous acid 629, 630, 640

I

Ilkovic equation 92
Incineration 26, 36-38, 47, 48, 50, 362, 373, 374, 471, 478, 480, 531, 698, 706
Indicator organisms 637, 660, 685
Indirect electrolysis 361, 398-401, 403, 405, 407, 409, 458
Inductive effect 364
Inner-cell process 417, 427
Instantaneous current efficiency 364, 365
Instrumental analysis 31
Intercellular coenzyme A 639
Interfacial capacitance 112
Iodine-based disinfectants 632
Ion chromatography 31, 201, 216, 270, 309
Ion exchange 44, 56, 109, 111, 209, 244, 277, 344, 361, 380, 394, 433, 434, 436-440, 443-446, 450, 452, 478, 490-493, 497, 638, 640, 703, 732
Ion-exchange chromatography 295
Ion-selective electrode 204, 319, 320, 321, 356, 358
Ionic current 58
Ionization potential 570
Irreversible processes 405
Isopropanol 401, 587
Isotope exchange 586, 619
Ispra Mark 13A process 706

K

Keratin solutions 520
Kerosene 401, 698

Index 771

L

Lactobacillus acidophilus 667
Lag time 668, 669, 672, 673
Landfill 21, 25, 37, 39, 44, 48, 49, 380
Laser flash photolysis 563, 564, 567, 571, 616
Laser-induced photochemical oxidative destruction 713, 714
Lazy river concept 599
Lead 1, 2, 7, 8, 13, 21-30, 33, 38, 47, 49, 51, 135, 153, 164, 165, 178, 180, 181, 192, 195, 196, 202, 218, 219, 226, 243, 244, 248, 249, 270, 286, 292, 293, 295, 299, 302, 303, 322, 329, 334, 335, 337, 342- 346, 349-353, 355, 369, 371, 379, 383, 420, 429, 432, 436, 451, 468, 469, 471, 472, 480, 505, 559, 635, 636, 728
Leaded gasoline 25, 26, 180
Leaking underground storage tank 702, 734
Lekkerkerk 2, 11
Levich equation 224
Limiting current 76, 78, 92, 104, 105, 221, 222, 262, 385, 386, 390, 393, 475, 477, 569
Linear sweep voltammetry 122, 144, 171, 173
Lock-in amplifier 122
Love Canal 2, 11
Luggin capillary 68

M

Malic enzyme 597, 623
Marcus theory 573, 574
Marine biofouling 639, 642, 677

Mass transport 57, 62, 63, 90, 99-103, 111, 160, 215, 224, 273, 291, 306, 326, 382, 387, 433, 469, 473, 474, 543, 596, 599, 691, 695, 714
Matrix exchange 343
Meat extract 375
Meat packing wastes 401
Membrane-assisted processes 432, 433, 435, 437, 439, 441, 443, 445, 447
Mercury 2, 7, 17, 25-28, 47, 49, 51, 91, 92, 95, 100, 133, 148, 159, 164-166, 181, 187, 191-194, 196, 198, 202, 215, 224, 231, 233, 234, 236, 240, 241, 244, 245, 256, 277-279, 281, 283-285, 287-290, 290, 298, 299, 304, 307, 323, 324, 328, 329, 332, 335-338, 341-345, 347-352, 377, 378, 473, 527, 533, 537
Metabolites 253, 254, 268, 269, 326, 337, 339
Metal felts 389, 395
Metal ion intake 24
Metal oxide electrodes 151, 403, 404
Metal screens 395
Metal speciation 17, 51, 204, 291, 312
Metal wool 389, 393, 394, 477
Metallated tetraphenylporphyrins 597
Metalloids 31, 167

Metals
69, 99, 123, 128, 151, 155, 192, 226, 231-233, 235, 236, 248, 250, 253, 268, 270, 277, 278, 280, 283, 287, 288, 290,

Metals 292, 298, 302, 303, 306, 307, 317, 323, 325, 329, 332, 341-346, 348-352, 354, 370, 380, 383, 384, 386, 388, 389, 391, 395, 398, 415, 416, 420-422, 433, 439, 442, 447, 450, 473, 475, 476, 478, 480, 482, 485, 489, 496, 503, 544, 608, 613, 637, 641, 692, 705, 721, 722

Methanol 137, 147, 149, 150, 259, 265, 374, 375, 420, 422, 444, 484, 486

Methyl isocyanate 2
Methyl isoborneol 637
Micro-interfaces 577
Microbial inactivation 628, 633, 682
Microbial treatment 36, 39, 41
Microwave conduction 516, 518, 520, 544, 571, 585
Mineralization 368, 584, 588, 590-592, 596, 605, 607, 630
Mixed potential 69, 81, 83, 123, 538, 555, 557, 571
Mobility 22, 63, 438, 452, 556
Most probable number 668
Moving bed electrode 397
Mucidone 637
Multiple-hit theory 650

N

N-type semiconductor 117, 118, 119, 120, 121, 536, 538, 608
Nanosized particles 240, 542
Natural organic matter 626, 635
Nephelometric turbidity unit 653
Nernst equation 70, 71, 73, 79, 84, 98, 383, 389
NiO 544, 613
Nitrate 7, 27, 29, 40, 43, 52, 53,

Nitrate 175, 177, 178, 195, 217-219, 233, 244, 279, 303, 321, 322, 335, 350, 352, 378, 379, 396, 443, 446, 447, 463, 472, 478, 490, 495, 592, 622, 640, 683

Nitro derivatives 51, 137, 373
Nitrogen-containing organics 137, 144, 145
Nitroxide radicals 513
NO_x removal 428

O

O-chlorobenzoic acid 378, 472
Odor removal 632, 636
Oil slicks 551, 607
Optical band-gap 116
Organic acids 152, 401, 597
Organo arsenicals 181, 183
Organometallic compounds 140, 155, 180
Outer-cell process 417
Overpotential 137, 158, 166, 175, 178, 231, 236, 241, 245, 277, 370, 371, 376, 383, 387, 407, 408, 412, 420, 442, 500, 505, 507, 517, 524, 526, 527, 531, 534, 535, 536, 538, 640
Oxidant generation 680
Oxychloride species 379
Oxygen-containing organics 148
Oxynitrogen ions 378
Ozonation 633, 636, 660, 663-665, 667, 674, 676
Ozone 6-10, 42, 46, 50,179, 201, 340, 371, 405, 406, 407, 415, 429, 481, 502, 503, 521-531, 595, 596,
Ozone 600-604, 627, 632, 633, 635, 636, 641, 652, 659, 664-667, 674-676, 681
Ozone depletion 6
Ozone layer 8, 10, 46

P

P-type semiconductor
116, 117, 118,
455, 683, 684
Parabolic trough concentrators 598
Parathion 183, 196, 259, 262, 263, 266, 306, 338, 352
PbO_2 154, 163, 165, 371, 373, 374, 376, 404, 407, 408, 641
Pentachlorophenol 254, 256, 378, 590, 606
Perhydroxyl radical 406, 601
Permselectivity 86, 209
Peroxide generation 512
Peroxo species 586
Peroxone 665, 666, 684
Pertechnate 379
Pesticides 6, 8, 11, 13, 14, 15, 17, 19, 22, 31, 40, 55, 253, 255-259, 262-266, 304, 306, 312, 338, 339, 373, 519, 527, 532, 603
Phage ms2 667, 684
Phenols 7, 20, 31, 40, 52, 137, 148, 153, 154, 191, 253-256, 265, 338, 506, 519, 532, 592, 602, 605, 627
Phenyl mercury compounds 181, 187
Phosphorus removal 416, 483
Photo-Kolbe decarboxylation 587, 588
Photo-reduction
500, 506, 536, 537, 538, 576
Photo-sinking 588, 621
Photoadsorption 541, 577, 588, 621
Photoassisted methods 499
Photocatalyst regeneration
542, 543, 597, 612, 615, 616, 623

Photocatalytic reactor design 597
Photocatalytic reactors 543, 596, 613, 622
Photocatalytic treatment 588, 592, 594, 595, 598, 600
Photochromic effect 561
Photocorrosion 541, 542, 569, 575
Photocurrent 116, 119, 122, 312-316, 506, 507, 548, 553, 563, 569, 573, 616
Photoelectric effect 120
Photoelectrochemical
phenomena 531
Photoelectrochemical cell 120, 553, 608
Photoelectrochemical
disinfection 50, 666, 667
Photoelectrowinning 538
Photoemission 120, 503
Photoinduced charge transfer 586
Photolytic ozonolysis 526
Photoreactivation 659, 660
Photovoltaic cells 120
Phthalocyanines 311, 387, 422, 474
Plug flow reactor 105
Point of zero charge 555, 594
Point source summation 648
Polarization 67, 68, 73, 74, 77, 78, 81, 82, 245, 312, 413, 415, 475, 491, 592
Polarography 88, 91, 92, 123, 131, 176, 180, 193, 194, 202, 204, 231, 247, 265, 294, 295, 298, 305, 306, 328, 340, 341, 346, 347, 350
Polarons 561
Poliovirus 627, 634, 657, 667, 675, 682
Polychlorinated biphenyls 7, 8, 15, 20-22, 53, 377, 495, 496, 605, 606

Polycyclic aromatic
    hydrocarbons 133, 137, 139
Polymers, electroactive 204, 208,
    216, 218, 219, 333, 343, 358
Porous graphite 391, 476
Potassium permanganate 627,
    633, 641
Potential distribution 390, 392,
    398, 475
Potentiometry 294, 295, 611
Potentiostatic control 191, 383, 561
Pourbaix diagrams
    57, 78, 79, 81, 155,
    156, 175
Powdered activated carbon 636
Prandtl layer 222
Precipitation 10, 29, 36, 41, 53,
    162, 171, 236, 297, 302, 321,
    345, 379, 380, 389, 410-412,
    435, 450, 452, 482, 483, 492,
    542, 596, 604
Propanol
    374, 375, 401, 403, 544, 612
Protein wastes 416
Proton exchange membrane
    407, 447, 481
Protozoan cysts 7, 35, 627,
    638, 639, 657, 658, 666, 676
Pulse voltammetry
    88, 92, 93, 95, 231,
    306, 307, 330, 352, 353
Pulsed amperometry 185
Pyrazole 147, 466
Pyrolysis 36-38, 235, 330, 390,
    463, 620
Pyruvic acid 597, 623

Q

Quantum yield 501, 507, 508, 543,
    545, 552, 553, 557, 562, 567,
    572, 575, 583, 593

Quenching 314, 316, 490, 570, 571

R

Radioactive wastes 6
Radiolytic methods 500
Radiolytic yield 644
Radon 6
Reactor configurations 385, 543,
    597, 599
Redox speciation 298
Reovirus 627
Residence time distribution 661
Residual disinfecting action 633
Reticulated vitreous carbon 100,
    221, 236, 325, 329, 331, 638
Reverse osmosis 44, 45, 380, 438,
    446
Reversible complexation 401
Reversible processes 386, 387, 399
Rhine River 2, 25
Ring hydroxylation 40, 527
Rotating cylinder electrodes 474
Rotating disk electrode 131, 192,
    224, 344
Ruthenium 265, 328, 332, 339,
    357, 379, 484, 490, 615, 641
Rutile 63, 679

S

Saccharomyces cerevisiae 667
Salt-splitting 441, 443
Scanning electrochemical
    microscopy 229, 326
Schottky diode 215
Sedimentation potential 449
Selective deposition 381, 384
Selenium 7, 27-29, 31, 52, 170,
    171, 175, 193, 194

Semiconductor 57, 59, 63, 69, 83,
    116-122, 124, 126, 455, 464,
    500, 531, 532, 534-536,
    538, 540-544, 546, 548-550,
    552, 555, 557, 558, 562,
    563, 567, 571, 574, 575,
    584, 587, 594-596,
    606- 611, 613-617, 622,
    623, 633, 674, 684
Semiconductor catalyst 46, 531
Semiconductor electrochemistry 117
Sewage 4, 6, 13, 14, 35, 294,
    401, 641, 681
Sewage sludge 401
Silver 190, 192, 195, 209, 210, 212,
    218, 219, 243, 271, 272, 287,
    289, 334, 349, 358, 378, 383,
    387, 389, 392, 394, 395,
    399- 401, 439, 475-478, 538,
    588, 595, 612, 615, 622,
    623, 625, 633, 638, 675, 680
Single crystals 328, 545, 612-614, 617
Size quantization 542, 548, 610, 613, 623
SnO_2 244, 310, 311, 355, 365, 370,
    375, 403, 404, 470, 479,
    542, 543
SO_2 removal 426
Solid polymer electrolytes 422, 447
Soluble metal oxide catalysts 403
Space-charge layer 118
Space-habitat environment 369
Spin trapping 587, 620
Spontaneous electron
    transfer 60, 456
Spouted bed electrode 397
Square-wave voltammetry 245
Standard hydrogen
    electrode 59, 61, 70, 541
Sterilization 625, 632, 633, 666,
    674, 677, 678, 681, 684

Streaming potential 449
Stripping analysis 97, 123, 125
Stripping voltammetry 88, 95, 96,
    193, 194, 198, 202, 231, 245,
    247, 249, 250, 276-279, 282,
    286, 287, 290, 295, 298,
    302-305, 329, 336, 337,
    341, 342, 344-353, 355
Sublethal stress 672, 673, 685
Sublethal UV damage repair 659
Sulfide 22, 42, 176, 184, 197, 210,
    212, 217, 218, 311, 321, 389,
    401, 419, 424, 425, 479,
    482-484, 487, 488, 533, 608,
    610, 612, 616, 617, 620, 623
Sulfite 42, 43, 69, 165, 175, 176,
    181, 187, 194, 198, 244, 249,
    335, 533, 608
Sulfone 184
Sulfur dioxide 175, 176, 194, 195,
    335, 340, 427, 489
Supercritical fluids 453, 466, 468
Surface charge 555, 560, 594, 611
Surface hydroxylation 585
Surface recombination 558, 561
Surface states 153, 329
Surface-enhanced Raman
    scattering 318, 358
Surfactant solutions 404
Surfactants 7, 20, 21, 31, 210, 218,
    280, 392, 532, 607
Survival ratio 648, 649
Synthetic detergents 416

T

Tafel plots 77
Tetraalkyl lead 180
Textile plant waste 520
Thin-layer cell 226, 227, 318, 564
Thiocyanate 334, 376, 567, 616
Thioethers 184

Thiol 184, 187, 304
Thiourea 159, 185, 186, 191, 256
THM formation potential 636, 664
Three-dimensional electrodes 111, 369, 382, 389, 391, 395, 396, 477, 638
Thymol blue 375
Time-domain reflectometry 562, 615
Titanium suboxides 463, 464
Total coliforms 631, 637, 650, 652, 653, 658, 667
Total dissolved solids 637
Total organic carbon 369, 371, 610, 635
Tributylphosphate 375
Trichloroethylene 7, 8, 15, 53, 137, 452, 506, 509, 519, 605, 621
Trihalomethanes 630, 635
Trioxane 375, 471
Triphenyl phosphine 183
Triphenyl phosphine oxide 183
Two-dimensional electrodes 112, 386

U

Ultrafiltration 45, 380, 636
Ultramicroelectrode 230, 231, 344
Ultrapure water sterilization 633
Underpotential deposition 98, 163, 171, 192, 195, 383
Urea 401, 612
Urotropin 375, 471
UV disinfection 644-649, 651-653, 657-661, 663, 665, 676, 680-683, 730-732
UV dose 647, 649-652, 656, 658
UV dose sensitivity 658

V

Vacuum scale 59

Vacuum UV region 508
Vibrio anguillarum 641
Vinyl chloride 2, 15, 47, 210, 453, 507
Viologen 244, 563, 565, 567, 569-571, 587, 597, 606, 615-617
Virological disinfection 658
Viruses 7, 14, 29, 30, 35, 36, 534, 626, 627, 635, 637-639, 657, 658, 675, 676, 681, 682, 684
Volatile organics 8, 529
Voltammetry 68, 78, 83, 88, 89, 91-96, 122, 123,131, 142, 144, 148, 157, 171, 173, 175, 182, 185, 188, 190, 193, 194, 197, 198, 202, 204, 220, 224, 231, 242, 243, 245, 247-250, 269, 276-279, 281, 282, 286, 287, 290, 294, 295, 298, 301-308, 323-326, 328-330, 332, 335-337, 341-350, 352-355, 357, 359, 484, 566
Vortex bed electrode 397

W

Wall-jet detector 207
Waste biomass 374
Water polishing 446
WO_3 310, 332, 542, 596, 609, 614

Z

Zero proton condition 594
Zeta potential 115, 555

TD
192.2
.R35
1997

DATE DUE

DE 20 02			